COEFF OF FRICTION
page 393

PLASTICS
TECHNOLOGY
HANDBOOK

PLASTICS ENGINEERING

Series Editor

Donald E. Hudgin

Professor
Clemson University
Clemson, South Carolina

1. Plastics Waste: Recovery of Economic Value, *Jacob Leidner*
2. Polyester Molding Compounds, *Robert Burns*
3. Carbon Black-Polymer Composites: The Physics of Electrically Conducting Composites, *edited by Enid Keil Sichel*
4. The Strength and Stiffness of Polymers, *edited by Anagnostis E. Zachariades and Roger S. Porter*
5. Selecting Thermoplastics for Engineering Applications, *Charles P. MacDermott*
6. Engineering with Rigid PVC: Processability and Applications, *edited by I. Luis Gomez*
7. Computer-Aided Design of Polymers and Composites, *D. H. Kaelble*
8. Engineering Thermoplastics: Properties and Applications, *edited by James M. Margolis*
9. Structural Foam: A Purchasing and Design Guide, *Bruce C. Wendle*
10. Plastics in Architecture: A Guide to Acrylic and Polycarbonate, *Ralph Montella*
11. Metal-Filled Polymers: Properties and Applications, *edited by Swapan K. Bhattacharya*
12. Plastics Technology Handbook, *Manas Chanda and Salil K. Roy*
13. Reaction Injection Molding Machinery and Processes, *F. Melvin Sweeney*
14. Practical Thermoforming: Principles and Applications, *John Florian*
15. Injection and Compression Molding Fundamentals, *edited by Avraam I. Isayev*

Additional Volumes in Preparation

PLASTICS TECHNOLOGY HANDBOOK

SECOND EDITION, REVISED AND EXPANDED

MANAS CHANDA

Department of Chemical Engineering
Indian Institute of Science
Bangalore, India

SALIL K. ROY

School of Building and Estate Management
National University of Singapore
Singapore

Marcel Dekker, Inc.   New York • Basel • Hong Kong

Library of Congress Cataloging-in-Publication Data

Chanda, Manas.
  Plastics technology handbook / Manas Chanda, Salil K. Roy. -- 2nd
ed., rev. and expanded.
      p.   cm. -- (Plastics engineering ; 26)
  Includes bibliographical references and index.
  ISBN 0-8247-8734-X
  1. Plastics.   I. Roy, Salil K.   II. Title.
III. Series: Plastics engineering (Marcel Dekker, Inc.) ; 26.
TA455.P5C46   1992
620.1'923--dc20   92-25551
                                                                      CIP

This book is printed on acid-free paper.

MARCEL DEKKER, INC.
270 Madison Avenue, New York, New York 10016

Current printing (last digit):
10 9 8 7 6 5 4 3 2 1

PRINTED IN THE UNITED STATES OF AMERICA

Preface to the Second Edition

The impetus to produce a second edition of *Plastics Technology Handbook* less than five years after publication of the first edition came from two main sources. One was the encouraging reader response and the other was the need to document recent developments in the field of plastics. This new edition, moreover, provided an opportunity to add several important topics and to revise each chapter, bringing the information up-to-date. In preparing the present edition the structure of the book has not been altered, and the straightforward style of presentation and lucidity of treatment that characterized the earlier edition have also been retained.

The significance of stabilizing agents and other additives in making plastics useful in many applications cannot be overemphasized. Discussion of this subject has therefore been greatly expanded and updated to include recent developments. Because the science and technology of processing elastomeric products differ significantly from those of plastics and fibers, a new, elaborate section has been added to Chapter 2 to deal with the compounding and processing of rubber products. A review of prepregging technology, which has gained much popularity in composite processing, is included in Chapter 2 as well. Recently, much attention has been paid to polymer blending as a simple tool to produce new products with tailored properties. This is done by blending common polymers, hence eliminating the need to add to existing inventories. A discussion of polyblending can be found in Chapter 2.

With plastics now found in every aspect of life and newer innovative plastic products constantly entering the market, a curious user must often wonder what polymer a product is made of. Polymer identification can often be performed using simple tests. This is highlighted in Chapter 3's extensive section on identification of plastics.

Most commercial demand for plastics is met by a small number of large-volume common polymers such as PE, PP, PS, and PVC and a few so-called engineering polymers possessing superior mechanical properties for engineering applications. There are, however, many polymers that have been developed to meet critical needs of specialized applications. Although their tonnage rating is very low, the importance of their use in modern technology cannot be overemphasized. Examples of such special-use polymers include conductive polymers, photoconductive polymers, piezoelectric polymers, photoresist polymers, polymeric ion-exchangers, and polymeric catalysts. Chapter 5 is a comprehensive discussion on polymers in special uses. In addition, Chapter 4 has been augmented with additional information.

With the diverse fields of technology undergoing rapid change both qualitatively and quantitatively, novel applications of both common polymers and specialty polymers are coming to light. New uses of polymers in such diverse areas as automotive, aerospace, packaging, agriculture, horticulture, domestic appliances, office equipment, communication, electronics and electrical applications are reviewed in Chapter 6.

It has been our pleasure since the publication of the first edition to receive from readers appreciative comments as well as suggestions for improvement, most of which have been implemented in this edition. Although it is not possible to thank everyone individually, we record our special gratitude to Professor G. L. Rempel, Chairman of the Department of Chemical Engineering, University of Waterloo, Waterloo, Ontario, Canada for his continuing assistance and support and to Dr. Stephan Jaenicke, Department of Chemistry, National University of Singapore, Singapore, who read the book critically and made extensive comments and suggestions for improvement. Finally, we acknowledge the tolerance shown by our wives, Mridula and Snigdha, and our daughters, Amrita and Rarita, who allowed us, though not ungrudgingly, to engage in this work so soon after the labor of love for the first edition.

Manas Chanda
Salil K. Roy

Preface to the First Edition

The title of the book requires that a definition of plastics be given, although it is very difficult and probably of little value to try to produce an accurate definition of the word *plastics*. Plastics are defined in the *Modern Plastics Encyclopedia* (1962) as a "large and varied group of materials which consist of or contain as an essential ingredient, a substance of high molecular weight which while solid in the unfinished state, at some stage in its manufacture is soft enough to be formed into various shapes, usually through the application, either singly *or* together, of heat and pressure." A more concise definition but one which requires clarification is "plastics materials are processable compositions based on macromolecules." With ever widening application of polymeric materials, the term *plastics*, however, has gained usage much beyond the original concept which classifies polymers into elastomers, plastics, and fibers, with plastics being less amorphous than the former and less crystalline and oriented than the latter. In view of this it is perhaps more appropriate to define plastics simply as "those materials which are considered to be plastics materials by common acceptance."

Plastics thus include by common usage all of the many thousands of grades of commercial materials, ranging in application from squeeze bottles, bread wrappers, baby pants, shoes, fabrics, paints, adhesives, rubbers, wire insulators, foams, greases, oils, and films to automobile and aircraft components, and missile and spacecraft bodies. In most cases (certainly with all synthetic materials) the macromolecules used in them are polymers, which are large molecules made by the joining together of many smaller ones.

Chapter 1 of this book provides a brief account of the molecular make-up and structural characteristics of these macromolecules and a short summary of the general methods of their preparation. It continues by showing how polymer properties are related to chemical structure and physical states of aggregation. A knowledge of this structure-property relation is essential for a fuller appreciation of the application potential of different polymers. This also leads to realization that the scope of application of polymers can be greatly extended since polymers can be modified and structural changes can be effected in a variety of ways, which is responsible, in a large measure, for the versatility and popularity of plastics.

Polymers are converted to a myriad of useful objects. Molding is one of the critical steps in this process. Molding processes vary depending on the type of plastics to be processed and also on the end products to be made. An outline of the more important of these molding processes is presented in Chapter 2.

Plastics materials differ very greatly from metals in respect of mechanical and other properties. These differences can be attributed to the molecular structural characteristics of the polymer base materials forming the plastics. The characteristic properties of plastics materials are responsible for their applications in many areas in preference to metals. Moreover, their unique properties make them the indispensable choice in many instances. The characteristic properties of polymeric materials are highlighted in Chapter 3, with broad classification into mechanical, electrical, optical, and thermal properties.

Polymers are ubiquitous, as they are used in a variety of forms including molded products such as radio cabinets, telephone sets, and thousands of other objects, wrapping, fibers, coatings, adhesives, and paints, and as components of composites. Instead of trying to know the different polymers used in this endless number of plastics products it is far easier to be acquainted with the various polymers and their characteristic features and properties so that their application areas can be self-evident. With this in view a large number of polymers have been reviewed in Chapter 4, highlighting their chemical nature, characteristic properties, and uses. The discussions in this chapter include industrial polymers of thermoplastic and thermosetting types, which are produced in substantial volumes and are more commonplace, as well as special polymers which can be described as tailor-made macromolecules with specific properties to satisfy critical needs in more sophisticated areas of application.

While SI units are being increasingly used in all branches of engineering, other systems of units, that is, the cgs system and the British (or fps) system, are still in common use. This is particularly true of plastics. In the present book, cgs and fps units have therefore been used, and in most places equivalent values are given in SI units. A suitable conversion table is also provided as an appendix.

This should assist the reader in gradual transition from the present use of mixed units to full use of SI units.

In writing a book of this kind, one accumulates indebtedness to a wide range of people, not the least to the authors of earlier publications in the field. Our faculty colleagues, innumerable students, and academic associates in other universities and colleges have provided much welcome stimulation and direct help. We are much indebted to all of them. We also acknowledge with gratitude the painstaking work of Mr. S. Sundaresh in typing the manuscript and the enthusiastic help of Mr. M. J. Venugopal in preparing the large body of artwork which conveys much of the message of the book. We would not be doing justice in thanking our wives, Mridula and Snigdha, and our daughters, Amrita and Rarita, in the limited space available. We are grateful for their forbearance and sacrifice.

Manas Chanda
Salil K. Roy

Contents

1

Characteristics of Polymers

WHAT IS A POLYMER?

A molecule has a group of atoms which have strong bonds among themselves but relatively weak bonds to adjacent molecules. Examples of small molecules are water (H_2O), methanol (CH_3OH), carbon dioxide, and so on. *Polymers* contain thousands to millions of atoms in a molecule which is large; they are also called *macromolecules*. Polymers are prepared by joining a large number of small molecules called *monomers*. Polymers can be thought of as big buildings, and monomers as the bricks that go into them.

Monomers are generally simple organic molecules containing a double bond or a minimum of two active functional groups. The presence of the double bond or active functional groups act as the driving force to add one monomer molecule upon the other repeatedly to make a polymer molecule. This process of transformation of monomer molecules to a polymer molecule is known as *polymerization*. For example, ethylene, the prototype monomer molecule, is very reactive because it has a double bond. Under the influence of heat, light, or chemical agents this bond becomes so activated that a chain reaction of self-addition of ethylene molecules is generated, resulting in the production of a high-molecular-weight material, almost identical in chemical composition to ethylene, known as *polyethylene*, the polymer of ethylene (Fig. 1.1).

The difference in behavior between ordinary organic compounds and polymeric materials is due mainly to the large size and shape of polymer molecules.

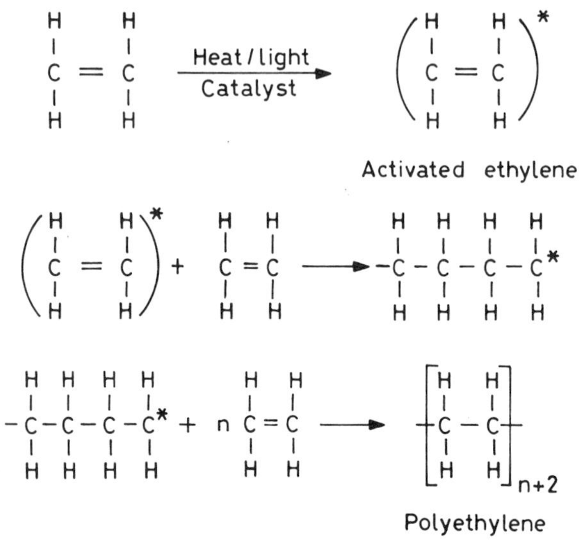

FIG. 1.1 Intermediate steps during formation of polyethylene.

Common organic materials such as alcohol, ether, chloroform, sugar, and so on, consist of small molecules having molecular weights usually less than 1000. The molecular weights of polymers, on the other hand, vary from 20,000 to hundreds of thousands.

The name *polymer* is derived from the Greek *poly* for many and *meros* for parts. A polymer molecule consists of a repetition of the unit called a *mer*. Mers are derived from *monomers*, which, as we have seen for ethylene, can link up or *polymerize* under certain conditions to form the polymer molecule. The number of mers, or more precisely the number of repetitions of the mer, in a polymer chain is called the *degree of polymerization* (DP). Since the minimum length or size of the molecule is not specified, a relatively small molecule composed of only, say, 3 mers might also be called a polymer. However, the term polymer is generally accepted to imply a molecule of large size (macromolecule). Accordingly, the lower-molecular-weight products with low DP should preferably be called *oligomers* (*oligo* = few) to distinguish them from polymers. Often the term *high polymer* is also used to emphasize that the polymer under consideration is of very high molecular weight.

Because of their large molecular size, polymers possess unique chemical and physical properties. These properties begin to appear when the polymer chain is of sufficient length—i.e., when the molecular weight exceeds a *threshold value*—and becomes more prominent as the size of the molecule increases. The dependence of the softening temperature of polyethylene on the degree of polymerization is shown in Figure 1.2a. The dimer of ethylene is

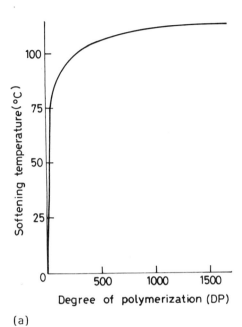

(a)

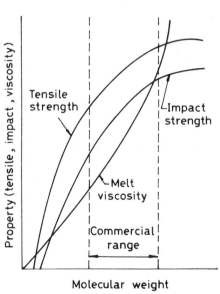

(b)

FIG. 1.2 Polymer properties versus polymer size. (a) Softening temperature of polyethylene. (b) Tensile strength, impact strength, and melt viscosity (schematic).

a gas, but oligomers with a DP of 3 or more (that is, C_6^- or higher paraffins) are liquids, with the liquid viscosity increasing with the chain length. Polyethylenes with DPs of about 30 are greaselike, and those with DPs around 50 are waxes. As the DP value exceeds 400 or the molecular weight exceeds about 10,000, polyethylenes become hard resins with softening points about 100°C. The increase in softening point with chain length in the higher-molecular-weight range is small. The relationship of such polymer properties as tensile strength, impact strength, and melt viscosity with molecular weight is indicated in Figure 1.2b. Note that the strength properties increase rapidly as first as the chain length increases and then level off, but the melt viscosity continues to increase rapidly. Polymers with very high molecular weights have superior mechanical properties but are difficult to process and fabricate due to their high melt viscosities. The range of molecular weights chosen for commercial polymers represents a compromise between maximum properties and processability.

MOLECULAR WEIGHT OF POLYMERS

In ordinary chemical compounds such as sucrose, all molecules are the same size and therefore have identical molecular weights (M). Such compounds are said to be *monodisperse*. In contrast, most polymers are *polydisperse*. Thus a polymer does not contain molecules of the same size and, therefore, does not have a single molecular weight. In fact, a polymer contains a large number of molecules—some big, some small. Thus there exists a variation in molecular size and weight, known as *molecular-weight distribution* (MWD), in every polymeric system, and this MWD determines to a certain extent the general behavior of polymers. Since a polymer consists of molecules of different sizes and weights, it is necessary to calculate an *average molecular weight* $(\overline{M})$ or an *average degree of polymerization* $(\overline{DP})$.

The molecular weights commonly used in the characterization of a polydisperse polymer are the *number average*, *weight average*, and *viscosity average*.

Consider a sample of a polydisperse polymer of total weight W in which $N =$ total number of moles; $N_i =$ number of moles of species i (comprising molecules of the same size); $n_i =$ mole fraction of species i; $W_i =$ weight of species i; $w_i =$ weight fraction of species i; $M_i =$ molecular weight of species i; $x_i =$ degree of polymerization of species i.

Number-Average Molecular Weight $(\overline{M}_n)$

From the definition of molecular weight as the weight of sample per mole, we obtain

$$\overline{M}_n = \frac{W}{N} = \frac{\sum N_i M_i}{N} = \sum n_i M_i \tag{1}$$

$$= \frac{\sum W_i}{\sum W_i/M_i} = \frac{\sum w_i}{\sum w_i/M_i} = \frac{1}{\sum w_i/M_i} \tag{2}$$

Dividing $\overline{M}_n$ by the mer weight M_0, we obtain a number average degree of polymerization, $\overline{DP}_n$, where

$$\overline{DP}_n = \frac{\overline{M}_n}{M_0} = \frac{\sum N_i x_i}{\sum N_i} \tag{3}$$

The quantity $\overline{M}_n$ is obtained by end-group analysis or by measuring a colligative property such as elevation of boiling point, depression of freezing point, or osmotic pressure [1,2].

Weight-Average Molecular Weight ($\overline{M}_w$)

Equation (1) indicates that in the computation of $\overline{M}_n$, the molecular weight of each species is weighted by the mole fraction of that species. Similarly, in the computation of weight-average molecular weight the molecular weight of each species is weighted by the weight fraction of that species:

$$\overline{M}_w = \sum w_i M_i = \frac{\sum W_i M_i}{\sum W_i} \tag{4}$$

$$= \frac{\sum N_i M_i^2}{\sum N_i M_i} \tag{5}$$

The weight-average degree of polymerization, $\overline{DP}_w$, is obtained by dividing $\overline{M}_w$ by the mer weight:

$$\overline{DP}_w = \frac{\overline{M}_w}{M_0} = \frac{\sum W_i x_i}{\sum W_i} \tag{6}$$

$\overline{M}_w$ can be determined by measuring light scattering of dilute polymer solutions [3,4]. $\overline{M}_w$ is always higher than $\overline{M}_n$. Thus for a polymer sample containing 50 mol % of a species of molecular weight 10,000 and 50 mol % of species of molecular weight 20,000, Eqs. (1) and (5) give $\overline{M}_n = 0.5 (10,000 + 20,000) = 15,000$ and $\overline{M}_w = [(10,000)^2 + (20,000)^2]/[10,000 + 20,000] \approx 17,000$.

Viscosity-Average Molecular Weight ($\overline{M}_v$)

The viscosity-average molecular weight is defined by the equation

$$\overline{M}_v = \left[\sum w_i M_i^a\right]^{1/a} = \left[\sum N_i M_i^{1+a} / \sum N_i M_i\right]^{1/a} \tag{7}$$

For $a = 1$, $\overline{M}_v = \overline{M}_w$, and for $a = -1$, $\overline{M}_v = \overline{M}_n$. $\overline{M}_v$ falls between $\overline{M}_w$ and $\overline{M}_n$, and for many polymers it is 10 to 20% below $\overline{M}_w$. $\overline{M}_v$ is calculated from the intrinsic viscosity [η] by the empirical relation

$$[\eta] = K\overline{M}_v^{\alpha} \tag{8}$$

where K and α are constants. [η] is derived from viscosity measurements by extrapolation to "zero" concentration [5,6].

In correlating polymer properties (such as reactivity) which depend more on the number of molecules in the sample than on the sizes of the molecules, $\overline{M}_n$ is a more useful parameter than $\overline{M}_w$ or $\overline{M}_v$. Conversely, for correlating polymer properties (such as viscosity) which are more sensitive to the size of the polymer molecules, $\overline{M}_w$ or $\overline{M}_v$ is more useful.

Because it is easy to determine, *melt index* often is used instead of molecular weight in routine characterization of polymers. It is defined as the mass rate of polymer flow through a specified capillary under controlled conditions of temperature and pressure. The index can often be related empirically to some average molecular weight, depending on the specific polymer. A lower melt index indicates a higher molecular weight, and vice versa.

Polydispersity Index

The ratio of weight-average molecular weight to number-average molecular weight is called the *dispersion* or *polydispersity index (I)* [7]. It is a measure of the width of the molecular-weight distribution curve (Fig. 1.3) and is used as such for characterization purposes. Normally I is between 1.5 and 2.5, but it may range to 15 or greater. The higher the value of I is, the greater is the spread of the molecular-weight distribution of the polymer. For a monodisperse system (e.g., pure chemicals), $I = 1$.

There is usually a molecular size for which a given polymer property will be optimum for a particular application. So a polymer sample containing the greatest number of molecules of that size will have the optimum property. Since samples with the same average molecular weight may possess different molecular-weight distributions, information regarding molecular-weight distribution is necessary for a proper choice of polymer for optimum performance [8]. A variety of fractionation techniques, such as fractional precipitation, precipitation chromatography, and gel permeation chromatography (GPC), based on properties such as solubility and permeability, which vary with molecular weight, may be used for separating polymers of narrow size ranges.

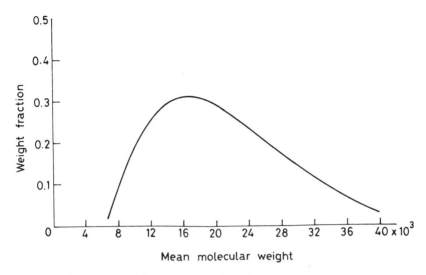

FIG. 1.3 Molecular-weight distribution of a polymer.

Example 1 A sample of poly(vinyl chloride) is composed according to the following fractional distribution (Fig. 1.3).

Wt. fraction	0.04	0.23	0.31	0.25	0.13	0.04
Mean mol. wt. $\times 10^{-3}$	7	11	16	23	31	39

(a) Compute $\overline{M}_n$, $\overline{M}_w$, $\overline{DP}_n$, and $\overline{DP}_w$.
(b) How many molecules per gram are there in the polymer?

Answer:
(a)

Wt. fraction (w_i)	Mean mol. wt. (M_i)	$w_i \times M_i$	w_i/M_i
0.04	7,000	280	0.57×10^{-5}
0.23	11,000	2,530	2.09×10^{-5}
0.31	16,000	4,960	1.94×10^{-5}
0.25	23,000	5,750	1.09×10^{-5}
0.13	31,000	4,030	0.42×10^{-5}
0.04	39,000	1,560	0.10×10^{-5}
Σ		19,110	6.21×10^{-5}

From Eq. (2)

$$\overline{M}_n = \frac{1}{6.21 \times 10^{-5}} = 16,100 \text{ g/mole}$$

From Eq. (4)

$$\overline{M}_w = 19,110 \text{ g/mole}$$

1 mer weight of vinyl chloride (C_2H_3Cl)

$$= (2)(12) + (3)(1) + 35.5 = 62.5 \text{ g/mer}$$

$$\overline{DP}_n = \frac{16,100 \text{ g/mole}}{62.5 \text{ g/mer}} = 258 \text{ mers/mole}$$

$$\overline{DP}_w = \frac{19,110 \text{ g/mole}}{62.5 \text{ g/mer}} = 306 \text{ mers/mole}$$

(b) Number of molecules per gram

$$= \sum \frac{w_i}{M_i} \text{ (Avogadro number)}$$

$$= (6.21 \times 10^{-5})(6.02 \times 10^{23})$$

$$= 3.74 \times 10^{19} \text{ molecules/g}$$

POLYMERIZATION PROCESSES

Polymerization processes can be broadly divided into *chain or addition polymerization* and *step polymerization* [9].

Chain or Addition Polymerization

In chain polymerization a simple, low-molecular-weight molecule possessing a double bond, referred to in this context as a monomer, is treated so that the double bond opens up and the resulting free valences join with those of other similar molecules to form a polymer chain. For example, vinyl chloride polymerizes to poly(vinyl chloride):

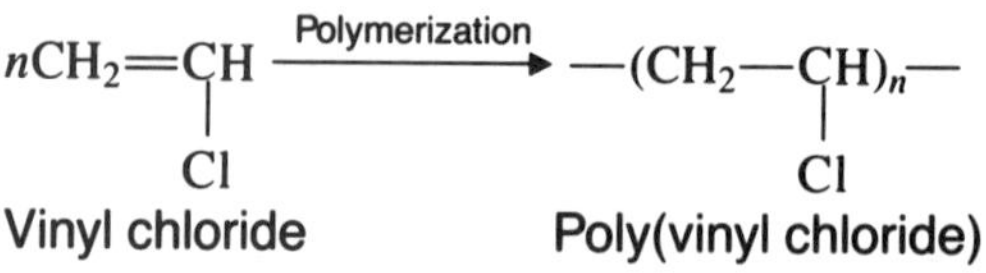

$$n\text{CH}_2\!\!=\!\!\underset{\substack{|\\ \text{Cl}}}{\text{CH}} \xrightarrow{\text{Polymerization}} -\!\!(\text{CH}_2\!-\!\!\underset{\substack{|\\ \text{Cl}}}{\text{CH}})_n\!\!- \tag{9}$$

Vinyl chloride   Poly(vinyl chloride)

It is evident that no side products are formed; consequently, the composition of the *mer* or repeat unit of the polymer ($-CH_2-CHCl-$) is identical to that of the monomer ($CH_2=CHCl$). The identical composition of the repeat unit of a polymer and its monomer(s) is, in most cases, an indication that the polymer is an *addition polymer* formed by chain polymerization process. The common addition polymers and the monomers from which they are produced are shown in Table 1.1.

Chain polymerization involves three processes: *chain initiation, chain propagation,* and *chain termination.* (A fourth process, *chain transfer,* may also be involved, but it may be regarded as a combination of chain termination and chain initiation.) Chain initiation occurs by an attack on the monomer molecule by a free radical, a cation, or an anion; accordingly, the chain polymerization processes are called *free-radical polymerization, cationic polymerization,* or *anionic polymerization.* A free radical is a reactive substance having an unpaired electron and is usually formed by the decomposition of a relatively unstable material called a *catalyst* or *initiator.* Benzoyl peroxide is a common free-radical initiator and can produce free radicals by thermal decomposition as

$$
\begin{array}{ccc}
O & O & O \\
\| & \| & \| \\
R-C-O-O-C-R & \xrightarrow{\text{Heat}} & R-C-O\cdot + R\cdot + CO_2
\end{array} \qquad (10)
$$

$(R = \text{Phenyl group})$
Initiator   Free radicals

Free radicals are, in general, very active because of the presence of unpaired electrons. A free-radical species can thus react to open the double bond of a vinyl monomer and add to one side of the broken bond, with the reactive center (unpaired electron) being transferred to the other side of the broken bond:

$$
\begin{array}{cccc}
O & H & O & H \\
\| & | & \| & | \\
R-C-O\cdot + CH_2=C & \longrightarrow & R-C-O-CH_2-C\cdot \\
& | & & | \\
& X & & X
\end{array} \qquad (11)
$$

Free   Vinyl
radical   monomer   New free radical

$(X = CH_3, C_6H_5, Cl, \text{etc.})$

The new species, which is also a free radical, is able to attack a second monomer molecule in a similar way, transferring its reactive center to the attacked molecule. The process is repeated, and the chain continues to grow as a large

TABLE 1.1 Typical Addition Polymers

Polymer name[a]	Monomer	Repeating unit
Polyethylene	$CH_2{=}CH_2$	$-CH_2-CH_2-$
Poly(vinyl chloride)	$CH_2{=}CHCl$	$-CH_2-\underset{\underset{Cl}{\vert}}{CH}-$
Polypropylene	$CH_2{=}CH-CH_3$	$-CH_2-\underset{\underset{CH_3}{\vert}}{CH}-$
Polystyrene	$CH_2{=}\underset{\underset{C_6H_5}{\vert}}{CH}$	$-CH_2-\underset{\underset{C_6H_5}{\vert}}{CH}-$
Polyacrylonitrile	$CH_2{=}CH-CN$	$-CH_2-\underset{\underset{CN}{\vert}}{CH}-$
Poly(vinyl acetate)	$CH_2{=}CH-O-\underset{\underset{O}{\Vert}}{C}-CH_3$	$-CH_2-\underset{\underset{O-\underset{\underset{O}{\Vert}}{C}-CH_3}{\vert}}{CH}-$
Poly(vinylidene chloride)	$CH_2{=}\underset{\underset{Cl}{\vert}}{\overset{\overset{Cl}{\vert}}{C}}$	$-CH_2-\underset{\underset{Cl}{\vert}}{\overset{\overset{Cl}{\vert}}{C}}-$
Polytetrafluoro-ethylene (Teflon)	$\underset{\underset{F}{\vert}\;\underset{F}{\vert}}{\overset{\overset{F}{\vert}\;\overset{F}{\vert}}{C{=}C}}$	$\underset{\underset{F}{\vert}\;\underset{F}{\vert}}{\overset{\overset{F}{\vert}\;\overset{F}{\vert}}{-C-C-}}$
Polychlorotrifluoro-ethylene(kel-F)	$\underset{\underset{F}{\vert}\;\underset{Cl}{\vert}}{\overset{\overset{F}{\vert}\;\overset{F}{\vert}}{C{=}C}}$	$\underset{\underset{F}{\vert}\;\underset{Cl}{\vert}}{\overset{\overset{F}{\vert}\;\overset{F}{\vert}}{-C-C-}}$
Polyisobutylene	$CH_2{=}\underset{\underset{CH_3}{\vert}}{\overset{\overset{CH_3}{\vert}}{C}}$	$-CH_2-\underset{\underset{CH_3}{\vert}}{\overset{\overset{CH_3}{\vert}}{C}}-$

TABLE 1.1 (continued)

Polymer name[a]	Monomer	Repeating unit
Poly(methyl acrylate)	$CH_2{=}CH{-}\overset{\|}{\underset{O}{C}}{-}OCH_3$	$-CH_2{-}\underset{\underset{O=C-OCH_3}{\|}}{CH}-$
Poly(methyl methacrylate)	$CH_2{=}\overset{\overset{CH_3}{\|}}{C}{-}\overset{\|}{\underset{O}{C}}{-}OCH_3$	$-CH_2{-}\overset{\overset{CH_3}{\|}}{\underset{\underset{O=C-OCH_3}{\|}}{C}}-$
Polyisoprene (natural rubber)	$CH_2{=}\overset{\underset{CH_3}{\diagdown}}{C}{-}CH{=}CH_2$	$-CH_2 \qquad CH_2-$ $\underset{CH_3 \diagup \qquad \diagdown H}{\diagdown C{=}C \diagup}$
Polybutadiene	$CH_2{=}CH{-}CH{=}CH_2$	$-CH_2{-}CH{=}CH{-}CH_2-$

[a]Addition polymers are commonly named by adding the name of the monomer on to the prefix 'poly' to form a single word. However when the monomer has a multiworded name, the name of the monomer after the prefix 'poly' is enclosed in parentheses.

number of monomer molecules are successively added to propagate the reactive center:

$$R{-}\overset{\|}{\underset{O}{C}}{-}O{-}CH_2{-}\overset{\overset{H}{\|}}{\underset{X}{C}}\cdot \xrightarrow[\substack{\text{Successive} \\ \text{addition} \\ \text{of } CH_2{=}CHX}]{} R{-}\overset{\|}{\underset{O}{C}}{-}O{+}CH_2{-}\overset{\overset{H}{\|}}{\underset{X}{C}})\cdot_m \qquad (12)$$

This process of *propagation* continues until another process intervenes and destroys the reactive center, resulting in the *termination* of the polymer growth. There may be several termination reactions depending on the type of the reactive center and the reaction conditions. For example, two growing radicals may combine to annihilate each other's growth activity and form an inactive polymer molecule; this is called termination by *combination or coupling*:

$$R-\underset{\underset{O}{\parallel}}{C}-O-(CH_2-\underset{\underset{X}{|}}{\overset{\overset{H}{|}}{C}})_m \cdot + \cdot (\underset{\underset{X}{|}}{\overset{\overset{H}{|}}{C}}-CH_2)_{\overline{n}}-O-\underset{\underset{O}{\parallel}}{C}-R$$

$$\downarrow$$

$$R-\underset{\underset{O}{\parallel}}{C}-O-(CH_2-\underset{\underset{X}{|}}{\overset{\overset{H}{|}}{C}})_{\overline{m}}(\underset{\underset{X}{|}}{\overset{\overset{H}{|}}{C}}-CH_2)_{\overline{n}}-O-\underset{\underset{O}{\parallel}}{C}-R \qquad (13)$$

Inactive polymer molecule

A second termination mechanism is *disproportionation*, shown by the following equation:

$$R-\underset{\underset{O}{\parallel}}{C}-O-(CH_2-\underset{\underset{X}{|}}{\overset{\overset{H}{|}}{C}})_m \cdot + \cdot (\underset{\underset{X}{|}}{\overset{\overset{H}{|}}{C}}-CH_2)_{\overline{n}}O-\underset{\underset{O}{\parallel}}{C}-R$$

$$(14)$$

$$R-\underset{\underset{O}{\parallel}}{C}-O-(CH_2-\underset{\underset{X}{|}}{\overset{\overset{H}{|}}{C}})_{\overline{m-1}} CH_2-\underset{\underset{X}{|}}{\overset{\overset{H}{|}}{C}}-H + \underset{\underset{X}{|}}{\overset{\overset{H}{|}}{C}}=CH(\underset{\underset{X}{|}}{\overset{\overset{H}{|}}{C}}-CH_2)_{\overline{n-1}}O-\underset{\underset{O}{\parallel}}{C}-R$$

In chain polymerizations initiated by free radicals, as in the previous example, the reactive center, located at the growing end of a molecule, is a free radical. As mentioned previously, chain polymerizations may also be initiated by ionic systems. In such cases the reactive center is ionic—a carbonium ion (in cationic initiation) or a carbanion (in anionic initiation). Regardless of the chain initiation mechanism—free radical, cationic, or anionic—once a reactive center is produced it adds many monomer molecules in a chain reaction and grows quite large extremely rapidly, usually within a few seconds or less. However, the relative slowness of the initiation stage causes the overall rate of conversion of monomer to be slow: the conversion of all monomers to polymers in most polymerizations requires at least 30 min, sometimes hours. Evidently, at any time during a chain polymerization process the reaction mixture will consist almost entirely of monomers and high polymers, with very few products at intermediate stages of growth (i.e., actively growing chains).

Step Polymerization

Step polymerization occurs by stepwise reaction between functional groups of reactants. The reaction leads successively from monomer to dimer, trimer, tetramer, pentamer, and so on, until finally a polymer molecule with large DP is formed. Note, however, that reactions occur at random between the intermediates (e.g., dimers, trimers, etc.) and the monomer as well as among the intermediates themselves. In other words, reactions of both types

$$n\text{-mer} + \text{monomer} \longrightarrow (n + 1)\text{-mer}$$

and

$$n\text{-mer} + m\text{-mer} \longrightarrow (n + m)\text{-mer}$$

occur equally. Thus, at any stage the product consists of molecules of varying sizes, giving a range of molecular weights. The average molecular weight builds up slowly in the step polymerization process, and a high-molecular-weight product is formed only after a sufficiently long reaction time when the conversion is more than 98%. In contrast, polymerization by chain mechanism proceeds very fast, a full-sized polymer molecule being formed almost instantaneously after a chain is initiated; the polymer size is thus independent of reaction time.

Most (but not all) of the step polymerization processes involve polycondensation (repeated condensation) reactions. Consequently, the terms "step polymerization" and "condensation polymerization" are often used synonymously. In a condensation reaction between two molecules, each molecule loses one atom or a group of atoms at the reacting end, which leads to the formation of a covalent bond between the two, while the eliminated atoms bond with each other to form small molecules such as water—hence the term *condensation reactions*. Consider, for example, the synthesis of a polyamide, i.e., a polymer with amide (—CONH—) as the characteristic linkage. If we start with, say, hexamethylenediamine and adipic acid as reactants, the first step in the formation of the polymer (nylon) is the reaction

$$H_2N\text{—}(CH_2)_6\text{—}NH_2 + HO\text{—}\overset{\displaystyle O}{\underset{\displaystyle \|}{C}}\text{—}(CH_2)_4\text{—}\overset{\displaystyle O}{\underset{\displaystyle \|}{C}}\text{—}OH$$

Hexamethylenediamine   Adipic acid

$$H_2N\text{---}(CH_2)_6\text{---}NHC\text{---}(CH_2)_4\text{---}C\text{---}OH + H_2O \qquad (15)$$
$$\underset{O}{\overset{\|}{}} \qquad \underset{O}{\overset{\|}{}}$$
$$\text{Monoamide}$$

The reaction continues step by step to give the polyamide nylon-6,6.* The overall reaction may thus be represented as

$$n H_2N\text{---}(CH_2)_6\text{---}NH_2 + n HO\text{---}C\text{---}(CH_2)_4\text{---}C\text{---}OH$$
$$\underset{O}{\overset{\|}{}} \qquad \underset{O}{\overset{\|}{}}$$

$$H\text{---}[NH\text{---}(CH_2)_6\text{---}NH\text{---}C\text{---}(CH_2)_4\text{---}C]_n\text{---}OH + (2n - 1)H_2O$$
$$\underset{O}{\overset{\|}{}} \qquad \underset{O}{\overset{\|}{}} \qquad (16)$$
$$\text{Poly(hexamethyleneadipamide)}$$

We see that the composition of the repeating unit (enclosed in brackets) equals that of two monomer molecules minus two molecules of water. Thus a condensation polymer may be defined as one whose synthesis involves elimination of small molecules or whose repeating unit lacks certain atoms present in the monomer(s). Some condensation polymers along with their characteristic linkages, repeating units, and condensation reactions by which they can be synthesized are shown in Table 1.2. With the development of polymer science and the synthesis of newer polymers, the previous definition of condensation polymer is inadequate. For example, in polyurethanes (Table 1.2), which are classified as condensation polymers, the repeating unit has the same net composition as the two monomers (i.e., a diol and a diisocyanate), which react without eliminating any small molecule. To overcome such problems, chemists have introduced a definition which describes condensation polymers as consisting of structural units joined by internal functional groups such as ester (—C—O—), amide
$$\underset{O}{\overset{\|}{}}$$
(—C—NH—), urethane —O—C—NH—), sulfide (—S—), ether (—O—),
$$\underset{O}{\overset{\|}{}} \qquad\qquad \underset{O}{\overset{\|}{}}$$

Nylon is the trade name for the polyamides from unsubstituted, nonbranched aliphatic monomers. Two numbers are added on to the word *nylon*, the first number indicating the number of carbon atoms in the diamine portion of the polyamide and the second number the number of carbon atoms in the diacyl portion.

TABLE 1.2 Typical Condensation Polymers

Polymer type	Characteristic linkage	Polymerization reaction*
Polyamide	$-\underset{\parallel O}{C}-NH-$	$nH_2N-R-NH_2 + nHO\underset{\parallel O}{C}-R'-\underset{\parallel O}{C}OH \longrightarrow H(NH-R-NH\underset{\parallel O}{C}-R'-\underset{\parallel O}{C})_n OH + (2n-1)H_2O$
		$nH_2N-R-NH_2 + nCl-\underset{\parallel O}{C}-R'-\underset{\parallel O}{C}-Cl \longrightarrow H(NH-R-NH\underset{\parallel O}{C}-R'-\underset{\parallel O}{C})_n Cl + (2n-1)HCl$
		$nH_2N-R-\underset{\parallel O}{C}OH \longrightarrow H(NH-R-\underset{\parallel O}{C})_n OH + (n-1)H_2O$
Polyester	$-\underset{\parallel O}{C}-O-$	$nHO-R-OH + nHO\underset{\parallel O}{C}-R'-\underset{\parallel O}{C}OH \longrightarrow HO(R-O\underset{\parallel O}{C}-R'-\underset{\parallel O}{C}O)_n H + (2n-1)H_2O$
		$nHO-R-OH + nR''O\underset{\parallel O}{C}-R'-\underset{\parallel O}{C}OR'' \longrightarrow HO(R-O\underset{\parallel O}{C}-R'-\underset{\parallel O}{C}O)_n R'' + (2n-1)R''OH$
		$nHO-R-\underset{\parallel O}{C}OH \longrightarrow HO(R-\underset{\parallel O}{C}O)_n H + (n-1)H_2O$
Polyurethane	$-O-\underset{\parallel O}{C}-NH-$	$nHO-R-OH + nOCN-R'-NCO \longrightarrow H(O-R-O\underset{\parallel O}{C}NH-R-NH\underset{\parallel O}{C})_{n-1}O-R-O\underset{\parallel O}{C}NH-R-NCO$

TABLE 1.2 (continued)

Polycarbonate	$nHO-\langle\bigcirc\rangle-R-\langle\bigcirc\rangle-OH + nCl-\underset{\underset{O}{\|}}{C}-Cl \longrightarrow H\left[O-\langle\bigcirc\rangle-R-\langle\bigcirc\rangle-O-\underset{\underset{O}{\|}}{C}\right]_n Cl + (2n-1)HCl$
Polysulphide	$nCl-R-Cl + nNa_2S_x \longrightarrow (R-S_x)_n + 2n\ NaCl$
Polysiloxane	$nHO-\underset{\underset{R}{\|}}{\overset{\overset{R}{\|}}{Si}}-OH \longrightarrow HO(\underset{\underset{R}{\|}}{\overset{\overset{R}{\|}}{Si}}-O)_n H + (n-1)H_2O$
Phenol-formaldehyde	$n\underset{OH}{\langle\bigcirc\rangle}-CH_2- + nCH_2=O \longrightarrow \left[\underset{OH}{\langle\bigcirc\rangle}CH_2-\underset{OH}{\langle\bigcirc\rangle}CH_2\right]_{n-1}-OH + (n-1)H_2O$
Urea-formaldehyde	$-NH-CH_2-\quad nH_2N-\underset{\underset{O}{\|}}{C}-NH_2 + n\underset{\underset{O}{\|}}{C}H_2 \longrightarrow (NH-\underset{\underset{O}{\|}}{C}-NH-CH_2)_n + nH_2O$
Melamine-formaldehyde	$-NH-CH_2-$

Melamine-formaldehyde:

$$n\ H_2N-C\overset{N=C-NH_2}{\underset{N-C}{\underset{\|}{\overset{\|}{\langle}}}}_{NH_2} + n\ \underset{\underset{O}{\|}}{C}H_2 \longrightarrow \left[HN-C\overset{N=C-NH-CH_2}{\underset{N-C}{\underset{\|}{\overset{\|}{\langle}}}}_{NH_2}\right]_n + nH_2O$$

*R, R′ and R″ represent aliphatic or aromatic grouping. The repeating unit of the polymer chain is enclosed in parentheses.

*R,R′ and R″ represent aliphatic or aromatic grouping. The repeating unit of the polymer chain is enclosed in parentheses.

carbonate (—O—C—O—), and sulfone (—S—) linkages. A polymer satisfy-

$$\text{carbonate } (-\text{O}-\overset{\text{O}}{\underset{\text{O}}{\overset{\|}{\text{C}}}}-\text{O}-), \text{ and sulfone } (-\overset{\text{O}}{\underset{\text{O}}{\overset{\|}{\text{S}}}}-) \text{ linkages.}$$

ing either or both of these definitions is classified as a condensation polymer. Phenol-formaldehyde, for example, satisfies the first definition but not the second.

Copolymerization

All the polymers we have considered so far contain only one type of repeating unit or mer in the chain. Polymers can also be synthesized by the aforesaid processes with more than one type of mer in the chain. Such polymers are called *copolymers*. They are produced by polymerizing a mixture of monomers (*copolymerization*) [10] or by special methods. Copolymers can be of different types, depending on the monomers used and the specific method of synthesis. The copolymer with a relatively random distribution of the different mers in its structure is referred to as a *random copolymer*. Representing, say, two different mers by A and B, a random copolymer can be depicted as

ABBABBBAABBAABAAABBA

Three other copolymer structures [11] are known: alternating, block, and graft copolymer structures (Fig. 1.4). In the *alternating copolymer* the two mers alternate in a regular fashion along the polymer chain:

ABABABABABABABABABAB

A *block copolymer* is a linear copolymer with one or more long uninterrupted sequences of each mer in the chain:

AAAAAAAAAABBBBBBBBBB

A *graft copolymer*, on the other hand, is a branched copolymer with a backbone of one type of mer to which are attached one or more side chains of another mer.

AAAAAAAAAAAAAAAA
   B
   B
   B
   B
   B

Copolymerization, which may be compared to alloying in metallurgy, is very useful for synthesizing polymers with the required combination of properties. For example, polystyrene is brittle, and polybutadiene is flexible; therefore

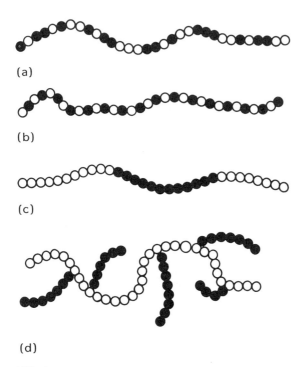

(a)

(b)

(c)

(d)

FIG. 1.4 Copolymer arrangements: (a) Two different types of mers (denoted by open and filled circles) are randomly placed. (b) The mers are alternately arranged. (c) A block copolymer. (d) A graft copolymer.

copolymers of styrene and butadiene should be more flexible than polystyrene but tougher than polybutadiene. The general-purpose rubber GRS (or SBR), the first practical synthetic rubber, is a copolymer of styrene and butadiene.

CONFIGURATIONS OF POLYMER MOLECULES

The long threadlike shape of polymer molecules induces generally a random arrangement leading to inter- and intramolecular entanglement, somewhat like a bowl of mee. A typical molecule of polyethylene, for example, might be represented by a cylindrical chain with a length of 50,000 Å and a diameter of less than 5 Å. This is similar to a rope that is 45 m long and 4.5 mm in diameter. A molecule such as this can easily get knotted and entangled with surrounding molecules. In the structure some parts of the molecular chains can be more ordered than others. The ordered regions are termed *micelles* or *crystallites*. These regions are embedded in the unordered or amorphous matrix (Fig. 1.5). Polymer

FIG. 1.5 Fringed-micelle picture of polymer.

molecules may extend through several micelles where polymer chain segments are precisely aligned together to undergo crystallization, whereas the chain segments between the micelles are unordered, forming the amorphous matrix. This theory is known as the *fringed-micelle theory*: the greater the degree of crystallinity of the polymer, the larger the volume fraction of micelles or crystallites in the polymer matrix.

It is quite logical that if we want a molecule to go into some kind of ordered, repetitive pattern, then its structure must also have a regularly repeating pattern. The degree of crystallinity of the polymer will thus increase with the linearity and steric regularity of the molecules and also with interchain attractive forces.

Linear polymers are found in nature, or they may be formed by polymerization of simple monomers. When monosubstituted ethylene monomers ($CH_2=CHR$) polymerize, the addition reaction may be head-to-tail, head-to-head/tail-to-tail, or a random mixture of the two:

$$-CH_2-\underset{\underset{R}{|}}{C}H-CH_2-\underset{\underset{R}{|}}{C}H-CH_2-\underset{\underset{R}{|}}{C}H-CH_2-\underset{\underset{R}{|}}{C}H-$$

Head-to-tail

$$-\underset{\underset{R}{|}}{C}H-\underset{\underset{R}{|}}{C}H-CH_2-CH_2-\underset{\underset{R}{|}}{C}H-\underset{\underset{R}{|}}{C}H-CH_2-CH_2-$$

Head-to-head/tail-to-tail

$$—CH_2—CH—CH—CH_2—CH—CH_2—CH—CH_2—CH_2—$$
$$\qquad\quad | \qquad | \qquad\qquad | \qquad\qquad |$$
$$\qquad\quad R \qquad R \qquad\quad R \qquad\qquad R$$

Random

The head-to-tail configuration is preferred almost to the exclusion of the other two. An important reason for this is steric hindrance, which favors head-to-tail reaction, especially if R is bulky.

Another aspect of stereoregularity is tacticity. Figure 1.6 shows a polymer chain in which all of the chain carbons are in the same plane. Three configurations can be obtained: A polymer molecule is *isotactic* if all the substituted groups lie on the same side of the main chain. In a *syndiotactic* polymer mole-

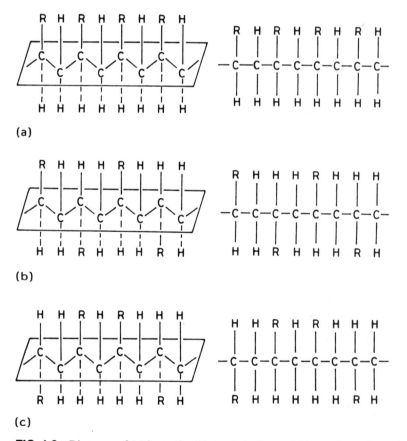

FIG. 1.6 Diagrams of (a) isotactic, (b) syndiotactic, and (c) atactic configuration in a vinyl polymer. The corresponding Fischer projections are shown on the right.

TABLE 1.3 Properties of Polypropylene Stereoisomers

Property	Stereoisomers		
	Isotactic	Syndiotactic	Atactic
Appearance	Hard solid	Hard solid	Soft rubbery
Melting temperature (°C)	175	131	<100
Density (g/cc)	0.90–0.92	0.89–0.91	0.86–0.89
Tensile strength [psi (N/m^2)]	5000 (3.4 × 10^7)	—	—
Solubility	Insoluble in most organic solvents	Soluble in ether and aliphatic hydrocarbon	Soluble in common organic solvents
Crystallinity (%)	<70	—	—
Glass transition temperature (°C)	0 to −35	—	−11 to −35

cule the substituted groups regularly alternate from one side to the other. The molecule is *atactic* if the positioning of substituted groups is random.

The relative arrangement of groups and atoms in successive monomer units in a polymer chain not only affects the crystallinity but also induces completely different properties in polymers. One example of this effect is found in polypropylene stereoisomers. (The three stereoisomers of polypropylene can be obtained by replacing R by CH_3 in Figure 1.6.) The structural difference results in profound variations in the properties of polypropylene isomers. As is evident from Table 1.3, the three polypropylene isomers appear to be three altogether different materials.

CONFORMATIONS OF A POLYMER MOLECULE

Just as a rope can be stretched, folded back on itself, curled up into a ball, entangled, knotted, and so forth, a polymer molecule can take on many conformations. Consider a molecule as a chain of N links, each link of length l_0 and attached to the preceding link by a rotating joint.

For a free rotating polymer chain the average conformation is characterized by the mean square distance $<r^2>$ the ends of the chain and is given by

$$<r^2> = Nl_0^2 \tag{17}$$

If we assume that adjacent links in the chain form fixed angles θ with each other but rotate freely about that angle, then

$$<r^2> = \frac{1 - \cos \theta}{1 + \cos \theta} Nl_0^2 = 2Nl_0^2 \tag{18}$$

since bonds in a tetrahedral carbon unit are at 109.5° to each other.

Thus, for a polyethylene molecule comprising, say, 40,000 freely rotating —CH_2— units, the end-to-end distance would only be $\sqrt{2N}l_0$ or approximately 375 Å, whereas the end-to-end distance, if fully extended, would be 50,000 Å.

Polymer Crystallinity

Metals invariably crystallize when they solidify under normal conditions. Polymers behave differently. Some solidify without crystallizing, whereas many others solidify without complete crystallization. Atoms in metals are almost spherical in shape, and we can view a metal crystal as a collection of individual balls arranged in a repeating long-range pattern. No polymer molecule has equi-axed shape with dimensions approximately equal in the three space directions.

Linear polymer molecules are characterized by their extreme elongation in one dimension. In polyethylene, for example, the molecules have strong covalent bonds along the molecular chain. The bonds between chains are much weaker but still present, so the chains tend to line up as shown in Figure 1.7.

A single molecule of polyethylene extends through many unit cells. Molecules are long, and that makes obtaining perfect crystallization difficult. Since molecules have different lengths, they do not end at the same position. Also the tendency to twist and turn makes perfect alignment difficult to achieve. The thickness of sheet-like single-crystal platelets of polyethylene is only about 100 Å, clearly not enough room for a polymer chain in that direction. On the other hand, it has been shown through electron diffraction data that the polymer chains are not laid out in the direction of the longer axes of these plates. The sheetlike crystals or *lamellae* consist of chains folded back and forth throughout the entire thickness of the platelets. Consequently, the surface of the crystal consists of hair-pin turns, as illustrated in Figure 1.8.

The folded-chain picture is well suited for highly crystalline polymers. Most polymers are, however, *semicrystalline*, having the characteristics of both crystalline solids and highly viscous liquids. This structure is revealed by x-ray and

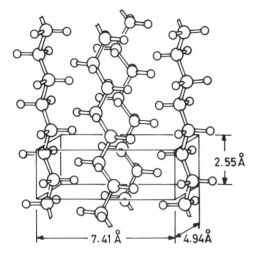

FIG. 1.7 Model of the packing of polymer chains in the crystal structure of polyethylene in which $a = 7.41$ Å, $b = 4.94$ Å, and successive pendant atoms are 2.55 Å apart along the chain axis.

electron diffraction patterns of polymers, which often show the sharp features characteristic of crystalline solids as well as the diffuse features typical of liquids.

For semicrystalline polymers it is often more advantageous to adopt the fringed-micelle picture (Fig. 1.5) or a model that uses the features of both

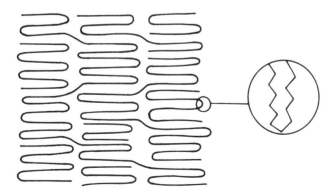

FIG. 1.8 Folded-chain lamellae of polyethylene.

the fringed-micelle and folded-chain concepts. Thus, as in the fringed-micelle concept, polymers may be pictured as two-phase systems composed of crystallites embedded in an amorphous matrix, whereas the structure of the crystallites may be that of the folded-chain lamella. The crystallites promote rigidity, hardness, and heat resistance, and the amorphous regions give rise to flexibility. A certain proportion of the amorphous region is indeed necessary to provide needed flexibility for absorbing impact shocks; a completely crystalline polymer is too brittle.

Determinants of Polymer Crystallinity

The extent to which polymer molecules will crystallize depends on their structures and on the magnitudes of the secondary bond forces among the polymer chains: the greater the structural regularity and symmetry of the polymer molecule and the stronger the secondary forces, the greater the tendency toward crystallization. We give a few examples:

1.   Linear polyethylene has essentially the best structure for chain packing. Its molecular structure is very simple and perfectly regular, and the small methylene groups fit easily into a crystal lattice. Linear polyethylene (high density) therefore crystallizes easily and to a high degree (over 90%) even though its secondary forces are small. Branching impairs the regularity of the structure and makes chain packing difficult. Branched polyethylene (low density) is thus only partially (50–60%) crystalline. Most of the differences in properties between low-density and high-density polyethylenes can be attributed to the higher crystallinity of the latter. Thus, linear polyethylenes have higher density than the branched material (density range of 0.95–0.97 g/cm^3 vs. 0.91–0.94 g/cm^3), higher melting point (typically 135° vs. 115°C), greater stiffness (modulus of 100,000 psi vs. 20,000 psi), greater tensile strength, greater hardness, and less permeability to gases and vapors.

2.   Substituents hanging off polymer chains lead to difficulties in packing and generally decrease the tendency toward crystallization. Moreover, crystallization does not take place easily when polymer molecules have a low degree of symmetry. Thus, polymers such as polystyrene, poly(methyl methacrylate), poly(vinyl acetate), etc., all of which have bulky side groups oriented at random with respect to the main carbon chain (in atactic polymers), show very poor crystallization tendencies and tend to have amorphous structures. However, crystallinity would result if the side groups could be arranged in a regular orientation. Indeed, this can be done by controlled polymerization with properly chosen catalysts.

3.   Copolymerization reduces the structural symmetry of a polymer. Thus it is a very effective method of decreasing the crystallization tendency of a polymer.

4.   Chain flexibility also affects the crystallizability of a polymer. Excessive flexibility in a polymer chain, as in natural rubber and polysiloxanes, gives rise to difficulty in chain packing, with the result that such polymers remain almost completely in the amorphous state. In the other extreme, excessive rigidity in polymers due to extensive cross-linking, as in thermosetting resins like phenol-formaldehyde and urea-formaldehyde, also results in an inability to crystallize.

5.   The presence of polar groups—such as amide, carboxyl, hydroxyl, chlorine, fluorine, and nitrile—along the polymer chains greatly increases the intermolecular or secondary attraction forces, which is favorable for crystallization. However, high secondary forces alone may not give rise to high crystallinity unless the chain segments are aligned. Mechanical stretching of the polymer makes this alignment easier. For example, nylon-6,6 (a polyamide) has less than the expected degree of crystallinity in the unstretched condition and is used as a plastic. Highly crystalline strong fibers are produced by stretching (cold-drawing) the polyamide polymer 400 to 500%. Mechanical stretching also makes it possible to develop a degree of order and crystallinity in several other thermoplastic resins that do not ordinarily crystallize. An unusual example of alignment and crystallization on stretching is rubber.

6.   The degree of crystallinity of polymeric materials is reduced by adding plasticizers. Crystallization in many synthetic resins is not always desirable because it makes shaping more difficult and reduces transparency by closely packing the polymer molecules. Therefore, plasticizers which cause partial neutralization of the intermolecular forces of attraction by coming between polymer molecules, thus enhancing flexibility and plasticity, are often added to the polymeric mass before shaping. The oldest example is *celluloid*, made by plasticizing nitrocellulose (ordinarily a crystalline material) with camphor. *Cellophane* (regenerated cellulose film produced by a *viscose* process) is plasticized with glycerine to prevent crystallization and loss of transparency. Polyvinyl chloride (PVC) is made flexible by adding plasticizers, such as dioctyl phthalate, for use as wire coating, upholstery, film, and tubing. The unplasticized rigid PVC is used for the production of pipe, sheet, and molded parts. The disadvantage of plasticizers is that they reduce the tensile strength and chemical resistance of the material.

STRUCTURAL SHAPE OF POLYMER MOLECULES

Polymers can be classified, based on the structural shape of polymer molecules, as linear, branched, or crosslinked. Schematic representations are given in Figure 1.9. *Linear* polymers have repeating units linked together in a continuous length (Fig. 1.9a). When branches protrude from the main polymer chain at *irregular* intervals, the polymer is termed a *branched* polymer. Branches may be

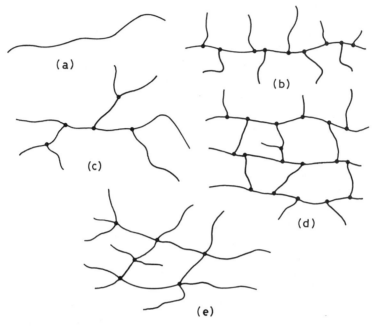

FIG. 1.9 Schematic representation of linear (a), branched (b and c), cross-linked (d and e) polymers. The branch points and junction points are indicated by heavy dots.

long or short, forming a comblike structure (Fig. 1.9b), or divergent (Fig. 1.9c), forming a dendritelike structure. [Regularly repeating side groups which are a part of the monomer structure are not considered as branches. Thus polypropylene is a linear polymer, as are polystyrene and poly(methyl methacrylate).]

Both linear and branched polymers are *thermoplastic*; that is, they can be softened and hardened reversibly by changing the temperature. Fabricating processes like injection molding, extrusion molding, casting, and blowing take advantage of this feature to shape thermoplastic resins. The rigidity of thermoplastic resins at low temperatures is attributed to the existence of secondary bond forces between the polymer chains. These bonds are destroyed at higher temperatures, thereby causing fluidity of the resin.

Polymers used as textile fibers are linear. However, they must satisfy two additional requirements: (1) high molecular weight and (2) a permanent orientation of the molecules parallel to the fiber axis. The molecules must have a high degree of order and/or strong secondary forces to permit orientation and crystallization. The chain orientation necessary to develop sufficient strength by crystallization is achieved by a process known as *cold drawing*, in which the initially formed filaments (unoriented or only slightly oriented) are drawn at a temperature above the glass transition temperature (discussed later), which is the temperature at which sufficient energy is available to the molecular segments to cause them to begin to rotate.

Elastomeric materials, like thermoplastic resins and fibers, are essentially linear polymers. But certain distinctive features in their molecular structure give rise to rubberlike elasticity. Elastomeric polymers have very long chain molecules occurring in randomly coiled arrangements in the unstressed condition. A large deformation is thus possible merely by reorienting the coiled molecules. When elongated, the molecular coils partially open up and become aligned more or less parallel to the direction of elongation. The aligned configuration represents a less probable state or a state of lower entropy than a random arrangement. The aligned polymer chains therefore have a tendency to return to their original randomly coiled state. The large deformability of elastomeric materials is due to the presence of a certain internal mobility that allows rearranging the chain orientation, the absence of which in linear chain plastic materials (at normal temperatures) constitutes the essential difference between the two groups.

Although the aforesaid requirements are necessary conditions for ensuring a large extent of deformability, the remarkable characteristic of the rubbery state—namely, nearly complete recovery—cannot be obtained without a permanent network structure, since permanent deformation rather than elastic recovery will occur. A small amount of cross-linkage is necessary to provide this essential network structure [12,13]. Natural rubber (polyisoprene), for example, simply flows like an extremely viscous liquid at room temperature if it is not cross-linked. Cross-linking is achieved by *vulcanization*, the process of introducing cross-links between long-chain molecules. In commercial vulcanization, sulfur cross-links are introduced into a rubber by heating raw rubber with sulfur (1–2% by weight) and accelerating agents. Sulfur reacts with the double-bonded carbon atoms to produce a network structure, as shown schematically in Figure 1.10. The amount of cross-linkage must be as small as possible to retain the structure; excessive cross-linkages will make the internal structure too stiff to permit even the required rearrangement of chain orientation during both deformation and recovery—in other words, it will destroy the rubbery state. An example of this is best furnished by ebonite, which is a rigid plastic made by vulcanizing natural rubber with large quantities of sulfur.

Example 2 (a) How much sulfur is required to fully cross-link natural rubber? (b) What is the sulfur content of vulcanized natural rubber that is 50% cross-linked? (Assume that each cross-link contains one sulfur atom).

Answer: Mer weight of isoprene (Fig. 1.10a):

$$C_5H_8 = (5)(12) + (8)(1) = 68 \text{ g/mer}$$

(a)

(b)

(c)

(c)

FIG. 1.10 Vulcanization of natural rubber with sulfur. (a) Linear polyisoprene (natural rubber). (b) Vulcanization produces cross-linking of chain molecules by means of sulfur atoms. (c) The effect of cross-linking is to introduce points of linkage or anchor points between chain molecules, restricting their slippage. The cross-links in elastomers are typically a few hundred carbon atoms apart.

From Fig. 1.10b, one sulfur atom, on the average, is required for cross-linking per mer of isoprene. Therefore,

(a) Amount of sulfur $= \dfrac{32}{68} \, 100$

$= 47 \text{ g}/100 \text{ g of raw rubber}$

(b) Sulfur content $= \dfrac{100 \, (0.5) \, (32)}{(0.5) \, (32) + 68}$

$= 19\%.$

Example 3 A rubber contains 60% butadiene, 30% isoprene, 5% sulfur, and 5% carbon black. What fraction of possible cross-links are joined by vulcanization? (Assume that all the sulfur is used in cross-linking.)

Answer:

1 mer weight of butadiene (C_4H_6)

$$= (4)\,(12) + (6)\,(1) = 54 \text{ g/mer}$$

1 mer weight of isoprene (C_5H_8)

$$= (5)\,(12) + (8)\,(1) = 68 \text{ g/mer}$$

1 atomic weight of sulfur $= 32$

Since, on the average, one sulfur atom per mer is required for cross-linking, we get

$$\text{fraction of cross-links} = \frac{5/32}{60/54 + 30/68}$$

$$= 0.101 \text{ or } 10.1\%$$

THERMAL TRANSITIONS IN POLYMERS

Amorphous polymers are in the glassy state and are usually transparent, stiff, hard, and somewhat brittle. They become more flexible and less brittle as a result of the onset of chain mobility. When heated to a characteristic temperature called the glass transition temperature (T_g), an abrupt change in volume and density occurs at T_g. The T_g temperatures depend to a certain extent on the rate of cooling; however, the T_g values are characteristic for each polymer.

When crystalline polymers are heated at relatively high temperatures, melting occurs, and these polymers become viscoelastic liquids at the melting points (T_m). The crystalline melting temperature and the glass transition temperature are shown schematically in Fig. 1.11. For a highly crystalline polymer (line ABCD) only crystalline melting would be observed. On the contrary, for a completely amorphous polymer (line ABEF) only glass transition would be observed. Most polymers are, however, semicrystalline and fall between these two extremes. They are therefore characterized by both types of transition temperatures: T_m is the melting temperature of the crystalline domains of a polymer sample, and T_g is the temperature at which the amorphous domains of the sample undergo glass transition.

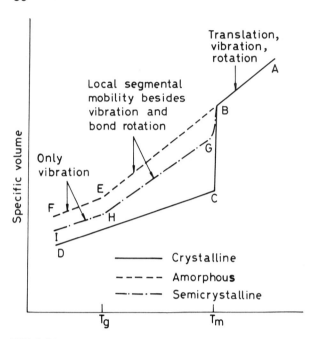

FIG 1.11 Determination of glass transition and crystalline melting temperatures by changes in specific volume (schematic). A: liquid region; B: liquid with some elastic response; BE: rubbery region; EF: glassy region; GH: crystallites in a rubber matrix; HI: crystallites in a glassy matrix; CD: crystalline solid.

The variation in specific volume with temperature for a semicrystalline polymer is represented by the curve ABGHI in Figure 1.11. When the melt of a polymer is cooled, the translational, rotational, and vibrational energies of the polymer molecules decrease, and when the total energies of the molecules have fallen to the point where the translational and rotational energies are essentially zero, crystallization is possible. Crystallization, however, occurs if certain symmetry requirements are met and the molecules are able to pack into an ordered arrangement. The temperature at which this occurs is T_m. For polymers T_m has a spread, generally of the order of 2 to 10°C. For semicrystalline polymers, crystallization occurs only over portions of polymer chains resulting in crystalline regions (crystallites) separated by amorphous regions. The energies of the molecules continue to decrease as the temperature decreases further below T_m until the temperature T_g is finally reached, at which the wriggling motions of the polymer chain segments in the amorphous regions stop due to cessation of bond rotations. Below T_g the amorphous regions are therefore no longer flexible and become hard and brittle like a glassy solid.

The T_g and T_m values for some polymers are shown in Table 1.4. In general, both T_g and T_m are affected in the same manner by considerations of polymer structure [14]. Thus, both T_g and T_m increase with higher molecular symmetry, structural rigidity, and secondary forces of polymer chains.

The T_g and T_m values of a polymer determine the temperature range in which it can be employed. Amorphous elastomeric polymers, for example, must be used at temperatures (region EB in Fig. 1.11) well above T_g to permit the high, local, segmental mobility required in such materials. Thus styrene-butadiene (25/75) copolymer ($T_g = -57°C$), polyisoprene ($T_g = -73°C$), and polyisobutylene ($T_g = -73°C$) can be used as rubbers at ambient temperatures. Amorphous structural polymers, such as polystyrene and poly(methyl methacrylate), depend on their glasslike rigidity below T_g for their utility; they should therefore have high T_g values so that under ambient conditions they are well below T_g.

TABLE 1.4 Glass Transition Temperatures (T_g) and Crystalline Melting Temperatures (T_m) of Polymers

Polymer	T_g (°C)	T_m (°C)
Polyethylene (high density)	− 115	137
Polyoxymethylene	− 85	181
Polyisoprene (natural rubber)	− 73	28
Polyisobutylene	− 73	44
Polypropylene	− 20	176
Poly(vinylidene chloride)	− 19	190
Poly(chlorotrifluoroethylene) (kel-F)	45	220
Poly(hexamethylene adipamide) (nylon-6,6)	53	265
Poly(ethylene terephthalate) (Terylene, Dacron)	69	265
Poly(vinyl chloride)	81	212
Polystyrene	100	240
Poly(methyl methacrylate) (Perspex, Lucite)	105	200
Cellulose triacetate	105	306
Polytetrafluoroethylene (Teflon)	127	327

Source: Data from J. Brandup and F. H. Immergut (eds.), *Polymer Handbook*, Interscience, New York, pp. III-32III-92 (1966).

Tough, leatherlike polymers are limited for use in the immediate vicinity of their T_g. Such behavior is observed in vinyl chloride-based plastics, which are used as substitutes for leather in automobile seat covers, travel luggage, and ladies' handbags. Highly crystalline fiber-forming polymers must be used at temperatures substantially below T_m (about 100°C), since changes in crystal structure can occur as T_m is approached. The T_m of a fiber must therefore be above 200°C to remain unaffected at use temperatures encountered in cleaning and ironing. (T_m should not, however, be excessively high—not more than 300°C; otherwise spinning of the fiber by melt-spinning processes may not be possible.) The T_g of a fiber, on the other hand, should have an intermediate value, because too high a value of T_g would interfere with the stretching operation as well as with ironing, and too low a value of T_g would not permit crease retention in fabrics. Nylon and terylene, as may be seen from Table 1.4, therefore have optimal values of T_m and T_g. Semicrystalline polymers with about 50% crystallinity are used at temperatures between T_g and T_m, since in this range the material exhibits moderate rigidity and a high degree of toughness, somewhat analogous to reinforced rubber. Branched polyethylene (low density), with $T_g = -120°C$ and $T_m = 115°C$, used at ambient temperatures is a typical example.

DESIGNING A POLYMER STRUCTURE FOR IMPROVED PROPERTIES

The three principles [15] applied to give strength and resistance to polymers are (1) crystallization, (2) cross-linking, and (3) increasing inherent stiffness of polymer molecules. Combinations of any two or all of the three strengthening principles have proved effective in achieving various properties with polymers (Fig. 1.12). For polymers composed of inherently flexible chains, crystallizaton and cross-linking are the only available means to enhance polymer properties. The factors affecting the crystallinity of polymers have been discussed previously, and the methods of introducing cross-links between polymer molecules are discussed later.

The third, and relatively new, strengthening principle is to increase chain stiffness. One possible way of stiffening a polymeric chain is to hang bulky side groups on the chain to restrict chain bending. For example, in polystyrene, benzene rings are attached to the carbon backbone of the chain; this causes stiffening of the molecule sufficient to make polystyrene a hard and rigid plastic with a softening temperature higher than that of polyethylene, even though the polymer is neither cross-linked nor crystalline. The method is advantageous because the absence of crystallinity makes the material completely transparent, and

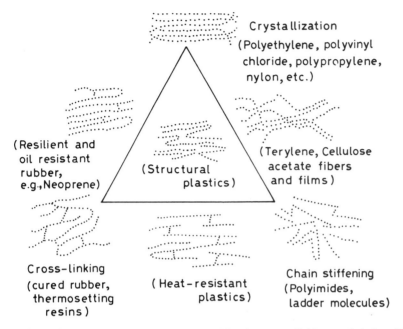

Crystallization
(Polyethylene, polyvinyl
chloride, polypropylene,
nylon, etc.)

(Resilient and
oil resistant
rubber,
e.g.,Neoprene)

(Structural
plastics)

(Terylene, Cellulose
acetate fibers
and films)

Cross-linking
(cured rubber,
thermosetting
resins)

(Heat-resistant
plastics)

Chain stiffening
(Polyimides,
ladder molecules)

FIG. 1.12 Three basic principles—crystallization, cross-linking, and chain stiffening--for making polymers strong and temperature resistant are represented at the three corners of the triangle. The sides and the center of the triangle indicate various combinations of the principles (after H. F. Mark [15]).

the absence of cross-linking makes it readily moldable. A similar example is poly(methyl methacrylate).

However, the disadvantage of attaching bulky side groups is that the material dissolves in solvents fairly easily and undergoes swelling, since the bulky side groups allow ready penetration by solvents and swelling agents. This problem can be eliminated by stiffening the backbone of the chain itself. One way to do this is to introduce rigid ring structures in the polymer chain. A classic example of such a polymer is *cellulose*, which is the structural framework of wood and the most abundant organic material in the world. Its chain molecule consisting of a string of ring-shaped condensed glucose molecules has an intrinsically stiff backbone. Cellulose therefore has a high tensile strength. In poly(ethylene terephthalate) fiber the chains are only moderately stiff, but the combination of chain stiffness and crystallization suffices to give the fiber high strength and a high melting point (265°C). The newer plastic *polycarbonate* containing benzene rings in the backbone of the polymer chain (Table 1.2) is so tough that it can withstand the blows of a hammer. *Ladder polymers* based on aromatic chains consist of double-stranded chains made up of benzene-type rings [16]. These hard polymers are completely unmeltable and insoluble.

The combination of all three principles has led to the development of new and interesting products with enhanced properties. Composites based on expoxy- and urethane-type polymers may be cited as an example. Thus, the stiff polymeric chains of epoxy and urethane types are cross-linked by curing reactions, and fillers are added to produce the equivalent of crystallization.

CROSS-LINKING OF POLYMER CHAINS

Cold flow can be prevented by cross-links between the individual polymer chains. The structure of polymer chains present in the cross-linked polymers is similar to the wire structure in a bedspring, and chain mobility, which permits one chain to slip by another (which is responsible for cold flow) is prevented. Natural rubber, for example, is a sticky product with no cross-linking, and its polymer chains undergo unrestricted slippage; the product has limited use. However, as we have seen, when natural rubber is heated with sulfur, cross-linking takes place (Fig. 1.10). Cross-linking by sulfur at about 5% of the possible sites gives rubber enough mechanical stability to be used in automobile tires but still enables it to retain flexibility. Introducing more sulfur introduces more cross-links and makes rubber inflexible and hard.

Cross-linking is an important process in polymer technology. A few common examples are illustrated in Figures 1.13–1.18. The cross-links are usually formed between polymer or prepolymer molecules by foreign atoms or mole-

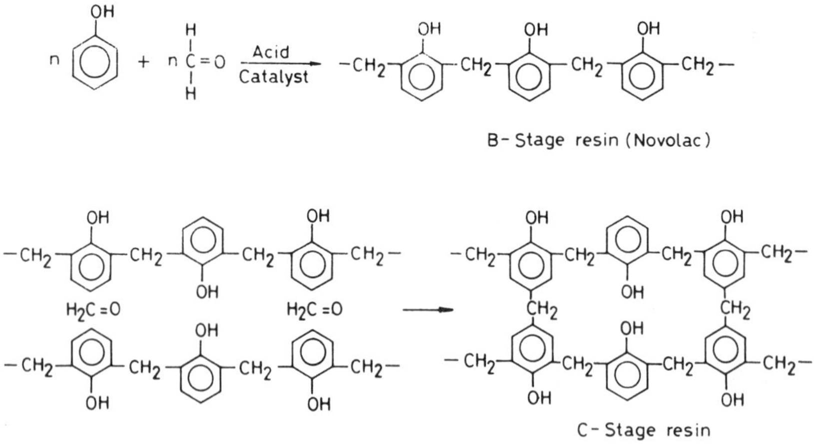

FIG. 1.13 Equations (idealized) for the production of phenolformaldehyde resins.

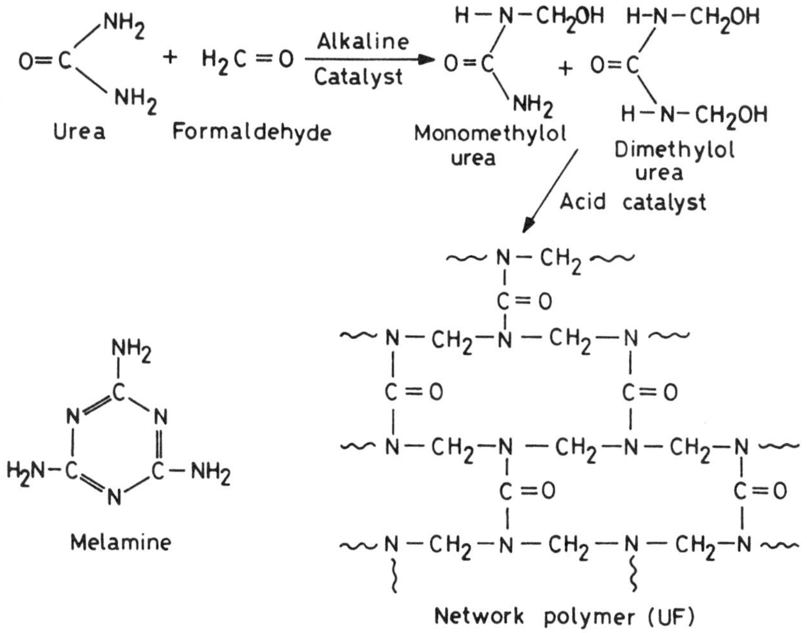

FIG. 1.14 The two important classes of amino-resins are the products of condensation reactions of urea and melamine with formaldehyde. Reactions for the formation of urea-formaldehyde amino resins (UF) are shown. Preparation of melamine-formaldehyde resins is similar.

cules [e.g., sulfur atoms in vulcanized rubber (Fig. 1.10) and styrene molecule in polyesters (Fig. 1.16)] or by small chain segments, as in phenolic and glyptal resins (Figs. 1.13 and 1.18).

A high degree of cross-linking gives rise to three-dimensional or space network polymers in which all polymer chains are linked to form one giant molecule. Thus, instead of being composed of discrete molecules, a piece of highly cross-linked polymer constitutes, essentially, just one molecule. At high degrees of cross-linking, polymers acquire rigidity, dimensional stability, and resistance to heat and chemicals [13]. Because of their network structure such polymers cannot be dissolved in solvents and cannot be softened by heat; strong heating only causes decomposition. Polymers or resins which are transformed into a cross-linked product, and thus take on a "set" on heating, are said to be of *thermosetting* type. Quite commonly, these materials are prepared, by intent, in only partially polymerized states (prepolymers), so that they may be deformed in the heated mold and then hardened by curing (cross-linking).

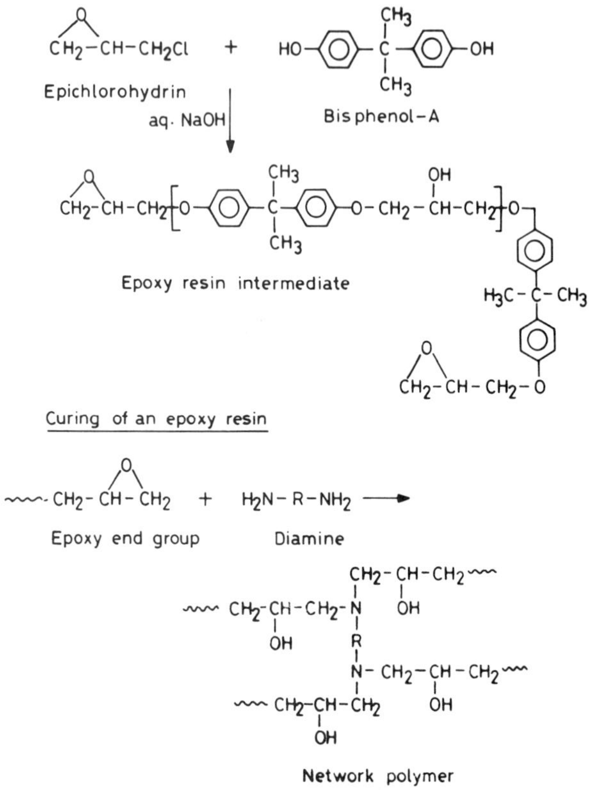

FIG. 1.15 Epoxy monomers and polymer and curing of epoxy resins. Polyamines such as diethylenetriamine ($H_2NC_2H_4NHC_2H_4NH_2$) are widely used for the production of network polymers by room temperature curing.

The most important thermosetting resins in current commercial applications are phenolic resins (Fig. 1.13), amino-resins (Fig. 1.14), epoxy resins (Fig. 1.15), unsaturated polyester resins (Fig. 1.16), urethane foams (Fig. 1.17), and the alkyds (Fig. 1.18). The conversion of an uncross-linked thermosetting resin into a cross-linked network is called *curing*. For curing, the resin is mixed with an appropriate hardener and heated. However, with some thermosetting systems (e.g., epoxies and polyesters), the cross-linked or network structure is formed even with little or no application of heat.

Aging of polymers is often accompanied by cross-linking due to the effect of the surroundings. Such cross-linking is undesirable because it greatly reduces the elasticity of the polymer, making it more brittle and hard. The well-known phenomenon of aging of polyethylene with loss of flexibility is due to cross-

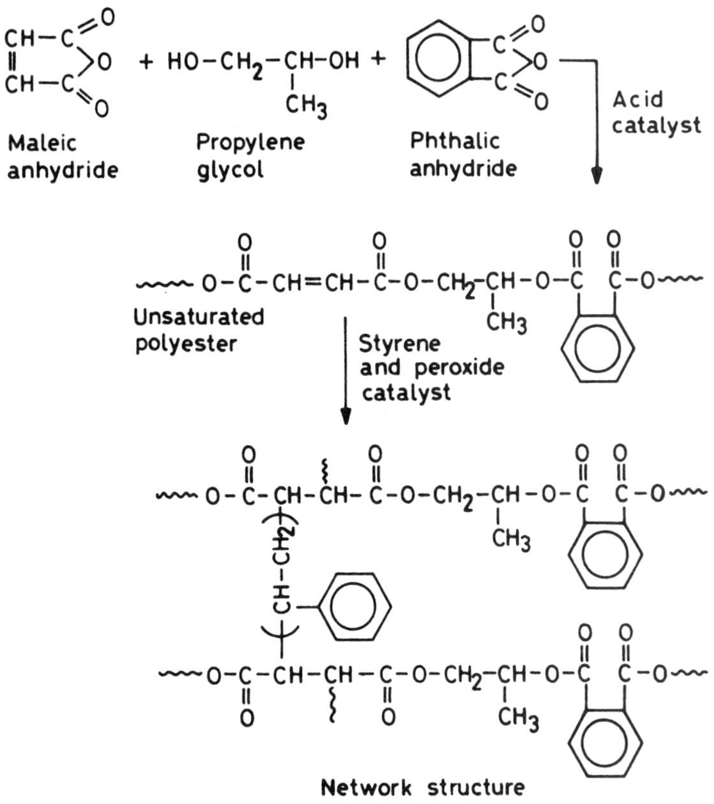

FIG. 1.16 Equations for preparation and curing of an unsaturated polyester resin. The presence of ethylenic unsaturation provides sites for cross-linking by a chain-reaction mechanism in the presence of styrene. Phthalic anhydride increases flexibility by increasing spacing of cross-links.

linking by oxygen under the catalytic action of sunlight (Fig. 1.19a). Cheap rubber undergoes a similar loss of flexibility with time due to oxidative cross-linking (Fig. 1.19b). This action may be discouraged by adding to the polymer an antioxidant, such as a phenolic compound, and an opaque filler, such as carbon black, to prevent entry of light.

SOLUBILITY BEHAVIOR OF POLYMERS

Knowledge of the solubility of various polymers in different solvents is important in assessing their chemical resistance and their application potentialities in

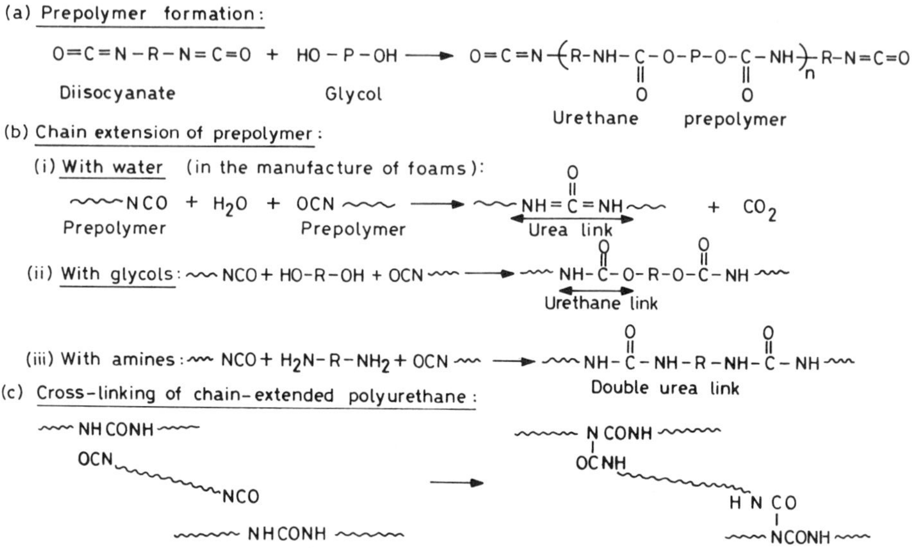
(a) Prepolymer formation:
O=C=N−R−N=C=O + HO−P−OH → O=C=N−(R−NH−C(=O)−O−P−O−C(=O)−NH)n−R−N=C=O
Diisocyanate
Glycol
Urethane prepolymer
(b) Chain extension of prepolymer:
(i) With water (in the manufacture of foams):
~NCO + H2O + OCN~ → ~NH−C(=O)−NH~ + CO2
Prepolymer
Prepolymer
Urea link
(ii) With glycols: ~NCO + HO−R−OH + OCN~ → ~NH−C(=O)−O−R−O−C(=O)−NH~
Urethane link
(iii) With amines: ~NCO + H2N−R−NH2 + OCN~ → ~NH−C(=O)−NH−R−NH−C(=O)−NH~
Double urea link
(c) Cross-linking of chain-extended polyurethane:
~NHCONH~
OCN
NCO
~NHCONH~
~N CONH~
OC NH
H N CO
~NCONH~

FIG. 1.17 Equations for preparations, chain extension, and curing of polyurethane.

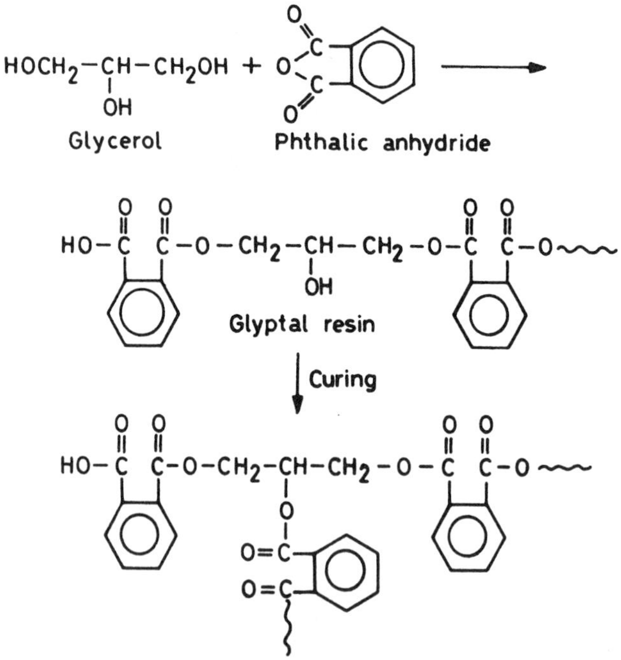
HOCH2−CH−CH2OH
OH
Glycerol
Phthalic anhydride
Glyptal resin
Curing

FIG. 1.18 Equation for preparation of network glyptal resin.

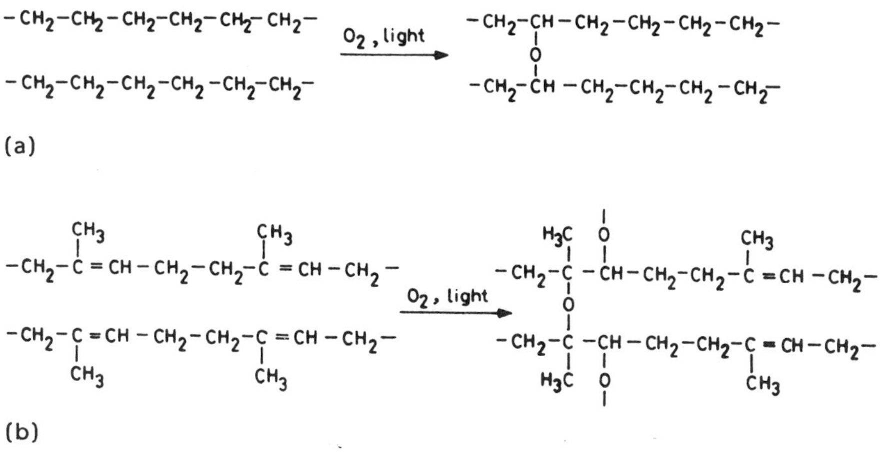

(a)

(b)

FIG. 1.19 Aging of (a) polyethylene and (b) natural rubber by oxidative cross-linking.

the fields of paints, spinning fibers, and casting films. Important also is the knowledge of the solubility of various materials, such as plasticizers and extenders in the polymer, especially since this has an important bearing on plastics formulation.

Because of the size and shape of the polymer molecules and other factors, the solubility relations in polymer systems are complex in comparison to those among low-molecular-weight compounds. Some empirical solubility rules have, nevertheless, been derived for applying to polymer systems, and it is also possible to make certain predictions about solubility characteristics of such systems.

The underlying reason that one material can act as a solvent for another is the compatibility of the materials—i.e., the ability of the molecules of the two materials to coexist without tending to separate. If we denote the force of attraction between the molecules of one material A by F_{AA}, that between the molecules of another material B by F_{BB}, and represent that between one A and one B molecule as F_{AB}, then the system will be compatible and a solution will result if $F_{AB} > F_{BB}$ and $F_{AB} > F_{AA}$. On the other hand, if F_{AA} or $F_{BB} > F_{AB}$, the system will be incompatible and the molecules will separate, forming two phases. In the absence of any specific interaction (e.g., hydrogen bonding) between solvent and solute, we can reasonably assume the intermolecular attraction forces between the dissimilar molecules to be approximately given by the geometric mean of the attraction forces of the corresponding pairs of similar molecules; that is, $F_{AB} = (F_{AA}F_{BB})^{1/2}$. Consequently, if F_{AA} and F_{BB} are equal, F_{AB} will also be similar and the materials should be soluble.

Solubility Parameter

A measure of the intermolecular attraction forces in a material is provided by the *cohesive energy*, which approximately equals the heat of vaporization (for liquids) or sublimation (for solids) per mole and so can be estimated from thermodynamic data. The cohesive energy per unit volume is called *cohesive energy density*, and the square root of this cohesive energy density is known as the *solubility parameter* (δ). The value for a solvent can thus be calculated from its latent heat of vaporization. However, since polymers, in general, cannot be vaporized without decomposition, this method of determination of the solubility parameter is obviously inapplicable to polymers. Consequently, other methods must be used. If a solvent can be found in which a given polymer will dissolve without producing any specific reaction or association, without any heat effect and volume change, then the solubility parameter of the polymer may be taken to be equal to that of the solvent. The value of δ for a given polymer can thus be estimated by testing a large number of solvents of known δ to find out which one conforms most closely to the aforesaid requirements.

Cross-linkage, if present, makes a polymer insoluble; the δ value in such a case can be obtained by finding the solvent that produces the maximum equilibrium swelling.

In contrast to this experimental method, which is laborious and time-consuming, the additive method of Small [17] for calculating the solubility parameter from a set of additive constants, F, called *molar attraction constants*, is very simple and accurate to the first decimal place, provided that hydrogen bonding between polymer molecules is insignificant. Values of molar attraction constants for the most common groups in organic molecules have been estimated by Small from the vapor pressure and heat of vaporization data for a number of simple molecules. These values of molar attraction constants, referring to 25°C, are shown in Table 1.5. The solubility parameter δ is given by the relationship

$$\delta = \frac{\rho}{M} \sum F \tag{19}$$

where $\sum F$ is the molar attraction constants summed over the groups present in the given compound; ρ and M are the density and the molecular weight of the compound. The method of computing the solubility parameter by using Small's table is illustrated in the following examples. [Small's method and constants yield values of δ in units of $(cal/cm^3)^{1/2}$. The SI value in $MPa^{1/2}$ may be obtained by multiplying by 2.04.]

Example 4 Dibutyl phthalate

Formula: $C_6H_4(COOCH_2CH_2CH_2CH_3)_2$
$M = 278$
$\rho = 1.04$ g/cm^3
1 phenylene at 658 = 658
2 CH$_3$ at 214 = 428
6 CH$_2$ at 133 = 798
2 COO   at 310 = 620
ΣF = 2504
$\delta = 1.04(2504/278) = 9.4$

Example 5 Polystyrene

Formula (for repeating unit): $-CH-CH_2-$
                                           $|$
                                       C_6H_5

M (for repeating unit) = 104
ρ (for polymer = 1.05 g/cm^3
1 phenyl at 735 = 735
1 CH$_2$ at 133 = 133
1 CH   at 28 = 28
ΣF = 896
$\Sigma F/M$ for the repeating unit = 896/104

The value of $\Sigma F/M$ for the whole polymer will evidently be the same as for the repeating unit. Hence

δ (for polymer) = 1.05 (896/104) = 9.05 (cal/cm^3)$^{1/2}$

Example 6 Poly(methyl methacrylate)

                                                      COOCH$_3$
                                                         $|$
Formula (for repeating unit): $-CH_2-C-$
                                                         $|$
                                                      CH$_3$

M (for repeating unit) = 100
ρ (for polymer) = 1.19 gm/cm^3
2 CH$_3$ at 214 = 428
1 CH$_2$ at 133 = 133
1 COO   at 310 = 310
     $|$
1$-$C$-$ at -93 = -93
     $|$
ΣF = 778
δ (for polymer) = 1.19 (778/100) = 9.26 (cal/cm^3)$^{1/2}$

TABLE 1.5 Molar Attraction Constants, F, at 25°C

Group	$F \left[\dfrac{(\text{cal cm}^3)}{\text{mole}} \right]^{1/2}$	Group	$F \left[\dfrac{(\text{cal cm}^3)}{\text{mole}} \right]^{1/2}$
$-CH_3$	214	H	80–100
$-CH_2-$ (single bonded)	133	O (ethers)	70
$-CH<$	28	CO (ketones)	275
$\begin{array}{c} \vert \\ -C- \\ \vert \end{array}$	−93	COO (esters)	310
$CH_2=$	190	CN	410
$-CH=$ (double bonded)	111	Cl single	270
$>C=$	19	Cl twinned as in $>CCl_2$	260
$CH\equiv C-$	285	Cl triple as in $-CCl_3$	250
$-C\equiv C-$	222		

Phenyl	735	Br single	340
Phenylene (o,m,p)	658	I single	425
Naphthyl	1146	CF_2 (in fluorocarbons	150
Ring (5-membered)	105–115	CF_3 only	274
Ring (6-membered)	95–105	S (sulphides)	225
Conjugation	20–30	SH (thiols)	315
		ONO_2 (nitrates)	~440
		NO_2 (aliphatic)	~440
		PO_4 (organic)	~500

Source: P. A. Small. Some factors affecting the solubility of polymers, *J. Appl. Chem.*, *3*: 71 (1953).

The calculated values of Table 1.6 and all values of Table 1.7 are those derived according to Small's method. In applying this method to crystalline polymers, it is desirable to use a density value corresponding to the amorphous state. This "amorphous density" can be obtained from the data above the melting point by extrapolation, or it can be obtained from other sources. Further, this method is not applicable where hydrogen bonding is significant and hence cannot be used for calculating solubility parameters of alcohols, amines, carboxylic acids, or other strongly hydrogen-bonded compounds.

Prediction of Solubility

In practice, it is observed that when there is no specific interaction (such as hydrogen bonding) between dissimilar molecules and also no tendency on the part of one species to crystallize, two compounds will be soluble, if their solubility parameter values, in units of $(cal/cm^3)^{1/2}$, differ by no more than unity. Coupled with this criterion the values in Tables 1.6 and 1.7 permit prediction of solubility or insolubility of polymers in different solvents. Thus, for instance, it is obvious that cellulose acetate ($\delta = 10.6$) is soluble in acetone ($\delta = 10.0$) but not in methanol ($\delta = 14.5$) or toluene ($\delta = 8.9$). Natural rubber ($\delta = 8.3$) dissolves, as expected, in toluene and carbon tetrachloride ($\delta = 8.6$) but not in ethanol ($\delta = 12.7$). Poly-(methyl methacrylate), an amorphous thermoplastic having a solubility parameter of about 9.2, dissolves in several liquids, including ethyl acetate ($\delta = 9.0$), ethylene dichloride ($\delta = 9.8$), trichloroethylene ($\delta = 9.3$), chloroform ($\delta = 9.3$), and toluene. Polystyrene, a hydrocarbon polymer with $\delta = 9.1$, dissolves in several hydrocarbons with similar solubility parameters, such as benzene ($\delta = 9.2$) and toluene. Cellulose dinitrate having a value of 10.5 is soluble in cellosolve ($\delta = 9.9$) or an equimolar mixture of ethanol and ethyl ether ($\delta = 7.4$).

Specific Interaction

The solubility-parameter approach is applicable only when the given polymer is amorphous and there exists no hydrogen bonding or other forms of specific interaction between the polymer and the solvent. Various modifications [18,19] have been proposed to account for these cases. Thus, hydrogen bonding has been used as a second parameter, in addition to δ [18], and the dielectric constant or some related quantity as a third [19]; but a point of diminishing return is soon reached.

Crystallization is accompanied by a decrease of free energy, so there exists a free-energy barrier restricting the solution of crystalline polymers. If there can be some form of interaction between a crystalline polymer and a solvent and if that interaction is strong enough to overcome the free-energy barrier, then the polymer will go into solution. On the other hand, if the polymer is highly crystalline but incapable of specific interaction, then it is unlikely to dissolve. It is

TABLE 1.6 Solubility Parameter Values for Polymers

| Polymer | $\delta\,[(cal/cm^3)]^{1/2}$ | |
	Observed[a]	Calculated[b]
Polytetrafluoroethylene	—	6.2
Polyethylene	7.9	8.1
Polypropylene	—	7.9
Polyisobutylene	8.05	7.7
Styrene-butadiene rubber (with 25% styrene)	8.09–8.6	8.5
Polyisoprene (natural rubber)	7.9–8.35	8.2
Polybutadiene	8.4–8.6	8.4
Polystyrene	8.5–9.7	9.1
Poly(vinyl acetate)	—	9.4
Poly(methyl methacrylate)	9.0–9.5	9.3
Buna N (butadiene/ acrylonitrile 75:25)	9.38–9.5	9.2
Neoprene GN	8.18–9.25	9.4
Poly(vinyl chloride)	9.48–9.7	9.6
Bisphenol A polycarbonate	—	9.5
Poly(ethylene terephthalate)	—	10.7
Cellulose (di) nitrate	10.56	10.5
Cellulose (di) acetate	10.6	11.4
Nylon-6,6	13.6	—
Polyacrylonitrile	15.4	12.8

[a]Collected from various sources.
[b]From Ref. 17.
Conversion of units: SI value of δ in $MPa^{1/2}$ may be obtained by multiplying by 2.04.

TABLE 1.7 Solubility Parameter Values for Common
Solvents at 25°C

Solvent	$\delta\,[(cal/cm^3)]^{1/2}$
n-Hexane	7.3
n-Heptane	7.4
Diethyl ether	7.4
n-Octane	7.6
n-Decane	7.7
Methylcyclohexane	7.8
Turpentine	8.1
Camphor	8.5
2,2-Dichloropropane	8.2
n-Butyl acetate	8.3
Amyl acetate	8.5
Methyl isopropyl ketone	8.5
Carbon tetrachloride	8.6
Xylene	8.8
Toluene	8.9
1,2-Dichloropropane	9.0
Ethylacetate	9.0
Benzene	9.2
Diacetone alcohol	9.2
Chloroform	9.3
Tetrahydrofuran	9.3
Trichloroethane	9.3
Tetrachloroethane	9.4
Tetraline	9.5
Methyl chloride	9.7
Methylene chloride	9.7
Ethylene dichloride	9.8

TABLE 1.7 (continued)

Solvent	$\delta\,[(cal/cm^3)]^{1/2}$
Cyclohexanone	9.9
Cellosolve	9.9
Dioxane	9.9
Carbon disulfide	10.0
Acetone	10.0
sec-Butanol	10.8
Pyridine	10.9
n-Butanol	11.4
Cyclohexanol	11.4
Isopropanol	11.5
n-Propanol	11.9
Dimethylformamide	12.1
Ethanol	12.7
Cresol	13.3
Formic acid	13.5
Methanol	14.5
Phenol	14.5
Glycerol	16.5
Water	23.4

Data collected from different sources.
SI value of δ in $MPa^{1/2}$ may be obtained by multiplying by 2.04.

thus understandable that polyethylene and polytetrafluoroethylene have no solvents at room temperature. However, even if the interaction with a solvent is weak or absent, the dissolution of a crystalline polymer may yet be brought about by raising the temperature to overcome the energy barrier. Thus, although at room temperature there are no solvents for polyethylene and polytetrafluoroethylene, they do dissolve at higher temperatures. But if a highly crystalline polymer is heated above its melting point and dissolved in a solvent with which it has no specific interaction, then the polymer will tend to crystallize on subsequent cooling.

Low-density polyethylene dissolves in benzene at about 60°C. Even though at room temperature there is no solvent for polyethylene, solvents with solubility parameters close to that of polyethylene will, however, cause swelling; such swelling is more significant in low-density polymers.

Polypropylene is very similar to high-density polyethylene and has a similar solubility parameter; consequently, the two polymers tend to be swollen by the same liquids.

Solvent Type

When specific interaction comes into play in the dissolution process, depending on the solubility parameter alone for predicting solubility relations might be quite misleading, because solvents promoting different types of interactions can act differently, even if they happen to have similar solubility parameters. For instance, since poly(vinyl chloride) and the polycarbonate derived from bisphenol A have similar solubility parameters, we might expect them to exhibit similar solubility characteristics in relation to different solvents. But they do not. Thus, tetrahydrofuran and cyclohexanone are good solvents for PVC but not for the polycarbonate, whereas chloroform and methylene chloride are good solvents for the polycarbonate but not for PVC. This difference presumably resides in the fact that PVC is a proton donor, whereas the polycarbonate is a proton acceptor (though both are weak in their respective behavior). Consequently, solvents such as cyclohexanone and tetrahydrofuran, which are proton acceptors, can dissolve PVC much better than the polycarbonate, and the reverse is obviously true in the case of solvents such as methylene chloride and chloroform since they are proton donors.

From similar arguments we can say that a polymide such as nylon-6,6 will dissolve in proton donors, and a polymer like polyacrylonitrile will dissolve in proton acceptors. Being highly polar and crystalline, both polyamide and polyacrylonitrile can, however, be expected to dissolve only in strongly interacting solvents such as formic acid, glacial acetic acid, phenols, and cresols for the former and dimethylformamide and tetramethylene sulfone for the latter.

Plasticizer Solubility

Plasticizers cause partial neutralization of the intermolecular forces of attraction, thus producing greater freedom of movement between the polymeric macromolecules. This is reflected in an enhanced flexibility and plasticity and in reduced tensile strength and chemical resistance of the materials. Leaving aside exceptional cases, plasticizers are mostly liquids, which are, indeed, simply high-molecular-weight (hence high-boiling) solvents for the polymers—the use of high-molecular weight (of at least 300) liquids being necessitated by the basic requirements of plasticizer permanence that necessarily demands low vapor pressure and a low diffusion rate of the plasticizer within the polymer. Besides permanence, another basic requirement of a plasticizer is that it must be compatible with the polymer. In light of our foregoing discussions, the compatibility requirements, in turn, implies that the plasticizer must conform to certain demands: (1) Its solubility parameter should be close to that of the polymer. (2) It should have the ability to undergo some specific interaction with the polymer, especially if the polymer itself has a tendency to crystallize. (3) If it is a solid, it should not be crystalline when it is incapable of some specific interaction with the polymer.

Solubility parameters for some common plasticizers are shown in Table 1.8, which on comparison with Table 1.6 allows some predictions—and these are indeed found to be correct—that aromatic oils will plasticize natural rubber; tritolyl phosphate will plasticize poly(methyl methacrylate); dimethyl phthalate will plasticize cellulose diacetate; dibenzyl ether will plasticize poly(vinylidene chloride); and octyl phthalates, tritolyl phosphate, and dioctyl sebacate will plasticize poly(vinyl chloride). The fact that camphor is an effective plasticizer for cellulose nitrate, in sharp contrast to what would be expected from consideration of the solubility parameter alone, can be explained by the ability of the carbonyl group in the camphor to undergo specific interaction with some group in the cellulose nitrate molecule. Celluloid, the first synthetic plastics material, is the product of cellulose nitrate plasticized with camphor.

In PVC formulations the plasticizer is sometimes partially replaced by some other material called an *extender* (such as chlorinated waxes and refinery oils), which, in itself, is not a plasticizer but can be added, up to a certain concentration, to a polymer-plasticizer combination. To make a three-component mixture compatible, the solubility parameter of the mixture of plasticizer and extender should be within unity of that of the polymer. The solubility parameter δ_{mix} of a mixture of two liquids of solubility parameters δ_1 and δ_2, respectively, can be calculated from

$$\delta_{mix} = x_1\delta_1 + x_2\delta_2$$

where x_1 and x_2 are the fraction concentrations of the two liquids. Thus, if a polymer is plasticized with a liquid of relatively high solubility parameter, then the blend would evidently tolerate a relatively higher proportion of an extender with low solubility parameter. As an example, we may cite the PVC-tritolyl phosphate formulation, which, as is to be expected, can take up more of a low-solubility-parameter extender than can the PVC-dioctyl sebacate formulation.

TABLE 1.8 Solubility Parameter Values[a] for Some
Plasticizers at 25°C

Plasticizer	$\delta[(cal/cm^3)]^{1/2}$
Paraffinic oils	7.5
Aromatic oils	8.0
Camphor	8.5
Diisooctyl adipate	8.7
Dioctyl sebacate	8.7
Diisodecyl phthalate	8.8
Dibutyl sebacate	8.9
Diethylhexyl phthalate	8.9
Diisooctyl phthalate	8.9
Di-2-butoxyethyl phthalate	9.3
Dibutyl phthalate	9.4
Triphenyl phosphate	9.8
Tritolyl phosphate	9.8
Trixylyl phosphate	9.9
Dibenzyl ether	10.0
Triacetin	10.0
Dimethyl phthalate	10.5

[a]Values computed by Small's method [17].
SI value of δ in $MPa^{1/2}$ may be obtained by multiplying by
2.04.

Rate of Dissolution

Note that the foregoing considerations of the thermodynamics of polymer dissolution only give the criteria of what may be called thermodynamically good
solvents. But thermodynamically good solvents may not necessarily be kinetically good as well, since in contrast to the aforesaid thermodynamic considerations the rate of polymer dissolution depends primarily on the rapidity of
diffusion of the polymer and the solvent into one another. Solvent diffusion
through a polymer involves solvent molecules passing through voids and other
gaps between polymer molecules. The rate of diffusion therefore largely depends

on the size of the solvent molecules and the size of the gaps. Solvents that promote rapid solubility are usually small, compact molecules. Mixtures of a kinetically good solvent and a thermodynamically good solvent are often very powerful and rapid polymer solvents.

EFFECTS OF CORROSIVES ON POLYMERS

Polymers are resistant to electrochemical corrosion. When a polymer has a δ value similar to that of water, dissolution occurs. However, if the δ value of the polymer is lower, it is not attacked by water or other polar solvents. The carbon–carbon bonds in polymer backbones are not cleaved even by boiling water. When amide, ester, or urethane groups are present in the polymer backbone, they may be attacked by hot acids or alkalies, and hydrolysis may occur. When these functional groups are present as pendant groups on the polymer chain, the reaction will be similar. However, the tendency for such attack is reduced when alkyl groups are present on the carbon atom attached to the functional group. Thus, poly(methyl methacrylate) but not poly(methyl acrylate) is resistant to acid or alkaline hydrolysis. Atoms or groups with strong carbon bonds, such as fluorine, chlorine, and ether groups, are resistant to attack by aqueous acids and alkalies. The chemical resistance of various plastics is summarized in Table 1.9.

THERMAL STABILITY AND FLAME RETARDATION

When polymers are heated above 200°C, the kinetic energy of molecules approaches the carbon–carbon bond energy, and bond dissociation occurs. The process is the reverse of polymerization. Some polymers give a quantitative yield of monomers when heated. Poly(methyl methacrylate) gives a quantitative yield of methyl methacrylate at elevated temperatures. Some polymers decompose into fragments that are larger than the corresponding monomers. Polyolefins usually behave this way. For linear or branched polymers, however, cleavage of one bond is required for the production of lower-molecular-weight products. For polymers with laddered or double-stranded chain structure, lower-molecular-weight products cannot be produced until two distinct bonds are cleaved.

Plastics can be grouped into eight temperature–time zones, depending on the temperature at which they retain 50% mechanical or physical properties when heated for different periods in air. These temperature–time zones are shown in Figure 1.20, and the materials falling in the different zones are listed in Table 1.10. The materials in zone 6 and above can compete with metals in high-performance applications because they perform in the same temperature range. Most polymers in zone 6 and above do not burn, but they may char and may be consumed very slowly in direct flames.

TABLE 1.9 Effects of Corrosive Environments on Plastics[a]

Plastic	Environment							
	Aliphatic solvents	Aromatic solvents	Chlorinated solvents	Esters and ketones	Weak bases	Strong bases	Strong acids	Strong oxidants
Acetal	A B	A B	A B	A B	A A	A B	E E	E E
Acrylic	B C	E E	E E	E E	A C	B E	D E	E E
Acrylonitrile-butadiene-styrene (ABS)	A E	D E	E E	E E	A C	A C	B E	D E
Cellulose acetate	A B	A C	A D	E E	A C	C E	C E	C E
Cellulose acetate butyrate	A C	D E	D E	E E	B D	C E	C E	C E
Cellulose acetate propionate	A C	D E	D E	E E	A B	C E	C E	C E
Epoxy (glass fiber filled)	A B	A B	A C	B C	A A	B C	B C	D D
Furan (asbestos filled)	A A	A A	A A	A A	B B	B B	A A	E E
Melamine	A A	A A	A A	A B	B C	B C	B C	B C
Phenolic (asbestos filled)	A A	A A	A A	C C	A C	D E	A A	D E

Polymer										
Polyamide	A A	A A	A B	A A	A B	A A	A B	B C	E E	E E
Polybenzimidazole	A A	A A	A A	A A	A A	A A	A A	A B	A A	A C
Polycarbonate	A A	A A	E E	E E	E E	E E	E E	E E	A A	A A
Poly(chlorotrifluoroethylene)	A A	A A	C D	A A	A A	A A	A A	A A	A A	A A
Polyester (glass fiber filled)	A B	A C	B D	C C	B C	B D	C E	B B	B B	B C
Polyethylene	C E	C E	D E	D E	A E	C C	E E	A A	A A	A A
Polypropylene	A D	B D	B D	A C	A C	A C	D E	A A	A A	A D
Polysulfone	A A	D D	E E	C D	A D	A D	E E	A D	A A	A A
Polystyrene	D E	D E	E E	D E	A E	E E	D E	D D	D D	D E
Poly(tetrafluoroethylene)	A A	A A	A A	A A	A A	A A	A A	A A	A A	A A
Polyurethane	A D	C D	D E	B C	A C	B C	D D	A C	A A	A D
Poly(vinyl chloride)	A E	D E	E E	D E	A E	D E	E E	A E	B B	B E
Silicone	B C	D D	D E	B D	A B	B D	D E	D D	D D	D E
Urea	A C	A C	A C	A B	A C	A B	B C	C B	B B	B C

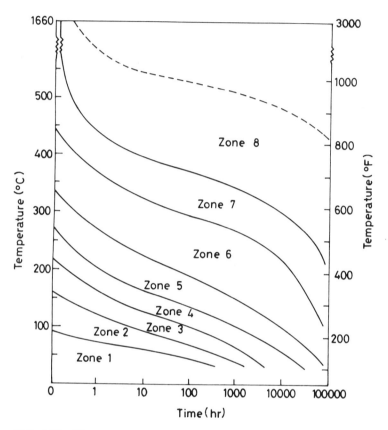

FIG. 1.20 Time-temperature zones indicating thermal stability of plastics (see Table 1.10). [Adapted from *Plastics World, 26*(3): 30 (1968).]

Rotation of chain segments is more difficult in polymers with cyclic rings in their chains. As a result, such polymers are stiffer and more resistant to deformation and have higher melting points and higher glass transition temperatures. Thermal stability is improved by the presence of aromatic and heterocyclic rings in polymer chains. This general approach has resulted in a number of new commercially available polymers with improved high-temperature properties. Polyimides and polybenzimidazoles are important examples of such high-temperature polymers. They can be exposed for a short time to temperatures as high as 1112°F (600°C).

Polyimides are synthesized [20] by the reactions of dianhydrides with diamines in a two-stage process (Fig. 1.21). In the first stage, the two materials form an intermediate poly(amic acid). Processing is accomplished after this

TABLE 1.10 Plastics Retaining 50% Mechanical or Physical Properties at Temperatures in Air

Zone 1

Acrylic
Cellulose acetate
Cellulose acetate butyrate
Cellulose acetate propionate
Cellulose nitrate
Polyallomer
Polyethylene, low density
Polystyrene
Poly(vinyl acetate)
Poly(vinyl alcohol)
Poly(vinyl butyral)
Poly(vinyl chloride)
Styrene-acrylonitrile
Styrene-butadiene
Urea-formaldehyde

Zone 2

Acetal
Acrylonitrile-butadiene-styrene
  copolymer
Ethyl cellulose
Ethylene-vinyl acetate copolymer
Furan
Ionomer
Polyamides
Polycarbonate
Polyethylene, high density
Polyethylene,cross-linked
Poly(ethylene terephthalate)
Polypropylene
Poly(vinylidene chloride)
Polyurethane

Zone 3

Poly(monochlorotrifluoroethylene)

Zone 4

Alkyd
Fluorinated ethylene propylene
  copolymer
Melamine formaldehyde
Phenol-furfural
Polyphenylene oxide
Polysulfone

Zone 5

Diallyl phthalate
Epoxy
Phenol-formaldehyde
Polyester
Poly(tetrafluoroethylene)

Zone 6

Polybenzimidazole
Polyphenylene
Silicone

Zone 7

Polyamide-imide
Polyimide

Zone 8

High-performance plastics being developed using intrinsically rigid linear macromolecules rather than the conventional crystallization and cross-linking

Source: Plastics World 26: 30 (Nov. 3, 1968).

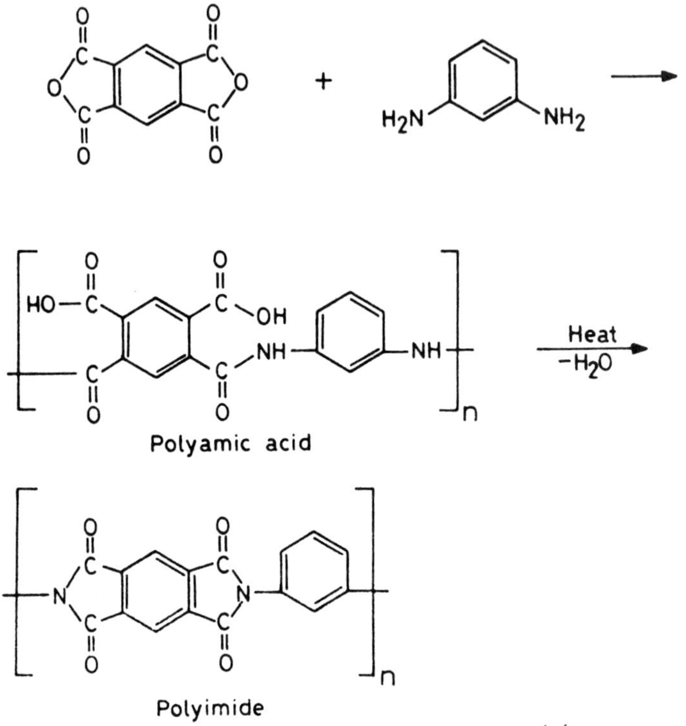

FIG. 1.21   Synthesis of polyimide from pyromellitic anhydride and *m*-phenylenediamine.

stage, since the polymer at this point is still soluble and fusible. The poly(amic acid) is formed into the desired physical form of the final polymer product (e.g., film, fiber, laminate, coating, etc.), and then the second stage of the reaction is carried out, in which the poly(amic acid) is cyclized in the solid state to the polyimide by heating at moderately high temperatures (above 150°C). The polymer after the second stage of the process is insoluble and infusible. Instead of *m*-phenylenediamine, shown in Figure 1.21, other diamines have also been used to synthesize polyimides—for example, *p*-phenylenediamine, *p,p*′-diamino-diphenyl ether, and *p,p*′-diaminodiphenyl sulfide.

Polyimides have been used up to 100 hr at 600°F (315.6°C). They have been used as a transparent head cover for fire fighters and as excellent flame- and high-temperature-resistant foams, which can be used as structurally stable insulation materials at high temperatures. Polyimide/glass fiber composites have many uses in the electronics and aerospace industries and in other high-performance applications.

Polybenzimidazoles are synthesized [21] by the reactions of aromatic diacids and aromatic tetraamines (Fig. 1.22). Polybenzimidazoles are mostly used as fibers in parachutes, for reentry vehicles, and so on. Composites of polybenzimidazoles with glass fibers have excellent basic strength and high-temperature performance. These composites have been extensively used in nose fairings, aircraft leading edges, reentry nose cones, radomes, and deicer ducts.

A weak link in the polybenzimidazole structure is the imino hydrogen. When this hydrogen is replaced by a phenyl group as in *N*-phenyl polybenzimidazole, a dramatic increase in high-temperature properties in oxidizing atmospheres is obtained.

In polymers such as polyphenylenes there are a few aliphatic linkages and many aromatic rings which account for their improved heat resistance. Several polyphenylenes are shown in Figure 1.23. Poly(phenylene oxide) has excellent dimensional stability at elevated temperatures. Repeated steam autoclaving does not deteriorate its properties. Poly(phenylene sulfide) is completely nonflammable. It is used in the form of composites with both asbestos and glass fibers.

Aliphatic linkage is completely eliminated in ladder polymers such as polybenzimidazopyrrolones, commonly called "pyrrones" (Fig. 1.24). Such polymers are highly stable in air. They do not burn or melt when heated but form carbon char without much weight loss. They are potentially the ultimate in heat- and flame-resistant materials.

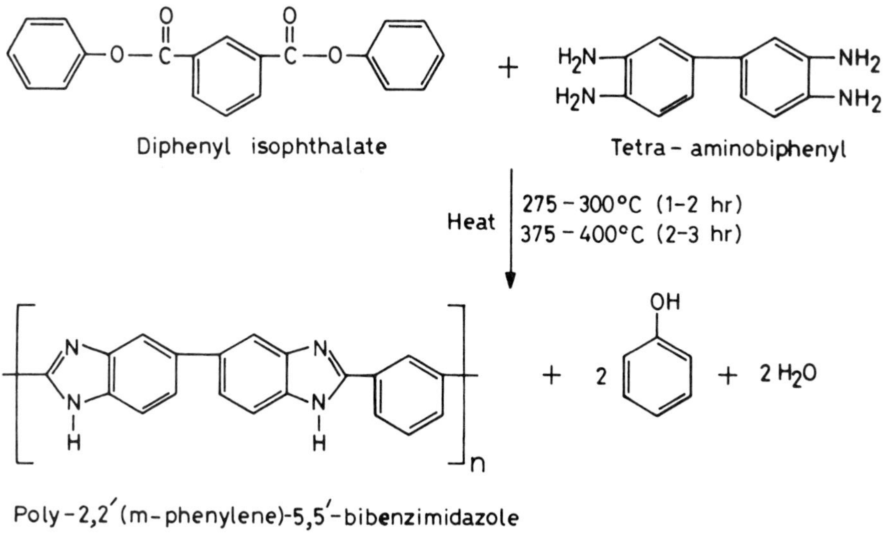

Diphenyl isophthalate   Tetra - aminobiphenyl

Heat | 275 – 300 °C (1–2 hr)
     | 375 – 400 °C (2–3 hr)

Poly-2,2′ (m-phenylene)-5,5′-bibenzimidazole

FIG. 1.22 Synthesis of polybenzimidazole by condensation polymerization.

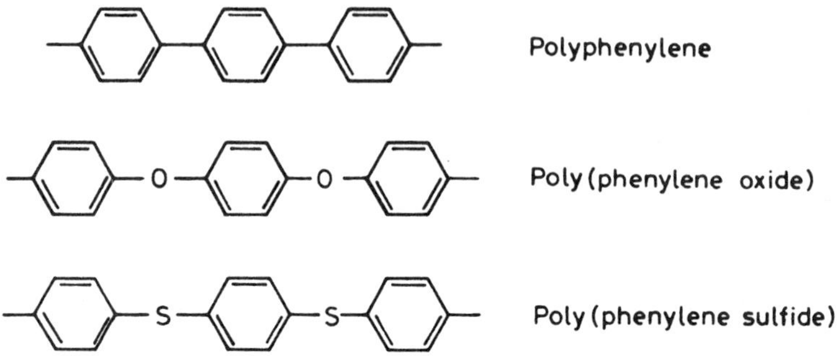

Polyphenylene

Poly(phenylene oxide)

Poly(phenylene sulfide)

FIG. 1.23 Structures of some polyphenylenes.

Another way to get good heat resistance is to use inorganic material as backbone chain as in silicone polymers. Here organic radicals are attached to silicone atoms on an inorganic silicon-oxygen structure. Presence of silicon-oxygen links gives such materials outstanding heat resistance. A silicone polymer has the structure

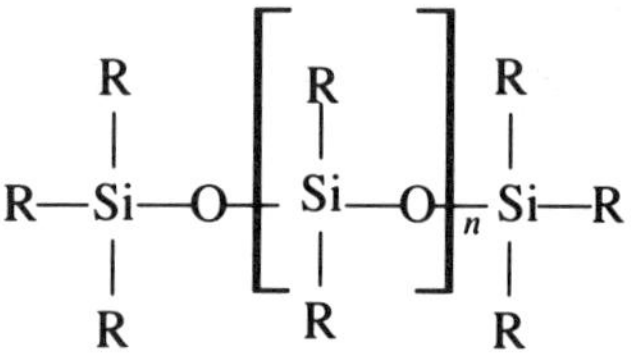

When n is a small number, the structure is that of a silicone oil, whereas silicon rubbers have high values of n. When the ratio R/Si is lower than 2, cross-linked polymers are obtained. Properties of silicone polymers are greatly affected by the type of organic radicals present. For a given chain length, a methyl silicone can be an oily liquid, but a phenyl silicone is a hard and brittle resin.

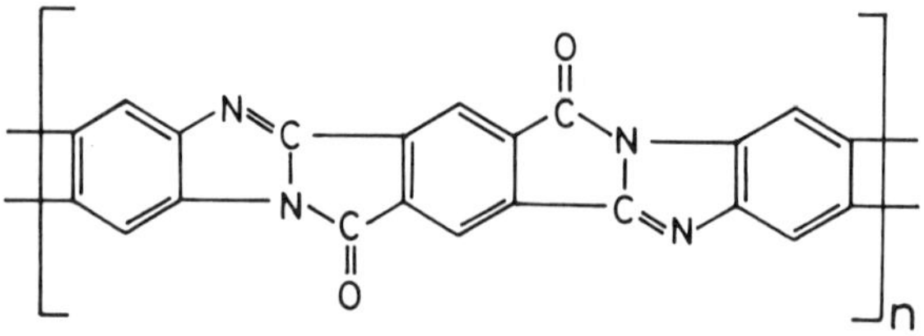

FIG. 1.24 Typical structure of a polybenzimidazopyrrolone.

Ablation

When subjected briefly to very high temperatures, some polymers, such as phenolics, can undergo rapid decomposition to gases and porous char, thereby dissipating the heat and leaving a protective thermal barrier on the substrate. This sacrificial loss of material accompanied by transfer of energy is known as *ablation*. Interaction of a high-energy environment (2500 to 5000°C) with the exposed ablative material results in sacrificial erosion of some amount of the surface material, and the attendant energy absorption controls the surface temperature and greatly restricts the flow of heat into the substrate. Most notable applications of ablative materials are in protecting space vehicles during reentry into the earth's atmosphere, protecting missile nose cones subjected to aerodynamic heating during hypersonic flight in the atmosphere, insulating sections of rocket motors from hot propulsion gases, resisting the intense radiant heat of thermonuclear blasts, and providing thermal protection for structural materials exposed to very high temperatures.

Polymers have been used as ablative materials for a combination of reasons [22]. Some of their advantages [22] are (1) high heat absorption and dissipation per unit mass expended, which may range from several hundred to several thousand calories per gram of ablative material; (2) automatic control of surface temperature by self-regulating ablative degradation; (3) excellent thermal insulation; (4) tailored performance by varying the individual material component and composition of ablative systems; (5) design simplicity and ease of fabrication; (6) light weight and low cost.

Polymer ablatives, however, can be used only for transitory periods of a few minutes or less at very high temperatures and heat load. Moreover, the sacrificial loss of surface material during ablation causes dimensional changes which must be predicted and incorporated into the design.

An ablative material should have a high heat of ablation, which measures the ability of the material to absorb and dissipate energy per unit mass. It should also possess good strength even after charring, since any sloughing off of chunks of material which have not decomposed or vaporized represents poor usage of the ablative system. Figure 1.25 profiles the various stages of heating, charring, and melting in two phenolic composites, one reinforced with glass and the other with nylon. The glass-reinforced system appears to be mechanically superior because it produces a molten glass surface, whereas the nylon-reinforced system may have higher thermal efficiency.

Flame Retardation [23]

All thermoset plastics are self-extinguishing. Among thermoplastics, nylon, polyphenylene oxide, polysulfone, polycarbonate, poly(vinyl chloride), chlorinated polyether, poly(chlorotrifluoroethylene) and fluorocarbon polymers have

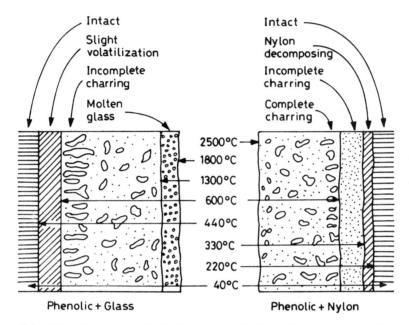

FIG. 1.25 Temperature distribution in two plastics during steady state ablation [after D. L. Schmidt: *Mod. Plastics, 37*: 131 (Nov. 1960), 147 (Dec. 1960)].

self-extinguishing properties. The burning characteristics of some polymers are summarized in Table 1.11. Halogenation enhances the flame retardancy of polymers. Thus when the chlorine content of PVC, which usually has 56% chlorine, is increased, as in cholorinated PVC (61% chlorine), the oxygen index increases from 43 to 60. (The oxygen index is a minimum percent of oxygen in a N_2/O_2 gas mixture necessary to support combustion. A material with an oxygen index greater than 27 usually passes the vertical burn test.)

A discussion on flame-retardant additives is given in a later section on polymer compounding.

DETERIORATION OF POLYMERS [24–26]

Deterioration of polymers is manifested in loss of strength, loss of transparency, warpage, cracks, erosion, and so on. Hydrophilic materials such as nylon or cellulose acetate can undergo swelling at high humidity or shrinkage due to low humidity. Degradation can occur due to imposition of energy in the form of heat, mechanical action, ultrasonic and sonic energy, radiation such as gamma

TABLE 1.11 Burning Characteristics and Burn Rates of Some Polymers

Polymer	Burning characteristics	Burn rate[a] (cm/min)
Polyethylene	Melts, drips	0.8–3.0
Polypropylene	Melts, drips	1.8–4.0
Poly(vinyl chloride)	Difficult to ignite, white smoke	Self-extinguishing
Poly(tetrafluoroethylene)	Melts, chars, bubbles	Nonburning
Fluorinated ethylene propylene copolymer	Does not ignite	Nonburning
Polybutylene	Burns	2.5
Acetal	Burns, bluish flame	1.3–2.8
Cellulose acetate	Burns, yellow flame sooty smoke	1.3–7.6
Cellulose propionate	Burns, drips	1.3–3.0
Cellulose acetate butyrate	Burns, drips	0.8–4.3
Acrylonitrile-butadiene-styrene (general purpose)	Burns	2.5–5.1
Styrene-acrylonitrile	Melts, chars, bubbles	1.0–4.0
Polystyrene	Softens, bubbles, black smoke	1.3–6.3
Acrylic	Burns slowly, drips	1.4–4.0
Nylons	Burns, drips, froths	Self-extinguishing
Phenylene oxide		Self-extinguishing
Polysulfone		Self-extinguishing
Chlorinated polyether		Self-extinguishing
Polyimide		Nonburning

[a]ASTM D-635 test procedure.
Source: Adapted from *Flame Retardancy of Polymeric Materials* (W. C. Kuryla and A. J. Papa, eds.), Marcel Dekker, New York, Vol. 3, 1973.

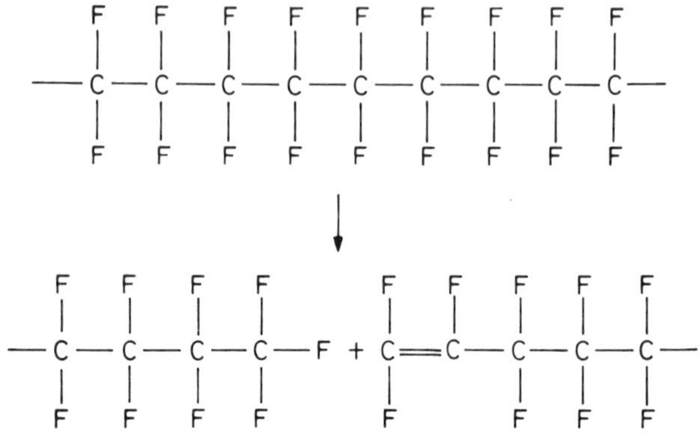

FIG. 1.26 Scission of polytetrafluoroethylene by irradiation.

rays, x-rays, visible light, ultraviolet light, infrared, and electrical action in the form of dielectric effects. Deterioration can occur by chemical effects such as oxidation, ozone attack, hydrolysis, attacks by solvents and detergents, cracking due to swelling of the plasticizer, hardening due to loss of the plasticizer, migration of plasticizers from layer to layer, crazing and cracking, delamination or debonding, void formation, etc.

Linear or thermoplastic polymers may deteriorate by scission. Ultraviolet light and neutrons can easily break a C—C bond of a vinyl-type polymer, producing smaller molecules (Fig. 1.26). As a result of breaking the bigger molecules into smaller molecules, strength and viscosity are reduced. Radiation can also induce branching. A photon or a neutron can supply the activation energy necessary to cause branching. Branching of polyethylene by a photon is shown in Figure 1.27.

Chemical Deterioration

Cross-linking is another way that linear or thermoplastic polymers may deteriorate. Two well-known examples are aging of polyethylene and natural rubber, with loss of flexibility due to cross-linking by oxygen under the catalytic action of sunlight (Fig. 1.19). Vulcanized rubber has only 5 to 20% of its possible positions anchored by sulfur cross-links. Over time there may be further cross-links of oxygen by the air, and the polymer may thus gradually lose its deformability and elasticity.

Because the oxygen molecule is a biradical, its reaction with a polymer usually results in a chain reaction involving free radicals. Thus, a polymer present in air or an atmosphere rich in oxygen can lose an H atom by reacting with O_2 to form a free radical, which then reacts with another oxygen molecule to form

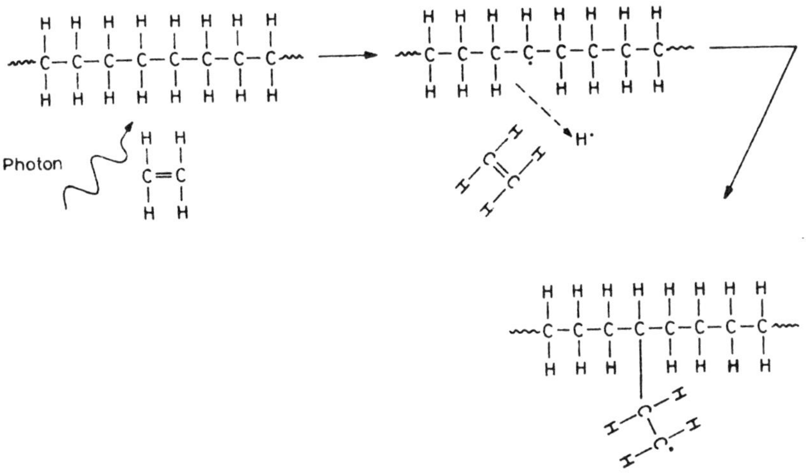

FIG. 1.27 Branching by irradiation with ultraviolet light. A neutron can produce the same effect.

a peroxy free radical. The latter, in turn, reacts with another unit of a polymer chain to produce a hydroperoxide and another free radical.

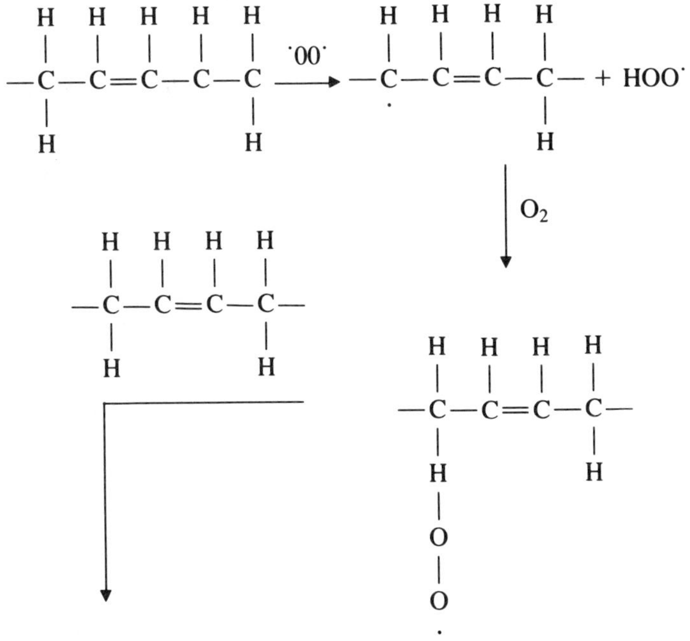

$$\begin{array}{ccccc} H & H & H & H & H \\ | & | & | & | & | \\ -C & -C & =C & -C & -C \\ | & & & & | \\ H & & & & H \end{array} \xrightarrow{\cdot OO\cdot} \begin{array}{cccc} H & H & H & H \\ | & | & | & | \\ -C & -C & =C & -C- \\ \cdot & & & | \\ & & & H \end{array} + HOO\cdot \qquad (20)$$

$$\downarrow O_2$$

$$\begin{array}{cccc} H & H & H & H \\ | & | & | & | \\ -C & -C & =C & -C- \\ | & & & | \\ H & & & H \end{array} \qquad (21)$$

$$\begin{array}{cccc} H & H & H & H \\ | & | & | & | \\ -C & -C & =C & -C- \\ | & & & | \\ H & & & H \\ | & & & \\ O & & & \\ | & & & \\ O & & & \\ \cdot & & & \end{array}$$

```
   H   H   H   H        H   H   H   H
   |   |   |   |        |   |   |   |
  —C—C=C—C  +  —C—C=C—C—                          (22)
   |       |        |           |
   O       H        ·           H
   |
   O
   |
   H
```

Free radicals can combine together and form a stable molecule.

```
                                    H   H   H   H
                                    |   |   |   |
                                   —C—C=C—C—            (23)
      H   H   H   H   H              |
      |   |   |   |   |              H
   2—C—C=C—C—C  ——→
      ·           |                  H
                  H                  |
                                   —C—C=C—C
                                    |   |   |   |
                                    H   H   H   H
```

Termination of the chain reaction can also occur with the formation of a peroxide which may be transitory.

```
   H   H   H   H                    H   H   H   H
   |   |   |   |                    |   |   |   |
  —C—C=C—C—                         C—C=C—C—
   ·           |                    |           |
               H                    O           H
                                    |
   ·                                |
   O
   |                    ——→
   O           H                    O           H
   |           |                    |           |
  —C—C=C—C—                        —C—C=C—C—            (24)
   |   |   |   |                    |   |   |   |
   H   H   H   H                    H   H   H   H
```

Inhibitors or antioxidants can also stop the process.

$$
\begin{array}{ccccc}
& H & H & H & H \\
& | & | & | & | \\
-& C & -C = C & -C & + RH \\
& \cdot & & & | \\
& & & & H
\end{array}
\longrightarrow
\begin{array}{cccc}
H & H & H & H \\
| & | & | & | \\
-C & -C = C & -C & - + R^{\cdot} \\
| & & & | \\
H & & & H
\end{array}
\qquad (25)
$$

In this way, the chain is broken, and an inactive radical $(R^{\cdot})$ is yielded. Aromatic amines or phenolic compounds can act as inhibitors.

However, new radicals can also be formed from the peroxy groups

$$
\begin{array}{cccc}
H & H & H & H \\
| & | & | & | \\
-C & -C = C & -C & - \\
| & & & | \\
O & & & H \\
| & & & \\
O & & & \\
| & & & \\
H & & &
\end{array}
\longrightarrow
\begin{array}{cccc}
H & H & H & H \\
| & | & | & | \\
-C & -C = C & -C & - + HO^{\cdot} \\
| & & & | \\
O & & & H \\
\cdot & & &
\end{array}
\qquad (26)
$$

$$
\begin{array}{cccc}
H & H & H & H \\
| & | & | & | \\
-C & -C = C & -C & - \\
| & & & | \\
O & & & H \\
| & & & \\
O & & & H \\
| & & & | \\
-C & -C = C & -C & -\\
| & | & | & | \\
H & H & H & H
\end{array}
\longrightarrow
2-
\begin{array}{cccc}
H & H & H & H \\
| & | & | & | \\
C & -C = C & -C & -\\
| & & & | \\
O & & & H \\
\cdot & & &
\end{array}
\qquad (27)
$$

Thus for each original reaction with oxygen there can be numerous propagation reactions. The peroxides may also cleave to give aldehydes, ketones, acids, or alcohols. Due to this type of molecular cleavage, the produce becomes soft with lower average molecular weight.

$$\begin{array}{c} \underset{|}{\overset{H}{C}} - \underset{|}{\overset{H}{C}} = \underset{|}{\overset{H}{C}} - \underset{|}{\overset{H}{C}} - \longrightarrow \text{aldehyde, ketone, acid, or alcohol} \qquad (28) \\ \underset{|}{O} \qquad H \\ O \\ | \\ H \end{array}$$

Ozone provides a very reactive source of oxygen because it breaks down to an oxygen molecule and a single reactive oxygen.

$$O_3 \longrightarrow O_2 + O^{\cdot} \qquad (29)$$

The temperature resistance of various plastics in air has been summarized in Figure 1.20 and Table 1.10.

Polymers containing hydrolyzable groups or which have hydrolyzable groups introduced by oxidation are susceptible to water attack. Hydrolyzable groups such as esters, amides, nitriles, acetals, and certain ketones can react with water and cause deterioration of the polymer. The dielectric constant, power factor, insulation resistance, and water absorption are most affected by hydrolysis. For polyesters, polyamides, cellulose, and cellulose ethers and esters, the hydrolyzable groups are weak links in the chain, and hydrolysis of such polymers can cause serious loss of strength. A summary of water absorption characteristics of common plastics and rubbers is presented in Table 1.12.

Microbiological Deterioration

Microbiological deterioration of plastics has been studied by numerous workers by pure-culture tests, soil burial, humidity cabinet tests, and electrical properties tests. The subject of moisture versus fungi as agents of damage to plastics is somewhat controversial. Many believe that moisture is of paramount importance and that fungi are only secondary in causing damage. On the other hand, there are many who believe that fungi alone are capable of accounting for electrical breakdown of some plastic materials under certain conditions. Table 1.13 provides a qualitative assessment of the resistance of plastics and rubbers to attack by microorganisms. The controversial assessment in certain cases may be due to discrepancies in reported data arising from differences in measurement technique, materials used, average molecular weights of materials, fillers, impurities in polymers, etc.

TABLE 1.12 Water Absorption Characteristics of Plastics and Rubbers

Material	Water abs. (%) (24 hr on sample 3.2 mm thick)
Phenol-formaldehyde resin cast (no filler)	0.3-0.4
Phenol-formaldehyde resin molded (wood-flour filler)	0.3-1.0
Phenol-formaldehyde resin molded (mineral filler)	0.01-0.3
Phenol-furfural resin (wood-flour filler)	0.2-0.6
Phenol-furfural resin (mineral filler)	0.2-1.0
Urea-formaldehyde resin (cellulose filler)	0.4-0.8
Melamine-formaldehyde resin (cellulose filler)	0.1-0.6
Melamine-formaldehyde resin (asbestos filler)	0.08-0.14
Ethyl cellulose	0.8-1.8
Cellulose acetate (molding)	1.9-6.5
Cellulose acetate (high acetyl)	2.2-3.1
Cellulose acetate-butyrate	1.1-2.2
Cellulose nitrate	1.0-2.0
Casein plastics	7-14
Poly(vinyl chloride) (plasticized)	0.1-0.6
Poly(vinylidene chloride) (molding)	<0.1
Poly(vinyl chloride acetate) (rigid)	0.07-0.08
Poly(vinyl chloride acetate) (flexible)	0.40-0.65
Poly(vinyl formal)	0.6-1.3
Poly(vinyl butyral)	1.0-2.0
Allyl resins (cast)	0.3-0.44
Polyester resins (rigid)	0.15-0.60

TABLE 1.12   (continued)

Material	Water abs. (%) (24 hr on sample 3.2 mm thick)
Polyester resins (flexible)	0.1–2.4
Poly(methyl methacrylate)	0.3–0.4
Polyethylene	<0.01
Polypropylene	0.01–0.1
Polystyrene	0.03–0.05
Polytetrafluoroethylene	0.00
Nylon (molded)	1.5
Rubbers (extruded)	0.4
Chlorinated rubber	0.1–0.3

Source: Data mainly from Ref. 24.

STABILIZATION OF POLYMERS [27–31]

We have seen that polymers tend to undergo degradation such as chain scission, depolymerization, cross-linking, oxidation, and so on. These changes can be effected by various environmental factors such as heat, light, radiation, oxygen, and water. Various stabilizing ingredients are added to plastics to prevent or minimize the degradative effects. These ingredients act either by interfering with degradative processes or by minimizing the cause of degradation.

Antioxidants and Related Compounds

Oxidation, as we have noted, is a free-radical chain process. The most useful stabilizing agents will therefore be those which combine with free radicals, as shown by Eq. (25), to give a stable species incapable of further reaction. These stabilizing agents are called *antioxidants*. They are the most frequently employed ingredients in plastics, fibers, rubbers, and adhesives. Stabilization is also achieved in some polymer systems by the use of additives which moderate the degradation reaction.

Two large, basic groups of antioxidants are normally distinguished: (1) chain terminating or *primary antioxidants*, and (2) hydroperoxide decomposers or *secondary antioxidants*, frequently called *synergists*.

TABLE 1.13 Resistance of Plastics and Rubbers to Attack by Microorganisms

	Resistance
Plastics	
Poly(methyl methacrylate)	Good
Polyacrylonitrile (orlon)	Good
Acrylonitrile-vinylchloride copolymer (Dynel)	Good
Cellulose acetate	Good, poor
Cellulose acetate butyrate	Good
Cellulose acetate propionate	Good
Cellulose nitrate	Poor
Ethyl cellulose	Good
Acetate rayon	Good
Viscose rayon	Poor
Phenol-formaldehyde	Good
Melamine-formaldehyde	Good, poor
Urea-formaldehyde	Good
Nylon	Good
Ethylene glycol terephthalate (Terylene)	Good
Polyethylene	Good, questionable
Polytetrafluoroethylene (Teflon)	Good
Polymonochlorotrifluoroethylene	Good
Polystyrene	Good
Poly(vinyl chloride)	Good, questionable
Poly(vinyl acetate)	Poor
Poly(vinyl butyral)	Good

TABLE 1.13   (continued)

	Resistance
Plastics (continued)	
Glyptal resins (alkyd resins)	Poor, moderate
Silicone resins	Good
Rubbers	
Pure natural rubber (caoutchouc)	Attacked
Natural rubber vulcanizate	Attacked
Crude sheet	Attacked
Pale crepe, not compounded	Attacked
Pale crepe, compounded	Resistant, attacked
Smoked sheet, not compounded	Attacked
Smoked sheet, compounded	Resistant, attacked
Reclaimed rubber	Attacked
Chlorinated rubber	Resistant
Neoprene, compounded	Resistant
GR-S, butadiene-styrene, compounded	Resistant, attacked
Hycar OR, butadiene-acrylonitrile compounded	Resistant, attacked
Buna N, butadiene-acrylonitrile, compounded	Attacked
GR-I(butyl), isobutylene-isoprene, compounded	Resistant, attacked
Thiokol, organic polysulfide, uncured	Attacked
Thiokol, organic polysulfide, vulcanized	Resistant
Thiokol, organic polysulfide, sheets for gasoline tank lining	Attacked
Silicone rubber	Resistant

Source: Adapted from Ref. 24.

The majority of primary antioxidants are sterically hindered phenols or secondary aromatic amines. They are capable of undergoing fast reaction with peroxy radicals and so are often called radical scavengers. For hindered phenols, for example, such reactions may be represented by the following scheme:

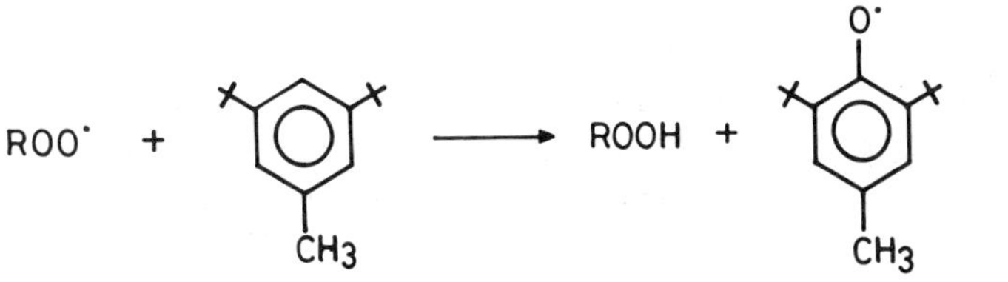

$$\text{ROO}^{\cdot} \; + \; \text{(hindered phenol)} \longrightarrow \text{ROOH} \; + \; \text{(phenoxy radical)} \tag{30}$$

$$\text{(phenoxy radical)} \longleftrightarrow \text{(cyclohexadienone radical)} \; + \; \text{ROO}^{\cdot} \longrightarrow \text{(peroxycyclohexadienone)} \tag{31}$$

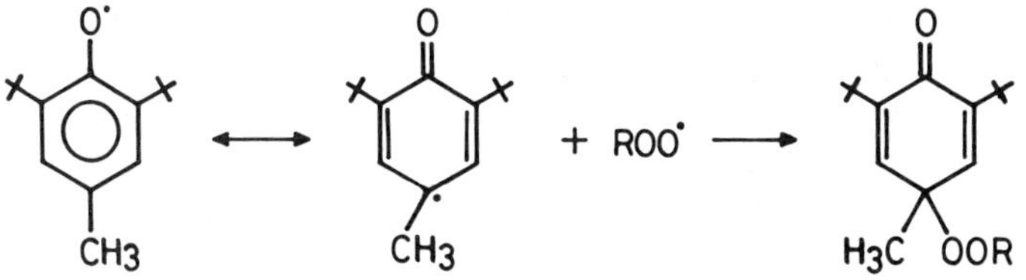

(† = tert butyl group)

Stabilization is achieved by the fact that reaction (30) competes with peroxy radical reactions in polymer degradation, such as reaction (22), transforming the reactive peroxy radical into a much less reactive phenoxy radical, which, in turn, is capable of reacting with a second peroxy radical according to reaction (31).

The phenoxy radicals formed in reaction (30) do not initiate new radical chains at the normal temperatures of use and testing, but such propagation reactions become possible at high temperatures. The thermal stability of peroxycyclohexadienones [formed by reaction (31)] is also limited, and their decomposition leads to new reaction chains. The effectiveness of sterically hindered phenols thus decreases with increasing temperature.

The secondary antioxidants are usually sulfur compounds (mostly thioethers and esters of thiodipropionic acid) or triesters of phosphorous acid (phosphites). A remedy for many types of discoloration in plastics is often the use of a phosphite or a thioether. Both have the ability to react with hydroperoxides, as those formed in reaction (30), to yield nonradical products, following heterolytic mechanisms. The reaction of phosphites to phosphates is an example:

$$\text{P(OR}')_3 + \text{ROOH} \longrightarrow \text{OP(OR}')_3 + \text{ROH} \tag{32}$$

Thioethers react with hydroperoxides in a first stage to yield sulfoxides and alcohols:

$$S\begin{subarray}{l} R_1 \\ \\ R_2 \end{subarray} + ROOH \longrightarrow OS\begin{subarray}{l} R_1 \\ \\ R_2 \end{subarray} + ROH \qquad (33)$$

Sulfoxides themselves yield, on further oxidation, even more powerful hydroperoxide decomposers than the original sulfides, in that they are able to destroy several equivalents of hydroperoxides. This catalytic effect is explained by the intermediate occurrence of sulfenic acids and sulfur dioxide. The fact that the phenomenon of synergism, which is defined as a cooperative action such that the total effect is greater than the sum of two or more individual effects taken independently, is often observed when primary and secondary antioxidants are combined has been explained with the concept of the simultaneous occurrence of the radical reactions [e.g., Eq. (30)] and the nonradical hydroperoxide decomposition [e.g., Eqs. (32) and (33)].

The preceding synergistic effect may be considered as an example of *heterosynergism*, which arises from the cooperating effect of two or more antioxidants acting by different mechanisms. *Homosynergism*, on the other hand, involves two compounds of unequal activity that are operating by the same mechanism, for example, a combination of two different chain-breaking antioxidants that normally function by donation of hydrogen to a peroxy free radical. The most likely mechanism of synergism in this case would involve the transfer of hydrogen from one inhibitor molecule to the radical formed in the reaction of the other inhibitor with a peroxy radical [see Eq. (30)], thus regenerating the latter inhibitor and prolonging its effectiveness.

Chemical Structures of Antioxidants

While sterically hindered phenols and secondary aromatic amines are the two main chemical classes of antioxidants for thermoplastics, the greatest diversity is, however, found in the class of the sterically hindered phenols. Most of these phenols are commercially available as relatively pure chemicals. Although aromatic amines are often more powerful antioxidants than phenols, their application is limited to vulcanized elastomers since the staining properties of aromatic amines prohibit their use in most thermoplastics.

Sterically hindered phenols of commercial importance may be further classified, according to their structure, into (1) alkylphenols, (2) alkylidene-bisphenols, (3) thiobisphenols, (4) hydroxybenzyl compounds, (5) acylamino-phenols, and (6) hydroxyphenyl propionates.

An important goal of antioxidant research has been to provide polyphenols of high molecular weight and low volatility. Most of the commercial sterically hindered phenols have molecular weights in the range 300–600 and above 600. Figure 1.28 depicts structural formulas of the important sterically hindered phenols used as antioxidants.

(a) Alkylphenols, e.g. :

(b) Alkylidene-bisphenols, e.g. :

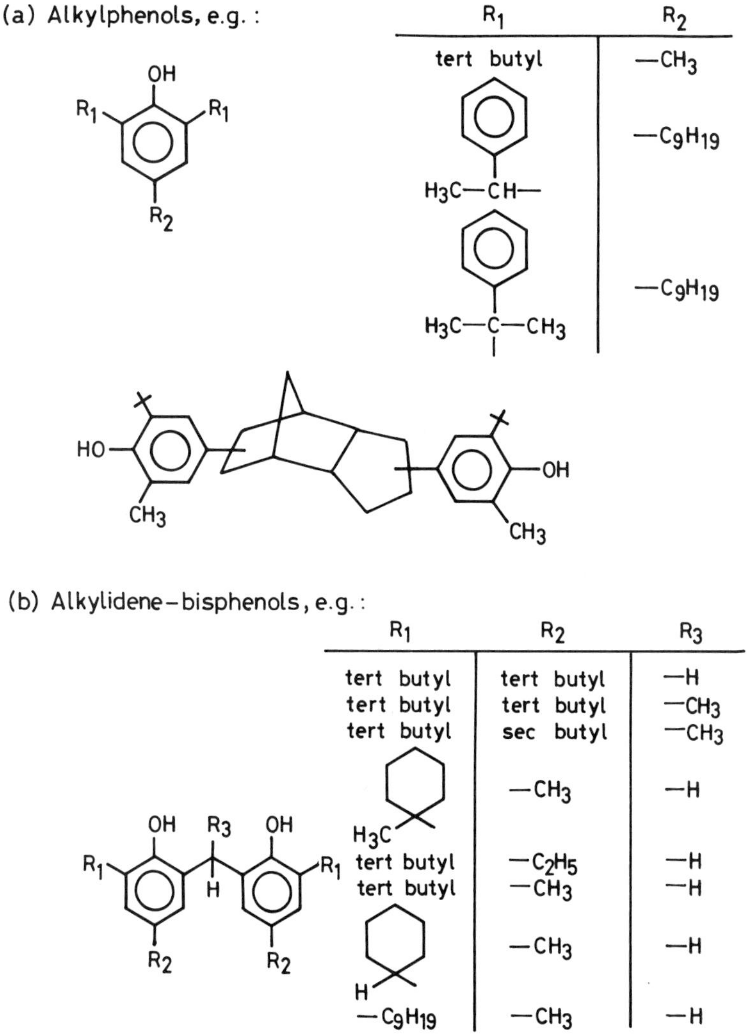

FIG. 1.28 Structural formulas of important sterically hindered phenols used as antioxidants for thermoplastics.

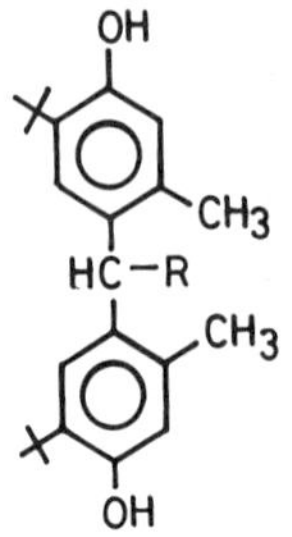

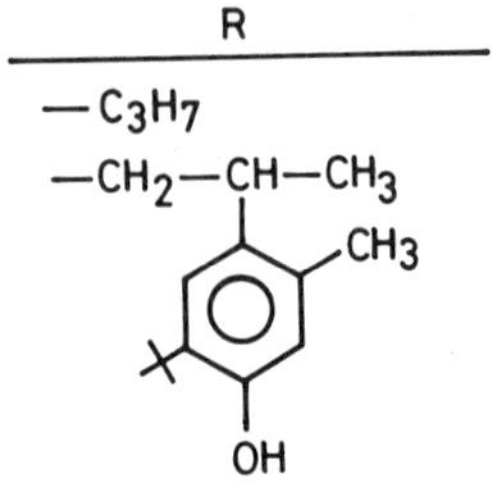

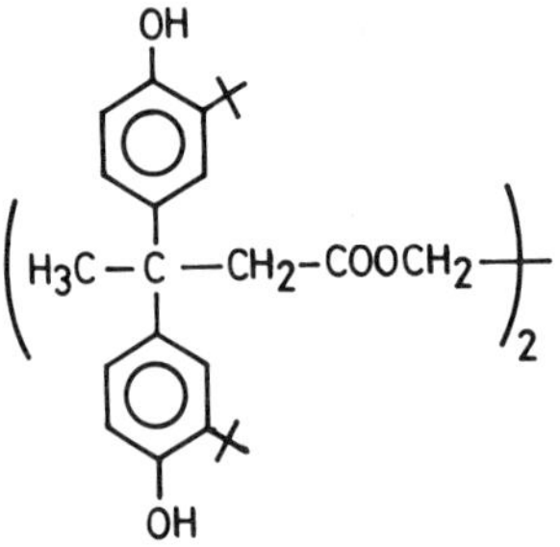

(c) Thiobisphenols, e.g. :

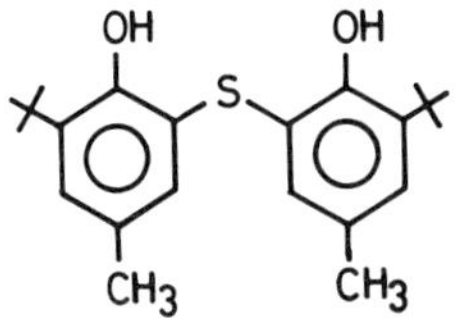

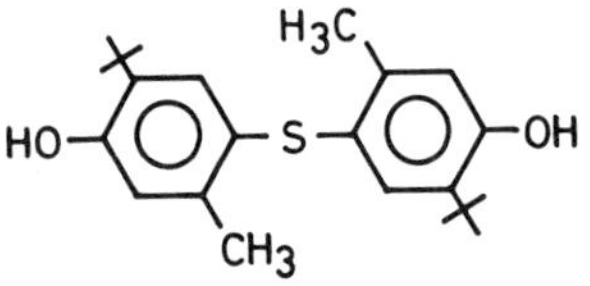

(d) Hydroxybenzyl compounds, e.g. :

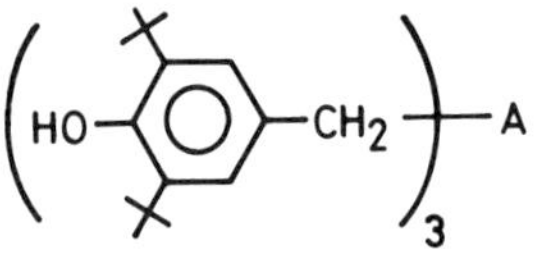

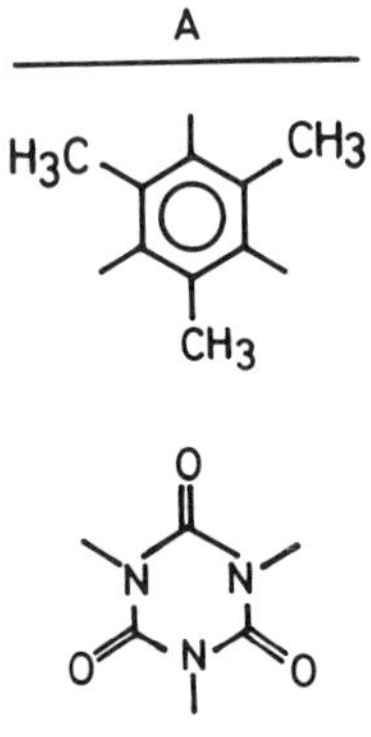

FIG. 1.28   (continued)

(e) Aminophenols, e.g.:

	R_1	R_2
	$-H$	$-COC_{11}H_{23}$
	$-H$	$-COC_{17}H_{35}$
	tert. butyl	(triazine with SC_8H_{17} groups)

(f) Hydroxyphenylpropionates e.g.:

n	A
1	$-OC_{18}H_{37}$
2	$-O(CH_2)_6O-$
2	$-NH-(CH_2)_6-NH-$
2	$-O(CH_2)_2-S-(CH_2)_2O-$
3	(isocyanurate ring structure)

The majority of secondary antioxidants are esters of thiodipropionic acid with fatty alcohols and triesters of phosphorus acid. Their chemical structures are shown in Fig. 1.29. Phosphites are more sensitive towards hydrolysis than esters of carboxylic acids, and this point has to be taken into account when storing and using phosphites. Aromatic phosphites are often preferred since they are inherently more resistant to hydrolysis than aliphatic phosphites.

(a) Thioethers. e.g. :

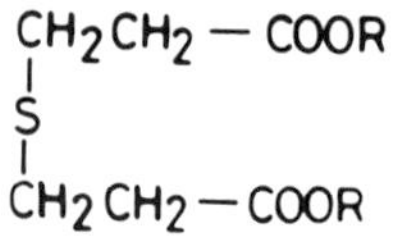

CH$_2$CH$_2$ — COOR
|
S
|
CH$_2$CH$_2$ — COOR

R
— C$_{12}$H$_{25}$
— C$_{13}$H$_{27}$
— C$_{18}$H$_{37}$

(b) Phosphites and phosphonites, e.g. :

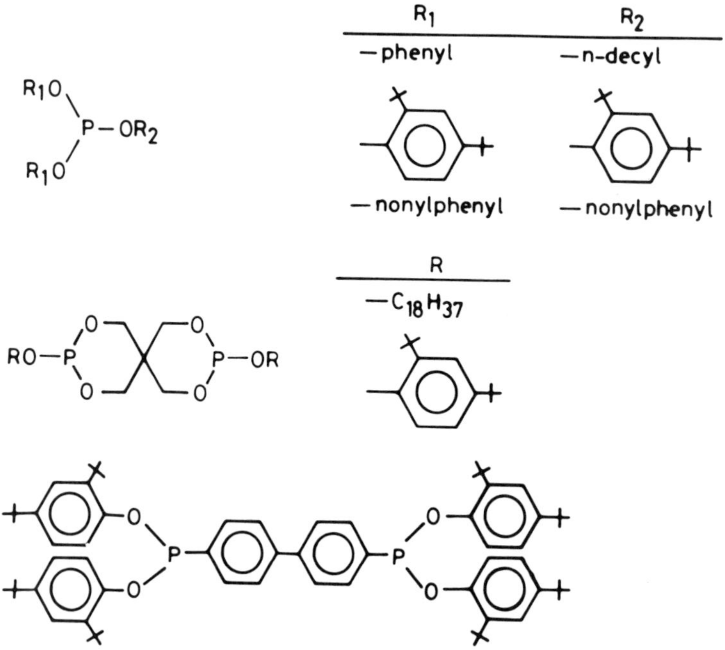

R$_1$O
  \
   P—OR$_2$
  /
R$_1$O

R$_1$	R$_2$
—phenyl	—n-decyl
—nonylphenyl	—nonylphenyl

RO—P

R
—C$_{18}$H$_{37}$

(c) Zinc dibutyldithiocarbamate

$$(C_4H_9)_2 \; N-\overset{\overset{\text{S}}{\|}}{C}-S-Zn-S-\overset{\overset{\text{S}}{\|}}{C}-N(C_4H_9)_2$$

FIG. 1.29 Structural formulas of some commercial secondary antioxidants for thermoplastics.

An index of trade names, manufacturers, and suppliers based on a choice of representative antioxidants for thermoplastics is given in Appendix 2A. Detailed lists are found in references [29] and [30].

Stabilization of Selected Polymers

Two important considerations in selecting antioxidants for polymers are toxicity and color formation. Thus, for use in food wrapping, any antioxidant or additive must be approved by the appropriate government agency. An antioxidant widely used in food products and food wrapping is butylated hydroxytoluene (BHT) (see Fig. 1.28a: R_1 = *tert*-butyl and R_2 = $-CH_3$). It is also added in small amounts to unsaturated raw rubbers before shipping to protect them during storage.

Materials that are not effective inhibitors when used alone may nevertheless be able to function as synergists by reacting with an oxidized form of an antioxidant to regenerate it and thus prolong its effectiveness. For example, carbon black forms an effective combination with thiols, disulfides, and elemental sulfur, even though these substances may be almost completely ineffective alone under comparable conditions. Synergistic combinations from a variety of chain terminators and sulfur compounds have been reported. A widely used combination of stabilizers for polyolefins is 2,6-di-*tert*-butyl-4-methylphenol (BHT) (Fig. 1.28a) with dilauryl thiodipropionate (DLTDP):

$$(H_{25}C_{12} \; O-\overset{\displaystyle O}{\overset{\displaystyle \|}{C}}-CH_2 \; CH_2)_2 \; S$$

This combination is of particular interest because both components are among the small group of stabilizers approved by the U.S. Food and Drug Administration for use in packaging materials for food products.

Other synergistic combinations have been reported involving free-radical chain terminators used with either ultraviolet absorbers or metal deactivators as the preventive antioxidants. The differences in mechanism of action of these different types of stabilizers permits them to act independently but cooperatively to provide greater protection than would be predicted by the sum of their separate effects.

Polypropylene

Because of the presence of tertiary carbon atoms occurring alternately on the chain backbone, propylene is particularly susceptible to oxidation at elevated temperatures. Since polypropylene is normally processed at temperatures between 220 and 280°C, it will degrade under these conditions (to form lower-molecular-weight products) unless it is sufficiently stabilized before it reaches the processor. The antioxidants are added at least partially during the manufacturing process and at the least during pelletizing. Antioxidant systems in technical use are composed of processing stabilizers, long-term heat stabilizers, calcium- or zinc-stearate, and synergists if necessary.

Typical processing stabilizers for polypropylene are butylated hydroxytoluene (BHT) as the primary antioxidant and phosphites and phosphonates as secondary antioxidants. Examples of the latter that are commonly used are: tetrakis-(2,4-di-*tert*-butyl-phenyl)-4,4′-bisphenylylenediphosphonite, distearyl-pentaerythrityl-disphosphonite, tris-(nonylphenyl)-phosphite, tris-(2,4-di*tert*-butyl-phenyl)-phosphite and bis(2,4-di-*tert*-butyl-phenyl)-pentaerythrityl-diphosphite. In commercial polypropylenes, phosphorous compounds are always used together with a sterically hindered phenol. The compounds are commonly added in concentrations between 0.05 and 0.25%.

The most important long-term heat stabilizers for polypropylene are phenols of medium (300 to 600) and especially high (600 to 1200) molecular weight, which are frequently used together with thioethers as synergists, e.g., dilauryl thiodipropionate (DLTDP), or distearyl thiodipropionate (DSTDP), or dioctadecyl disulfide.

Polyethylene

High-density polyethylenes (HDPE) are less sensitive to oxidation than polypropylene, so lower stabilizer concentrations are generally sufficient. As in polypropylene, antioxidants can be added during a suitable stage of manufacture or during pelletizing. The antioxidants in technical use are the same as for polypropylene. Phenols of medium- and high-molecular weight are also active as long-term heat stabilizers. Concentrations between 0.03 and 0.15% are usual.

Low-density polyethylene (LDPE) is extensively used for the manufacture of films. During processing, which is carried out at temperatures of approximately 200°C, cross-linking, and thus formation of gel, can occur through oxidation if the polymer is not stabilized. Such gel particles are visible in the film as agglomerates, known as "fish eyes" or "arrow heads." The processing stabilizers used in LDPE consist of systems commonly used for polypropylene, namely, combinations of a phosphite or phosphonite and a long term heat stabilizer (hindered phenol) in overall concentrations up to 0.1%. Concentrations seldom exceed 0.1%, since the compatibility of any additive in LDPE is considerably lower than in any other polyolefins.

For primary cable insulation, cable jackets, and pipes manufactured from LDPE, medium to high-molecular-weight grades are used and are frequently cross-linked after processing. Long-term heat stabilization is of primary importance in these applications, since lifetimes of up to 50 years are required (usually at elevated temperatures with short-time peaks up to 100°C for cables). The antioxidants have to be extremely compatible and resistant to extraction. Some typical antioxidants customary for insulation are as follows:

For power cable insulation. Pentaerythrityltetrakis-3-(3,5-di-*tert*-butyl-4-hydroxyphenyl)-propionate, polymeric 2,2,4-trimethyl-1,2-dihydroquinoline, 4,4'-thiobis-(3-methyl-6-*tert*-butyl-phenol), 2,2'-thiodiethyl-bis-3-(3,5-di-*tert*-butyl-4-hydroxyphenyl)-propionate, and distearyl thiodipropionate are used as synergists.

For communication cable insulation. 2,2'-Thiobis-(4-methyl-6-*tert*-butyl-phenol), 4,4'-thiobis-(3-methyl-6-*tert*-butyl-phenol), 2,2'-methylenebis-(4-methyl-6-α-methylcyclohexyl-phenol), 1,1,3-tris-(5-*tert*-butyl-4-hydroxy-2-methylphenyl)-butane, 2,2'-thiodiethyl-bis-3-(3,5-di-*tert*-butyl-4-hydroxy-phenyl)-propionate, pentaerythrityl-tetrakis-3-(3,5-di-*tert*-butyl-4-hydroxy-phenyl)-propionate, and dilauryl thiodipropionate are used as synergists.

The simultaneous use of metal deactivators has become increasingly important for cable insulation. This is discussed later.

Polystyrene

Unmodified crystal polystyrene is relatively stable under oxidative conditions, so that for many applications the addition of an antioxidant is not required. Nevertheless, repeated processing may lead to oxidative damage of the material, leading to an increase of melt flow index and to embrittlement of the material. Stabilization is effected by the addition of octadecyl-3-(3,5-di-*tert*-butyl-4-hydroxy-phenyl)-propionate at concentrations of up to 0.15%, if necessary in combination with phosphites or phosphonites to improve color.

Compared to unmodified crystal polystyrene, impact polystyrene consisting of copolymers of styrene and butadiene are more sensitive to oxidation. This sensitivity is a consequence of the double bonds in polybutadiene component and manifests itself in yellowing and the loss of mechanical properties of the polymer. In impact polystyrene, the following antioxidants or their mixture are used in total concentrations of 0.1 to 0.25%: BHT, octadecyl-3-(3,5-di-*tert*-butyl-4-hydroxyphenyl)-propionate, 1,1,3-tris-(5-*tert*-butyl-4-hydroxy-2-methylphenyl)-butane, and dilauryl thiodipropionate.

Acrylonitrile-Butadiene-Styrene Copolymers

Like impact polystyrene, acrylonitrile-butadiene-styrene copolymers (ABS) are sensitive to oxidation caused by the unsaturation of the elastomeric component. The processes for the manufacture of ABS require the drying (at 100 to 150°C) of powdery polymers that are extremely sensitive to oxidation. Thus, antioxidants have to be added before the coagulation step, normally in emulsified form, although sometimes in solution. The primary antioxidants are frequently used together with a synergist. Primary antioxidants commonly used for ABS are BHT, 2,2'-methylenebis-(4-ethyl or methyl-6-*tert*-butyl-phenol), 2,2'-methyl enebis-(4-methyl-6-cyclohexyl-phenol), 2,2'-methylenebis-(4-methyl-6-nonyl-phenol), octadecyl-3-(3,5-di-*tert*-butyl-4-hydroxyphenyl)-propionate, and 1,1,3-tris(-5-*tert*-butyl-4-hydroxy-2-methylphenyl)-butane. Important synergists are tris-(nonyl-phenyl)-phosphite and dilauryl thiodipropionate. These antioxidants are either liquids or show comparatively low melting points, which is an important prerequisite for the formation of stable emulsions.

Polycarbonate

Thermoxidative degradation of polycarbonate manifests itself in yellowing that is readily seen because of the transparency of polycarbonate. For this reason, stabilization against discoloration is considered to be important. Severe requirements are also imposed on the nonvolatility and thermostability of the stabilizers for polycarbonate, because the processing temperatures are extraordinarily high (about 320°C). The stabilizers are generally added during the pelletizing step.

Yellowing of polycarbonate during processing is retarded by the addition of phosphites or phosphonites. They are used in concentrations of 0.05 to 0.15%, possibly in combination with an epoxy compound as acid acceptor. The addition of these stabilizers not only decreases the yellowness index but also inhibits the rise of the melt flow index and negative influence on impact strength of processing.

The effective processing stabilizers are not suitable, however, for preventing the aging effect of long-term use. For long-term heat stabilization, a sterically hindered phenolic antioxidant is added. An effective antioxidant is octadecyl-3-(3,5-di-*tert*-butyl-4-hydroxyphenyl)-propionate.

Nylons

Degradation of nylons due to processing and aging causes discoloration and loss of mechanical properties, although not at the same rate. For example, yellowing is observed already after short periods of oven aging at 165°C, but tensile strength and elongation are hardly affected during the same time period. Discoloration of polyamides during processing can be suppressed to a certain extent by the addition of phosphites, e.g., tris-(2,4-di-*tert*-butyl-phenyl)-phosphite, in concentrations of 0.2 to 0.4%.

The stabilization of nylons is mainly a matter of long-term stabilization. Three main groups of stabilizers have become known: (1) copper salts, especially in combination with halogen and/or phosphorus compounds (e.g., copper acetate + potassium iodide/phosphoric acid), (2) aromatic amines (e.g., *N*,*N'*-dinaphthyl-*p*-phenylenediamine or *N*-phenyl-*N'*-cyclohexyl-p-phenylenediamine), and (3) hindered phenols.

Copper-halogen systems are effective in very low concentrations (10 to 50 ppm of copper; approximately 1000 ppm of halogen), although they are extracted quite easily with water and cause discoloration of the substrate. Next to the copper-halogen systems, aromatic amines are the most effective stabilizers. They are used in rather high concentrations, from 0.5 to 2%. However, since they have strongly discoloring properties they are used mainly for technical articles, which tolerate discoloration.

Hindered phenols do not show the above-mentioned disadvantages. They are the stabilizers of choice whenever good oxidation stability has to be coupled with good color stability and, possibly, food approval of the end article. The most important hindered phenols in use are the following: *N,N'*-hexamethylenebis-3-(3,5-di-*tert*-butyl-4-hydroxyphenyl)-propionamide,1,1,3-tris-(5-*tert*-butyl-4-hydroxy-2-methylphenyl-butane, 1,3,5-tris-(3,5-di-*tert*-butyl-4-hydroxybenzyl)-mesitylene, and BHT. The antioxidants can be added already during polycondensation, the normal concentration ranging from 0.3 to 0.7%.

Thermoplastic Elastomers

Styrene-based thermoplastic elastomers (see Chapter 4) are sensitive to oxidation since they contain unsaturated soft segments. These elastomers are manufactured by solution polymerization processes in aliphatic hydrocarbons. In order to prevent autoxidation during the finishing steps (stripping, drying), which manifests itself by a rise in melt flow index and discoloration of the raw polymer, antioxidant is added to the polymer solution before finishing. Hence the antioxidant has to be soluble in the polymerization solvent.

A number of hindered phenols are used in practice in a total concentration of approximately 0.5%. Examples of primary antioxidants used are BHT, 1,3,5-tris-(3,5-di-*tert*-butyl-4-hydroxybenzyl)-mesitylene, and octadecyl-3-(3,5-di-*tert*-butyl-4-hydroxyphenyl)-propionate. Tris-(nonylphenyl)-phosphite is used as a synergist. BHT may, however, be partially lost during finishing and drying. The phenols of higher molecular weight are used as they have low volatility and have the further advantage of protecting the material also during processing and end use.

Thermoplastic polyester elastomers (see Chapter 4) contain readily oxidizable polyether soft segments that make stabilization necessary. Essentially two antioxidants, namely, 4,4'-di(α, α-dimethylbenzyl)-diphenylamine and *N,N'*-hexamethylenebis-3-(3,5-di-*tert*-butyl-4-hydroxyphenyl)-propionamide are in use in concentrations of up to 1%. The antioxidants can be added during pelletizing, or even better, already during polycondensation. When added during polycondensation, *N,N'*-hexamethylenebis-3-(3,5-di-*tert*-butyl-4-hydroxyphenyl)-propionamide is partly chemically bound to the polymer because of its amide structure. The antioxidant thus becomes highly stable to extraction.

The oxidative stability of thermoplastic polyurethane elastomers is determined by the length and the structure of the linear polyether or polyester soft segments. The stability of polyether urethanes against autoxidation is distinctly lower compared to that of polyester urethanes, but the latter are less stable to hydrolysis. Antioxidants may be used in polyurethanes for stabilization against loss of mechanical properties and against discoloration in injection molding grades and as gasfading inhibitors in elastomeric fibers.

The antioxidants used in thermoplastic polyurethane elastomers are hindered phenols, e.g., BHT, octadecyl-3-(3,5-di-*tert*-butyl-4-hydroxyphenyl)-propionate, and pentaerythrityl-tetrakis-3-(3,5-di-*tert*-butyl-4-hydroxyphenyl)-propionate, and aromatic amines, e.g., 4,4′-di-*tert*-octyl-diphenylamine, as well as their combinations. Aromatic amines can, however, be employed only in very limited concentrations (250 to 550 ppm at most) because of their discoloring properties.

Polyacetal

Polyoxymethylenes have a marked tendency to undergo thermal depolymerization with loss of formaldehyde. To prevent thermal depolymerization, polyoxymethylenes are structurally modified, the two possibilities being acetylation to block the reactivity of the end groups or copolymerization with cyclic ethers, e.g., ethylene oxide. Polyacetals are also sensitive towards autoxidation, which invariably leads to depolymerization as a result of chain scission. The formaldehyde released by depolymerization is very likely to be oxidized to formic acid, which can catalyze further depolymerization.

The stabilizer systems for polyacetals are invariably composed of a hindered phenol with a costabilizer. The hindered phenols in use are 2,2′-methylenebis-(4-methyl-6-*tert*-butyl-phenol), 1,6-hexamethylenebis-3-(3,5-di-*tert*-butyl-4-hydroxyphenyl)-propionate, and pentaerythrityl-tetrakis-3-(3,5-di-*tert*-butyl-4-hydroxyphenyl)-propionate. A large number of nitrogen-containing organic compounds have been described as costabilizers for polyacetals, e.g., dicyandiamide, melamine, terpolyamides, urea, and hydrazine derivatives. The effectiveness of these compounds is based on their ability to react with formaldehyde and to neutralize acids, especially formic acid, formed by oxidation. In addition to nitrogen compounds, salts of long-chain fatty acids (e.g., calcium stearate, calcium ricinoleate, or calcium citrate) are also used as acid acceptors. The practical concentrations are 0.1–0.5% for the phenolic antioxidant and 0.1–1.0% for the costabilizer.

Poly(Vinyl Chloride)

Poly(vinyl chloride)(PVC) is relatively unstable under heat and light. The first physical manifestation of degradation is a change in the PVC color, which on heating, changes from the initial water-white to pale yellow, orange, brown, and finally black. Further degradation causes adverse changes in mechanical and electrical properties.

The most widely accepted mechanism for PVC degradation is one based on a free-radical chain. Thermal initiation probably involves loss of a chlorine atom adjacent to some structural abnormality, such as terminal unsaturation, which reduces the stability of the C-Cl bond [28]. The chlorine radical thus formed abstracts a hydrogen to form HCl, and the resulting chain radical then reacts to form a chain unsaturation with regeneration of another chlorine radical.

$$Cl\cdot + \text{ⱳ}CH_2\text{—}CH\text{ⱳ} \xrightarrow{-HCl} \text{ⱳ}\overset{\cdot}{C}H\text{—}CH\text{ⱳ} \longrightarrow \text{ⱳ}CH\text{=}CH\text{ⱳ} + Cl\cdot$$
$$\qquad\qquad\;\; | \qquad\qquad\qquad\quad |$$
$$\qquad\qquad\;\; Cl \qquad\qquad\qquad\quad Cl$$

Thus, as hydrogen chloride is removed, polyene structures are formed:

$$\text{ⱳ}CH_2\text{—}CH\text{—}CH_2\text{—}CH\text{—}CH_2\text{—}CH\text{ⱳ} \xrightarrow{-HCl}$$
$$\qquad\;\; | \qquad\quad | \qquad\quad |$$
$$\qquad\;\; Cl \qquad\quad Cl \qquad\quad Cl$$
$$\qquad\qquad\qquad\qquad\qquad \text{ⱳ}CH\text{=}CH\text{—}CH\text{=}CH\text{—}CH\text{=}CH\text{ⱳ}$$

The reaction can also be initiated by ultraviolet light. In the presence of oxygen the reactions are accelerated (as evidenced by the acceleration of color formation), and ketonic structures are formed in the chain.

Stabilizers are almost invariably added to PVC to improve its heat and light stability. The species found effective in stabilizing PVC are those that are able to absorb or neutralize HCl, react with free radicals, react with double bonds, or neutralize other species that might accelerate degradation. Lead compounds, such as basic lead carbonate and tribasic lead sulfate, and metal soaps of barium, cadmium, lead, zinc, and calcium are used as stabilizers. Obviously, they can react with HCl. Epoxy plasticizers aid in stabilizing the resin. Another group of stabilizers are the organotin compounds, which find application because of their resistance to sulfur and because they can yield crystal-clear compounds.

Rubber

In stabilizing rubber, antioxidants are classified as staining or nonstaining, depending on whether they develop color in use. In carbon-black-loaded rubber tire a staining material is not harmful. For white rubber goods, however, it is important that the additives used be colorless and that they stay colorless while protecting against oxidation. Some antioxidants that find use in rubber are phenyl-β-naphthylamine (PBNA), which is staining, and 2,2'-thiobis (6-*tert*-butyl-para-cresol) (see Fig. 1.28c), which is nonstaining.

It has been established that ozone formation is a general phenomenon that occurs when free radicals can combine with oxygen. Rubber products made from natural rubber, styrene-butadiene rubber (SBR), or nitrile rubber, when stretched moderately and exposed to low concentrations of ozone, crack rapidly and sometimes disastrously. The cracking is caused by cuts that appear on the surface and may penetrate deeply, causing serious damage. These cuts appear only when the rubber is under tension during exposure to ozone and if they cross the lines of tension at an angle of 90°. Since the cracking problem is associated only with elastomers containing some degree of residual unsaturation, the process is considered to be related to the attack of ozone on the unsaturation. Unlike molecular oxygen (O_2), ozone appears to add directly to the double bond, which is often followed by chain scission. Use of diene rubbers in an environment containing ozone, such as an automobile tire in some urban locations, therefore calls for stabilization against ozone.

The effect of ozone is both greatly delayed and reduced when various antiozonants, such as *N,N'*-di-alkyl-*p*-phenylene-diamines (I) or similar compounds, are incorporated in the rubber. They may react directly with ozone or with the ozone-olefin reaction products in such a way as to prevent chain scission. SBR is much more receptive to protection than either the nitrile or natural rubbers. Of the latter two the nitrile is more readily inhibited. If a wax is also added in small quantities with the antiozonant, the retardation of the attack of ozone is in certain instances enhanced several fold.

$R_1, R_2 = Alkyl$

(I)

It is believed that waxes function by providing an unreactive layer on the rubber surface. Since, unlike oxygen, ozone reacts only at the surface and does not diffuse into the sample, a surface layer of relatively unreactive wax presents an impervious surface to ozone. The wax thus needs to bloom (i.e., exude to the surface) in order to afford protection. A great deal of attention has therefore been paid to the factors affecting migratory aptitude of wax. Waxes with good protective power have been found to have melting points between 65 and 72°C, refractive indices in the range of 1.432 to 1.438, and branching to the extent of 30 to 50% side chains [28].

Of the two general categories of waxes—paraffin and microcrystalline—the latter are more strongly held to the surface. However, the use of waxes alone to provide protection against ozone attack is rather well restricted to static conditions of service. Whenever constant flexing is present, even the more strongly held microcrystalline waxes flake off and protection is lost. Combinations of waxes and chemical antiozonants are therefore used to provide protection under both static and dynamic conditions of service. In fact, it is felt that waxes can aid in the diffusion of chemical antiozonant to the rubber surface.

METAL DEACTIVATORS

As described previously, thermooxidative degradation of polyolefins proceeds by a typical free-radical chain mechanism in which hydroperoxides are key intermediates because of their thermally-induced homolytic decomposition to free radicals, which in turn initiate new oxidation chains. However, since the monomolecular homolytic decomposition of hydroperoxides into free radicals require relatively high activation energies, this process becomes effective only at temperatures in the range of 120°C and higher. But in the presence of certain metal ions, hydroperoxides can undergo catalytic decomposition even at room temperature by a redox reaction to radical products. Two oxidation-reduction reactions can be involved depending on the metal and its state of oxidation:

$$ROOH + M^{n+} \longrightarrow RO\cdot + M^{(n+1)+} + OH^- \tag{34}$$

$$ROOH + M^{(n+1)+} \longrightarrow ROO\cdot + M^n + H^+ \tag{35}$$

$$2ROOH \xrightarrow{M^{n+}/M^{(n+1)+}} RO\cdot + ROO\cdot + H_2O \tag{36}$$

The electron transfer producing free radicals as shown above is preceded by the formation of unstable coordination complexes of the metal ions with alkyl hydroperoxides. The relative importance of reactions (34) and (35) depends upon the relative strength of the metal ion as an oxidizing or reducing agent. When the metal ion has two valence states of comparable stability, both reactions (34) and (35) will occur, and a trace amount of the metal can convert a large amount of peroxide to free radicals according to the sum of the two reactions [reaction (36)]. This is true of compounds of metals such as Fe, Co, Mn, Cu, Ce, and V, commonly called transition metals.

The presence of the above-mentioned metal ions increases the decomposition rate of hydroperoxides and the overall oxidation rate in the autoxidation of a hydrocarbon to such an extent that even in the presence of antioxidants, the induction period of oxygen uptake is drastically shortened. In such a case, sterically hindered phenols or aromatic amines even at rather high concentrations, do not retard the oxidation rate satisfactorily. A much more efficient inhibition is then achieved by using metal deactivators, together with anti-oxidants. Metal deactivators are also known as *copper inhibitors*, because, in practice, the copper-catalyzed oxidation of polyolefins is by far of greatest importance. This is due to the fact that polyolefins are the preferred insulation material for communication wire and power cables, which generally contain copper conductors.

The function of a metal deactivator is to form an inactive complex with the catalytically active metal species. Specially suited for this purpose are chelating agents, which can form metal complexes of high thermal stability. The general feature of chelating agents is that they may contain several ligand atoms like N, O, S, P, alone or in combination with functional groups such as hydroxyl, carboxyl, or carbamide groups. The chelating agent *N,N'*-bis-(*o*-hydroxybenzal) oxalyldihydrazide is a good example for the above mentioned structural possibilities. It forms a soluble complex (II) with the first mole of a copper salt by binding at the phenolic group, and a second mole of copper salt binds at amide nitrogens to form an insoluble complex.

(II)

Besides their main function to retard efficiently the metal-catalyzed oxidation of polyolefins, metal deactivators have to possess a number of ancillary properties to be useful in actual service. These include sufficient solubility or ease of dispersion, high extraction resistance, and low volatility and sufficient thermal stability under processing and service conditions.

Within the last two decades, a number of chemical structures have been proposed as metal deactivators for polyolefins. These include carboxylic acid amides of aromatic mono- and di-carboxylic acids and *N*-substituted derivatives such as *N,N'*-diphenyloxamide, cyclic amides such as barbituric acid, hydrazones and bishydrazones of aromatic aldehydes such as benzaldehyde and salicylaldehyde or of *o*-hydroxy-arylketones, hydrazides of aliphatic and aromatic mono- and di-carboxylic acids as well as *N*-acylated derivatives thereof, bis-acylated hydrazine derivatives, polyhydrazides, and phosphorus acid ester of a thiobisphenol. An index of trade names and suppliers of a few commercial metal deactivators is given in Appendix 2B.

Though there are metals other than copper (such as iron, manganese and cobalt) that can accelerate thermal oxidation of polyolefins and related polymers such as EPDM, in practice, however, the inhibition of copper-catalyzed degradation of polyolefins is of paramount importance because of the steadily increasing use of polyolefin insulation over copper conductors. Among polyolefins, polyethylene is still the most common primary insulation material for wire and cable. In the United States, high-density polyethylene and ethylene-propylene copolymers are used in substantial amounts for communications wire insulation.

For stabilization of polyolefins in contact with copper, it is often mandatory to combine a metal deactivator with an antioxidant. Metal deactivators in actual use are essentially N,N'-bis-[3-(3′,5′-di-*tert*-butyl-4′-hydroxy-phenyl)propionyl]-hydrazine and N,N'-dibenzaloxalyldihydrazide. The latter compound requires predispersion in a masterbatch because of its insolubility in polyolefins. This is not needed, however, for the former compound, which at commonly used concentrations is molecularly dispersed in polyolefins after processing. The required additive concentrations range from 0.05 to 0.5% depending on the polymer, the nature of the insulation (solid, cellular), whether the cable is petrolatum filled, and on service conditions.

Increasingly, communication cables are filled with petrolatum to improve waterproofing, whereas the insulation may be solid or cellular. Both trends, namely, petrolatum filling and cellular insulation, exert an influence on the oxidative stability of such cables. Thermal stability of high density polyethylene decreases by about 35% in changing from solid to cellular insulation. Moreover, in contact with petrolatum the stability of solid polyethylene decreases by 35% and that of cellular polyethylene decreases by 10 to 40%. Even in the most adverse instance, i.e., cellular insulation in contact with petrolatum, elevated concentrations of antioxidant and metal deactivator make it possible to achieve a high level of stability.

LIGHT STABILIZERS

Most polymers are affected by exposure to light, particularly the ultraviolet (UV) portion of sun's spectrum between 300 and 400 nm. (Fortunately, the earth's atmosphere filters out most of the light waves shorter than 300 nm.) Light contributes very actively to polymer degradation, especially when oxygen is present, which is a normal situation for most plastic materials. For example, when rubber is exposed to UV radiation at 45°C, it oxidizes three times as fast as the dark at 70°C. Light and oxygen induce degradation reactions in plastics that may not only discolor them but also exert a detrimental influence on numerous mechanical and physical properties. The inhibition of these degradation reactions is essential or else, the applications of many plastics would be drastically reduced. The inhibition can be achieved through addition of special chemicals, light stabilizers, or UV stabilizers, which are capable of interfering with the physical and chemical processes of light-induced degradation.

Carbon black is used as a stabilizer in a limited number of formulations where color is not a criterion. It not only absorbs light, but it can also react with free-radical species that might be formed. The weathering properties of polyethylene are improved by the incorporation of carbon blacks (at 2–3% concentration). Weather-resistant wire and cable insulation, pipe for outdoor applications, films for mulching, and water conservation in ponds and canals may be made from polyethylene containing carbon black.

In spite of the fact that carbon black and other pigments may also protect the plastics from the effects of light, we shall consider in the following only those organic and organo-matallic compounds that impart only slight discoloration or no discoloration at all to the plastics to be stabilized.

The most important classes of light stabilizers from a practical point of view are 2-hydroxybenzophenones, 2-hydroxyphenylbenzotriazoles, hindered amines, and organic nickel compounds. In addition, salicylates, *p*-hydroxy-benzoates, resorcinol monobenzoates, cinnamate derivatives, and oxanilides are also used. Light stabilizers are generally used in concentrations between 0.05 and 2%, the upper limit of this range being employed only exceptionally. The sterically hindered amines represent the latest development in the field. They have outperformed the previously available light stabilizer classes in numerous synthetic resins.

Light Stabilizer Classes

The designation of the light stabilizer classes is based on mechanisms of UV stabilization. Free radicals are formed in polymers exposed to light, as a consequence of the excitation of absorbing functionalities in the polymer. This is a function of the energy of the light and of the structure of the polymer molecule. Since in the presence of oxygen, the polymer will simultaneously oxidize (photooxidation), it is often difficult to distinguish the pure photochemical processes from the thermal processes (oxidation), which are then superimposed. Some of the fundamental mechanisms involved have been evaluated [31]. The mechanism developed initially for the thermal oxidation of rubber can be applied to other substrates and also to photooxidation. The scheme shown below illustrates a possible reaction sequence to photooxidation:

Chain initiation

$$\left.\begin{array}{l} \text{Hydroperoxides ROOH} \\ \text{Carbonyl compounds} > C=O \\ \text{Catalyst residues (Ti, } \ldots) \\ \text{Charge transfer complexes (RH, } O_2) \end{array}\right\} \xrightarrow[\text{M}^{n+}/\text{M}^{(n+1)+}]{\Delta,\ h\nu} \begin{array}{l} \text{Free radicals} \\ (R\cdot, RO\cdot, \\ HO\cdot, HO_2^{\cdot}, \ldots) \end{array}$$

$$(37)$$

Chain propagation

$$R\cdot + O_2 \longrightarrow RO_2^{\cdot} \tag{38}$$

$$RO_2^{\cdot} + RH \longrightarrow RO_2H + R\cdot \tag{39}$$

Chain branching

$$ROOH \xrightarrow[\text{M}^{n+}/\text{M}^{(n+1)+}]{\Delta,\ \text{or}\ h\nu} RO\cdot + HO\cdot \tag{40}$$

$$2ROOH \xrightarrow[\text{M}^{n+}/\text{M}^{(n+1)+}]{\Delta,\ \text{or}\ h\nu} RO_2^{\cdot} + RO\cdot + H_2O \tag{41}$$

$$RO\cdot + RH \longrightarrow ROH + R\cdot \tag{42}$$

$$HO\cdot + RH \longrightarrow H_2O + R\cdot \tag{43}$$

Chain termination

$$\left.\begin{array}{l} R\cdot + R\cdot \longrightarrow \\ R\cdot + RO_2^{\cdot} \longrightarrow \\ RO_2^{\cdot} + RO_2^{\cdot} \longrightarrow \end{array}\right\} \text{nonradical products} \tag{44}$$

In the light of the above photooxidation scheme there are four possibilities of protection against UV light. These are based on (1) prevention of UV light absorption or reduction of the amount of light absorbed by the chromophores; (2) reduction of the initiation rate through deactivation of the excited states of the chromophoric groups: (3) intervention in the photooxidative degradation process by transformation of hydroperoxides into more stable compounds, without generation of free radicals, before hydroperoxides undergo photolytic cleavage; and (4) scavenging of the free radicals as soon as possible after their formation, either as alkyl radicals or as peroxy radicals.

According to the four possibilities of UV protection mechanism described above, the light stabilizer classes can be designated as (1) UV absorbers, (2) quenchers of excited states, (3) hydroperoxide decomposers, and (4) free radical scavengers. It must be mentioned, however, that this classification is a simplification and that some compounds may be active in more than one way and often do so.

UV Absorbers

The protection mechanism of UV absorbers is based essentially on absorption of the harmful UV radiation and its dissipation in a manner that does not lead to photosensitization, i.e., conversion to energy corresponding to high wavelengths or dissipation as heat. Besides having a very high absorption themselves, these compounds must be stable to light, i.e., capable of absorbing radiative energy without undergoing decomposition.

Hydroxybenzophenones and hydroxyphenyl benzotriazoles are the most extensively studied UV absorbers. Though the main absorptions of 2-hydroxybenzophenone are situated in the uninteresting wavelength domain around 260 nm, substituents such as hydroxy and alkoxy groups push this absorption towards longer wavelengths, between 300 and 400 nm, and at the same time total absorption in the UV increases. The hydroxybenzophenone type UV absorbers (III) are essentially derived from 2,4-dihydroxybenzophenone (III, R = H). Through choice of adequate alkyl group R in the alkoxy groups it is possible to optimize the protective power and the compatibility with the plastics to be stabilized.

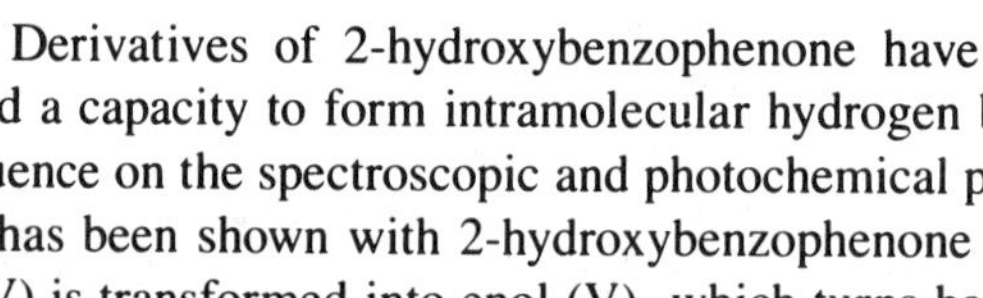

$$R = H, CH_3 \text{ to } C_{12}H_{25}$$

(III)

Derivatives of 2-hydroxybenzophenone have highly conjugated structures and a capacity to form intramolecular hydrogen bonds that exert a decisive influence on the spectroscopic and photochemical properties of these compounds. It has been shown with 2-hydroxybenzophenone (IV) that on exposure to light (IV) is transformed into enol (V), which turns back into its initial form (IV) on losing thermal energy to the medium [reaction (45)]:

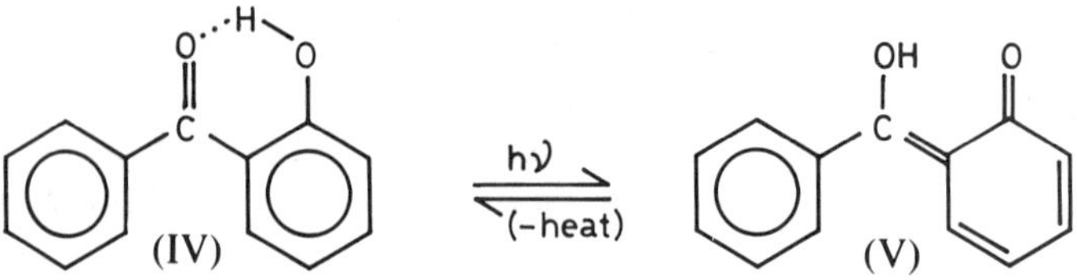

(45)

The light energy consumed by the UV absorber corresponds to the quantity of energy needed to break the hydrogen bond. This explanation is supported by the fact that compounds that cannot lead to the formation of intramolecular hydrogen bonds (benzophenone or 2-methoxybenzophenone) do not absorb in the UV wavelength range.

Hydroxyphenyl benzotriazoles have the structure (VI) where X is H or Cl (chlorine shifts the absorption to longer wavelengths), R_1 is H or branched alkyl, and R_2 is CH_3 to C_8H_{17} linear and branched alkyl (R_1 and R_2 increase the affinity to polymers). Some technically important materials in this class are

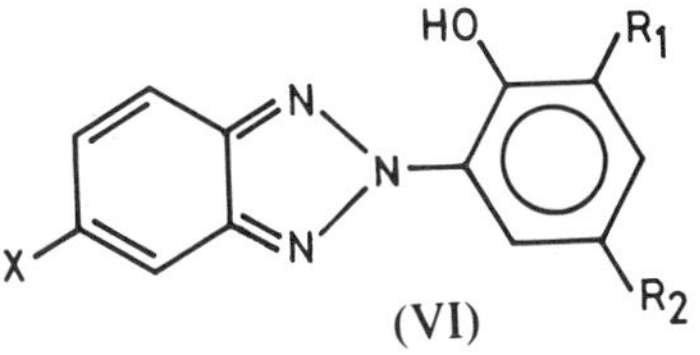

(VI)

2-(2'-hydroxy-5'-methyl-phenyl)-benzotriazole, 2-(2'-hydroxy-3'-5'-di-*tert*-butyl-phenyl)-benzotriazole, and 2-(2'-hydroxy-3',5'-di-*tert*-butyl-phenyl)-5-chlorobenzotriazole. In comparison with 2-hydroxybenzophenones, the 2-(2'-hydroxyphenyl) benzotriazoles have higher molar extinction coefficients and steeper absorption edges towards 400 nm.

The exact mechanism of light absorption by hydroxybenzotriazoles is not known. However, the formation of intramolecular hydrogen bond, as in (VII), and of zwitter ions having a quinoid structure, as in (VIII), may be responsible for the transformation of light radiation energy into chemical modifications:

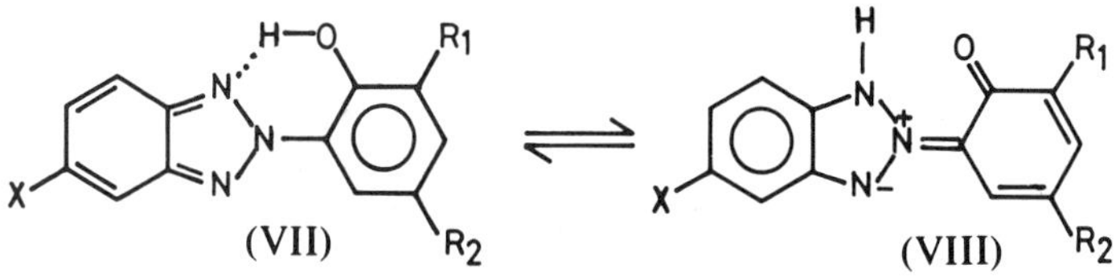

(VII)   (VIII)   (46)

It may be noted that the tendency to form chelated rings by the creation of hydrogen bonds between hydroxide and carbonyl groups [as in (IV)] or groups containing nitrogen [as in (VII)] is a characteristic property of all UV absorbers.

A fundamental disadvantage of UV absorbers is the fact that they need a certain absorption depth (sample thickness) for good protection of a plastic.

Therefore, the protection of thin section articles, such as films and fibers, with UV absorbers alone is only moderate.

Quenchers

Quenchers (Q) are light stabilizers that are able to take over energy absorbed from light radiation by the chromophores (K) present in a plastic material and thus prevent polymer degradation. The energy absorbed by quenchers can be dissipated either as heat [reaction (49)] or as fluorescent or phosphorescent radiation [reaction (50)]:

$$K + h\nu \longrightarrow K^* \tag{47}$$

$$K + Q \longrightarrow K + Q^* \tag{48}$$

$$Q^* \longrightarrow Q + \text{heat} \tag{49}$$

$$Q^* \longrightarrow Q + h\nu' \tag{50}$$

For energy transfer to occur from the excited chromophore K^* (donor) to the quencher Q (acceptor), the latter must have lower energy states than the donor. The transfer can take place by two processes: (1) long-range energy transfer or Förster mechanism and (2) contact, or collisional, or exchange energy transfer.

The Förster mechanism is based on a dipole-dipole interaction and is usually observed in the quenching of excited states. It has been considered as a possible stabilization mechanism for typical UV absorbers with extinction coefficients greater than 10,000. The distance between chromophore and quencher in this process may be as large as 5 or even 10 nm, provided there is a strong overlap between the emission spectrum of the chromophore and the absorption spectrum of the quencher.

However, for an efficient transfer to take place in the contact or exchange energy transfer process, the distance between quencher and chromophore must not exceed 1.5 nm. From calculations based on the assumption of random distribution of both stabilizer and sensitizer in the polymer, it is thus concluded that exchange energy transfer cannot contribute significantly to stabilization. This would not apply, however, if some kind of association between sensitizer and stabilizer takes place (for example, through hydrogen bonding).

Considering the dominant role of hydroperoxides in polyolefin photooxidation [cf. Eqs. (40) and (47)], quenching of excited —OOH groups would contribute significantly to stabilization. However, since the —OOH excited state is dissociative, i.e., its lifetime is limited to one vibration of the O—O bond, contact energy transfer during this very short time (about 10^{-3} s) appears highly unlikely if the —OOH group is not already associated with the quencher. The quenching action being thus independent of the thickness of the samples, quenchers are specifically useful for the stabilization of thin section articles such as films and fibers.

Metallic complexes that act as excited state quenchers are used to stabilize polymers, mainly polyolefins. They are nickel and cobalt compounds corresponding to the following structure:

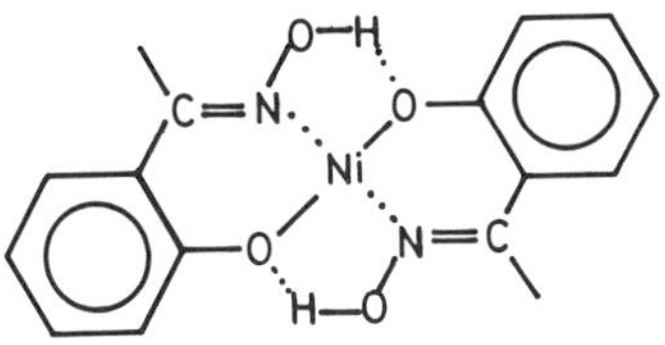

(IX)

Metallic complexes based on Ni, Co, and substituted phenols, thiophenols, dithiocarbamates, dithiophosphinates, or phosphates are used. Typical representatives are nickel-di-butyldithiocarbamate, *n*-butylamin-nickel-2,2′-thio-bis-(4-*tert*-octyl-phenolate), nickel-bis-[2,2′-thio-bis-(4-*tert*-octyl-phenolate)] and nickel-(0-ethyl-3,5-di-*tert*-butyl-4-hydroxy-benzyl) - phosphonate. But their use is not as widespread as for other UV absorbers because they tend to be green.

Hydroperoxide Decomposers

Since hydroperoxides play a determining role in the photooxidative degradation of polymers, decomposition of hydroperoxides into more stable compounds, before the hydroperoxides undergo photolytic cleavage, would be expected to provide an effective means of UV protection. Metal complexes of sulfur-containing compounds such as dialkyldithiocarbamates (X), dialkyldithiophosphates (XI) and thiobisphenolates (XII) are very efficient hydroperoxide decomposers even if used in almost catalytic quantities. Besides reducing the hydroperoxide content of pre-oxidized polymer films, they also can act as very efficient UV stabilizers. This explains the fact that an improvement in UV stability is often observed on combining UV absorbers with phosphite or nickel compounds.

Free-Radical Scavengers

Besides the absorption of harmful radiation by UV absorbers, the deactivation of excited states by quenchers, and the decomposition of hydroperoxides by some phosphorus and/or sulfur containing compounds, the scavenging of free-radical intermediates is another possibility of photostabilization, analogous to that used for stability against thermooxidative degradation. It has been shown that compounds (XII), (XIII), (XIV), and (XV) are effective

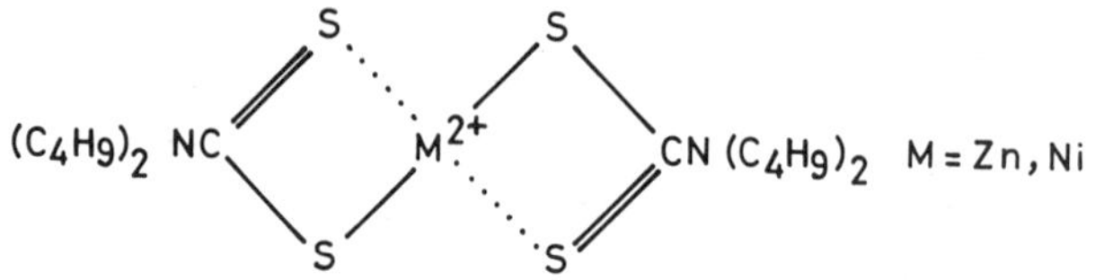

$(C_4H_9)_2$ NC$\underset{S}{\overset{S}{\diagup}}M^{2+}\underset{S}{\overset{S}{\diagdown}}$CN $(C_4H_9)_2$ M = Zn, Ni

(X)

$(i-C_3H_7)_2$N$\diagdown$$\underset{(i-C_3H_7)_2N}{}P\underset{S}{\overset{S}{\diagup}}Ni\underset{S}{\overset{S}{\diagdown}}P\diagup$$\overset{N(i-C_3H_7)_2}{}$

(XI)

Ni······NH$_2$C$_4$H$_9$

R = tert. octyl

(XII)

radical scavengers. The radicals generated by homolytic cleavage of hydro-peroxide:

$$PPOOH \longrightarrow PPO^{\cdot} + {}^{\cdot}OH \tag{51}$$

may be removed by reactions, such as

$$HO^{\cdot} + InH \longrightarrow [InH . . . {}^{\cdot}OH] \tag{52}$$

$$PPO^{\cdot} + InH \longrightarrow PPOH + In^{\cdot} \tag{53}$$

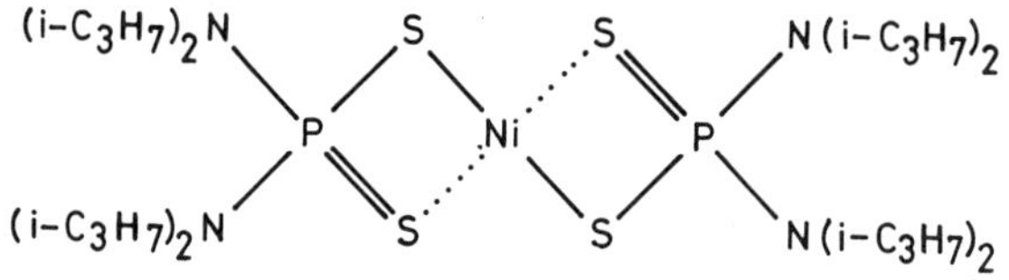

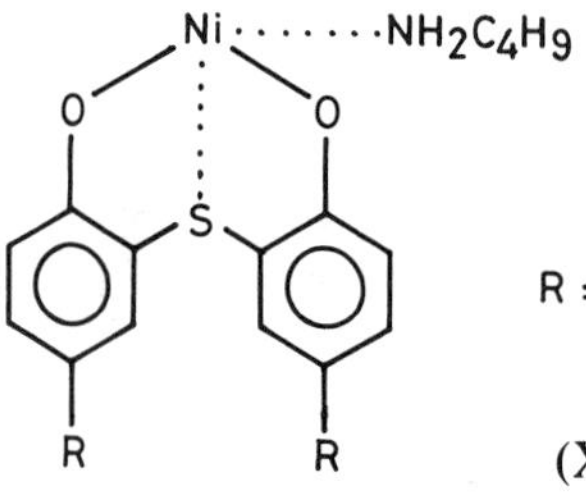

with a radical scavenger InH.

The latest development in the field of light stabilizers for plastics is represented by sterically hindered amine-type light stabilizers (HALS). A typical such compound is bis-(2,2,6,6-tetramethyl-4-piperidyl)-sebacate (XVI). Since it does not absorb any light above 250 nm, it cannot be considered a UV absorber or a quencher of excited states. This has been confirmed in polypropylene through luminescence measurements.

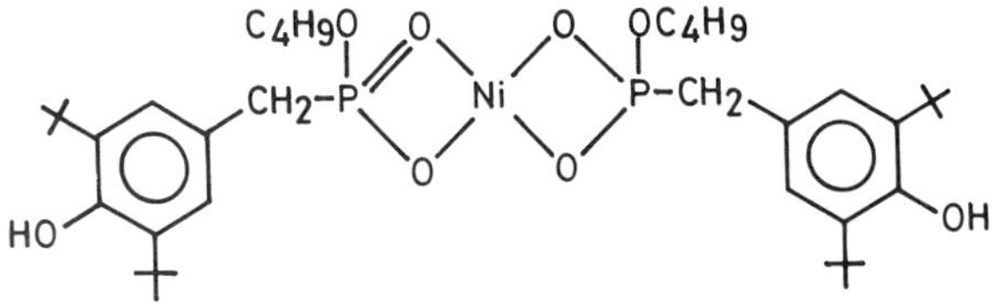

($-\!\!\!+$ = tert butyl) (XIII)

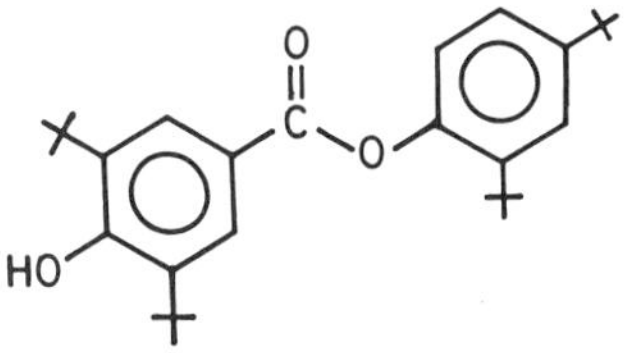

(XIV) 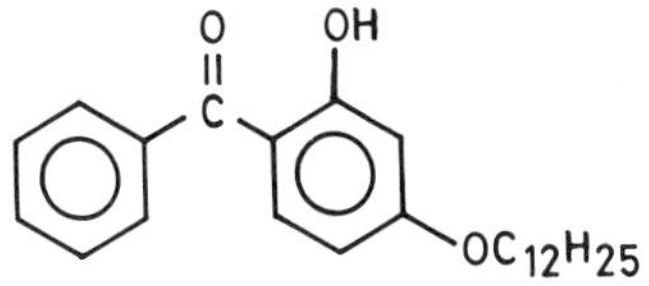　　　　　　(XV)

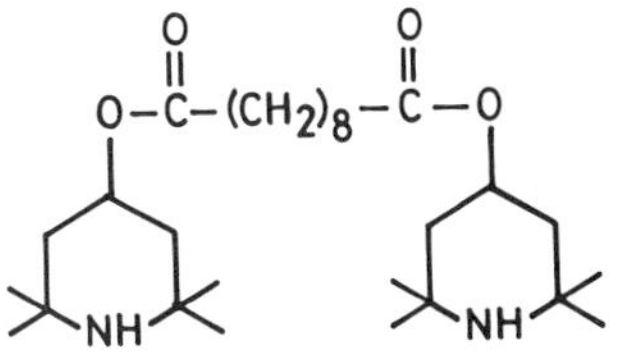

(XVI)

A low-molecular weight HALS such as (XVI), denoted henceforth as HALS-I, has the disadvantage of relative volatility and limited migration and extraction resistance, which are undesirable in special plastics applications (for example, in fine fibers and tapes). For such applications, it is advantageous to use polymeric sterically hindered amines such as poly-(*N*-β-hydroxyethyl-2,2,6,6-tetramethyl-4-hydroxypiperidyl succinate) represented by (XVII) and a more complex polymeric hindered amine represented by (XVIII). In later discussions, they will be designated as HALS-II and HALS-III, respectively. Though they do not reach completely the performance of the low-molecular weight HALS-I, they are nevertheless superior to the other common light stabilizers used at several-fold higher concentrations.

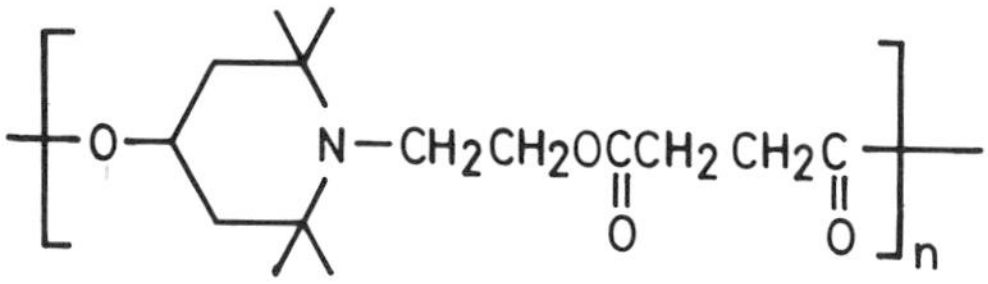

(XVII)

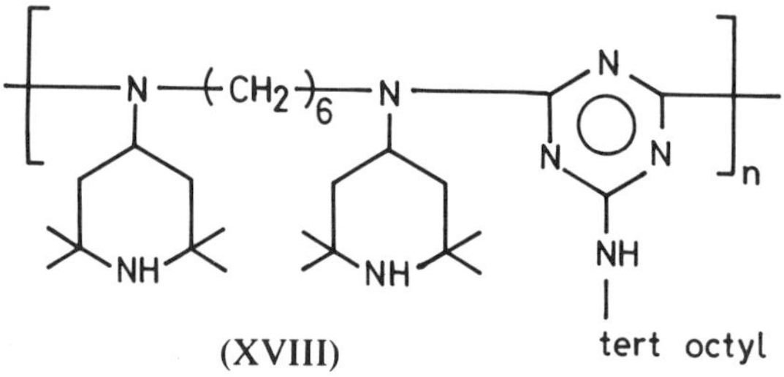

(XVIII)   tert octyl

The protection mechanisms of HALS, known so far mostly from studies with model systems, can be summarized as follows: From ESR measurements it is concluded that, under photooxidative conditions, HALS are converted, at least in part, to the corresponding nitroxyl radicals (XIX). The latter, through reaction (54), are thought to be the true radical trapping species.

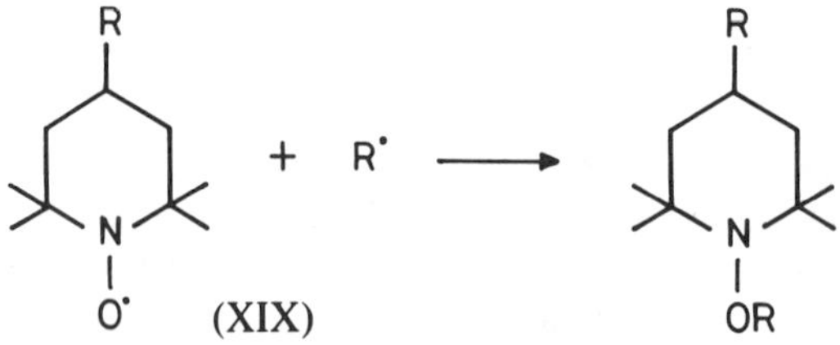

(54)

Another explanation of the UV protection mechanism of HALS involves the hydroxylamine ethers (XX) formed in reaction (54). There is indirect evidence that (XX) can react very quickly with peroxy radicals, thereby regenerating nitroxyl radicals [reaction (55)]. Reactions (54) and (55), which constitute the "Denisov cycle," result in an overall slowdown of the usual chain oxidation reactions (38) and (39).

$$\text{(XX)} + RO_2^{\bullet} \longrightarrow \text{(XIX)} + RO_2R \qquad (55)$$

Nitroxyl radicals may also react with polymer radical to form hydroxylamine [reaction (56)], and the latter can react with peroxy radicals and hydroperoxides according to reactions (57) and (58):

$$\text{>NO}^{\bullet} + R^{\bullet} \longrightarrow \text{>NOH} + \text{>C}{=}\overset{|}{\text{C}}{-}R' \qquad (56)$$

$$\text{>NOH} + RO_2^{\bullet} \longrightarrow \text{>NO}^{\bullet} + RO_2H \qquad (57)$$

$$\text{>NOH} + ROOH \longrightarrow \text{>NO}^{\bullet} + H_2O + RO^{\bullet} \qquad (58)$$

The formation of associations between HALS and hydroperoxides followed by reaction of these associations with peroxy radicals [reactions (59) and (60)] represents another possibility of retarding photooxidation.

$$\text{>NH} + HOOR \rightleftharpoons [\text{>NH} \ldots HOOR] \qquad (59)$$

$$[\text{>NH} \ldots HOOR] + RO_2^{\bullet} \longrightarrow ROOR + H_2O + \text{>NO}^{\bullet} \qquad (60)$$

The hydroperoxides associated with HALS may undergo photolysis producing hydroxy and alkoxy radicals in close proximity to the amines. The radicals may then abstract a hydrogen atom from the amine and form hydroxylamine and hydroxylamine ethers [reaction (61)]:

$$
>\!\!NH \ldots HOOR \rightarrow \left[\begin{array}{c} \cdot OH \\[2mm] >\!\!N\!\!-\!\!H \\[2mm] \cdot OR \end{array} \right] \begin{array}{l} \rightarrow >\!\!NOH + ROH \\[4mm] \rightarrow >\!\!NOR + H_2O \end{array} \qquad (61)
$$

Still other mechanisms have been postulated, e.g., the interaction of HALS with α, β-unsaturated carbonyl compounds and the formation of charge-transfer complexes between HALS and peroxy radicals. However, despite extensive publications in the field, a complete knowledge of the processes occurring in polymer photooxidation in the presence of HALS is still not available.

LIGHT STABILIZERS FOR SELECTED PLASTICS

In choosing a light stabilizer for a given plastic, several factors, in addition to its protective power, play an important role. In this respect, one may cite physical form (liquid/solid, melting point), thermal stability, possible interaction with other additives and fillers that may eventually lead to discoloration of the substrate, volatility, toxicity (food packaging), and, above all, compatibility with the plastics material considered. An additive can be considered as compatible if, during a long period of time, no blooming or turbidity is observed at room temperature and at the elevated temperatures that may be encountered during projected use of the plastic material.

Among the light stabilizer classes available commercially, only a few may be used in a broad range of plastics. Thus, nickel compounds are used almost exclusively in polyolefins, whereas poly(vinyl chloride) is stabilized with UV absorbers only.

A list of trade names and manufacturers of selected light stabilizers is given in Appendix 2C.

Polypropylene

For the light stabilization of polypropylene, representatives of the following stabilizer classes are mainly used: 2-(2′-hydroxyphenyl)-benzotriazoles, 2-hydroxy-4-alkoxybenzophenones, nickel-containing light stabilizers, 3,5-di-*tert*-butyl-4-hydroxybenzoates, as well as sterically hindered amines (HALS). Nickel-containing light stabilizers are used exclusively in thin sections, such as films and tapes, whereas all other classes may be used in thin as well as thick sections, though UV absorbers have only limited effectiveness in thin sections. Nickel-containing additives are also used as "dyesites" because they allow the dying and printing of polypropylene fibers with dyestuffs susceptible to complexation with metals.

Polyethylene

As all polyolefins, polyethylene is sensitive to UV radiation, although less than polypropylene. For outdoor use polyethylene needs special stabilization against UV light. The light stabilizers for polyethylene are in principle the same as for polypropylene. On accelerated weathering, HALS show much better performance in HDPE tapes than UV absorbers, despite the latter being used in much higher concentrations. The comparison between HALS is, however, in favor of the polymeric HALS-III, which has the same performance when added at a concentration of 0.05% as HALS-I and HALS-II at 0.1%.

Among the numerous commercial light stabilizers only a few are suitable for low density polyethylene (LDPE). This is mainly due to the fact that most light stabilizers are not sufficiently compatible at levels of concentration necessary for the required protection and so they bloom more or less rapidly. Initially, UV absorbers of the benzophenone and benzotriazole types were used to protect LDPE materials. With the development of nickel-quenchers, significant improvement in the light stability of LDPE films has been achieved. For reasons of economy, however, combinations of nickel-quenchers with UV absorbers are mainly used. The service life of LDPE films may be increased by raising the concentrations of light stabilizers. Additive contents close to 2% may be found in greenhouse films thought to last up to 3 years outdoors.

A further improvement in UV stability of LDPE was expected with the development of HALS. However, LDPE compatibility of HALS available in the early years was insufficient, resulting in relatively poor performance on outdoor weathering. It was only with the development of polymeric HALS that these difficulties were overcome.

Tests have shown that HALS-II is significantly superior to UV absorbers and Ni-stabilizers so that the same performance can be achieved with much smaller concentrations. However, use of combinations of the polymeric HALS-II with a UV absorber leads to a significant improvement of the efficiency in comparison with the HALS used alone at the same concentration as the combination. A further boost of the performance can be achieved through use of the polymeric HALS-III. For example, the performance of HALS-II can be reached by using HALS-III at about half the concentration. The superiority of HALS-III becomes even more pronounced in films of thickness below 200 μm.

In linear low density polyethylene (LLDPE) also, the polymeric HALS-II and HALS-III show much better performance than other commercial light stabilizers. Blooming is observed with low molecular weight HALS-I, similar to that found with LDPE.

In more polar substances such as ethylene-vinyl acetate copolymers (EVA), the low molecular weight HALS-I can be used. However, in this substrate too, the polymeric HALS-II and HALS-III are significantly superior to the low molecular weight HALS

Styrenic Polymers

Double bonds are not regarded as the chromophores responsible for initiation of photooxidation in polystyrenes because they absorb below 300 nm. However, peroxide groups in the polymer chain, resulting from copolymerization of oxygen with styrene, are definitely photolabile. Moreover, oxidation products such as aromatic ketones of the acetophenone type, which have been detected by emission spectroscopy, are formed during processing of styrene polymers at high temperatures. Aromatic ketones (AK) in the triplet state are able to abstract hydrogen from polystyrene (PSH) [reaction (62)]:

$$AK \longrightarrow AK*$$

$$AK* + PSH \longrightarrow AKH^\bullet + PS^\bullet \tag{62}$$

This raction is considered the most important initiation mechanism for styrene photooxidation in the presence of aromatic ketones.

Styrenic plastics such as acrylonitrile/butadiene/styrene graft copolymers (ABS) and impact-resistant polystyrenes are very sensitive towards oxidation, mainly because of their butadiene content. Degradation on weathering starts at the surface and results in rapid loss of mechanical properties such as impact strength.

Because of the lack of efficient light stabilizers, ABS has not been used outdoors on a large scale. However, by combining two light stabilizers with different protection mechanisms, e.g., a UV absorber of the benzotriazole class and the sterically hindered amine HALS-I, it is possible to achieve good stabilization even in ABS. This is a case of synergism in which the UV absorber protects the deeper layers, while HALS-I assures surface protection. At the same time discoloration of the ABS polymer is also reduced significantly. The same holds for polystyrene and styrene-acrylonitrile copolymers (SAN), and the best protection is obtained with HALS/UV absorber combinations. Light stabilization is necessary for articles of these polymers for which UV exposure can be expected (e.g., covers for fluorescent lights).

Poly(Vinyl Chloride)

"Pure" poly(vinyl chloride) (PVC) does not absorb any light above 220 nm. Different functional groups and structural irregularities that may arise during polymerization and processing have thus been considered as possible initiating chromophores. They include irregularities in the polymer chain as well as hydroperoxides, carbonyl groups, and double bonds.

Thermal stabilizers used in PVC also confer some degree of light stability. Ba/Cd salts and organic tin carboxylates, for example, confer already some UV stability to PVC on outdoor exposure. However, for transparent and translucent PVC articles requiring high UV stability, the light stability conferred by thermal stabilizers is not sufficient. The addition of light stabilizers in such cases is therefore mandatory. So far, UV absorbers yield the best results in practical use. HALS-I has almost no effect.

Polycarbonate

Bisphenol-A polycarbonate (PC) absorbs UV light below 360 nm but its absorption is intense only below 300 nm. Insufficient light stability of PC on outdoor use is manifested by yellowing, which increases rapidly.

Studies indicate that on absorption of UV light, PC undergoes photo-Fries rearrangement, which gives first a phenyl salicylate, and after absorption of a second photon and subsequent rearrangement, it gives 2,2'-dihydroxybenzophenone groups (Fig. 1.30). The absorption of these groups reaches into the visible region, and PC yellowing has been essentially attributed to them. In addition to the Fries reaction, the formation of O_2-charge transfer complexes in PC, similar to those found in polyolefins and leading to the formation of hydroperoxides, is considered to contribute significantly to photooxidation in the early stages.

Among the stabilizer classes, only UV absorbers are in use for stabilization of PC. In choosing a UV absorber, intrinsic performance, volatility, adequate thermal stability at the elevated processing temperatures (about 320°C), and effect on initial color of the PC articles should be considered. Benzotriazole, oxanilide, and cinnamate-type UV absorbers are effective photostabilizers for PC with benzotriazoles giving the best performance among the three types.

Polyacrylates

Poly(methyl methacrylate) (PMMA) is highly transparent in the UV region, and thus much more light stable than other thermoplastics. UV absorbers may therefore be used to confer a UV filter effect to PMMA articles. PMMA window

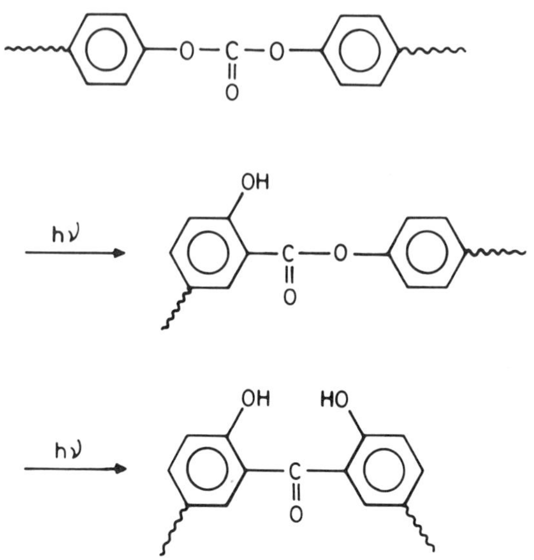

FIG. 1.30 Photo-Fries rearrangement on polycarbonate.

panes for solar protection containing 0.05 to 0.2% 2-(2'-hydroxy-5'-methylphenyl)-benzotriazole are well-known examples of this application. The rear lights of motor cars, electric signs, and covers for fluorescent lights are some applications for which PMMA is UV-stabilized. The excellent light-stabilizing performance of HALS is also found with PMMA.

Polyacetal

Polyacetal is markedly unstable towards light because even UV radiation of wavelengths as high as 365 nm may initiate its degradation. Polyacetal cannot therefore be used outdoors if it does not contain any light stabilizers. Even after a short weathering, surface crazes and pronounced chalking are observed.

Carbon black (0.5 to 3%) is a good stabilizer for polyacetal when sample color is not important. Other possibilities for stabilization are the use of 2-hydroxybenzophenone and, especially, 2-hydroxyphenylbenzotriazole-type UV absorbers. Stabilization with HALS/UV absorber is superior to that with UV absorber alone.

Polyurethanes

Light stability of polyurethanes depends to a large extent on their chemical structure, and both components (i.e., isocyanate and polyol) have an influence. Polyurethanes based on aliphatic isocyanates and polyester diols show the best light stability if yellowing is considered, whereas polyurethanes based on aromatic isocyanates and polyether diols are worst in this respect.

Light stabilizers are used mainly in the coatings industry (textile coatings, synthetic leather). In addition to some UV absorbers of the 2-(2'-hydroxyphenyl)-benzotriazole type, HALS used alone or in combination with benzotriazoles are especially effective stabilizers.

Polyamides

The absorption of aliphatic polyamides in the short wavelength region of sunlight is attributed largely to the presence of impurities. Direct chain scission at wavelengths below 300 nm and photosensitized oxidation above 300 nm have been considered a long time ago as responsible for photooxidation. Antioxidants used in polyamides often confer good light stability as well. However, enhanced performance is obtained with the combination of an antioxidant with a light stabilizer. The sterically hindered amines are significantly superior. For example, molded polyamide samples stabilized with HALS (0.5%) are found to exhibit approximately twice the light stability of the samples stabilized with a phenolic antioxidant (0.5%) or combination of the latter with a UV absorber (0.5%).

DIFFUSION AND PERMEABILITY

There are many instances where diffusion and permeation of a gas, vapor, or liquid through a plastics material is of considerable importance in the processing and usage of the material. For example, dissolution of a polymer in a solvent occurs through diffusion of the solvent molecules into the polymer, which thus swells and eventually disintegrates. In a plastisol process the gelation of PVC paste, which is a suspension of PVC particles in a plasticizer such as tritolyl phosphate, involves diffusion of the plasticizer into the polymer mass, resulting in a rise of the paste viscosity. Diffusion processes are involved in the production of cellulose acetate film by casting from solution, as casting requires removal of the solvent. Diffusion also plays a part in plastics molding. For example, lubricants in plastics compositions are required to diffuse out of the compound during processing to provide lubrication at the interface of the compound and the mold. Incompatibility with the base polymer is therefore an important criterion in the choice of lubricants in such cases.

Permeability of gases and vapors through a film is an important consideration in many applications of polymers. A high permeability is sometimes desirable. For example, in fruit-packaging applications of plastics film it is desirable to have high permeability of carbon dioxide. On the other hand, for making inner tube and tubeless tires, or in a child's balloon, the polymer used must have low air permeability.

Diffusion

Diffusion occurs as a result of natural processes that tend to equalize the concentration gradient of a given species in a given environment. A quantitative relation between the concentration gradient and the amount of one material transported through another is given by Fick's first law:

$$dm = -D \frac{dc}{dx} A \, dt \tag{63}$$

where dm is the number of grams of the diffusing material crossing area A (cm^2) of the other material in time dt (sec), D is the diffusion coefficient (cm^2/sec) whose value depends on the diffusing species and the material in which diffusion occurs, and dc/dx is the concentration gradient, where the units of x and c are centimeters and grams per cubic centimeter.

Diffusion in polymers occurs by the molecules of the diffusing species passing through voids and other gaps between the polymer molecules. The diffusion rate will therefore heavily depend on the molecular size of the diffusing species and on the size of the gaps. Thus, if two solvents have similar solubility parameters, the one with smaller molecules will diffuse faster in a polymer. On the other hand, the size of the gaps in the polymer depend to a large extent on the physical state of the polymer—that is, whether it is crystalline, rubbery, or glassy. Crystalline structures have an ordered arrangement of molecules and a high degree of molecular packing. The crystalline regions in a polymer can thus be considered as almost impermeable, and diffusion can occur only in amorphous regions or through regions of imperfection: hence, the more crystalline the polymer, the greater will be its tendency to resist diffusion. Amorphous polymers, as noted earlier, exist in the rubbery state above the glass transition temperature and in the glassy state below this temperature. In the rubbery state there is an appreciable "free volume" in the polymer mass, and molecular segments also have considerable mobility, which makes it highly probable that a molecular segment will at some stage move out of the way of the diffusing molecule, thus contributing to a faster diffusion rate. In the glassy state, however, the molecular segments cease to have mobility, and there is also a reduction in free volume or voids in the polymer mass, both of which lower the rate of diffusion. Thus, diffusion rates will be highest in rubbery polymers and lowest in crystalline polymers.

Permeability

Permeation of gas, vapor, or liquid through a polymer film consists of three steps: (1) a solution of permeating molecules in the polymer, (2) diffusion through the polymer due to concentration gradient, and (3) emergence of permeating molecules at the outer surface. Permeability is therefore the product of solubility and diffusion; so where the solubility obeys Henry's law one may write

$$P = DS \qquad (64)$$

where P is the permeability, D is the diffusion coefficient, and S is the solubility coefficient.

Hence, factors which contribute to greater solubility and higher diffusivity will also contribute to greater permeability. Thus, a hydrocarbon polymer like polyethylene should be more permeable to hydrocarbons than to liquids of similar molecular size but of different solubility parameter, and smaller hydrocarbons should have higher permeability. The permeabilities of a number of polymers to the common atmospheric gases [32,33], including water vapor, are given in Table 1.14.

It appears that regardless of the film material involved, oxygen permeates about four times as fast as nitrogen, and carbon dioxide about 25 times as fast. The fact that the ratios of the permeabilities for all gases, apart from water vapor, are remarkably constant, provided there is no interaction between the film material and the diffusing gas, leads one to express the permeability as the product of three factors [31]: one determined by the nature of the polymer film, one determined by the nature of the gas, and one accounting for the interaction between the gas and the film; i.e.,

$$P_{i,\,k} = F_i G_k H_{i,\,k} \qquad (65)$$

where $P_{i,k}$ is the permeability for the system polymer i and gas k; F_i, G_k and $H_{i,k}$ are the factors associated with the film, gas, and interaction respectively. When $H_{i,k} \approx 1$, there is little or no interaction, and as the degree of interaction increases, $H_{i,k}$ becomes larger. With little or no interaction, Eq. (65) becomes

$$P_{i,\,k} = F_i G_k \qquad (66)$$

It then appears that the ratio of the permeability of two gases (k,l) in the same polymer (i) will be the same as the ratio between the two G factors

$$\frac{P_{i,k}}{P_{i,l}} = \frac{G_k}{G_l} \qquad (67)$$

If the G value of one of the gases, usually nitrogen, is taken as unity, G values of other gases and F values of different polymers can be calculated from Eqs. (67) and (66). These values are reliable for gases but not for water vapor. Some F and G values for polymers are given in Table 1.15. Evidently, the F values correspond to the first column of Table 1.14, since G for N_2 is 1. The G values for O_2 and CO_2 represent the averages of P_{O_2}/P_{N_2} and P_{CO_2}/P_{N_2} in columns 5 and 6, respectively.

TABLE 1.14 Permeability Data for Various Polymers

Polymer	Permeability $(P \times 10^{10} \; cm^3/cm^2/mm/sec/cm \; Hg)$			
	N_2 (30°C)	O_2 (30°C)	CO_2 (30°C)	H_2O (25°C, 90% R.H)
Poly(vinylidene chloride) (Saran)	0.0094	0.053	0.29	14
Polychlorotrifluoro- ethylene	0.03	0.10	0.72	2.9
Poly(ethylene terephtha- late) (Mylar A)	0.05	0.22	1.53	1,300
Rubber hydrochloride (Pliofilm ND)	0.08	0.30	1.7	240
Polyamide (Nylon 6)	0.10	0.38	1.6	7,000
Poly(vinyl chloride) (unplasticized)	0.40	1.20	10	1,560
Cellulose acetate	2.8	7.8	68	75,000
Polyethylene (d = 0.954–0.960)	2.7	10.6	35	130
Polyethylene (d = 0.922)	19	55	352	800
Polystyrene	2.9	11	88	12,000
Polypropylene (d = 0.910)	—	23	92	680
Butyl rubber	3.12	13.0	51.8	—
Polybutadiene	64.5	191	1,380	—
Natural rubber	80.8	233	1,310	—

Source: References 31 and 32.

Ratios (to N_2 permeability as 1.0)

P_{O_2}/P_{N_2}	P_{CO_2}/P_{N_2}	P_{H_2O}/P_{N_2}	Nature of Polymer
5.6	31	1,400	Crystalline
3.3	24	97	Crystalline
4.4	31	26,000	Crystalline
3.8	21	3,000	Crystalline
3.8	16	70,000	Crystalline
3.0	25	3,900	Semicrystalline
2.8	24	2,680	Glassy
3.9	13	48	Crystalline
2.9	19	42	Semicrystalline
3.8	30	4,100	Glassy
—	—	—	Crystalline
4.1	16.2	—	Rubbery
3.0	21.4	—	Rubbery
2.9	16.2	—	Rubbery

POLYMER COMPOUNDING [29, 34, 35]

In many commercial plastics, the polymer is only one of several constituents, and it is not necessarily the most important one. Such systems are made by polymer *compounding*—a term used for the mixing of polymer with other ingredients. These other ingredients or supplementary agents are collectively referred to as *additives*. They may include chemicals to act as plasticizing agents, various

TABLE 1.15 F and G Constants for Polymers and Gases

Polymer	F	Gas	G
Poly(vinylidene chloride) (Saran)	0.0094	N_2	1.0
Poly(chlorotrifluoroethylene)	0.03		
Poly(ethylene terephthalate)	0.05	O_2	3.8
Rubber hydrochloride (Pliofilm)	0.08	H_2S	21.9
Nylon 6	0.1	CO_2	24.2
Cellulose acetate (+15% plasticizer)	5		
Polyethylene (d = 0.922)	19		
Ethyl cellulose (plasticized)	84		
Natural rubber	80.8		
Butyl rubber	3.12		
Nitrile rubber	2.35		
Polychloroprene	11.8		
Polybutadiene	64.5		

Source: Reference 31.

types of filling agents, stabilizing agents, antistatic agents, colorants, flame re-tardants, and other ingredients added to impart certain specific properties to the final product.

The properties of compounded plastics may often be vastly different from those of the base polymers used in them. A typical example is SBR, which is the largest-volume synthetic rubber today. The styrene-butadiene copolymer is a material that does not extrude smoothly, degrades rapidly on exposure to warm air, and has a tensile strength of only about 500 psi (3.4×10^6 N/m²). However, proper compounding changes this polymer to a smooth-processing, heat-stable rubber with a tensile strength of over 3000 psi (20×10^6 N/m²). Since all prop-erties cannot be optimized at once, compounders have developed thousands of specialized recipes to optimize one or more of the desirable properties for par-ticular applications (tires, fan belts, girdles, tarpaulins, electrical insulation, etc.). In SBR the compounding ingredients can be (1) reinforcing fillers, such as carbon black and silica, which improve tensile strength or tear strength; (2) inert fillers and pigments, such as clay, talc, and calcium carbonate, which make the polymer easier to mold or extrude and also lower the cost; (3) plasticizers and extenders, such as mineral oils, fatty acids, and esters; (4) antioxidants, basically amines or phenols, which stop the chain propagation in oxidation; and (5) curatives, such as sulfur for unsaturated polymers and peroxides for saturated polymers, which are essential to form the network of cross-links that ensure elasticity rather than flow.

Polymer applications which generally involve extensive compounding are rubbers, thermosets, adhesives, and coatings. Fibers and thermoplastic polymers (with the exception of PVC) are generally not compounded to any significant extent. Fibers, however, involve complex after-treatment processes leading to the final product. PVC, which by itself is a rigid solid, owes much of its versatility in applications to compounding with plasticizers. The plasticizer content varies widely with the end use of the product but is typically about 30% by weight.

Of the compounding ingredients, fillers and plasticizers are more important in terms of quantities used. Other additives used in smaller quantities are antioxidants, stabilizers, colorants, flame retardants, etc. The ingredients used as antioxidants and light stabilizers, and their effects have been discussed previously. Fillers, plasticizers and flame retardants are described next.

Fillers

Fillers play a crucial role in the manufacture of plastics. Alone many plastics are virtually useless, but they are converted into highly useful products by combining them with fillers. For example, phenolic and amine resins are almost always used in combination with substances like wood flour, pure cellulose, powdered mica, and asbestos. Glass fiber is used as a filler for fiber-reinforced composites with epoxy or polyester resins.

Another extremely important example is the use of carbon filler for rubber. Rubber would be of little value in modern industry were it not for the fact that the filler carbon greatly enhances its mechanical properties like tensile strength, stiffness, tear resistance, and abrasion resistance. Enhancement of these properties is called *reinforcement*, and the fillers which produce the strengthening effect are known as *reinforcing fillers*. Other fillers may not appreciably increase strength, but they may improve other properties of the polymer, thus making it easier to mold, which reduces cost.

Fillers used in plastics can be divided into two types: particulate and fibrous. Typical fillers in these two categories and the improvements they bring about are summarized in Table 1.16. In some instances they are added to perform one or more prime functions or to provide special properties. Asbestos, for example, provides high-temperature resistance and improves dimensional stability. Mica improves the electrical and thermal properties of all compounds. Glass fibers produce high strength. Carbon black is the only important reinforcing filler for most elastomers. It also imparts UV resistance and coloring.

TABLE 1.16 Some Fillers and Their Effects on Plastics

Fillers	Chemical resistance	Heat resistance	Dimensional stability	Tensile strength	Stiffness	Impact strength	Hardness	Lubricity	Electrical insulation	Electrical conductivity	Thermal conductivity	Moisture resistance	Processability	Recommended for use in[a]
Alpha cellulose			+	+					+					S
Alumina	+	+	+											S/P
Aluminum powder										+	+			S
Asbestos	+	+	+		+	+	+		+					S/P
Calcium carbonate		+	+		+		+						+	S/P
Calcium silicate		+	+		+		+							S
Carbon black		+	+		+					+	+		+	S/P
Carbon fiber										+	+			S
Cellulose			+	+	+	+	+		+					S/P
Cotton (macerated/chopped fibers)			+	+	+	+	+		+					S
Fibrous glass	+	+	+	+	+	+	+		+			+		S/P

Material	1	2	3	4	5	6	7	8	9	10	11	12	13	Type
Graphite			+	+		+	+		+	+	+	+	+	S/P
Jute							+	+	+					S
Kaolin	+	+				+	+		+		+	+	+	S/P
Kaolin (calcined)	+	+			+		+		+		+	+	+	S/P
Mica	+	+			+	+	+		+		+	+	+	S/P
Molybdenum disulfide	+	+							+	+				P
Nylon (macerated/chopped fibers)	+				+	+	+	+			+	+	+	S/P
Acrylic fiber (Orlon)	+	+			+	+	+	+	+	+	+	+	+	S/P
Rayon					+	+	+	+	+	+				S
Silica, amorphous	+	+			+									S/P
TFE-fluorocarbon						+			+		+			S/P
Talc	+	+			+	+	+		+		+	+	+	S/P
Wood flour					+					+	+			S

aP = in thermoplastics only; S = in thermosets only; S/P = in both thermoplastics and thermosets.

Beryllium oxide-filled resins gain high conductivity without loss of electrical properties. Metal particles have been used as fillers for plastics to improve or impart certain properties. Thus, aluminum has been used for applications ranging from making a decorative finish to improving thermal conductivity. Copper particles are used in plastics to provide electrical conductivity. Lead is used because it dampens vibrations, acts as barrier to gamma-radiation, and has high density.

Of the new space-age products used as reinforcing fillers, carbon fibers and boron fibers have the greatest potential for use in high-strength advanced composites. Carbon fibers made by pyrolizing organic fibers such as rayon and polyacrylonitrile have tensile strengths approaching 3.3×10^5 psi (2.3×10^9 N/m^2). Boron fibers made by depositing boron from a BCl_3—H_2 mixture onto tungsten wire have tensile strengths approaching 5×10^5 psi (3.5×10^9 N/m^2). The specific strengths and specific moduli of polymer composites made with these fibers are far above those attainable in monolithic structural materials such as high-strength aluminum, steel, and titanium. These composites can thus lead to significant weight savings in actual applications. A relatively recent addition to the high-performance fiber field is the organic polymeric fiber Kevlar-49, developed by Du Pont. It has a higher specific strength than glass, boron, or carbon. Furthermore, Kevlar, a polyamide, is cheap, having one-sixth the price of acrylic-based graphite fibers.

The extensive range of fillers and reinforcing agents used nowadays indicates the importance that these materials have attained. The main difference between inert and reinforcing fillers lies in the fact that modulus of elasticity and stiffness are increased to a greater or less extent by all fillers, including the spherical types such as chalk or glass spheres, whereas tensile strength can be appreciably improved only by a fiber reinforcement. Heat deflection temperature, i.e. stiffness at elevated temperatures, cannot be increased by spherical additives to the same extent as by fiber reinforcement. On the other hand, fillers in flake form, such as talc or mica, likewise produce a marked improvement in the heat deflection temperature but lead to a decrease in tensile strength and to elongation at break. The influences of common fillers on the properties of polyolefins are compared in Table 1.17.

PLASTICIZERS

Plasticizers are organic substances of low volatility that are added to plastics compounds to improve their flexibility, extensibility, and processability. They increase flow and thermoplasticity of plastic materials by decreasing the viscosity of polymer melts, the glass transition temperature (T_g) the melting temperature (T_m), and the elasticity modulus of finished products.

TABLE 1.17 Influence of Fillers on the Properties of Polyolefins [29]

Material	Filler content (% by wt.)	Density (g/cm³)	Elongation (%)	Tensile strength (N/mm²)	Modulus of elasticity (N/mm²)	Ball indentation (N/mm²)	Heat deflection temperature (1.85 N/mm²) (°C)	Melt index 190/5 (g/10 min)
Low density polyethylene	-	0.92	⊊500	10	210	16	35	1.7
Chalk	40	1.26	220	16	900	-	-	0.2
glass fiber	30	1.11	65	24	1200	33	-	1.6
Asbestos	30	1.13	20	20	670	22	-	2.3
Talc	30	1.14	40	16	600	30	-	1.1
Mica	30	1.16	46	13	440	27	-	1.2
Glass spheres (<50 µm)	30	1.11	73	10	290	19	-	2.6
Quartz powder	30	1.14	77	12	400	23	-	2.5
High density polyethylene	-	0.95	>550	27	1400	46	50	22
Glass fiber	30	1.16	2	60	7800	68	108	6
Chalk	30	1.17	9	20	1900	48	-	18
Wood flour	30	-	6	28	2100	-	-	-
Kaolin	40	1.24	11	26	2000	-	-	2.5
Asbestos	40	-	10	30	2100	-	77	0.7
Polypropylene	-	0.90	620	31	1600	75	62	1.0
Short glass fiber	30	1.20	5	49	7500	100	120	6.0
Asbestos	40	1.28	7	45	5000	84	96	4.0
Talc	40	1.21	11	30	3100	83	95	3.0
Mica	30	-	-	30	6900	-	-	-
Chalk	40	1.23	180	25	3400	63	73	2.5
Wood flour	40	-	8	20	2200	-	-	-

Plasticizers are particularly used for polymers that are in a glassy state at room temperature. These rigid polymers become flexible by strong interactions between plasticizer molecules and chain units, which lower their brittle-tough transition or brittleness temperature (T_b) (the temperature at which a sample breaks when struck) and their T_g value, and extend the temperature range for their rubbery or viscoelastic state behavior (see Fig. 1.11).

Mutual miscibility between plasticizers and polymers is an important criterion from a practical point of view. If a polymer is soluble in a plasticizer at a high concentration of the polymer, the plasticizer is said to be a *primary plasticizer*. Primary plasticizers should gel the polymer rapidly in the normal processing temperature range and should not exude from the plasticized material. *Secondary plasticizers*, on the other hand, have lower gelation capacity and limited compatibility with the polymer. In this case, two phases are present after plasticization process—one phase where the polymer is only slightly plasticized, and one phase where it is completely plasticized. Polymers plasticized with secondary plasticizers do not, therefore, deform homogeneously when stressed as compared to primary plasticizers. The deformation appears only in the plasticizer-rich phase and the mechanical properties of the system are poor. Unlike primary plasticizers, secondary plasticizers cannot be used alone and are usually employed in combination with a primary plasticizer.

Plasticizer properties are determined by their chemical structure because they are affected by the polarity and flexibility of molecules. The polarity and flexibility of plasticizer molecules determine their interaction with polymer segments. Plasticizers used in practice contain polar and nonpolar groups, and their ratio determines the miscibility of a plasticizer with a given polymer.

Plasticizers for PVC can be divided into two main groups [33] according to their nonpolar part. The first group consists of plasticizers having polar groups attached to aromatic rings and is termed the *polar aromatic group*. Plasticizers such as phthalic acid esters and tricresyl phosphate belong to this group. An important characteristic of these substances is the presence of the polarizable aromatic ring. It has been suggested that they behave like dipolar molecules and form a link between chlorine atoms belonging to two polymer chains or to two segments of the same chain, as shown in Fig. 1.31(a). Plasticizers belonging to this group are introduced easily into the polymer matrix. They are characterized by ability to produce gelation rapidly and have a temperature of polymer-plasticizer miscibility that is low enough for practical use. These plasticizers are therefore called *solvent-type* plasticizers, and their kerosene extraction (bleeding) index is very low. They are, however, not recommended for cold-resistant materials.

The second group consists of plasticizers having polar groups attached to aliphatic chains and is called the *polar aliphatic group*. Examples are aliphatic alcohols and acids or alkyl esters of phosphoric acid (such as trioctyl phosphate).

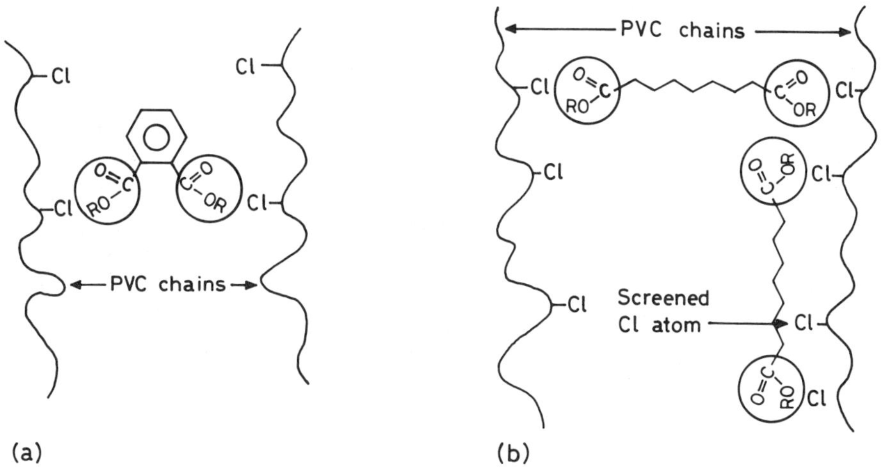

FIG. 1.31 Action of (a) a polar aromatic plasticizer and (b) a polar aliphatic plasticizer on PVC chains.

Their polar groups interact with polar sites on polymer molecules, but since their aliphatic part is rather bulky and flexible other polar sites on the polymer chain may be screened by plasticizer molecules. This reduces the extent of intermolecular interactions between neighboring polymer chains, as shown in Fig. 1.31b.

Polar aliphatic plasticizers mix less well with polymers than do polar aromatics and, consequently, may exude (bloom) from the plasticized polymer more easily. Their polymer miscibility temperature is higher than that for the first group. These plasticizers are called *oil-type* plasticizers, and their kerosene extraction index is high. Their plasticization action is, however, more pronounced than that of polar aromatic plasticizers at the same molar concentration. Moreover, since the aliphatic portions of the molecules retain their flexibility over a large temperature range, these plasticizers give a better elasticity to finished products at low temperature, as compared to polar aromatic plasticizers, and allow the production of better cold-resistant materials. In PVC they also cause less coloration under heat exposure.

In practice plasticizers usually belong to an intermediate group. Mixtures of solvents belonging to the two groups discussed above are used as plasticizers to meet the requirements for applications of the plasticized material.

Plasticizers can also be divided into groups according to their chemical structure to highlight their special characteristics. Several important plasticizers in each group (with their standard abbreviations) are cited below.

Phthalic Acid Esters

Di(2-ethyl hexyl) phthalate (DOP) and diisooctyl phthalate (DIOP) are largely used for PVC and copolymers of vinyl chloride and vinyl acetate as they have an affinity to these polymers, produce good solvation, and maintain good flexibility of finished products at low temperature. The use of *n*-octyl-*n*-decyl phthalate in the production of plastics materials also allows good flexibility and ductility at low temperature. Diisodecyl phthalate (DIDP), octyl decyl phthalate (ODP), and dicapryl phthalate (DCP) have a lower solvency and are therefore used in stable PVC pastes. Butyl octyl phthalate (BOP), butyl decyl phthalate (BDP, and butyl benzyl phthalate (BBP) have a good solvency and are used to adjust melt viscosity and fusion time in the production of high-quality foams. They are highly valued for use in expandable plasticized PVC.

Dibutyl phthalate (DBP) is not convenient for PVC plasticization because of its relatively high volatility. It is a good gelling agent for PVC and vinyl chloride-vinyl acetate copolymer (PVCA) and so is sometimes used as a secondary plasticizer in plasticizer mixtures to improve solvation. DBP is mainly used for cellulose-based varnishes and for adhesives. It has a high dissolving capacity for cellulose nitrate (CN).

Dimethyl phthalate (DMP) also has high dissolving capacity for CN. It has good compatibility with cellulose esters and are used in celluloid made from CN and plastic compounds or films made from other cellulosic polymers, cellulose acetate (CA), cellulose acetate-butyrate (CAB), cellulose acetate-propionate (CAP), and cellulose propionate (CP). It is light stable but highly volatile. Diethyl phthalate (DEP) possesses properties similar to DMP and is slightly less volatile.

Phosphoric Acid Esters

Tricresyl phosphate (TCP), trioctyl phosphate, diphenyl 2-ethylhexyl phosphate, and tri(2-ethylhexyl) phosphate (TOP) are used as plasticizers as well as flame retardants. They have a low volatility, resist oil extraction well, and are usually combined with other plasticizers. TCP is a good flame retardant plasticizer for technical PVC, PVAC, NC, and CAB. It is used for PVC articles especially in electrical insulation, but it is not recommended for elastic materials to be used at low temperature. Trioctyl phosphate is a better choice in low temperature applications but it offers a lower resistance to kerosene and oil extraction. Diphenyl 2-ethylhexyl phosphate is a good gelling agent for PVC. It also can be used for materials designed for low temperature applications. TOP gels NC, PVCA, and PVC. It has markedly higher volatility than DOP, and it gives plastisols of low viscosity.

Fatty Acid Esters

The esters of aliphatic dicarboxylic acids, mainly adipic, azelaic, and sebacic acid, are used as plasticizers for PVC and PVCA. Di-2-ethylhexyl adipate (DOA), benzyl butyl adipate, di-2-ethylhexyl azelate (DOZ), and di-2-ethylhexyl sebacate (DOS) are good examples. They give the polymer outstanding low-temperature resistance and are distinguished by their high light stability. Another characteristic is their low viscosity, which is valuable in the manufacture of PVC plastisols. The solvating action of these esters on PVC at room temperature is weak. This also has a favorable effect on the initial viscosity and storage stability of plastisols, which is observed even when, for example, plasticizer mixtures of DOP and an aliphatic dicarboxylic ester with DOP contents of up to 80% are used. Such combinations of plasticizers help to improve gelation.

DOS exceeds the low-temperature resistance of all other products in the group. It has the least sensitivity to water and has a relatively low volatility. DOS is used most often because of these properties and because of its high plasticization efficiency.

Monoisopropyl citrate, stearyl citrate, triethyl citrate (TEC), butyl and octyl maleates, and fumarates are other important plasticizers for the preparation of stable PVC pastes and of low-temperature-resistant PVC products. Triethyl citrate and triethyl acetylcitrate are among the few plasticizers to have good solvency for cellulose acetate. In comparison with diethyl phthalate, with which they are in competition in this application, there are slight but not serious differences in volatility and water sensitivity. The interest in citrate esters is due to a favorable assessment of their physiological properties. They are intended for plastic components used for packaging of food products.

Butyl and octyl maleates, being unsaturated, are used for copolymerization with vinyl chloride, vinyl acetate, and styrene to provide internal plasticization.

In the production of poly(vinyl butyral), which is used as an adhesive interlayer film between glass plates for safety glass, triethyleneglycol di(2-ethyl) butyrate is an extremely valuable plasticizer that has a proven record of success over many years. It has the required light stability and a plasticizing effect tailored to the requirements of poly(vinyl butyral) that gives the films suitable adhesion to glass ensuring good splinter adhesion over the temperature range from -40 to $+70°C$.

The ethylhexanoic esters of triethylene glycol and polyethylene glycol are good plasticizers for cellulose acetate butyrate (CAB). They, however, have only limited compatibility with poly(vinyl butyral).

Through the action of peracids on esters of unsaturated fatty acids, oxygen adds to the double bond forming an epoxide group. For example, epoxidized oleates have the following structure:

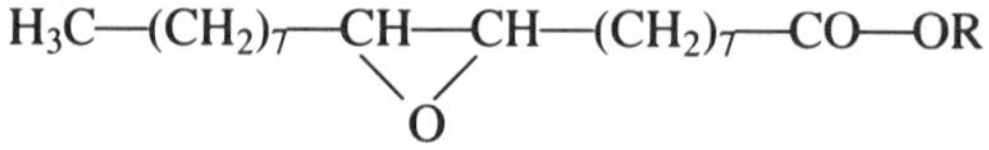

$$H_3C-(CH_2)_7-CH-CH-(CH_2)_7-CO-OR$$
$$\diagdown \diagup$$
$$O$$

Since the epoxide group reacts with acids, epoxidized fatty acid esters have become popular as plasticizers for PVC with a good stabilizing effect. Such compounds are not only able to bind hydrogen chloride according to reaction:

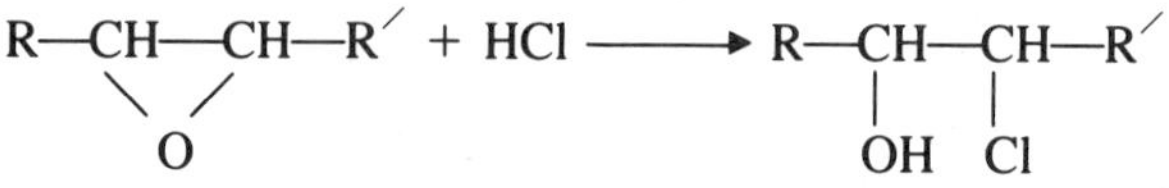

$$R-CH-CH-R' + HCl \longrightarrow R-CH-CH-R'$$
$$\diagdown \diagup \qquad\qquad\qquad | \quad |$$
$$O \qquad\qquad\qquad\qquad OH \quad Cl$$

but they are also able to substitute labile chlorine atoms in PVC under the catalytic influence of metal ions:

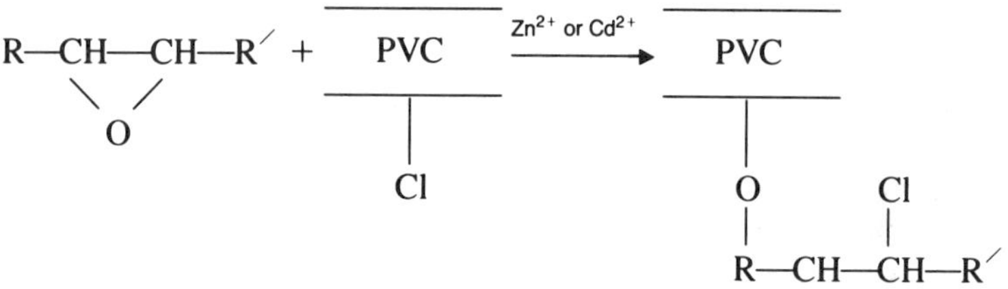

$$R-CH-CH-R' + \overline{\quad PVC \quad} \xrightarrow{Zn^{2+} \text{ or } Cd^{2+}} \overline{\quad PVC \quad}$$

with Cl on first PVC; products O and Cl, and R—CH—CH—R′.

Polymeric Plasticizers

Polymeric plasticizers can be divided into two main types. Oligomers or polymers of molecular weight ranging from 600 to 8000 belong to the first group. They include poly(ethylene glycol) and polyesters. High-molecular plasticizers comprise the second group.

Poly(ethylene glycol) is used to plasticize proteins, casein, gelatin, and poly(vinyl alcohol). Polyester plasticizers are condensation products of dicarboxylic acids with simple alcohols corresponding to the following two general formulas:

Ac−G—AcD−G—Ac
Al−AcD—G−AcD—Al

where

Ac = monocarboxylic acid
G   = glycol
AcD = dicarboxylic acid
Al = monofunctional alcohol

In practice, esters from adipic, sebacic, and phthalic acids are frequently used as polyester plasticizers. The value of n may vary from 3 to 40 for adipates and from 3 to 35 for sebacates. Polyester plasticizers are seldom used alone. They are used in combination with monomeric plasticizers to reduce the volatility of the mixed solvents. They offer a higher resistance to plasticizer migration and to extraction by kerosene, oils, water, and surfactants. Polyester plasticizers are used specially in PVC-based blends and in nitrocellulose varnishes.

Ethylene-vinyl acetate (EVA) polymers (containing 65–70% by weight of vinyl acetate) are of industrial interest as high-molecular weight plasticizers for PVC, mainly because of their low cost. A polymeric plasticizer PB-3041 available from Du Pont allows the preparation of a highly permanent plasticized PVC formulation. It is believed to be a terpolymer of ethylene, vinyl acetate, and carbon monoxide. Also, butylene terephthalate-tetrahydrofuran block copolymers, with the trade name of Hytrel (Du Pont), are used as excellent permanent plasticizers of PVC.

Miscellaneous Plasticizers

Hydrocarbons and chlorinated hydrocarbons (chloroparaffins) belong to the secondary plasticizer type.

Aromatic and aliphatic hydrocarbons are used as extenders, particularly in the manufacture of PVC plastisols that must maintain as stable a viscosity as possible for relatively long periods of time (dip and rotational molding). The petroleum industry offers imprecisely defined products with an aliphatic-aromatic structure for use as extenders. They are added to plastisols in small amounts as viscosity regulators. Dibenzyl toluene also serves the same purpose. A disadvantage of this type of extenders is the risk of exudation if excessive quantities are used.

The normal liquid chlorinated paraffins used as plasticizers for PVC have viscosities ranging from 100 to 40,000 mPa·s at 20°C. Products with chlorine contents ranging from 30 to 70% are on the market. Compatibility with PVC increases with increasing chlorine content but the plasticizing effect is reduced. The low viscosity products (chlorine content 30 to 40%) are used as secondary plasticizers for PVC. They have a stabilizing effect on viscosity in plastisols. Chlorinated paraffins can be used up to a maximum of 25% of the total plasticizer content of the PVC plastisol without the risk of exudation. As chlorine-containing substances, these plasticizers also have a flame-retarding effect.

At a chlorine content of 50%, chlorinated paraffins are sufficiently compatible with PVC to be used alone, i.e., as primary plasticizers, in semi-rigid compounds under certain circumstances. Products with extremely good compatibility, however, have low plasticizing action, requiring larger amounts to be used.

Among other plasticizers, *o*- and *p*-toluene-sulfonamides are used to improve the processability of urea and melamine resins, and cellulose-based adhesives.

For the same reason, *N*-ethyl-*o*-toluenesulfonamide and *N*-ethyl-*p*-toluenesulfonamide are used for polyamides, casein, and zein.

Diglycol benzoates are liquid plasticizers rather like benzyl butyl phthalate (BBP) in their plasticizing action in PVC. In the manufacture of PVC floor coverings, the benzoates offer the advantage of a highly soil resistant surface. In this respect, they are even superior to BBP, which itself performs well here.

Pentaerythritol esters can be used instead of phthalic acid esters when low volatility is one of the major factors in the choice of plasticizer for PVC. Pentaerythritol triacrylate can be used to plasticize PVC and to produce cross-links under UV radiation. Such plasticized cross-linked PVC has some application in the cable industry.

Commonly used plasticizers for some polymers beside PVC are listed below:

Cellulosics: Dimethyl phthalate (DMP), diethyl phthalate (DEP), dibutyl phthalate (DBP), dimethylcyclohexyl phthalate, dimethyl glycol phthalate, trichloroethyl phosphate, cresyldiphenyl phosphate, some glycolates, some sulfonamides.

Poly(vinyl butyrals):   Triethyleneglycol di(2-ethyl)butyrate, triethyleneglycol propionate, some adipates, some phosphates.

Polyurethanes: Dibutyl phthalate (DBP), diethyleneglycol dibenzoate (DEGB), and most of the plasticizers used for PVC are preferred for their good plasticization action.

ANTISTATIC AGENTS

By virtue of their chemical constitution, most plastics are powerful insulators, a property that makes them useful for many electrical applications. A disadvantage of the insulation property, however, is the accumulation of static electricity, which is not discharged fast enough due to the low surface conductivity of most plastics—a difference between plastics and metals. On plastics and other nonmetallic materials, frictional contact is necessary to generate static charges. On immobile objects, for example, static electricity may build up simply by friction with the ambient air. Electrostatic charges can also be produced on the surfaces of polymeric materials in the course of processing operations such as extrusion, calendering, and rolling up of plastic sheets or films.

The superficial electrical potential generated by friction may reach values up to a few tens of kilovolts, and this presents serious difficulties for practical applications and to users. Spark formation through heavy charging with subsequent ignition of dust or solvent/air mixtures have been a cause of many destructive explosions. In the service life of plastics products, static charging can give rise to many other troublesome effects, as for example, interference with the sound reproduction of records by dust particles picked up by electrostatic charges, production delays due to clinging of adjacent films or sheets, lump formation during pneumatic transportation, and static build-up on people passing over synthetic fiber carpeting or plastic floor-coverings, with a subsequent ''shock'' as the charge flows off, usually when the person touches a door handle.

There are many ways to eliminate surface electrostatic charges, for example, by increasing the humidity or the conductivity of the surrounding atmosphere, or by increasing the electric conductance of materials with the use of electroconducting carbon blacks, powdered metals, or antistatic agents.

Electroconducting carbon blacks are largely utilized to increase the electric conductivity of organic polymers. The electric conductivity of carbon blacks depends, inter alia, on the capacity to form branched structures in the polymer matrix, and on the size and size distribution of carbon black particles. The branched and tentacular structures of carbon in the polymer matrix are responsible for the electric conductivity, as is the case for lamp, acetylene, and furnace carbon blacks. The specific resistance of the carbon particles decreases with their size and then increases with further diminution of the size. A wide size distribution is believed to favor the formation of branched structures contributing to greater conductivity.

In spite of the effectiveness of some carbon blacks in reducing surface charges on plastics materials, the use of antistatic agents have increased steadily. The simplest antistatic agent is water. It is adsorbed on the surface of objects exposed to a humid atmosphere, and it forms a thin electroconducting layer with impurities adsorbed from the air. Such a layer is even formed on the surface of hydrophobic plastics, probably because of the existence of a thin layer of dirt.

Antistatic agents commonly used are substances that are added to plastics molding formulations or to the surface of molded articles in order to minimize the accumulation of static electricity. In general terms, antistatic agents can be divided according to the method of application into external and internal agents.

External Antistatic Agents

External antistatic agents are applied to the surface of the molded article from aqueous or alcoholic solution by wetting, spraying, or soaking the plastic object in solution, followed by drying at room temperature or under hot blown air. Their concentration varies between 0.1 and 2% by weight. Almost all surface active compounds are effective, as well as numerous hygroscopic substances such as glycerin, polyols, and polyglycols, which lack the surface activity feature. The most important external antistatic agents from the practical point of view are quaternary ammonium salts and phosphoric acid derivatives. The advantage of external agents lies in the fact that, in their performance, the properties of compatibility with the polymer and controlled migration in the polymer, which play an important part in the performance of internal antistatic agents described later, are not of any consideration.

It is generally assumed that surface-active molecules accumulate on the surface and are oriented such that the hydrophobic part containing the hydrocarbon chain extends into the plastic, and the hydrophilic part points outwards where it is able to adsorb water on the surface. The phase boundary angle between water and plastic is reduced by the surface active antistatic agents thus allowing the adsorbed water to be uniformly distributed on the plastic surface. A water film forms on the surface thus increasing the conductivity by means of an ion conduction process. This also explains why surface conductivity, and hence the antistatic action, are found to increase with increasing atmospheric humidity.

For the ion conduction in the surface film, a conductivity mechanism that is similar to protonic conductivity in water has been suggested:

$$
\begin{array}{ccccccc}
& \overset{\displaystyle H}{\overset{\displaystyle |}{}} & & & & \overset{\displaystyle H}{\overset{\displaystyle |}{}} & \\
H^{+}\ldots O{-}H\ldots O{-}H\ldots O{-}H & \rightleftharpoons & H{-}O\ldots H{-}O\ldots H{-}O\ldots H^{+} \\
\underset{\displaystyle H}{\overset{\displaystyle |}{}} \quad\quad \underset{\displaystyle H}{\overset{\displaystyle |}{}} & & \underset{\displaystyle H}{\overset{\displaystyle |}{}} \quad\quad \underset{\displaystyle H}{\overset{\displaystyle |}{}}
\end{array}
$$

This mechanism is based on a comparison of the conductivities of substances of different chemical structure. For instance, primary amines are efficient as antistatic agents but secondary amines are not. The conductivity of tertiary amines, on the other hand, depends on the nature of *N*-hydroxyalkyl substituents. Among amides, only *N,N*-disubstituted derivatives and mainly those having two hydroxyalkyl substituents are effective. The presence of many OH groups in the molecule makes the efficiency of the antistatic agent less dependent on the humidity. Antistatic agents bearing OH or NH_2 groups are able to associate in chain form via hydrogen bonding and display antistatic activity even at low atmospheric humidity, unlike compounds that are able to form only intramolecular hydrogen bonds.

Many antistatic agents also show hygroscopic properties, thereby intensifying the attraction of water to the surface. At constant atmospheric humidity, a hygroscopic compound attracts more water to the surface and so increases antistatic effectiveness.

Internal Antistatic Agents

Internal antistatic agents are incorporated in the polymer mass as additives, either before or during the molding process. However, to be functional, they must be only partially miscible with the polymer so that they migrate slowly to the surface of the plastics material. Their action therefore appears after a few hours and even a few days after compounding, depending on the mutual miscibility of the agent and the polymer. The concentration of internal antistatic agents in plastics varies from 0.1 to 10% by weight.

Interfacial antistatic agents are all of interfacially active character, the molecule being composed of a hydrophilic and a hydrophobic component. The hydrophobic part confers a certain compatibility with the particular polymer and is responsible for anchoring the molecule on the surface, while the hydrophilic part takes care of the binding and exchange of water on the surface.

Chemical Composition of Antistatic Agents

The selection of an antistatic agent for a given polymer is based on its chemical composition. Accordingly, antistatic agents can be divided in many groups, as presented below.

Antistatic Agents Containing Nitrogen

Antistatic agents containing nitrogen are mainly amines, amides, and their derivatives, such as amine salts, and addition compounds between oxiranes and aminoalcohols. Pyrrolidone, triazol, and polyamine derivatives also belong to this group. A few representative examples belonging to this group are:

Ethoxylated amines of the type
$$R-N\begin{cases} (CH_2CH_2O)_xH \\ (CH_2CH_2O)_xH \end{cases}$$

Amine oxides of the type
$$R_2-\overset{\overset{\displaystyle R_1}{\mid}}{\underset{\underset{\displaystyle R_3}{\mid}}{N}}\longrightarrow O$$

Fatty acid polyglycolamides of the type
$$R-\overset{\overset{\displaystyle O}{\parallel}}{C}-NH(CH_2CH_2O)_xH$$

The following two commercial products are typical examples: (1) Alacstat C-2 (Alcolac Chemical Corp., U.S.A.). This is a N,N-bis(2-hydroxyethyl)-alkylamine used for polyolefins at 0.1% by weight. (2) Catanac 477 (American Cyanamid Co., U.S.A.). It is N-(3-dodecyloxy-2-hydroxypropyl)ethanolamine ($C_{12}H_{25}OCH_2CHOHCH_2NHCH_2CH_2OH$) used for linear polyethylene (0.15%

by weight), for polystyrene (1.5% by weight), and for polypropylene (1% by weight). These compounds are not recommended for PVC.

Other compounds containing nitrogen and used as antistatic agents are quaternary ammoniun salts, quaternized amines, quaternized heterocycles obtained from imidazoline and pyridine, and amides of quaternized fatty acids. Some of these compounds can be utilized for PVC for example:

$$
\text{Ammonium salts} \quad \left(\begin{array}{c} R_2 \\ | \\ R_3—N—R_1 \\ | \\ R_4 \end{array} \right)^{+} \quad X^-
$$

$$
\text{Quaternized ethoxylated amine} \quad \left(\begin{array}{c} R_1 \\ | \\ R_2—N—(CH_2CH_2O)_xH \\ | \\ R_3 \end{array} \right)^{+} \quad X^-
$$

$$
\text{Amide of quaternized fatty acid} \quad R_1—\overset{\displaystyle O}{\overset{\displaystyle \|}{C}}—NH—R
$$

Some typical examples of commercial products are Catanac SN, Catanac 609, Catanac LS, and Catanac SP, all of American Cyanamid Co. Catanac SN is stearamidopropyl-dimethyl-β-hydroxyethylammonium nitrate:

$$
[C_{17}H_{35}CONHCH_2CH_2CH_2—\overset{\displaystyle CH_3}{\overset{\displaystyle |}{\underset{\displaystyle |}{\underset{\displaystyle CH_3}{N}}}}—CH_2CH_2OH]^{+}NO_3^{-}
$$

It is available as a 50% by weight solution in a water-isopropanol mixture (pH 4–6). It begins to decompose at 180°C and decomposes very rapidly at 250°C. It is soluble in water, acetone, alcohols, and in a series of polar solvents. It is used as an internal and external antistatic agent for rigid PVC, impact-PS, polyacrylates, ABS terpolymers, paper, and textiles. For applying externally, the concentration of the solution is reduced to 1–10% by weight. The solution can be applied on the surface of plastic products by rubbing them with a flannel cloth impregnated with the solution. For application as an internal antistatic agent, the commercial higher-concentration solution is directly added to the polymer mixture during the milling or mixing process.

Catanac 609 is *N,N*-bis(2-hydroexyethyl)-*N*-(3'-dodecyloxy-2'-hydroxy-propyl) methylammonium methylsulfate:

$$
[C_{12}H_{25}OCH_2CHOHCH_2\text{—}\overset{\displaystyle CH_2CH_2OH}{\underset{\displaystyle CH_2CH_2OH}{\overset{|}{\underset{|}{N}}}}\text{—}CH_3]^+ \ CH_3SO_4^-
$$

It is supplied mainly as a 50% by weight solution in a water-propanol mixture. It is applied as an external antistatic agent at 2% by weight concentration (pH 4–6) and is recommended for phonograph records and other products made of PVC and its copolymers.

Catanac LS is 3-lauramidopropyl-trimethylammonium methylsulfate:

$$
[C_{11}H_{23}CONHCH_2CH_2CH_2\text{—}\overset{\displaystyle CH_3}{\underset{\displaystyle CH_3}{\overset{|}{\underset{|}{N}}}}\text{—}CH_3]^+ \ CH_3SO_4^-
$$

It exists in a crystalline form with a melting point of 99–103°C.

Catanac SP is stearamidopropyl-dimethyl-β-hydroxyethyl-ammonium dihydrogenphosphate:

$$
[C_{17}H_{35}CONHCH_2CH_2CH_2\text{—}\overset{\displaystyle CH_3}{\underset{\displaystyle CH_3}{\overset{|}{\underset{|}{N}}}}\text{—}CH_2CH_2OH]^+ \ H_2PO_4^-
$$

supplied as a 35% by weight solution in a water-isopropanol mixture (pH 6–8). It begins to decompose slightly at 200°C, and at 250°C its decomposition is rapid. It is soluble in water, acetone and alcohols, and is not corrosive to metals even during prolonged contact. Catanac SP can be used as an internal and external antistatic agent at 1–3% by weight.

Antistatic Agents Containing Phosphorus

Antistatic agents that contain phosphorus can be used for all polymers, although they are recommended mainly for PVC with which they also function as plasticizers. They are phosphoric acid derivatives, phosphine oxides, triphosphoric acid derivatives, and substituted phosphoric amides. Typical examples are the following:

Phosphoric acid esters $O{=}P(OR)_3$

Ethoxylated alcohols and phosphoric acid esters $O{=}P[O(CH_2CH_2O)_x{-}R]_3$

Ammonium salts of phosphoric acid esters

$$\left(O{=}P{\begin{array}{c} {\diagup}O{-}R_1 \\ {-}O{-}R_2 \\ {\diagdown}O \end{array}} \right)^{-} \left({\begin{array}{c} H{\diagdown}\quad{\diagup}R_3 \\ N \\ H{\diagup}\quad{\diagdown}R_4 \end{array}} \right)^{+}$$

Antistatic Agents Containing Sulfur

Antistatic agents that contain sulfur include compounds such as sulfates, sulfonates, derivatives of aminosulfonic acids, condensation products of ethyleneoxide and sulfonamides, and sulfonates of alkylbenzimidazoles, dithiocarbamides, and sulfides.

Sulfur-containing antistatic agents are recommended mainly for PVC and PS because they do not interfere with heat stabilizers. They are not suitable for PMMA, polyolefins, and polyamides. As examples, there are alkylpolyglycolether sulfates, $RO(CH_2CH_2O)_xSO_3{}^-Me^+$, which are marketed under the trade name Statexan HA (Bayer).

Betaine-Type Antistatic Agents

Betaines are trialkyl derivatives of glycine, which exist as dipolar ions of the formula $R_3{}^+NCH_2CO_2{}^-$. Typical examples of betaine-type antistatic agents are stearylbetaine and dodecyldimethyl-ethanesulfobetaine. They are used mainly for polyolefins.

Nonionic Antistatic Agents

Nonionic antistatic agents are nonionizable, interfacially active compounds of low polarity. The hydrophilic portion of the molecule is usually represented by hydroxyl groups and the hydrophobic portion by organic groups. This group of antistatic agents includes the following: polyhydroxy derivatives of glycine, sugars, and fatty acids, sometimes modified by addition of oxiranes, most of the products being nontoxic and hygienically acceptable but having low antistatic effects; heavy alcohol derivatives such as alkylpolyglycol ethers, which are recommended for polyolefins and PVC; alkylphenolpolyglycol ethers having good heat stability, which are used mainly for polyolefins; polyglycol ethers obtained from the reaction of glycols with oxiranes, which are used for polyolefins; and fatty acid polyglycol esters $RCO_2(CH_2CH_2O)_xH$, which are used for many types of plastics.

Nonionic antistatic agents are supplied for the most part in liquid form or as waxes with a low softening region. The low polarity of this class makes its members ideal internal antistatic agents for polyethylene and polypropylene.

A number of substances that cannot be included in any of the groups above are also good antistatic agents. They are silicone copolymers, organotin derivatives, oxazoline derivatives, organoboron derivatives, and perfluorated surfactants. Some of them show excellent heat stability.

FLAME RETARDANTS

Efforts to develop flame-retarding plastics materials have been focussing on the increasing use of thermoplastics. As a result, flame-retarding formulations are available today for all thermoplastics, which strongly reduce the probability of their burning in the initiating phase of a fire. The possibility of making plastics flame-retardant increases their scope and range of application. (It must be noted, however, that in the case of fire, the effectiveness of flame retardants depends on the period of time and intensity of the fire. Even a product containing the most effective flame-retardants cannot resist a strong and long-lasting fire.)

Flame retardants are defined as chemical compounds that modify pyrolysis reactions of polymers or oxidation reactions implied in combustion by slowing them down or by inhibiting them. The combustion of thermoplastics (see Fire-Resistant Polymers in Chapter 5) is initiated by heating the material up to its decomposition point through heat supplied by radiation, flame, or convection. During combustion, free radicals are formed by pyrolysis and, besides noncombustible gases such as carbon dioxide and water, combustible gases are formed—mainly hydrocarbons, hydrogen, and carbon monoxide. The radicals combine with oxygen and combustible gases in a radical chain reaction, thereby causing heat release and further decomposition of the plastics material. For continued combustion it is necessary to have sufficient oxygen as well as combustible gaseous compounds.

The combustion will be slowed down or inhibited if free radicals, which are evolved by pyrolysis, are blocked. For example, it is presumed that the following reactions take place when thermoplastics containing organobromine compounds as flame retardants are used:

Exothermic
propagation $\qquad HO^{\cdot} + CO = CO_2 + H^{\cdot}$ $\qquad\qquad$ (68)

Chain branching $H^{\cdot} + O_2 = HO^{\cdot} + O^{\cdot}$ $\qquad\qquad$ (69)

Chain transfer $O^{\cdot} + HBr = HO^{\cdot} + Br^{\cdot}$ $\qquad\qquad$ (70)

Inhibition $\qquad HO^{\cdot} + HBr = H_2O + Br^{\cdot}$ $\qquad\qquad$ (71)

Regeneration $\qquad RH^{\cdot} + Br^{\cdot} = R^{\cdot} + HBr$ $\qquad\qquad$ (72)

The radical chain reaction is interrupted when the highly reactive HO· radical, which plays a key role in the combustion process, is replaced by the less reactive Br· radical [reaction (71)]. While Eqs. (70) and (71) show reactions with HBr (which is released during combustion), experiments have shown that halogen-containing organic compounds may also react as undecomposed molecules.

Flame retardants, in view of the above discussion, can be defined as chemical compounds that modify pyrolysis reactions of polymers or oxidation reactions implied in the combustion by slowing these reactions down or by inhibiting them. In practice, mixtures of flame retardants are often used to combine different types of retardancy effects. In spite of a few thousand references on flame retardants, only a small number of compounds are produced commercially. They are mainly chlorine, bromine, phosphorus, antimony, aluminum, and boron-containing compounds. Trade names and manufacturers of some selected flame retardants are given in Appendix 2D. Most flame retarding agents may be divided into three groups: (1) halogen compounds, (2) phosphorus compounds, and (3) halogen-antimony synergetic mixtures.

Halogen Compounds

As flame retardants, bromine compounds show superiority over chlorine compounds. For example, in case of combustion of *n*-heptane/air mixture, the minimum concentrations (in vol. %) found necessary for the extinction of flame are 5.2% for dibromomethane and 17.5% for trichloromethane. The efficiency of halogen-containing agents depends on the strength of the carbon-halogen bond. Thus the relative efficiency of aliphatic halides decreases in the following order:

Aliphatic bromides > aliphatic chlorides = aromatic
bromides > aromatic chlorides

It is also known that aromatic bromides have a higher decomposition temperature (250–300°C) than aliphatic bromides (200–250°C).

Bromine-containing compounds on a weight basis are at least twice as effective as chlorine-containing ones. Because of the smaller quantities of bromine compounds needed, their use hardly influences the mechanical properties of the base resins and reduces markedly the hydrogen halide content of the combustion gases. Aliphatic, cycloaliphatic, as well as aromatic and aromatic-aliphatic bromine compounds are used as flame retardants.

Bromine compounds with exclusively aromatic-bound bromine are produced in large quantities. Occupying the first place is tetrabromobisphenol-A, which is employed as a reactive flame retardant in polycarbonates and epoxy resins. Similarly, tetrabromophthalic anhydride has been used in the production of flame-retarded unsaturated polyester resins.

Brominated diphenyls and brominated diphenyl ethers–for instance, octabromodiphenyl ether (melting point 200–290°C) and decabromodiphenyl ether (melting point 290–310°C)—have gained great significance. Possessing excellent heat stability, they are excellent flame retardants for those thermoplastics that have to be processed at high temperatures, such as linear polyesters and ABS.

Aromatic-aliphatic bromine compounds, like the bis(dibromopropyl)-ether of tetrabromobisphenol A, or bromoethylated and aromatic ring-brominated compounds, such as 1,4-bis-(bromoethyl)-tetrabromobenzene, combine the high heat stability of aromatic-bound bromine with the outstanding flame retardancy of aliphatic-bound bromine. They are used mainly as flame retardants for polyolefins, including polyolefin fibers.

Polymeric aromatic bromine compounds, such as poly(tribromostyrene), poly(pentabromobenzyl acrylate) or poly(dibromophenylene oxide) are suitable for application in polyamides, ABS, and polyesters.

Oligomeric bromine-containing ethers on the basis of tetrabromoxylene dihalides and tetrabromobisphenol-A (molecular weight between 3000 and 7000) are useful flame retardants for polypropylene, polystyrene, and ABS.

The following cycloaliphatic compounds are suitable as flame retardants in polyolefins: hexabromocyclododecane and bromine-containing Diels-Alder reaction products based on hexachlorocyclopentadiene.

Phosphorus Compounds

Whereas halogen acids, which are released during the combustion of thermoplastics containing organic halogen compounds, interrupt the chain reaction [e.g., reaction (71)], the flame-retarding action of phosphorus compounds is not yet fully understood. It is supposed, however, that in case of fire, phosphorus compounds facilitate polymer decomposition, whereby the very stable poly(meta-phosphoric acid) formed creates an insulating and protecting surface layer between the polymer and the flame. The phosphoric acid also reacts with the polymer and produces a char layer protecting the surface.

Phosphorus-containing flame retardants may be divided according to the oxidation state of the element into phosphates $[(RO)_3PO]$, phosphites $[(RO)_3P]$, phosphonites $[(RO)_2PR']$, phosphinates $[(RO)R'_2PO]$, phosphines $[R_3P]$, phosphine oxides $[R_3PO]$, and phosphonium salts $[R_4PX]$. The halogenated phosphorus-containing compounds represent a very important group of flame retardants with high efficiency. A few examples are tris-(tribromoneopentyl)-phosphite, tris-(dibromopropyl)-phosphate, tris-(dibromophenyl)-phosphate, and tris-(tribromophenyl)-phosphate. A similar flame retardancy may also be obtained by the combinations of phosphorus-containing compounds with various halogen derivatives. These combinations are relatively universal in their action.

Phosphorus-containing flame retardants are suitable for polar polymers such as PVC, but for polyolefins their action is not sufficient. In this case they are used together with Sb_2O_3 and with halogenated compounds. Alkyl-substituted aryl phosphates are incorporated into plasticized PVC and modified PPO (Noryl) to a great extent; for other plastics they are less important. Quaternary phosphonium compounds are recommended as flame retardants for ABS and polyolefins.

Halogen-Antimony Synergetic Mixtures

Antimony oxide is an important component of flame retarded thermoplastics containing halogen compounds. Used alone, the flame-retardant effect of antimony oxide is not sufficient, but the effect increases significantly as halogen is added to the system. The basis of this antimony-halogen *synergism* is probably the formation of volatile flame-retardant SbX_3 (see below), where X is Cl or Br. Antimony-halogen combinations have been widely used in polymer for flame retardation, and the interaction of antimony (mostly as antimony oxide) with halogenated polymers or polymers containing halogenated additives represents the classic case of flame-retardant synergism. Experimental results indicate that the best Sb:X ratio is 1 to 3. More effective are ternary synergetic mixtures such as Sb_2O_3-pentabromotoluene-chloroparaffin.

The antimony oxide-alkyl (aryl) halide system functions on heating according to the following mechanism (shown for chloride):

$$RCl \rightarrow HCl + R' \, CH{=}CH_2 \tag{73}$$

$$Sb_2O_3 + 6HCl \rightarrow 2SbCl_3 + 3H_2O \tag{74}$$

$$Sb_2O_3 + 2HCl \rightarrow 2SbOCl + H_2O \tag{75}$$

Volatile $SbCl_3$ reduces the formation of radicals in the flame and also affects the oxidation process strongly, thus terminating the combustion. As with phosphorus compounds, it promotes carbonization of the polymer. It is believed that SbOCl and $SbCl_3$ already function as dehydrogenation agents in the solid phase of the polymer in flame. At 240–280°C, SbOCl is transformed to higher oxychlorides and $SbCl_3$, and at temperatures around 500°C the higher antimony oxychlorides are converted back into antimony oxide by releasing again $SbCl_3$:

$$5SbOCl \xrightarrow{\ 240-280°\,C\ } Sb_4O_5Cl_2 + SbCl_3 \tag{76}$$

$$4Sb_4O_5Cl \xrightarrow{\ 410-475°\,C\ } 5Sb_3O_4Cl + SbCl_3 \tag{77}$$

$$3Sb_3O_4Cl \xrightarrow{\ 475-565°\,C\ } 4Sb_2O_3 + SbCl_3 \tag{78}$$

The flame retardation is also attributed to the consumption of thermal energy by these endothermic reactions, to the inhibition reactions in the flame, and to the release of hydrochloric acid directly in the flame. The pigmentation effect of mixtures containing Sb_2O_3 is a disadvantage that limits their use for light-colored and transparent products.

The demands on a flame retardant and thermoplastics formulated with such agents are manifold. The flame retardant should provide a durable flame-retarding effect by the addition of only small quantities of the additive; it should be as cheap as possible and the manner of incorporation should be easy; it should be nontoxic and should not produce fire effluents with increased toxicity; it should not decompose at the processing temperatures; it should not volatilize and smell; and the mechanical, optical, and physical properties of the thermoplastics should be affected as little as possible. It is understandable that these far-ranging demands cannot be satisfied by only one flame retardant and for all thermoplastics with their manifold applications. One is therefore forced to seek the optimum flame retarding formulation for each thermoplastic and the specific application. Examples of typical formulations of several flame-retarded thermoplastics are given in Appendix 3.

SMOKE SUPPRESSANTS

In recent years, the hazards of smoke and toxic gas production by burning have received much greater recognition. Since burning of a large amount of polymer represents a fuel rich combustion, it produces large amounts of toxic carbon monoxide and carbon-rich black smoke, causing death and inhibiting escape from the burning environment. A comparison of smoke production from burning different polymers is given in Table 1.18.

The problem of smoke suppression has not yet been solved satisfactorily, because the demands on flame retardancy and reduced smoke development are principally of antagonistic nature. Most of the developmental work on smoke suppressants has been carried out in the field of PVC. For this polymer, highly dispersed aluminum trihydrate is widely used as a flame retardant and as a smoke suppressant. Antimony trioxide-metal borates (barium borate, calcium borate, and zinc borate) are good smoke reducers. Zinc borate of the formula $2ZnO:3B_2O_3:3.5H_2O$ has shown the best effect with regard to reduced flammability and reduced smoke development of PVC. It is possible to reduce smoke development of rigid PVC significantly by adding rather small quantities of molybdic oxide. For example, 2% of molybdic oxide reduces the smoke density to about 35%, but does not decrease the limiting oxygen index value of rigid PVC.

Several methods have been proposed for the determination of the smoke density. Reproducible values are obtained with the laboratory method worked out by

TABLE 1.18   Smoke Emission on Burning (NBS Smoke Chamber, Flaming Condition)

Polymer	Maximum smoke density (D_{max})
ABS	800
Poly(vinyl chloride)	520
Polystyrene	475
Polycarbonate	215
Polyamide-imide	169
Polyarylate	109
Polytetrafluoroethylene	95
Polyethersulfone	50
Polyether ether ketone	35

the National Bureau of Standards, whereby a specimen of 7.6 × 7.6 cm is heated by radiation. The smoke density is then measured optically.

Colorants

A wide variety of inorganic and organic materials are added to polymers to impart color. For transparent colored plastics materials, oil-soluble dyes or organic pigments (such as phthalocyanines) having small particle size and refractive index near that of the plastic are used. Others, including inorganic pigments, impart opaque color to the plastic. Some of the common colorants for plastics, among many others, are barium sulfate and titanium dioxide (white), ultramarine blues, phthalocyanine blues and greens, chrome greens, molybdate oranges, chrome yellows, cadmium reds and yellows, quinacridone reds and magentas, and carbon black. Flake aluminum is added for a silver metallic appearance, and lead carbonate or mica for pearlescence.

The principal requirements of a colorant are that it have a high covering power-cost ratio and that it withstand processing and service conditions of the plastic. The colorant must not catalyze oxidation of the polymer nor adversely affect its desirable properties. Colorants are normally added to the powdered plastic and mixed by tumbling and compounding on a hot roll or in an extruder.

TOXICITY OF PLASTICS

With the expanding use of plastics in all walks of life, including clothing, food packaging, and medical and paramedical applications, attention has been focused on the potential toxic liability of these man-made materials.

Toxic chemicals can enter the body in various ways, particularly by skin absorption, inhalation, and swallowing. Although some chemicals may have an almost universal effect on humans, others may attack few persons. The monomers used in the synthesis of many of the polymers are unsaturated compounds with reactive groups such as vinyl, styrene, acrylic, epoxy, and ethylene imine groups. Such compounds can irritate the mucous membranes of the eyes and respiratory tract and sensitize the skin. They are suspected of inducing chemical lesions and carcinogenic and mutagenic effects. It was long ago that carcinogenic properties of the monomers vinyl chloride and chloroprene were reported, and there may be more trouble ahead in this respect. However, the monomers used in the production of polyester or polyamide resins are usually less reactive and may be expected to be less harmful.

Although many monomers are harmful chemicals, the polymers synthesized from them are usually harmless macromolecues. But then one has to take into account that the polymers may still contain small amounts of residual monomer and catalyst used in the polymerization process. Moreover, polymers are rarely used as such but are compounded, as we saw earlier, with various additives such as plasticizers, stabilizers, curing agents, etc., for processing into plastic goods. Being relatively smaller in molecular size, the residual monomer, catalyst residues, and the additives can migrate from the plastic body into the environment. They may eventually migrate into food products packed in plastic containers, or they may interact with the biological substrate when plastics materials are used as parts of tissue and organs implanted into humans. The toxic potential of thermodegradation and combustion products of plastics when these materials are burned, either deliberately or by accident, is also an important consideration in view of the widening use of plastics as structural and decorative lining material in buildings, vehicles, and aircraft.

Plastic Devices in Pharmacy and Medicine

Within the past two decades there has been an increase in the use of a variety of plastic materials in the pharmaceutical, medical, dental, and biochemical engineering fields. Such plastic devices can be classified according to use into five basic groups: (1) collection and administrative devices—e.g., catheters, blood transfusion sets, dialyzing units, injection devices; (2) storage devices—e.g., bags for blood and blood products, containers for drug products, nutritional products, diagnostic agents; (3) protective and supportive devices—e.g., protective clothes, braces, films; (4) implants having contact with mucosal tissue—e.g., dentures, contact lenses, intrauterine devices; and (5) permanent implants—e.g., orthopedic implants, heart valves, vascular grafts, artificial organs.

The toxic potential of plastic devices becomes more relevant in those applications which involve long periods of contact with the substrate. For short periods of contact, however, a great number of plastic materials manufactured today as medical and paramedical devices will produce little or no irritant response.

With polyethylene, polypropylene, Teflon, Dacron, polycarbonate, and certain types of silicone rubbers, the migration of additives from the material is so small that a biological response cannot be detected. Besides, many of these materials contain additives only in extremely low concentrations, some perhaps having only a stabilizing agent in concentrations of less than 0.1 wt. %. Incidence of tissue response increases when the plastic materials require greater concentrations of various types of additives. It should be stressed, however, that the toxic effects of a material will depend on the intrinsic toxicity of the additives and their ability to migrate from the material to the tissue in contact with it. It is important to know precisely the toxic potential of each of the ingredients in the final polymerized and formulated material to be used in an implantable or storage device.

Packaging

Many pharmaceutical manufacturers have now taken to packaging their drug product in plastic containers. The plastics used most in these applications are those manufactured from polyethylene and polypropylene, though other materials have also been used. Since both polyethylene and polypropylene contain extremely small amounts of additives (mostly as antioxidants and antistatic agents), the possibility of their release in sufficient concentrations to endanger the patient is negligible.

Tubings and Blood Bag Assemblies

Plastic tubings are used for many medical applications, such as catheters, parts of dialysis and administration devices, and other items requiring clear, flexible tubings. The most successful tubing is PVC, which has been plasticized to give it the desired flexibility. Plasticized PVC is also the chief material used in America for making bags for blood and blood products. In Europe polyethylene and polypropylene have been mainly used. The plasticizers used most for flexible PVC are the esters of phthalic acid; one that is mostly employed from this group is di-2-ethylhexyl phthalate. Long-term feeding studies have demonstrated the nontoxic nature of the plasticizer under the experimental conditions used. Organotin compounds, which are one of the best groups of stabilizers for vinyl polymers, have been used to stabilize PVC, but, in general, their toxicity has decreased their use in medical applications.

Implants

Man-made materials have been used for making implants to save human life. It is possible that these materials, depending on their specific nature and the site of implant, will degrade with long periods of contact and release polymer fragments in the body. These in turn may elicit one or more biological responses, including carcinogenesis. However, no well-substantiated evidence has been reported showing that man-made materials have caused cancer in humans, although the real answer will be available only when these materials have been implanted for periods of time exceeding 20 to 30 years.

Perhaps the most commonly used implantable material is silicone rubber. This material, if properly prepared by the manufacturer, does not cause local toxic response. Various types of epoxy polymers and polyurethane materials also have found one or more medical applications.

Adhesives and Dental Materials

In most medical and paramedical applications of plastics, the materials used are those already produced by the manufacturers in a polymerized or formulated form. Certain surgical and dental applications, however, require that the material be polymerized or formulated just prior to use. Surgical cements and adhesives, a host of dental filling materials, materials for dentures, cavity liners, and protective coatings for tooth surfaces are in this category.

Cyanoacrylates have become very useful as tissue adhesives in surgical applications, because they polymerize rapidly in contact with moisture and create an extremely tenacious film. Methyl cyanoacrylate, used initially, has now fallen out of favor because of its toxic properties. The butyl and heptyl analogs are, however, quite satisfactory and do not produce objectionable tissue response. They also degrade at a much slower rate than the methyl compound in a biological environment.

Many restorative materials in dentistry are made from one of the acrylic monomers, the most used being the methyl ester of methacrylic acid. A liquid component containing the acrylic monomer and stabilizers is mixed with a powder containing polymerized acrylic and several different types of additives to form the final solid material, which, before setting, can be fashioned into any shape. These types of systems are used by the dentist in his daily practice. Since the dental materials prepared in this way by mixing components may have residual monomers or reactive agents that have not been removed, these can migrate to tissue and cause tissue response. In clinical practice, however, similar responses may not always be noted.

Toxicity of Plastic Combustion Products

Fires involving plastics produce not only smoke but also other pyrolysis and combustion products. Since most of the polymeric materials contain carbon,

carbon monoxide is one of the products generated from the heating and burning of these materials. Depending on the material, the temperature, and the presence or absence of oxygen, other harmful gases may also evolve. These include HCl, HCN, NO_2, SO_2, and fluorinated gases. Presently, various flame-retarding agents are added to plastics to reduce their combustion properties. When heated or subjected to flame, these agents can change the composition of the degradation products and may produce toxic responses not originally anticipated [37].

Toxicity Testing

Ideally, materials for medical and paramedical applications should be tested or evaluated at three levels: (1) on the ingredients used to make the basic resin, (2) on the final plastic or elastomeric material, and (3) on the final device. Organizations such as the Association for the Advancement of Medical Instrumentation, the U.S.A. Standard Institute, and the American Society for Testing and Materials (F4 Committee) have developed toxicity testing programs for materials used in medical applications. The American Dental Association has recommended standard procedures for biological evaluation of dental materials [38].

REFERENCES

1. W. R. Krigbaum and P. J. Flory, Treatment of osmotic pressure data, *J. Polymer Sci., 9*: 503 (1952).
2. F. W. Billmeyer, Jr., Recent advances in determining polymer molecular weights and sizes, *Appl. Polymer Symposia, 10*: 1 (1969).
3. P. Debye, Molecular weight determination by light scattering, *J. Phys. and Coll. Chem., 51*: 18 (1947).
4. B. A. Brice, M. Halwer, and R. Speiser, Photoelectric light scattering photometer for determining high molecular weights, *J. Opt. Soc. Am., 40*: 768 (1950).
5. P. F. Onyon, Viscometry, *Techniques of Polymer Characterization* (P. W. Allen, ed.), Butterworths, London, Chap. 6 (1951).
6. G. Mayerhoff, The viscometric determination of molecular weight of polymer (in German), *Fortschr. Hochpolym. Forsch., 3*: 59 (1961).
7. F. W. Billmeyer, Jr., Characterization of molecular weight distribution in high polymers, *J. Polymer Sci., C8*: 161 (1965).
8. H. W. McCormick, F. M. Brower, and L. Kin, The effect of molecular weight distribution on the physical properties of polystyrene, *J. Polymer Sci., 39*: 87 (1959).
9. H. Mark and G. S. Whitby (eds.), *Collected Papers of Wallace Hume Carothers on High Polymeric Substances*, Interscience, New York (1940).
10. R. W. Lenz, Applied polymer reaction kinetics, *Ind. Eng. Chem., 62(2)*: 54 (1970).
11. F. T. Wall, The structure of copolymers, *J. Am. Chem. Soc., 66*: 2050 (1944).
12. P. J. Flory, Network structure and the elastic properties of vulcanized rubber, *Chem. Revs., 35*: 51 (1944).

13. L. C. Nielsen, Cross-linking—effect on physical properties of polymers, *J. Macromol. Sci. Revs. Macromol. Chem., C3*: 69 (1969).
14. R. F. Boyer, The relation of transition temperatures to chemical structure in high polymers, *Rubber Chem. Tech., 36*: 1303 (1963).
15. H. F. Mark, The nature of polymeric materials, *Sci. American, 217 (3)*, Sept. 19: 148, 156 (1967).
16. C. G. Overberger and J. A. Moore, Ladder polymers, *Fortschr. Hochpolym. Forsch. 7*: 113 (1970).
17. P. A. Small, Some factors affecting the solubility of polymers, *J. Appl. Chem., 3*: 71 (1953).
18. H. Burrell and B. Immergut, Solubility parameter values, *Polymer Handbook* (J. Brandrup and E. H. Immergut, eds., with the collaboration of H. G. Elias) Interscience, New York, pp. IV–341–IV–368 (1966).
19. J. D. Crowley, G. S. Teague, Jr., and J. W. Lowe, Jr., A three-dimensional approach to solubility, *J. Paint Technol., 38*: 269 (1966).
20. H. Lee, D. Stoffey, and K. Neville, *New Linear Polymers*, McGraw-Hill, New York (1967).
21. *Celanese R and D. Product Information: Polybenzimidazoles in Ion Exchange Applications*, Celanese Fiber Operations, Charlotte, North Carolina.
22. D. L. Schmidt, *Mod. Plastics, 37*: 131 (Nov. 1960), 147 (Dec. 1960).
23. W. C. Kuryla and A. J. Papa (eds.), *Flame Retardancy of Polymeric Materials*, Marcel Dekker, New York, Vol. 3, 1973.
24. A. Lightbody, M. E. Roberts, and C. J. Wessel, Preservation of plastics and rubber, *Deterioration of Materials* (G. A. Greathouse and C. J. Wessel, eds.), Van Nostrand Reinhold, New York.
25. S. H. Pinner (ed.), *Weathering and Degradation* of Plastics, Columbine, London, 1966.
26. D. V. Rosato and R. T. Schwartz (eds.), *Environmental Effects on Polymeric Materials*, Interscience, New York (1968).
27. L. Bateman (ed.), *The Chemistry and Physics of Rubber-like Substances*, John Wiley, New York, Chap. 12 (1963).
28. W. L. Hawkins, *Polymer Stabilization*, John Wiley, New York (1972).
29. R. Gächter and H. Müller (eds.), *Plastics Additives Handbook*, Hanser Publishers, Munich/New York (1987).
30. T. J. Henman, *World Index of Polyolefin Stabilizers*, Kogan Page Ltd., London (1982).
31. B. Ranby and J. F. Rabek, *Photodegradation, Photooxidation and Photostabilization of Polymers*, Wiley Interscience, London (1975).
32. V. T. Stannett and M. Szwarc, Permeability of Polymer films to gases—simple relation, *J. Polymer Sci., 16*: 89 (1955).
33. F. A. Paine, Packaging with flexible materials, *J. Roy. Inst. Chem., 86*: 263 (1962).
34. R. B. Seymour, *Additives for Plastics*, Academic Press, New York, vol. 2 (1978).
35. J. Stepek and H. Daoust, *Additives for Plastics*, Springer-Verlag, New York (1982).
36. H. J. Sanders, Artificial organs—Parts I and II, *Chem. and Eng. News, 32* (April 5, 1971); *68* (April 12, 1971).

37. J. Autian, Toxicologic aspects of flammability and combustion of polymer materials, *J. Fire Flammability, 1*: 239 (1970).
38. American Dental Association: Recommended standard procedures for biological evaluation of dental materials, *J. Am. Dent. Assoc., 84*: 382 (1972).

2

Fabrication Processes

TYPES OF PROCESSES

As indicated in Chapter 1, the family of polymers is extraordinarily large and varied. There are, however, some fairly broad and basic approaches that can be followed when designing or fabricating a product out of polymers or, more commonly, polymers compounded with other ingredients. The type of fabrication process to be adopted depends on the properties and characteristics of the polymer and on the shape and form of the final product.

In the broad classification of plastics there are two generally accepted categories: *thermoplastic* resins and *thermosetting* resins.

Thermoplastic resins consist of long polymer molecules, each of which may or may not have side chains or groups. The side chains or groups, if present, are not linked to other polymer molecules (i.e., are not cross-linked). Thermoplastic resins, usually obtained as a granular polymer, can therefore be repeatedly melted or solidified by heating or cooling. Heat softens or melts the material so that it can be formed; subsequent cooling then hardens or solidifies the material in the given shape. No chemical change usually takes place during this shaping process.

In thermosetting resins the reactive groups of the molecules form cross-links between the molecules during the fabrication process. The cross-linked or "cured" material cannot be softened by heating. Thermoset materials are usually supplied as a partially polymerized molding compound or as a liquid monomer-polymer mixture. In this uncured condition they can be shaped with or without pressure and polymerized to the cured state with chemicals or heat.

With the progress of technology the demarcation between thermoplastic and thermoset processing has become less distinct. For thermosets processes have been developed which make use of the economic processing characteristics of thermoplastics. For example, cross-linked polyethylene wire coating is made by extruding the thermoplastic polyethylene, which is then cross-linked (either chemically or by irradiation) to form what is actually a thermoset material that cannot be melted again by heating. More recently, modified machinery and molding compositions have become available to provide the economics of thermoplastic processing to thermosetting materials. Injection molding of phenolics and other thermosetting materials are such examples. Nevertheless, it is still a widespread practice in industry to distinguish between thermoplastic and thermosetting resins. Examples of common thermoplastics are ABS, PVC, SAN, acetals, acrylics, cellulosics, polyethylenes, polypropylenes, polystyrenes, polycarbonates, polyesters, nylons, and fluoropolymers. Some of the more common thermosets are phenolics, ureas, melamines, epoxies, alkyds, polyesters, silicones, and urethanes.

Compression and transfer molding are the most common methods of processing thermosetting plastics. For thermoplastics, the more important processing techniques are extrusion, injection, blow molding, and calendering; other processes are thermoforming, slush molding, and spinning.

COMPRESSION MOLDING [1–3]

Compression molding is the most common method by which thermosetting plastics are molded. In this method the plastic, in the form of powder, pellet, or disc, is dried by heating and then further heated to near the curing temperature; this heated charge is loaded directly into the mold cavity. The temperature of the mold cavity is held at 150 to 200°C, depending on the material. The mold is then partially closed, and the plastic, which is liquefied by the heat and the exerted pressure, flows into the recess of the mold. At this stage the mold is fully closed, and the flow and cure of the plastic are complete. Finally, the mold is opened, and the completely cured molded part is ejected.

Compression-molding equipment consists of a matched mold, a means of heating the plastic and the mold, and some method of exerting force on the mold halves. For severe molding conditions molds are usually made of hardened steel. Brass, mild steel, or plastics are used as mold materials for less severe molding conditions or short-run products.

In compression molding a pressure of 2250 psi (158 kg/cm^2) to 3000 psi (211 kg/cm^2) is suitable for phenolic materials. The lower pressure is adequate only for an easy-flow material and a simple uncomplicated shallow molded shape.

For a medium-flow material and where there are average-sized recesses, cores, shapes, and pins in the molding cavity, a pressure of 3000 psi (211 kg/cm^2) or above is required. For molding urea and melamine materials, pressures of approximately one and one-half times that needed for phenolic material are necessary.

The time required to harden thermosetting materials is commonly referred to as the *cure time*. Depending on the type of molding material, preheating temperature, and the thickness of the molded article, the cure time may range from seconds to several minutes.

In compression molding of thermosets the mold remains hot throughout the entire cycle; as soon as a molded part is ejected, a new charge of molding powder can be introduced. On the other hand, unlike thermosets, thermoplastics must be cooled to harden. So before a molded part is ejected, the entire mold must be cooled, and as a result, the process of compression molding is quite slow with thermoplastics. Compression molding is thus commonly used for thermosetting plastics such as phenolics, urea, melamine, and alkyds; it is not ordinarily used for thermoplastics. However, in special cases, such as when extreme accuracy is needed, thermoplastics are also compression molded. One example is the phonograph records of vinyl and styrene thermoplastics; extreme accuracy is needed for proper sound reproduction. Compression molding is ideal for such products as electrical switch gear and other electrical parts, plastic dinnerware, radio and television cabinets, furniture drawers, buttons, knobs, handles, etc.

Compression molds can be divided into hand molds, semiautomatic molds, and automatic molds. The design of any of these molds must allow venting to provide for escape of steam, gas, or air produced during the operation. After the initial application of pressure the usual practice is to open the mold slightly to release the gases. This procedure is known as *breathing*.

Hand molds are used primarily for experimental runs, for small production items, or for molding articles which, because of complexity of shape, require dismantling of mold sections to release them. *Semiautomatic molds* consist of units mounted firmly on the top and bottom platens of the press. The operation of the press closes and opens the mold and actuates the ejector system for removal of the molded article. However, an operator must load the molding material, actuate press controls for the molding sequence, and remove the ejected piece from the mold. This method is widely used. *Fully automatic molds* are specially designed for adaptation to a completely automatic press. The entire operation cycle, including loading and unloading of the mold, is performed automatically, and all molding operations are accurately controlled. Thermosetting polymers can be molded at rates up to 450 cycles/hr. Tooling must be of the highest standard to meet the exacting demands of high-speed production. Automatic molds offer the most economical method for long production runs because labor costs are kept to a minimum.

The three common types of mold designs are *open flash*, *fully positive*, and *semipositive*.

Open Flash [4]

In an open flash mold a slight excess of molding powder is loaded into the mold cavity (Fig. 2.1a). On closing the top and bottom platens, the excess material is forced out and flash is formed. The flash blocks the plastic remaining in the

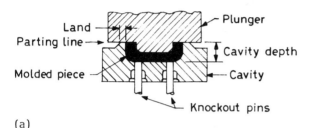

(a)

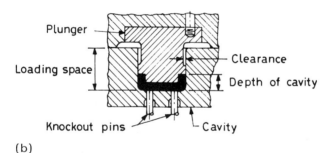

(b)

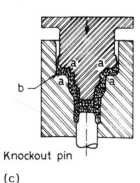

Knockout pin

(c)

FIG. 2.1 Compression molds. (a) A simple flash mold. (b) A positive mold. Knockout pins could extend through plunger instead of through cavity. (c) Semipositive mold as it appears in partly closed position before it becomes positive. Material trapped in area b escapes upward. (d) Semipositive mold in closed position. When corners at a pass, mold is positive.

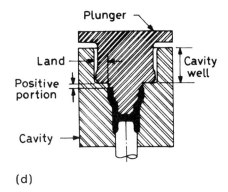

(d)

cavity and causes the mold plunger to exert pressure on it. Gas or air can be trapped by closing the mold too quickly, and finely powdered material can be splashed out of the mold. However, if closing is done carefully, the open flash mold is a simple one, giving very good results.

Since the only pressure on the material remaining in the flash mold when it is closed results from the high viscosity of the melt which did not allow it to escape, only resins having high melt viscosities can be molded by this process. Since most rubbers have high melt viscosities, the flash mold is widely used for producing gaskets and grommets, tub and flash stoppers, shoe heels, door mats, and many other items.

Because of lower pressure exerted on the plastic in the flash molds, the molded products are usually less dense than when made using other molds. Moreover, because of the excess material loading needed, the process is somewhat wasteful as far as raw materials are concerned. However, the process has the advantage that the molds are cheap, and very slight labor costs are necessary in weighing out the powder.

Fully Positive [4]

In the fully positive mold (Fig. 2.1b) no allowance is made for placing excess powder in the cavity. If excess powder is loaded, the mold will not close; an insufficient charge will result in reduced thickness of the molded article. A correctly measured charge must therefore be used with this mold—it is a disadvantage of the positive mold. Another disadvantage is that the gases liberated during the chemical curing reaction are trapped inside and may show as blisters on the molded surface. Excessive wear on the sliding fit surface on the top and bottom forces and the difficulty of ejecting the molding are other reasons for discarding this type of mold. The mold is used on a small scale for molding thermosets, laminated plastics, and certain rubber components.

Semipositive [4]

The semipositive mold (Fig. 2.1c and d) combines certain features of the open flash and fully positive molds and makes allowance for excess powder and flash. It is also possible to get both horizontal and vertical flash. Semipositive molds are more expensive to manufacture and maintain than the other types, but they are much better from an applications point of view. Satisfactory operation of semipositive molds is obtained by having clearance (0.025 mm/25 mm of diameter) between the plunger (top force) and the cavity. Moreover, the mold is given a 2 to 3° taper on each side. This allows the flash to flow on and the entrapped gases to escape along with it, thereby producing a clean, blemish-free mold component.

TRANSFER MOLDING [1–6]

In transfer molding, the thermosetting molding powder is placed in a chamber or pot outside the molding cavity and subjected to heat and pressure to liquefy it. When liquid enough to start flowing, the material is forced by the pressure into the molding cavity, either by a direct *sprue* or through a system of *runners* and *gates*. The material sets hard to the cavity shape after a certain time (cure time) has elapsed. When the mold is disassembled, the molded part is pushed out of the mold by *ejector pins*, which operate automatically.

Figure 2.2 shows the molding cycle of *pot-type transfer* molding, and Figure 2.3 shows *plunger-type transfer molding* (sometimes called *auxiliary raw transfer molding*). The taper of the sprue in pot-type transfer is such that, when the mold is opened, the sprue remains attached to the disc of material left in the pot, known as *cull*, and is thus pulled away from the molded part, whereas the latter is lifted out of the cavity by the ejector pins (Fig. 2.2c). In plunger-type transfer molding, on the other hand, the cull and the sprue remain with the molded piece when the mold is opened (Fig. 2.3c).

Another variation of transfer molding is screw transfer molding (Fig. 2.4). In this process the molding material is preheated and plasticized in a screw chamber and dropped into the pot of an inverted plunger mold. The preheated molding material is then transferred into the mold cavity by the same method as shown in Figure 2.3. The screw-transfer-molding technique is well suited to fully automatic operation. The optimum temperature of a phenolic mold charge is 240 ± 20°F (115 ± 11°C), the same as that for pot-transfer and plunger molding techniques.

For transfer molding, generally pressures of three times the magnitude of those required for compression molding are required. For example, usually a pressure of 9000 psi (632 kg/cm^2) and upward is required for phenolic molding material (the pressure referred to here is that applied to the powder material in the transfer chamber).

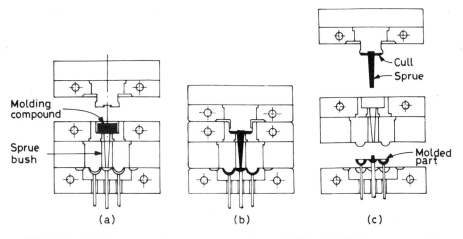

FIG. 2.2 Molding cycle of a pot-type transfer mold [6]. (a) Molding compound is placed in the transfer pot and then (b) forced under pressure when hot through an orifice and into a closed mold. (c) When the mold opens, the sprue remains with the cull in the pot, and the molded part is lifted out of the cavity by ejector pins.

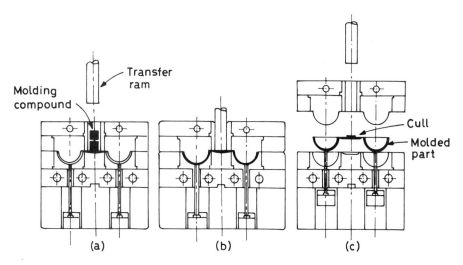

FIG. 2.3 Molding cycle of a plunger-type transfer mold [6]. An auxiliary ram exerts pressure on the heat-softened material in the pot (a) and forces it into the mold (b). When the mold is opened, the cull and sprue remain with the molded piece (c).

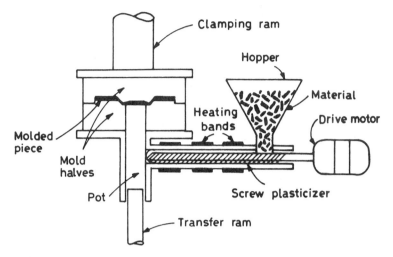

FIG. 2.4   Drawing of a screw-transfer molding machine [6].

The principle of transferring the liquefied thermosetting material from the transfer chamber into the molding cavity is similar to that of the injection molding of thermoplastics (described later). Therefore the same principle must be employed for working out the maximum area which can be molded—that is, the projected area of the molding multiplied by the pressure generated by the material inside the cavity must be less than the force holding the two halves together. Otherwise, the molding cavity plates will open as the closing force is overcome.

Transfer molding has an advantage over compression molding in that the molding powder is fluid when it enters the mold cavity. The process therefore enables production of intricate parts and molding around thin pins and metal inserts (such as an electrical lug). Thus, by transfer molding, metal inserts can be molded into the component in predetermined positions held by thin pins, which would, however, bend or break under compression-molding conditions. Typical articles made by the transfer molding process are terminal-block insulators with many metal inserts and intricate shapes, such as cups and caps for cosmetic bottles.

Ejection of Molding

Ejection of a molded plastic article from a mold can be achieved by using ejector pins, sleeves, or stripper plates. Ejector pins are the most commonly used method because they can be easily fitted and replaced. The ejector pins must be located in a position where they will eject the article efficiently without causing distortion of the part. They are worked by a common ejector plate or a bar located under the mold, and operated by a central hydraulic ejector ram. The ejector pins are fitted either to the bottom force or to the top force depending on whether it is necessary for the molding to remain in the bottom half of the female part or on the top half of the male part of the tool. The pins are usually constructed of a hardened steel to avoid wear.

Heating System

Heating is extremely important in plastics molding operations because the tool and auxiliary parts must be heated to the required temperature, depending on the powder being molded, and the temperature must be maintained throughout the molding cycle. The molds are heated by steam, hot water, gas, or electricity (resistance heaters, band heaters, low-voltage heaters, and induction heaters). Steam heating is preferred for compression and transfer molding, although electricity is also used because it is cleaner and has low installation costs. The main disadvantage of the latter method is that the heating is not fully even, and there is tendency to form hot spots.

Types of Presses

Presses used for compression and transfer molding of thermosets can be of many shapes and designs, but they can be broadly classified as hand, mechanical, or hydraulic types. *Hand* presses have relatively lower capacity, ranging from 10 to 100 tons, whereas hydraulic presses have considerably higher capacity (500 tons). *Hydraulic* presses may be of the upstroke or downstroke varieties. In the simple upstroke press, pressure can be applied fairly quickly, but the return is slow. In the downstroke press fitted with a prefilling tank, this disadvantage of the upstroke press is removed, and a higher pressure is maintained by prefilling with liquid from a tank.

The basic principles of hydraulics are used in the presses. Water or oil is used as the main fluid. Water is cheap but rusts moving parts. Oil is more expensive but it does not corrode and it does lubricate moving parts. The main disadvantage of oil is that it tends to form sludge due to oxidation with air.

The drive for the presses is provided by single pumps or by central pumping stations, and accumulators are used for storing energy to meet instantaneous pressure demand in excess of the pump delivery. The usual accumulator consists of a single-acting plunger working in a cylinder. The two main types of accumulators used are the *weight-loaded type* and the *air-loaded type*. The weight-loaded type is heavy and therefore not very portable. There is also an initial pressure surge on opening the valve. The pressure-surge problem is overcome in the air- or gas (nitrogen)-loaded accumulator. This type is more portable but suffers a small pressure loss during the molding cycle.

Preheating

To cut down cycle times and to improve the finished product of compression molding and transfer molding, the processes of preheating and preforming are commonly used. With preheating, relatively thick sections can be molded without porosity. Other advantages of the technique include improved flow of resin, lower molding pressures, reduced mold shrinkage, and reduced flash.

Preheating methods are convection, infrared, radio frequency, and steam. Thermostatically controlled gas or electrically heated ovens are inexpensive methods of heating. The quickest, and possibly the most efficient, method is radio-frequency heating, but it is also the most expensive. Preheaters are located adjacent to the molding press and are manually operated for each cycle.

Preforming

Preforming refers to the process of compressing the molding powder into the shape of the mold before placing it in the mold or to *pelleting*, which consists of compacting the molding powder into pellets of uniform size and approximately known weight. Preforming has many advantages, which include avoiding waste, reduction in bulk factor, rapid loading of charge, and less pressure than uncompacted material. Preformers are basically compacting presses. These presses may be mechanical, hydraulic, pneumatic, or rotary cam machines.

Flash Removal

Although mold design takes into consideration the fact that flash must be reduced to a minimum, it still occurs to some extent on the molded parts. It is thus necessary to remove the flash subsequent to molding. This removal is most often accomplished with tumbling machines. These machines tumble molded parts against each other to break off the flash. The simplest tumbling machines are merely wire baskets driven by an electric motor with a pulley belt. In more elaborate machines blasting of molded parts is also performed during the tumbling operation.

INJECTION MOLDING OF THERMOPLASTICS [7–9]

Injection molding is the most important molding method for thermoplastics. It is based on the ability of thermoplastic materials to be softened by heat and to harden when cooled. The process thus consists essentially of softening the material in a heated cylinder and injecting it under pressure into the mold cavity, where it hardens by cooling. Each step is carried out in a separate zone of the same apparatus in a cyclic operation.

A diagram of a typical injection-molding machine is shown in Figure 2.5. Granular material (the plastic resin) falls from the hopper into the barrel when

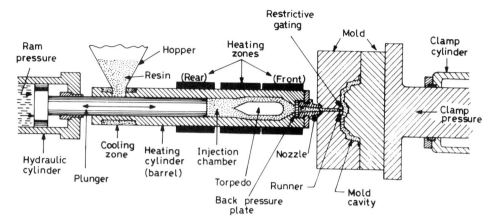

FIG. 2.5 Cross section of a typical plunger injection-molding machine [10].

the plunger is withdrawn. The plunger then pushes the material into the heating zone, where it is heated and softened (plasticized or plasticated). Rapid heating takes place due to spreading of the polymer into a thin film around a *torpedo*. The already molten polymer displaced by this new material is pushed forward through the nozzle, which is in intimate contact with the mold. The molten polymer flows through the sprue opening in the die, down the runner, past the gate, and into the mold cavity. The mold is held tightly closed by the clamping action of the press platen. The molten polymer is thus forced into all parts of the mold cavities, giving a perfect reproduction of the mold. The material in the mold must be cooled under pressure below T_m or T_g before the mold is opened and the molded part is ejected. The plunger is then withdrawn, a fresh charge of material drops down, the mold is closed under a locking force, and the entire cycle is repeated. Mold pressures of 8000 to 40,000 psi (562 to 2812 kg/cm^2) and cycle times as low as 15 sec are achieved on some machines.

Note that the feed mechanism of the injection molding machine is activated by the plunger stroke. The function of the torpedo in the heating zone is to spread the polymer melt into thin film in close contact with the heated cylinder walls. The fins, which keep the torpedo centered, also conduct heat from the cylinder walls to the torpedo, although in some machines the torpedo is heated separately.

Injection-molding machines are rated by their capacity to mold polystyrene in a single shot. Thus a 2-oz machine can melt and push 2 oz of general-purpose polystyrene into a mold in one shot. This capacity is determined by a number of factors such as plunger diameter, plunger travel, and heating capacity.

The main components of an injection-molding machine are (1) the injection unit which melts the molding material and forces it into the mold; (2) the clamping unit which opens the mold and closes it under pressure; (3) the mold used; and (4) the machine controls.

Types of Injection Units

Injection-molding machines are known by the type of injection unit used in them. The oldest type is the single-stage plunger unit (Fig. 2.5) described above. As the plastics industry developed, another type of plunger machine appeared, known as a two-stage plunger (Fig. 2.6a). It has two plunger units set one on top of the other. The upper one, also known as a preplasticizer, plasticizes the molding material and feeds it to the cylinder containing the second plunger, which operates mainly as a shooting plunger, and pushes the plasticized material through the nozzle into the mold.

Later, another variation of the two-stage plunger unit appeared, in which the first plunger stage was replaced by a rotating screw in a cylinder (Fig. 2.6b). The screw increases the heat transfer at the walls and also does considerable heating by converting mechanical energy into heat. Another advantage of the screw is its mixing and homogenizing action. The screw feeds the melt into the second plunger unit, where the injection ram pushes it forward into the mold.

Although the single-stage plunger units (Fig. 2.5) are inherently simple, the limited heating rate has caused a decline in popularity: they have been mostly supplanted by the *reciprocating screw-type machines*. In these machines (Fig. 2.7) the plunger and torpedo (or spreader) that are the key components of plunger-type machines are replaced by a rotating screw that moves back and forth like a plunger within the heating cylinder. As the screw rotates, the flights pick up the feed of granular material dropping from the hopper and force it along

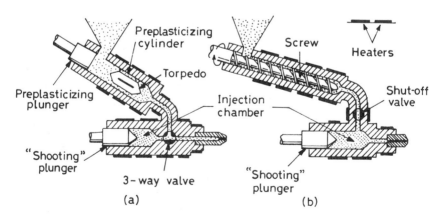

FIG. 2.6 Schematic drawings of (a) a plunger-type preplasticizer and (b) a screw-type preplasticizer atop a plunger-type injection-molding machine [10].

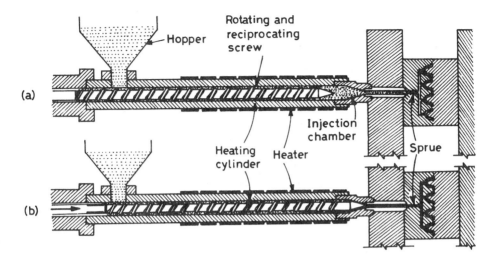

FIG. 2.7 Cross section of a typical screw-injection molding machine, showing the screw (a) in the retracted position and (b) in the forward position [10].

the heated wall of the barrel, thereby increasing the rate of heat transfer and also generating considerable heat by its mechanical work. The screw, moreover, promotes mixing and homogenization of the plastic material. As the molten plastic comes off the end of the screw, the screw moves back to permit the melt to accumulate. At the proper time the screw is pushed forward without rotation, acting just like a plunger and forcing the melt through the nozzle into the mold. The size of the charge per shot is regulated by the back travel of the screw. The heating and homogenization of the plastics material are controlled by the screw rotation speed and wall temperatures.

Clamping Units

The clamping unit keeps the mold closed while plasticized material is injected into it and opens the mold when the molded article is ejected. The pressure produced by the injection plunger in the cylinder is transmitted through the column of plasticized material and then through the nozzle into the mold. The unlocking force, that is, the force which tends to open the mold, is given by the product of the injection pressure and the projected area of the molding. Obviously, the clamping force must be greater than the unlocking force to hold the mold halves closed during injection.

Several techniques can be used for the clamping unit: (1) hydraulic clamps, in which the hydraulic cylinder operates on the movable parts of the mold to open and close it (Fig. 2.8); (2) toggle or mechanical clamps, in which the hydraulic

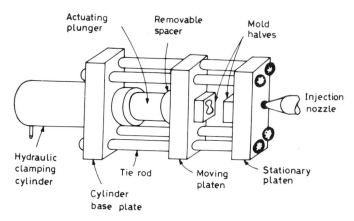

FIG. 2.8   Hydraulic clamping unit [6].

cylinder operates through a toggle linkage to open and close the mold (Fig. 2.9); and (3) various types of hydraulic mechanical clamps that combine features of (1) and (2).

Clamps are usually built as horizontal units, with injection taking place through the center of the stationary platen, although vertical clamp presses are also available for special jobs.

Molds

The mold is probably the most important element of a molding machine. Although the primary purpose of the mold is to determine the shape of the molded part, it performs several other jobs. It conducts the hot plasticized material from the heating cylinder to the cavity, vents the entrapped air or gas, cools the part until it is rigid, and ejects the part without leaving marks or causing damage. The mold design, construction, the craftsmanship largely determine the quality of the part and its manufacturing cost. The injection mold is normally described by a variety of criteria, including (1) number of cavities in the mold; (2) material of construction, e.g., steel, stainless steel, hardened steel, beryllium copper, chrome-plated aluminum, and epoxy steel; (3) parting line, e.g., regular, irregular, two-plate mold, and three-plate mold; (4) method of manufacture, e.g., machining, hobbing, casting, pressure casting, electroplating, and spark erosion; (5) runner system, e.g., hot runner and insulated runner; (6) gating type, e.g., edge, restricted (pinpoint), submarine, sprue, ring, diaphragm, tab, flash, fan, and multiple; and (7) method of ejection, e.g., knockout pins, stripper ring, stripper plate, unscrewing cam, removable insert, hydraulic core pull, and pneumatic core pull.

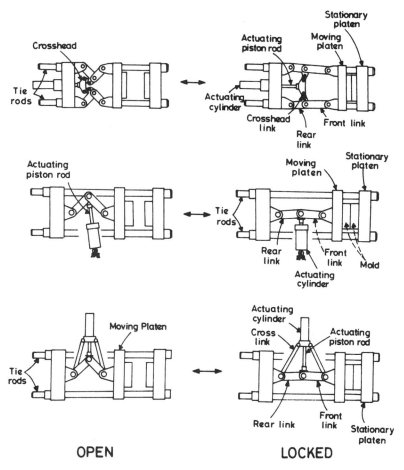

FIG. 2.9 Several types of toggle clamps [6].

Mold Designs

Molds used for injection molding of thermoplastic resins are usually flash molds, because in injection molding, as in transfer molding, no extra loading space is needed. However, there are many variations of this basic type of mold design.

The design most commonly used for all types of materials is the two-plate design (Fig. 2.10). The cavities are set in one plate, the plungers in the second plate. The sprue bushing is incorporated in that plate mounted to the stationary half of the mold. With this arrangement it is possible to use a direct center gate that leads either into a single-cavity mold or into a runner system for

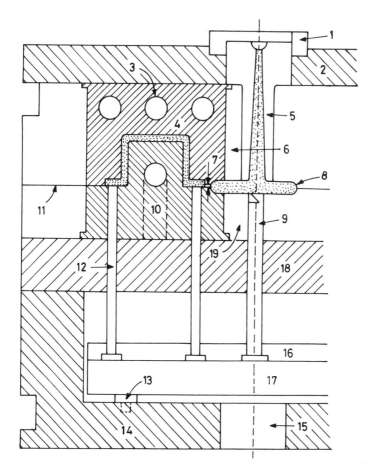

FIG. 2.10 A two-plate injection-mold design: (1) locating ring; (2) clamping plate; (3) water channels; (4) cavity; (5) sprue bushing; (6) cavity retainer; (7) gate; (8) full round runner; (9) sprue puller pin; (10) plunger; (11) parting line; (12) ejector pin; (13) stop pin; (14) ejector housing; (15) press ejector clearance; (16) pin plate; (17) ejector bar; (18) support plate; (19) plunger retainer.

a multi-cavity mold. The plungers are ejector assembly and, in most cases, the runner system belong to the moving half of the mold. This is the basic design of an injection mold, though many variations have been developed to meet specific requirements.

A three-plate mold design (Fig. 2.11) features a third, movable, plate which contains the cavities, thereby permitting center or offset gating into each cavity for multicavity operation. When the mold is opened, it provides two open-

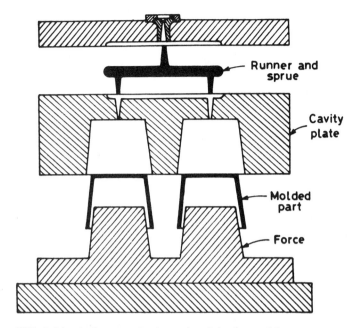

FIG. 2.11 A diagram of a three-plate injection mold.

ings, one for ejection of the molded part and the other for removal of the runner and sprue.

Moldings with inserts or threads or coring that cannot be formed by the normal functioning of the press require installation of separate or *loose details* or *cores* in the mold. These loose members are ejected with the molding. They must be separated from the molding and reinstalled in the mold after every cycle. Duplicate sets are therefore used for efficient operation.

Hydraulic or pneumatic cylinders may be mounted on the mold to actuate *horizontal coring* members. It is possible to mold angular coring, without the need for costly loose details, by adding *angular core pins* engaged in sliding mold members. Several methods may be used for unscrewing internal or external threads on molded parts: For high production rates *automatic unscrewing* may be done at relatively low cost by the use of rack-and-gear mechanism actuated by a double-acting hydraulic long-stroke cylinder. Other methods of unscrewing involve the use of an electric gear-motor drive or friction-mold wipers actuated by double-acting cylinders. Parts with interior undercuts can be made in a mold which has provision for angular movement of the core, the movement being actuated by the ejector bar that frees the metal core from the molding.

Number of Mold Cavities

Use of multiple mold cavities permits greater increase in output speeds. However, the greater complexity of the mold also increases significantly the manufacturing cost. Note that in a single-cavity mold the limiting factor is the cooling time of the molding, but with more cavities in the mold the plasticizing capacity of the machine tends to be the limiting factor. Cycle times therefore do not increase prorate with the number of cavities. There can be no clear-cut answer to the question of optimum number of mold cavities, since it depends on factors such as the complexity of the molding, the size and type of the machine, cycle time, and the number of moldings required. If a fairly accurate estimate can be made of the costs and cycle times for molds with each possible number of cavities and a cost of running the machine (with material) is assumed, a break-even quantity of the number of moldings per hour can be calculated and compared with the total production required.

Runners

The channels through which the plasticized material enters the gate areas of the mold cavities are called *runners*. Normally, runners are either full round or trapezoidal in cross section. Round cross section offers the least resistance to the flow of material but requires a duplicate machining operation in the mold, since both plates must be cut at the parting line. In three-plate mold designs, however, trapezoidal runners are preferred, since sliding movements are required across the parting-line runner face.

One can see from Figure 2.11 that a three-plate mold operation necessitates removal of the runner and sprue system, which must be reground, and the material reused. It is possible, however, to eliminate the runner system completely by keeping the material in a fluid state. This mold is called a *hot-runner mold* (Fig. 2.12). The material is kept fluid by the hot-runner manifold, which is heated with electric cartridges. The advantage of a hot-runner mold is that in a long-running job it is the most economical way of molding—there is no regrinding, with its attendant cost of handling and loss of material, and the mold runs automatically, eliminating variations caused by operators. A hot-runner mold also presents certain difficulties: It takes considerably longer to become operational, and in multicavity molds balancing the gate and the flow and preventing drooling are difficult. These difficulties are partially overcome in an insulated-runner mold, which is a cross between a hot-runner mold and a three-plate mold and has no runner system to regrind. An insulated-runner mold is more difficult to start and operate than a three-plate mold, but it is considerably easier than a hot-runner mold.

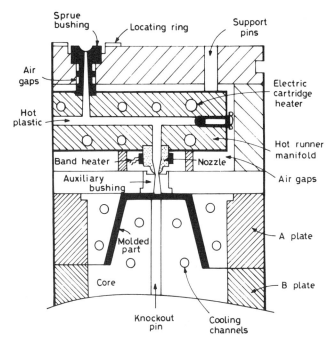

FIG. 2.12 Schematic drawing of a hot-runner mold [6].

Gating

The gate provides the connection between the runner and the mold cavity. It must permit enough material to flow into the mold to fill out the cavity. The type of the gate and its size and location in the mold strongly affect the molding process and the quality of the molded part. There are two types of gates: large and restricted. Restricted (pinpointed) gates are usually circular in cross section and for most thermoplastics do not exceed 0.060 in. in diameter. The apparent viscosity of a thermoplastic is a function of shear rate—the viscosity decreases as the shear rate and, hence, the velocity increases. The use of the restricted gate is therefore advantageous, because the velocity of the plastic melt increases as it is forced through the small opening; in addition, some of the kinetic energy is transformed into heat, raising the local temperature of the plastic and thus further reducing its viscosity. The passage through a restricted area also results in higher mixing.

The most common type of gate is the *edge gate* (Fig. 2.13a), where the part is gated either as a restricted or larger gate at some point on the edge. The edge gate is easy to construct and often is the only practical way of gating. It can be

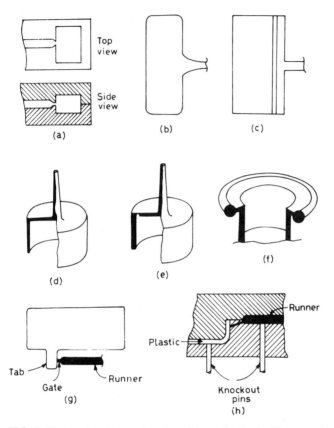

FIG. 2.13 Gating design: (a) edge; (b) fan; (c) flash; (d) sprue; (e) diaphragm; (f) ring; (g) tab; (h) Submarine.

fanned out for large parts or when there is a special reason. Then it is called a *fan gate* (Fig. 2.13b). When it is required to orient the flow pattern in one direction, a *flash gate* (Fig. 2.13c) may be used. It involves extending the fan gate over the full length of the part but keeping it very thin.

The most common gate for single-cavity molds is the *sprue gate* (Fig. 2.13d). It feeds directly from the nozzle of the machine into the molded part. The pressure loss is therefore a minimum. But the sprue gate has the disadvantages of the lack of a cold slug, the high stress concentration around the gate area, and the need for gate removal. A *diaphragm gate* (Fig. 2.13e) has, in addition to the sprue, a circular area leading from the sprue to the piece. This type of gate is suitable for gating hollow tubes. The diaphragm eliminates stress concentration around the gate because the whole area is removed, but the cleaning of this gate is more difficult than for a sprue gate. *Ring gates* (Fig. 2.13f) accomplish the same purpose as gating internally in a hollow tube, but from the outside.

When the gate leads directly into the part, there may be surface imperfection due to jetting. This may be overcome by extending a tab from the part into which the gate is cut. This procedure is called *tab gating* (Fig. 2.13g). The tab has to be removed as a secondary operation.

A *submarine gate* (Fig. 2.13h) is one that goes through the steel of the cavity. It is very often used in automatic molds.

Venting

When the melted plastic fills the mold, it displaces the air. The displaced air must be removed quickly, or it may ignite the plastic and cause a characteristic burn, or it may restrict the flow of the melt into the mold cavity, resulting in incomplete filling. For venting the air from the cavity, slots can be milled, usually opposite the gate. The slots usually range from 0.001 to 0.002 in. deep and from 3/8 to 1 in. wide. Additional venting is provided by the clearance between knockout pins and their holes. Note that the gate location is directly related to the consideration of proper venting.

Parting Line

If one were inside a closed mold and looking outside, the mating junction of the mold cavities would appear as a line. It also appears as a line on the molded piece and is called the *parting line*. A piece may have several parting lines. The selection of the parting line in mold design is influenced by the type of mold, number of cavities, shape of the piece, tapers, method of ejection, method of fabrication, venting, wall thickness, location and type of gating, inserts, post-molding operations, and aesthetic considerations.

Cooling

The mold for thermoplastics receives the molten plastic in its cavity and cools it to solidity to the point of ejection. The mold is provided with cooling channels. The mold temperature is controlled by regulating the temperature of the cooling fluid and its rate of flow through the channels. Proper cooling or coolant circulation is essential for uniform repetitive mold cycling.

The functioning of the mold and the quality of the molded part depend largely on the location of the cooling channel. Since the rate of heat transfer is reduced drastically by the interface of two metal pieces, no matter how well they fit, cooling channels should be located in cavities and cores themselves rather than only in the supporting plates. The cooling channels should be spaced evenly to prevent uneven temperatures on the mold surface. They should be as close to the plastic surface as possible, taking into account the strength of the mold material. The channels are connected to permit a uniform flow of the cooling or heating medium, and they are thermostatically controlled to maintain a given temperature.

Another important factor in mold temperature control is the material the mold is made from. Beryllium copper has a high thermal conductivity, about twice that of steel and four times that of stainless steel. A beryllium copper cavity should thus cool about four times as fast as a stainless steel one. A mold made of beryllium copper would therefore run significantly faster than one of stainless steel.

Ejection

Once the molded part has cooled sufficiently in the cavity, it has to be ejected. This is done mechanically by KO (knockout) pins, KO sleeves, stripper plates, stripper rings or compressed air, used either singly or in combination. The most frequent problem in new molds is with ejection. Because there is no mathematical way of predicting the amount of ejection force needed, it is entirely a matter of experience.

Since ejection involves overcoming the forces of adhesion between the mold and the plastic, the area provided for the knockout is an important factor. If the area is too small, the knockout force will be concentrated, resulting in severe stresses on the part. As a result, the part may fail immediately or in later service. In materials such as ABS and high-impact polystyrene, the severe stresses can also discolor the plastic.

Sticking in a mold makes ejection difficult. Sticking is often related to the elasticity of steel and is called *packing*. When injection pressure is applied to the molten plastic and forces it into the mold, the steel deforms; when the pressure is relieved, the steel retracts, acting as a clamp on the plastic. Packing is often eliminated by reducing the injection pressure and/or the injection forward time. Packing is a common problem in multicavity molds and is caused by unequal filling. Thus, if a cavity seals off without filling, the material intended for that cavity is forced into other cavities, causing overfilling.

Standard Mold Bases

Standardization of mold bases for injection molding, which was unknown prior to 1940, was an important factor in the development of efficient mold making. Standard mold bases were pioneered by the D-M-E Co., Michigan, to provide the mold maker with a mold base at lower cost and with much higher quality than if the base were manufactured by the mold maker. Replacement parts, such as locating ring and sprue bushings, loader pins and bushings, KO pins and push-back pins of high quality are also available to the molder. Since these parts

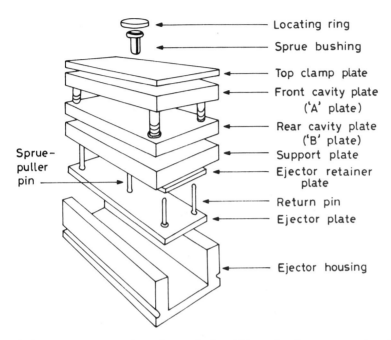

Locating ring

Sprue bushing

Top clamp plate

Front cavity plate
('A' plate)

Rear cavity plate
('B' plate)

Sprue–puller pin

Support plate

Ejector retainer plate

Return pin

Ejector plate

Ejector housing

FIG. 2.14 Exploded view of a standard mold base showing component parts.

are common for many molds, they can be stocked by the molder in the plant and thus down time is minimized. An exploded view of the components of a standard injection-mold base assembly is shown in Figure 2.14.

INJECTION MOLDING OF THERMOSETTING RESINS [6, 11–13]

Prior to 1940, no serious effort was made to adapt injection-molding techniques to thermosetting resins, but progress has since been rapid. Before 1950, three methods were developed—jet molding, straight injection molding, and offset molding. However, these techniques have only historical interest since they never really achieved significant commercial penetration.

Jet molding involved the application of a special nozzle to a standard injection-molding machine. Straight injection molding (i.e., molding in a straight line) was developed by altering standard plunger-type injection-molding machines to accommodate the thermosetting character of the resins. A jacketed cylinder, smoothly tapering to the orifices in the specially designed nozzles was incorporated, and the torpedos or spreaders of standard injection machines were eliminated. Offset injection molding may be described as a transfer-molding operation carried out with an altered horizontal-injection-molding machine in which the injection plunger displaces the heat-softened material into the heated transfer pot and from there into runners and cavities where rapid curing takes place. Offset molds are generally used when parts with inserts are to be made. The advantages claimed for the offset process are fast cycles, large capacities, and low plunger pressure.

Screw-Injection Molding of Thermosetting Resins

The foregoing three machines are basically plunger-type machines. But in the late 1960s shortly after the development of screw-transfer machines, the concept of screw-injection molding of thermosets, also known as *direct screw transfer* (or DST), was introduced. The potential of this technique for low-cost, high-volume production of molded thermoset parts was quickly recognized, and today screw-injection machines are available in all clamp tonnages up to 1200 tons and shot sizes up to 10 lb. Coupled with this, there has been a new series of thermosetting molding materials developed specifically for injection molding. These materials have long life at moderate temperatures (approximately 200°F), which permits plastication in the screw barrel, and react (cure) very rapidly when the temperature is raised to 350 to 400°F (177 to 204°C), resulting in reduced cycle time.

A typical arrangement for a direct screw-transfer injection-molding machine for thermosets is shown in Figure 2.15. The machine has two sections mounted

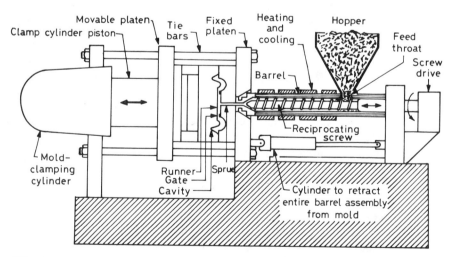

FIG. 2.15 Schematic of a direct screw-transfer molding machine for thermosets [6].

on a common base. One section constitutes the plasticizing and injection unit, which includes the feed hopper, the heated barrel that encloses the screw, the hydraulic cylinder which pushes the screw forward to inject the plasticized material into the mold, and a motor to rotate the screw. The other section clamps and holds the mold halves together under pressure during the injection of the hot plastic melt into the mold.

The thermosetting material (in granular or pellet form) suitable for injection molding is fed from the hopper into the barrel and is then moved forward by the rotation of the screw. During its passage, the material receives conductive heat from the wall of the heated barrel and frictional heat from the rotation of the screw. For thermosetting materials, the screw used is a zero-compression-ratio screw—i.e., the depths of flights of the screw at the feed-zone end and at the nozzle end are the same. By comparison, the screws used in thermoplastic molding machines have compression ratios such that the depth of flight at the feed end is one and one-half to five times that at the nozzle end. This difference in screw configuration is a major difference between thermoplastic- and thermosetting-molding machines.

As the material moves forward in the barrel due to rotation of the screw, it changes in consistency from a solid to a semifluid, and as it starts accumulating at the nozzle end, it exerts a backward pressure on the screw. This back pressure is thus used as a processing variable. The screw stops turning when the required amount of material—the charge—has accumulated at the nozzle end of the barrel, as sensed by a limit switch. (The charge is the exact volume of material required to fill the sprue, runners, and cavities of the mold.) The screw is then moved forward like a plunger by hydraulic pressure (up to 20,000 psi) to force the hot plastic melt through the sprue of the mold and into the runner system, gates, and mold cavities. The nozzle temperature is controlled to maintain a proper balance between a hot mold (350–400°F), and a relatively cool barrel (150–200°F).

Molded-in inserts are commonly used with thermosetting materials. However, since the screw-injection process is automatic, it is desirable to use post-assembled inserts rather than molded-in inserts because molded-in inserts require that the mold be held open each cycle to place the inserts. A delay in the manual placement disrupts an automatic cyclic operation, affecting both the production rate and the product quality.

Tolerances of parts made by injection molding of thermosetting materials are comparable to those produced by the compression and transfer methods described earlier. Tolerances achieved are as low as ±0.001 in./in., although ordinarily tolerances of ±0.003 to 0.005 in./in. are economically practical.

Thermosetting materials used in screw-injection molding are modified from conventional thermosetting compounds. These modifications are necessary to provide the working time-temperature relationship required for screw plasticating. The most commonly used injection-molding thermosetting materials are the phenolics. Other thermosetting materials often molded by the screw-injection process include melamine, urea, polyester, alkyd, and diallyl phthalate (DAP).

Since the mid-1970s the injection molding of glass-fiber-reinforced thermosetting polyesters gained increasing importance as better materials (e.g., low-shrinkage resins, pelletized forms of polyester/glass, etc.), equipment, and tooling became available. Injection-molded reinforced thermoset plastics have thus made inroads in such markets as switch housings, fuse blocks, distributor caps, power-tool housings, office machines, etc. Bulk molding compounds (BMC), which are puttylike FRP (fibrous glass-reinforced plastic) materials, are injection molded to make substitutes of various metal die castings.

For injection molding, FRP should have some specific characteristics. For example, it must flow easily at lower-than-mold temperatures without curing and without separating into resin, glass, and filler components, and it should cure rapidly when in place at mold temperature. A traditional FRP material shrinks about 0.003 in./in. during molding, but low-shrink FRP materials used for injection molding shrink as little as 0.000 to 0.0005 in./in. Combined with proper tooling, these materials thus permit production of pieces with dimensional tolerances of ±0.0005 in./in.

Proper design of parts for injection molding requires an understanding of the flow characteristics of material within the mold. In this respect, injection-molded parts of thermosets are more like transfer-molded parts than to compression-molded parts. Wall-section uniformity is an important consideration in part design. Cross sections should be as uniform as possible, within the dictates of part requirements, since molding cycles, and therefore costs, depend on the cure time of the thickest section. (For thermoplastics, however, it is the cooling time that is critical.) A rule of thumb for estimating cycle times for a 1/4-in. wall section is 30 sec for injection-molded thermosets (compared to 45 sec for thermoplastics). As a guideline for part design, a good working average for wall thickness is 1/8 to 3/16 in., with a minimum of 1/16 in.

STANDARDS FOR TOLERANCES ON MOLDED ARTICLES [6]

Molding materials shrink after they are taken out of the mold. The extent of shrinkage depends on the character of the material and the final mold temperature. A significantly large dimensional change takes place in cold-mold articles, because the curing or hardening is accomplished outside of the mold. Moreover, many of the conventional thermosetting and thermoplastic materials continue to shrink for many months after molding. Such data on normal material shrinkage after molding are commonly supplied by material suppliers.

The mold shrinkage described should not, however, be confused with tolerance. *Tolerance* is the variation in mold shrinkage rather than the shrinkage itself. Dimensional tolerances in a molded article are the allowable variations, plus and minus, from a nominal or mean dimension. Charts were developed by the Society of Plastics Industry, Inc. [12] to help obtain the magnitude of practical tolerances on the dimensions of articles molded from a variety of thermosetting and thermoplastic materials, including phenolics, melamine and urea, epoxy, diallyl phthalate, alkyd, polyethylene, polypropylene, ABS, acetal, acrylic, polystyrene, polycarbonate, and flexible vinyl. The charts were based on data obtained from representative suppliers and molders of plastics who responded to a questionnaire based on a hypothetical molded article, of which a cross section is shown in the charts. The shape of the article represents a variety of tolerance problems on diameter, length, depth, and thickness. The charts present standards of tolerances for articles of this shape molded from different plastics materials. Tables 2.1 to 2.4 illustrate these standards for two typical thermosets and two thermoplastics. The tolerances are based on a 1/8-in. wall

TABLE 2.1 Standards for Tolerances: General Purpose Phenolic [6]

Drawing Code	Dimensions (Inches)		
A = Diameter	0.000 / 1.000		
B = Depth	3.000		
C = Height	5.000		
	6 000 to 12.000 for each additional inch add (inches)	Comm. + .002	Fine ± .001
D = Bottom wall		.008	.005
E = Side Wall		.005	.003
F = Hole Size Diameter	0.000 to 0.125	.002	.001
	0.125 to 0.250	.002	.001
	0.250 to 0.500	.003	.002
	0.500 & Over	.003	.002
G = Hole Size Depth	0.000 to 0.250	.004	.002
	0.250 to 0.500	.004	.002
	0 500 to 1 000	.005	.003
Draft Allowance per side		1°	1/2°

Note:
These tolerances do not include allowance for aging characteristics of material

TABLE 2.2 Standards for Tolerances: Epoxy [6]

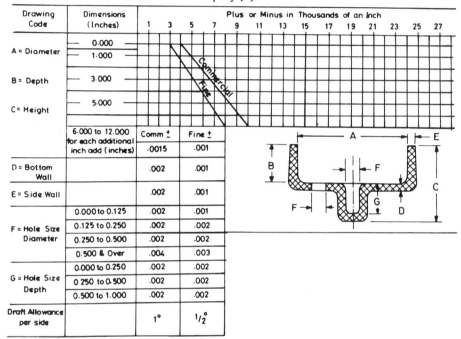

Drawing Code	Dimensions (Inches)	Plus or Minus in Thousands of an Inch
A = Diameter	0.000 — 1.000	
B = Depth	3.000	
C = Height	5.000	

	6.000 to 12.000 for each additional inch add (inches)	Comm ±	Fine ±
		.0015	.001
D = Bottom Wall		.002	.001
E = Side Wall		.002	.001
F = Hole Size Diameter	0.000 to 0.125	.002	.001
	0.125 to 0.250	.002	.002
	0.250 to 0.500	.002	.002
	0.500 & Over	.004	.003
G = Hole Size Depth	0.000 to 0.250	.002	.002
	0.250 to 0.500	.002	.002
	0.500 to 1.000	.002	.002
Draft Allowance per side		1°	1/2°

Note:

These tolerances do not include allowance
for aging characteristics of material.

section in each case. The *commercial* values shown in the charts represent common production tolerances obtainable under average manufacturing conditions. The *fine* values represent the narrowest possible tolerances obtainable under close supervision and control of production and, hence, at a greater cost. Note that the charts are not to be taken as offering hard-and-fast rules applicable to all conditions. They can best be used as a basis for establishing standards for individual molded articles.

EXTRUSION [14–16]

The *extrusion* process is basically designed to continuously convert a soft material into a particular form. An oversimplified analogy may be a household meat grinder. However, unlike the extrudate from a meat grinder, plastic extrudates generally approach truly continuous formation. Like the usual meat

TABLE 2.3 Standards for Tolerances: Low Density Polyethylene [6]

Drawing Code	Dimensions (Inches)	Plus or Minus in Thousands of an Inch														
		1	3	5	7	9	11	13	15	17	19	21	23	25	27
A = Diameter	0.000														
	1.000														
B = Depth	3.000														
C = Height	5.000														

		Comm ±	Fine ±
	6.000 to 12.000 for each additional inch add (inches)	.005	.004
D = Bottom Wall		.005	.004
E = Side Wall		.005	.004
F = Hole Size Diameter	0.000 to 0.125	.003	.002
	0.125 to 0.250	.004	.003
	0.250 to 0.500	.005	.004
	0.500 & Over	.006	.005
G = Hole Size Depth	0.000 to 0.250	.003	.003
	0.250 to 0.500	.004	.004
	0.500 to 1.000	.006	.005
Draft Allowance per side		2°	1°

Note:
These tolerances do not include allowance for aging characteristics of material.

grinder, the extruder (Fig. 2.16) is essentially a screw conveyor. It carries the cold plastic material (in granular or powdered form) forward by the action of the screw, squeezes it, and, with heat from external heaters and the friction of viscous flow, changes it to a molten stream. As it does this, it develops pressure on the material, which is highest right before the molten plastic enters the die. The screen pack, consisting of a number of fine or coarse mesh gauzes supported on a breaker plate and placed between the screw and the die, filter out dirt and unfused polymer lumps. The pressure on the molten plastic forces it through an adapter and into the die, which dictates the shape of the final extrudate. A die with a round opening as shown in Figure 2.16, produces pipe; a square die opening produces a square profile, etc. Other continuous shapes, such as the film, sheet, rods, tubing, and filaments, can be produced with appropriate dies. Extruders are also used to apply insulation and jacketing to wire and cable and to coat substrates such as paper, cloth, and foil.

When thermoplastic polymers are extruded, it is necessary to cool the extrudate below T_m or T_g to impart dimensional stability. This cooling can often be

TABLE 2.4 Standards for Tolerances: ABS [6]

Drawing Code	Dimensions (Inches)	Plus or Minus in Thousands of an Inch
A = Diameter	0.000 — 1.000	
B = Depth	3.000	
C = Height	5.000	

Drawing Code	Dimensions (Inches)	Comm. ±	Fine ±
	6.000 to 12.000 for each additional inch add (inches)	.003	.002
D = Bottom Wall		.004	.002
E = Side Wall		.003	.002
F = Hole Size Diameter	0.000 to 0.125	.002	.001
	0.125 to 0.250	.002	.001
	0.250 to 0.500	.003	.002
	0.500 & Over	.004	.002
G = Hole Size Depth	0.000 to 0.250	.003	.002
	0.250 to 0.500	.004	.002
	0.500 to 1.000	.005	.003
Draft Allowance per side		2°	1°

Note:
These tolerances do not include allowance for aging characteristics of material.

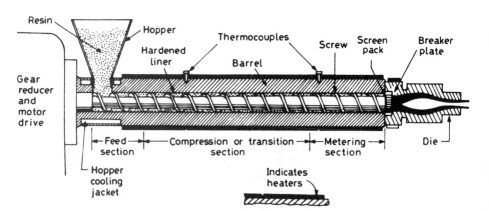

FIG. 2.16 Scheme for a typical single-screw extruder showing extruding pipe.

done simply by running the product through a tank of water, by spraying cold water, or, even more simply, by air cooling. When rubber is extruded, dimensional stability results from cross-linking (vulcanization). Interestingly, rubber extrusion for wire coating was the first application of the screw extruder in polymer processing.

Extruders have several other applications in polymer processing: in the blow-molding process they are used to make hollow objects such as bottles; in the blown-film process they are used for making wide films; they are also used for compounding plastics (i.e., adding various ingredients to a resin mix) and for converting plastics into the pellet shape commonly used in processing. In this last operation specialized equipment, such as the die plate-cutter assembly, is installed in place of the die, and an extrusion-type screw is used to provide plasticated melt for various injection-molding processes.

Extruder Capacity

Standard sizes of single-screw extruders are 1½, 2, 2½, 3¼, 3½, 4½, 6, and 8 in., which denote the inside diameter of the barrel. As a rough guide, extruder capacity Q_e, in pounds per hour, can be calculated from the barrel diameter D_b, in inches, by the empirical relation [16]

$$Q_e = 16 \, D_b^{2.2} \tag{1}$$

Another estimate of extruder capacity can be made by realizing that most of the energy needed to melt the thermoplastic stems from the mechanical work, whereas the barrel heaters serve mainly to insulate the material. If we allow an efficiency from drive to screw of about 80%, the capacity Q_e (lb/hr) can be approximately related to the power supplied H_p (horsepower), the heat capacity of the material C_p [Btu/lb °F], and the temperature rise from feed to extrudate ΔT (°F) by

$$Q_e = 1.9 \times 10^3 H_p / C_p \Delta T \tag{2}$$

Equation 2 is obviously not exact since the heat of melting and other thermal effects have been ignored. Equation (2) coupled with Eq. (1) enables one to obtain an estimate of ΔT. Thus, for processing of poly(methyl methacrylate), for which C_p is about 0.6 Btu/lb °F, in a 2-in. extruder run by a 10-hp motor, Eq. (1) gives $Q_e = 74$ lb/hr, and Eq. (2) indicates that $\Delta T \simeq 430°F$. In practice, a ΔT of 350°F is usually adequate for this polymer.

Extruder Design and Operation

The most important component of any extruder is the screw. It is often impossible to extrude satisfactorily one material by using a screw designed for another material. Therefore screw designs vary with each material.

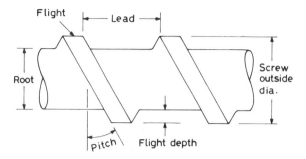

FIG. 2.17 Detail of screw.

Typical Screw Construction

The screw consists of a steel cylinder with a helical channel cut into it (Fig. 2.17). The helical ridge formed by the machining of the channel is called the *flight*, and the distance between the flights is called the *lead*. The lead is usually constant for the length of the screw in single-screw machines. The helix angle is called *pitch*. Helix angles of the screw are usually chosen to optimize the feeding characteristics. An angle of 17.5° is typical, though it can be varied between 12 and 20°. The screw outside diameter is generally just a few thousandths of an inch less than the inside diameter (ID) of the barrel. The minimal clearance between screw and barrel (ID) prevents excessive buildup of resin on the inside barrel wall and thus maximizes heat transfer.

The screw may be solid or cored. *Coring* is used for steam heating or, more often, for water cooling. Coring can be for the entire length of the screw or for a portion of it, depending on the particular application. Full-length coring of the screw is used where large amounts of heat are to be removed. The screw is cored only in the initial portions at the hopper end when the objective is to keep the feed zone cooler for resins which tend to soften easily. Screws are often fabricated from 4140 alloy steel, but other materials are also used. The screw flights are usually hardened by flame-hardening techniques or inset with a wear-resistant alloy (e.g., Stellite 6).

Screw Zones

Screws are characterized by their *length-diameter ratio* (commonly written as *L/D* ratios). *L/D* ratios most commonly used for single-screw extruders range from 15:1 to 30:1. Ratios of 20:1 and 24:1 are common for thermoplastics, whereas lower values are used for rubbers. A long barrel gives a more homogeneous extrudate, which is particularly desirable when pigmented materials are handled. Screws are also characterized by their *compression ratios*—the ratio of the flight depth of the screw at the hopper end to the flight depth at the die end. Compression ratios of single-screw extruders are usually 2:1 to 5:1.

The screw is usually divided into three sections, namely, feed, compression, and metering (Fig. 2.16). One of the basic parameters in screw design involves the ratio of lengths between the feed, compression (or transition), and metering sections of the screw. Each section has its own special rate. The feed section picks up the powder, pellets, or beads from under the hopper mouth and conveys them forward in the solid state to the compression section. The feed section is deep flighted so that it functions in supplying enough material to prevent starving the forward sections.

The gradually diminishing flight depth in the compression section causes volume compression of the melting granules. The volume compression results in the trapped air being forced back through the feed section instead of being carried forward with the resin, thus ensuring an extrudate free from porosity. Another consequence of volume compression is an increase in the shearing action on the melt, which is caused by the relative motion of the screw surfaces with respect to the barrel wall. The increased shearing action produces good mixing and generates frictional heat, which increases fluidity of the melt and leads to a more uniform temperature distribution in the molten extrudate. The resin should be fully melted into a reasonably uniform melt by the time it enters the final section of the screw, known as the *metering section*. The function of the metering section is to force the molten polymer through the die at a steady rate and to iron out pulsations. For many screw designs the compression ratio is 3 to 5; i.e., the flight depth in the metering section is one-third to one-fifth that in the feed section.

Motor Drive

The motor employed for driving the screw of an extruder should be of more than adequate power required for its normal needs. Variable screw speeds are considered essential. Either variable-speed motors or constant-speed motors with variable-speed equipment, such as hydraulic systems, step-change gear boxes, and infinitely variable-speed gear boxes may be used. Thrust bearings of robust construction are essential because of the very high back pressure generated in an extruder and the trend towards higher screw speeds. Overload protection in the form of an automatic cutout should be fitted.

Heating

Heat to melt the polymer granules is supplied by external heaters or by frictional heat generated by the compression and shearing action of the screw on the polymer. Frictional heat is considerable, and in modern high-speed screw extruders it supplies most of the heat required for steady running. External heaters serve only to insulate the material and to prevent the machine from stalling at the start of the run when the material is cold. The external heater may be an oil, steam, or electrical type. Electrical heating is most popular because it is compact, clean, and accurately controlled. Induction heating is also used because it gives quicker heating with less variation and facilitates efficient cooling.

The barrel is usually divided into three or four heating zones; the temperature is lowest at the feed end and highest at the die end. The temperature of each zone is controlled by carefully balancing heating and cooling. Cooling is done automatically by either air or water. (The screw is also cored for heating and cooling.) The screw is cooled where the maximum amount of compounding is required, because this improves the quality of the extrudate.

Screw Design

The screw we have described is a simple continuous-flight screw with constant pitch. The more sophisticated screw designs include flow disrupters or mixing sections (Fig. 2.18). These mixer screws have mixing sections which are designed as mechanical means to break up and rearrange the laminar flow of the melt within the flight channel, which results in more thorough melt mixing and more uniform heat distribution in the metering section of the screw. Mixer screws have also been used to mix dissimilar materials (e.g., resin and additives or simply dissimilar resins) and to improve extrudate uniformity at higher screw speeds (>100 rpm). A few typical mixing section designs are shown in Figure 2.19. The fluted-mixing-section-barrier-type design (Fig. 2.19a) has proved to be especially applicable for extrusion of polyolefins. For some mixing problems, such as pigment mixing during extrusion, it is convenient to use rings (Fig. 2.19b) or mixing pins (Fig. 2.19c) and sometimes parallel interrupted mixing flights having wide pitch angles (Fig. 2.19d).

A later development in extruder design has been the use of venting or degassing zones to remove any volatile constituents from the melt before it is extruded through the die. This can be achieved by placing an obstruction in the barrel (the

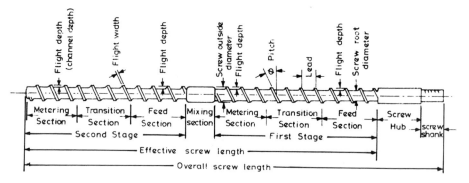

FIG. 2.18 Single-flight, two-stage extrusion screw with mixing section.

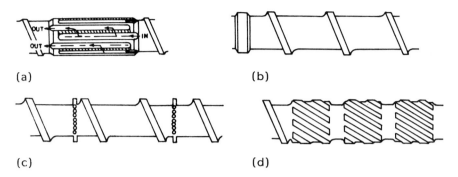

FIG. 2.19 Mixing section designs [6]: (a) fluted-mixing-section-barrier type; (b) ring-barrier type; (c) mixing pins; (d) parallel interrupted mixing flights.

reverse flights in Fig. 2.20) and by using a valved bypass section to step down the pressure developed in the first stage to atmospheric pressure for venting. In effect, two screws are used in series and separated by the degassing or venting zone. Degassing may also be achieved by having a deeper thread in the screw in the degassing section than in the final section of the first screw, so the polymer melt suddenly finds itself in an increased volume and hence is at a lower pressure. The volatile vapors released from the melt are vented through a hold in the top of the extruder barrel or through a hollow core of the screw by way of a hole drilled in the trailing edge of one of the flights in the degassing zone. A vacuum is sometimes applied to assist in the extraction of the vapor. Design and operation must be suitably controlled to minimize plugging of the vent (which, as noted above, is basically an open area) or the possibility of the melt escaping from this area.

Many variations are possible in screw design to accommodate a wide range of polymers and applications. So many parameters are involved, including such

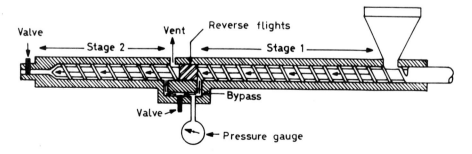

FIG. 2.20 A two-stage vented extruder with a valved bypass [17].

variables as screw geometry, materials characteristics, operating conditions, etc., that the industry now uses computerized screw design, which permits analysis of the variables by using mathematical models to derive optimum design of a screw for a given application.

Various screw designs have been recommended by the industry for extrusion of different plastics. For polyethylene, for example, the screw should be long with an L/D of at least 16:1 or 30:1 to provide a large area for heat transfer and plastication. A constant-pitch, decreasing-channel-depth, metering-type polyethylene screw or constant-pitch, constant-channel-depth, metering-type nylon screw with a compression ratio between 3 to 1 and 4 to 1 (Fig. 2.21) is recommended for polyethylene extrusion, the former being preferable for film extension and extrusion coating. Nylon-6, 6 melts at approximately 260°C (500°F). Therefore, an extruder with an L/D of at least 16:1 is necessary. A screw with a compression ratio of 4:1 is recommended.

Multiple-Screw Extruders

Multiple-screw extruders (that is, extruders with more than a single screw) were developed largely as a compounding device for uniformly blending plasticizers, fillers, pigments, stabilizers, etc., into the polymer. Subsequently, the multiple-screw extruders also found use in the processing of plastics.

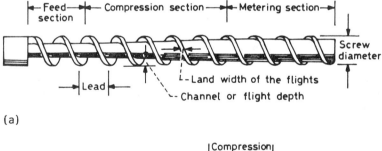

(a)

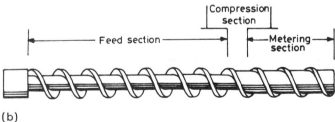

(b)

FIG. 2.21 (a) Constant pitch, decreasing-channel-depth metering type polyethylene screw. (b) Constant pitch, constant-channel-depth metering type nylon screw (not to scale) [6].

Multiple-screw extruders differ significantly from single-screw extruders in mode of operation. In a single-screw machine, friction between the resin and the rotating screw makes the resin rotate with the screw, and the friction between the rotating resin and the barrel pushes the material forward, and this also generates heat. Increasing the screw speed and/or screw diameter to achieve a higher output rate in a single-screw extruder will therefore result in a higher buildup of frictional heat and higher temperatures. In contrast, in twin-screw extruders with intermeshing screws the relative motion of the flight of one screw inside the channel of the other pushes the material forward almost as if the machine were a positive-displacement gear pump which conveys the material with very low friction. In twin-screw extruders, heat is therefore controlled independently from an outside source and is not influenced by screw speed. This fact becomes especially important when processing a heat-sensitive plastic like PVC. Multiple-screw extruders are therefore gaining wide acceptance for processing vinyls, although they are more expensive than single-screw machines. For the same reason, multiple-screw extruders have found a major use in the production of high-quality rigid PVC pipe of large diameter.

Several types of multiple-screw machines are available, including intermeshing corotating screws (in which the screws rotate in the same direction, and the flight of one screw moves inside the channel of the other), intermeshing counterrotating screws (in which the screws rotate in opposite directions), and nonintermeshing counterrotating screws. Multiple-screw extruders can involve either two screws (twin-screw design) or four screws. A typical four-screw extruder is a two-stage machine, in which a twin-screw plasticating section feeds into a twin-screw discharge section located directly below it. The multiple screws are generally sized on output rates (lb/hr) rather than on L/D ratios or barrel diameters.

Blown-Film Extrusion [15,16]

The blown-film technique is widely used in the manufacture of polyethylene and other plastic films. A typical setup is shown in Figure 2.22. In this case the molten polymer from the extruder head enters the die, where it flows round a mandrel and emerges through a ring-shaped opening in the form of a tube. The tube is expanded into a *bubble* of the required diameter by the pressure of internal air admitted through the center of the mandrel. The air contained in the bubble cannot escape because it is sealed by the die at one end and by the nip (or pinch) rolls at the other, so it acts like a permanent shaping mandrel once it has been injected. An even pressure of air is maintained to ensure uniform thickness of the film bubble.

The film bubble is cooled below the softening point of the polymer by blowing air on it from a cooling ring placed round the die. When the polymer, such

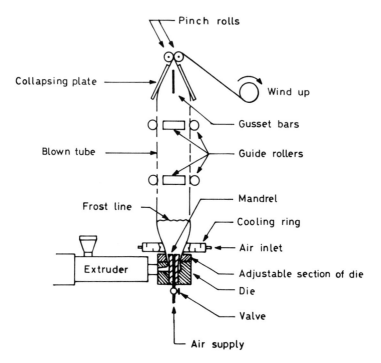

FIG. 2.22 Typical blown-film extrusion setup.

as polyethylene, cools below the softening point, the crystalline material is cloudy compared with the clear amorphous melt. The transition line which coincides with this transformation is therefore called the *frost line*.

The ratio of bubble diameter to die diameter is called the *blowup* ratio. It may range as high as 4 or 5, but 2.5 is a more typical figure. Molecular orientation occurs in the film in the hoop direction during blowup, and orientation in the machine direction, that is, in the direction of the extrudate flow from the die, can be induced by tension from the pinch rolls. The film bubble after solidification (at frost line) moves upward through guiding devices into a set of pinch rolls which flatten it. It can then be slit, gusseted, and surface-treated in line. (Vertical extrusion, shown in Figure 2.22, is most common, although horizontal techniques have been successfully used.)

Blown-film extrusion is an extremely complex subject, and a number of problems are associated with the production of good-quality film. Among the likely defects are variations in film thickness, surface imperfections (such as "fish eyes," "orange peel," haze), wrinkling, and low tensile strength. The factors affecting them are also numerous. "Fish eyes" occur due to imperfect mixing in the extruder or due to contamination of the molten polymer. Both factors are controlled by the screen pack.

The blown-film technique has several advantages: the relative ease of changing film width and caliber by controlling the volume of air in the bubble and the speed of the screw; the elimination of the end effects (e.g., edge bead trim and nonuniform temperature that result from flat film extrusion); and the capability of biaxial orientation (i.e., orientation both in the hoop direction and in the machine direction), which results in nearly equal physical properties in both directions, thereby giving a film of maximum toughness.

After extrusion, blown-film is often slit and wound up as flat film, which is often much wider than anything produced by slot-die extrusion. Thus, blown-films of diameters 7 ft or more have been produced, giving flat films of widths up to 24 ft. One example is reported [14] of a 10-in. extruder with 5-ft diameter and a blowup ratio of 2.5, producing 1100 lb/hr of polyethylene film, which when collapsed and slit is 40 ft wide. Films in thicknesses of 0.004 to 0.008 in. are readily produced by the blown-film process. Polyethylene films of such large widths and small thicknesses find extensive uses in agriculture, horticulture, and building.

Flat Film or Sheet Extrusion

In the flat-film process the polymer melt is extruded through a slot die (T-shaped or "coat hanger" die), which may be as wide as 10 ft. The die has relatively thick wall sections on the final lands (as compared to the extrusion coating die) to minimize deflection of the lips from internal melt pressure. The die opening (for polyethylene) may be 0.015 to 0.030 in.—even for films that are less than 0.003 in. thick. The reason is that the speed of various driven rolls used for taking up the film is high enough to draw down the film with a concurrent thinning. (By definition, the term *film* is used for material less than 0.010 in. thick, and *sheet* for that which is thicker.) Typical dies for film or sheet extrusion are shown in Figure 2.23.

Following extrusion, the film may be chilled below T_m or T_g by passing it through a water bath or over two or more chrome-plated chill rollers which have been cored for water cooling. A schematic drawing of a *chill-roll* (also called *cast-film*) operation is shown in Figure 2.24. The polymer melt extruded as a web from the die is made dimensionally stable by contacting several chill rolls before being pulled by the powered carrier rolls and wound up. The chrome-plated surface of the first roll is highly polished so that the product obtained is of extremely high gloss and clarity.

In flat-film extrusion (particularly at high takeoff rates), there is a relatively high orientation of the film in the machine direction (i.e., the direction of the extrudate flow) and a very low one in the traverse direction.

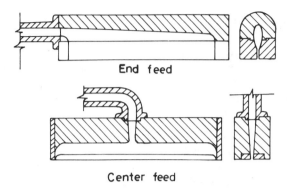

End feed

Center feed

FIG. 2.23 A flat die of the type shown here is required for extruding film through a water bath.

Biaxially oriented film can be produced by a flat-film extrusion by using a tenter (Fig. 2.25). Polystyrene, for example, is first extruded through a slit die at about 190°C and cooled to about 120°C by passing between rolls. Inside a temperature-controlled box the moving sheet, rewarmed to 130°C, is grasped on either side by tenterhooks which exert a drawing tension (longitudinal stretching) as well as a widening tension (lateral stretching). Stretch ratios of 3:1 to 4:1 in both directions are commonly employed for biaxially oriented polystyrene film. Biaxial stretching leads to polymers of improved tensile strength. Commercially available oriented polystyrene film has a tensile strength of 10,000 to 12,000 psi (703 to 843 kg/cm^2), compared to 6,000 to 8,000 psi (422 to 562 kg/cm^2) for unstretched material.

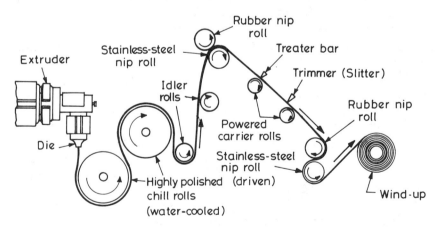

FIG. 2.24 Sketch of chill-roll film extrusion [10].

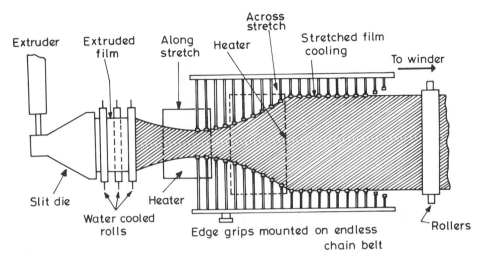

FIG. 2.25   Plax process for manufacture of biaxially stretched polystyrene film.

Biaxial orientation effects are important in the manufacture of films and sheet. Biaxially stretched polypropylene, poly(ethyleneterephthalate) (e.g., Melinex) and poly(vinylidene chloride) (Saran) produced by flat-film extrusion and tentering are strong films of high clarity. In biaxial orientation, molecules are randomly oriented in two dimensions just as fibers would be in a random mat; the orientation-induced crystallization produces structures which do not interfere with the light waves. With polyethylene, biaxial orientation often can be achieved in blown-film extrusion.

Pipe or Tube Extrusion

The die used for the extrusion of pipe or tubing consists of a die body with a tapered mandrel and an outer die ring which control the dimensions of the inner and outer diameters, respectively. Since this process involves thicker walls than are involved in blown-film extrusion, it is advantageous to cool the extrudate by circulating water through the mandrel (Fig. 2.26) as well as by running the extrudate through a water bath.

The extrusion of rubber tubing, however, differs from thermoplastic tubing. For thermoplastic tubing, dimensional stability results from cooling below T_g or T_m, but rubber tubing gains dimensional stability due to a cross-linking reaction at a temperature above that in the extruder. The high melt viscosity of the rubber being extruded ensures a constant shape during the cross-linking.

A complication encountered in the extrusion of continuous shapes is *die swell*. Die swell is the swelling of the polymer when the elastic energy stored in

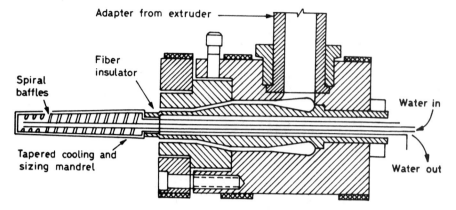

FIG. 2.26 An extrusion die fitted with a tapered cooling and sizing mandrel for use in producing either pipe or tubing [17].

capillary flow is relaxed on leaving the die. The extrusion of flat sheet or pipe is not sensitive to die swell, since the shape remains symmetrical even though the dimensions of the extrudate differ from those of the die. Unsymmetrical cross sections may, however, be distorted.

Wire and Cable Coverings

The covering or coating of wire and cable in continuous lengths with insulating plastics is an important application of extrusion, and large quantities of resin are used annually for this purpose. This application represented one of the first uses of extruders for rubber about 100 years ago. The wire and cable coating process resembles the process used for pipe extrusion (Fig. 2.26) with the difference that the conductor (which may be a single metal strand, a multiple strand, or even a bundle of previously individually insulated wires) to be covered is drawn through the mandrel on a continuous basis (Fig. 2.27). For thermoplastics such as polyethylene, nylon, and plasticized PVC, the coating is hardened by cooling below T_m or T_g by passing through a water trough. Rubber coatings, on the other hand, are to be cross-linked by heating subsequent to extrusion.

Extrusion Coating

Many substrates, including paper, paperboard, cellulose film, fiberboard, metal foils, or transparent films are coated with resins by direct extrusion. The resins most commonly used are the polyolefins, such as polyethylene, polypropylene, ionomer, and ethylene-vinyl acetate copolymers. Nylon, PVC, and polyester are used to a lesser extent. Often combinations of these resins and substrates are

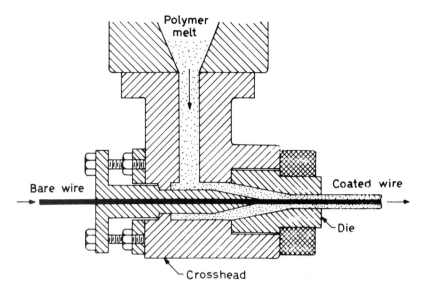

FIG. 2.27 Crosshead used for wire coating.

used to provide a multilayer structure. [A related technique, called *extrusion laminating*, involves two or more substrates, such as paper and aluminum foil, combined by using a plastic film, (e.g., polyethylene) as the adhesive and as a moisture barrier.] Coatings are applied in thicknesses of about 0.2 to 15 mils, the common average being 0.5 to 2 mils, and the substrates range in thickness from 0.5 to more than 24 mils.

The equipment used for extrusion coating is similar to that used for the extrusion of flat film. Figure 2.28 shows a typical extrusion coating setup. The thin molten film from the extruder is pulled down into the nip between a chill roll and a pressure roll situated directly below the die. The pressure between these two rolls forces the film on to the substrate while the substrate, moving at a speed faster than the extruded film, draws the film to the required thickness. The molten film is cooled by the water-cooled, chromium-plated chill roll. The pressure roll is also metallic but is covered with a rubber sleeve, usually neoprene or silicone rubber. After trimming, the coated material is wound up on conventional windup equipment.

Profile Extrusion

Profile extrusion is similar to pipe extrusion (Fig. 2.26) except that the sizing mandrel is obviously not necessary. A die plate, in which an orifice of appropriate geometry has been cut, is placed on the face of the normal die assembly.

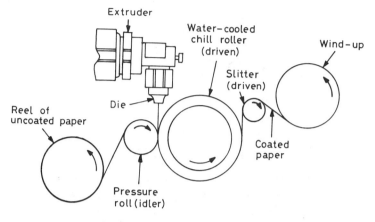

FIG. 2.28   Sketch of paper coating for extrusion process.

The molten polymer is subjected to surface drag as it passes through the die, resulting in reduced flow through the thinner sections of the orifice. This effect is countered by altering the shape of the orifice, but often this results in a wide difference in the orifice shape from the desired extrusion profile. Some examples are shown in Figure 2.29.

BLOW MOLDING [18]

Basically, blow molding is intended for use in manufacturing hollow plastic products, such as bottles and other containers. However, the process is also used

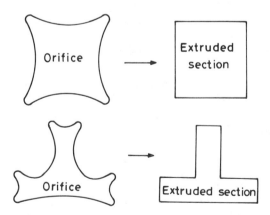

FIG. 2.29   Relationships between extruder die orifice and extruded section.

for the production of toys, automobile parts, accessories, and many engineering components. The principles used in blow molding are essentially similar to those used in the production of glass bottles. Although there are considerable differences in the processes available for blow molding, the basic steps are the same: (1) melt the plastic; (2) form the molten plastic into a *parison* (a tubelike shape of molten plastic); (3) seal the ends of the parison except for one area through which the blowing air can enter; (4) inflate the parison to assume the shape of the mold in which it is placed; (5) cool the blow-molded part; (6) eject the blow-molded part; (7) trim flash if necessary.

Two basic processes of blow molding are extrusion blow molding and injection blow molding. These processes differ in the way in which the parison is made. The extrusion process utilizes an unsupported parison, whereas the injection process utilizes a parison supported on a metal core. The extrusion blow-molding process by far accounts for the largest percentage of blow-molded objects produced today. The injection process is, however, gaining acceptance.

Although any thermoplastic can be blow-molded, polyethylene products made by this technique are predominant. Polyethylene squeeze bottles form a large percentage of all blow-molded products.

Extrusion Blow Molding

Extrusion blow molding consists basically of the extrusion of a predetermined length of parison (hollow tube of molten plastic) into a split die, which is then closed, sealing both ends of the parison. Compressed air is introduced (through a blowing tube) into the parison, which blows up to fit the internal contours of the mold. As the polymer surface meets the cold metal wall of the mold, it is cooled rapidly below T_g or T_m. When the product is dimensionally stable, the mold is opened, the product is ejected, a new parison is introduced, and the cycle is repeated. The process affords high production rates.

In continuous extrusion blow molding, a molten parison is produced continuously from a screw extruder. The molds are mounted and moved. In one instance the mold sets are carried on the periphery of a rotating vertical wheel (Fig. 2.30a), in another on a rotating horizontal table (Fig. 2.30b). Such rotary machines are best suited for long runs and large-volume applications.

In the ram extrusion method the parison is formed in a cyclic manner by forcing a charge out from an accumulated molten mass, as in the preplasticizer injection-molding machine. The transport arm cuts and holds the parison and lowers it into the waiting mold, where shaping under air pressure takes place (Fig. 2.31).

A variation of the blow-extrusion process which is particularly suitable for heat-sensitive resins such as PVC is the *cold preform molding*. The parison is produced by normal extrusion and cooled and stored until needed. The

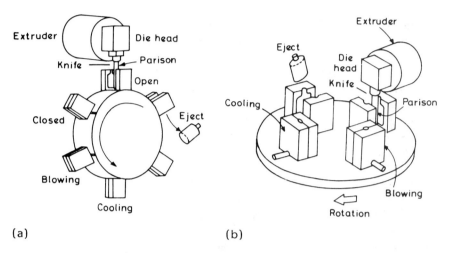

FIG. 2.30 Continuous extrusion blow molding [6]. (a) Continuous vertical rotation of a wheel carrying mold sets on the periphery. (b) Rotating horizontal table carrying mold sets.

required length of tubing is then reheated and blown to shape in a cold mold, as in conventional blow molding. Since, unlike in the conventional process, the extruder is not coupled directly to the blow-molding machine, there is less chance of a stoppage occurring, with consequent risk of holdup and degradation of the resin remaining in the extruder barrel. There is also less chance of the occurrence of "dead" pockets and consequent degradation of resin in the

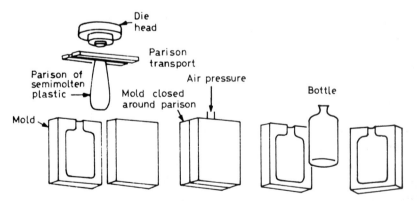

FIG. 2.31 Continuous extrusion blow molding with parison transfer. The transport arm cuts the extruded parison from the die head and lowers it into the waiting mold.

straight-through die used in this process than in the usual crosshead used with a conventional machine.

Injection Blow Molding

In this process the parison is injection molded rather than extruded. In one system, for example, the parison is formed as a thick-walled tube around a blowing stick in a conventional injection-molding machine. The parison is then transferred to a second, or blowing, mold in which the parison is inflated to the shape of the mold by passing compressed air down the blowing stick. The sequence is shown in Figure 2.32. Injection blow molding is relatively slow and is more restricted in choice of molding materials as compared to extrusion blow molding.

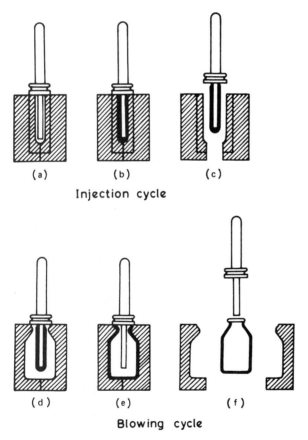

FIG. 2.32 Sketch of injection blow molding process.

The injection process, however, affords good control of neck and wall thicknesses of the molded object. With this process it is also easier to produce unsymmetrical moldings.

CALENDERING [19]

Calendering is the leading method for producing vinyl film, sheet, and coatings. In this process continuous sheet is made by passing a heat-softened material between two or more rolls. Calendering was originally developed for processing rubber, but it is now widely used for producing thermoplastic films, sheets, and coatings. A major portion of thermoplastics calendered is accounted for by flexible (plasticized) PVC. Most plasticized PVC film and sheet, ranging from the 3-mil film for baby pants to the 0.10-in "vinyl" tile for floor coverings, is calendered.

The calendering process consists of feeding a softened mass into the nip between two rolls where it is squeezed into a sheet, which then passes round the remaining rolls. The processed material thus emerges as a continuous sheet, the thickness of which is governed by the gap between the last pair of rolls. The surface quality of the sheet develops on the last roll and may be glossy, matt, or embossed. After leaving the calender, the sheet is passed over a number of cooling rolls and then through a beta-ray thickness gage before being wound up.

The plastics mass fed to the calender may be simply a heat-softened material, as in the case of, say, polyethylene, or a rough sheet, as in the case of PVC. The polymer (PVC) is blended with stabilizers, plasticizers, etc., in ribbon blenders, gelated at 120 to 160°C for about 5 to 10 min in a Banbury mixer, and the gelated lumps are made into a rough sheet on a two-roll mill before being fed to the calender.

Calenders may consist of two, three, four, or five hollow rolls arranged for steam heating or water cooling and are characterized by the number of rolls and their arrangement. Some arrangements are shown in Figure 2.33. Thick sections of rubber can be made by applying one layer of polymer upon a previous layer (*double plying*) (Fig. 2.33b). Calenders can be used for applying rubber or plastics to fabrics (Fig. 2.33c). Fabric or paper is fed through the last two rolls of the calender so that the resin film is pressed into the surface of the web. For profiling, the plastic material is fed to the nip of the calender, where the material assumes the form of a sheet, which is then progressively pulled through two subsequent banks to resurface each of the two sides (Fig. 2.33d). For thermoplastics the cooling of the sheet can be accomplished on the rolls with good control over dimensions. For rubber, cross-linking can be carried out with good control over dimensions, with the support of the rolls. Despite the simple appearance of the

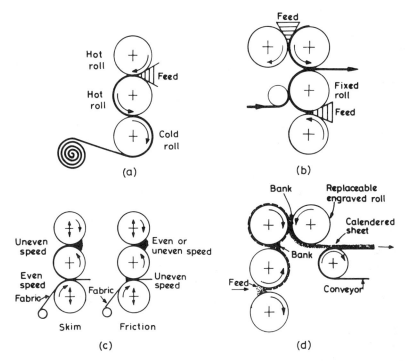

FIG. 2.33 Typical arrangements of calender rolls: (a) single-ply sheeting; (b) double-ply sheeting; (c) applying rubber to fabrics; (d) profiling with four-roll engraving calender. [After G. G. Winspear (ed.), *Vanderbilt Rubber Handbook*, R. T. Vanderbilt Co., p. 392 (1958)].

calender compared to the extruder, the close tolerances involved and other mechanical problems make for the high cost of a calendering unit.

SPINNING OF FIBERS [20–22]

The term *spinning*, as used with natural fibers, refers to the twisting of short fibers into continuous lengths. In the modern synthetic fiber industry, however, the term is used for any process of producing continuous lengths by any means. (A few other terms used in the fiber industry should also be defined. A *fiber* may be defined as a unit of matter having a length at least 100 times its width or diameter. An individual strand of continuous length is called a *filament*. Twisting together filaments into a strand gives continuous filament *yarn*. If the filaments are assembled in a loose bundle, we have *tow* or *roving*. These can be chopped into small lengths (an inch to several inches long), referred to as *staple*.

Spun yarn is made by twisting lengths of staple into a single continuous strand, and *cord* is formed by twisting together two or more yarns.)

The primary fabrication process in the production of synthetic fibers is the *spinning*—i.e., the formation—of filaments. In every case the polymer is either melted or dissolved in a solvent and is put in filament form by forcing through a die, called *spinneret*, having a multiplicity of holes. Spinnerets for rayon spinning, for example, have as many as 10,000 holes in a 15-cm-diameter platinum disc, and those for textile yarns may have 10 to 120 holes; industrial yarns such as tire cord might be spun from spinnerets with up to 720 holes. Three major categories of spinning processes are *melt*, *dry*, and *wet spinning* [20]. The features of the three processes are shown in Figure 2.34, and the typical cross sections of the fibers produced by them are shown in Figure 2.35.

Melt Spinning

In *melt spinning*, which is the same as melt extrusion, the polymer is heated and the viscous melt is pumped through a spinneret. An inert atmosphere is provided in the melting chamber before the pump. Special pumps are used to operate in the temperature range necessary to produce a manageable melt (230 to 315°C). For nylon, for example, a gear pump is used to feed the melt to the spinneret (Fig. 2.34a). For a polymer with high melt viscosity such as polypropylene, a screw extruder is used to feed a heated spinneret. Dimensional stability of the fiber is obtained by cooling under tension.

Dry Spinning

In *dry spinning*, a polymer dissolved in a solvent is fed to a spinneret by using a pump. Dimensional stability of the fiber is achieved by evaporation of the solvent (Fig. 2.34b), which involves diffusion. The skin which forms first on the fiber by evaporation from the surface gradually collapses and wrinkles as more solvent diffuses out and the diameter decreases. The cross section of a dry-spun fiber thus has an irregularly lobed appearance (Fig. 2.35). Recovery of the solvent used for dissolving the polymer is important to the economics of the process. Cellulose acetate dissolved in acetone and polyacrylonitrile dissolved in dimethylformamide are two typical examples.

Wet Spinning

Wet spinning also involves pumping a solution of the polymer to the spinneret. However, unlike dry spinning, dimensional stability is achieved by precipitating the polymer in a nonsolvent (Fig. 2.34c). For example, polyacrylonitrile in dimethylformamide can be precipitated by passing a jet of the solution through a bath of water, which is miscible with the solvent but coagulates the polymer. For

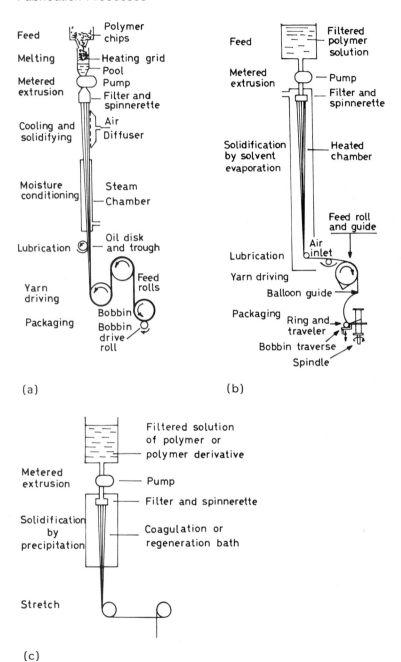

FIG. 2.34 Schematic of the three principal types of fiber spinning [22]: (a) melt spinning; (b) dry spinning; (c) wet spinning.

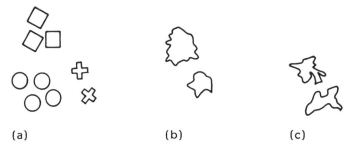

(a)   (b)   (c)

FIG. 2.35 Typical cross-sections of fibers produced by different spinning processes: (a) melt-spun nylon from various shaped orifices; (b) dry-spun cellulose acetate from round orifice; (c) wet-spun viscose rayon from round orifice.

wet-spinning cellulose triacetate a mixture methylene chloride and alcohol can be used to dissolve the polymer, and a toluene bath can be used for precipitation of the polymer. In some cases the precipitation can also involve a chemical reaction. An important example is viscose rayon, which is made by regenerating cellulose from a solution of cellulose xanthate in dilute alkali.

$$R{-}OH \xrightarrow{\text{CS}_2,\ \text{NaOH}} R{-}O{-}\overset{\displaystyle S}{\overset{\|}{C}}{-}S^-\ Na^+ + H_2O$$

cellulose xanthate

$$H_2O \Big|\ \begin{matrix} H_2SO_4 \\ NaHSO_4 \end{matrix}$$

$$R{-}OH + CS_2 + Na^+\ (HSO_4{}^-)$$

cellulose

If a slot die rather than a spinneret is used, the foregoing process would yield cellulose film (cellophane) instead of fiber.

Cold Drawing of Fibers

Almost all synthetic fibers are subjected to a drawing (stretching) operation to orient the crystalline structure in the direction of the fiber axis. Drawing orients crystallites in the direction of the stretch so that the modulus in that direction is increased and elongation at break is decreased. Usually the drawing is carried out at a temperature between T_g and T_m of the fiber. Thus, polyethylene ($T_g = -115°C$) can be drawn at room temperature, whereas nylon-6,6 ($T_g = 53°C$)

should be heated or humidified to be drawn. T_g is depressed by the presence of moisture, which acts as a plasticizer. The drawing is accomplished by winding the yarn around a wheel or drum driven at a faster surface velocity than a preceding one.

THERMOFORMING [23]

When heated, thermoplastic sheet becomes as soft as a sheet of rubber, and it can then be stretched to any given shape. This principle is utilized in thermoforming processes which may be divided into three main types: (a) vacuum forming, (2) pressure forming (blow forming), and (3) mechanical forming (e.g., matched metal forming), depending on the means used to stretch the heat-softened sheet.

Since fully cured thermoset sheets cannot be resoftened, forming is not applicable to them. Common materials subjected to thermoforming are thermoplastics such as polystyrene, cellulose acetate, cellulose acetate butyrate, PVC, ABS, poly(methyl methacrylate), low- and high-density polyethylene, and polypropylene. The bulk of the forming is done with extruded sheets, although cast, calendered, or laminated sheets can also be formed.

In general, thermoforming techniques are best suited for producing moldings of large area and very thin-walled moldings, or where only short runs are required. Thermoformed articles include refrigerator and freezer door liners complete with formed-in compartments for eggs, butter, and bottles of various types, television masks, dishwasher housings, washing machine covers, various automobile parts (instrument panels, arm rests, ceilings, and door panels), large patterned diffusers in the lighting industry, displays in advertising, various parts in aircraft industry (windshields, interior panels, arm rests, serving trays, etc.), various housings (typewriters, dictaphones, and duplicating machines), toys, transparent packages, and much more.

Vacuum Forming

In *vacuum forming*, the thermoplastic sheet can be clamped or simply held against the rim of a mold and then heated until it becomes soft. The soft sheet is then sealed at the rim, and the air from the mold cavity is removed by a suction pump so that the sheet is forced to take the contours of the mold by the atmospheric pressure above the sheet (Fig. 2.36a). The vacuum in the mold cavity is maintained until the part cools and becomes rigid.

Straight cavity forming is not well adapted to forming a cup or box shape because as the sheet, drawn by vacuum, continues to fill out the mold and solidify, most of the stock is used up before it reaches the periphery of the base, with the result that this part becomes relatively thin and weak. This difficulty is

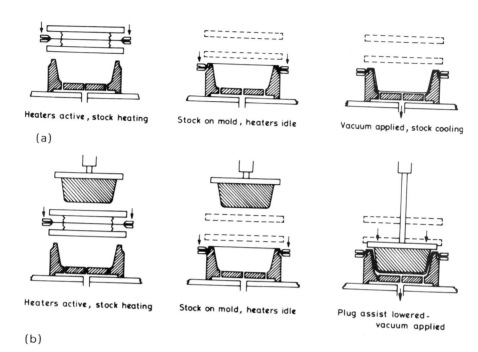

Heaters active, stock heating   Stock on mold, heaters idle   Vacuum applied, stock cooling

(a)

Heaters active, stock heating   Stock on mold, heaters idle   Plug assist lowered -
                                                                              vacuum applied

(b)

FIG. 2.36   (a) Vacuum forming. (b) Plug-assist forming using vacuum.

alleviated and uniformity of distribution in such shapes is promoted if the *plug assist* is used (Fig. 2.36b). The plug assist is any type of mechanical helper which carries extra stock toward an area where the part would otherwise be too thin. Plug-assist techniques are adaptable both to vacuum-forming and pressure-forming techniques. The system shown in Figure 2.36b is thus known as *plug-assist vacuum forming*.

Pressure Forming

Pressure forming is the reverse of vacuum forming. The plastic sheet is clamped, heated until it becomes soft, and sealed between a pressure head and the rim of a mold. By applying air pressure (Fig. 2.37), one forces the sheet to take the contours of the mold. Exhaust holes in the mold allow the trapped air to escape. After the part cools and becomes rigid, the pressure is released and the part is removed. As compared to vacuum forming, pressure forming affords a faster production cycle, greater part definition, and greater dimensional control.

A variation of vacuum forming or pressure forming, called *free forming* or *free blowing*, is used with acrylic sheeting to produce parts that require superior

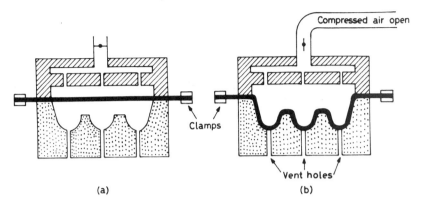

FIG. 2.37 Pressure forming: (a) heated sheet is clamped over mold cavity; (b) compressed air pressure forces the sheet into the mold.

optical quality (e.g., aircraft canopies). In this process the periphery is defined mechanically by clamping, but no mold is used, and the depth of draw or height is governed only by the vacuum or compressed air applied.

Mechanical Forming

Various mechanical techniques have been developed for thermoforming that use neither air pressure nor vacuum. Typical of these is *matched mold forming* (Fig. 2.38). A male mold is mounted on the top or bottom platen, and a matched female mold is mounted on the other. The plastic sheet, held by a clamping frame, is heated to the proper forming temperature, and the mold is then closed, forcing the plastic to the contours of both the male and the female molds. The molds are held in place until the plastic cools and attains dimensional stability, the latter facilitated by internal cooling of the mold. The matched mold technique affords excellent reproduction of mold detail and dimensional accuracy.

CASTING PROCESSES

There are two basic types of casting used in plastics industry: simple casting and plastisol casting.

Simple Casting

In *simple casting*, the liquid is simply poured into the mold without applying any force and allowed to solidify. Catalysts that cause the liquid to set are often added. The resin can be a natural liquid or a granular solid liquefied by heat.

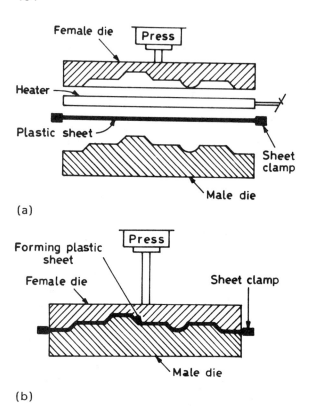

FIG. 2.38 Matched mold forming: (a) heating; (b) forming.

After the liquid resin is poured into the closed mold, the air bubbles are removed and the resin is allowed to cure either at room temperature or in an oven at low heat. When completely cured, the mold is split apart and the finished casting is removed. In the production of simple shapes such as rods, tubes, etc., usually a two-piece metal mold with an entry hole for pouring in the liquid resin is used. For making flat-cast acrylic plexiglass or lucite sheets, two pieces of polished plate glass separated by a gasket with the edge sealed and one corner open are usually used as a mold.

Both thermosets and thermoplastics may be cast. Acrylics, polystyrene, polyesters, phenolics, and epoxies are commonly used for casting.

Plastisol Casting [24]

Plastisol casting, commonly used to manufacture hollow articles, is based on the fact that plastisol in fluid form is solidified as it comes in contact with a

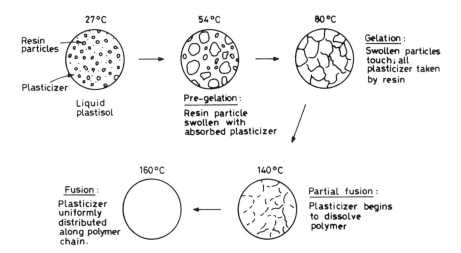

FIG. 2.39 Various changes in a plastisol system in the transformation from a liquid dispersion to a homogeneous solid.

heated surface. A plastisol is a suspension of PVC in a liquid plasticizer to produce a fluid mixture that may range in viscosity from a pourable liquid to a heavy paste. This fluid may be sprayed onto a surface, poured into a mold, spread onto a substrate, etc. The plastisol is converted to a homogeneous solid ("vinyl") product through exposure to heat [e.g., 350°F (176°C)], depending on the resin type and plasticizer type and level. The heat causes the suspended resin to undergo fusion—that is, dissolution in the plasticizer (Fig. 2.39)—so that on cooling, a flexible vinyl product is formed with little or no shrinkage. The product possesses all the excellent qualities of vinyl plastics.

Dispersion-grade PVC resins are used in plastisols. These resins are of fine particle size (0.1–2 micron in diameter), as compared to suspension type resins (commonly 75–200 micron in diameter) used in calender and extrusion processing. A plastisol is formed by simply mixing the dispersion-grade resin into the plasticizer with sufficient shearing action to ensure a reasonable dispersion of the resin particles. (PVC plasticizers are usually monomeric phthalate esters, the most important of them being the octyl esters based on 2-ethylhexyl alcohol and isooctyl alcohol, namely dioctyl phthalate and diisooctyl phthalate, respectively.) The ease with which virtually all plastisol resins mix with plasticizer to form a smooth stable dispersion/paste is due to the fine particle size and the emulsifier coating on the resin particles. (The emulsifier coating aids the wetting of each particle by the plasticizer phase.)

The liquid nature of the plastisol system is the key to its ready application. The plastisol may be spread onto a cloth, paper, or metal substrate, or otherwise cast or slushed into a mold. After coating or molding, heat is applied, which causes the PVC resin particles to dissolve in the plasticizer and form a cohesive mass, which is, in effect, a solid solution of polymer in plasticizer. The various changes a plastisol system goes through in the transformation from a liquid dispersion to a homogeneous solid are schematically shown in Figure 2.39. At 280°F (138°C) the molecules of plasticizer begin to enter between the polymer units, and fusion begins. If the plastisol were cooled after being brought to this temperature, it would give a cohesive mass with a minimum of physical strength. Full fusion occurs and full strength is accomplished when the plastisol is brought to approximately 325°F (163°C) before cooling. The optimum fusion temperature, however, depends on resin type and plasticizer type.

For coating applications it is common practice to add solvent (diluent) to a plastisol to bring down viscosity. This mixture is referred to as *organosol*. It may be applied by various coating methods to form a film on a substrate and then is heated to bring about fusion, as in the case of plastisol.

Unlike coating applications, there are some applications where it is desirable to have an infinite viscosity at low shear stress. For such applications, a plastisol can be gelled by adding a metallic soap (such as aluminum stearate) or finely divided filler as a gelling agent to produce a *plastigel*. A plastigel can be cold molded, placed on a pan, and heated to fusion without flow. The whole operation is like baking cookies.

A *rigidsol* is a plastisol of such formulation that it becomes a rigid, rather than a flexible, solid when fused. A very rigid product can be obtained when the plasticizer can be polymerized during or right after fusion. For example, a rigidsol can be made from 100 parts of PVC resin, 100 parts of triethylene glycol dimethacrylate (network forming plasticizer) and 1 part of di-*tert*-butyl peroxide (initiator). This mixture has a viscosity of only 3 poises compared with 25 poises for phthalate-based plastisol. However, after being heated for 10 min at 350°F (176°C), the resin solvates and the plasticizer polymerizes to a network structure, forming a hard, rigid, glassy solid with a flexural modulus of over 2.5×10^5 psi (1.76×10^4 kg/cm^2) at room temperature.

Three important variations of the plastisol casting are *dip casting*, *slush casting*, and *rotational casting*.

Dip Casting

A heated mold is dipped into liquid plastisol (Fig. 2.40a) and then drawn at a given rate. The solidified plastisol (with mold) is then cured in an oven at 350 to 400°F (176 to 204°C). After it cools, the plastic is stripped from the mold. Items with intricate shapes such as transparent ladies' overshoes, flexible gloves, etc., can be made by this process.

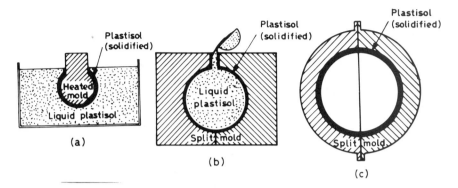

FIG. 2.40 Plastisol casting processes: (a) dip casting; (b) slush casting; (c) rotational casting.

The dipping process is also used for coating metal objects with vinyl plastic. For example, wire dish drainers, coat hangers, and other industrial and household metal items can be coated with a thick layer of flexible vinyl plastic by simply dipping in plastisol and applying fusion.

Slush Casting

Slush casting is similar to slip casting (drain) of ceramics. The liquid is poured into a preheated hollow metal mold, which is the shape of the outside of the object to be made (Fig. 2.40b). The plastisol in immediate contact with the walls of the hot mold solidifies. The thickness of the cast is governed by the time of stay in the mold. After the desired time of casting is finished, the excess liquid is poured out and the solidified plastisol with the mold is kept in an oven at 350 to 400°F (176 to 204°C). The mold is then opened to remove the plastic part, which now bears on its outer side the pattern of the inner side of the metal mold. Slush molding is used for hollow, open articles. Squeezable dolls or parts of dolls and boot socks are molded this way.

Rotational Casting

In rotational casting a predetermined amount of liquid plastisol is placed in a heated, closed, two-piece mold. The liquid is uniformly distributed against the walls of the mold in a thin uniform layer (Fig. 2.40c) by rotating the mold in two planes. The solidified plastisol in the mold is cured in an oven; the mold is then opened, and the part is removed. This method is used to make completely enclosed hollow objects. Doll parts, plastic fruits, squeeze bulbs, toilet floats, etc. can be made by rotational casting of plastisols.

REINFORCING PROCESSES [25]

A reinforced plastic (RP) consists of a polymeric resin strengthened by the properties of a reinforcing material. Glass is one of the most commonly used RP reinforcement materials and is used in various forms, such as fibers, fabrics, chopped strands, mats, or rovings.

Continuous-strand glass fiber gives unidirectional reinforcement. Glass fabric reinforces the object in two directions. Chopped glass strands (approximately 50 mm long) give a random reinforcement. Glass mats cost less than fabric and give random reinforcement. Woven glass roving gives high strength and costs less than glass fabrics. Other common reinforcing materials are asbestos (mainly used in phenolics and silicones) and paper (mainly used with phenolics and melamines). Several special reinforcing materials are high-silica glass, quartz, and graphite, which are used to meet unusual temperature demands. Synthetic fibers also find many special uses for RP.

Resins used for RP are thermosetting resins such as unsaturated polyesters, epoxy, phenolic, melamine, and silicone. Unsaturated polyesters, which are the leading resin for reinforced plastics, are comparatively low in cost and can be cured at ordinary temperatures. They are used primarily with fibrous glass for radars, aircraft parts, motor components, cars and commercial vehicles, boat hulls, building panels, and ducting.

Several methods are employed to make reinforced plastics. Although each method has the characteristics of either molding or casting, the processes may be described as (1) hand layup, (2) spray-up, (3) matched molding, (4) vacuum-bag molding, (5) pressure-bag molding, (6) continuous pultrusions, (7) filament winding and prepreg molding.

Hand Layup or Contact Molding

A mold is first treated with a release agent (such as wax or silicone-mold release), and a coating of the liquid resin (usually polyester or epoxy) is brushed, rolled, or sprayed on the surface of the mold. Fiberglass cloth or mat is impregnated with resin and placed over the mold. Air bubbles are removed, and the mat or cloth is worked into intimate contact with the mold surface by squeegees, by rollers, or by hand (Fig. 2.41a). Additional layers of glass cloth or mat are added, if necessary, to build up the desired thickness. The resin hardens due to curing, as a result of the catalyst or hardener that was added to the resin just prior to its use. Curing occurs at room temperature, though it may be speeded up by heat. Ideally, any trimming should be carried out before the curing is complete, because the material will still be sufficiently soft for knives or shears. After curing, special cutting wheels may be needed for trimming.

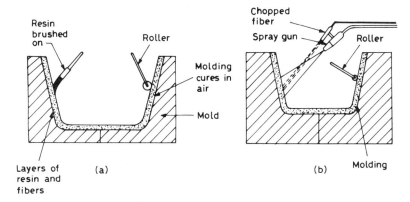

FIG. 2.41 (a) Basic hand layup method. (b) Spray-up technique.

Lowest-cost molds such as simple plaster, concrete, or wood are used in this process, since pressures are low and little strength is required. However, dimensional accuracy of the molded part is relatively low and, moreover, maximum strength is not developed in the process because the ratio of resin to filler is relatively high.

The hand layup process can be used for fabricating boat hulls, automobile bodies, swimming pools, chemical tanks, ducts, tubes, sheets, and housings, and for building, machinery, and autobody repairs.

Spray-Up

A release agent is first applied on the mold surface, and measured amounts of resin, catalyst, premoter, and reinforcing material are sprayed with a multi-headed spray gun (Fig. 2.41b). The spray guns used for this work are different from those used for spraying glazes, enamels, or paints. They usually consist of two or three nozzles, and each nozzle is used to spray a different material. One type, for example, sprays resin and promoter from one nozzle, resin and catalyst from another, and chopped glass fibers from a third. The spray is directed on the mold to build up a uniform layer of desired thickness on the mold surface. The resin sets rapidly only when both catalyst and promoter are present. This method is particularly suitable for large bodies, tank linings, pools, roofs, etc.

Matched Metal Molding

Matched metal molding is used when the manufacture of articles of close tolerances and a high rate of production are required. Possible methods are preform molding, sheet molding, and dough molding. In *preform molding* the reinforcing material in mat or fiber form is preformed to the approximate shape and placed

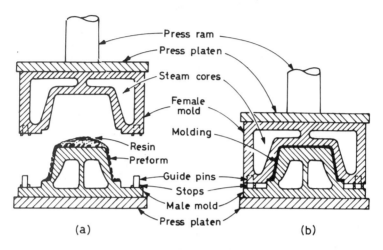

FIG. 2.42   Matched metal molding. (a) Before closing of die. (b) After closing of die.

on one-half of the mold, which was coated previously with a release agent. The resin is then added to the preform, the second half of the mold (also coated previously with a release agent) is placed on the first half, and the two halves of the mold are then pressed together and heated (Fig. 2.42). The resin flows, impregnates the preform, and becomes hard. The cured part is removed by opening the mold. Because pressures of up to 200 psi (14 kg/cm^2) can be exerted upon the material to be molded, a higher ratio of glass to resin may be used, resulting in a stronger product. The cure time in the mold depends on the temperature, varying typically from 10 min at 175°F (80°C) to only 1 min at 300°F (150°C). The cure cycle can thus be very short, and a high production rate is possible.

The molding of *sheet-molding compounds* (SMC) and *dough-molding compounds* (DMC) is done "dry"—i.e., it is not necessary to pour on resins. SMC, also called *prepreg*, is basically a polyester resin mixture (containing catalyst and pigment) reinforced with chopped strand mat or chopped roving and formed into a pliable sheet that can be handled easily, cut to shape, and placed between the halves of the heated mold. The application of pressure then forces the sheet to take up the contours of the mold. DMC is a doughlike mixture of chopped strands with resin, catalyst, and pigment. The charge of dough, also called *premix*, may be placed in the lower half of the heated mold, although it is generally wise to preform it to the approximate shape of the cavity. When the mold is closed and pressure is applied, DMC flows readily to all sections of the cavity. Curing generally takes a couple of minutes for mold temperatures from 250 to 320°F (120 to 160°C). This method is used for the production of switch gear, trays, housings, and structural and functional components.

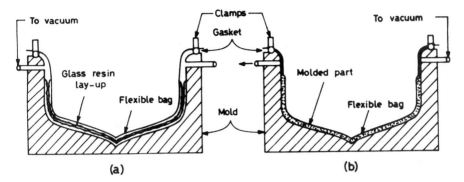

FIG. 2.43 Vacuum-bag molding. (a) Before vacuum applied. (b) After vacuum applied.

Vacuum-Bag Molding

In vacuum-bag molding the reinforcement and the resin mixed with catalyst are placed in a mold, as in the hand layup method, and an airtight flexible bag (frequently rubber) is placed over it. As air is exhausted from the bag, atmospheric air forces the bag against the mold (Fig. 2.43). The resin and reinforcement mix now takes the contours of the mold. If the bag is placed in an autoclave or pressure chamber, higher pressure can be obtained on the surface. After the resin hardens, the vacuum is destroyed, the bag opened and removed, and the molded part obtained. The technique has been used to make automobile body, aircraft component, and prototype molds.

Pressure-Bag Molding

In pressure-bag molding the reinforcement and the resin mixed with catalyst are placed in a mold, and a flexible bag is placed over the wet layup after a separating sheet (such as cellophane) is laid down. The bag is then inflated with an air pressure of 20 to 50 psi (1.4 to 3.5 kg/cm^2). The resin and reinforcement follow the contours of the mold (Fig. 2.44). After the part is hardened, the bag is deflated and the part is removed. The technique has been used to make radomes, small cases, and helmets.

Continuous Pultrusion

Continuous strands, in the form of roving or other forms of reinforcement, are impregnated with liquid resin in a resin bath and pulled through a long, heated steel die which shapes the product and controls the resin content. The final cure is effected in an oven through which the stock is drawn by a suitable pulling device. The method produces shapes with high unidirectional strength (e.g.,

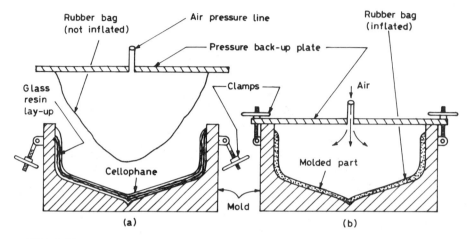

FIG. 2.44 Pressure-bag molding. (a) During layup. (b) During curing.

I-beams, rods, and shafts). Polyesters account for 90% of pultrusion resin, epoxies for the balance.

Filament Winding

In the filament-winding method, continuous strands of glass fiber are used in such a way as to achieve maximum utilization of the fiber strength. In a typical process rovings or single strands are fed from a creel through a bath of resin and wound on a suitably designed rotating mandrel. Arranging for the resin-impregnated fibers to traverse the mandrel at a controlled and predetermined (programmed) manner (Fig. 2.45) makes it possible to lay down the fibers in any desired fashion to give maximum strengths in the direction required. When the right number of layers have been applied, curing is done at room temperature or in an oven. For open-ended structures, such as cylinders or conical shapes, mandrel design is comparatively simple, either cored or solid steel or aluminum being ordinarily used for the purpose. For structures with integrally wound end closures, such as pressure vessels, careful consideration must be given to mandrel design and selection of mandrel material. A sand-poly(vinyl alcohol) combination, which disintegrates readily in hot water, is an excellent choice for diameters up to 5 ft (1.5 m). Thus, a mandrel made of sand with water-soluble poly(vinyl alcohol) as a binder can be decomposed with water to recover the filament-wound part. Other mandrel materials include low-melting alloys, eutectic salts, soluble plasters, frangible or breakout plasters, and inflatables.

Because of high glass content, filament-wound parts have the highest strength-to-weight ratio of any reinforced thermoset. The process is thus highly

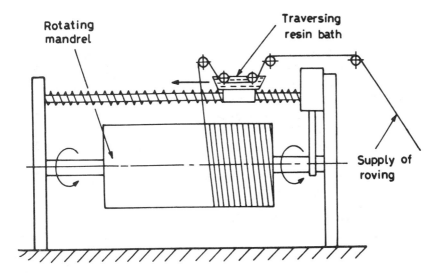

FIG. 2.45   Sketch of filament winding.

suited to pressure vessels where reinforcement in the highly stressed hoop direction is important. Pipe installation, storage tanks, large rocker motor cases, interstage shrouds, high-pressure gas bottles, etc., are some of the products made by filament winding. The main limitation on the process is that it can only be used for fabricating objects which have some degree of symmetry about a central axis.

Prepreg Molding

Optimal strength and stiffness of continuous fiber-reinforced polymeric composites is obtained through controlled orientation of the continuous fibers. One means to achieve this is by prepreg molding. In this process, unidirectionally oriented layers of fibers are pre-impregnated with the matrix resin and cured to an intermediate stage of polymerization (B-stage). When desired, this pre-impregnated composite precursor, called a prepreg, can be laid up in the required directions in a mold for quick conversion into end components through the use of hot curing techniques. Prepregs can thus be described as preengineered laminating materials for the manufacture of fiber-reinforced composites with controlled orientation of fibers.

For the designer, a precisely controlled ply of prepreg represents a building block, with well-defined mechanical properties from which a structure can be developed with confidence. For the fabricator, on the other hand, prepregs provide a single, easy-to-handle component that can be applied immediately to the lay-up of the part to be manufactured, be it aircraft wing skin or fishing rod tube. The prepreg has the desired handleability already built in to suit the lay-up and curing process being utilized, thus improving efficiency and consistency.

Prepregs have been used since the late 1940s, but they have only achieved wide prominence and recognition since the development of the higher performance reinforcing fibers, carbon and kevlar. The quantum leap in properties provided by these new fibers generated a strong development effort by prepreg manufacturers and there is no doubt that new developments made in this area have been as significant, if not as evident, as that of the introduction of the new fibers.

The use of prepreg in the manufacture of a composite components offers several advantages over the conventional wet lay-up formulations:

1. Being a readily formulated material, prepreg minimizes the material's knowledge required by a component manufacturer. The cumbersome process of stocking various resins, hardeners, and reinforcements is avoided.
2. With prepreg a good degree of alignment in the required directions with the correct amount of resin is easily achieved.
3. Prepreg offers a greater design freedom due to simplicity of cutting irregular shapes.
4. Material wastage is virtually eliminated as offcuts of prepregs can be used as random molding compounds.
5. Automated mass-production techniques can be used for prepreg molding, and the quality of molded product is reproducible.
6. Toxic chemical effects on personnel using prepregs are minimized or eliminated.

A flowchart showing the key stages in the fabrication of composite structures from raw materials with an intermediate prepregging step is given in Fig. 2.46. It can be seen that there are two basic constituents to prepreg—the reinforcing fiber and the resin system. All of the advanced reinforcing fibers are available in continuous form, generally with a fixed filament diameter. The number of filaments that the supplier arranges into a 'bundle' (or yarn) varies widely, and is an important determinant of the ability to weave fabric and make prepregs to a given thickness. In the majority of cases, the yarn is treated with a size to protect it from abrasion during the weaving or prepregging process. Often, the size is chosen such that it is compatible with the intended resin system.

Resin systems have developed into extremely complex multi-ingredient formulations in an effort to ensure the maximum property benefit from the fiber. Normally there are four methods of impregnation: (1) solution dip; (2) solution spray; (3) direct hot-melt coat; and (4) film calendering.

The solution dip and solution spray impregnation techniques work with a matrix resin dissolved in a volatile carrier. The low viscosity of the resin

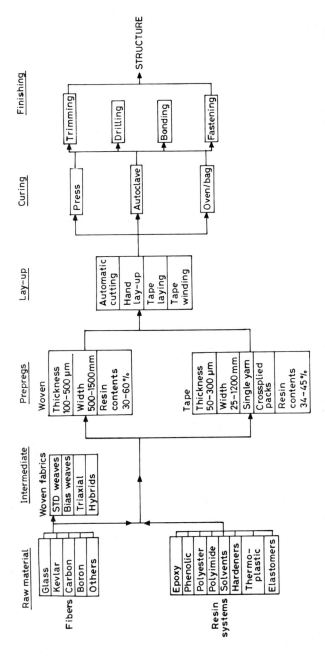

FIG. 2.46 Flow chart showing key stages in the fabrication of composite structures from raw materials by prepreg molding [26].

solution allows good penetration of the reinforcing fiber bundles with resin. In solution dipping, the fiber, in yarn or fabric form, is passed through the resin solution and picks up an amount of solids dependent upon the speed of through-put and the solids level. With solution spraying, on the other hand, the required amount of resin formulation is metered directly onto the fiber. In both cases, the impregnated fiber is then put through a heat cycle to remove the solvent and 'advance' the chemical reaction in the resin to give the correct degree of tack.

Direct hot melt can be performed in a variety of ways. In one method, the reinforcing fiber web is dipped into a melt resin bath. A doctor blade, scraper bar, or metering roller controls the resin content. Alternately, the melt resin is first applied to a release paper, the thickness of the resin being determined by a doctor blade. The melt resin on the release paper is then brought into contact with a collimated fiber bundle and pressed into it in a heated impregnation zone.

In film calendering, which is a variation of the above method, the resin formulation is cast into a film from either hot-melt or solution and then stored. Thereafter in a separate process, the reinforcing fiber is sandwiched between two films and calendered so that the film is worked into the fiber.

The decision of which method to use is dependent upon several factors. Hot melt film processes are faster and cheaper, but certain resin formulations cannot be handled in this way, and hence solution methods have to be used. The solution dip method is often preferred for fabrics as the need to squeeze hot-melt and film into the interstices of the fabric can cause distortion of the weave pattern.

In a process based on a biconstituent tow impregnation concept, the polymeric matrix is introduced in fibrous form and a comingled tow of polymer and reinforcing fibers are fed into a heated impregnation zone (Fig. 2.47). In this method, it may be possible to effect better wetout of the reinforcing fibers with the matrix polymer, especially with high viscosity thermoplastic matrix polymers, through intimate comingling.

The machines necessary to accomplish the above prepregging procedures are many and varied. There are three distinct aspects to quality control: raw material screening, on-line control, and batch testing. All three are obviously important, but the first two are more critical.

Fibers and base resins are supplied against certificates of conformance and often property test certificates. On-line control during the manufacture of prepreg revolves around the correct ratio of fiber to resin. This is done by a traversing Beta-gauge, which scans the dry fiber (either unidirectional or fabric) and then the impregnated fiber and provides a continuous real-time plot of the ratio across the width of the material. This can be linked back to the resin system application point for continuous adjustment.

Batch testing is carried out to verify prepreg properties, such as resin content, volatile level, and flow. The resin 'advancement' (chemical reaction) is monitored via a DSC (Differential Scanning Calorimeter) and the formulation

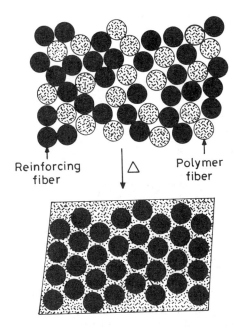

FIG. 2.47 Biconstituent tow impregnation using a co-mingled tow of polymer and re-inforcing fibers [26].

consistency by testing the Tg via DSC or DMA (Dynamic Mechanical Analyzer). The laminate properties are also determined. All are documented and quoted on a Release Certificate.

Commercial prepregs are available with different trade names. Fibredux 914 of Ciba-Geigy is a modified epoxy resin preimpregnated into unidirectional fibers of carbon (HM or HT), glass (E type and R type), or aramid (Kevlar 49) producing prepregs that, when cured to form fiber reinforced composite components, exhibit very high strength retention between −60°C to +180°C operational temperatures.

'Scotchply' brand reinforced plastics of Industrial Specialties Division of 3M Company are structural-grade thermosetting molding materials, consisting of unidirectional nonwoven glass fibers embedded in an epoxy resin matrix. The product is available both in prepreg form in widths up to 48 inches (1.22 m) and in flat sheet stock in sizes up to 48 inches (1.22 m) × 72 inches (1.83 m). The cured product is claimed to possess extraordinary fatigue life, no notch sensitivity, high ultimate strength, and superior corrosion resistance. The high ultimate strength and fatigue resistance of Scotchply laminate and relatively low modulus, make it about six times more efficient for spring use. The range of application of the reinforced plastic includes vibratory springs, sonar housings, landing gear, picker blades, snowmobile track reinforcement, helicopter blades, seaplane pontoons, missile casing, and archery bow laminate.

FOAMING PROCESSES

Plastics can be foamed in a variety of ways. The foamed plastics, also referred to as cellular or expanded plastics, have several inherent features which combine to make them economically important. Thus, a foamed plastic is a good heat insulator by virtue of the low conductivity of the gas (usually air) contained in the system, has a higher ratio of flexural modulus to density than when unfoamed, has greater load-bearing capacity per unit weight, and has considerably greater energy-storing or energy-dissipating capacity than the unfoamed material. Foamed plastics are therefore used in the making of insulation, as core materials for load-bearing structures, as packaging materials used in product protection during shipping, and as cushioning materials for furniture, bedding, and upholstery.

Among those plastics which are commercially produced in cellular form are polyurethane, PVC, polystyrene, polyethylene, polypropylene, epoxy, phenol-formaldehyde, urea-formaldehyde, ABS, cellulose acetate, styrene-acrylonitrile, silicone, and ionomers. However, note that it is possible today to produce virtually every thermoplastic and thermoset material in cellular form. In general, the basic properties of the respective polymers are present in the cellular products except, of course, those changed by conversion to the cellular form.

Foamed plastics can be classified according to the nature of cells in them into *closed-cell type* and *open-cell type*. In a closed-cell foam each individual cell, more or less spherical in shape, is completely closed in by a wall of plastic, whereas in an open-cell foam individual cells are interconnecting, as in a sponge. Closed-cell foams are usually produced in processes where some pressure is maintained during the cell formation stage. Free expansion during cell formation typically produces open-cell foams. Most foaming processes, however, produce both kinds. A closed-cell foam makes a better buoy or life jacket because the cells do not fill with liquid. In cushioning applications, however, it is desirable to have compression to cause air to flow from cell to cell and thereby dissipate energy, so the open-cell type is more suitable. Foamed plastics can be produced in a wide range of densities—from 0.1 lb/ft^3 (0.0016 g/cm^3) to 60 lb/ft^3 (0.96 g/cm^3)—and can be made flexible, semirigid, or rigid.

Obtained forms of foamed plastics are blocks, sheets, slabs, boards, molded products, and extruded shapes. These plastics can also be sprayed onto substrates to form coatings, foamed in place between walls (i.e., poured into the empty space in liquid form and allowed to foam), or used as a core in mechanical structures. It has also become possible to process foamed plastics by conventional processing machines like extruders and injection-molding machines.

Foaming of plastics can be done in a variety of ways. Most of them typically involve creating gases to make foam during the foaming cycle. Common foaming processes are the following.

1. Air is whipped into a dispersion or solution of the plastic, which is then hardened by heat or catalytic action or both.
2. A low-boiling liquid is incorporated in the plastic mix and volatilized by heat.
3. Carbon dioxide gas is produced within the plastic mass by chemical reaction.
4. A gas, such as nitrogen, is dissolved in the plastic melt under pressure and allowed to expand by reducing the pressure as the melt is extruded.
5. A gas, such as nitrogen, is generated within the plastic mass by thermal decomposition of a chemical blowing agent.
6. Microscopically small hollow beads of resin or even glass (e.g., microballoons) are embedded in a resin matrix.

Examples of these processes now follow.

Polystyrene Foams

Polystyrene, widely used in injection and extrusion molding, is also extensively used in the manufacture of plastic foams for a variety of applications. Polystyrene produces light, rigid, closed-cell plastic foams having low thermal conductivity and excellent water resistance, meeting the requirements of low-temperature insulation and buoyancy applications. Two types of low-density polystyrene foams are available to the fabricator, molder, or user: (1) extruded polystyrene foam and (2) expandable polystyrene for molded foam.

Extruded Polystyrene Foam

This material is manufactured as billets and boards by extruding molten polystyrene containing a blowing agent (nitrogen gas or chemical blowing agent) under elevated temperature and pressure into the atmosphere where the mass expands and solidifies into a rigid foam. Many sizes of extruded foam are available, some as large as 10 in. × 24 in. × 9 ft. The billets and boards can be used directly or cut into different forms. One of the largest markets for extruded polystyrene in the form of boards is in low-temperature insulation (e.g., truck bodies, railroad cars, refrigerated pipelines, and low-temperature storage tanks for such things as liquefied natural gas). Another growing market for extruded polystyrene boards is residential insulation. Such boards are also used as the core material for structural sandwich panels, used prominently in the construction of recreational vehicles.

Expandable Polystyrene

Expandable polystyrene is produced in the form of free-flowing pellets or beads containing a blowing agent. Thus, pellets chopped from an ordinary melt extruder or beads produced by suspension polymerization are impregnated with a hydrocarbon such as pentane. Below 70°F (21°C) the vapor pressure of the pentane dissolved in the polymer is low enough to permit storage of the impregnated material (expandable polystyrene) in a closed container at ordinary temperature and pressure. Even so, manufacturers do not recommend storing for more than a few months.

The expandable polystyrene beads may be used in a tubular blow-extrusion process (Fig. 2.48) to produce polystyrene foam sheet, which can subsequently be formed into containers, such as egg cartons and cold-drink cups, by thermoforming techniques.

Expandable polystyrene beads are often molded in two separate steps: (1) *preexpansion* or prefoaming of the expandable beads by heat, and (2) further expansion and fusion of the preexpanded beads by heat in the enclosed space of a shaping mold.

Steam heat is used for preexpansion in an agitated drum with a residence time of a few minutes. As the beads expand, a rotating agitator prevents them from fusing together, and the preexpanded beads, being lighter, are forced to the top of the drum and out the discharge chute. They are then collected in storage bins for aging prior to molding. The usual lower limit of bulk density for bead pre-

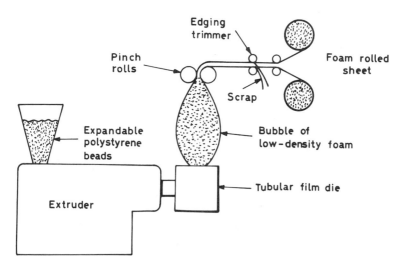

FIG. 2.48 Tubular blow extrusion of production of low-density polystyrene foam sheet. [From F. H. Collins, *SPE J.*, *16*, 705 (July 1960).]

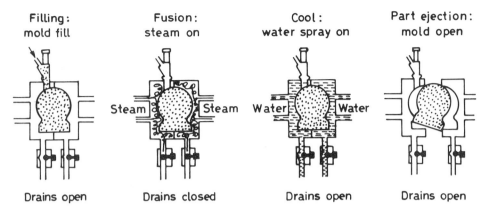

FIG. 2.49 Molding of preexpanded (prefoamed) polystyrene beads. [Adapted from *PELASPAN Expandable Polystyrene*, Form 171-414, Dow Chemical Co. (1966).]

expansion is 1.0 lb/ft^3 (0.016 g/cm^3), compared to the virgin bead bulk density of about 35 lb/ft^3 (0.56 g/cm^3).

Molding of preexpanded (prefoamed) beads requires exposing them to heat in a confined space. In a typical operation (Fig. 2.49) prefoamed beads are loaded into the mold cavity, the mold is closed, and steam is injected into the mold jacket. The prefoamed beads expand further and fuse together as the temperature exceeds T_g. The mold is cooled by water spray before removing the molded article. Packages shaped to fit their contents (small sailboats, toys, drinking cups, etc.) are made in this way. Special machines have been designed to produce thin-walled polystyrene foam cups. Very small beads at a prefoamed density of approximately 4 to 5 lb/ft^3 (0.06 to 0.08 g/cm^3) are used, which allow easy flow into the molding cavity and produce a cup having the proper stiffness for handling.

Polystyrene Foams

Polyurethane foams, also known as urethane foams or U-foams, are prepared by reacting hydroxyl-terminated compounds called *polyols* with an isocyanate (see Fig. 1.17). Isocyanates in use today include toluene diisocyanate, known as TDI, crude methylenebis(4-phenyl-isocyanate), known as MDI, and different types of blends, such as TDI/crude MDI. Polyols, the other major ingredient of the urethane foam, are active hydrogen-containing compounds, usually polyester diols and polyether diols.

It is possible to prepare many different types of foams by simply changing the molecular weight of the polyol, since it is the molecular backbone formed by the reaction between isocyanate and polyol that supplies the reactive sites for cross-linking (Fig. 1.17), which in turn largely determines whether a given foam will be flexible, semirigid, or rigid. In general, high-molecular-weight polyols with low functionality produce a structure with a low amount of cross-linking and, hence, a flexible foam. On the other hand, low-molecular-weight polyols of high functionality produce a structure with a high degree of cross-linking and, consequently, a rigid foam. Of course, the formulation can be varied to produce any degree of flexibility or rigidity within these two extremes.

The reactions by which urethane foam are produced can be carried out in a single stage (*one-shot process*) or in a sequence of several stages (*prepolymer process* and *quasi-prepolymer process*). These variations led to 27 basic types of products or processes, all of which have been used commercially. In the one-shot process all of the ingredients—isocyanate, polyol, blowing agent, catalyst, additives, etc.—are mixed simultaneously, and the mixture is allowed to foam. In the prepolymer method (Fig. 1.17), a portion of the polyol is reacted with an excess of isocyanate to yield a prepolymer having isocyanate end groups. The prepolymer is then mixed with additional polyol, catalyst, and other additives to cause foaming. The quasi-prepolymer process is intermediate between the prepolymer and one-shot processes.

Flexible Polyurethane Foams

The major interest in flexible polyurethane foams is for cushions and other upholstery materials. Principal market outlets include furniture cushioning, carpet underlay, bedding, automotive seating, crash pads for automobiles, and packaging. The density range of flexible foams is usually 1 to 6 lb/ft^3 (0.016 to 0.096 g/cm^3). The foam is made in continuous loaves several feet in width and height and then sliced into slabs of desired thickness.

One-Shot Process

The bulk of the flexible polyurethane foam is now being manufactured by the one-shot process using polyether-type polyols because they generally produce foams of better cushioning characteristics. The main components of a one-shot formulation are polyol, isocyanate, catalyst, surfactant, and blowing agent. Today the bulk of the polyether polyols used for flexible foams are propylene oxide polymers. The polymers prepared by polymerizing the oxide in the presence of propylene glycol as an initiator and a caustic catalyst are diols having the general structure

$$CH_3\!-\!CH\!-\!CH_2\!-\!O\!-\!(CH_2\!-\!CH\!-\!O)_{\overline{n}}CH_2\!-\!CH\!-\!CH_3$$
$$\qquad\ \ |\qquad\qquad\qquad\quad\ |\qquad\qquad\qquad |$$
$$\qquad\ \ OH\qquad\qquad\qquad\ CH_3\qquad\qquad OH$$

The polyethers made by polymerizing propylene oxide using trimethyl propane, 1,2,6-hexanetriol, or glycerol as initiator are triols of the following general type.

$$HO\text{---}(C_3H_6O)_{\overline{n}}CH_2\text{---}CH(OH)\text{---}CH_2\text{---}(C_3H_6O)_{\overline{n}}OH$$

The higher hydroxyl content of these polyethers leads to foams of better load-bearing characteristics. Molecular weight in the range 3000 to 3500 is found to give the best combination of properties.

The second largest component in the foam formulation is the isocyanate. The most suitable and most commonly used isocyanate is 80:20 TDI—i.e., 80:20 mixture of tolylene-2,4-diisocyanate and tolylene-2,6-diisocyanate.

One-shot processes require sufficiently powerful catalysts to catalyze both the gas evolution and chain extension reaction (Fig. 1.17). Use of varying combinations of an organometallic tin catalyst (such as dibutyltin dilaurate and stannous octoate) with a tertiary amine (such as alkyl morpholines and triethylamine), makes it possible to obtain highly active systems in which foaming and cross-linking reactions could be properly balanced.

The surface active agent is an essential ingredient in formulations. It facilitates the dispersion of water in the hydrophobic resin by decreasing the surface tension of the system. In addition, it also aids nucleation, stabilizes the foam, and regulates the cell size and uniformity. A wide range of surfactants, both ionic and nonionic, have been used at various times. Commonly used among them are the water-soluble polyether siloxanes.

Water is an important additive in urethane foam formulation. The water reacts with isocyanate to produce carbon dioxide and urea bridges (Fig. 1.17). An additional amount of isocyanate corresponding to the water present must therefore be incorporated in the foaming mix. The more water that is present, the more gas that is evolved and the greater number of active urea points for cross-linking. This results in foams of lower density but higher degree of cross-linking. So when soft foams are required, a volatile liquid such as trichloromonofluoromethane (bp 23.8°C) may be incorporated as a blowing agent. This liquid will volatilize during the exothermic urethane reaction and will increase the total gas in the foaming system, thereby decreasing the density, but it will not increase the degree of cross-linking. However, where it is desired to increase the cross-link density independently of the isocyanate-water reaction, polyvalent alcohols, such as glycerol and pentaerythritol, and various amines may be added as additional cross-linking agents. A typical formulation of one-shot urethane foam system is shown in Table 2.5.

Most foam is produced in block form from Henecke-type machines (Fig. 2.50) or some modification of them. In this process [27], several streams of the ingredients are fed to a mixing head which oscillates in a horizontal plane. In a typical process, four streams may be fed to the mixing head: e.g., polyol and fluorocarbon (if any); isocyanate; water, amine, and silicone; and

TABLE 2.5 Urethane One-Shot Foam Formulation

Ingredient	Parts by weight
Poly(propylene oxide), mol. wt. 2000 and 2 OH/molecule	35.5
Poly(propylene oxide) initiated with trifunctional alcohol, mol. wt. 3000 and 3 OH/molecule	35.5
Toluene diisocyanate (80:20 TDI)	26.0
Dibutyltin dilaurate	0.3
Triethylamine	0.05
Water	1.85
Surfactant (silicone)	0.60
Trichloromonofluoromethane (CCl_3F)	12.0

Final density of foam = 1.4 lb/ft^3 or 0.022 g/cm^3
(2.0 lb/ft^3 or 0.032 g/cm^3 if CCl_3F is omitted)

Source: Reference 28.

tin catalyst. The reaction is carried out with slightly warmed components. Foaming is generally complete within a minute of the mixture emerging from the mixing head. The emergent reacting mixture runs into a trough, which is moving backward at right angles to the direction of traverse of the reciprocating mixing head. In this way the whole trough is covered with the foaming mass.

Other developments of one-shot flexible foam systems include *direct molding*, where the mixture is fed into a mold cavity (with or without inserts such as springs, frames, etc.) and cured by heat. In a typical application, molds would be filled and closed, then heated rapidly to 300 to 400°F (149 to 204°C) to develop maximum properties. A good deal of flexible urethane foam is now being made by the *cold-cure* technique. This involves more reactive polyols and isocyanates in special foaming formulations which would cure in a reasonable time to their maximum physical properties without the need for additional heat over and above that supplied by the exothermic reaction of the foaming process. Cold-cure foaming is used in the production of what is known as *high-resilient foams* having high *sag factor* (i.e., ratio of the load needed to compress foam by 65% to the load needed to compress foam by 25%), which is most important to

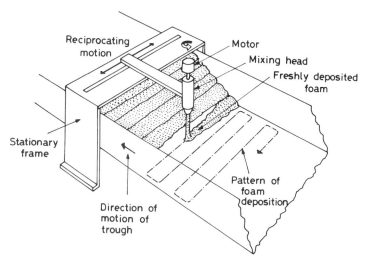

FIG. 2.50 Henecke-type machine for production of methane foam in block form by one-shot process [26].

cushioning characteristics. True cold-cure foams will produce a sag factor of 3 to 3.2, compared to 2 to 2.6 for hot-cured foams.

Prepolymer Process

In the prepolymer process the polyol is reacted with an excess of isocyanate to give an isocyanate-terminated prepolymer which is reasonably stable and has less handling hazards than free isocyanate. If water, catalysts, and other ingredients are added to the product, a foam will result. For better load-bearing and cushioning properties, a low-molecular-weight triol, such as glycerol and trimethylolpropane, is added to the polyol before it reacts with the isocyanate. The triol provides a site for chain branching.

Although the two-step prepolymer process is less important than the one-shot process, it has the advantage of low exotherms, greater flexibility in design of compounds, and reduced handling hazards.

Quasi-Prepolymer Process

In the quasi-prepolymer process a prepolymer of low molecular weight and hence low viscosity is formed by reacting a polyol with a large excess of isocynate. This prepolymer, which has a large number of free isocyanate groups, is then reacted at the time of foaming with additional hydroxy compound, water, and catalyst to product the foam. The additional hydroxy compound, which may be a polyol or a simple molecule such as glycerol or ethylene glycol, also functions as a viscosity depressant. The system thus has the advantage of having low-viscosity components, compared to the prepolymer process, but there are problems with high exotherms and a high free-isocyanate content. Quasi-propolymer systems based on polyester polyols and polyether polyols are becoming important in shoe soling, the former being most wear resistant and the latter the easiest to process.

Rigid and Semirigid Foams

The flexible foams discussed in previous sections have polymer structures with low degrees of cross-linking. Semirigid and rigid forms of urethane are products having higher degrees of cross-linking. Thus, if polyols of higher functionality—i.e., more hydroxyl groups per molecule—are used, less flexible products may be obtained, and in the case of polyol with a sufficiently high functionality, rigid foams will result.

The normal density range for rigid and semirigid foams is about 1 to 3 lb/ft^3 (0.016 to 0.048 g/cm^3). Some packaging applications, however, use densities down to 0.5 lb/ft^3 (0.008 g/cm^3); for furniture applications densities can go as high as 20 to 60 lb/ft^3 (0.32 to 0.96 g/cm^3), thus approaching solids. At densities of from 2 lb/ft^3 (0.032 g/cm^3) to 12 lb/ft^3 (0.19 g/cm^3) or more, these foams combine the best of structural and insulating properties.

Semirigid (or semiflexible) foams are characterized by low resilience and high energy-absorbing characteristics. They have thus found prime outlet in the automotive industry for applications like safety padding, arm rests, horn buttons, etc. These foams are cold cured and involve special polymeric isocyanates. They are usually applied behind vinyl or ABS skins. In cold curing, the liquid ingredients are simply poured into a mold in which vinyl or ABS skins and metal inserts for attachments have been laid. The liquid foams and fills the cavity, bonding to the skin and inserts. Formulations and processing techniques are also available to produce self-skinning semirigid foam in which the foam comes out of the mold with a continuous skin of the same material.

Rigid urethane foams have outstanding thermal insulation properties and have proved to be far superior to any other polymeric foam in this respect. Besides, these rigid foams have excellent compressive strength, outstanding buoyancy (flotation) characteristics, good dimensional stability, low water absorption, and the ability to accept nails or screws like wood. Because of these characteristics, rigid foams have found ready acceptance for such applications as insulation, refrigeration, packaging, construction, marine uses, and transportation.

For such diverse applications several processes are now available to produce rigid urethane foam. These include foam-in-place (or pour-in-place), spraying, molding, slab, and laminates (i.e., foam cores with integral skins produced as a single unit). One-shot techniques can be used without difficulty, although in most systems the reaction is slower than with the flexible foam, and conditions of manufacture are less critical. Prepolymer and quasi-prepolymer systems were also developed in the United States for rigid and semirigid foams, largely to reduce the hazards involved in handling TDI where there are severe ventilation problems.

In the *foam-in-place process* a liquid urethane chemical mixture containing a fluorocarbon blowing agent is simply poured into a cavity or metered in by machine. The liquid flows to the bottom of the cavity and foams up, filling all cracks and corners and forming a strong seamless core with good adhesion to the inside of the walls that form the cavity. The cavity, of course, can be any space, from the space between two walls of a refrigerator to that between the top and bottom hull of a boat. However, if the cavity is the interior of a closed mold, the process is known as *molding*.

Rigid urethane foam can be applied by *spraying* with a two-component spray gun and a urethane system in which all reactants are incorporated either in the polyol or in the isocyanate. The spraying process can be used for applying rigid foam to the inside of building panels, for insulating cold-storage rooms, for insulating railroad cars, etc.

Rigid urethane foam is made in the form of *slab* stock by the one-shot technique. As in the Henecke process (Fig. 2.50), the reactants are metered separately into a mixing head where they are mixed and deposited onto a conveyor. The mixing head oscillates in a horizontal plane to insure an even deposition. Since the foaming urethane can structurally bond itself to most substrates, it is possible (by metering the liquid urethane mixture directly onto the surface skin) to produce board stock with integral skins already attached to the surface of the foam. Sandwich-construction building panels are made by this technique.

Foamed Rubber

Although foamed rubber and foamed urethanes have many similar properties, the processes by which they are made differ radically. In a simple process a solution of soap is added to natural rubber latex so that a froth will result on beating. Antioxidants, cross-linking agents, and a foam stabilizer are added as aqueous dispersions. Foaming is done by combined agitation and aeration with automatic mixing and foaming machines. The stabilizer is usually sodium silicofluoride (Na_2SiF_6). The salt hydrolyzes, yielding a silica gel which increases the viscosity of the aqueous phase and prevents the foam from collapsing. A typical cross-linking agent is a combination of sulfur and the zinc salt of mercaptobenzothiazole (accelerator). Cross-linking (curing or vulcanization) with this agent takes place in 30 min at 100°C. When making a large article such as a mattress, a metal mold may be filled with the foamed latex and heated by

TABLE 2.6 Foamed Rubber Formulation

Ingredient	Parts by weight
Styrene-butadiene latex (65% solids)	123
Natural rubber latex (60% solids)	33
Potassium oleate	0.75
Sulfur	2.25
Accelerators:	
Zinc diethyldithiocarbamate	0.75
Zinc salt of mercaptobenzothiazole	1.0
Trimene base (reaction product of ethyl chloride, formaldehyde, and ammonia)	0.8
Antioxidant (phenolic)	0.75
Zinc oxide	3.0
Na_2SiF_6	2.5

Source: Reference 29.

steam at atmospheric pressure. After removing the foamed rubber article from the mold, it may be dewatered by compressing it between rolls or by centrifuging and by drying with hot air in a tunnel dryer. In foamed rubber formulation a part of the natural rubber latex can be replaced by a synthetic rubber latex. One such combination is shown in Table 2.6.

Vinyl Foams

Plastisols are the most widely used route to flexible expanded vinyl products. Varied types of upholstery, garment fabrics, and flooring products are made from coatings with expandable plastisol compounds. Dolls, gaskets, resilient covers for tool handles, etc., are produced from such compounds by molding. Plastisols are expanded by four general processes: (1) physical frothing, (2) chemical frothing, (3) chemical foaming, and (4) incorporating gas into plastic held under high pressure with subsequent expansion (pressure sponge).

Froths are the result of incorporating gas into liquid plastisol, which is then subjected to a heat cycle, giving cellular products that are highly *open* in structure. The reason is that many bubble walls thin out and break due to thermal expansion during the heat cycle. *Foams* result from expansion of thermoplastic plastisol hot *melt* (after fusion) at or near atmospheric pressure. Since the bubbles do not have to withstand much, if any, thermal expansion, the result is a highly unicellular or *closed-cell* structure.

The chemical used as a *frothing agent* should have sufficiently low decomposition temperature, or else an appropriate plastisol formulation should be employed so that the plastisol does not gel before frothing takes place. After frothing, the plastisol is fused into a homogeneous plasticized mass. Chemically frothed plastisols are used predominantly in molding applications. Slipper soles, for example, are molded from them.

Many *chemical blowing (foaming) agents* have been developed for cellular elastomers and plastics, which generally speaking, are organic nitrogen compounds that are stable at normal storage and mixing temperatures but undergo decomposition with gas evolution at reasonably well-defined temperatures. Three important characteristics of a chemical blowing agent are the decomposition temperature, the volume of gas generated by unit weight, and the nature of the decomposition products. Since nitrogen is an inert, odorless, nontoxic gas, nitrogen-producing organic substances are preferred as blowing agents. Several examples of blowing agents [30] especially recommended for vinyl plastisols are shown in Table 2.7; in each case the gas generated is nitrogen.

To produce uniform cells, the blowing agent must be uniformly dispersed or dissolved in the plastisol and uniformly nucleated. It should decompose rapidly and smoothly over a narrow temperature range corresponding to the attainment of a high viscosity or gelation of the plastisol system. The gelation involves solvation of the resin in plasticizer at 300 to 400°F (149 to 204°C), the temperature depending on the ingredients employed in the plastisol. The foam quality is largely determined by the matching of the decomposition of the blowing agent to the gelation of the polymer system. If gelation occurs before gas evolution, large holes or fissures may form. On the other hand, if gas evolution occurs too soon before gelation, the cells may collapse, giving a coarse, weak, and spongy product. Among the blowing agents listed in Table 2.7, azobisformamide (ABFA) is the most widely used for vinyls because it fulfills the requirements efficiently. ABFA decomposition can also be adjusted through proper choice of metal organic activators so that the gas evolution occurs over a narrow range within the wide range given in Table 2.7.

Closed-cell foams result when the decomposition and gelation are carried out in a closed mold almost filled with plastisol. After the heating cycle, the material is cooled in the mold under pressure until it is dimensionally stable. The mold is then opened, and the free article is again subjected to heat (below the previous molding temperature) for final expansion. Protective padding, life jackets, buoys, and floats are some items made by this process.

The blowing agents given in Table 2.7 can be used to make foamed rubber. A stable network in this product results from the cross-linking reaction

TABLE 2.7 Commercial Blowing (Foaming) Agents

Chemical type	Decomposition temperature in air (°C)	Decomposition position range in plastics (°C)	Gas yield (mL/g)
Azo compounds			
Azobisformamide (Azodicarbonamide)	195–200	160–200	220
Azobisisobutyronitrile	115	90–115	130
Diazoaminobenzene	103	95–100	115
N-Nitroso compounds			
N,N'-Dimethyl-N,N'-dinitrosoterephthalimide	105	90–105	126
N,N'-Dinitrosopentamethylenetetramine	195	130–190	265
Sulfonyl hydrazides			
Benzenesulfonylhydrazide	>95	95–100	130
Toluene-(4)-sulfonyl hydrazide	103	100–106	120
Benzene-1,3-disulfonyl hydrazide	146	115–130	85
4,4'-oxybis(benzenesulfonylhydrazide)	150	120–140	125

Source: Reference 30.

(vulcanization), which thus corresponds to the step of fusion in the case of plastisols. Some thermoplastics also can be foamed by thermal decomposition of blowing agents even though they do not undergo an increase in dimensional stability at an elevated temperature. In this case the viscosity of the melt is high enough to slow down the collapse of gas bubbles so that when the polymer is cooled below its T_m a reasonably uniform cell structure can be built in. Cellular polyethylene is made in this way.

RUBBER COMPOUNDING AND PROCESSING TECHNOLOGY [31–34]
Compounding Ingredients

No rubber becomes technically useful if its molecules are not cross-linked, at least partially, by a process known as curing or vulcanization. For natural rubber and many synthetic rubbers, particularly the diene rubbers, the curing agent most commonly used is sulfur. But sulfur curing takes place at technically viable rates only at a relatively high temperature ($>140°C$) and, moreover, if sulfur alone is used, optimum curing requires use of a fairly high dose of sulfur, typically 8–10 parts per hundred parts of rubber (phr), and heating for nearly 8 h at 140°C. Sulfur dose has been substantially lowered, however, with the advent of organic accelerators. Thus, incorporation of only 0.2–2.0 phr of accelerator allows reduction of sulfur dose from 8–10 phr to 0.5–3 phr and effective curing is achieved in a time scale of a few minutes to nearly an hour depending on temperature (100–140°C) and type of the selected accelerator. The low sulfur dose required in the accelerated sulfur vulcanization has not only eliminated bloom (migration of unreacted sulfur to the surface of the vulcanizate), which was a common feature of the earlier nonaccelerated technology, but also has led to the production of vulcanizates of greatly improved physical properties and good resistance to heat and aging.

The selection of the accelerator depends largely on the nature of the rubber taken, the design of the product, and the processing conditions. It is important to adopt a vulcanizing system that not only gives a rapid and effective cross-linking at the desired vulcanizing temperatures but also resists premature vulcanization (scorching) at somewhat lower temperatures encountered in such operations as mixing, extrusion, calendering, and otherwise shaping the rubber before final cross-linking. This may require the use of delayed-action type accelerators as exemplified by sulfenamides. Other principal types of accelerators with different properties are guanidines, thiazoles, dithiocarbamates, thiurams, and xanthates (Table 2.8). Accelerators are more appropriately classified according to the speed of curing induced in their presence in NR systems. In the order of increasing speed of curing, they are classified as *slow*, *medium*, *semiultra*, and *ultra* accelerators.

TABLE 2.8 Accelerator for Sulfur Vulcanization of Rubbers

Accelerator type and formula	Chemical name	Accelerator activity
Guanidines		
	Diphenyl guanidine (DPG)	Medium accelerator
	Triphenyl guanidine (TPG)	Slow accelerator
Thiazoles		
	Mercaptobenzothiazole (MBT)	Semi-ultra accelerator
	Mercaptobenzothiazyl disulfide (MBTS)*	Semi-ultra (delayed action) accelerator
Sulfenamides		
	N-Cyclohexyl benzothiazyl sulfenamide (CBS)	Semi-ultra (delayed action) accelerator
	N-Oxydiethylenebenzothiazyl sulfenamide (NOBS) or 2-Morpholinothiobenzothiazole (MBS)	Semi-ultra (delayed action) accelerator
	N-t-Butylbenzothiazyl sulfenamide (TBBS)	Semi-ultra (delayed action) accelerator

TABLE 2.8 Continued

Accelerator type and formula	Chemical name	Accelerator activity
Dithiocarbamates		

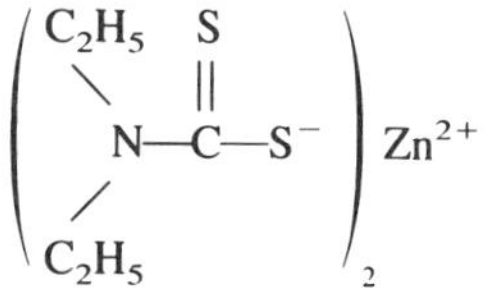

Zinc diethyl dithiocarbamate (ZDC) — Ultra accelerator

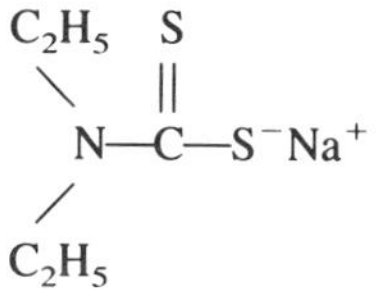

Sodium diethyl dithiocarbamate (SDC) — Ultra accelerator, water soluble (used for latex)

Thiuram sulfides

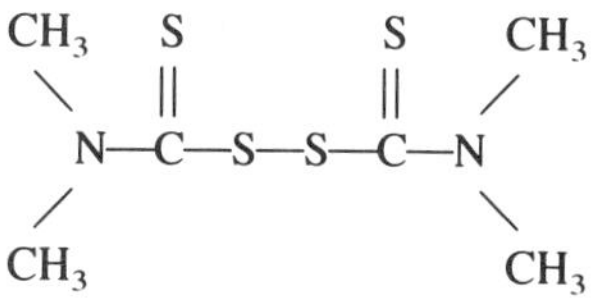

Tetramethyl thiuram disulfide (TMTD, TMT)* — Ultra accelerator

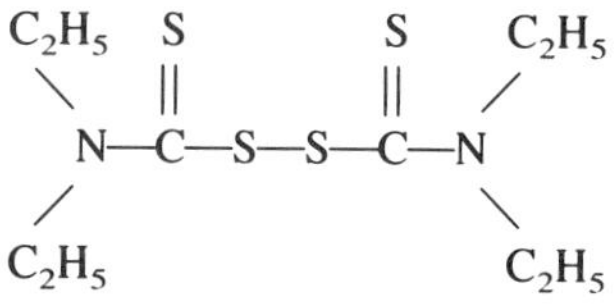

Tetraethyl thiuram disulfide (TETD, TET)* — Ultra accelerator

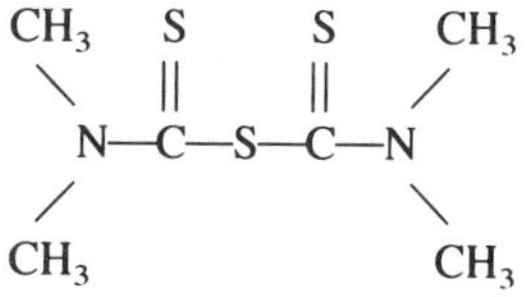

Tetramethyl thiuram monosulfide (TMTM) — Ultra accelerator

Xanthates

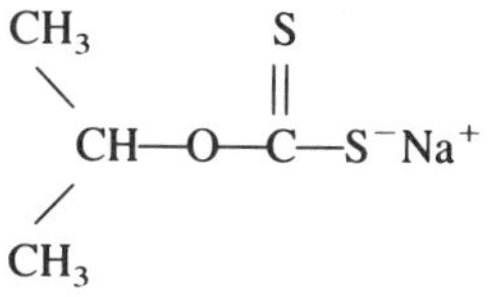

Sodium isopropyl xanthate (SIX) — Ultra accelerator, water soluble (suited for latex)

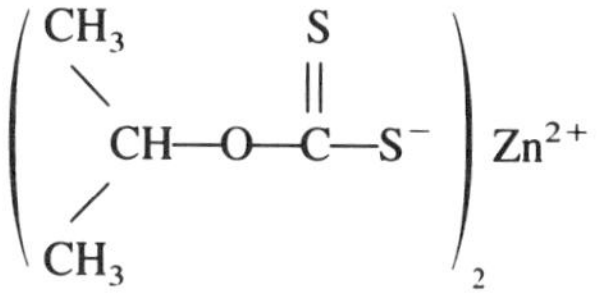

Zinc isopropyl xanthate (ZIX) — Ultra accelerator

*Sulfur donors

The problem of scorching or premature vulcanization is very acute with ultra or fast accelerators. Rubber stocks are usually bad conductors of heat and therefore flow of heat to the interior of a vulcanizing stock from outside is very slow. As a result, in thick items the outer layers may reach a state of overcuring before the core or interior layers begin to cure. For such thick items, a slow accelerator (Table 2.8) is most suitable.

For butyl and EPDM rubbers, which have very limited unsaturations, slow accelerators are, however, unsuitable and, fast accelerators should be used at high temperatures for good curing at convenient rates. Since butyl rubber is characterized by *reversion*, a phenomenon of decrease of tensile strength and modulus with time of cure after reaching a maximum, duration of heating at curing temperatures must be carefully controlled, and prolonged heating must be avoided.

For rubbers with higher degree of unsaturations, an ideal accelerator is one that is stable during mixing, processing, and storage of the mix, but that reacts and decomposes sharply at the high vulcanization temperature to effect fast curing. These requirements or demands are closely fulfilled by the delayed-action accelerators typical examples of which are given in Table 2.8.

The spectacular effects of modern organic accelerators in sulfur vulcanization of rubber are observed only in the presence of some other additives known as accelerator *activators*. They are usually two-component systems comprising a metal oxide and a fatty acid. The primary requirement for satisfactory activation of accelerator is good dispersibility or solubility of the activators in rubber. Oxides of bivalent metals such as zinc, calcium, magnesium, lead, and cadmium act as activators in combination with stearic acid. A combination of zinc oxide and stearic acid is almost universally used. (Where a high degree of transparency is required, the activator may be a fatty acid salt such as zinc stearate.) Besides speeding up the rate of curing, activators also bring about improvements in physical properties of vulcanizates. This is highlighted by the data given in Table 2.9.

A sulfur-curing system thus has basically four components: a sulfur vulcanizing agent, an accelerator (sometimes combinations of accelerators), a metal oxide, and a fatty acid. In addition, in order to improve the resistance to scorching, a *prevulcanization inhibitor* such as N-cyclohexylthiophthalimide may be incorporated without adverse effects on either the rate of cure or physical properties of the vulcanizate.

The level of accelerator used varies from polymer to polymer. Some typical curing systems for the diene rubbers (NR, SBR, and NBR) and for two olefin rubbers (IIR and EPDM—see Appendix 1B for abbreviations) are given in Table 2.10.

In addition to the components of the vulcanization system, several other additives are commonly used with diene rubbers. Rubbers in general, and diene

TABLE 2.9 Effect of Activator on Vulcanization

Time of cure, min	Tensile strength, psi (MPa)	
	ZnO, phr	
	0.0	5.0
15	100 (0.7)	2300 (16)
30	400 (2.8)	2900 (20)
60	1050 (7.2)	2900 (20)
90	1300 (9.0)	2900 (20)

Base compound: NR (pale creep) 100; sulfur 3; mercaptobenzothiazole (MBT) 0.5.
Temperature of vulcanization 142°C.

rubbers in particular, are blended with many more additives than is common for most thermoplastics with the possible exceptions of PVC. The major additional classes of additives are:

1. Antidegradants (antioxidants and antiozonants)
2. Processing aids (peptizers, plasticizers, softeners and extenders, tackifiers, etc.)
3. Fillers
4. Pigments
5. Others (retarders, blowing agents)

The use of antioxidants and antiozonants has already been described in Chapter 1.

TABLE 2.10 Components of Sulfur Vulcanization Systems

Additive[a] (phr)	Rubber				
	NR	SBR	NBR	IIR	EPDM
Sulfur	2.5	2.0	1.5	2.0	1.5
Zinc oxide	5.0	5.0	5.0	3.0	5.0
Stearic acid	2.0	2.0	1.0	2.0	1.0
TBBS	0.6	1.0	—	—	—
MBTS	—	—	1.0	0.5	—
MBT	—	—	—	—	1.5
TMTD	—	—	0.1	1.0	0.5

[a]See Table 2.8 for accelerator abbreviations.

Processing Aids

Peptizers are added to rubber at the beginning of mastication (see later) and are used to increase the efficiency of mastication. They act chemically and effectively at temperatures greater than 65°C and hasten the rate of breakdown of rubber chains during mastication. Common peptizers are zinc thiobenzoate, zinc-2-benzamidothiophenate, thio-β-naphthol, etc. Processing aids other than the peptizer and compounding ingredients (additives) are added after the rubber attains the desired plasticity on mastication.

Common process aids, besides the peptizer, are pine tar, mineral oil, wax, factice, coumarone-indene resins, petroleum resins, rosin derivatives, and polyterpenes. Their main effect is to make rubber soft and tacky to facilitate uniform mixing, particularly when high loading of carbon black or other fillers is to be used.

Factice (vulcanized oil) is a soft material made by treating drying or semi-drying vegetable oils with sulfur monochloride (cold or white factice) or by heating the oils with sulfur at 140–160°C (hot or brown factice). The use of factice (5–30 phr) allows efficient mixing and dispersion of powdery ingredients and gives a better rubber mix for the purpose of extrusion.

Ester plasticizers (phthalates and phosphates) that are used to plasticize PVC (see Chapter 1) are also used as process aids, particularly with NBR and CR. Polymerizable plasticizers such as ethylene glycol dimethacrylate are particularly useful for peroxide curing rubbers. They act as plasticizers or tackifiers during mixing and undergo polymerization by peroxide initiation during cure.

The diene hydrocarbon rubbers are often blended with hydrocarbon oils. The oils decrease polymer viscosity and reduce hardness and low temperature brittle point of the cured product. They are thus closely analogous to the plasticizers used with thermoplastics but are generally known as *softeners*. Three main types of softeners are distinguished: aliphatic, aromatic, and naphthenic. The naphthenics are preferred for general all-round properties.

Natural rubbers exhibit the phenomenon known as *tack*. Thus when two clean surfaces of masticated rubber are brought into contact the two surfaces strongly adhere to each other, which is a consequence of interpenetration of molecular ends followed by crystallization. Amorphous rubbers such as SBR do not display such tack and it is necessary to add *tackifiers* such as rosin derivatives and polyterpenes.

Fillers

The principles of use of inert fillers, pigments, and blowing agents generally follow those described in Chapter 1. Major fillers used in the rubber industry are classified as (1) nonblack fillers such as china clay, whiting, magnesium carbonate, hydrated alumina, anhydrous, and hydrated silicas and silicates including those in the form of ground mineral such as slate powder, talc, or french chalk, and (2) carbon blacks.

Rather peculiar to the rubber industry is the use of the fine particle size reinforcing fillers, particularly carbon black. Fillers may be used from 50 phr to as high as 100–120 phr or even higher proportions. Their use improves such properties as modulus, tear strength, abrasion resistance, and hardness. They are essential with amorphous rubbers such as SBR and polybutadiene that has little strength without them. They are less essential with strain-crystallizing rubbers such as NR for many applications but are important in the manufacture of tires and related products.

Carbon blacks are essentially elemental carbon and are produced by thermal decomposition or partial combustion of liquid or gaseous hydrocarbons to carbon and hydrogen. The principal types, according to their method of production, are *channel black, furnace black*, and *thermal black*.

Thermal black is made from natural gas by the *thermatomic process* in which methane is cracked over hot bricks at a temperature of 1600°F (871°C) to form amorphous carbon and hydrogen. Thermal black consists of relatively coarse particles and is used principally as a pigment. A few grades (FT and MT referring to fine thermal and medium thermal) are also used in the rubber industry.

Most of the carbon black used in the rubber industry is made by the furnace process (furnace black), that is, by burning natural gas or vaporized aromatic hydrocarbon oil in a closed furnace with about 50% of the air required for complete combustion. Furnace black produced from natural gas has an intermediate particle size, while that produced from oil can be made in wide range of controlled particle sizes and is particularly suitable for reinforcing rubbers. Quite a variety of grades of furnace blacks are available, e.g., fine furnace black (FF), high modulus (HMF), high elongation (HEF), reinforcing (RF), semireinforcing (SRF), high abrasion (HAF), super abrasion (SAF), intermediate super abrasion (ISAF), fast extruding (FEF), general purpose (GPF), easy processing (EPF), conducting (CF), and super conducting furnace black (SCF).

Channel black is characterized by lower pH, higher volatile content, and high surface area. It has the smallest particle size of any industrial material. A few grades of channel blacks (HPC, MPC, or EPC corresponding to hard, medium, or easy processing channel) are used in the rubber industry.

For carbon-black fillers, structure, particle size, particle porosity, and overall physico-chemical nature of particle surface are important factors in deciding cure rate and degree of reinforcement attainable. The pH of the carbon black has a profound influence. Acidic blacks (channel blacks) tend to retard the curing process while alkaline blacks (furnace blacks) produce a rate-enhancing effect in relation to curing, and may even give rise to scorching.

Another important factor is the particle size of the carbon black filler. The smaller the particle size, the higher the reinforcement, but the poorer the processability because of the longer time needed for dispersion and the greater heat produced during mixing. Blacks of the smallest particle size are thus unsuitable for use in rubber compounding.

For carbon black fillers the term *structure* is used to represent the clustering together and entanglement of fine carbon particles into long chains and three-dimensional aggregates. *High-structure blacks* produce high-modulus vulcanizates as high shear forces applied during mixing break the agglomerates down to many active free radical sites, which bind the rubber molecules, thereby leading to greater reinforcement. In *nonstructure* blacks the aggregates are almost nonexistent.

Most of the nonblack fillers used in rubber compounds are of nonreinforcing types. They are added for various objectives, the most important being cost reduction. Precipitated silica (hydrated), containing about 10–12% water with average particle size ranging from 10–40 nm, produce effective reinforcements and are widely used in translucent and colored products. Finely ground magnesium carbonate and aluminum silicate also induce good reinforcing effects. Precipitated calcium carbonate and activated calcium carbonate (obtained by treating calcium carbonate with a stearate) are used as semireinforcing fillers.

Short fibers of cotton, rayon, or nylon may be added to rubber to enhance modulus and tear and abrasion resistance of the vulcanizates. Some resins such as 'high styrene resins' and novolac-type phenolic resins mixed with hexamethylene tetramine may also be used as reinforcing fillers or additives. Whereas SBR has a styrene content of about 23.5% and is rubbery, styrene-butadiene copolymers containing about 50% styrene are leatherlike whilst with 70% styrene the materials are more rigid thermoplastics but with low softening points. Both of these copolymers are known in the rubber industry as *high styrene resins* and are usually blended with a hydrocarbon rubber such as NR and SBR. Such blends have found use in shoe soles, car wash brushes and other moldings, but in recent years have suffered increasing competition from conventional thermoplastics and thermoplastic rubbers.

Mastication and Mixing

A deficiency of natural rubber, compared with the synthetics, is its very high molecular weight, which makes mixing of compounding ingredients and subsequent processing by extrusion and other shaping operations difficult. For natural rubber it is thus absolutely necessary, while for synthetic rubbers it is helpful, to subject the stock to a process of breakdown of the molecular chains prior to compounding. This is effected by subjecting the rubber to high mechanical work (shearing action), a process commonly known as *mastication*. Mastication and mixing are conveniently done using two-roll mills and internal mixers. The oxygen in air plays a critical role during mastication.

Rubber that has been masticated is more soft and flows more readily than the unmasticated material. Mastication also allows preparation of solutions of high solids content because of the much lower solution viscosity of the degraded rubber. Rubber is also rendered tacky by mastication, which means that the uncured rubber sticks to itself readily so that articles of suitable thickness can be built up from layers of masticated rubber or rubberized fabric without the use of a solvent.

Open Mill

The mainstays of the rubber industry for over 70 years has been the two-roll (open) mill and the Banbury (internal) mixer. Roll mills were first used for rubber mixing over 120 years ago. The plastics and adhesives industries later adopted these tools.

The two-roll mill (Fig. 2.51) consists of two opposite-rotating rollers placed close to one another with the roll axes parallel and horizontal, so that a relatively small gap or nip (adjustable) between the cylindrical surface exists. The speeds of the two rolls are usually different, the front roll having a slower speed. For natural rubber mixing, a friction ratio of 1:1.2 for the front to back roll may be used. For some synthetic rubbers or highly filled NR mixes, friction ratios close to 1.0 produce good results.

The nip is adjusted so that when pieces of rubber are placed between the rolls, they are deformed by shearing action and squeezed through the nip to which they are returned by the operator. On repeated passage through the nip, the separate pieces of rubber soon join into a single mass to form a band around the front roll and a moving "bank" above the nip. As the rolls keep on rotating, the operator uses a knife to cut through the band on the front roll, removes

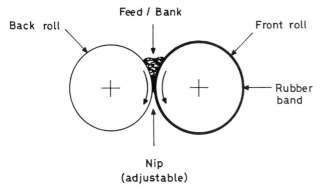

FIG. 2.51 Section showing the features of a two-roll open mill.

the mass in parts from time to time, and places it in a new position to ensure uniform treatment and mixing through the nip.

With most rubbers other than natural rubber, addition of different compounding ingredients may be normally started soon after a uniform band is formed on the roll and a bank is obtained. For natural rubber, however, milling is usually continued to masticate the rubber to the desired plasticity, and mixing of compounding ingredients is started only after the adequate mastication. Mixing is effected by adding the different ingredients onto the bank. They are gradually dispersed into the rubber, which is cut at intervals, rolled over, and recycled through the nip of the moving rolls to produce uniform mixing. Since the rate of mastication is a function of temperature, time and temperature of mastication have to be controlled or kept uniform from batch to batch in order to get the desired uniform products from different batches.

Roll mills vary greatly in size from very small laboratory machines with rollers of about 1 in. in diameter and driven by fractional horsepower motors to very large mills with rollers of nearly 3 ft in diameter and 7 or 8 ft in length and driven by motors over 100 hp.

Internal Batch Mixers

Internal batch mixers are widely used in the rubber industry. They are also used for processing plastics such as vinyl, polyolefins, ABS, and polystyrene, along with thermosets such as melamines and ureas because they can hold materials at a constant temperature.

The principle of internal batch mixing was first introduced in 1916 with the development of the Banbury mixer (Fig. 2.52a). A Banbury-type internal mixer essentially consists of a cylindrical chamber or shell within which materials to be mixed are deformed by rotating blades or rotors with protrusions. The mixer is provided with a feed door and hopper at the top and a discharge door at the bottom. In most cases a shell actually consists of two adjacent cylindrical shells, partly intersecting each other. In each shell there is a rotor describing an arc concentric with the shell section. The ridge between the two cylindrical chamber sections helps force mixing, and the acute convergence of the rotors and the chamber walls imparts shearing. As the rubber or mix is worked and sheared between the two rotors and between each rotor and the body of the casing, mastication takes place over the wide area, unlike in a open mill where it is restricted only in the area of the nip between the two rolls.

The rotor blade of the Banbury mixer is pear shaped, but the projection is spiral along the axis and the two spirals interlock and rotate in opposite directions (Fig. 2.52b). The interaction of rotor blades between themselves, in addition to producing shearing action, causes folding or "shuffling" of the mass, which is further accentuated by the helical arrangement of the blade along the axis of the rotor, thereby imparting motion to the mass in the third, or axial,

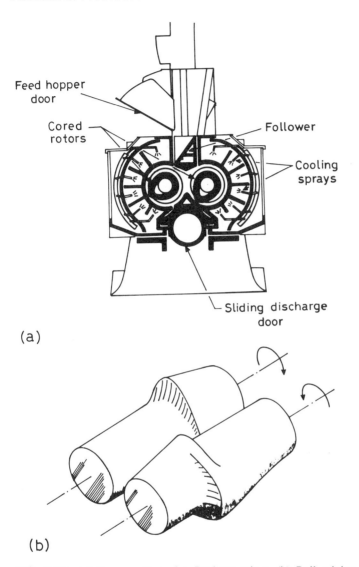

FIG. 2.52 (a) Cross-section of a Banbury mixer. (b) Roll mixing blades in Banbury mixer.

direction. This combination of intensive working produces a highly homogeneous mix.

An important and novel feature of the Banbury mixer is a vertical ram to press the mass into contact with the two rotors. Rubbers, fillers, and other ingredients are charged through the feed hopper and then held in the mixing chamber under the pressure of the hydraulic (or manual) ram. As a result, incorporation of solids is more rapid. Both the cored rotors and the walls of the mixing chamber can be cooled or heated by circulating fluid. Because of the large power consumption of such a machine (up to 500 hp) the cylinder walls are usually water-cooled by sprays (Fig. 2.52a).

Other major machines in use in the rubber industry are the Shaw intermix and the Baker-Perkins shear-mix. Both the Banbury-type and the intermix mixers have passed through modifications and refinements at different points in time. Mechanical feeding and direct oil injection in measured doses into the mixing chamber through a separate oil injection port are notable features of modern internal mixers. Higher rotor speeds and, in the Banbury type, the higher ram pressures, are used for speedy output. In Banbury-type mixers, the rotors run at different speeds, while in intermix mixers the rotor speeds are equal but the kneading action between the thicker portion of one rotor and the thinner portion of the other produces a frictional effect.

The operation of internal mixers is power intensive, and a given job is performed at a much higher speed and in a much shorter time than on a two-roll open mill. However, the major vulcanizing agent (such as sulfur) is often added later on a two-roll mill so as to eliminate possibilities of scorching. And even if this is not practiced, the mix, after being discharged from the internal mixer, is usually passed through a two-roll mill in order to convert it from irregular lumps to a sheet form for convenience in subsequent processing.

Reclaimed Rubber

The use of reclaimed rubber in a fresh rubber mix not only amounts to waste utilization but offers some processing and economic advantages to make it highly valued in rubber compounding. Though waste vulcanized rubber is normally not processable, application of heat and chemical agents to ground vulcanized waste rubber leads to substantial depolymerization whereby conversion of the rubber to a soft, plastic processable state is effected. Rubber so regenerated for reuse is commonly known as *reclaimed rubber* or simply as *reclaim*. Reclaimed rubber can be easily revulcanized.

Worn-out tires and scraps and trimmings of other vulcanized products constitute the raw material for reclaimed rubber. Therefore a good reclaiming process must not only turn the rubber soft and plastic but also must remove reinforcing cords and fabrics that may be present. There are a number of commercial processes [33] for rubber regeneration: (1) alkali digestion process, (2) neutral or zinc chloride digestion process, (3) heater or pan process, and (4) Reclaimator process.

Tires are most commonly reclaimed by digestion processes. For processes (1) and (2), debeaded tires and scraps, cut into pieces, are ground with two-roll mills or other devices developed for the purpose. Two-roll mills generally used for grinding tires turn at a ratio of about 1:3, thus providing the shearing action necessary to rip the tire apart. The rubber chunks are screened, and the larger material is recycled until the desired size is reached. The ground rubber is then mixed with a peptizer, softener, and heavy naphtha, and charged into spherical autoclaves with requisite quantities of water containing caustic soda for process (1) or zinc chloride for process (2). The textile is destroyed and mostly lost in the digestion process. Steam pressure and also the amount of air or oxygen in the autoclave greatly influence the period necessary for rubber reclaiming. On completion of the process, the pressure is released, the contents of the autoclave discharged into water, centrifuged, pressed to squeeze out water, and dried. The material is finally processed through a two-roll mill during which mineral fillers and oils may be added to give a product of required specific gravity and oil extension.

Butyl and natural rubber tubes and other fiber-free scrap rubbers are reclaimed by means of the heater or pan process. Brass tube fittings and other metal are removed from the scrap. The scrap is mechanically ground, mixed with reclaiming agents, loaded into pans or devulcanizing boats, and autoclaved at steam pressures of 10–14 atm (1.03–1.40 MPa) for 3–8 h. The reclaim is finally processed much the same way as in the digester process.

The Reclaimator process is more attractive than the above processes. The Reclaimator is essentially a high-pressure extruder that devulcanizes fiber-free rubber continuously. Ground scrap is mechanically treated in hammer mills to remove the textile material, mixed with reclaiming oils and other materials, and then fed into the Reclaimator. High pressure and shear between the rubber mixture and the extruder barrel walls effectively reclaim the rubber mixture. Devulcanization occurs at 175–205°C in a few minutes and turns the rubber into reclaim that issues from the machine continuously. The whole regeneration, which is a dry process, may be completed in about 30 min.

The reclaiming oils and chemicals are complex wood and petroleum derivatives that swell the rubber and provide access for breaking the rubber bonds with heat, pressure, chemicals, and mechanical shearing. Approximately 2–4 parts of oil are used per 100 parts of scrap rubber. Some examples of reclaiming oils include monocyclic and mixed terpenes, i.e., pine-tar products, saturated polymerized petroleum hydrocarbons, aryl disulfides in petroleum oil, cycloparaffinic hydrocarbons, and alkyl aryl polyether alcohols.

Reclaimed rubber contains all the fillers present in the original scrap or waste rubber. It shows very good aging characteristics and is characterized by less heat development during mixing and processing as compared to fresh rubber. The use of reclaim in a fresh rubber mix is advantageous not on consideration of physical and mechanical properties but essentially for smooth processing and reduced cost.

Some Major Rubber Products

The most important application of rubbers is in the transport sector, with tires and related products consuming nearly 70% of the rubber produced. Next in importance is the application in belting for making flat conveyor and (power) transmission belts and V-belts (for power transmission), and in the hose industry for making different hoses. Rubber also has a large outlet in cellular and microcellular products. Other important and special applications of rubber are in the areas of adhesives, coated fabrics, rainwear, footwear, pipes and tubing, wire insulation, cables and sheaths, tank lining for chemical plants and oil storage, gaskets and diaphragms, rubber mats, rubber rollers, sports goods, toys and balloons, and a wide variety of molded mechanical and miscellaneous products. Formulations of a few selected rubber compounds are given in Appendix 4.

Tires

Tire technology is a very specialized area, and a tire designer is faced with the difficult task of trying to satisfy all the needs of the vehicle manufacturer, the prime factors of consideration, however, being safety and tread life.

Figure 2.53 shows the constructions of a standard bias (diagonal) ply tire and a radial ply tire. The major components of a tire are: bead, carcass, sidewall, and tread. In terms of material composition, a tire on an average contains nearly 50% of its weight in actual rubber; for oil extended rubbers (typically containing 25 parts of aromatic or cycloparaffinic oils to 75 parts of rubber), it is less. The remainder includes carbon black, textile cord, and other compounding ingredients plus the beads.

The *bead* is constructed from a number of turns or coils of high tensile steel wire coated with copper and brass to ensure good adhesion of the rubber coating applied on it. The beads function as rigid, practically inextensible units that retain the inflated tire on the rim.

The *carcass* forms the backbone of the tire. The main part of the carcass is the tire cord. The cords consist of textile threads twisted together. Rayon, nylon, and polyester cords are widely used. Steel cords are also used. While the practice of laying rubber-impregnated cords in position is still followed, the use of woven fabric is more widespread. To promote a good bond with rubber, the fiber or cord is treated with an adhesive composition such as water-soluble resorcinol-formaldehyde resin and aqueous emulsion of a copolymer of butadiene, styrene, and vinyl pyridine (70:15:15). The resin-coated cord or fabric is dried and coated with a rubber compound by calendering, whereby each cord is isolated from its neighbor. For conventional tires, the rubber-coated fabric is then cut to a predetermined width and bias angle. The bias-cut plies are joined end-to-end into a continuous length and batched into roll form, interleaved with a textile lining to prevent self-adhesion.

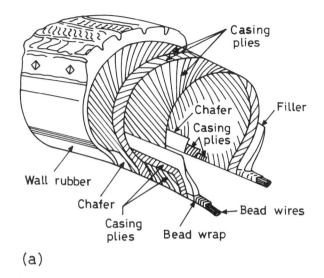

(a)

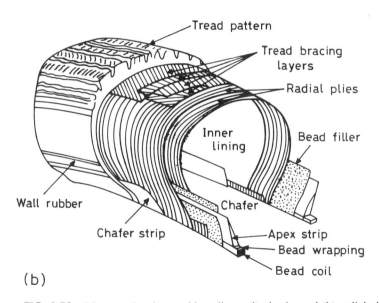

(b)

FIG. 2.53 Diagram showing (a) bias (diagonal) ply tire and (b) radial ply tire.

The *sidewall* is a layer of extruded rubber compound that protects the carcass framework from weathering and from damage from chafing. Together with the tread and overlapping it in the buttress region, the sidewall forms the outermost layer of the tire. It is the most highly strained tire component and is susceptible to two types of failure—flex cracking and ozone cracking.

The *tread* is also formed by extrusion, but different rubber compounds are used for sidewall and tread (see Appendix 4). When making the sidewall and the tread in two separate extrusion lines, it is useful to take the extruded sidewall to the second extruder, which would deliver the tread to the sidewall. To simplify the tire-building operation, the sidewall and the tread may also be produced as a single unit by simultaneous extrusion of the two compounds in a single band— the tread over the sidewall—with two extruders arranged head-to-head with a common double die. The band is rapidly cooled to avoid scorching and then cut to the appropriate length. The tread receives its characteristic pattern from the mold when the subsequently built tire is vulcanized.

While building the tire, a layer of specially compounded cushion rubber may be used to keep heat development on flexing to a minimum and achieve better adhesion between the tread and the carcass. One or more layers of fabric, known as breakers or bracing layers, may be placed below the cushion. The bracing layers raise the modulus of the tread zone and level out local blows to the tread as it contacts the road.

The tire building is carried out on a flat drum, which is rotated at a controlled speed. The plies of rubber-coated cord fabric are placed in position, one over the other, and rolled down as the drum rotates, the inner lining being placed next to the drum and the ends of the drum being flanged to suit the bead configuration of the tire. The plies of rubber-coated textile are assembled in three basic constructions—bias (diagonal), radial, and bias belted.

Bias tires have an even number of plies with cords at an angle of 30–38° from the tread center line. Passenger-car bias tires commonly have two or four plies, with six for heavy duty service. Truck tires are often built with six to twelve plies, although the larger earthmover types may contain thirty or more.

In the radial-ply tire, one or two plies are set at an angle of 90° from the center line and a *breaker* or *belt* of rubber-coated wire or textile is added under the tread. This construction gives a different tread-road interaction, resulting in a decreased rate of wear. The sidewall is thin and very flexible. The riding and steering qualities are noticeably different from those of a bias-ply tire and require different suspension systems.

The bias-belted tire, on the other hand, has much of the tread wear and traction advantage of the radial tire, but the shift from bias to bias-belted tires requires less radical change in vehicle suspension systems and in tire building machines. These features make the bias-belted tire attractive to both automobile and tire manufacturers.

When the tire building on the drum is complete, the drum is collapsed and the uncured ("green") tire is removed. The cylindrical shape of the uncured tire obtained from the building drum is transformed into a toroidal shape in the mold resulting in a circumferential stretch of the order of 60%. For shaping and curing, the uncured tire is pressed against the inner face of the heated mold by an internal bag or bladder made of a pre-cured heat-resistant rubber and inflated by a high-pressure steam or circulating hot water. Different types of molding presses are in use. In the Bag-O-Matic type of press, the uncured tire is placed over a special type of bag or bladder, and the operation of shaping, curing, and ejection of the cured tire are accomplished by an automatic sequence of machine operation.

Pneumatic tires require an air container or *inner tube* of rubber inside the tire. The manufacture of inner tubes is done essentially in three steps: extrusion, cutting into length, and insertion of valve and vulcanization. The ends of the cut length of the tube are joined after insertion of the valve prior to vulcanization.

Tubeless tires have, instead of an inner tube, an *inner liner*, which is a layer of rubber cured inside the casing to contain the air, and a *chafer* around the bead contoured to form an airtight seal with the rim.

Belting and Hoses

Uses of flat conveyor and (power) transmission belts and V-belts (for power transmission) are to be found in almost all major industries. V-belts of different types cover applications ranging from fan belts for automobiles, belts for low-power drives for domestic, laboratory, and light industrial applications, and high-power belts for large industrial applications.

Textile cords or fabric and even steel cords constitute an important part of all rubber belting and hoses. The various types of cords used in the tire industry are also in use in the belting industry.

Essential steps in making *conveyor belts* are: (1) drying the fabric; (2) frictioning of the hot fabric with a rubber compound and topping to give additional rubber between plies and the outer ply and cover, using a three- or four-roll calender; (3) belt building; and (4) vulcanization.

The process of belt building essentially consists of cutting, laying, and folding the frictioned and coated fabric to give the desired number of plies. The construction may be *straight-laid*, *folded-jacket*, or *graded-ply* types (Fig. 2.54), with the joints in neighboring plies being staggered to eliminate weakness and failure. Finally, the cover coat is applied by calendering.

Vulcanization of conveyor belts may be carried out in sections using press cure or continuously by means of a Rotocure equipment. In press cure, vulcanization is done by heating in long presses, the belt being moved between successive cures by a length less than the length of the press platens. Since in this

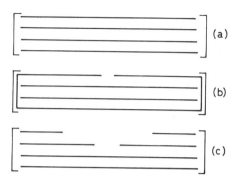

FIG. 2.54 Construction of conveyor belts: (a) straight laid, (b) folded jacket, and (c) graded ply.

process the end or overlap zones receive an additional cure, it is desirable to minimize damage or weakness due to overcure by using flat cure mixes and allowing water cooling at each end section of the platen. In each step of vulcanization, the section of the belt to be vulcanized is gripped and stretched hydraulically to minimize or eliminate elongation during use. The difficulties of press cure may, however, be avoided by adopting continuous vulcanization with a Rotocure equipment in which the actual curing operation is carried out between an internally steam heated cylinder and a heated steel band. Rotocure is also useful for the vulcanization of transmission belts and rubber sheeting.

The most common feature of V-belts is their having a cross-section of the shape of a regular trapezium with the unparallel sides at an angle of 40° (Fig. 2.55). The V-belts usually consist of five sections: (1) the top section known as the *tension section*, (2) the bottom section, called the *compression section*, (3) the cord section located at the neutral zone, (4) the *cushion section* on either side of the cord section, and (5) one or two layers of rubberized fabric, called the jacket section covering the whole assembly.

Three different rubber compounds are required for use in the above construction of a V-belt: (1) a base compound, which is the major constituent on a weight basis, (2) a soft and resilient cushion compound required for surrounding and protecting the reinforcing cord assembly, and (3) a friction compound used for rubberizing the fabric casing of the belt (see Appendix 4).

Relatively short length V-belts are built layer by layer on rotatable collapsible drum formers. The separate belts are then cut out with knives and transferred to a skiving machine that imparts the desired V-shape. The fabric jacket is then applied, and the belts are vulcanized in open steam using multi-cavity ring molds for smaller belts. Long belts are built similarly on V-groove sheave pul-

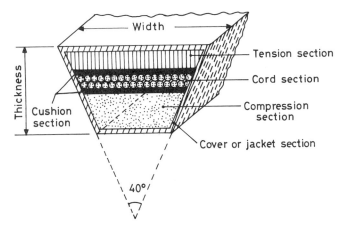

FIG. 2.55 Cross-section showing V-belt construction.

leys using weftless cord fabric in place of individually wound cord. They are vulcanized endlessly by molding in a hydraulic press under controlled tension.

A rubber *hose* has three concentric layers along the length. While the innermost part consisting of a rubber lining or tube is required to resist the action of the material that would pass through, the top layer is meant to play the role of a protective layer to resist weathering, oils, chemicals, abrasion, etc. Between the inner lining and the outer cover is given a layer of reinforcement of textile yarn or steel wire applied by spiralling, knitting, braiding, or circular loom weaving. A cut woven fabric wrapped straight or on the bias may also be used to reinforce the inner lining or tube.

Essentially, the process of hose building consists of extruding the lining or tube, braiding or spiralling the textile around the cooled tube, and applying an outer cover of rubber to the reinforced hose using a cross-head extruder. Several methods are employed for vulcanization. In one process, the built hose is passed through a lead press or lead extruder to give a layer of lead cover to the hose. The hose is then wound on a drum, filled with water or air, and the ends are sealed. The whole assembly is then heated to achieve vulcanization. The water or air inside expands, and the vulcanizing hose is pressed against the lead acting as the mold. On completion of cure, the sealed ends are cut open, the lead cover is removed by slitting lengthwise in a stripping machine, and the cured hose is coiled up.

Cellular Rubber Products

Cellular rubber may be described as an assembly of a multitude of cells distributed in a rubber matrix more or less uniformly. The cells may be interconnected (open cells) as in a sponge or separate (closed cells). Foam rubber made from a liquid starting material such as latex, described earlier, is of open-cell type. Cellular products made from solid rubber are commonly called sponge (open cell structure) and expanded rubber (closed cell structure).

The technology of making cellular products from solid rubber is solely dependent on the incorporation of a blowing agent, usually a gas such as nitrogen or a chemical blowing agent, into the rubber compound. The most widely used chemical blowing agent for this application is dinitroso pentamethylene tetramine (DNPT). The curing is carried out either freely using hot air or steam or in a mold that is only partially filled with the molding compound. Synthetic rubbers, particularly SBR, are preferred as they allow precise control over level of viscosity required for obtaining consistent product quality. The sponge and expanded rubber products include carpet backing, sheets, profiles, and moldings.

The development of microcellular rubber has brought a revolution in footwear technology. Microcellular rubber with an extremely fine noncommunicating cell structure and very comfortable wearing properties, is the lightest form of soling that can be produced. Density of soling as low as 0.3 g/cm^3 may be obtained with a high dose (8–10 phr) of DNPT at a curing temperature of 140–150°C. For common solings, the density normally varies between 0.5 and 0.8 g/cm^3.

High hardness and improved abrasion resistance along with low density, desired in microcellular soling, can be achieved by using SBR and high styrene resins with NR in right proportions. Higher proportions of high styrene resins give products of higher hardness and abrasion resistance and lower density. Silicious fillers such as precipitated silica and aluminum or calcium silicate also give higher hardness, abrasion resistance, and split tear strength. Microcellular crumbs can be used in considerable quantity along with china clay and whiting to reduce the product cost. Higher proportions of stearic acid (5–10 phr) are normally used in microcellular compounds in order to bring down the decomposition temperature of DNPT type blowing agents (see Appendix 4). Post-cure oven stabilization of the microcellular sheets, typically at 100°C for 4 h, reduces the delayed shrinkage after cure to a minimum.

MISCELLANEOUS PROCESSING TECHNIQUES
Coating Processes

A *coating* is a thin layer of material used to protect or decorate a substrate. Most often, a coating is intended to remain bonded to the surface permanently, although there are strippable coatings which are used only to afford temporary protection. An example of the latter type is the strippable hot-melt coating with ethyl cellulose as the binder [35], which is used to protect metal pieces such as drill bits or other tools and gears from corrosion and mechanical abrasion during shipping and handling.

Two of the principal methods of coating substrates with a polymer, namely, *extrusion coating* and *calendering* have already been dealt with in this chapter. Other methods of coating continuous webs include the use of dip, knife, brush, and spray. *Dip coating*, as applied to PVC, has already been described in a previous section on plastisols. In *knife coating* the coating is applied either by passing the web over a roll running partly immersed in the coating material or by running the coating material onto the face of the web while the thickness of the coating is controlled by a sharply profiled bar (or knife). This technique, also referred to as *spreading*, is used extensively for coating fabrics with PVC. The PVC is prepared in the form of a paste, and more than one layer is usually applied, each layer being gelated by means of heat before the next layer is added.

Lacquers are a class of coatings in which film formation results from mere evaporation of the solvent(s). The term "lacquer" usually connotes a solution of a polymer. Mixtures of solvents and *diluents* (extenders which may not be good solvents when used alone) are usually needed to achieve a proper balance of volatility and compatibility and a smooth coherent film or drying. Some familiar examples of lacquers are the spray cans of touch-up paint sold to the auto owner. These are mostly pigmented acrylic resins in solvents together with a very volatile solvent [usually dichlorodifluoromethane (CCl_2F_2)] which acts as a propellant. A typical lacquer formulation for coating steel surfaces contains polymer, pigment, plasticizer (nonvolatile solvent), and volatile solvents.

Latex paints or *emulsion paints* are another class of coatings which form films by loss of a liquid and deposition of a polymer layer. The paints are composed of two dispersions: (1) a resin dispersion, which is either a latex formed by emulsion polymerization or a resin in emulsion form, and (2) a dispersion of colorants, fillers, extenders, etc., obtained by milling the dry ingredients into water. The two dispersions are blended to produce an emulsion paint. Surfactants and protective colloids are added to stabilize the product. Emulsion paints are characterized by the fact that the binder (polymer) is in a water-dispersed form, whereas in a solvent paint it is in solution form. In emulsion systems the external water phase controls the viscosity, and the molecular weight of the polymer in the internal phase does not affect it, so polymers of high molecular weight are readily utilized in these systems. This is an advantage of emulsion paints. The minimum temperature at which the latex particles will coalesce to form a continuous layer depends mainly on the T_g. The T_g of a latex paint polymer is therefore adjusted by copolymerization or plasticization to a suitable range. The three principal polymer latexes used in emulsion paints are styrene-butadiene copolymer, poly(vinyl acetate), and acrylic resin.

Although the term "paint" has been used for latex-based systems as well as many others, traditionally it refers to one of the oldest coating systems known— that of a pigment combined with a drying oil, usually a solvent. Drying oils (e.g., linseed, tung), by virtue of their multiple unsaturation, behave like polyfunctional monomers which can polymerize ("dry") to produce film by a combination of oxidation and free-radical propagation. Oil-soluble metallic soaps are used to catalyze the oxidation process.

Combinations of resins with drying oils yield *oleoresinous varnishes*, whereas addition of a pigment to a varnish yields an *enamel*. The combination of hard, wear-resistant resin with softer, resilient, drying-oil films can be designed to give products with a wide range of durability, gloss, and hardness. Another route to obtaining a balanced combination of these properties is the *alkyd resin*, formed from *alcohols* and *acids* (and hence the "alkyd"). Alkyds are actually a type of polyester resin and are produced by direct fusion of glycerol, phthalic anhydride, and drying oil at 410 to 450°F (210 to 232°C). The process involves an esterification reaction of the alcohol and the anhydride and transesterification of the drying oil. A common mode of operation today is thus to start with the free fatty acids from the drying oil rather than with the triglycerides.

Fluidized Bed Coating

Fluidized bed coating is essentially an adaptation of dip coating and designed to be used with plastics in the form of a powder of fine particle size. It is applied for coating metallic objects with plastics. The uniqueness of the process lies in the fact that both thermoplastics and thermosetting resins can be used for the coating. Uniform coating of thicknesses form 0.005 to 0.080 in. (0.13 to 2.00 mm) can be built on many substrates such as aluminum, carbon steel, brass, and expanded metal. The coating is usually applied for electrical insulation and to enhance the corrosion resistance and chemical resistance of metallic parts. Polymers often used for coating are polyethylene, PVC, nylon, and cellulosics.

A metal object to be coated is heated to well above the fusion point of the resin and dipped into a bed of powdered resin fluidized by the passage of air through a porous plate. The bed is not heated; only the surface of the object to be coated is hot. A layer of powder adheres to the hot surface and melts to form a continuous coating. After removing from the bed, the object may be heated further to ensure the integrity of the coating. Products coated by this method include valve bodies used in chemical industries, electrical transformer casings, covers and laminations, steel pipe, appliance parts, and hardware.

Spray Coating

Spray coating is especially useful for articles that are too large for dip coating or fluidized bed coating. The process consists of blowing out fine polymer powders through a specially designed burner nozzle, which is usually flame heated by means of acetylene or some similar gas, or it can be heated electrically. Compressed air or oxygen is used as the propelling force for blowing the polymer powder.

Powder Molding of Thermoplastics

Static (Sinter) Molding

The process is often used with polyethylene and is limited to making open-ended containers. The mold which represents the exterior shape of the product is filled with powder. The filled mold is heated in an oven, causing the powder to melt and thus creating a wall of plastics on the inner surface of the mold. After a specific time to build the required wall thickness, the excess powder is dumped from the mold and the mold is returned to the oven to smooth the inner wall. The mold is then cooled, and the product is removed. The product is strain free, unlike pressure-molded products. In polyethylene this is especially significant if the product is used to contain oxidizing acids.

Rotational Molding

Rotational molding (popularly known as *rotomolding*) is best suited for large, hollow products, requiring complicated curves, uniform wall thickness, a good finish, and stress-free strength. It has been used for a variety of products such as car and truck body components (including an entire car body), industrial containers, telephone booths, portable outhouses, garbage cans, ice buckets, appliance housings, light globes, toys, and boat hulls. The process is applicable to most thermoplastics and has also been adapted for possible use with thermosets.

Essentially four steps are involved in rotational molding: loading, melting and shaping, cooling, and unloading. In the loading stage a predetermined weight of powdered plastic is charged into a hollow mold. The mold halves are closed, and the loaded mold is moved into a hot oven where it is caused to simultaneously rotate in two perpendicular planes. A 4:1 ratio of rotation speeds on minor and major axes is generally used for symmetrically shaped objects, but wide variability of ratios is necessary for objects having complicated configurations. On most units the heating is done by air or by a liquid of high specific heat. The temperature in the oven may range from 500 to 900°F (260 to 482°C), depending on the material and the product. As the molds continue to rotate in the oven, the polymer melts and forms a homogeneous layer of molten plastic, distributed evenly on the mold cavity walls through gravitational force (centrifugal force is not a factor). The molds are then moved, while still rotating, from the oven into the cooling chamber, where the molds and contents are cooled by forced cold air, water fog, or water spray. Finally, the molds are opened, and the parts removed. (Rotational molding of plastisol, described earlier, is similar to that described here. However, in this case the plastic is charged in the form of liquid dispersion, which gels in the cavity of the rotating mold during the heating cycle in the oven.)

The most common rotational molding machine in use today is the carousel-type machine. These are three-arm machines consisting of three cantilevered

Charging area

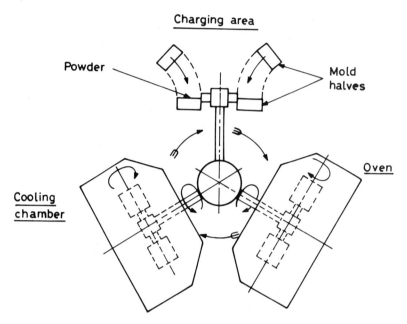

FIG. 2.56 Basics of continuous-type three-arm rotational molding machine.

arms or mold spindles extending from a rotating hub in the center of the unit. In operation, individual arms are simultaneously involved in different phases (loading, heating, and cooling) in three stations so that no arms are idle at any time (Fig. 2.56).

Modern rotational-molding machines enable large parts to be molded (e.g., 500-lb car bodies and 500-gal industrial containers). For producing small parts an arm of a carousel-type machine may hold as many as 96 cavities.

Centrifugal Casting

Centrifugal casting is generally used for making large tube forms. It consists of rotating a heated tube mold which is charged uniformly with powdered thermoplastic along its length. When a tubular molten layer of the desired thickness builds up on the mold surface, the heat source is removed and the mold is cooled. The mold, however, continues to rotate during cooling, thus maintaining uniform wall thickness of the tube. Upon completion of the cooling cycle, the plastic tube, which has shrunk away from the mold surface, is removed, and the process is repeated. Usual tube sizes molded by the process range from 6 to 30 in. in diameter and up to 96 in. in length.

Welding

Often it is necessary to join two or more components of plastics to produce a particular setup or to repair a broken part. For some thermoplastics solvent welding is applicable. The process uses solvents which dissolve the plastic to provide molecular interlocking and then evaporate. Normally it requires close-fitting joints. The more common method of joining plastics, however, is to use heat, with or without pressure. Various heat welding processes are available. Those processes in common commercial use are described here.

Hot-Gas Welding

Hot-gas welding, which bears a superficial resemblance to welding of metals with an oxyacetylene flame, is particularly useful for joining thermoplastic sheets in the fabrication of chemical plant items, such as tanks and ducting. The sheets to be joined are cleaned, beveled, and placed side by side so that the two beveled edges form a V-shaped channel. The tip of a filler rod (of the same plastic) is placed in the channel, and both it and the adjacent area of the sheets are heated with a hot-gas stream (200–400°C) directed from an electrically heated hot-gas nozzle (Fig. 2.57a), which melts the plastics. The plastics then fuse and unite the two sheets. The hot gas may be air in PVC welding, but for polyethylene an inert gas such as nitrogen must be used to prevent oxidation of the plastics during welding.

Fusion Welding

Fusion or hot-tool welding is accomplished with an electrically heated hot plate or a heated tool (usually of metal), which is used to bring the two plastic surfaces to be joined to the required temperature. The *polyfusion process* for joining plastic pipes by means of injection-molded couplings is an example of this type

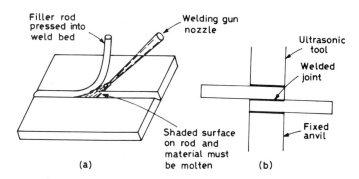

FIG. 2.57 Welding of plastics: (a) hot gas welding; (b) ultrasonic contact welding.

of welding. The tool for this process is so shaped that one side of it fits over the pipe while the other side fits into the coupling. The tool is heated and used to soften the outside wall of the pipe and the inside wall of the coupling. The pipe and coupling are firmly pressed together and held until the joint cools to achieve the maximum strength of the weld. The tool is chrome plated to prevent the plastic sticking to its surfaces.

Friction Welding

In friction or spin-welding of thermoplastics one of the two pieces to be joined is fixed in the chuck of a modified lathe and rotated at high speed while the other piece is held against it until frictional heat causes the polymer to flow. The chuck is stopped, and the two pieces are allowed to cool under pressure. The process is limited to objects having a circular configuration. Typical examples are dual-colored knobs, molded hemispheres, and injection-molded bottle halves.

High-Frequency Welding

Dielectric or high-frequency welding can be used for joining those thermoplastics which have high dielectric-loss characteristics, including cellulose acetate, ABS, and PVC. Obviously, polyethylene, polypropylene, and polystyrene cannot be welded by this method. The device used for high-frequency welding is essentially a radio transmitter operated at frequencies between 27 and 40 MHz. The energy obtained from the transmitter is directed to electrodes of the welding apparatus. The high-frequency field causes the molecules in the plastic to vibrate and rub against each other very fast, which creates frictional heat sufficient to melt the interfaces and produce a weld.

Ultrasonic Welding

In ultrasonic welding the molecules of the plastic to be welded are sufficiently disturbed by the application of ultrahigh-frequency mechanical energy to create frictional heat, thereby causing the plastics to melt and join quickly and firmly. The machinery for ultrasonic welding consists of an electronic device which generates electrical energy at 20/50 kHz/sec and a transducer (either magnetostrictive or piezoelectric) to convert the electrical energy to mechanical energy. In the contact-welding method (Fig. 2.57b) the ultrasonic force from the transducer is transmitted to the objects (to be welded) through a tool or "horn," generally made of titanium. The amplitude of the motion of the horn is from 0.0005 to 0.005 in. (0.013 to 0.13 mm) depending on the design. The method is generally used for welding thin or less rigid thermoplastics, such as films, or sheets of polyethylene, plasticized PVC, and others having low stiffness.

DECORATION OF PLASTICS

Plastics products can be decorated in various ways, which include, more commonly, screen printing, hot stamping, heat-transfer printing, in-mold decoration, painting, molded-in lettering, two-color molding, electroplating, and vacuum metallizing. Most of these processes require bonding other media, such as inks, enamels, and other materials to the plastics to be decorated. Some plastics, notably polyolefins and acetals, are, however, highly resistant to bonding and need separate treatment to activate the surface. Commonly used treatment processes are flame treatment, electronic treatments such as corona discharge and plasma discharge, and chemical treatment.

In *flame treatment*, plastic objects such as bottles, film, etc., are passed through an oxidizing gas flame. Momentary contact with the flame causes oxidation of the surface, which makes it receptive to material used in decorating the product.

In the *corona discharge process* the plastic film to be treated is allowed to pass over an insulated metal drum beneath conductors charged with a high voltage. When the electron discharge ("corona") between the charged conductors and the drum strikes the intervening film surface, oxidation occurs and makes the surface receptive to coatings. Molded products are also treated in a similar manner, often by fully automatic machinery.

In the *plasma process*, air at low pressure is passed through an electric discharge where it is partially dissociated into the plasma state and then expanded into a closed vacuum chamber containing the plastic object to be treated. The plasma reacting with the surfaces of the plastic alters their physicochemical characteristics in a manner that affords excellent adhesion to surface coatings. The process can be used for batch processing of plastics products, including films which may be unreeled in the vacuum chamber for treatment.

Acetal resin products are surface treated by a chemical process consisting of subjecting the product to a short acid dip that results in an etched surface receptive to paint.

Screen Process Printing

Screen process printing (see also Chapter 5) requires making a stenciled screen of the image to be printed on the plastic surface. The printing consists of forcing ink or paint through the interstices of the screen with a squeegee.

The screen is usually made of nylon or other synthetic material. Stainless steel or other metallic screens are also used. An art copy positive of the image is light-exposed to a photosensitive stencil film secured to the screen. The clear or unexposed areas harden, and the exposed areas wash out on subsequent treatment in a developer bath. Open and closed areas corresponding to the art copy are thus obtained on the screen. After fixing in a hardener and drying, the screen becomes ready for use.

Polyolefins such as polyethylene and polypropylene must be surface treated before being printed. The most effective way is in an integrated machine where surface treatment takes place right before printing. A time lapse will mean that the treatment will lose some of its effect. Three methods are used: flame treatment, corona discharge, and chemical treatment. Flame treatment is considered the most practical and most widely used.

Screen printing is performed manually for small lots at the rate of 200 to 600 pieces per hour. Large-volume products, such as detergent bottles, are printed by automatic screening at rates approaching 2000 to 3000 impressions per hour, depending on the nature of the product. Printing of multicolor impressions by this process requires the use of a screen for each color and adequate drying time for each color before the next color is applied. Split screens enable simultaneous printing of several colors. However, split screens can only be used if colors are not mixed in the design.

Hot Stamping

Hot stamping is one of the original methods of decorating plastics materials. The process uses a leaf or foil, which is usually an acetate or cellophane carrier film containing a thermoplastic color coat. When a heated die presses the foil against the part to be decorated, the color coating is released and adheres to the part. The operation may be manual, semiautomatic, or fully automatic.

Heat-Transfer Printing

Heat-transfer printing is an important method of single- and multi-color decoration of films, molded products, and blown containers. The process consists of transferring the ink image from a carrier stock onto the receiving surface by heat and pressure. The carrier stock consists of a paper support, a fusible release coat, and a thermoplastic ink image. The ink becomes tacky when heated by a hot platen. When it is then pressed against the receiving plastic surface, the ink image bonds to the surface. The ink also carries with it some of the fusible release coat, which thus provides the image with a glossy protective coating. In actual practice the ink transfer is effected by means of a heated rubber roll to insure a full line contact of the ink image with the variable product surface. The rubber roll is specially designed to suit the product configuration, whether round, concave, convex, or oval.

In-Mold Decoration

In-mold decoration is obtained as an integral part of the product. It therefore produces one of the most durable and permanent decorations. High-quality melamine dinnerware is decorated by this method, and so are a host of other household and hardware plastics goods.

Thermosetting plastics are decorated with a two-stage process. For melamine products, for example, the mold is loaded with the molding powder in the usual manner and closed. It is opened after a partial cure, and the decorative "foil" or overlay is placed in position. The mold is then closed again, and the curing cycle is completed. The overlay consists of a cellulose sheet having printed decoration and covered with a thin layer of partially cured clear melamine resin. During the molding cycle the overlay is fused to the product and becomes a part of the molding. The process is relatively inexpensive, especially when a multicolor decoration is required.

For in-mold decoration of thermoplastic products, single-stage process is used. The foil or overlay is thus placed in the mold cavity prior to the injection of the polymer. It is held in place in the mold by its inherent static charge. Shifting is prevented during molding by inducing an additional charge by passing the wand of an electronic static charging unit over the foil after it is properly positioned. The overlay, in all cases, is a printed or decorated film (0.003–0.005 in. thick) of the same polymer. Thus, polystyrene film is used for a polystyrene product, and polypropylene film for a polypropylene product. A similar procedure may also be used for decorating blow-molded products.

Painting

Many plastic products are decorated by painting (e.g., television cabinets, electric-iron handles, toys). Although all plastics can be decorated by painting, they require special consideration of the resin-solvent system of the paint to achieve adhesion, adequate covering, and chemical resistance. Paint is applied by manual or automatic spraying and dipping. Three-dimensional effects can be achieved by placing multicolor painting intaglio designs on the undersurface of clear plastic products.

Molded-In Lettering

Molded letters, figures, and decorated designs are often used on molded parts to identify and to decorate the product. Letters may be raised or depressed, or they may face up from a depressed panel. Raised lettering is by far the cheaper because it requires engraved letters on the mold surface which can be cut or stamped into the cavity of the mold. Depressed lettering, on the other hand, requires raised mold letters to create the depressed letters on the product. Such molds are more expensive to make. If mold cavities are hobbed, the hob is engraved with the depressed letters so that the hob forms raised letters in the mold cavity. The molded part thus becomes a replica of the hob and contains depressed letters. The depressed letters are filled with paint, and the surface is wiped, which produces an attractive appearance.

Raised or depressed lettering usually does not exceed 0.03 in. (0.76 mm). For lettering which must be applied to the parting line or side walls, it is desirable to provide sufficient draft for the letters to "draw" out the mold.

Molded-in lettering offers unusual decorative possibilities for transparent thermoplastics materials. Thus lettering may be cut in on the underside of the product, and contrasting colors may be used for the background and lettering to produce an attractive three-dimensional effect.

Two-Color Molding

Two-color molding is an injection-molding process where two colors are successively molded. The most common examples of such two-color or double-shot molded products are typewriter and business machine keys. The shell is first molded with one color and then transferred to another cavity where the second color is injected.

Electroplating

Since plastics are nonconductors of electricity, electroplating requires that the surface be properly conditioned and sensitized. The surface of the plastic part after conditioning should be hydrophilic (readily wetted by water). The hydrophilic part is sensitized by absorbing on the surface a material (such as acidified stannous chloride) which is readily oxidized. The oxidation of the sensitizer in the subsequent nucleation step (typically, treatment with an acidified palladium chloride solution) serves to deposit a catalytic film on the plastic surface. When the catalytic surface is next introduced into an electroless plating bath, the reaction between the metal salt and the reducing agent, both present in the bath, takes place only on this surface and deposits a layer of the metal. Detailed information on plating procedures are usually obtained from the resin manufacturers and suppliers to the plating industry.

Electroplating of plastic products provides the high-quality appearance and wear resistance of metal combined with the light weight and corrosion resistance of plastics. Physical properties such as tensile strength, flexural strength, heat-deflection temperature, etc., are also enhanced in the plated plastics.

Plating is done on many plastics, including phenolic, urea, ABS, acetal, and polycarbonate. Many automotive, appliance, and hardware uses of plated plastics include knobs, instrument cluster panels, bezel, speaker grilles, and nameplates. In marine searchlights zinc has been replaced by chrome-plated ABS plastics to gain lighter weight, greater corrosion resistance, and lower cost. An advantage of plastics plating is that, unlike metal die castings which require buffing in most cases after plating, plastics do not ordinarily require this extra expensive operation. The use of plated plastics also affords the possibility of obtaining attractive texture contrasts.

Vacuum Metallizing

Vacuum metallizing is a process whereby a bright thin film of metal is deposited on the surface of a molded product or film under a high vacuum. The metal may be gold, silver, or most generally, aluminum. The process produces a somewhat delicate surface compared to electroplating.

Small clips of the metal to be deposited are attached to a filament. When the filament is heated electrically, the clips melt and coat the filament. An increased supply of electrical energy then causes vaporization of this metal coating, and plating of the plastic product takes place. To minimize surface defects and enhance the adhesion of the metal coating, manufacturers initially give the plastics parts a lacquer base coat and dry in an oven. The lacquered parts are secured to a rack fitted with filaments, to which are fastened clips of metal to be vaporized. The vaporization and deposition are accomplished at high vacuum (about 1/2 micron). The axles supporting the part holding the fixtures are moved so as to rotate the parts during the plating cycle to promote uniform deposition. The thickness of the coating produced is about 5×10^{-6} in. (127 nm). After the deposition is completed, the parts are removed and dipped or sprayed with top-coat lacquer to protect the metal film from abrasion. Color tones, such as gold, copper, brass, etc., may be added to this coating if desired.

Vacuum metallizing of polymer films, such as cellulose acetate, butyrate, and Mylar, is performed in essentially the same way. Film rolls are unreeled and rewound during the deposition process to metallize the desired surface. A protective abrasion-resistant coating is then applied to the metallized surface in an automatic coating machine.

REFERENCES

1. *Modern Plastics Encyclopedia*, McGraw-Hill, New York, Vols. 42–45 (1964–1967).
2. N. M. Bikales (ed.), *Molding of Plastics*, Interscience, New York (1971).
3. J. A. Butler, *Compression and Transfer Molding*, Plastics Institute Monograph, Iliffe, London (1964).
4. E. W. Vaill, *Mod. Plastics, 40* (1A, Encycl. Issue): 767 (Sept. 1962).
5. A. L. Maiocco, Transfer molding, past, present and future, *SPE Tech. Papers, 10*: XIV–4 (1964).
6. J. Frados, *Plastics Engineering Handbook*, SPJ, Van Nostrand Reinhold, New York (1976).
7. I. I. Rubin, *Injection Molding of Plastics*, John Wiley, New York (1973).

8. J. Brown, *Injection Molding of Plastic Components*, McGraw-Hill, New York (1979).
9. J. B. Dym, *Injection Molds and Molding*, Van Nostrand Reinhold, New York (1979).
10. *Petrothene: A Processing Guide*, 3rd ed., U.S. Industrial Chemicals Co., New York (1965).
11. L. J. Zukov, Injection molding of thermosets, *SPE J.*, *13* (10): 1057 (1963).
12. Y. Morita, Screw injection molding of thermosets, *SPE Tech. Papers, 12*: XIV–5 (1966).
13. J. C. O'Brien, Business of injection-molding thermosets, *Plast. Eng., 32*(2), 23(Feb. 1976).
14. G. Schenkel, *Plastics Extrusion Technology and Theory*, American Elsevier, New York (1966).
15. N. M. Bikales, *Extrusion and Other Plastics Operations*, Interscience, New York (1971).
16. E. G. Fisher, *Extrusion of Plastics*, Newnes-Butterworth, London (1976).
17. R. T. Van Ness, G. R. De Hoff, and R. M. Bonner, *Mod. Plastics, 45* (14A, Encyl. Issue); 672 (Oct. 1968).
18. E. G. Fisher, *Blow Molding of Plastics*, Iliffe, London (1971).
19. R. A. Elden and A. D. Swan, *Calendering of Plastics*, Plastics Institute Monograph, Iliffe, London (1971).
20. H. F. Mark, S. M. Atlas, and E. Cernia (eds.), *Man-made Fibers: Science and Technology*, Interscience, New York, 3 Vols. (1967–1968).
21. R. W. Moncrieff, *Man-Made Fibers*, John Wiley, New York (1963).
22. J. L. Riley, *Polymer Processes* (C. E. Schildknecht, ed.), Interscience, New York, Chap. XVIII (1956).
23. R. L. Butzko, *Plastics Sheet Forming*, Van Nostrand Reinhold, New York (1958).
24. H. A. Sarvetnick (ed.), *Plastisols and Organosols*, Van Nostrand Reinhold, New York (1972).
25. G. Lubin (ed.), *Handbook of Composites*, Van Nostrand Reinhold, New York (1982).
26. W. J. Lee, J. C. Seferis, and D. C. Bonner, Prepreg processing science, *SAMPE Quarterly, 17*(2), 58(Jan. 1986).
27. L. N. Phillips and D. B. V. Parker, *Polyurethanes-Chemistry, Technology and Properties*, Iliffe, London, 1964.
28. *One-step Urethane Foams*, Bull. F40487, Union Carbide Corp. (1959).
29. H. J. Stern, *Rubber: Natural and Synthetic*, Palmerton, New York (1967).
30. H. R. Lasman, *Mod. Plastics, 45* (1A, Encycl. Issue): 368 (Sept. 1967).
31. P. Schidrowitz and T. R. Dawson (eds.), *History of the Rubber Industry*, IRI, London (1952).
32. W. Hoffman, *Vulcanization and Vulcanizing Agents*, Maclaren, London (1967).
33. A. Nourry (ed.), *Reclaimed Rubber, Its Developments, Applications and Future*, Maclaren, London (1962).
34. P. Ghosh, *Polymer Science and Technology of Plastics and Rubbers*, Tata McGraw-Hill, New Delhi (1990).
35. D. H. Parker, *Principles of Surface Coating Technology*, John Wiley, New York (1965).

3

Plastics Properties and Testing

INTRODUCTION

There are two stages in the process of becoming familiar with plastics. The first is rather general and involves an introduction to the unique molecular structures of polymers, their physical states, and transitions which have marked influence on their behavior. These have been dealt with in Chapter 1. The second stage, which will be treated in this chapter, is more specific in that it involves a study of the specific properties of plastics which dictate their applications. Besides the relative ease of molding and fabrication, many plastics offer a range of important advantages in terms of high strength/weight ratio, toughness, corrosion and abrasion resistance, low friction, and excellent electrical resistance. These qualities have made plastics acceptable as materials for a wide variety of engineering applications. It is important therefore that an engineer be aware of the performance characteristics and significant properties of plastics.

In this chapter plastics have been generally dealt with in respect to broad categories of properties, namely, mechanical, electrical, thermal, and optical. In this treatment the most characteristic features of plastic materials have been highlighted.

An important facet of materials development and proper materials selection is testing and standardization. The latter part of this chapter is therefore devoted to this aspect. It presents schematically (in simplified form) a number of standard test methods for plastics, highlighting the principles of the tests and the properties measured by them.

MECHANICAL PROPERTIES [1–4] *see page 778*

Several unfamiliar aspects of material behavior of plastics need to be appreciated, the most important probably being that, in contrast to most metals at room temperature, the properties of plastics are time dependent. Then superimposed on this aspect are the effects of the level of stress, the temperature of the material, and its structure (such as molecular weight, molecular orientation, and density). For example, with polypropylene an increase in temperature from 20 to 60°C may typically cause a 50% decrease in the allowable design stress. In addition, for each 0.001 g/cm^3 change in density of this material there is a corresponding 4% change in design stress. The material, moreover, will have enhanced strength in the direction of molecular alignment (that is, in the direction of flow in the mold) and less in the transverse direction. Because of the influence of so many additional factors on the behavior of plastics, properties (such as modulus) quoted as a single value will be applicable only for the conditions at which they are measured. Properties measured as single values following standard test procedures are therefore useful only as a means of quality control. They would be useless as far as design is concerned, because to design a plastic component it is necessary to have complete information, at the relevant service temperature, on the time-dependent behavior (viscoelastic behavior) of the material over the whole range of stresses to be experienced by the component.

Stress-Strain Behavior

The stress-strain behavior of plastics measured at a constant rate of loading provides a basis for quality control and comparative evaluation of various plastics. The diagram shown in Figure 3.1a is most typical of that obtained in tension for a constant rate of loading. For compression and shear the behavior is quite similar except that the magnitude and the extent to which the curve is followed are different.

In the diagram, load per unit cross section (*stress*) is plotted against deformation expressed as a fraction of the original dimension (*strain*). Even for different materials the nature of the curves will be similar, but they will differ in (1) the numerical values obtained and (2) how far the course of the typical curve is followed before failure occurs. Cellulose acetate and many other thermoplastics may follow the typical curve for almost its entire course. Thermosets like phenolics, on the other hand, have cross-linked molecules, and only a limited amount of intermolecular slippage can occur. As a result, they undergo fracture at low strains, and the stress-strain curve is followed no further than to some point below the knee, such as point 1.

Ultimate strength, elongation, and elastic modulus (Young's modulus) can be obtained from the stress-strain study (Fig. 3.1a). For determining *Young's modulus* (E) the slope of the initial tangent, i.e., the steepest portion of the curve, is measured. Other moduli values are also used for plastics (see Fig. 3.1b).

The appearance of a permanent set is said to mark a *yield point*, which indicates the upper limit of usefulness for any material. Unlike some metals, in particular, the ferrous alloys, the *drop-of-beam* effect and a sharp knee in the stress-strain diagram are not exhibited by plastics. An arbitrary yield point is usually assigned to them. Typical of these arbitrary values is the 0.2% or the 1%

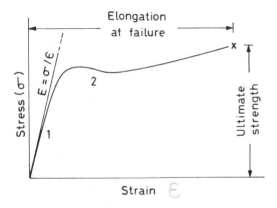

(a)

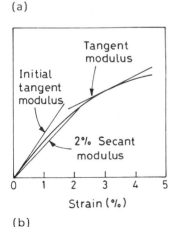

(b)

FIG. 3.1 (a) Nominal stress-strain diagram. (b) Typical moduli values quoted for plastics.

offset yield stress (Fig. 3.2a). Alternatively, a yield stress can be defined as that at which the ratio of total stress to total strain is some selected amount, say 50% or 70% of the elastic modulus (Fig. 3.2b). In the first case the yield stress is conveniently located graphically by offsetting to the right the stated amount of 0.2% (or 1%) and drawing a line paralleling that drawn for the elastic modulus. The point at which this line intersects the observed stress-strain line defines the yield stress. In the second case also the point of intersection of the line drawn with a slope of 0.7E, for instance, with the observed stress-strain line determines the yield stress.

Up to point 1 in Figure 3.1a, the material behaves as an elastic solid, and the deformation is recoverable. This deformation, which is small, is associated with the bending or stretching of the inter-atomic bonds between atoms of the polymer molecules (Fig. 3.3a). This type of deformation is nearly instantaneous and recoverable. There is no permanent displacement of the molecules relative to each other.

Between points 1 and 2 in Figure 3.1a, deformations have been associated with a straightening out of a kinked or coiled portion of the molecular chains, if loaded in tension. (For compression the reverse is true.) This can occur without intermolecular slippage. The deformation is recoverable ultimately but not instantaneously and hence is analogous to that of a nonlinear spring. Although the deformation occurs at stresses exceeding the stress at the proportional limit, there is no permanent change in intermolecular arrangement. This kind of deformation, characterized by recoverability and nonlinearity, is very pronounced in the rubber state. The greatest extension that is recoverable marks the elastic limit for the material. Beyond this point extensions occur by displacement of molecules with respect to each other (Fig. 3.3c), as in Newtonian flow of a

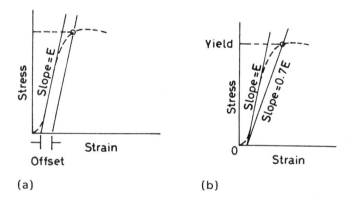

FIG. 3.2 Location of a yield value.

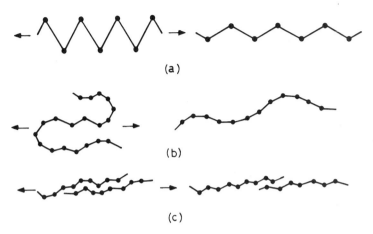

FIG. 3.3 Deformation in plastics. (a) Stretching of polymer molecule. (b) Straightening out of a coiled molecular chain. (c) Intermolecular slippage.

liquid. The displaced molecules have no tendency to slip back to their original positions, therefore these deformations are permanent and not recoverable.

Poisson's ratio is a measure of the reduction in the cross section accompanying stretching and is the ratio of the transverse strain (a contraction for tensile stress) to longitudinal strain (elongation). Poisson's ratio for many of the more brittle plastics such as polystyrene, the acrylics, and the thermoset materials is about 0.3; for the more flexible plasticized materials, such as cellulose acetate, the value is somewhat higher, about 0.45. Poisson's ratio for rubber is 0.5 (characteristic of a liquid); it decreases to 0.4 for vulcanized rubber and to about 0.3 for ebonite. Poisson's ratio varies not only with the nature of the material but also with the magnitude of the strain for a given material. All values cited here are for zero strain.

Strain energy per unit volume is represented as the area under the stress-strain curve. It is another property that measures the ability of a material to withstand rough treatment and is related to toughness of the material. The stress-strain diagram thus serves as a basis for classification of plastics. *Strong* materials have higher ultimate strength than *weak* materials. *Hard* or unyielding materials have a higher modulus of elasticity (steeper initial slope) than *soft* materials. *Tough* materials have high elongations with large strain energy per unit volume. Stress-strain curves for type cases are shown in Figure 3.4.

It must be emphasized that the type behavior shown in Figure 3.4 depends not only on the material but also very definitely on conditions under which the test is made. For example, the bouncing putty silicone is puttylike under slow rates of loading (type curve a) but behaves as an elastic solid under high rates of

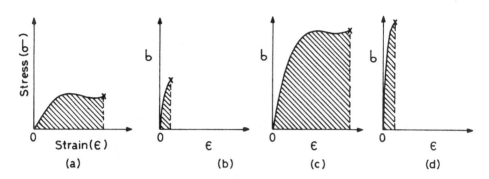

FIG. 3.4 Classification of plastics on the basis of stress-strain diagram. (a) Soft and weak. (b) Weak and brittle. (c) Strong and tough. (d) Hard and strong.

impact (type curve b or d). Figure 3.5a shows that at high extension rates (>1 mm/sec) unplasticized PVC is almost brittle with a relatively high modulus and strength. At low extension rates (<0.05 mm/sec) the same material exhibits a lower modulus and a high ductility, because at low extension rates the polymer molecular chains have time to align themselves under the influence of the applied load. Thus the material is able to flow at the same rate as it is being strained. (The interesting phenomenon observed in some plastics is known as *cold drawing*.) Further examples of the effect of conditions on the behavior of plastics are illustrated by the stress-strain curves for plasticized cellulose acetate when determined at different temperatures (Fig. 3.5b). Thus the material is hard and strong at low temperatures, relatively tough at ordinary temperatures, and

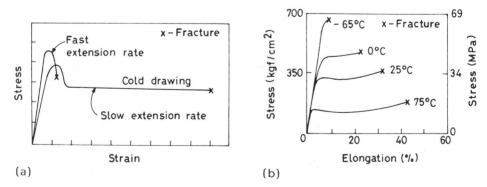

FIG. 3.5 (a) Typical tensile behavior of unplasticized PVC. (b) Stress-strain curves of cellulose acetate at different temperatures.

soft and weak at higher temperatures. This behavior may be attributed to variable molecular slippage effects associated with plasticizer action.

Viscoelastic Behavior of Plastics

In a perfectly elastic (Hookean) material the stress, σ, is directly proportional to the strain, ϵ. For uniaxial stress and strain the relationship may be written as

$$\sigma = \text{constant} \times \epsilon \tag{1}$$

where the constant is referred to as the *modulus of elasticity*.

In a perfectly viscous (Newtonian) liquid the shear stress, τ, is directly proportional to the rate of strain, $\dot{\gamma}$, and the relationship may be written as

$$\tau = \text{constant} \times \dot{\gamma} \tag{2}$$

where the constant is referred to as the *viscosity*.

Polymeric materials exhibit stress-strain behavior which falls somewhere between these two ideal cases; hence, they are termed *viscoelastic*. In a viscoelastic material the stress is a function of both strain and time and so may be described by an equation of the form

$$\sigma = f(\epsilon, t) \tag{3}$$

This equation represents *nonlinear viscoelastic* behavior. For simplicity of analysis it is often reduced to the form

$$\sigma = \epsilon f(t) \tag{4}$$

which represents *linear viscoelasticity*. It means that in a tensile test on linear viscoelastic material, for a fixed value of elapsed time, the stress will be directly proportional to strain.

The most characteristic features of viscoelastic materials are that they exhibit time-dependent deformation or strain when subjected to a constant stress (*creep*) and a time-dependent stress when subjected to a constant strain (*relaxation*). Viscoelastic materials also have the ability to recover when the applied stress is removed. To a first approximation, this *recovery* can be considered as a reversal of creep.

Creep Behavior

Except for a few exceptions like lead, metals generally exhibit creep at higher temperatures. Plastics and rubbers, however, possess very temperature-sensitive creep behavior; they exhibit significant creep even at room temperature. In creep tests a constant load or stress is applied to the material, and the variation of deformation or strain with time is recorded. A typical creep curve plotted from such a creep test is shown in Figure 3.6a. The figure shows that there is

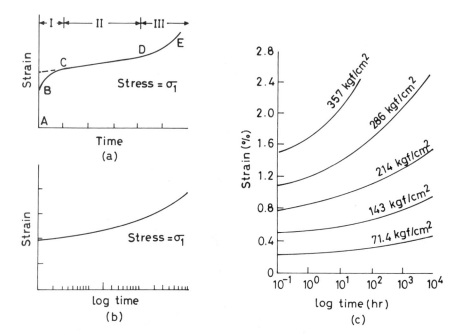

FIG. 3.6 Typical creep curve (a) with linear time scale and (b) with logarithmic time scale. (c) Family of creep curves for poly(methyl methacrylate) at 20°C (1 kgf/cm^2 = 0.098 MPa).

typically an almost instantaneous elastic strain AB followed by a time-dependent strain, which occurs in three stages: *primary* or *transient creep* BC (stage 1), *secondary* or *steady-state creep* CD (stage II), and *tertiary* or *accelerated creep* DE (stage III).

The primary creep has a rapidly decreasing strain rate. It is essentially similar in mechanism to retarded elasticity and, as such, is recoverable if the stress is removed. The secondary or steady-state creep is essentially viscous in character and is therefore nonrecoverable. The strain rate during this state is commonly referred to as the *creep rate*. It determines the useful life of the material. Tertiary creep occurs at an accelerated rate because of an increase in the true stress due to necking of the specimen.

Normally a logarithmic time scale is used to plot the creep curve, as shown in Figure 3.6b, so that the time dependence of strain after long periods can be included. If a material is linearly viscoelastic [Eq. (4)], then at any selected time each line in a family of creep curves (with equally spaced stress levels) should be offset along the strain axis by the same amount. Although this type of behavior may be observed for plastics at low strains and short times, in most cases the behavior is nonlinear, as indicated in Figure 3.6c.

Plastics generally exhibit high rates of creep under relatively low stresses and temperatures, which limits their use for structural purposes. Creep behavior varies widely from one polymer to another; thermoset polymers, in general, are much more creep resistant than thermoplastic polymers. The steady-state creep in plastics and rubbers (often referred to as *cold flow*) is due to viscous flow, and increases continuously. Clearly, the material cannot continue to get larger indefinitely, and eventually fracture will occur. This behavior is referred to as *creep rupture. Creep strength* of these materials is defined as the maximum stress which may be applied for a specified time without causing fracture. The creep strength of plastics is considerably increased by adding fillers and other reinforcing materials, such as glass fibers and glass cloth, since they reduce the rate of flow.

The creep and recovery of plastics can be simulated by an appropriate combination of elementary mechanical models for ideal elastic and ideal viscous deformations. Although there are no discrete molecular structures which behave like individual elements of the models, they nevertheless aid in understanding the responses of plastics materials.

Maxwell Model

The Maxwell model consists of a spring and dashpot connected in series (Fig. 3.7a). When a load is applied, the elastic displacement of the spring occurs immediately and is followed by the viscous flow of liquid in the dashpot which requires time. After the load is removed, the elastic displacement is recovered immediately, but the viscous displacement is not recovered.

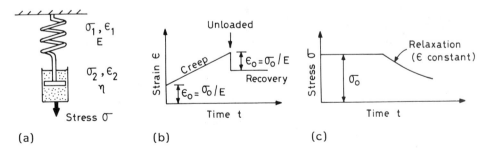

FIG. 3.7 (a) The Maxwell model. (b) and (c) Responses of the model under time-dependent modes of deformation.

Stress-Strain Relation

The spring is the elastic component of the response and obeys the relation

$$\sigma_1 = E \epsilon_1 \tag{5}$$

σ_1 and ϵ_1 are the stress and strain, respectively, and E is a constant.

The dashpot (consisting of a piston loosely fitting in a cylindrical vessel containing a liquid) accounts for the viscous component of the response. In this case the stress σ_2, is proportional to the rate of strain $\dot{\epsilon}_2$; i.e.,

$$\sigma_2 = \eta \dot{\epsilon}_2 \tag{6}$$

where η is a material constant called the *coefficient of viscous traction*.

The total strain, ϵ, of the model under a given stress, σ, is distributed between the spring and the dashpot elements:

$$\epsilon = \epsilon_1 + \epsilon_2 \tag{7}$$

From Eq. (7) the rate of total displacement with time is

$$\dot{\epsilon} = \dot{\epsilon}_1 + \dot{\epsilon}_2 \tag{8}$$

and from Eqs. (5) to (7),

$$\dot{\epsilon} = \frac{1}{E} \dot{\sigma}_1 + \frac{1}{\eta} \sigma_2 \tag{9}$$

But both elements are subjected to the entire stress, σ,

$$\sigma = \sigma_1 = \sigma_2 \tag{10}$$

Therefore Eq. (9) can be written as

$$\dot{\epsilon} = \frac{1}{E} \dot{\sigma} + \frac{1}{\eta} \sigma \tag{11}$$

which is the governing equation of the Maxwell model. It is interesting to consider the responses that this model predicts under three common time-dependent modes of deformation.

(i) *Creep.* If a constant stress σ_O is applied, then Eq. (11) becomes

$$\dot{\epsilon} = \frac{1}{\eta} \sigma_O \tag{12}$$

which indicates a constant rate of increase of strain with time—i.e., steady-state creep (Fig. 3.7a).

(ii)*Relaxation.* If the strain is held constant, then Eq. (11) becomes

$$0 = \frac{1}{\epsilon} \dot{\sigma} + \frac{1}{\eta} \sigma$$

Solving this differential equation with the initial condition $\sigma = \sigma_0$ at $t = t_0$ gives

$$\sigma = \sigma_0 \exp\left(\frac{-E}{\eta}\right)t \tag{13}$$

This indicates an exponential decay of stress with time (Fig. 3.7c).

(iii) *Recovery*. When the initial stress, σ_0, is removed, there is an instantaneous recovery of the elastic strain, ϵ_0, and then, as shown by Eq. (11), the strain rate is zero, so there is no further recovery (Fig. 3.7b).

This model therefore gives an acceptable first approximation to the relaxation behavior of an actual material, but it is inadequate for predicting creep and recovery behavior.

Kelvin or Voigt Model

In the Kelvin or Voigt model the spring and dashpot elements are connected in parallel, as shown in Figure 3.8a. This model roughly approximates the behavior of rubber. When the load is applied at zero time, the elastic deformation cannot occur immediately because the rate of flow is limited by the dashpot. Displacements continue until the strain equals the elastic deformation of the spring and it resists further movement. On removal of the load the spring recovers the displacement by reversing the flow through the dashpot, and ultimately there is no strain. The mathematical relations are derived next.

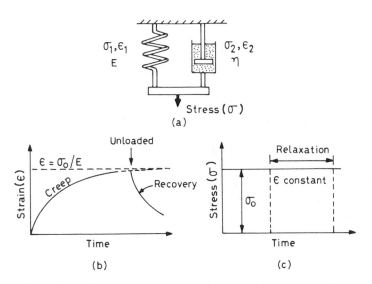

FIG. 3.8 (a) The Kelvin or Voigt model. (b) and (c) Responses of the model under time-dependent modes of deformation.

Stress-Strain Relation

Since the two elements are connected in parallel, the total stress will be distributed between them, but any deformation will take place equally and simultaneously in both of them; that is,

$$\sigma = \sigma_1 + \sigma_2 \tag{14}$$

$$\epsilon = \epsilon_1 + \epsilon_2 \tag{15}$$

From Eqs. (5), (6), and (14),

$$\sigma = E\epsilon_1 + \eta\dot{\epsilon}_2$$

or, using Eq. (15),

$$\sigma = E\epsilon + \eta\dot{\epsilon} \tag{16}$$

which is the governing equation for the Kelvin (or Voigt) model. Its predictions for the common time-dependent deformations are derived next.

(i) *Creep*. If a constant stress, σ_0, is applied, Eq. (16) becomes

$$\sigma_0 = E\epsilon + \eta\dot{\epsilon} \tag{17}$$

The differential equation may be solved for the total strain, ϵ, to give

$$\epsilon = \frac{\sigma_0}{E}\left[1 - \exp\left(\frac{-E}{\eta}\right)t\right] \tag{18}$$

This equation indicates that the deformation does not appear instantaneously on application of stress, but it increases gradually, attaining asymptotically its maximum value $\epsilon = \sigma_0/E$ at infinite time (Fig. 3.8b).

(ii) *Relaxation*. If the strain is held constant, then Eq. (16) becomes

$$\sigma = E\epsilon$$

Therefore, the stress is constant, and the predicted response is that of an elastic material—that is, no relaxation.

(iii) *Recovery*. If the stress is removed, Eq. (16) becomes

$$O = E\epsilon + \eta\dot{\epsilon}$$

This differential equation may be solved with the initial condition $\epsilon = \epsilon_0$ at the time of stress removal to give

$$\epsilon = \epsilon_0 \exp\left(\frac{-E}{\eta}\right)t \tag{19}$$

This equation represents an exponential recovery of strain which, as a comparison with Eq. (18) shows, is a reversal of the predicted creep.

The Kelvin (or Voigt) model therefore gives an acceptable first approximation to creep and recovery behavior but does not predict relaxation. By comparison, the previous model (Maxwell model) could account for relaxation but was poor in relation to creep and recovery. It is evident therefore that a better simulation of viscoelastic materials may be achieved by combining the two models.

Four-Element Model

A combination of the previous two models comprising four elements is shown in Figure 3.9a. The total strain is

$$\epsilon = \epsilon_1 + \epsilon_2 + \epsilon_k \tag{20}$$

where ϵ_k is the strain response of the Kelvin model. From Eqs. (5), (6), and (18),

$$\epsilon = \frac{\sigma_0}{E_1} + \frac{\sigma_0 t}{\eta_1} + \frac{\sigma_0}{E_2}\left[1 - \exp\left(\frac{-E_2}{\eta_2} \right)t \right] \tag{21}$$

Thus the strain rate is

$$\dot{\epsilon} = \frac{\sigma_0}{\eta_1} + \frac{\sigma_0}{\eta_2}\exp\left(-\frac{E_2}{\eta_2} \right)t \tag{22}$$

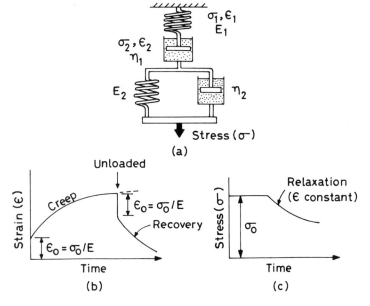

FIG. 3.9 (a) Four-element model. (b) and (c) Responses of the model under time-dependent modes of deformation.

The response of this model to creep, relaxation, and recovery situations is thus the sum of the effects described previously for the Maxwell and Kelvin models and is illustrated in Figure 3.9b. Though the model is not a true representation of the complex viscoelastic response of polymeric materials, it is nonetheless an acceptable approximation to the actual behavior. The simulation becomes better as more and more elements are added to the model, but the mathematics also becomes more complex.

Zener Model

Another model, attributed to Zener, consists of three elements connected in series and parallel, as illustrated in Figure 3.10, and known as the *standard linear solid*. Following the procedure already given, we derive the governing equation of this model:

$$\eta_3 \dot{\sigma} + E_1 \sigma = \eta_3 (E_1 + E_2)\dot{\epsilon} + E_1 E_2 \epsilon \tag{23}$$

This equation may be written in the form

$$a_1 \dot{\sigma} + a_0 \sigma = b_1 \dot{\epsilon} + b_0 \epsilon \tag{24}$$

where a_1, a_0, b_1, and b_0 are all material constants. A more general form of Eq. (24) is

$$a_n \frac{\partial^n \sigma}{\partial t^n} + a_{n-1} \frac{\partial^{n-1} \sigma}{\partial t^{n-1}} + \cdots + a_0 \sigma = b_m \frac{\partial^m \epsilon}{\partial t^m} + \cdots + b_0 \epsilon \tag{25}$$

The modern theory of viscoelasticity favors this type of equation. The models described earlier are special cases of this equation.

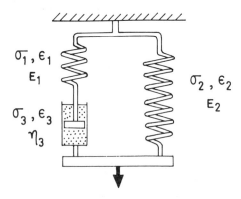

FIG. 3.10 The standard linear solid.

Superposition Principle

Each of the creep curves in Figure 3.6c depicts the strain response of a material under a constant stress. However, in service, materials are often subjected to a complex sequence of stresses or stress histories, and obviously it is not practical to obtain experimental creep data for all combinations of loading. In such cases a theoretical model can be very useful for describing the response of a material to a given loading pattern. The most commonly used model is the *Boltzmann superposition principle*, which proposes that for a linear viscoelastic material the entire loading history contributes to the strain response, and the latter is simply given by the algebraic sum of the strains due to each step in the load. The principle may be expressed as follows. If an equation for the strain is obtained as a function of time under a constant stress, then the modulus as a function of time may be expressed as

$$E(t) = \frac{\sigma}{\epsilon(t)} \tag{26}$$

Thus if the applied stress is σ_0 at zero time, the creep strain at any time, t, will be given by

$$\epsilon(t) = \frac{1}{E(t)} \sigma_0 \tag{27}$$

On the other hand, if the stress, σ_0, was applied at zero time and an additional stress, σ_1, at time u, the Boltzmann superposition principle says that the total strain at time t is the algebraic sum of two independent responses; that is,

$$\epsilon(t) = \frac{1}{E(t)} \sigma_0 + \frac{1}{E(t-u)} \sigma_1 \tag{28}$$

For any series of stress increments this equation can be generalized to

$$\epsilon(t) = \sum_{u=-\infty}^{u=t} \sigma_i \frac{1}{E(t-u)} \tag{29}$$

The lower limit of the summation is taken as $-\infty$ since the entire stress history contributes to the response.

As an illustration, for a series of step changes in stress as in Figure 3.11a, the strain response predicted by the model is shown schematically in Figure 3.11b. The time-dependent strain response (creep curve) due to the stress σ_0 applied at zero time is predicted by Eq. (26) with $\sigma = \sigma_0$. When a second stress, σ_1, is added to σ_0, the new curve will be obtained, as illustrated in Figure 3.11b, by adding the creep due to σ_1 to the anticipated creep due to σ_0. Removal of all stress at a subsequent time u_2 is then equivalent to removing the creep stain due

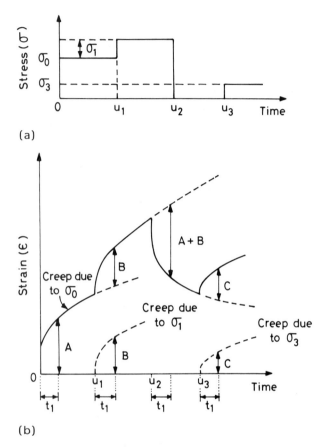

FIG. 3.11 (a) Stress history. (b) Predicted strain response using Boltzmann's superposition principle.

to σ_0 and σ_1, independently, as shown in Figure 3.11b. The procedure is repeated in a similar way for other stress changes.

To take into account a continuous loading cycle, we can further generalize Eq. (29) to

$$\epsilon(t) = \int_{-\infty}^{t} \frac{1}{E(t-u)} \frac{d\sigma(u)}{du} \, du \tag{30}$$

In the same way the stress response to a complex strain history may be derived as

$$\sigma(t) = \int_{-\infty}^{t} E(t - u) \frac{d\epsilon(u)}{du} \, du \tag{31}$$

When the stress history has been defined mathematically, substitution in Eq. (30) and integration within limits gives the strain at the given time. The stress at a given time is similarly obtained from Eq. (31).

Isometric and Isochronous Curves

Isometric curves are obtained by plotting stress vs. time for a constant strain; *isochronous curves* are obtained by plotting stress vs. strain for a constant time of loading. These curves may be obtained from the creep curves by taking a constant-strain section and a constant-time section, respectively, through the creep curves and replotting the data, as shown in Figure 3.12.

An isometric curve provides an indication of the relaxation of stress in the material when the strain is kept constant. Since stress relaxation is a less com-

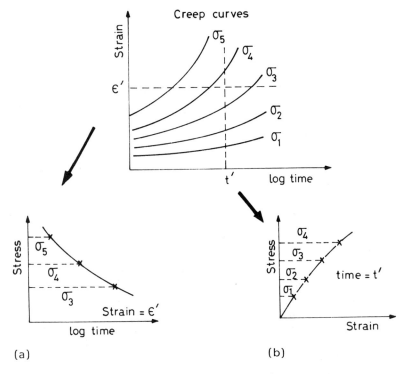

FIG. 3.12 (a)Isometric and (b) isochronous curves from creep curves.

mon experimental procedure than creep testing, an isometric curve, derived like the preceding curves from creep curves, is often used as a good approximation of this property.

Isochronous curves, on the other hand, are more advantageously obtained by direct experiments because they are less time consuming and require less specimen preparation than creep testing. The experiments actually involve a series of mini creep and recovery tests on the material. Thus a stress is applied to a specimen of the material, and the strain is recorded after a time *t* (typically 100 sec). The stress is then removed, and the material is allowed to recover, usually for a time 4*t*. A higher stress is then applied to the same specimen, and the strain is recorded after time *t*; the stress is then removed, and the material is allowed to recover. This procedure is repeated until there are sufficient points to plot the isochronous curve.

Note that the isochronous test method is quite similar to that of a conventional incremental loading tensile test and differs only in that the presence of creep is recognized and the "memory" of the material for its stress history is overcome by the recovery periods. Isochronous data are often presented on log-log scales because this provides a more precise indication of the nonlinearity of the data by yielding a straight-line plot of slope less than unity.

Pseudoelastic Design Method

Due to the viscoelastic nature of plastics, deformations depend on such factors as the time under load and the temperature. Therefore the classical equations available for the design of structural components, such as springs, beams, plates, and cylinders, and derived under the assumptions that (1) the modulus is constant and (2) the strains are small and independent of loading rate or history and are immediately reversible, cannot be used indiscriminately. For example, classical equations are derived using the relation

Stress = constant × strain

where the constant is the modulus. From the nature of the creep curves shown in Figure 3.12a, it is clear that the modulus of a plastic is not constant. Several approaches have been developed to allow for this fact, and some of them also give very accurate results; but mathematically they are quite complex, and this has limited their use. However, one method that has been widely accepted is the *pseudoelastic design method*. In this method appropriate values are chosen for the time-dependent properties, such as modulus, and substituted into the classical equations. The method has been found to give sufficiently accurate results, provided that the value of the modulus is chosen judiciously, taking into account the service life of the component and the limiting strain of the plastic. Unfortunately, however, there is no straightforward method for finding the limiting strain of a plastic. The value may differ for various plastics and even for the same plastic in different applications. The value is often arbitrarily chosen, although several methods have been suggested for arriving at an appropriate value. One method is to draw a secant modulus which is 0.85 of the initial tangent modulus and to note the strain at which this intersects the stress-strain curve (see Fig. 3.1b). But this method may be too restrictive for many plastics, particularly those which are highly crystalline. In most situations the maximum allowable strain is therefore decided in consultations between designer and product manufacturer.

Once an appropriate value for the maximum strain is chosen, design methods based on creep curves and the classical equations are quite straightforward, as shown in the following examples.

Example 1 A plastic beam, 200 mm long and simply supported at each end, is subjected to a point load of 10 kg at its mid-span. If the width of the beam is 14 mm, calculate a suitable depth so that the central deflection does not exceed 5 mm in a service life of 20,000 hr. The creep curves for the material at the service temperature of 20°C are shown in Figure 3.13a. The maximum permissible strain in this material is assumed to be 1%.

Answer: The linear elastic equation for the central deflection, δ, of the beam is

$$\delta = \frac{PL^3}{48EI}$$

where

P = load at mid-span
L = length of beam
E = modulus of beam material
I = second moment of area of beam cross section

The second moment of area is

$$I = \frac{bd^3}{12} = \frac{14d^3}{12} \text{ mm}^4$$

So from the expression for δ,

$$d^3 = \frac{PL^3}{56E\delta}$$

The only unknown on the right side is E. For plastic this is time dependent, but a suitable value corresponding to the maximum permissible strain may be

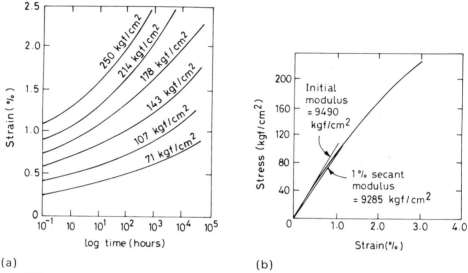

(a)   (b)

FIG. 3.13 (a) Creep curves for material used in illustrative examples. (b) Isochronous curve at 20,000 hr service life (1 kgf/cm² = 0.098 MPa).

obtained by referring to the creep curves in Figure 3.13a. A constant-time section across these curves at 20,000 hr gives the isochronous curve shown in Figure 3.13b. Since the maximum strain is recommended as 1%, a secant modulus may be taken at this value. It is 9285 kgf/cm² (= 92.85 kgf/mm²). Using this value in the above equation gives

$$d^3 = \frac{10(200)^3}{56 \times 92.85 \times 5}$$

$$d = 14.5 \text{ mm}$$

Example 2 A thin-wall plastic pipe of diameter 150 mm is subjected to an internal pressure of 8 kgf/cm² at 20°C. It is suggested that the service life of the pipe should be 20,000 hr with a maximum strain of 2%. The creep curves for the plastic material are shown in Figure 3.13a. Calculate a suitable wall thickness for the pipe.

Answer: The hoop stress, σ, in a thin-wall pipe of diameter d and thickness h, subjected to an internal pressure, P, is given by

$$\sigma = \frac{Pd}{2h} \quad \text{so} \quad h = \frac{Pd}{2\sigma}$$

A suitable design stress may be obtained from the creep curves in Figure 3.13a. By referring to the 20,000-hr isochronous curve (Fig. 3.13b) derived from these curves, the design stress at 2% strain is obtained as 167.7 kg/cm^2. (Note that a similar result could have been obtained by plotting a 2% isometric curve from the creep curves and reading the design stress at a service life of 20,000 hr.) Substituting the design stress into the equation for h gives

$$h = \frac{8 \times 150}{2 \times 167.7} = 3.58 \text{ mm}$$

It may be seen from the creep curves (Fig. 3.13a) that when the pipe is first pressurized, the strain is less than 1%. Then as the material creeps, the strain increases steadily to reach its limit of 2% at 20,000 hr.

In both examples it has been assumed that the service temperature is 20°C. If this is not the case, then creep curves at the appropriate temperature should be used. However, if none are available, a linear extrapolation between available temperatures may be sufficient for most purposes. Again, for some materials like nylon the moisture content of the material has a significant effect on its creep behavior. In such a case creep curves are normally available for the material in both wet and dry states, and appropriate data should be used, depending on the service conditions.

Effect of Temperature

Many attempts have been made to obtain mathematical expressions which describe the time and temperature dependence of the strength of plastics. Since for many plastics at constant temperature a plot of stress, σ, against the logarithm of time to failure (creep rupture), t, is approximately linear, one of the expressions most commonly used is

$$t = Ae^{-B\sigma} \tag{32}$$

where A and B are constants. In reality, however, they depend on factors such as material structure and on temperature.

The most successful attempts to include the effects of temperature in a relatively simple expression have been made by Zhurkov and Bueche, who used an equation of the form [5]

$$t = t_0 \exp\left(\frac{U_0^{-\gamma\sigma}}{RT}\right) \tag{33}$$

where t_0 is a constant which has approximately the same value for most plastics, U_0 is the activation energy of the fracture process, γ is a coefficient which depends on the structure of the material, R is the molar gas constant, and T is the absolute temperature.

A series of creep rupture tests on a given material at a fixed temperature would permit the values for U_0 and γ for the material to be determined from this expression. The times to failure at other stresses and temperatures could then be predicted.

The relative effects of temperature rises on different plastic materials depend on the structure of each material and, particularly, whether it is crystalline or amorphous. If a plastic is largely amorphous (e.g., polymethyl methacrylate, polystyrene), then it is the glass transition temperature (T_g) which will determine the maximum service temperature, since above T_g the material passes into the rubbery region (see Fig. 1.11). On the other hand, in plastics which have a high degree of crystallinity (e.g., polyethylene, polypropylene), the amorphous regions are small, so T_g is only of secondary importance. For them it is the melting temperature which will limit the maximum service temperature. The lowest service temperatures which can be used are normally limited by the brittleness introduced into the material. The behavior of plastics materials at room temperature is related to their respective T_g values. This aspect has been dealt with in Chapter 1.

Plastics Fractures

The principal causes of fracture of a plastic part are the prolonged action of a steady stress (*creep rupture*), the application of a stress in a very short period of time (*impact*), and the continuous application of a cyclically varying stress (*fatigue*). In all cases the process of failure will be accelerated if the plastic is in an aggressive environment.

Two basic types of fracture under mechanical stresses are recognized: *brittle fracture* and *ductile fracture*. These terms refer to the type of deformation that precedes fracture. Brittle fractures re potentially more dangerous because there occurs no observable deformation of the material. In a ductile failure, on the other hand, large nonrecoverable deformations occur before rupture actually takes place and serve as a valuable warning. A material thus absorbs more energy when it fractures in a ductile fashion than in a brittle fashion. In polymeric materials fracture may be ductile or brittle, depending on several variables, the most important of which are the straining rate, the stress system, and the temperature. Both types of failures may thus be observed in the one material, depending on the service conditions.

Impact Behavior of Plastics

Tests of brittleness make use of impact tests. The main causes of brittle failure in materials have been found to be (1) triaxiality of stress, (2) high strain rates, and (3) low temperatures. Test methods developed for determining the impact behavior of materials thus involve striking a notched bar with a pendulum. This is the most convenient way of subjecting the material to triaxiality of stress (at the notch tip) and a high strain rate so as to promote brittle failures.

The standard test methods are the *Charpy* and *Izod* tests, which employ the pendulum principle (Fig. 3.14a). The test procedures are illustrated in Figure 3.14b and 3.14c. The specimen has a standard notch machined in it and is placed with the notch on the tension side. In the Charpy test the specimen is supported as a simple beam and is loaded at the midpoint (Fig. 3.14b). In the Izod test it is supported as a cantilever and is loaded at its end (Fig. 3.14c). The standard energy absorbed in breaking the specimen is recorded.

The results of impact tests are often scattered, even with the most careful test procedure. A normal practice in such cases is to quote the median strength rather than the average strength, because the median is more representative of the bulk of the results if there is a wide scatter with a few very high or very low results. Impact strengths are normally expressed as

$$\text{Impact} = \frac{\text{Energy absorbed to break}}{\text{Area at notch section}}$$

$$(\text{ft} - \text{lbf/in}^2, \text{ cm} - \text{kgf/cm}^2, \text{ or } \text{J/m}^2)$$

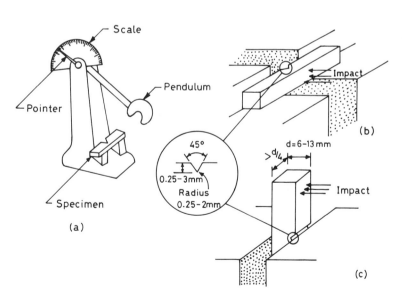

FIG. 3.14 Impact test. (a) Schematic diagram of Charpy impact testing machine. (b) Arrangement of Charpy impact specimen. (c) Mounting of Izod impact specimen.

Occasionally, the less satisfactory term of energy to break per unit width may be quoted in units of ft − lbf/in, cm − kgf/cm or J/m.

The choice of notch depth and tip radius will affect the impact strength observed. A sharp notch is usually taken as a 0.25-mm radius, a blunt notch as a 2-mm radius. The typical variation of impact strength with notch-tip variation for several thermoplastics is presented in Figure 3.15. It is evident that the use of a sharp notch may even rank plastic materials in an order different from that obtained by using a blunt notch. This fact may be explained by considering the total energy absorbed to break the specimen as consisting of energy necessary for crack initiation and for crack propagation. When the sharp notch (0.25-mm radius) is used, it may be assumed that the energy necessary to initiate the crack is small, and the main contribution to the impact strength is the propagation energy. On this basis Figure 3.15 would suggest that high-density polyethylene and ABS have relatively high crack-propagation energies, whereas materials such as PVC, nylon, polystyrene, and acrylics have low values. The large improvement

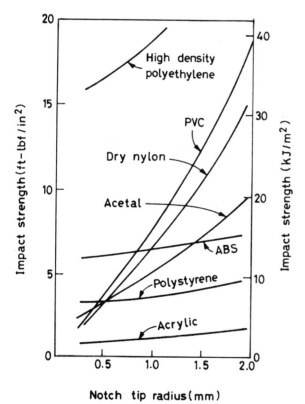

FIG. 3.15 Variation of impact strength with notch radius for several thermoplastics.

in impact strength observed for PVC and nylon when a blunt notch is used would imply that their crack-initiation energies are high. On the other hand, the smaller improvement in the impact strength of ABS with a blunt notch would suggest that the crack-initiation energy is low. Thus the benefit derived from using rounded corners would be much less for ABS than for materials such as nylon or PVC.

Temperature has a pronounced effect on the impact strength of plastics. In common with metals, many plastic materials exhibit a transition from ductile behavior to brittle as the temperature is reduced. The variation of impact strength with temperature for several common thermoplastics is shown in Figure 3.16. The ranking of the materials with regard to impact strength is seen to be influenced by the test temperature. Thus, at room temperature (approximately 20°C) polypropylene is superior to acetal; at subzero temperatures (e.g., −20°C) polypropylene does not perform as well as acetal. This comparison pertains to

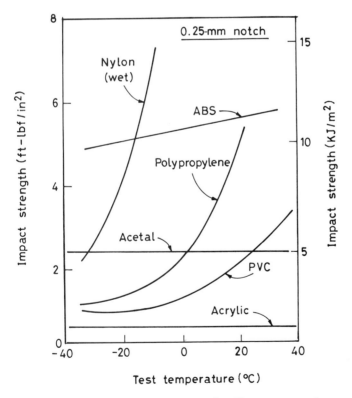

FIG. 3.16 Variation of impact strength with temperature for several thermoplastics with sharp notch.

impact behavior measured with a sharp (0.25-mm) notch. Note that notch sharpness can influence the impact strength variation with temperature quite significantly. Figure 3.17 shows that when a blunt (2-mm) notch is used, there is indeed very little difference between acetal and polypropylene at 20°C, whereas at −20°C acetal is much superior to polypropylene.

It may be seen from Figures 3.16 and 3.17 that some plastics undergo a change from ductile or tough (high impact strength) to brittle (low impact strength) behavior over a relatively narrow temperature change. This allows a temperature for ductile-brittle transition to be cited. In other plastic materials this transition is much more gradual, so it is not possible to cite a single value for transition temperature. It is common to quote in such cases a brittleness temperature, $T_B(\frac{1}{4})$. This temperature is defined as the value at which the impact strength of the material with a sharp notch (¼-mm tip radius) is $10 kJ/m^2$ (4.7 ft-lbf/in²). When quoted, it provides an indication of the temperature above which there should be no problem of brittle failure. However, it does not mean that a material should never be used below its $T_B(\frac{1}{4})$, because this temperature, by definition, refers only to the impact behavior with a sharp notch. When the

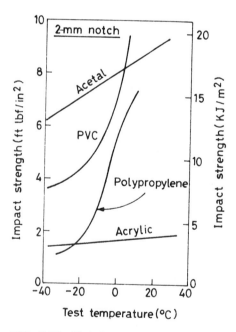

FIG. 3.17 Variation of impact strength with temperature for several thermoplastics with blunt notch.

material is unnotched or has a blunt notch, it may still have satisfactory impact behavior well below $T_B(\frac{1}{4})$.

Other environmental factors besides temperature may also affect impact behavior. For example, if the material is in the vicinity of a fluid which attacks it, then the crack-initiation energies may be reduced, resulting in lower impact strength. Some materials, particularly nylon, are significantly affected by water, as illustrated in Figure 3.18. The absorption of water produces a spectacular improvement in the impact behavior of nylon.

Note that the method of making the plastic sample and the test specimen can have significant effect on the measured values of the properties of the material. Test specimens may be molded directly or machined from samples which have been compression molded, injection molded, or extruded. Each processing method involves a range of variables, such as melt temperature, mold or die temperature, and shear rate, which influence the properties of the material. Fabrication defects can affect impact behavior: for example, internal voids, inclusion, and additives, such as pigments, which can produce stress concentrations within the material. The surface finish of the specimen may also affect impact behavior. All of this accounts for the large variation usually observed in the results of testing one material processed and/or fabricated in different ways. It also emphasizes the point that if design data are needed for a particular application, then the test specimen must match as closely as possible the component to be designed.

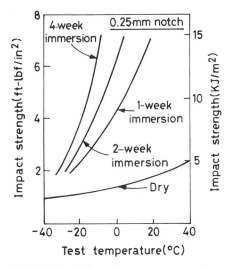

FIG. 3.18 Effect of water content on impact strength of nylon.

TABLE 3.1 Impact Behavior of Common Thermoplastics Over a Range of Temperatures

Plastic material	Temperature (°C)							
	−20	−10	0	10	20	30	40	50
Polyethylene (low density)	A	A	A	A	A	A	A	A
Polyethylene (high density)	B	B	B	B	B	B	B	B
Polypropylene	C	C	C	C	B	B	B	B
Polystyrene	C	C	C	C	C	C	C	C
Poly(methyl methacrylate)	C	C	C	C	C	C	C	C
ABS	B	B	B	B	B	B	A	A
Acetal	B	B	B	B	B	B	B	B
Teflon	B	A	A	A	A	A	A	A
PVC (rigid)	B	B	B	B	B	B	A	A
Polycarbonate	B	B	B	B	A	A	A	A
Poly(phenylene oxide)	B	B	B	B	B	B	A	A
Poly(ethylene terephthalate)	B	B	B	B	B	B	B	B
Nylon (dry)	B	B	B	B	B	B	B	B
Nylon (wet)	B	B	B	A	A	A	A	A
Glass-filled nylon (dry)	C	C	C	C	C	C	C	B
Polysulfone	B	B	B	B	B	B	B	B

A = Tough (specimens do not break completely even when sharply notched); B = notch brittle; C = brittle even when unnotched.
Source: Data from Reference 6. *page 336*

In some applications impact properties of plastics may not be critical, and only a general knowledge of their impact behavior is needed. In these circumstances the information provided in Table 3.1 would be adequate. The table lists the impact behavior of a number of commonly used thermoplastics over a range of temperatures in three broad categories [6].

Fatigue of Plastics

A material subject to alternating stresses over long periods may fracture at stresses much below its maximum strength under static loading (tensile strength)

due to the phenomenon called *fatigue*. Fatigue has been recognized as one of the major causes of fracture in metals. Although plastics are susceptible to a wider range of failure mechanisms, it is likely that fatigue still plays an important part in plastics failure.

For metals the fatigue process is generally well understood and is divided into three stages: *crack initiation*, *crack growth*, and *fracture*. Repeated cyclic stressing causes highly localized yielding and strain hardening. The gradual reduction of ductility in the strain-hardened areas finally results in the initiation of submicroscopic cracks. Cracks may also be initiated under relatively low stresses at surface imperfections, such as those from machining marks, fabrication flaws, and surface damage. The cyclic action of the load causes the crack to grow until it is so large that the remainder of the cross section cannot support the load and a break suddenly occurs. The fatigue failure is thus of a brittle type and particularly serious, because there is no visual warning that a failure is imminent. The fatigue theory of metals is well developed, but the fatigue theory of polymers is not. The completely different molecular structure of polymers means that there is unlikely to be a similar type of crack initiation process as in metals, though it is possible that once a crack is initiated the subsequent phase of propagation and failure may be similar.

Fatigue cracks may develop in plastics in several ways. If the plastic article has been machined, surface flaws capable of propagation may be introduced. However, if the article has been molded, it is more probable that fatigue cracks will develop from within the bulk of the material. In a crystalline polymer the initiation of cracks capable of propagation may occur through slip of molecules. In addition to acting as a path for crack propagation, the boundaries of spherulites (see Chapter 1), being areas of weakness, may thus develop cracks during straining. In amorphous polymers cracks may develop at the voids formed during viscous flow.

A number of features are peculiar to plastics, which make their fatigue behavior a complex subject not simply analyzed. Included are viscoelastic behavior, inherent damping, and low thermal conductivity. Consider, for example, a sample of plastic subjected to a cyclic stress of fixed amplitude. Because of the high damping and low thermal conductivity of the material, some of the input energy will be dissipated in each cycle and appear as heat. The temperature of the material will therefore rise, and eventually a stage will be reached when the heat transfer to the surroundings equals the heat generation in the material. The temperature of the material will stabilize at this point until a conventional metal-type fatigue failure occurs.

If, in the next test, the stress amplitude is increased to a higher value, the material temperature will rise further and stabilize, followed again by a metal-type fatigue failure. In Figure 3.19, where the stress amplitude has been plotted against the logarithm of the number of cycles to failure, failures of this type have been labeled as *fatigue failures*. This pattern will be repeated at higher stress amplitudes until a point is reached when the temperature rise no longer stabilizes but continues to rise, resulting in a short-term thermal softening failure in the material. At stress amplitudes above this crossover point there will be thermal failures in an even shorter time. Failures of this type have been labeled as *thermal failures* in Figure 3.19. The fatigue curves in Figure 3.19 thus have two distinct regimes—one for the long-term conventional fatigue failures, and one for the relatively short-term thermal softening failures.

The frequency of the cyclic stress would be expected to have a pronounced effect on the fatigue behavior of plastics, a lower frequency promoting the conventional-type fatigue failure rather than thermal softening failure. Thus it is evident from Figure 3.19 that if the frequency of cycling is reduced, then stress amplitudes which would have produced thermal softening failures at a higher frequency may now result in temperature stabilization and eventually fatigue failure. Therefore, if fatigue failures are required at relatively high stresses, the frequency of cycling must be reduced. Normally, fatigue failures at one frequency on the extrapolated curve fall from the fatigue failures at the previous frequency (Fig. 3.19). As the cyclic stress amplitude is further reduced in some

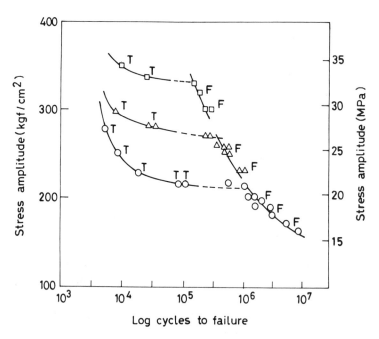

FIG. 3.19 Typical fatigue behavior of a thermoplastic at several frequencies. F, fatigue failure; T, thermal failure. ○, 5.0 Hz; △, 1.67 Hz; ⊓, 0.5 Hz. (Adapted from Ref. 4.)

plastics, the frequency remaining constant, the fatigue failure curve becomes almost horizontal at large values of the number of stress cycles (*N*). The stress amplitude at which this leveling off occurs is clearly important for design purposes and is known as the *fatigue limit*. For plastics in which fatigue failure continues to occur even at relatively low stress amplitudes, it is necessary to define an *endurance limit*—that is, the stress amplitude which would not cause fatigue failure up to an acceptably large value of *N*.

Hardness

Hardness of a material may be determined in several ways: (1) resistance to indentation, (2) rebound efficiency, and (3) resistance to scratching. The first method is the most commonly used technique for plastics. Numerous test methods are available for measuring the resistance of a material to indentation, but they differ only in detail. Basically they all use the size of an indent produced by a hardened steel or diamond indentor in the material as an indication of its hardness—the smaller the indent produced, the harder the material, and so the greater the hardness number. Hardness tests are simple, quick, and nondestructive, which account for their wide use for quality control purposes.

Indentation Hardness

The test methods used for plastics are similar to those used for metals. The main difference is that because plastics are viscoelastic allowance must be made for the creep and the time-dependent recovery which occurs as a result of the applied indentation load.

Brinell Hardness Number

A hardened steel ball 10 mm in diameter is pressed into the flat surface of the test specimen under load of 500 kg for 30 sec. The load is then removed, and the diameter of the indent produced is measured (Fig. 3.20). The Brinell hardness number (BHN) is then defined as

$$\text{BHN} = \frac{\text{Load applied to indentor (kgf)}}{\text{Contact area of indentation (mm}^2)}$$

$$= \frac{2P}{\pi D(D - \sqrt{D^2 - d^2})} \tag{34}$$

where *D* is the diameter of the ball and *d* is the diameter of the indent. Tables are available to convert the diameter of the indent into BHN.

Although the units of Brinell hardness are kgf/mm^2, it is quoted only as a number. A disadvantage of the Brinell hardness test when used for plastics is

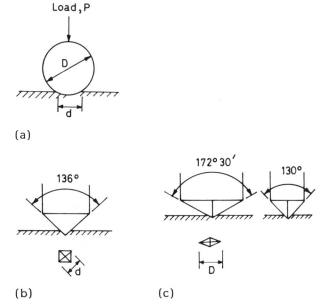

FIG. 3.20 Indentation hardness tests. (a) Brinell test. (b) Vickers test. (c) Knoop test.

that the edge of the indent is usually not well defined. This problem is overcome in the following test.

Vickers Hardness Number

The Vickers hardness test differs from the Brinell test in that the indentor is a diamond (square-based) pyramid (Fig. 3.20) having an apex angle of 136°. If the average diagonal of the indent is d, the hardness number is calculated from

$$\text{Vickers hardness number} = \frac{\text{Load applied to indentor (kgf)}}{\text{Contact area of indentation (mm}^2)}$$

$$= 1.854\left(\frac{P}{d^2}\right) \tag{35}$$

As in the Brinell test, tables are available to convert the average diagonal into the Vickers number.

Knoop Hardness Number

The indentor used in the Knoop hardness is a diamond pyramid, but the lengths of the two diagonals, as shown in Figure 3.20, are different. If the long diagonal of the indent is measured as D, the hardness number is obtained from

$$\text{Knoop hardness number} = 14.23\left(\frac{P}{D^2}\right) \tag{36}$$

Time-dependent recovery of the indentation in plastics is a problem common to all three tests. To overcome this problem, allow a fixed time before making measurements on the indent.

Rockwell Hardness Number

The Rockwell test differs from the other three tests because the depth of the indent rather than its surface area is taken as a measure of hardness. A hardened steel ball is used as the indentor. A major advantage of the Rockwell test is that no visual measurement of the indentation is necessary, and the depth of the indent is read directly as a hardness value on the scale. The test involves three steps, as shown in Figure 3.21. A minor load of 10 kg is applied on the steel ball, and the scale pointer is set to zero within 10 sec of applying the load. In addition to this minor load, a major load is applied for 15 sec. A further 15 sec after removal of the major load (with the minor load still on the ball), the hardness value is read off the scale. Since creep and recovery effects can influence readings, it is essential to follow a defined time cycle for the test.

Several Rockwell scales (Table 3.2) are used, depending on the hardness of the material under test (Table 3.3). The scale letter is quoted along with the hardness number, e.g., Rockwell R60. Scales R and L are used for low-hardness materials, and scales M and E when the hardness value is high. When the hardness number exceeds 115 on any scale, the sensitivity is lost, so another scale should be used.

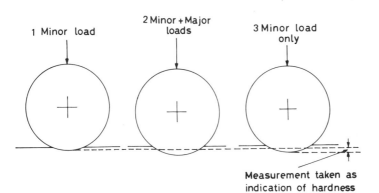

FIG. 3.21 Stages in Rockwell hardness test: 1, minor load; 2, minor and major loads; 3, minor load only.

TABLE 3.2 Rockwell Hardness Scales

Scale	Major load (kg)	Dia. of indentor (in.)
R	60	1/2
L	60	1/4
M	100	1/4
E	100	1/8

Barcol Hardness

The Barcol hardness tester is a hand-operated hardness measuring device. Its general construction is shown in Figure 3.22. With the back support leg placed on the surface, the instrument is gripped in the hand and its measuring head is pressed firmly and steadily onto the surface until the instrument rests on the stop ring. The depth of penetration of the spring-loaded indentor is transferred by a lever system to an indicating dial, which is calibrated from 0 to 100 to indicate increasing hardness. To allow for creep, one normally takes readings after 10 sec. The indentor in the Barcol tester Model No. 934–1 is a truncated steel cone having an included angle of 26° with a flat tip of 0.157 mm (0.0062 in.) in

TABLE 3.3 Choice of Hardness Test Methods Based on Modulus Range of Plastics

	Material	Test method
Low modulus	Rubber	Shore A or BS 903
	Plasticized PVC	Shore A or BS 2782
	Low-density polyethylene	Shore D
	Medium-density polyethylene	Shore D
	High-density polyethylene	Shore D
	Polypropylene	Rockwell R
	Toughened polystyrene	Rockwell R
	ABS	Rockwell R
	Polystyrene	Rockwell M
	Poly(methyl methacrylate)	Rockwell M
High modulus		

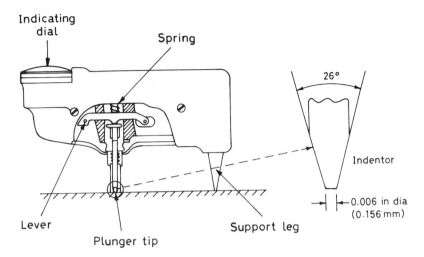

FIG. 3.22 General construction of Barcol hardness tester.

diameter. The values obtained using this instrument are found to correlate well to Rockwell values on the M scale. This instrument is used for metals and plastics. Two other models, No. 935 and No. 936, are used for plastics and very soft materials, respectively.

Durometer Hardness

A durometer is an instrument for measuring hardness by pressing a needle-like indentor into the specimen. Operationally, a durometer resembles the Barcol tester in that the instrument is pressed onto the sample surface until it reaches a stop ring. This forces the indentor into the material, and a system of levers transforms the depth of penetration into a pointer movement on an indicating dial, which is calibrated from 0 to 100. The two most common types of durometers used for plastics are the Shore Type A and Shore Type D. They differ in the spring force and the geometry of the indentor, as shown in Figure 3.23. Due to creep, readings should be taken after a fixed time interval, often chosen as 10 sec. Typical hardness values of some of the common plastics measured by different test methods are shown in Table 3.4.

Rebound Hardness

The energy absorbed when an object strikes a surface is related to the hardness of the surface: the harder the surface, the less the energy absorbed, and the greater the rebound height of the object after impact. Several methods have been

Type A   Type D

822g   10 lbf

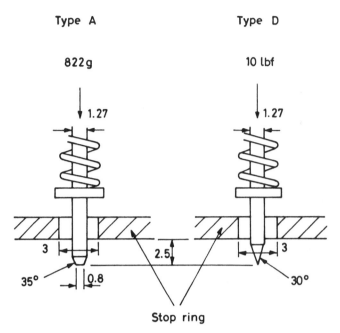

FIG. 3.23 Two types of Shore durometer.

developed to measure hardness in this way. The most common method uses a *Shore scleroscope*, in which the hardness is determined from the rebound height after the impact of a diamond cone dropped onto the surface of the test piece. Typical values of Scleroscope hardness together with the Rockwell M values (in parentheses) for some common plastics are as follows: PMMA 99 (M 102), LDPE 45 (M 25), polystyrene 70 (M 83), and PVC 75 (M 60).

Scratch Hardness

Basically, scratch hardness is a measure of the resistance the test sample has to being scratched by other materials. The most common way of qualifying this property is by means of the Mohs scale. On this scale various materials are classified from 1 to 10. The materials used, as shown in Figure 3.24, range from talc (1) to diamond (10). Each material on the scale can scratch the materials that have a lower Mohs number; however, the Mohs scale is not of much value for classifying plastic materials, because most common plastics fall in the 2 to 3 Mohs range. However, the basic technique of scratch hardness may be used to establish the relative merits of different plastic materials from their ability to scratch one another.

TABLE 3.4 Some Typical Hardness Values for Plastics

Material	Brinell	Vickers	Knoop	Rockwell	Barcol	Shore D
High-density polyethylene	4	2		R40		70
Polypropylene	7	6		R100		74
Polystyrene	25	7	17	M83	76	74
Poly(methyl methacrylate)	20	5	16	M102	80	90
Poly(vinyl chloride)	11	9		M60		80
Poly(vinyl chloride-co-vinyl acetate)	20	5	14	M75		
Polycarbonate		7		M70	70	60
Nylon		5	15	M75		80
Cellulose acetate	12	4	12	M64		70

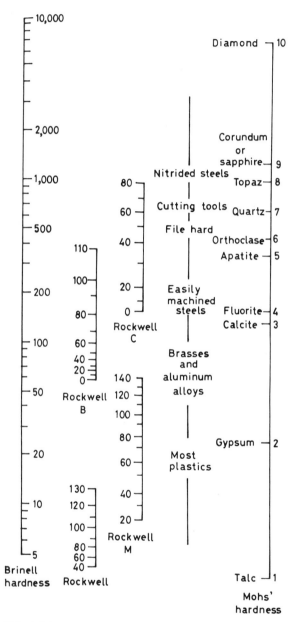

FIG. 3.24 Comparison of hardness scales (approximate).

Scratch hardness is particularly important in plastics used for their optical properties and is usually determined by some sort of *mar-resistance* test. In one type of test a specimen is subjected to an abrasive treatment by allowing exposure to a controlled stream of abrasive, and its gloss (specular reflection) is measured before and after the treatment. In some tests the light transmission property of the plastic is measured before and after marring.

Stress Corrosion Cracking of Polymers [7, 8]

Stress corrosion cracking of polymers occurs in a corrosive environment and also under stress. This kind of crack starts at the surface and proceeds at right angles to the direction of stress. The amount of stress necessary to cause stress corrosion cracking is much lower than the normal fracture stress, although there is a minimum stress below which no stress corrosion cracking occurs.

The stress corrosion resistance of polymers depends on the magnitude of the stress, the nature of the environment, the temperature, and the molecular weight of the specimen. Ozone cracking is a typical example of stress corrosion cracking of polymers. The critical energy for crack propagation (τ_c) in ozone cracking varies very little from one polymer to another and is about 100 erg/cm^2 (0.1 J/m^2). This value is much lower than the τ_c values for mechanical fracture, which are about 10^7 erg/cm^2 (10^4 J/m^2). In ozone cracking very little energy is dissipated in plastic or viscoelastic deformations at the propagating crack, and that is why τ_c is about the same as the true surface energy. The only energy supplied to the crack is that necessary to provide for the fresh surfaces due to propagation of the crack, because in ozone cracking chemical bonds at the crack tip are broken by chemical reaction, so no high stress is necessary at the tip.

The critical energy τ_c is about 4000 erg/cm^2 for PMMA in methylated spirits at room temperature, but the value is lower in benzene and higher in petroleum ether. Thus τ_c in this case is much higher than the true surface energy but still much lower than that for mechanical crack propagation.

REINFORCED PLASTICS [9–11]

The modulus and strength of plastics can be increased significantly by means of reinforcement. A reinforced plastic consists of two main components—a matrix, which may be either a thermoplastic or thermosetting resin, and a reinforcing filler, which is mostly used in the form of fibers (but particles, for example glass spheres, are also used). The greater tensile strength and stiffness of fibers as compared with the polymer matrix is utilized in producing such composites. In general, the fibers are the load-carrying members, and the main role of the matrix is to transmit the load to the fibers, to protect their surface, and to raise the energy for crack propagation, thus preventing a brittle-type fracture. The strength of the fiber-reinforced plastics is determined by the strength of the fiber and by the nature and strength of the bond between the fibers and the matrix.

Types of Reinforcement

The reinforcing filler usually takes the form of fibers, since it is in this form that the maximum strengthening of the composite is attained. A wide range of amorphous and crystalline materials can be used as reinforcing fibers, including glass, carbon, asbestos, boron, silicon carbide, and more recently, synthetic polymers (e.g., Kevlar fibers from aromatic polyamides). Some typical properties of these reinforcing fibers are given in Table 3.5.

Glass is relatively inexpensive, and in fiber form it is the principal form of reinforcement used in plastics. The earliest successful glass reinforcement had a low-alkali calcium-alumina borosilicate composition (E glass) developed specifically for electrical insulation systems. Although glasses of other compositions were developed subsequently for other applications, no commercial glass better than E glass has been found for plastics reinforcement. However, certain special glasses having extra high-strength properties or modulus have been produced in small quantities for specific applications (e.g., aerospace technology).

Glass fibers are usually treated with finishes. The function of a finish is to secure good wetting and to provide a bond between the inorganic glass and the organic resin. The most important finishes are based on silane compounds— e.g., vinyltrichlorosilane or vinyltriethoxysilane.

Types of Matrix

The matrix in reinforced plastics may be either a thermosetting or thermoplastic resin. The major thermosetting resins used in conjunction with glass-fiber reinforcement are unsaturated polyester resins and, to a lesser extent, epoxy resins. These resins have the advantage that they can be cured (cross-linked) at room temperature, and no volatiles are liberated during curing.

Among thermoplastic resins used as the matrix in reinforced plastics, the largest tonnage group is the polyolefins, followed by nylon, polystyrene, thermoplastic polyesters, acetal, polycarbonate, and polysulfone. The choice of any thermoplastic is dictated by the type of application, the service environment, and the cost.

Analysis of Reinforced Plastics

Fibers exert their effect by restraining the deformation of the matrix while the latter transfers the external loading to the fibers by shear at the interface. The resultant stress distributions in the fiber and the matrix tend to be complex. Theoretical analysis becomes further complicated because fiber length, diameter,

TABLE 3.5 Typical Properties of Reinforcing Fibers

Fiber	Density (g/cm^3)	Tensile strength		Tensile modulus	
		10^4 kgf/cm^2	GPa	10^5 kgf/cm^2	GPa
E Glass	2.54-2.56	3.5-3.7	3.4-3.6	7.1-7.7	70-76
Carbon	1.75-2.0	2.1-2.8	2.1-2.8	24.5-40.8	240-400
Asbestos	2.5-3.3	2.1-3.6	2.1-3.5	14.3-19.4	140-190
Boron	2.6	3.0-3.6	3.0-3.5	40.8-45.9	400-450
Silicon carbide	3.2-3.4	3.0-3.7	3.0-3.6	46.9-50.0	460-490
Kevlar-49	1.45	3.0-3.7	3.0-3.6	13.2	130

Source: Data from Reference 4.

and orientation are all factors. A simplified analysis follows for two types of fiber reinforcement commonly used, namely, (1) continuous fibers and (2) discontinuous fibers.

Continuous Fibers

We will examine what happens when a load is applied to an ideal fiber composite in which the matrix material is reinforced by fibers which are uniform, continuous, and arranged uniaxially, as shown in Figure 3.25a. Let us assume that the fibers are gripped firmly by the matrix so that there is no slippage at the fiber-matrix interface and both phases act as a unit. Under these conditions the strains in the matrix and in the fiber under a load are the same (Fig. 3.25b), and the total load is shared by the fiber and the matrix:

$$P_c = P_m + P_f \tag{37}$$

where P is the load and the subscripts c, m, and f refer, respectively, to composite, matrix and fiber.

Since the load $P = \sigma A$, Eq. (37), expressed in terms of stresses (σ) and cross-sectional areas (A), becomes

$$\sigma_c A_c = \sigma_m A_m + \sigma_f A_f \tag{38}$$

Rearranging gives

$$\sigma_c = \sigma_m \left(\frac{A_m}{A_c} \right) + \sigma_f \left(\frac{A_f}{A_c} \right) \tag{39}$$

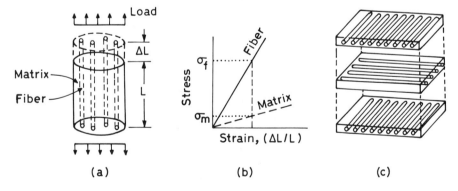

FIG. 3.25 (a) Continuous-fiber reinforced composite under tensile load. (b) Isostrain assumption in a composite. (c) Arrangement of fibers in a cross-plied laminate.

Since the fibers run throughout the length of the specimen, the ratio A_m/A_c can be replaced by the volume fraction $\phi_m = V_m/V_c$, and similarly A_f/A_c by ϕ_f. Equation (39) thus becomes

$$\sigma_c = \sigma_m \phi_m + \sigma_f \phi_f \tag{40}$$

Equation (40) represents the rule of mixture for stresses. It is valid only for the linear elastic region of the stress-strain curve (see Fig. 3.1). Since $\phi_m + \phi_f = 1$, we can write

$$\sigma_c = \sigma_m (1 - \phi_f) + \sigma_f \phi_f \tag{41}$$

Since the strains on the components are equal,

$$\epsilon_c = \epsilon_m = \epsilon_f \tag{42}$$

Equation (40) can now be rewritten to give the rule of mixture for moduli

$$E_c \epsilon_c = E_m \epsilon_m \phi_m + E_f \epsilon_f \phi_f$$

i.e.,

$$E_c = E_m \phi_m + E_f \phi_f \tag{43}$$

Equation (42) also affords a comparison of loads carried by the fiber and the matrix. Thus for elastic deformation Eq. (42) can be rewritten as

$$\frac{\sigma_c}{E_c} = \frac{\sigma_m}{E_m} = \frac{\sigma_f}{E_f}$$

or

$$\frac{P_f}{P_m} = \frac{E_f}{E_m}\left(\frac{A_f}{A_m}\right) = \frac{E_f}{E_m}\left(\frac{\phi_f}{\phi_m}\right) \tag{44}$$

Because the modulus of fibers is usually much higher than that of the matrix, the load on a composite will therefore be carried mostly by its fiber component (see Example 3). However, a critical volume fraction of fibers (ϕ_{crit}) is required to realize matrix reinforcement. Thus for Eqs. (41) and (44) to be valid, $\phi_f > \phi_{crit}$.

The efficiency of reinforcement is related to the fiber direction in the composite and to the direction of the applied stress. The maximum strength and modulus are realized in a composite along the direction of the fiber. However, if the load is applied at 90° to the filament direction, tensile failure occurs at very low stresses, and this transverse strength is not much different than the matrix strength. To counteract this situation, one uses cross-plied laminates having alternate layers of unidirectional fibers rotated at 90°, as shown in Figure 3.25c.

(A more isotropic composite results if 45° plies are also inserted.) The stress-strain behavior for several types of fiber reinforcement is compared in Figure 3.26.

As already noted, if the load is applied perpendicularly to the longitudinal direction of the fibers, the fibers exert a relatively small effect. The strains in the fibers and the matrix are then different, because they act independently, and the total deformation is thus equal to the sum of the deformations of the two phases

$$V_c\epsilon_c = V_m\epsilon_m + V_f\epsilon_f \qquad (45)$$

Dividing Eq. (45) by V_c and applying Hooke's law, since the stress is constant, give

$$\frac{\sigma}{E_c} = \frac{\sigma\phi_m}{E_m} + \frac{\sigma\phi_f}{E_f} \qquad (46)$$

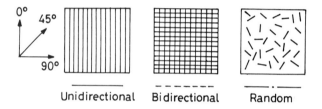

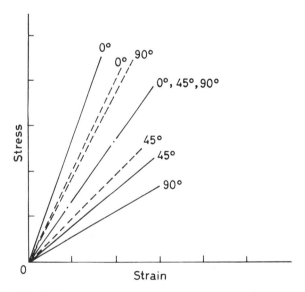

FIG. 3.26 Stress-strain behavior for several types of fiber reinforcement.

Dividing by σ and rearranging, we get

$$E_c = \frac{E_m E_f}{E_m \phi_f + E_f \phi_m} \tag{47}$$

The fiber composite thus has a lower modulus in transverse loading than in longitudinal loading.

Example 3 A unidirectional fiber composite is made by using 75% by weight of E glass continuous fibers (sp. gr. 2.4) having a Young's modulus of 7×10^5 kg/cm^2 (68.6 GPa), practical fracture strength of 2×10^4 kg/cm^2 (1.9 GPa), and an epoxy resin (sp. gr. 1.2) whose modulus and tensile strength, on curing, are found to be 10^5 kg/cm^2 (9.8 GPa) and 6×10^2 (58.8 MPa), respectively. Estimate the modulus of the composite, its tensile strength, and the fractional load carried by the fiber under tensile loading. What will be the value of the modulus of the composite under transverse loading?

Answer: Volume fraction of glass fibers (ϕ_f)

$$= \frac{0.75/2.4}{0.75/2.4 + 0.25/1.2} = 0.60$$

$$\phi_m = 1 - 0.6 = 0.4$$

From Eq. (43),

$$E_c = 0.4(10^5) + 0.6(7 \times 10^5)$$

$$= 4.6 \times 10^5 \text{ kg/cm}^2 \text{ (45 GPa)}$$

From Eq. (40),

$$\sigma_c = 0.4(6 \times 10^2) + 0.6(2 \times 10^4)$$

$$= 1.22 \times 10^4 \text{ kg/cm}^2 \text{ (1.2 GPa)}$$

Equation (44), on rearranging, gives

$$\frac{P_f}{P_c} = \frac{E_f}{E_c} \phi_f$$

$$= \frac{7 \times 10^5}{4.6 \times 10^5} \times 0.6 = 0.91$$

Thus, nearly 90% of the load is carried by the fiber, and the weakness of the plastic matrix is relatively unimportant.

For transverse loading, from Eq. (47),

$$E_c = \frac{10^5(7 \times 10^5)}{0.6 \times 10^5 + 0.4(7 \times 10^5)}$$

$$= 2 \times 10^5 \text{ kg/cm}^2 \text{ (19.6 GPa)}$$

Equations (41) and (43) apply to ideal fiber composites having uniaxial arrangement of fibers. In practice, however, not all the fibers are aligned in the direction of the load. This practice reduces the efficiency of the reinforcement, so Eqs. (41) and (43) are modified to the forms

$$\sigma_c = \sigma_m(1 - \phi_f) + k_1\sigma_f\phi_f \tag{48}$$

$$E_c = E_m(1 - \phi_f) + k_2E_f\phi_f \tag{49}$$

If the fibers are bidirectional (see Fig. 3.26), then the strength and modulus factors, k_1 and k_2, are about 0.3 and 0.65, respectively.

Discontinuous Fibers

If the fibers are discontinuous, the bond between the fiber and the matrix is broken at the fiber's ends, which thus carry less stress than the middle part of the fiber. The stress in a discontinuous fiber therefore varies along its length. A useful approximation pictures the stress as being zero at the end of the filler and as reaching the maximum stress in the fiber at a distance from the end (Fig. 3.27a). The length over which the load is transferred to the fiber is called the *transfer length*. As the stress on the composite is increased, the maximum fiber stress as well as the transfer length increase, as shown in Figure 3.27a, until a limit is reached, because the transfer regions from the two ends meet at the middle of the fiber (and so no further transfer of stress can take place), or because the fiber fractures. For the latter objective to be reached, so as to attain the maximum strength of the composite, the fiber length must be greater than a minimum value called the *critical fiber length*, l_c.

Consider a fiber of length l embedded in a polymer matrix, as shown in Figure 3.27b. One can then write, equating the tensile load on the fiber with the shear load on the interface,

$$\frac{\sigma\pi d^2}{4} = \tau\pi dl \tag{50}$$

where σ is the applied stress, d is the fiber diameter, and τ is the shear stress at the interface.

The critical fiber length, l_c can be derived from a similar force balance for an embedded length of $l_c/2$. Thus,

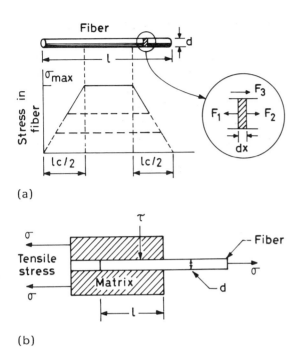

(a)

(b)

FIG. 3.27 Composite reinforced with discontinuous fibers (a) A total length l_c at the two ends of a fiber carries less than the maximum stress. (b) Interfacial strength of the matrix fiber.

$$l_c = \frac{\sigma_{ff} d}{2\tau_i}$$

and

$$\frac{l_c}{d} = \frac{\sigma_{ff}}{2\tau_i} \tag{51}$$

where σ_{ff} is the fiber strength and τ_i is the shear strength of the interface or the matrix, whichever is smaller.

So if the composite is to fail through tensile fracture of the fiber rather than shear failure due to matrix flow at the interface between the fiber and the matrix, the ratio l_c/d, known as the *critical aspect ratio*, must be exceeded, or, in other words, for a given diameter of fiber, d, the critical fiber length, l_c, must be exceeded. If the fiber length is less than l_c, the matrix will flow around the fiber, and maximum transfer of stress from matrix to fiber will not occur. Using Eq. (51), we can estimate the value of l_c/d from the values of σ_{ff} and τ_i, and vice versa. Typical values of l_c/d for glass fiber and carbon fiber in an epoxy resin matrix are 30– 100 and 70, respectively.

If the fibers are discontinuous, then, since the stress is zero at the end of the fiber, the average stress in the fibers will be less than the value $\sigma_{f\ max}$, which it would have achieved if the fibers had been continuous over the whole length of the matrix. The value of the average stress will depend on the stress distribution in the end portions of the fibers and also on their lengths. If the stress distributions are assumed to be as shown in Figure 3.27a, then the average stress in the fibers may be obtained as follows.

Considering a differential section of the fiber as shown in Figure 3.27a, we obtain

$$F_1 = \sigma_f \frac{\pi d^2}{4}$$

$$F_2 = \left(\sigma_f + \frac{d\sigma_f}{dx} dx \right) \frac{\pi d^2}{4}$$

$$F_3 = \tau \pi d\ dx$$

For equilibrium,

$$F_1 = F_2 + F_3$$

so

$$\sigma_f \frac{\pi d^2}{4} = \left(\sigma_f + \frac{d\sigma_f}{dx} dx \right) \frac{\pi d^2}{4} dx + \tau \pi d\ dx$$

$$\frac{d}{4} d\sigma_f = -\tau\ dx \tag{52}$$

Integrating gives

$$\frac{d}{4} \int_0^{\sigma_f} d\sigma_f = -\int_{1/2}^x \tau\ dx$$

$$\sigma_f = \frac{4\tau(1/2 - x)}{d} \tag{53}$$

Three cases may now be considered.

Fiber Length Less Than l_c (Fig. 3.28a)

In this case the peak stress occurs at $x = 0$. So from Eq. (53),

$$\sigma_f = \frac{2\tau l}{d}$$

The average fiber stress is obtained by dividing the area of the stress-fiber length diagram by the fiber length; that is,

$$\bar{\sigma}_f = \frac{(1/2)2\tau l/d}{l} = \frac{\tau l}{d}$$

The stress, σ_c, in the composite is now obtained from Eq. (48)

$$\sigma_c = \sigma_m(1 - \phi_f) + \frac{\tau l k_1}{d}\phi_f \tag{54}$$

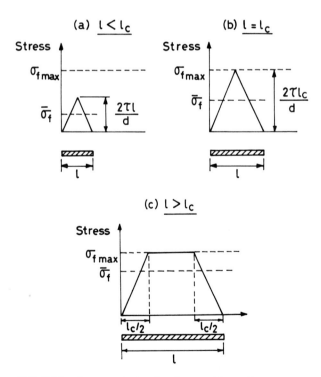

FIG. 3.28 Stress variation for short and long fibers.

Fiber Length Equal To l_c (Fig. 3.28b)

In this case the peak value of stress occurs at $x = 0$ and is equal to the maximum fiber stress.

So

$$\sigma_f = \sigma_{f\,max} = \frac{2\tau l_c}{d} \tag{55}$$

Average fiber stress $= \overline{\sigma}_f = \frac{1}{2} \frac{l_c(2\tau l_c/d)}{l_c}$

i.e.,

$$\overline{\sigma}_f = \frac{\tau l_c}{d}$$

So from Eq. (48),

$$\sigma_c = \sigma_m(1 - \phi_f) + k_1 \left(\frac{\tau l_c}{d}\right) \phi_f \tag{56}$$

Fiber Length Greater than l_c (Fig. 3.28c)

(i) For $l/2 > x > (l - l_c)/2$,

$$\sigma_f = \frac{4\tau}{d} \left(\frac{1}{2}l - x\right)$$

(ii) For $(l - l_c)/2 > x > 0$,

$$\sigma_f = \text{constant} = \sigma_{f\,max} = \frac{2\tau l_c}{d}$$

The average fiber stress, from the area under the stress-fiber length graph, is

$$\overline{\sigma}_f = \frac{(l_c/2)\sigma_{f\,max} + (l - l_c)\sigma_{f\,max}}{l}$$

$$= \left(1 - \frac{l_c}{2l}\right) \sigma_{f\,max} \tag{57}$$

So from Eq. (48),

$$\sigma_c = \sigma_m(1 - \phi_f) + k_1\phi_f \left(1 - \frac{l_c}{2l}\right) \sigma_{f\,max} \tag{58}$$

It is evident from Eq. (57) that to get the average fiber stress as close as possible to the maximum fiber stress, the fibers must be considerably longer than the critical length. At the critical length the average fiber stress is only half of the maximum fiber stress, i.e., the value achieved in continuous fibers.

Equations such as Eq. (58) give satisfactory agreement with the measured values of strength and modulus for polyester composites reinforced with chopped strands of glass fibers. These strength and modulus values are only about 20 to 25% of those achieved by reinforcement with continuous fibers, because in a random arrangement of short fibers only a small percentage of the fibers is aligned along the line of action of the applied stress. However, even if all the short fibers were aligned in the direction of the stress, the properties would still be inferior compared to the continuous-fiber case, because, as we have seen, short fibers carry less stress than continuous fibers.

Example 4   Calculate the maximum and average fiber stresses for glass fibers of diameter 15 μm and length 2 mm embedded in a polymer matrix. The matrix exerts a shear stress of 40 kgf/cm^2 (3.9 MPa) at the interface, and the critical aspect ratio of the fiber is 50.

Answer:

$$l_c = 50 \times 15 \times 10^{-3} = 0.75 \text{ mm}$$

Since $l > l_c$, then

$$\sigma_{max} = \frac{2\tau l_c}{d} = 2 \times 40 \times 50$$

$$= 4 \times 10^3 \text{ kgf/cm}^2 \ (= 392 \text{ MPa})$$

Also

$$\overline{\sigma}_f = \left(1 - \frac{l_c}{2l}\right) \sigma_{f\,max}$$

$$= 1 - \frac{0.75}{2 \times 2} (4 \times 10^3)$$

$$= 3.25 \times 10^3 \text{ kgf/cm}^2 \ (= 318 \text{ MPa})$$

Deformation Behavior of Fiber-Reinforced Plastic

As we have seen, the presence of fibers in the matrix has the effect of stiffening and strengthening it. The tensile deformation behavior of fiber-reinforced composites depends largely on the direction of the applied stress in relation to the orientation of the fibers, as illustrated in Figure 3.26. The maximum strength and modulus are achieved with unidirectional fiber reinforcement when the stress is aligned with the fibers (0°), but there is no enhancement of matrix properties when the stress is applied perpendicular to the fibers. With random orientation of fibers the properties of the composite are approximately the same in all directions, but the strength and modulus are somewhat less than for the continuous-fiber reinforcement.

In many applications the stiffness of a material is just as important as its strength. In tension the stiffness per unit length is given the product EA, where E is the modulus and A is the cross-sectional area. When the material is subjected to flexure, the stiffness per unit length is a function of the product EI, where I is the second moment of area of cross section (see Example 1). Therefore the stiffness in both tension and flexure increases as the modulus of the material increases, and the advantages of fiber reinforcement thus become immediately apparent, considering the very high modulus values for fibers.

Fracture of Fiber-Reinforced Plastics

Although the presence of the reinforcing fibers enhances the strength and modulus properties of the base material, they also cause a complex distribution of stress in the materials. For example, even under simple tensile loading, a triaxial stress system is set up since the presence of the fiber restricts the lateral contraction of the matrix. This system increases the possibility of brittle failure in the material. The type of fracture which occurs depends on the loading conditions and fiber matrix bonding.

Tension

With continuous-fiber reinforcement it is necessary to break the fibers before overall fracture can occur. The two different types of fracture which can occur in tension are shown in Figure 3.29. It is interesting to note that when an individual fiber in a continuous-fiber composite breaks, it does not cease to contribute to the strength of the material, because the broken fiber then behaves like a long "short fiber" and will still be supporting part of the external load at sections remote from the broken end. In short-fiber composites, however, fiber breakage is not an essential prerequisite to complete composite fracture, especially when the interfacial bond is weak, because the fibers may then be simply pulled out of the matrix as the crack propagates through the latter.

Compression

In compression the strength of glass-fiber reinforced plastics is usually less than in tension. Under compressive loading, shear stresses are set up in the matrix

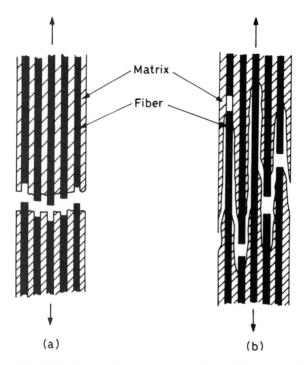

FIG. 3.29 Typical fracture modes in fiber-reinforced plastics. (a) Fracture due to strong interfacial bond. (b) Jagged fracture due to weak interfacial bond.

parallel to the fibers. The fiber aligned in the loading direction thus promote shear deformation. Short-fiber reinforcement may therefore have advantages over continuous fibers in compressive loading because in the former not all the fibers can be aligned, so the fibers which are inclined to the loading plane will resist shear deformation. If the matrix-fiber bond is weak, debonding may occur, causing longitudinal cracks in the composite and buckling failure of the continuous fibers.

Flexure or Shear

In flexure or shear, as in the previous case of compression, plastics reinforced with short fibers are probably better than those with continuous fibers, because in the former with random orientation of fibers at least some of the fibers will be correctly aligned to resist the shear deformation. However, with continuous-fiber reinforcement if the shear stresses are on planes perpendicular to the continuous fibers, then the fibers will offer resistance to shear deformation. Since high volume fraction (ϕ_f) can be achieved with continuous fibers, this resistance can be substantial.

Fatigue Behavior of Reinforced Plastics

Like unreinforced plastics, reinforced plastics are also susceptible to fatigue. There is, however, no general rule concerning whether glass reinforcement enhances the fatigue endurance of the base material. In some cases the unreinforced plastic exhibits greater fatigue endurance than the reinforced material; in other cases, the converse is true.

In short-fiber glass-reinforced plastics, cracks may develop relatively easily at the interface close to the ends of the fibers. The cracks may then propagate through the matrix and destroy its integrity long before fracture of the composite takes place. In many glass-reinforced plastics subjected to cyclic tensile stresses, debonding may occur after a few cycles even at modest stress levels. Debonding may be followed by resin cracks at higher stresses, leading eventually to the separation of the component due to intensive localized damage. In other modes of loading, e.g., flexure or torsion, the fatigue endurance of glass-reinforced plastics is even worse than in tension. In most cases the fatigue endurance is reduced by the presence of moisture.

Plastics reinforced with carbon and boron, which have higher tensile moduli than glass, are stiffer than glass-reinforced plastics and are found to be less vulnerable to fatigue.

Impact Behavior of Reinforced Plastics

Although it might be expected that a combination of brittle reinforcing fibers and a brittle matrix (e.g., epoxy or polyester resins) would have low impact strength, this is not the case, and the impact strength of most glass-reinforced plastics is many times greater than the impact strengths of the fibers or the matrix. For example, polyester composites with chopped-strand mat have impact strengths from 45 to 70 ft-lbf/in² (94 to 147 kJ/m², whereas a typical impact strength for polyester resin is only 1 ft-lbf/in² (2.1 kJ/m²).

The significant improvement in impact strength by reinforcement is explained by the energy required to cause debonding and to overcome friction in pulling the fibers out of the matrix. It follows from this that impact strengths would be higher if the bond between the fiber and the matrix is relatively weak, because if the interfacial bond is very strong, impact failure will occur by propagation of cracks across the matrix and fibers requiring very little energy. It is also found that in short-fiber-reinforced plastics the impact strength is maximum when the fiber length has the critical value. The requirements for maximum impact strength (i.e., short fiber and relatively weak interfacial bond) are thus seen to be contrary to those for maximum tensile strength (long fibers and strong bond).

The structure of a reinforced plastic material should therefore be tailored in accordance with the service conditions to be encountered by the material.

ELECTRICAL PROPERTIES

Conduction of electric current in materials is the result of the movement of charged particles (ions and electrons) under an applied electric field. In metals, the outermost (valence) electrons, which are loosely bound to the nucleus, are responsible for the high electronic conductivity commonly observed. In plastic insulators, on the other hand, the outermost electrons of the atoms are shared only with adjacent atoms in homopolar (covalent) bonding. In an electric field the firmly bound charges in such a material can be displaced by only limited amounts from their equilibrium positions; the extent of displacement depends on the natures of the electric field and of the restoring forces opposing displacement. The existence of these restoring forces gives plastics the ability to store electrical energy as potential energy. The amount of energy that can thus be stored per unit volume of the insulating material depends on the geometric arrangement and the nature and numerical density of the displaced charge carriers. The permittivity property of a material refers to its ability to store energy in this way to that of vacuum.

The displacement of charge in an insulator is accompanied by energy losses due to viscous or frictionlike forces opposing the displacement. The displacement of charge is thus unable to immediately follow changes in the applied electric field. The extent to which the charge displacement lags behind the change in the applied field is a measure of the friction-type energy losses occurring in the material. Since an oscillating electric field can be considered as a rotating vector, this *lag* is conveniently measured as a *loss angle*. Dipole orientation and ionic conduction, along with electronic and atomic polarizations, play an important part in permittivity and power loss in materials.

The usefulness of an insulator or dielectric ultimately depends on its ability to act as a separator for points across which a potential difference exists. This ability depends on the dielectric strength of the material, which is defined as the maximum voltage gradient that the material withstands before failure or loss of the material's insulating properties occurs.

Besides permittivity (dielectric constant), dielectric losses, and dielectric strength, another property used to define the dielectric behavior of a material is the insulation resistance, i.e., the resistance offered by the material to the passage of electric current. This property may be important in almost all applications of insulators.

The resistivity (i.e., reciprocal of conductivity) of a plastic material with a perfect structure would tend to be infinite at low electric fields. However, the various types of defects which occur in plastics may act as sources of electrons or ions which are free to contribute to the conductivity or that can be thermally activated to do so. These defects may be impurities, discontinuities in the structure, and interfaces between crystallites and between crystalline and amorphous phases. Common plastics therefore have finite, though very high, resistivities from 10^8 to 10^{20} ohm-cm. These resistivity values qualify them as electrical insulators.

Polymeric materials have also been produced which have relatively large conductivities and behave in some cases like semiconductors and even photoconductors [12]. For example, polyphenylacetylene, polyaminoquinones, and polyacenequinone radical polymers have been reported with resistivities from 10^3 to 10^8 ohm-cm. It has been suggested that the conductivity in these organic semiconductors is due to the existence of large numbers of unpaired electrons, which are "free" within a given molecule and contribute to the conduction current by "hopping" (tunneling) from one molecule to an adjacent one.

Dielectric Strength

Dielectric strength is calculated as the maximum voltage gradient that an insulator can withstand before puncture or failure occurs It is expressed as volts (V) per unit of thickness, usually per mil (1 mil = $\frac{1}{1000}$ in.).

Puncture of an insulator under an applied voltage gradient results from small electric leakage currents which pass through the insulator due to the presence of various types of defects in the material. (Note that only a perfect insulator would be completely free from such leakage currents.) The leakage currents warm the material locally, causing the passage of a greater current and greater localized warming of the material, eventually leading to the failure of the material. The failure may be a simple puncture in the area where material has volatilized and escaped, or it may be a conducting carbonized path (*tracking*) that short circuits the electrodes. It is obvious from the cause of dielectric failure that the measured values of dielectric strength will depend on the magnitude of the applied electric field and on the time of exposure to the field. Since the probability of a flaw and a local leakage current leading ultimately to failure increases with the thickness of the sample, dielectric strength will also be expected to depend on the sample thickness.

The measurement of dielectric strength (Fig. 3.30a) is usually carried out either by the short-time method or by the step-by-step method. In the former method the voltage is increased continuously at a uniform rate (500 V/sec) until failure occurs. Typically, a $\frac{1}{8}$-in. thick specimen requiring a voltage of about 50,000 V for dielectric failure will thus involve a testing period of 100 sec or so. In step-by-step testing, definite voltages are applied to the sample for a definite time (one minute), starting with a value that is half of that obtained by short-

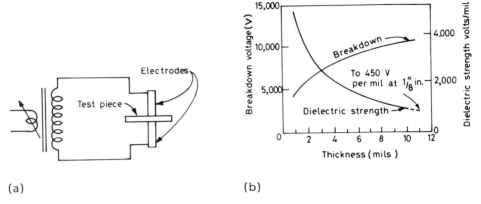

FIG. 3.30 (a) Dielectric strength test. (b) Dependence of dielectric strength on thickness of sample.

time testing, with equal increments of 2000 V until failure occurs. Since step-by-step testing provides longer exposure to the electric field, dielectric strength values obtained by this method are lower than those obtained by the short-time test. Conditions of step-by-step testing correspond more nearly with those met in service. Even so, service failure almost invariably occurs at voltages below the measured dielectric strength. It is thus necessary to employ a proper safety factor to provide for the discrepancy between test and service conditions.

Increase in thickness increases the voltage required to give the same voltage gradient, but the probability of a flaw and a local leakage current leading ultimately to failure also increases. The breakdown voltage increases proportionally less than thickness increases, and as a result the dielectric strength of a material decreases with the thickness of specimen (Fig. 3.30b). For this reason, testing of insulation plastic should be done with approximately the thickness in which it is to be used in service.

It is seen from Figure 3.30b that the dielectric strength increases rapidly with decreasing thickness of the sample. A rule of thumb is that the dielectric strength varies inversely with the 0.4 power of the thickness. For example, if the dielectric strength of poly(vinyl chloride) plastic is 375 V/mil in a thickness of 0.075 inch, it would be $375(75/15)^{0.4}$ or about 700 V/mil in foils only 15 mils thick. The fact that thin foils may have proportionally higher dielectric strength is utilized in the insulation between layers of transformer turns.

The dielectric strength of an insulation material usually decreases with increase in temperature and is approximately inversely proportional to the absolute temperature. But the converse is not observed, and below room temperature dielectric strength is substantially independent of temperature change.

Mechanical loading has a pronounced effect on dielectric strength. Since a mechanical stress may introduce internal flaws which serve as leakage paths, mechanically loaded insulators may show substantially reduced values of dielectric strength. Reductions up to 90% have been observed.

Dielectric strength of an insulating material is influenced by the fabrication detail. For example, flow lines in a compression molding or weld lines in an injection molding may serve as paths of least resistance for leakage currents, thus reducing the dielectric strength. Even nearly invisible minute flaws in a plastic insulator may reduce the dielectric strength to one-third its normal value.

Insulation Resistance

The resistance offered by an insulating material to the electric current is the composite effect of volume and surface resistances, which always act in parallel. *Volume resistance* is the resistance to leakage of the electric current through the body of the material. It depends largely on the nature of the material. But *surface resistance*, which is the resistance to leakage along the surface of a material, is largely a function of surface finish and cleanliness. Surface resistance is reduced by oil or moisture on the surface and by surface roughness. On the other hand, a very smooth or polished surface gives greater surface resistance.

A three-electrode system, as shown in Figure 3.31, is used for measurement of insulation resistance. In this way the surface and volume leakage currents are separated. The applied voltage must be well below the dielectric strength of the

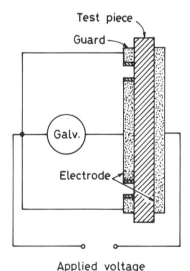

FIG. 3.31 Insulation resistance test.

material. Thus, in practice, a voltage gradient less than 30 V/mil is applied. From the applied voltage and the leakage current, the leakage resistance is computed. Since the measured value depends, among other things, on the time during which the voltage is applied, it is essential to follow a standardized technique, including preconditioning of the specimen to obtain consistent results.

The insulation resistance of a dielectric is represented by its *volume resistivity* and *surface resistivity*. The volume resistivity (also known as *specific volume resistance*) is defined as the resistance between two electrodes covering opposite faces of a centimeter cube. The range of volume resistivities of different materials including plastics is shown in Figure 3.32. Values for plastics range from approximately 10^{10} ohm-cm for a typical cellulose acetate to about 10^{19} ohm-cm for a high-performance polystyrene. The surface resistivity (also known as *specific surface resistance*) is defined as the resistance measured between the opposite *edges* of the surface of a material having an area of 1 cm^2. It ranges from 10^{10} ohm for cellulose acetate to 10^{14} ohm for polystyrene.

The insulation resistance of most plastic insulating materials is affected by temperature and the relative humidity of the atmosphere. The insulation resistance falls off appreciably with an increase in temperature or humidity. Even polystyrene, which has very high insulation resistance at room temperature, becomes generally unsatisfactory above 80°C (176°F). Under these conditions polymers like polytetrafluoroethylene and polychlorotrifluoroethylene are more suitable. Plastics that have high water resistance are relatively less affected by high humidities.

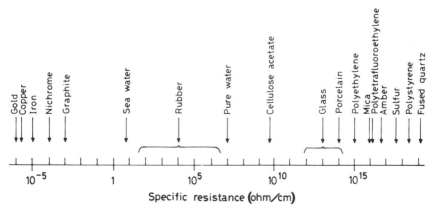

FIG. 3.32   The resistivity spectrum.

Arc Resistance

The arc resistance of a plastic is its ability to withstand the action of an electric arc tending to form a conducting path across the surface. In applications where the material is subject to arcing, such as switches, contact bushes, and circuit breakers, resistance to arc is an important requirement. Arcing tends to produce a conducting carbonized path on the surface. The arc resistance of an insulator may be defined as the time in seconds that an arc may play across the surface without burning a conducting path. A schematic of an arc-resistance test is shown in Figure 3.33.

Plastics that carbonize easily (such as phenolics) have relatively poor arc resistance. On the other hand, there are plastics (such as methacrylates) that do not carbonize, although they would decompose and give off combustible gases. There would thus be no failure in the usual sense. Special arc-resistant formulations involving noncarbonizing mineral fillers are useful for certain applications. But when service conditions are severe in this respect, ceramics ought to be used, because they generally have much better arc resistance than organic plastics.

Related to arc resistance is ozone resistance. This gas is found in the atmosphere around high-voltage equipment. Ignition cable insulation, for example, should be ozone resistant. Natural rubber is easily attacked and deteriorated by ozone. Fortunately, most synthetic resins have good ozone resistance and are satisfactory from this point of view.

Dielectric Constant

The effect of a dielectric material in increasing the charge storing capacity of a capacitor can be understood by considering the parallel-plate type sketched in

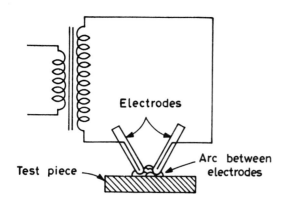

FIG. 3.33 Arc-resistance test.

Figure 3.34. If a voltage V is applied across two metal plates, each of area A m^2, separated by a distance, d m, and held parallel to each other in vacuum, the electric field established between the plates (Fig. 3.34a) is

$$E = \frac{-V}{d} \tag{59}$$

The charge density, Q_0/A, where Q_0 is the total charge produced on the surface area A of each plate, is directly proportional to the electric field.

$$\frac{Q_0}{A} = -\epsilon_0 E = \epsilon_0 \frac{V}{d} \tag{60}$$

or

$$Q_0 = \frac{A\epsilon_0}{d} V = C_0 V \tag{61}$$

The proportionality constant, ϵ_0, is called the *dielectric constant* (or *permittivity*) of a vacuum. It has units of

$$\epsilon_0 = \frac{Q_0/A}{V/d} = \frac{\text{coul}/\text{m}^2}{V/m} = \frac{\text{coul}}{V/m} = \frac{\text{farad}}{m}$$

and a value of 8.854×10^{-12} farad/m.

The quantity C_0 in Eq. (61) is the *capacitance* of a capacitor (condenser) with a vacuum between its plates. It can be defined as the ratio of the charge on either of the plates to the potential differences between the plates.

Now if a sheet of a dielectric material is inserted between the plates of a capacitor (Fig. 3.34b), an increased charge appears on the plates for the same voltage, due to *polarization* of the dielectric. The applied field E causes polarizaiton of the entire volume of the dielectric and thus gives rise to induced charges, or bound charges, Q', at its surface, represented by the ends of the dipole chains. These induced charges may be pictured as neutralizing equal charges of opposite signs on the metal plates. If one assumes, for instance, that the induced charge $-Q'$ neutralizes an equal positive charge in the upper plate of the capacitor (Fig. 3.34b), the total charge stored in the presence of the dielectric is $Q = Q_0 + Q'$.

The ratio of the total charge Q to the free charge Q_0 (which is not neutralized by polarization) is called the *relative dielectric constant* or *relative permittivity*, ϵ_r, and is characteristic of the dielectric material.

$$\epsilon_r = \frac{\text{Total charge}}{\text{Free charge}} = \frac{Q}{Q_0} \tag{62}$$

Obviously, ϵ_r is always greater than unity and has no dimensions. For most materials ϵ_r exceeds 2 (Table 3.6).

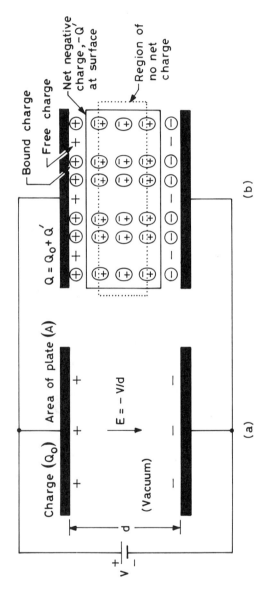

FIG. 3.34 Schematic illustration of the effect of dielectric material in increasing the charge storing capacity of a capacitor.

TABLE 3.6 Dielectric Properties of Electrical Insulators

	ε_r	tan δ		Dielectric strength[a]
Material	60 Hz	60 Hz	10^6 Hz	(V/mil)
Ceramics				
Porcelain	6	0.010	—	—
Alumina	9.6	—	<0.0005	200–300
Zircon	9.2	0.035	0.001	60–290
Soda-lime	7	0.1	0.01	—
Fused silica	4	0.001	0.0001	—
Mica	7	—	0.0002	3000–6000 (1–3 mil specimen)
Polymers				
Polyethylene	2.3	<0.0005	—	450–1000
Polystyrene	2.5	0.0002	0.0003	300–1000
Polyvinyl chloride	7	0.1	—	300–1000
Nylon-6,6	4	0.02	0.03	300–400
Teflon	2.1	<0.0001	—	400

[a]Specimen thickness ⅛ in.

Dividing both the numerator and denominator of Eq. (62) by the applied voltage V and applying the definition of C from Eq. (61), we obtain

$$\epsilon_r = \frac{\epsilon}{\epsilon_0} = \frac{C}{C_0} \tag{63}$$

The relative dielectric constant or relative permittivity is thus defined as the ratio of the capacitance of a condenser with the given material as the dielectric to that of the same condenser without the dielectric. (The dielectric constant for air is 1.0006. It is usually taken as unity—that is, the same as a vacuum—and the relative dielectric constant is referred to simply as the *dielectric constant*.)

The dielectric susceptibility, χ, is defined as

$$\chi = \epsilon_r - 1 = \frac{\epsilon - \epsilon_0}{\epsilon_0} \tag{64}$$

It thus represents the part of the total dielectric constant which is a consequence of the material. From Eq. (62),

$$\chi = \frac{\text{Bound charge}}{\text{Free charge}}$$

The magnitude of the induced or bound charge Q' per unit area is the polarization, P, which has the same units as charge density.

Therefore,

$$P = \frac{Q'}{A} = \frac{Q - Q_0}{A} \tag{65}$$

Substituting from Eqs. (60), (62), and (64), we obtain

$$P = \chi\epsilon_0 E = \epsilon_0(\epsilon_r - 1)E \tag{66}$$

This equation, as we shall see later, provides a link between the permittivity, which is a macroscopic, measurable property of a dielectric, and the atomic or molecular mechanisms in the dielectric which give rise to this property.

Polarization and Dipole Moment

In terms of the wave-mechanical picture, an atom may be looked upon as consisting of a positively charged nucleus surrounded by a negatively charged cloud, which is made up of contributions from electrons in various orbitals. Since the centers of positive and negative charges are coincident (see Fig. 3.35a), the net dipole moment of the atom is zero. If an electric field is applied, however, the electron cloud will be attracted by the positive plate and the nucleus by the negative plate, with the result that there will occur a small displacement of the center of gravity of the negative charge relative to that of the positive charge (Fig. 3.35a). This phenomenon is described by the statement that the field has *induced an electric dipole* in the atom; and the atom is said to have suffered *electronic polarization*—"electronic" because it arises from the displacement of the electron cloud relative to the nucleus.

The electric dipole moment of two equal but opposite charges, $+q$ and $-q$, at a distance r apart is defined as qr. For the atomic model of Figure 3.35a, it can be shown by balancing the opposite forces of the electric field and the coulombic attraction between the nucleus and the center of the electron cloud that the dipole moment, μ, induced in an atom by the field E is

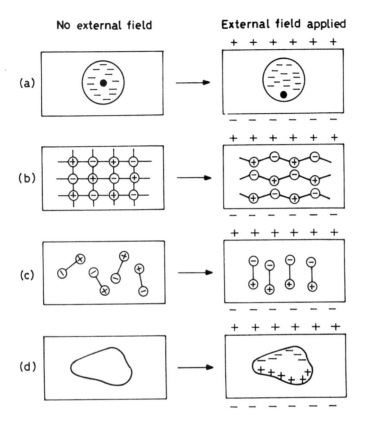

FIG. 3.35 Schematic illustrations of polarization mechanisms. (a) Electronic displacement. (b) Ionic displacement. (c) Dipole orientation. (d) Space charge.

$$\mu = (4\pi\epsilon_0 R^3)E = \alpha_e E \tag{67}$$

where α_e is a constant, called the electronic *polarizability* of the atom. It is proportional to the volume of the electron cloud, R^3. Thus the polarizability increases as the atoms become larger.

On the macroscopic scale we have earlier defined the polarization, P, to represent the bound charges induced per unit area on the surface of the material. Therefore, if we take unit areas on opposite faces of a cube separated by a distance d, the dipole moment due to unit area will be

$$\mu = Pd$$

For $d = 1$ (i.e., for unit volume) $\mu = P$. The polarization P is thus identical with the dipole moment per unit volume (check the units: $\text{coul}/\text{m}^2 = \text{coul} \cdot \text{m}/\text{m}^3$).

A dipole moment may arise through a variety of mechanisms, any or all of which may thus contribute to the value of P. The total polarization may be represented as a sum of individual polarizations, each arising from one particular mechanism (Fig. 3.35) or, more appropriately, as an integrated sum of all the individual dipole moments per unit volume.

$$P = P_e + P_i + P_o + P_s$$

$$= \frac{\sum \mu_e + \sum \mu_i + \sum \mu_o + \sum \mu_s}{V} \tag{68}$$

Electronic polarization, P_e, as discussed previously, arises from electron displacement within atoms or ions in an electric field; it occurs in all dielectrics. Similarly, displacements of ions and atoms within molecules and crystal structures (Fig. 3.35b) under an applied electric field give rise to *ionic* (or *atomic*) *polarization*, P_i. *Orientation polarization*, P_o, arises when asymmetric (polar) molecules having permanent dipole moments are present, since they become preferentially oriented by an electric field (Fig. 3.35c). *Interfacial* (or *space charge*) *polarization*, P_s, is the result of the presence of higher conductivity phases in the insulating matrix of a dielectric, causing localized accumulation of charge under the influence of an electric field (Fig. 3.35d).

Any or all of these mechanisms may be operative in any material to contribute to its polarization. A question to be discussed now is, which of the mechanisms are important in any given dielectric? The answer lies in studying the frequency dependence of the dielectric constant.

Dielectric Constant vs. Frequency

Let us consider first a single dipole in an electric field. Given time, the dipole will line up with its axis parallel to the field (Fig. 3.35c). If now the field is reversed, the dipole will turn 180° to again lie parallel to the field, but it will take a finite time; so if the frequency of the field reversal increases, a point will be reached when the dipole cannot keep up with the field, and the alteration of the dipole direction lags behind that of the field. For an assembly of dipoles in a dielectric, this condition results in an apparent reduction in the dielectric constant of the material. As the frequency of the field continues to increase, at some stage the dipoles will barely have started to move before the field reverses. Beyond this frequency, called the *relaxation frequency*, the dipoles make virtually no contribution to the polarization of the dielectric.

We may now consider the various mechanisms and predict, in a general way, the relaxation frequency for each one. Electrons with their extremely small mass have little inertia and can follow alterations of the electric field up to very high frequencies. Relaxation of electronic polarization is not observed until about

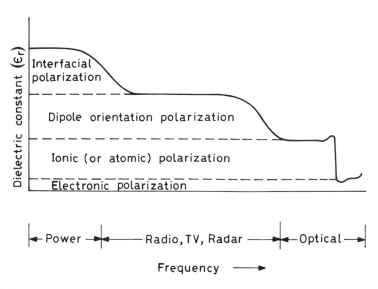

FIG. 3.36 Dielectric constant vs. frequency.

10^6 Hz (ultraviolet region). Atoms or ions vibrate with thermal energy, and the frequencies of these vibrations correspond to the infrared frequencies of the electromagnetic spectrum. The relaxation frequencies for ionic polarization are thus in the infrared range. Molecules or groups of atoms (ions) behaving as permanent dipoles may have considerable inertia, so relaxation frequencies for orientation polarization may be expected to occur at relatively smaller frequencies, as in the radio-frequency range. Since the alternation of interfacial polarization requires a whole body of charge to be moved through a resistive material, the process may be slow. The relaxation frequency for this mechanism is thus low, occurring at about 10^3 Hz.

Figure 3.36 shows a curve of the variation of the dielectric constant (relative permittivity) with frequency for a hypothetical solid dielectric having all four mechanisms of polarization. Note that except at high frequencies the electronic mechanism makes a relatively low contribution to permittivity. However, in the optical range of frequencies, only this mechanism and the ionic mechanism operate; they therefore strongly influence the optical properties of materials.

Dielectric Constant vs. Temperature

Liquids have higher dielectric constants than solids because dipole orientation is easier in the former. The effect is shown schematically in Figure 3.37a. After the

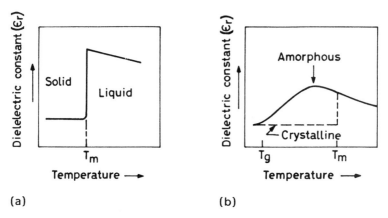

FIG. 3.37 Variation of dielectric constant with temperature (schematic). (a) Crystalline material. (b) Amorphous polymer. A crystalline polymer containing polar group would behave as shown by dashed lines.

abrupt change due to melting, the dielectric constant decreases as the temperature is increased, which is due to the higher atomic or molecular mobility and thermal collisions tending to destroy the orientation of dipoles.

Figure 3.37b shows the schematic variations of dielectric constants with temperature for amorphous solids, such as glasses and many polymers. Above the glass transition temperature (T_g), atoms and molecules have some freedom of movement, which allows orientation of permanent dipoles with the field, thereby increasing the dielectric constant. Since the polar groups which contribute to orientation polarization are not identically situated in an amorphous matrix, the dielectric constant changes over a temperature range rather than abruptly at a single temperature as in a crystalline material (cf. Fig. 3.37a). The decrease in the dielectric constant after melting is again due to greater molecular mobility and thermal collisions.

Dielectric Losses

The behavior of a dielectric under an applied load has much in common with that of a material subjected to mechanical loading. The displacements of atoms and molecules within a material, when a mechanical force is applied, do not occur instantaneously but lag behind the force, resulting in elastic aftereffect and energy dissipation by mechanical hysteresis under an alternating force. Similarly, in dielectrics the lag of polarization behind the applied field produces energy dissipation by electrical hysteresis in an alternating field (Fig. 3.38b). Such energy losses are related to the *internal dipole* friction. The rotation of dipoles

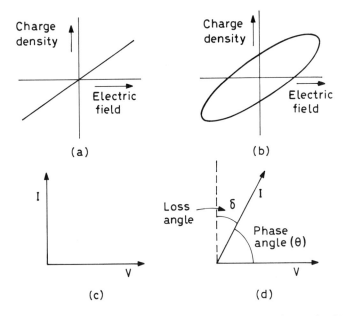

FIG. 3.38 Charge density vs. electric field. (a) Loss-free cycle. (b) High-loss cycle. (c) Phase shift in a perfect capacitor. (d) Phase shift in a real capacitor.

with the field is opposed by the internal friction of the material, and the energy required to maintain this rotation contributes to the power loss in dielectrics.

Besides electrical hysteresis, leakage currents also contribute to dielectric losses. Leakage currents occur mainly by ionic conduction through the dielectric material and are usually negligible except at high temperatures. There are various ways of measuring energy losses by a dielectric. A fundamental property of a capacitor is that if an alternating voltage is applied across it in a vacuum, the current that flows to and from it due to its successive charging and discharging is 90° out of phase with the voltage (Fig. 3.38c), and no energy is lost. However, in real capacitors containing a dielectric, the lag of polarization causes a phase shift of the current (Fig. 3.38). The phase shift angle, δ is called the *loss angle*, and its tangent (tan δ) is referred to as the *loss tangent* or *dissipation factor*. (An ideal dielectric would have a phase angle of 90°, and hence the loss angle would be zero.) The sine of the loss angle (sin δ) or the cosine of the phase angle (cos θ) is termed the *power factor*. In electrical applications the *power loss (PL)* is defined as the rate of energy loss per unit volume and is derived to be

$$PL = \frac{\omega E^2 \epsilon_o}{2} \epsilon_r \tan \delta \qquad (69)$$

where E is the electric field and ω is the angular velocity; $\epsilon_r \tan \delta$ is called the *loss factor*.

For most materials δ is very small; consequently, the three measures of energy dissipation—δ, $\tan \delta$, and $\sin \delta$ ($= \cos \theta$)—are all approximately the same. Since these values vary somewhat with frequency, the frequency must be specified (see Table 3.6).

It is evident from this discussion that the power loss and heat dissipation in a dielectric will be aided by a high dielectric constant, high dissipation factor, and high frequency. Therefore, for satisfactory performance electrical insulating materials should have a *low* dielectric constant and a *low* dissipation factor but a *high* dielectric strength (Table 3.6) and a *high* insulation resistance. Polyethylene and polystyrene with their exceptionally low dissipation factors (<0.0005) and low dielectric constant (2.3–2.5) are the most suitable for high-frequency applications, as in television and radar. For dielectrics used in capacitors, however, a high dielectric constant is desirable.

Dielectric Losses of Polar Polymers

When a polymer having polar groups (e.g., polymethyl methacrylate, polyvinyl chloride) is placed in an electric field, the polar groups behaving as dipoles tend to orient themselves in response to the field (Fig. 3.39a). In an alternating field the friction of the dipoles rotating with the field contributes to the dielectric loss. This loss is small at low frequencies where the polar groups are able to

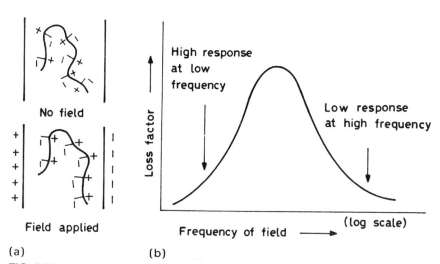

(a)   (b)

FIG. 3.39 (a) Orientation of a polar polymer molecule in an electric field. (b) Dielectric respone of a polar polymer in an alternating electric field.

respond easily to the field, and also at high frequencies where the polar groups are unable to change their alignment with the field. The loss is maximum in the transitional region where the polymer is passing from high response at low frequency to low response at high frequency (Fig. 3.39b).

Since the friction of dipoles in an alternating field produces heat, polar polymers can be heated by the application of radio-frequency field in fabrication processes (see high-frequency welding, Chapter 2).

OPTICAL PROPERTIES [13]

Optical characteristics of plastics include color, clarity, general appearance, and more directly measurable properties, such as index of refraction. For optical applications, however, other properties, including dimensional stability, scratch resistance, temperature limitation, weatherability, and water absorption, must be considered.

Optical Clarity

Most resins by nature are clear and transparent. They can be colored by dyes and will become opaque as pigments or fillers are added. Polystyrene and poly-(methyl methacrylate) are well known for their optical clarity, which even exceeds that of most glass. Optical clarity is a measure of the light transmitting ability of the material. It depends, among other things, on the length of the light path, which can be quantitatively expressed by the *Lambert–Beer law*, or $\log (I/I_0) = -AL$, where I/I_0 is the fraction of light transmitted, L is the path length, and A is the *absorptivity* of the material at that wavelength. The absorptivity describes the effect of path length and has the dimension of reciprocal length.

As an example, an absorptivity of 0.60 cm^{-1} corresponds, by the log relation, to a 25.2% transmission through a thickness of 1 cm and 88.0% transmission through a thickness of 0.1 cm. These figures, however, do not take into account the loss due to reflections that occur at air–plastic interfaces. One method of estimating this loss at each surface is the *Fresnel* relation, $(n - 1)^2/(n + 1)^2$, where n is the refractive index against air. For example, if $n = 1.500$, then according to this relation, there is a 4.0% reflection loss at each surface. The apparent transmissions in the example thus become

$0.96 \times 0.252 \times 0.96 = 0.232$ or 23.2% for a 1-cm sheet

and

$0.96 \times 0.88 \times 0.96 = 0.810$ or 81.0% for a 0.1-cm sheet

Perfect optical clarity for this material would thus correspond to a net transmission of $0.96 \times 1 \times 0.96$ or about 92%.

Transparent colored materials are obtained by adding a dye to a water white resin. A color results when a dye removes part of the visible light traveling through the piece. The red color, for example, is produced by a dye which absorbs the blue, green, and yellow components of the light and transmits the red unchanged (Fig. 3.40). However, for any dye to be effective it must be soluble in the plastic, and it is best incorporated in the plastic before molding. Fluorescent dyes absorb radiant energy at one wavelength, perhaps in the ultraviolet, and emit it as less energetic but more visible radiation.

Cast phenolics, allyls, cellulosics, and many other clear plastics show a natural tendency to absorb in the blue and to be yellowish. Ultraviolet photodecomposition is largely responsible for the development of yellowish color in plastics exposed to sunlight. Incorporation of an invisible ultraviolet-absorbing material such as phenyl salicylate or dihydroxybenzophenone in the plastic greatly reduces photodecomposition. Addition of a blue or green tint into the plastic can also mask the yellow color. In addition to yellowing, plastics show darkening due to outdoor exposure as the transmission curve shifts downward, and there is pronounced absorption at wavelengths shorter than about 5000 A (Fig. 3.41).

The optical clarity of a plastics specimen is measured by the lack of optical haze. *Haze* is arbitrarily defined as the fraction of transmitted light which is deviated 2 1/2° or more from an incident beam by forward scattering. When the haze value is greater than 30%, the material is considered to be translucent. The forward scattering of the light beam, which is responsible for haze, occurs at internal interfaces such as caused by a dust particle, bubble, particles of a filler or pigment, or by density changes.

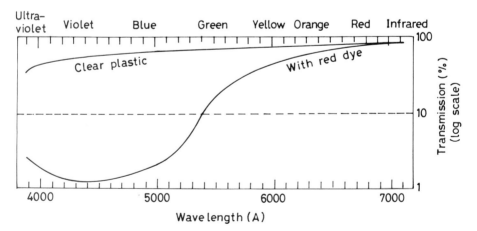

FIG. 3.40 Light transmission diagram for plastic.

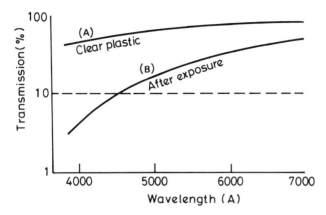

FIG. 3.41   Yellowing of plastics on exposure.

Due to scattering at interfaces, a crystalline plastic with myriads of crystallite regions bounded by interfaces is translucent. Crystalline polyethylene is thus translucent at room temperature, but, on warming, the crystallites disappear and the material becomes transparent. It can thus be inferred that plastics which are transparent at room temperature, such as polystyrene or poly(methyl methacrylate), are of the noncrystalline type and without fillers. The effect of interfaces due to fillers depends to a large extent on the difference in the indices of refraction of plastic and filler. Thus transparent glass-filled polyester panels are obtained if the indices of refraction of glass and resin are identical and the glass is surface treated to enable the resin to wet the glass completely.

Color Assessment

In the plastics industry color is mostly assessed by visual examination. The problem with visual evaluation is the differences between individual observers. However, by using a predetermined light source at fixed distance and viewing a sample together with a standard by an experienced observer, a pretty good idea can be gotten about the color of a plastic sample. If the sample and standard do not match, we must know the difference. The problem is handled by use of one or more *limit standards* in addition to the target standard. Depending on the amount of difference that can be tolerated, the number of limit standards as well as their degree of difference from the target is determined. In some elaborate systems the limit standards are set up in three directions from the target standard: the high and low limits for lightness and darkness (value), hue, and saturation (chroma).

Lightness or value describes the extent to which an object appears to reflect the incident light: the darker the color, the less an object reflects the light falling on it. Lighter colors reflect a larger fraction of the incident light. Hue describes whether an object appears red, yellow, blue, green, or some intermediate shade like orange. *Saturation* or *chroma* is the amount of chromatic response or the amount of color or hue present.

Munsell Color-Order System [14]

In the Munsell system, lightness or value is represented from top to bottom, hue by a hue circle, and saturation or chroma by the distance from the scale of grays making up the center column (Fig. 3.42)

Brightness or value varies from black at the bottom with a value of 0 to white at the top with a value of 10. The hue circle is made up of 10 hues—five principal hues (red, yellow, green, blue, and purple) and five intermediate hues. Each of the 10 hues is divided into 10 steps. At the center of the hue circle is the gray pole, and the colors increase in saturation as the radius of the hue circle increases from the gray pole. The most saturated colors occur at the periphery of this hue circle.

The Munsell system is illustrated by the 1500 or so samples of the *Munsell Book of Color*. Each page represents a constant hue, and on each page samples

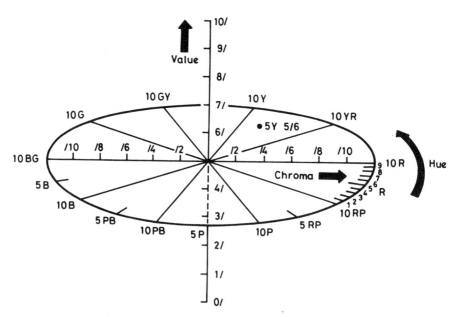

FIG. 3.42 Diagram of Munsell color-order system. (From Ref. 14.)

are represented by value in the vertical direction and chroma in the horizontal direction. Each sample carries a notation denoting its position. The notation for a color of hue 5 YR, value 5, chroma 6, is 5 YR 5/6—a yellowish red. The notation for a color midway between 2.5 YR 5/6 and 5 YR 6/8 is 3.75 YR 5.5/7.

Instrumental Methods of Color Measurement

The instrumental color measurement has its roots in the Commission Internationale l'Eclairage (CIE) order system. In this system every light stimulus is expressed by three numbers, X, Y, Z, called the *tristimulus values*. The spectral power distribution of the light source, the spectral reflectance of the sample, and the spectral response of the eye in the form of color-matching functions are the three important components in specifying color. These three components must be multiplied and integrated over the visible wavelength region to calculate the tristimulus values, as shown in Figure 3.43.

The CIE tristimulus values can be used directly to see if two colors match, but most often they are used in "maps" of color space called *chromaticity diagrams*. Hue and saturation are evaluated as

$$x = \frac{X}{x + y + z} \qquad\qquad y = \frac{Y}{x + y + z}$$

The third dimension, lightness, is included by adding the lightness scale, Y, perpendicular to the two-dimensional chromaticity diagram defined by x and y. A typical chromaticity diagram with hue, saturation, and lightness coordinates is shown in Figure 3.44. Thus color can be described by tristimulus values (X, Y, Z) or chromaticity coordinates (x, y) and lightness (Y).

Index of Refraction

The index of refraction for any transparent material is the ratio of the sine of the angle of an incident ray to the sine of the angle of refraction (Fig. 3.45). It also corresponds to the ratio of the speed of light in a vacuum (or air, closely) to its speed in the material. The refractive index values of several common plastics are compared with those of other materials in Table 3.7.

The refraction property makes possible the focusing action of a lens. Plastic lenses have the advantage that they are light in weight (about half as heavy as glass) and practically nonshattering. But they have the disadvantage of low scratch resistance and comparatively poor dimensional stability for temperature changes. Where tolerances are less critical, plastic lenses can be mass produced by virtue of the moldability of plastics; these lenses are quite satisfactory for inexpensive camera view finders, for example, but in applications where

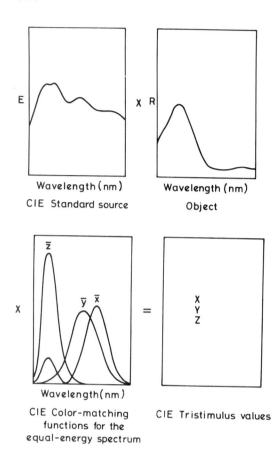

FIG. 3.43 Calculation of tristimulus values from the three important components for specifying color. (From Ref. 15.)

exacting tolerances are required, such as cameras, periscopes, or similar high-resolution devices, molded plastic lenses have not been suitable. If plastic lenses are to be used in these applications, they should be ground and polished in much the same manner as in glass, though, of course, it is easier to do so.

The refractive index, n, of an isotropic material is given by the Lorentz–Lorenz relation

$$\frac{n^2 - 1}{n^2 + 2} = \frac{4}{3}\pi P$$

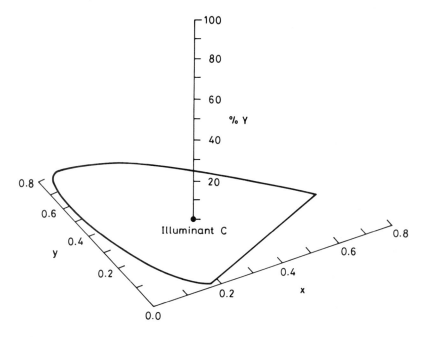

FIG. 3.44 Chromaticity diagram with hue, saturation, and lightness coordinates. (From Ref. 15.)

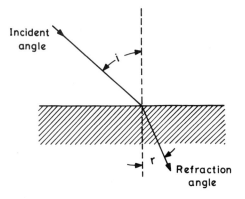

FIG. 3.45 Refraction of light. Index of refraction: $n = (\sin i)/(\sin r)$.

TABLE 3.7 Index of Refraction

Material	Index of refraction (n)
Air	1.00
Water	1.33
Cellulose acetate	1.48
Poly(methyl methacrylate)	1.49
Common glass	1.52
Poly(vinyl chloride) copolymer	1.53
Polystyrene	1.59
Flint glass	1.65
Diamond	2.42

where P is the *optical polarizability* per unit volume of material. For a pure substance it is more convenient to write this equation in the form

$$\frac{(n^2 - 1)M}{(n^2 + 2)\rho} = \frac{4}{3} \pi N_A \alpha = R \tag{70}$$

where M is the molecular weight of the substance, ρ is the density, N_A is Avogardro's number, α is the *optical polarizability* of a single molecule of the substance, and R is the *molar refraction*. The quantity α represents the amount of polarization of the molecule per unit electric field caused by the alternating electric field associated with the light ray passing through the material. The polarization may be regarded as due to the shift of the center of charge of the more loosely bound electrons relative to the nucleus (see Fig. 3.35).

The refractive index varies with wavelength of light and is measured by the *optical dispersion*—that is, the difference in the refractive indexes for different wavelengths. It is responsible for the spectrum-separating ability of a prism in a spectroscope. Most plastics have relatively low optical dispersion. This property makes them more suitable for eyeglasses and large lenses for projection television.

Piped Lighting Effect

The difference in indexes of refraction for air and for a transparent solid (plastic or glass) is responsible for the ability of the latter, used as a rod or plate, to bend

or pipe light around a curve. Bending can be explained as follows. It is evident
from Figure 3.45 that a light beam in air cannot enter a material with $i > 90°$.
Since $\sin 90° = 1$, the maximum angle of refraction is that for which the sine is
$1/n$, where n is the index of refraction. For poly(methyl methacrylate) $n = 1.49$,
and the maximum angle of refraction becomes $\sin^{-1}(1/1.49)$ or $42°$ approximately. Obviously a light beam refracted at a greater angle cannot have come
from the air, but it must have been reflected internally. A plastic plate with small
curvature therefore gives internal reflections and can thus bend a light beam, as
shown in Figure 3.46. Consideration of geometric optics shows the minimum
radius of curvature for this piping of light to correspond to $1/(n - 1)$ times the
thickness.

The light-bending ability of transparent materials is made use of in several
novel applications (e.g., illuminating sensitive tissues by distant lamps and edge
lighting of signs). In the former, illumination from a distant lamp is piped by the
plastic, and the heat effects of the lamp are largely eliminated by low thermal
conductivity of the plastic. In the latter application, light entering the edge of a
transparent sheet is carried by internal reflections between polished surfaces until it reaches a roughened area from where it escapes with luminescent glow.
Note that glass might also serve well in place of plastics in these applications.
However, the choice of plastics brings with it the advantages that it is nonshattering, lightweight, and easy to fabricate.

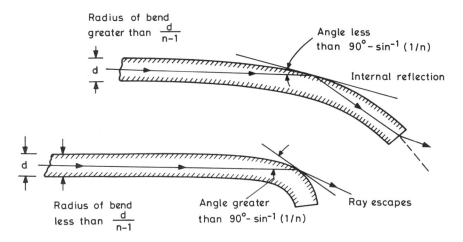

FIG. 3.46 Principle of piped light.

Stress-Optical Characteristics

An important phenomenon observed in amorphous plastics (also observed in optical glass) is the development of optical anisotropy due to stress. The stress-optical characteristic of transparent plastics is the basis of the important technique of photoelasticity by which stress and strain in complicated shapes, for which no analytical solution is readily available, can be determined experimentally and simply. The amount of strain in various parts of a transparent plastic model of a machine part, subjected to loads simulating those in actual operation, is determined by measuring the anisotropy with polarized light, therefore from this measurement the distribution of stress and strain in the actual metal part can be deduced. Very complicated shapes, including complete structures such as bridges or aircraft wings, have been successfully analyzed in this manner, employing plastic models. The same stress-optical characteristic also permits examination of locked-in stresses in a molded plastic part. The part is examined under *polarized* light, and the amount of stress is indicated by the number of *fringes* or *rings* that become visible. Illumination with white light gives colorful patterns involving all the colors of the spectrum. Monochromatic light is, however, used for stress analysis because it permits more precise measurements.

In general, a single ray of light entering an anisotropic transparent material, such as crystals of sufficiently low symmetry and strained or drawn polymers, is propagated as two separate rays which travel with different velocities and are polarized differently. Both the velocities and the state of polarization vary with the direction of propagation. This phenomenon is known as *birefringence* or *double refraction*, and the material is said to be *birefringent*. The stress-optical characteristic of plastics arises from this phenomenon of birefringence, induced by strains due to applied stress. The applied stress produces different densities along different axes. The stress-optical coefficient of most plastics is about 1000 psi (70 kgf/cm^2 or 6.9 MPa) per inch (2.54 cm) thickness per fringe. This means that a 1-in. thick part when illuminated with monochromatic light and viewed through a polarizing filter will show a dark fringe or ring for a stress of 1000 psi. This sensitivity is many times higher than that shown by glass and makes transparent plastics useful in photoelastic stress analysis.

In one variation of the process the deforming stress is applied to a warm, soft plastic model which is then quenched to room temperature. The resulting locked-in stress may then be analyzed at leisure or, perhaps more conveniently, by cutting sections from the model and examining them separately. For this application, however, the plastic should have stress-optical stability. In this regard, unplasticized transparent plastics such as poly(methyl methacrylate), polystyrene, and cast poly(allyl phthalate) are superior to plasticized materials such as cellulose acetate.

THERMAL PROPERTIES [16]

The useful thermal properties of plastics include specific heat, thermal expansion, thermal conductivity, and thermal softening.

Specific Heat

For most plastics the specific heat value (calories per gram per °C) lies between 0.3 to 0.4. On a weight basis this value is much higher than that of most metals. Both iron and copper, for example, have specific heats of about 0.1 at ordinary temperatures. However, on a volume basis, the specific heats of plastics are lower than those of common metals, because of the substantially lower density of plastics.

Determination of precise values of specific heats of plastics is difficult since temperature changes of the material may be accompanied by chemical transformations, such as traces of precuring and, perhaps, some drying and evaporation. Approximate values of specific heat of a resin, per monomer unit, can be found from additive atomic values: e.g., carbon, 1.8; hydrogen, 2.3; silicon, 3.8; oxygen, 4.0; fluorine, 5.0; sulfur and phosphorus, 5.4; most other elements, 6.2. For example, for PVC we can calculate a specific heat of $2 \times 1.8 + 3 \times 2.3 + 6.2$ or 16.7. Since the molecular weight of the monomer is 62.5, the specific heat in calories per gram per °C is $16.7/62.5$ or 0.27, which agrees closely with the reported value of 0.28. The specific heats of mixtures and copolymers can be estimated in a similar way ($1 \text{ cal}/g-°C = 4.187 \times 10^3 \text{ J}/kg-K$).

Knowledge of specific heat of a material helps to determine the energy requirement for increasing its temperature. In compression molding and injection molding the theoretical heat requirement can be calculated as the sum of this direct heat and any latent heat of melting minus any energy released by chemical reaction. But this heat requirement is not large compared to the heat loss by radiation and conduction from the press. The proportion of losses is substantially lower for injection molding than for compression molding. It is nevertheless customary in both cases to establish the heating requirements as well as mold cooling requirements by a process of trial and error. In dielectric preheating, however, the specific heat of a molding powder is of more direct concern; this knowledge along with the amount of material and time enables one to calculate the required amount of radio-frequency power. For example, to raise the temperature of a 1-kg preform of specific heat 0.35 through 80°C in 1 min requires $1 \times 0.35 \times 80$ or 28 kcal/min or about 2 kW.

Thermal Expansion

Linear (thermoplastic) polymers have very high thermal expansion coefficients since they are weakly bonded materials and need less input of thermal energy to expand the structure. This applies to all polymers of the vinyl type which have expansion coefficients of about $90 \times 10^{-6}/°C$. Network (thermosetting) polymers having a three-dimensional framework of strong covalent bonds exhibit less thermal expansion and have expansion coefficients in the range of 30 to 70 $\times$ $10^{-6}/°C$. These may be compared with values of $11 \times 10^{-6}/°C$ for mild steel, $17 \times 10^{-6}/°C$ for ordinary brass, and less than $10 \times 10^{-6}/°C$ for ceramics. In spite of high thermal expansion, plastics do not easily undergo thermal cracking, because they also have very low elastic moduli and large strains do not induce high stresses.

One area where the high expansion of polymers plays a significant role is the molded dimensions of plastic parts. The linear mold shrinkage due to thermal contraction from molding to room temperature is usually about ½ to 1%. Polyethylene and certain other materials exhibit even a higher shrinkage. As a result, plastic parts with close tolerances are difficult to make. It is also possible during the molding operation that different parts are not all at a uniform temperature. This may lead to differential shrinkage during cooling and produce *warping*, locked-up internal stresses, and weakening of the part.

Warping may be prevented by a suitable mounting. Internal stresses in plastics that are more subject to creep and cold flow tend to relieve themselves slowly; warming the part accelerates the process. Internal stresses in polystyrene and other brittle plastics can be removed by an annealing process similar to the one used in glass manufacture. It involves a controlled heating-cooling cycle, and for plastics it is obtained by immersing the parts in a liquid held at the proper temperature, followed by very slow cooling. Polystyrene for example can be heated in water at 80°C and then slowly cooled to 65°C. It is then cooled in undisturbed air.

The relatively high thermal expansion of plastics poses a problem in the use of molded metal inserts which are sometimes required for electrical contacts, screw thread mountings, or increased strength. To minimize stresses due to inserts, manufacturers usually use plastics with low coefficient of expansion. Phenolics and ureas are commonly used because their coefficient of expansion are among the lowest of the common plastics. For many plastics the use of molded metal inserts is not satisfactory because of the excessive stress produced. As an example, polystyrene with brass inserts on cooling from the molding temperature of 160 to 20°C produces a strain of $(70 - 17) \times 10^{-6} \times (160 - 20)$ or 0.0074. For an elastic modulus of 0.46×10^6 psi (3.2 GPa), the internal stresses become 3400 psi (23 MPa). The presence of this much internal stress renders the part useless, since the tensile strength of polystyrene is only 3600 psi (25 MPa). However, by use of an appropriate filler the thermal expansion of plastics and, hence, the internal stresses can be reduced. Addition of 11% by weight of aluminum oxides in polystyrene gives a mixture whose thermal expansion is identical to that of brass. Brass inserts in this alumina-polystyrene composite show no evidence of internal stresses and are quite satisfactory.

When synthetic resins are used as cements, the thermal expansion needs to be matched; this matching is done by judiciously choosing the filler. In the production of panels of aluminum alloy overlaid on phenolic laminate, a filler consisting of a glass fiber-starch mixture is used to reduce the expansion characteristic of the resin and to match its expansion exactly with that of aluminum. The aluminum panel then adheres completely and cannot be removed without tearing the metal or rupturing the phenolic core. In cases where two materials differ in expansion coefficients, the daily temperature variations set up stress cycles of real magnitude, and the effect is magnified at every variation in adhesion. Flexible resins such as cellulose nitrate may be able to withstand these stress cycles quite well, but the inherently brittle resins tend to fail, and, for them, adjustment of thermal expansion by use of an appropriate filler is the best solution.

There are cases where the high thermal expansion of plastics is used to advantage. One example is the shrink fitting of handles of cellulose nitrate on screwdrivers. Another example is in mold design where the shrinkage on cooling is sufficient to permit a small undercut.

Thermal Conductivity

Thermal conductivities of plastics are relatively low and approximate 0.0004 (cal-cm)/(°C-cm²-sec). The corresponding values are 0.95 for copper, 0.12 for cast iron, 0.002 for asbestos, 0.0008 for wood, and 0.0001 for cork (SI values in J/s-m-k are obtained by multiplying by 418.7). Because of their low thermal conductivities, plastics are used for handles to cooking utensils and for automobile steering wheels. The low thermal conductivity is also responsible for the pleasant feel of plastic parts. Quite hot or quite cold objects can be handled with less difficulty if they are made of plastic, since the thermal insulation afforded by the plastic prevents a continuous rush of heat energy to (or from) the hand.

Both thermal conductivity and temperature resistance of plastics have to be considered for their use at high temperatures. As an example consider a teapot handle; it must not deform even at 100°C. Therefore, common thermoplastics such as cellulose acetate are ruled out, but both a wood-flour-filled and an asbestos-filled phenolic might be considered.

The thermal conductivity of a mixture is nearly proportional to the volume percentage of each component. Wood-flour-filled phenolic has a higher thermal conductivity than the pure resin, but the conductivity of this composite is still low enough to justify its use as the handle to a teapot. This composite can also withstand temperatures up to 100°C sufficiently well to give the handle a reasonable service life. For parts subjected to higher temperatures, asbestos-filled phenolic is a better choice. It can be used as the insulating connection to an electric iron, for example.

Design of a handle determines to a great extent its service life. Quite often handles are found fastened directly to the hot object without regard for any temperature limitation; at the junction the plastic becomes brittle because of high temperature, and failure occurs. If the handle can be separated from the heated part and some cooling arrangement is included as part of the design, improved performance is to be expected.

Transition Temperatures and Temperature Limitations

Both first- and second-order transitions are observed in polymers. Melting and allotropic transformations are accompanied by latent-heat effects and are known as first-order transitions. During second-order transitions, changes in properties occur without any latent-heat effects. Below the second-order-transition temperature (glass transition temperature) a rubberlike material acts like a true solid (see Chapter 1). Above this temperature the fixed molecular structure is broken down partially by a combination of thermal expansion and thermal agitation. The glass transition temperature of polystyrene is 100°C; below 100°C polystyrene is hard and brittle, and above 100°C it is rubberlike and becomes easily deformed.

The heat-distortion temperature is defined as the temperature at which the midpoint of a beam ½ by ½ by 5 inches, supported ½-in. from each end, shows a net deflection of 0.01 in. when loaded centrally with 2.5 kg and heated at the specified rate of 2°C/min. Testing is also done at one-quarter of this load. For most materials the two heat-distortion temperatures are within 10°.

A large difference in these two temperatures indicates a material sensitive to temperature change, and for such materials at any elevated temperature stress should be minimum up to the heat-distortion temperature. A plastic is expected to maintain its shape under load, and hence this temperature represents an upper limiting point at which a plastic may be used.

Recommendations of upper temperature limits for plastics are usually based on general experience, although some consideration is given to high-load and low-load heat-distortion temperatures. The size and shape of the part as well as its molding conditions govern to a certain extent the maximum permissible temperature; service conditions such as temperature variations and humidity are also important.

As the service temperature is raised, thermoplastic materials become softer and thermosets become brittle with time because of oxidation of resin or scorching of a filler, such as wood flour. One should know the temperature limitations of plastics in order to specify and use them intelligently (see Chapter 1). Since a small change in temperature may make a great difference in properties, it is important to know precisely the temperature ranges to be encountered by the materials in service.

TESTING OF PLASTICS

Testing of plastic materials may be undertaken for several reasons, including quality assurance, obtaining typical value data, and assessment of a material with regard to its suitability for a given application. Tests for the latter case differ from those for the first two in philosophy as well as in technique.

Quality assurance tests and tests used to obtain typical value data require techniques that can be used either to ensure process uniformity or to compare materials under standard laboratory conditions. These tests are usually oversimplified and are limited only to these two aspects. Testing for end-use performance—that is, to establish the material's suitability for a given purpose—requires techniques that can be used to establish that a material or component fulfills its functions by retaining all significant properties throughout its expected service life. These techniques may be simply the standard test methods, modified when necessary to conform to the service conditions, or they may require development of special methods [17].

A situation peculiar to plastics is that physical properties of these materials cannot be presented as unchanging or typical values that describe a material's universal behavior. This aspect is often overlooked by engineers, which is evident from the fact that many of the tests as well as testing equipment used for plastics were originally adopted, with minor modification, from those used for steel, wood, concrete, and paper. It is, however, not hard to realize that materials of construction based on high polymers would require special techniques for characterization. The reason is that when plastics are subjected to mechanical stress or to electrical, optical, or thermal fields, the polymer molecules rearrange themselves at a rate varying with time, temperature, and other environmental conditions. Such molecular rearrangement gives rise to viscoelastic phenomena in mechanical tests, polarization phenomena in electric measurements, and cross-linking, photodegradation, microstructural changes, polarization, etc., due to optical excitation. Such phenomena are often overlooked by materials engineers in favor of obtaining oversimplified single-point data.

Test methods employed in physical testing can be divided into three classes [18]: fundamental tests, component tests, and hybrid tests. *Fundamental tests* yield data concerning basic physical properties. They provide a means of evaluating the physical properties independent of the design and dimensions of either the test specimen or the test equipment. *Component tests* yield data from a finished component part to allow prediction of the *behavior of the* component in service conditions. Although this test is the most desirable, in many cases a component part is usually not available or cannot be tested as such. *Practical tests* are special forms of component tests. A practical test is so designed that the boundary conditions of the test are identical to those in which the component would be exposed. They are thus valid for specific applications only. A *hybrid test*, on the other hand, is one that stands somewhere between the practical and the scientific. The test yields data from special specimen geometries chosen for obtaining general information supposedly simulating certain service conditions. Examples of hybrid tests are the Izod test, using test specimen of specific geometries, and scratch-resistance tests using scribers cut from arbitrary materials and of arbitrary geometry. Hybrid tests are acceptable in quality assurance testing where the measurement of variability is more important than the significance of the absolute result. However, for research purposes and to establish suitability, fundamental or component tests are to be preferred even though they are more time consuming and more expensive than a hybrid test.

Many organizations are engaged in the development of standards in various subject areas. The number exceeds 400 in the United States alone [19]. The American Society for Testing and Materials (ASTM) has issued approximately 4000 standard specifications and test methods which cover a wide and diverse range, including fundamental, component, practical, and hybrid methods required for the various needs of testing. These tests are so long established and their number is so large that their influence often extends beyond their domain of applicability. The use of single-point hybrid test results in general design calculations and predictions is a common example. Such single-point results are primarily intended as quality control or practical values and certainly have no applicability in design or predictive calculations. Such an application can sometimes lead to an erroneous conclusion.

Standard test methods by nature are arbitrary. The input excitation and boundary conditions of the test are specified so that the results (data) can be used for comparison or as a basis for trade or acceptance. But these conditions seldom match the actual use conditions encountered. The application of the test data in actual use may thus be misleading. However, it is not the test methods that are to be criticized for this but rather it is the failure of the experimenter to extend these measurements to cover the expected input functions, the boundary conditions, and the environment to which the material will be exposed as well as the material's expected service life. A greater awareness of the need for such extended measurements and more input from designers will hopefully lead to greater activity in the future for the development of standard design and performance tests for plastics.

Standard Properties of Plastics

Figures 3.47 to 3.70 illustrate schematically (in simplified form) the bases of some of the standard properties of plastics [20]. Where standard test methods

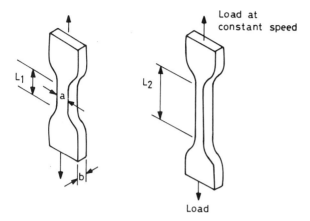

FIG. 3.47 Tests for tensile strength (stress at fracture of the specimen) and elongation (extension of materials under load) of plastics. The first provides a measure of the breaking strength of the material but is radically affected by the rate of loading and the ambient temperature. Tensile strength in $lbf/in.^2$ or kgf/cm^2 = load (lbf or kgf)$/a \times b(in.^2$ or cm^2). Standard test methods: BS 2782 method 301, ASTM D638, ISO R527. Elongation % = $(L_2 - L_1) \times 100/L_1$. Standard test methods: BS 2782 method 301, ASTM D638.

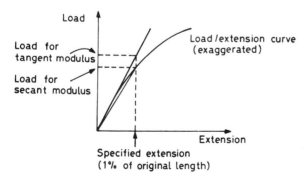

FIG. 3.48 Test for tangent and secant moduli of plastics. For tangent modulus, load on tangent to load-extension curve at specified extension is used for calculating the stress value, while for secant modulus load on secant to load-extension curve is used. The modulus is given by stress/strain. Standard test methods: BS 2782 method 302, ASTM D638, ISO R527.

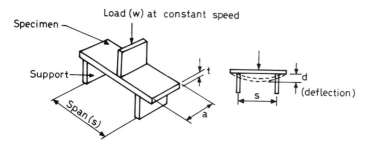

FIG. 3.49 Tests for flexural properties of plastics. Flexural strength = $3ws/2at$ (lbf/in^2 or kgf/cm^2). Standard test method: ASTM D790. Modulus in flexure = $s^3w/4at^3d$ (lbf/in.^{2} or kgf/cm^2). Standard test methods: BS 2782 method 302D, ASTM D790, ISO R178.

have been developed, these have been included. The principles of the different tests are shown, and the properties measured are indicated.

Note that the property values of plastics are highly dependent on the specimen preparation, equipment, and testing techniques used. For this reason it is essential to refer to the appropriate official standard test method when executing the work.

Following is a selective list of the many standards relating to plastics included in the *1982 Annual Book of ASTM Standards*, parts 35 and 36. In the serial designations prefixed to the titles, the number following the hyphen indicates the year of original issue or of adoption as standard or, in the case of revision, the

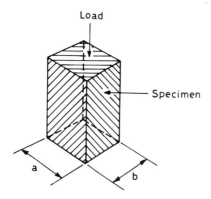

FIG. 3.50 Test for compressive strength of plastics. Compressive strength (lbf/in^2 or kgf/cm^2) = load (lbf or kgf)/$a \cdot b$ (in^2 or cm^2). Standard test methods: ASTM D795, BS 2782 method 303, ISO R604.

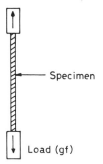

FIG. 3.51 Test for tenacity of filaments, cords, twines, etc. Tenacity (gf/denier) = breaking load (gf)/denier of specimen. (Denier of filaments, cords, twines, etc., is equal to the weight in grams of 9000 meters of the sample.)

year of last revision. Thus, standards adopted or revised during the year 1981 have 81 as their final number. Standards that have been simply reapproved without change are indicated by the year of last reapproval in parentheses, for example (1981).

D 149-81 Tests for dielectric breakdown voltage and dielectric strength of solid electrical insulating materials at commercial power frequencies (see Fig. 3.65)

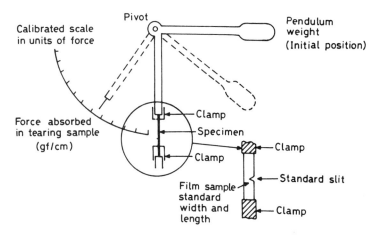

FIG. 3.52 Test for tear propagation resistance of plastic film and thin sheeting. Tear resistance (gf/cm/mil) = force required to tear sample (gf/cm)/thickness of film (mil). Standard test methods: ASTM D1922, BS 2782 method 308B.

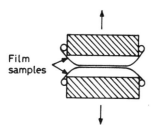

FIG. 3.53 Test for blocking of plastic film. Blocking force (lbf/in.^{2} or kgf/cm^2) = load (lbf or kgf)/initial area of films in contact (in.^{2} or cm^2). Standard test method: ASTM D1893.

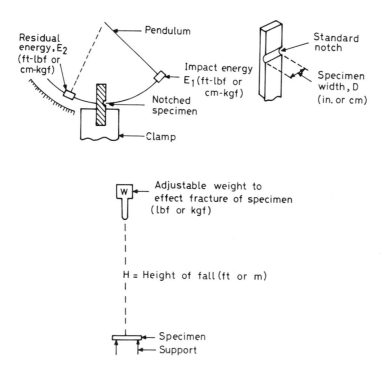

FIG. 3.54 Tests for impact resistance of plastics. Izod impact strength = $(E_1 - E_2)/D$ ft − lbf/in. of notch or cm − kgf/cm of notch. Standard test methods: ASTM D256; BS 2782 method 306A, ISO R180. Falling weight impact strength = $W \cdot H$ ft − lbf or m − kgf. (F_{50} is the energy required to fracture 50% of the specimens.) Standard test method: BS 2782 method 306B.

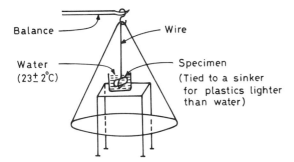

FIG. 3.55 Test for specific gravity and density of plastics. Sp. gr. $= a/(a - w - b)$ where a = wt. of specimen without wire, b = wt. of specimen completely immersed and of the wire partially immersed in water, and w = wt. of partially immersed wire. Standard test method: ASTM D792.

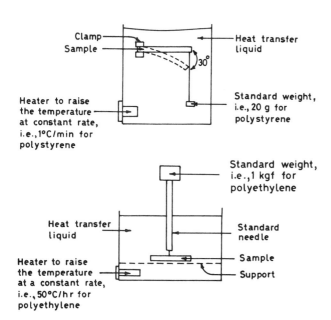

FIG. 3.56 Tests for softening temperatures of plastics. Cantilever softening point is the temperature at which the sample bends through 30°. Standard test method: BS 2782 method 102C. Vicat softening point is the temperature at which the needle penetrates 1 mm into the sample. Standard test methods: ASTM D1525, BS 2782 method 102D, ISO R306.

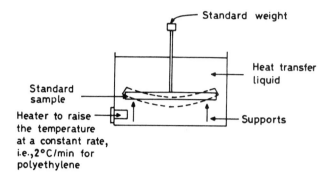

FIG. 3.57 Test for deflection temperature of plastics under flexural load. Heat distortion temperature is the temperature at which a sample deflects by 0.1 in. (2.5 mm). Two measurements are made and quoted: (a) with a stress of 66 lbf/in.^{2} (4.6 kgf/cm^2) and (b) with a stress of 264 lbf/in.^{2} (18.5 kgf/cm^2). Standard test methods: ASTM D648, BS 2782 method 102, ISO R75.

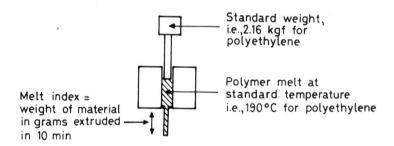

FIG. 3.58 Melt index of plastics. The test measures the rate of flow of polymer melt. It provides an indication of the ease of processing. Standard test methods: ASTM D1238, BS 2782 method 105C, ISO R292.

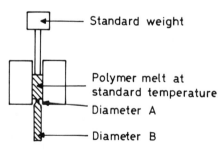

FIG. 3.59 Measurement of swelling of die extrudate. Die swell ratio $= B/A$.

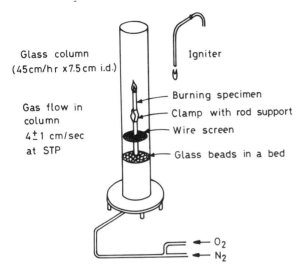

Glass column
(45cm/hr x 7.5 cm i.d.)

Gas flow in
column
4 ± 1 cm/sec
at STP

Igniter

Burning specimen

Clamp with rod support

Wire screen

Glass beads in a bed

O_2
N_2

FIG. 3.60 Determination of oxygen index. Oxygen index, $n\% = 100 \times O_2/(O_2 + N_2)$, where O_2 = volumetric flow rate of oxygen, cm^3/sec, at the minimum concentration necessary to support flaming combustion of a material initially at room temperature, and N_2 = corresponding flow rate of nitrogen, cm^3/sec. Standard test method: ASTM D2863.

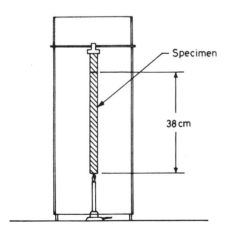

Specimen

38 cm

FIG. 3.61 Test for rate of burning. Burning rate = 38 $(cm)/t$ (min). Standard test method: ASTM D568.

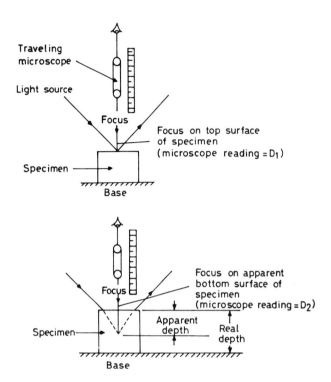

FIG. 3.62 Test for index of refraction of transparent plastics. Refractive index = real depth (measured with vernier)/apparent depth $(D_2 - D_1)$. Standard test method: ASTM D542, ISO R489.

D 150-81   Tests for AC loss characteristics and permittivity (dielectric constant) of solid electrical insulating materials (see Fig. 3.68)

D 256-81   Tests for impact resistance of plastics and electrical insulating materials (see Fig. 3.54)

D 257-78   Tests for DC resistance or conductance of insulating materials (see Fig. 3.67)

D 494-46 (1979)   Test for acetone extraction of phenolic molded or laminated products

D 495-73 (1979)   Test for high-voltage, low-current, dry arc resistance of solid electrical insulation

D 542-50 (1977)   Tests for index of refraction of transparent organic plastics (see Fig. 3.62)

D 543-67 (1978)   Test for resistance of plastics to chemical reagents

D 568-77   Test for rate of burning and/or extent and time of burning of flexible plastics in a vertical position (see Fig. 3.61)

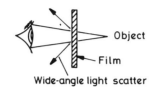

Wide-angle light scatter

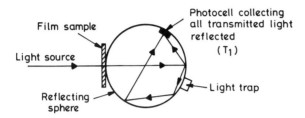

After T₁ is determined the sphere is rotated to measure T₂

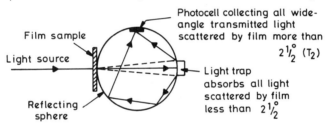

FIG. 3.63 Test for haze of transparent plastics. Haze, $\% = 100 \times T_2/T_1$. A low haze value is important for good short distance vision. Standard test method: ASTM D1003.

D 569-59 (1976)   Measuring the flow properties of thermoplastic molding materials

D 570-81   Test for water absorption of plastics

D 621-64 (1976)   Tests for deformation of plastics under load

D 635-81   Test for rate of burning and/or extent and time of burning of self-supporting plastics in a horizontal position

D 637-50 (1977)   Test for surface irregularities of flat transparent plastic sheets

D 638-82   Test for tensile properties of plastics

D 648-72 (1978)   Test for deflection temperature of plastics under flexural load (see Fig. 3.57)

D 671-71 (1978)   Test for flexural fatigue of plastics by constant-amplitude-of-force

D 673-70 (1982)   Test for mar resistance of plastics

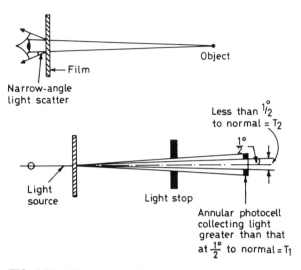

FIG. 3.64 Measurement of narrow-angle light-scattering property of plastic film. Clarity, $\% = 100 \times T_1/(T_1 + T_2)$.

D 695-80   Test for compressive properties of rigid plastics
D 696-79   Test for coefficient of linear thermal expansion of plastics
D 700-81   Spec. for phenolic molding compounds
D 702-81   Spec. for cast methacrylate plastics sheets, rods, tubes, and shapes
D 703-81   Spec. for polystyrene molding and extrusion materials
D 704-81   Spec. for melamine-formaldehyde molding compounds
D 705-81   Spec. for urea-formaldehyde molding compounds

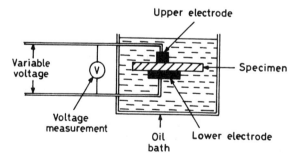

FIG. 3.65 Test for dielectric strength of solid insulating materials. Dielectric strength (V/mil) = maximum voltage before failure (V)/thickness of specimen (mil). Standard test methods: ASTM D149, BS 2782 method 201.

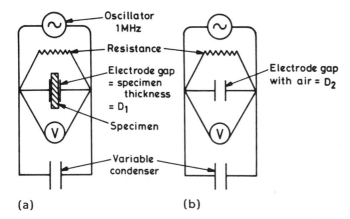

(a)   (b)

FIG. 3.66   Test for permittivity (dielectric constant) of insulating materials. (a) Position of maximum voltage obtained with sample by adjusting variable capacitor. Electrode gap = specimen thickness = D_1; (b) position of maximum voltage obtained with air by adjusting electrode gap to D_2 (variable capacitor remains as set in (a)). Dielectric constant = D_1 (in. or mm)/D_2 (in. or mm). Standard test methods: ASTM D150, BS 2782 method 207A.

D 706-81   Spec. for cellulose acetate molding and extrusion compounds
D 707-81   Spec. for cellulose acetate butyrate molding and extrusion compounds
D 729-81   Spec. for vinylidene chloride molding compounds
D 731-67 (1976)   Test for molding index of thermosetting molding powder
D 746-79   Test for brittleness temperature of plastics and elastomers by impact

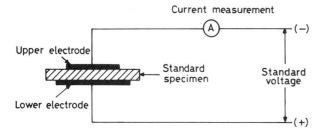

FIG. 3.67   Test for DC resistance of insulating materials. Electrical resistance of specimen (ohm) = applied voltage (*V*)/current measured (*A*). Volume resistivity (ohm − cm) = resistance of specimen (ohm) × arc of upper electrode (cm²)/specimen thickness (cm). Standard test methods: ASTM D257, BS 2782 method 202.

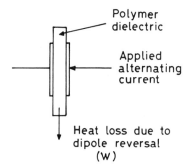

FIG. 3.68 Test for AC loss characteristics of solid insulating materials. Power factor = $W/V \times I$, where W = power loss in watts and $V \times I$ = effective sinusoidal voltage × current in volt-amperes. Standard test methods: ASTM D150, BS 2782 method 207A.

D 785-65 (1981)   Test for Rockwell hardness of plastics and electrical insulating materials (see Fig. 3.21)
D 788-81   Spec. for methacrylate molding and extrusion compounds
D 789-81   Spec. for nylon injection molding and extrusion materials
D 790-81   Tests for flexural properties of unreinforced and reinforced plastics and electrical insulating materials (see Fig. 3.49)
D 792-66 (1979)   Tests for specific gravity and density of plastics by displacement (see Fig. 3.55)

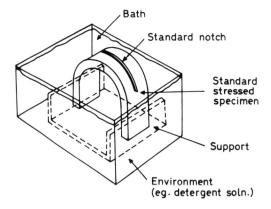

FIG. 3.69 Test for environmental stress cracking of ethylene plastics. Stressed cracking resistance (F_{50}) = time taken for 50% of the specimens to fail (h). Standard test method: ASTM D1693.

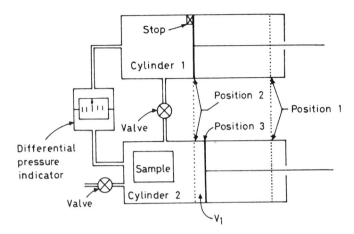

FIG. 3.70 Determination of open-cell content of rigid cellular plastics. Open cell content (approx.) = $100 (V - V_1)/V$, where V_1 = displacement volume of specimen, cm^3 and V = geometric volume of specimen, cm^3. Standard test method: ASTM D2856.

D 817-72 (1978) Testing cellulose acetate propionate and cellulose acetate butyrate

D 834-61 (1979) Tests for free ammonia in phenol-formaldehyde molded materials

D 864-52 (1978) Test for coefficient of cubical thermal expansion of plastics

D 882-81 Tests for tensile properties of thin plastic sheeting

D 953-80 Test for bearing strength of plastics

D 955-73 (1979) Measuring shrinkage from mold dimensions of molded plastics

D 1003-61 (1977) Test for haze and luminous transmittance of transparent plastics (see Fig. 3.63)

D 1004-78 Test for resistance of transparent plastics to abrasion

D 1045-80 Sampling and testing plasticizers used in plastics

D 1180-57 (1978) Test for bursting strength of round rigid plastic tubing

D 1201-81 Spec. for thermosetting polyester molding compounds

D 1238-79 Test for flow rates of thermoplastics by extrusion plastometer (see Fig. 3.58)

D 1242-56 (1981) Tests for resistance of plastic materials to abrasion

D 1248-81 Spec. for polyethylene plastics molding and extrusion materials

D 1299-55 (1979) Test for shrinkage of molded and laminated thermosetting plastics at elevated temperature

D 1303-55 (1979)   Test for total chlorine in vinyl chloride polymers and copolymers

D 1430-81   Spec. for polychlorotrifluoroethylene (PCTFE) plastics

D 1431-81   Spec. for styrene-acrylonitrile copolymer molding and extrusion materials

D 1433-77   Test for rate of burning and/or extent and time of burning of flexible thin plastic sheeting supported on a 45° incline

D 1434-75   Test for gas transmission rate of plastic film and sheeting

D 1435-75 (1979)   Rec. practice for outdoor weathering of plastics

D 1457-81   Spec. for PTFE molding and extrusion materials

D 1463-81   Spec. for biaxially oriented styrene plastics sheet

D 1494-60 (1980)   Test for diffuse light transmission factor of reinforced plastics panels

D 1525-76   Test for vicat softening temperature of plastics (see Fig. 3.56)

D 1562-81   Spec. for cellulose propionate molding and extrusion compounds

D 1593-81   Spec. for nonrigid vinyl chloride plastic sheeting

D 1602-60 (1980)   Test for bearing load of corrugated reinforced plastics panels

D 1603-76   Test for carbon black in olefin plastics

D 1621-73 (1979)   Test for compressive properties of rigid cellular plastics

D 1622-63 (1975)   Test for apparent density of rigid cellular plastics

D 1623-78   Test for tensile and tensile adhesion properties of rigid cellular plastics

D 1636-81   Spec. for allyl molding compounds

D 1637-61 (1976)   Test for tensile heat-distortion temperature of plastic sheeting

D 1638-74   Testing urethane foam isocyanate raw materials

D 1652-73 (1980)   Test for epoxy content of epoxy resins

D 1673-79   Tests for relative permittivity and dissipation factor of expanded cellular plastics used for electrical insulation

D 1693-70 (1980)   Test for environmental stress cracking of ethylene plastics (see Fig. 3.69)

D 1705-61 (1980)   Particle size analysis of powdered polymers and copolymers of vinyl chloride

D 1708-79   Test for tensile properties of plastics by use of microtensile specimens

D 1709-75 (1980)   Test for impact resistance of polyethylene film by the free-falling dart method

D 1746-70 (1978)   Test for transparency of plastic sheeting

D 1755-81   Spec. for poly(vinyl chloride) resins

D 1763-81   Spec. for epoxy resins

D 1784-81 Spec. for rigid poly(vinyl chloride) compounds and chlorinated poly(vinyl chloride) compounds

D 1788-81 Spec. for rigid acrylonitrile-butadiene-styrene (ABS) plastics

D 1789-65 (1977) Test for welding performance of poly(vinyl chloride) structures

D 1824-66 (1980) Test for apparent viscosity of plastisols and organosols at low shear rates by Brookfield viscometer

D 1893-67 (1978) Test for blocking of plastic film (see Fig. 3.53)

D 1894-78 Test for static and kinetic coefficients of friction of plastic film and sheeting

D 1895-69 (1979) Tests for apparent density, bulk factor, and pourability of plastic materials

D 1921-63 (1975) Tests for particle size (sieve analysis) of plastic materials

D 1922-67 (1978) Test for propagation tear resistance of plastic film and thin sheeting by pendulum method (see Fig. 3.52)

D 1925-70 (1977) Test for yellowness index of plastics

D 1927-81 Spec. for rigid poly(vinyl chloride) plastic sheet

D 1929-77 Test for ignition properties of plastics

D 1938-67 (1978) Test for tear propagation resistance of plastic film and thin sheeting by a single-tear method

D 2103-81 Spec. for polyethylene film and sheeting

D 2117-64 (1978) Test for melting point of semicrystalline polymers

D 2123-81 Spec. for rigid poly(vinyl chloride-vinyl acetate) plastic sheet

D 2124-70 (1979) Analysis of components in poly(vinyl chloride) compounds using an infrared spectrophotometric technique

D 2133-81 Spec. for acetal resin injection molding and extrusion materials

D 2146-80 Spec. for propylene plastic molding and extrusion materials

D 2222-66 (1978) Test for methanol extract of vinyl chloride resins

D 2236-81 Tests for dynamic mechanical properties of plastics by means of a torsional pendulum

D 2240-81 Test for rubber property, durometer hardness

D 2287-81 Spec. for nonrigid vinyl chloride polymer and copolymer molding and extrusion compounds

D 2393-80 Test for viscosity of epoxy resins and related components

D 2445-69 (1981) Test for thermal oxidative stability of propylene plastics

D 2457-70 (1977) Test for specular gloss of plastic films

D 2473-81 Spec. for polycarbonate plastic molding, extrusion, and casting materials

D 2474-81 Spec. for vinyl chloride copolymer resins

D 2530-81 Spec. for nonoriented propylene plastic film

D 2552-69 (1980) Test for environmental stress rupture of type III polyethylenes under constant tensile load

D 2566-79   Test for linear shrinkage of cured thermosetting casting resins during cure

D 2583-81   Test for indentation hardness of rigid plastics by means of a Barcol impressor (see Fig. 3.22)

D 2647-80   Spec. for cross-linkable ethylene plastics

D 2673-81   Spec. for oriented polypropylene film

D 2734-70 (1980)   Test for void content of reinforced plastics

D 2765-68 (1978)   Tests for degree of cross-linking in cross-linked ethylene plastics as determined by solvent extraction

D 2842-69 (1975)   Test for water absorption of rigid cellular plastics

D 2849-69 (1980)   Testing urethane foam polyol raw materials

D 2856-70 (1976)   Test for open-cell content of rigid cellular plastics by the air pycnometer (see Fig. 3.70)

D 2857-70 (1977)   Test for dilute solution viscosity of polymers

D 2863-77   Measuring the minimum oxygen concentration to support candle-like combustion of plastics (oxygen index) (see Fig. 3.60)

D 2923-70 (1977)   Test for rigidity of polyolefin film and sheeting

D 2951-71 (1977)   Test for thermal stress cracking resistance of types III and IV polyethylene plastics

D 2990-77   Test for tensile, compressive, and flexural creep and creep rupture of plastics

D 3039-76   Test for tensile properties of fiber-resin composites

D 3171-76   Test for fiber content of resin-matrix composites by matrix digestion

D 3351-74 (1980)   Test for gel count of plastic film

D 3354-74 (1979)   Test for blocking load of plastic film by the parallel-plate method

D 3355-74 (1980)   Test for fiber content of unidirectional fiber-resin composites by electrical resistivity

D 3418-82   Test for transition temperatures of polymers by thermal analysis

D 3536-76   Test for molecular-weight averages and molecular weight-distribution of polystyrene by liquid exclusion chromatography (gel permeation chromatography—GPC)

D 3576-77   Test for cell size of rigid cellular plastics

D 3592-77   Practice for determining molecular weight by vapor pressure osmometry

D 3713-78   Measuring response of solid plastics to ignition by a small flame

D 3749-78   Test for residual vinyl chloride monomer in poly(vinyl chloride) homopolymer resins by gas chromatographic headspace technique

D 3750-79   Practice for determination of number-average molecular weight of polymers by membrane osmometry

D 3835-79   Test for measuring the rheological properties of thermoplastics with a capillary rheometer

D 3881-81   Spec. for phenolic transfer- and injection-molding compounds

D 3985-81   Test for oxygen gas transmission rate through plastic film and sheeting using a coulometric sensor

D 4001-81   Practice for determination of weight-average molecular weight of polymers by light scattering

D 4019-81   Test for moisture in plastics by coulometry

D 4066-81   Spec. for nylon injection and extrusion materials

D 4093-82   Photoelastic measurements of birefringence and residual strains in transparent or translucent plastic materials

E 96-80   Tests for water vapor transmission of materials

E 252-78   Test for thickness of thin foil and film by weighing

E 831-81   Test for linear thermal expansion of solid materials by thermodilatometry

IDENTIFICATION OF COMMON PLASTICS

The first step in the identification of polymers is a critical visual examination. While the appearance of the sample may indicate whether it is essentially a raw polymer or a compounded and processed item, learning about its form, feel, odor, color, transparency or opacity, softness, stiffness, brittleness, bounce, and surface texture may be important in the process of the identification of the polymer. For example, polystyrene, the general purpose polymer, is transparent and brittle, and produces a characteristic metallic tinkle when objects molded from it are dropped or struck.

Identification of plastics is carried out by a systematic procedure: preliminary test, detection of elements, determination of characteristic values, and, finally, specific tests. For an exact identification, however, the test sample should first be purified so that it contains no additives (plasticizers, fillers, pigments, etc.) that may affect the results of an analysis. Purification is achieved by solvent extraction; either the material is dissolved out and polymer is obtained by reprecipitation or evaporation of the solvent, or the pure polymer remains as the insoluble residue. The solvent varies, and a general method cannot be given. However, for many materials, particularly for those in which additives do not interfere, the unpurified material can be investigated and qualitative preliminary tests used.

In general, the following series of investigations should be carried out for preliminary tests of the material: (1) behavior on heating in an ignition tube and in the flame; (2) qualitative detection of heteroelements such as N, S, and halogens; (3) determination of properties such as density, refractive index, and melting point or range; (4) determination of solubility as an aid to polymer identification; and (5) ash or sulfated ash determination as a test for inorganic additives (fillers, pigments, stabilizers).

Behavior on Heating and Ignition

Many polymers can be roughly identified by their behavior when carefully heated and ignited. The test must be carried out carefully on small quantities of material. Nitrocellulose and plastics containing this (e.g., celluloid) burn with explosive violence and other materials such as poly(vinyl chloride) or fluoro-hydrocarbons decompose with the evolution of poisonous or irritating vapors. Only small quantities of material should therefore be taken for heating tests. The heating should be done gently, because if the heating is too rapid or intense, the characteristic changes may not be observed.

In a typical procedure, a small piece or amount (0.1 g) of the test sample is placed on a cleaned glass or stainless steel spatula, previously heated to remove any traces of combustible or volatile materials, and then warmed gently over a small colorless bunsen flame until the sample begins to fume. The decomposing sample is removed from the flame and the nature of the fume or gas is examined with respect to color, odor, inflammability, and chemical identity including acidity, alkalinity, etc.

The sample is next moved to the hottest zone of the small bunsen flame and note is taken of the following: (1) if the material burns and if so, how readily; (2) the nature and color of the flame as the material burns; (3) whether the material is self-extinguishing or continues to burn after removal from the flame, and (4) the nature of the residue.

Observations on heating and ignition of some common polymers are listed in Table 3.8.

Tests for Characteristic Elements

The results of tests for characteristic elements such as nitrogen, sulfur and halogens may serve to roughly indicate the nature of the unknown material, including the nature of the base polymer and additives, if present. The following tests may be performed for qualitative detection of N, S, and halogens.

1. *Nitrogen.* About 50 mg of the test material is heated carefully in an ignition tube with twice the amount of freshly cut sodium, until the sodium melts. A further small amount of material is added and the tube is heated to red heat. It is then dropped carefully into water (20 ml) in a mortar. The solid is powdered and the solution filtered. A little freshly prepared ferrous sulfate is added to the filtrate and the latter boiled for 1 min. It is acidified with dilute hydrochloric acid, and 1–2 drops of ferric chloride solution added. A deep blue coloration or precipitate (Prussian blue) indicates nitrogen (if very little nitrogen is present, a

TABLE 3.8 Heating Tests of Some Common Polymers [21, 22]

Polymer	Color of flame	Odor of vapor	Other notable points
The material burns but self-extinguishes on removal from flame			
Poly(vinyl chloride)	Yellow-orange, green bordered	Resembles hydro-chloric acid and plasticizer (usually ester like)	Strongly acidic fumes (HCl), black residue
Poly(vinylidene chloride)	As above	Resembles hydro-chloric acid	As above
Polychloroprene	Yellow, smoky	As above	As above
Phenol-formaldehyde resin	Yellow, smoky	Phenol, formaldehyde	Very difficult to ignite, vapor reaction neutral
Melamine-formaldehyde resin	Pale yellow, light	Ammonia, amines (typically fish like), formaldehyde	Very difficult to ignite, vapor reaction alkaline
Urea-formaldehyde resin	As above	As above	As above
Nylons	Yellow-orange, blue edge	Resembling burnt hair	Melts sharply to clear, flowing liquid; melt can be drawn into a fiber; vapor reaction alkaline
Polycarbonate	Luminous, sooty	Phenolic	Melts and chars
Chlorinated rubber	Yellow, green-bordered	Acrid	Strongly acidic fumes, liberation of HCl; swollen, black residue
The material burns and continues burning on removal from flame			
Polybutadiene (BR)	Yellow, blue base, smoky	Disagreeable, sweet	Chars readily; vapor reaction neutral
Polyisoprene (NR, gutta percha, synthetic)	Yellow, sooty	Pungent, disagreeable, like burnt rubber	As above
Styrene-butadiene rubber (SBR)	Yellow, sooty	Pungent, fruity smell of styrene	As above
Nitrile rubber (NBR)	Yellow, sooty	Like burnt rubber/burnt hair	As above
Butyl rubber (IIR)	Practically smoke free candle like	Slightly like burnt paper	Melt does not char readily
Polysulfide rubber (polymer itself emits unpleasant, mercaptan like odor)	Smoke-free, bluish	Pungent; smell of H_2S	Yellow, acidic (SO_2) fumes

TABLE 3.8 Continued

Polymer	Color of flame	Odor of vapor	Other notable points
Cellulose (cotton, cellophane, viscose rayon, etc.)	Yellow	Burnt paper	Chars, burns without melting
Cellulose acetate	Yellow-green, sparks	Acetic acid, burnt paper	Melts, drips, burns rapidly, chars, acidic fumes
Cellulose acetate butyrate	Dark yellow (edges slightly blue), somewhat sooty, sparks	Acetic acid/butyric acid, burnt paper	Melts and forms drops which continue burning
Cellulose nitrate (plasticized with camphor)	Yellow	Camphor	Burns very fast, often with explosion
Methyl cellulose	Yellow, luminous	Burnt paper	Melts and chars
Ethyl cellulose	Pale yellow with blue-green base	Slightly sweet, burnt paper	Melts and chars
Polyacrylonitrile	Yellow	Resembling burnt hair	Dark residue; vapor reaction alkaline
Poly(vinyl acetate)	Yellow, luminous, sooty	Acetic acid	Sticky residue, acidic vapor
Poly(vinyl alochol)	Luminous, limited smoky	Unpleasant, charry smell	Burns in flame, self-extinguishing slowly on removal; black residue
Poly(vinyl butyral)	Bluish (yellow edge)	Like rancid butter	Melts, forms drops
Poly(vinyl acetal)	Purple edge	Acetic acid	Does not drip like poly(vinyl butyral)
Poly(vinyl formal)	Yellow-white	Slightly sweet	Does not drip like poly(vinyl butyral)
Polyethylene	Luminous (blue center)	Like paraffin wax (extinguished candles)	Melts, forms drops; droplets continue burning
Poly(-olefins) (PP, EPR, etc.)	As above	As above	As above
Polystyrene	Luminous, very sooty	Sweet (styrene)	Softens, easily ignited
Poly(methyl metyhacrylate)	Luminous, yellow (blue edge), slightly sooty, crackling	Sweet, fruity	Softens, chars slightly

TABLE 3.8 Continued

Polymer	Color of flame	Odor of vapor	Other notable points
Polyoxymethylene	Bluish flame	Formaldehyde	Melts and decomposes; vapor reaction neutral
Polyesters (alkyds, polyester fiber)	Yellow, luminous, sooty	Acrid; sweet aromatic	Melts, burns uniformly or chars giving acidic distillate and black residue

green-blue color is formed initially). This test sometimes fails with substances containing nitro-groups and with nitrogen-containing heterocyclic compounds.

2. *Sulfur*. The fusion is carried out as for nitrogen. The filtrate (2 ml) is acidified with acetic acid and a few drops of an aqueous solution of lead acetate is added. A black precipitate of lead sulfide indicates the presence of sulfur.

3. *Halogens*. The fusion is carried out as for nitrogen. The filtrate (2 ml) is acidified with dilute nitric acid, boiled in the fume cupboard to remove H_2S and HCN, and then a few drops of silver nitrate solution is added. A white precipitate, turning grey-blue, indicates chlorine, a yellowish precipitate indicates bromine and a yellow precipitate indicates iodine.

In Beilstein's test, which provides a simple means of detecting halogens, a minute quantity of the test material is placed on a copper wire (initially heated in a nonluminous bunsen flame until the flame appears colorless and then cooled) and heated in the outer part of the flame. Carbon burns first with a luminous flame. A subsequent green to blue-green color, produced by volatilized copper halide, indicates halogen (chlorine, bromine, iodine).

The different polymers may be classified into several groups according to the elements present as shown in Table 3.9. The focus of identification may be further narrowed down on the basis of other preliminary observations, e.g., fusibility or otherwise, melting point or range, heat distortion temperature, flame tests and tests for thermal degradation, and solubility or extractability in water or different organic solvents. The solubility behaviors of common polymers are compared in Table 3.10.

Specific Tests

When the observations and results of preliminary tests have been considered and most of the possible structures for the polymer base eliminated, an exact identification can be made by carrying out specific tests. Some specific tests for ready identification of specific polymers are described below.

TABLE 3.9 Classification of Most Common Polymers According to Elements Present

Group	Element found	Rubbers[a]	Plastics/fibers
Group 1	N	Nitrile (NBR), polyure-thane (ester/ether urethanes)	Cellulose nitrate, silk, polyamides, polyimides, polyurethanes, polyacrylonitrile, SAN and ABS resins, urea-formaldehyde resins, melamine-formaldehyde resins, etc.
Group 2	S	S-vulcanized diene rubbers (NR, IR, SBR, BR, CR, IIR, EPDM, NBR), polysulfide rubbers and chlorosulfonated poly-ethylenes.	Wool, polysulfones, ebonite, etc.
Group 3	Cl	Polychloroprene, chloro-sulfonated polyethyl-enes, etc.	Poly(vinyl chloride), poly(vinyl-idene chloride) and related copolymers, polychlorotrifluoro-ethylene, chlorinated or hydro-chlorinated rubber, etc.
Group 4	Absence of N, S, Cl etc.	Peroxide cross-linked or unvulcanized hydrocar-bon rubbers (NR, IR, SBR, BR, IIR, EPDM, EPR, etc.)	Petroleum resins, coumarone-indene resins, cellulose and cellulosics other than cellulose nitrate, polyesters, polyethers or acetal resins, polyolefins, polystyrene, poly(methyl meth-acrylate), poly(vinyl acetate/ alcohol), etc.

[a]See Appendix 1B for abbreviations.

1. *Tests for polystyrene and styrenic copolymers.* The test depends on the aromatic rings of the styrene units in the polymer chain. The polymer sample (0.1 g) is heated under reflux in concentrated nitric acid and the clear mixture is poured into water (25 ml) to yield a pale yellow precipitate which is then extracted with ether (2×25 ml). The ether extracts are combined and washed thoroughly with water (2×5 ml), then extracted with dilute sodium hydroxide solution (2×5 ml) and finally with water (5 ml). The alkaline extracts and the aqueous extract are combined and the nitro compounds present in the mixture are reduced by adding granulated zinc (1 g) and concentrated hydrochloric acid and warming gently. The solution is cooled, filtered and then diazotized with a dilute solution (5 ml) of sodium nitrite under ice-cooled condition. The solution is finally poured into excess of alkaline β-naphthol solution producing a deep red color.

TABLE 3.10   Solubility Behavior of Some Common Plastics [21]

Resin	Soluble in	Insoluble in
Polyolefins		
Polyethylene	Dichloroethylene, xylene, tetralin, decalin (boiling)	Alcohols, esters, halogenated hydrocarbons
Polypropylene	Chloroform, trichloroethylene, xylene, tetralin, decalin (boiling)	Alcohols, esters
Polyisobutylene	Ethers, petroleum ether	Alcohols, esters
Poly(vinyl chloride)	Dimethyl formamide, tetrahydrofuran, cyclohexanone	Alcohols, hydrocarbons, butyl acetate
Poly(vinylidene chloride)	Butyl acetate, dioxane, ketones, tetrahydrofuran	Alcohols, hydrocarbons
Polytetrafluoroethylene	Insoluble	All solvents
Polychlorotrifluoroethylene	o-Chlorobenzotrifluoride (above 120°C)	All solvents
Polystyrene	Benzene, methylene chloride, ethyl acetate	Alcohols, water
ABS	Chlorinated hydrocarbons, eg., p-dichlorobenzene	Alcohols, water
Polybutadiene	Benzene	Aliphatic hydrocarbons, alcohols, esters, ketones
Polyisoprene	Benzene	Alcohols, esters, ketones
Acrylics		
Polyacrylonitrile	Dimethylformamide and nitrophenol	Alcohols, esters, ketones, hydrocarbons
Polyacrylamide	Water	Alcohols, esters, hydrocarbons
Esters of polyacrylic acid	Aromatic hydrocarbons, esters, chlorinated hydrocarbons, ketones, tetrahydrofuran	Aliphatic hydrocarbons
Esters of polymethacrylic acid	Aromatic hydrocarbons, chlorinated hydrocarbons, esters, ketones, dioxane	Aliphatic hydrocarbons, alcohols, ethers
Poly(vinyl acetate)	Aromatic hydrocarbons, chlorinated hydrocarbons, ketones, methanol	Aliphatic hydrocarbons
Poly(vinyl alcohol)	Formamide, water	Aliphatic and aromatic hydrocarbons, alcohols, ethers, esters, ketones
Poly(vinyl acetals)	Esters, ketones, tetrahydrofuran, (butyrals in 9:1 chloroform-methanol mixture)	Aliphatic hydrocarbons, methanol

TABLE 3.10 Continued

Resin	Soluble in	Insoluble in
Poly(vinyl ethers)		
Methyl ether	Water, alcohol, benzene, chlorinated hydrocarbons, ethers, esters	Petroleum ether
Ethyl ether	Petroleum ether, benzene, chlorinated hydrocarbons, alcohols, ethers, esters, ketones	Water
Polyesters	Benzyl alcohol, nitrated hydro-carbons, phenols	Alcohols, esters, hydro-carbons
Polycarbonate	Chlorinated hydrocarbons, cyclohex-anone, dimethyl formamide, cresol	(Only swelling in usual solvents)
Nylon	Formic acid, phenols, trifluoro-ethanol	Alcohols, esters, hydro-carbons
Molded phenolic resins	Benzylamine (at 200°C)	All common solvents
Molded amino resins (urea, melamine)	Benzylamine (at 160°C)	All common solvents
Polyurethanes		
Noncross-linked	Methylene chloride, hot phenol, dimethylformamide	Petroleum ether, benzene, alcohols, ethers
Cross-linked	Dimethylformamide	Common solvents
Polyoxymethylene	Insoluble	All solvents
Poly(ethylene oxide)	Alcohols, chlorinated hydrocarbons, water	Petroleum ether
Epoxy resins		
Uncured	Alcohols, ketones, esters, dioxane, benzene, methylene chloride	Aliphatic hydrocarbons, water
Cured	Practically insoluble	
Cellulose, regenerated	Schweizer's reagent	Organic solvents
Cellulose ethers		
Methyl	Water, dil. sodium hydroxide	Organic solvents
Ethyl	Methanol, methylene chloride	Water, aliphatic and aromatic hydrocarbons
Cellulose esters (acetate, acetate-butyrate, propionate)	Ketones, esters	Aliphatic hydrocarbons
Cellulose nitrate	Esters (ethyl acetate, butyl acetate, etc.), ketones (acetone, methyl ethyl ketone, etc.), mixtures (eg. 80% methyl isobutyl ketone + 20% isopropyl alcohol, 80% butyl acetate + 20% isopropyl alcohol)	Aliphatic hydrocarbons (hexane, heptane, etc.) methyl alcohol, water
Natural rubber	Aromatic hydrocarbons, chlorinated hydrocarbons	Petroleum ether, alcohols ketones, esters

TABLE 3.10 Continued

Resin	Soluble in	Insoluble in
Chlorinated rubber	Esters, ketones, linseed oil (80-100°C), carbon tetrachloride, tetrahydrofuran	Aliphatic hydrocarbons
Styrene-butadiene rubber	Ethyl acetate, benzene, methylene chloride	Alcohols, water

2. *Test for poly(vinyl chloride)*. Specific tests for chlorine containing polymers are performed only when the presence of chlorine is confirmed by preliminary test. The simplest method of chlorine determination is the Beilstein test, described previously. Plasticizers are removed from the test material by ether extraction and the Beilstein test for chlorine is repeated to make certain that chlorine is still present. The material is then dissolved in tetrahydrofuran, filtered and the polymeric product reprecipitated by adding methanol. On addition of a few drops of a ~10% methanolic solution of sodium hydroxide to 2–5 ml of ~5% solution of the plasticizer-free material, the mixture changes with time from colorless to light yellow, yellow-brown, dark brown, and finally to black. In another test, poly(vinyl chloride) and vinyl chloride polymers readily form brown coloration and eventually dark brown precipitates when their pyridine solutions are boiled and treated with a few drops of methanolic sodium hydroxide (5%).

3. *Test for poly(vinylidene chloride)*. Poly(vinylidene chloride), when immersed in morpholine, develops a brown color both in the liquid and the partially swollen polymer. The change takes place faster if the mixture is warmed in a water bath.

4. *Test for poly(vinyl alcohol) and poly(vinyl acetate)*. When a small volume of iodine solution (0.1 g iodine and 1.0 g KI dissolved in 20 ml of a 1:1 alcohol-water mixture and diluted to 100 ml using 2N hydrochloric acid solution) is added to an equal volume of 0.25% neutral or acidic solution of poly(vinyl alcohol), a blue color develops almost immediately or on addition of a pinch of borax into the solution. Poly(vinyl acetate), however, turns deep brown on contact with the dilute iodine solution.

5. *Test for polyacrylonitrile and acrylonitrile copolymers*. When a strip of cupric acetate paper that is freshly moistened with a dilute solution of benzidine in dilute acetic acid is held in the pyrolytic vapors of polyacrylonitrile or its copolymers, a bright blue color develops readily. In another test, if the condensed pyrolyzate of polyacrylonitrile or its copolymers is made alkaline, boiled with a trace of ferrous sulfate, and then acidified, Prussian blue precipitate is readily produced.

6. *Test for phenolic resins.* The test material (dry) is heated in an ignition tube over a small flame. The mouth of the tube is covered with a filter paper, prepared by soaking it in an ethereal solution of 2,6-dibromoquinone-4-chloroimide and then drying it in air. After the material has been pyrolyzed for about a minute, the paper is removed and moistened with 1–2 drops of dilute ammonia solution. A blue color indicates phenols (care must be taken with plastics that contain substances yielding phenols on pyrolysis, e.g., phenyl and cresyl phosphates, cross-linked epoxide resins, etc.).

7. *Test for formaldehyde condensate resins.* Formaldehyde enters into the composition of several resins and polymers. Common examples are phenol-, urea-, and melamine-formaldehyde resins, polyoxymethylene and poly(vinyl formal). Formaldehyde is evolved when these are thermally treated or boiled with water in the presence of an acid (H_2SO_4). The aqueous extracts or acid distillates are treated with chromotropic acid (1,8-dihydroxynaphthalene-3,6-disulfonic acid). A few drops of 5% aqueous chromotropic acid solution are added to the aqueous test solution and then an excess of concentrated sulfuric acid is added, and the mixture preferably warmed to nearly 100°C for a few minutes. In the presence of formaldehyde, the solution turns violet/dark violet.

Phenol, urea, and melamine are also obtained as intermediates on acid hydrolysis of the corresponding formaldehyde condensate resins and appropriate tests for phenol, urea, and melamine may be employed for identification purposes.

a. *Phenol.* On addition to the extract (10 ml), approximately 0.5N potassium hydroxide (8–10 ml) and 2 ml of diazotized *p*-nitroaniline (5% sodium nitrite solution added to an ice-cold solution of 2–5 mg of *p*-nitroaniline dissolved in 500 ml of approximately 3% hydrochloric acid, until the solution becomes just colorless), a red or violet color develops indicating phenol. No differentiation between phenol and its homologs is possible by this test.

b. *Urea and melamine.* The aqueous extract is divided into two parts. (solutions 1 and 2). Solution 1 is made alkaline with dilute sodium hydroxide, and 1 ml of sodium hypochlorite solution is added. Solution 2 is treated with freshly prepared furfural reagent (5 drops of pure, freshly distilled furfural, 2 ml of acetone, 1 ml of concentrated hydrochloric acid, and 2 ml of water).

For urea, solution 1 remains colorless and solution 2 becomes orange to red after 3–5 hr. For melamine, solution 1 slowly becomes orange and solution 2 remains colorless.

Detection of urea with urease also provides a useful differentiation between urea and melamine resins. For this test, the powdered sample (0.25 g) is placed in a 100 ml Erlenmeyer flask and boiled with 5% sulphuric acid until the smell of formaldehyde has disappeared. The mixture is neutralized with sodium hydroxide (phenolphthalen as indicator). Then 1 drop of 1N sulphuric acid and 10 ml of 10% urease solution is added, a strip of red litmus paper is placed in the vapors and the flask is stoppered. The appearance of a blue color in the litmus paper after a short time indicates urea and thus the presence of urea resin.

8. *Test for cellulose esters.* Cellulose esters respond to the Molisch test for carbohydrates. The sample is dissolved in acetone and treated with 2–3 drops of a 2% ethanolic solution of α-naphthol; a volume of 2–2.5 ml of concentrated H_2SO_4 is so added as to form a lower layer. A red to red-brown ring at the interface of the liquids indicates cellulose (glucose). A green ring at the interface indicates nitrocellulose and differentiates it from other cellulose esters. A more sensitive test for nitrocellulose is provided by an intense blue color reaction when a drop of a solution of diphenylamine in concentrated H_2SO_4 (5% w/v) is added to the sample in the absence of other oxidizing agents.

9. *Test for polyamides.* When a strip of filter paper soaked in a fresh saturated solution of *o*-nitrobenzaldehyde in dilute sodium hydroxide is held over the pyrolytic vapors of polyamides containing adipic acid, a mauve-black color is readily developed. Pyrolytic vapors of polyamides from diacids other than adipic acid produce a grey color when tested similarly.

10. *Tests for natural rubber and synthetic rubbers.* Rubbers may be identified by testing the pyrolytic vapors from test samples. A strip of filter paper soaked in an ethereal solution containing *p*-dimethylaminobenzaldehyde (3%) and hydroquinone (0.05%) and then moistened with a 30% solution of trichloroacetic acid in isopropanol produces different color reactions in the presence of pyrolytic vapors of different rubbers. Vapors from natural rubber (NR) produce a deep blue or blue-violet color, and those from styrene-butadiene rubber (SBR) pyrolysis turn the paper green or blue with a distinct green tinge. Polyisobutylenes and butyl rubber resemble NR, and silicone rubbers resemble SBR, in this color reaction. Pyrolytic vapors from nitrile rubbers, on the other hand, give a brown or brown-yellow color and those from polychloroprenes (neoprene) turn the test paper grey with a yellow tinge.

In another test, pyrolytic vapors from polyisobutylenes and butyl rubbers produce a bright yellow color on filter paper freshly soaked in a solution obtained by dissolving yellow mercuric oxide (5 g) in concentrated sulfuric acid (15 ml) and water (85 ml) on boiling. Other rubbers may either produce little change or give an uncharacteristic brown color.

Chromic acid oxidation provides a simple test for polyisoprenes. About 0.1 g of sample is gently heated in a test tube in the presence of chromic acid solution (5 ml) prepared by dissolving chromium trioxide (2 g) in water (5 ml) and adding concentrated sulfuric acid (1.5 ml). 1,4-Polyisoprenes (the synthetic equivalent of natural rubber) yield acetic acid, which can be readily identified by its odor.

If the above test is positive, the sample may be further subjected to a modified Weber color test for polyisoprenes. For this test, an acetone-extracted polymer (0.05 g) is dissolved or suspended in carbon tetrachloride, treated with a little solution of bromine in carbon tetrachloride and heated in a water bath to remove excess bromine. The residue is then warmed with a little phenol. In the presence of polyisoprenes, the solid or the solution turns violet or purple and gives a purple solution with chloroform. The test is also positive for natural rubber and butyl rubber.

REFERENCES

1. R. M. Ogorkiewicz, *Engineering Properties of Thermoplastics*, John Wiley, New York (1970).
2. J. G. Williams, *Stress Analysis of Polymers*, Longmans, London (1973).
3. S. Turner, *Mechanical Testing of Plastics*, CRC Press, Cleveland, Ohio (1973).
4. R. J. Crawford, *Plastics Engineering*, Pergamon, London (1981).
5. G. M. Bartenev and Y. S. Zuyev, *Strength and Failure of Viscoelastic Materials*, Pergamon, London (1968).
6. P. I. Vincent, *Impact Tests and Service Performance of Thermoplastics*, Plastics Institute, London (1970).
7. M. Braden and A. N. Gent, Attack of ozone on stretched rubber vulcanizates, *J. Appl. Polymer Sci.*, 3: 90 (1960).
8. E. Gaube, *Kunststoffe*, Creep resistance and stress cracking of low pressure polyethylene, 49: 446 (1959).
9. L. Holloway, *Glass Reinforced Plastics in Construction*, Surrey University Press (1978).
10. M. O. W. Richardson, *Polymer Engineering Composites*, Applied Science Publishers, London (1977).
11. A. F. Johnston, *Engineering Design Properties of GRP*, British Plastics Federation, p. 215 (1979).
12. J. J. Brophy and J. W. Buttrey, *Organic Semiconductors*, Macmillan, New York (1962)
13. D. W. Saunders, Optical properties of high polymers, *Physics of Plastics* (P. D. Ritchie, ed.) Van Nostrand Reinhold, New York, p. 386 (1965).
14. F. W. Billmeyer and M. Saltzman, *Principles of Color Technology*, Interscience, New York (1966).
15. D. Osmer, SPE Annual Technical Conference, 1977.
16. M. Gordon, Thermal properties of high polymers, *Physics of Plastics* (P. D. Ritchie, ed.), Van Nostrand Reinhold, New York, p. 219 (1965).
17. D. W. Von Krevelen, *Properties of Polymers, Their Estimation and Correlation with Chemical Structure*, Elsevier, Amsterdam, New York (1976).
18. R. E. Evans, A Rationale for testing to establish suitability for an end-use application, *Physical Testing of Plastics*, ASTM Special Technical Publication 736, ASTM, Philadelphia (1981).
19. W. E. Brown, F. C. Frost, and P. E. Willard, *Testing of Polymers* (J. V. Schmitz, ed.), Interscience, New York, Vol. 1, p. 1 (1965).

20. A. B. Glanville, The Plastics Engineer's Data Book, Industrial Press, New York (1971).
21. A. Krause and A. Lange, *Introduction to Chemical Analysis of Plastics*, Iliffe Books Ltd., London (1969).
22. P. Ghosh, *Polymer Science and Technology of Plastics and Rubber*, Tata McGraw-Hill, New Delhi (1990).

4

Industrial Polymers

INTRODUCTION

The first completely synthetic plastic, phenol-formaldehyde, was introduced by L. H. Baekeland in 1909, nearly four decades after J. W. Hyatt had developed a semisynthetic plastic—cellulose nitrate. Both Hyatt and Baekeland invented their plastics by trial and error. Thus the step from the idea of macromolecules to the reality of producing them at will was still not made. It had to wait till the pioneering work of Hermann Staudinger, who, in 1924, proposed linear molecular structures for polystyrene and natural rubber. His work brought recognition to the fact that the macromolecules really are linear polymers. After this it did not take long for other materials to arrive. In 1927 poly(vinyl chloride) (PVC) and cellulose acetate were developed, and 1929 saw the introduction of urea-formaldehyde (UF) resins.

The production of nylon-6,6 (first synthesized by W. H. Carothers in 1935) was started by Du Pont in 1938, and the production of nylon-6 (perlon) by I. G. Farben began in 1939, using the caprolactam route to nylon developed by P. Schlack. The latter was the first example of ring-opening polymerization. The years prior to World War II saw the rapid commercial development of many important plastics, such as acrylics and poly(vinyl acetate) in 1936, polystyrene in 1938, melamine-formaldehyde (formica) in 1939, and polyethylene and polyester in 1941. The amazing scope of wartime applications accelerated the development and growth of polymers to meet the diverse needs of special materials in different fields of activity.

The development of new polymeric materials proceeded at an even faster pace after the war. Epoxies were developed in 1947, and acrylonitrile-butadiene-styrene (ABS) terpolymer in 1948. The polyurethanes, introduced in Germany in 1937, saw rapid development in the United States as the technology became available after the war. The discovery of Ziegler–Natta catalysts in the 1950s brought about the development of linear polyethylene and stereoregular polypropylene. These years also saw the emergence of acetal, polyethylene terephthalate, polycarbonate, and a host of new copolymers. The next two decades saw the commercial development of a number of highly temperature-resistant materials, which included poly(phenylene oxide) (PPO), polysulfones, polyimides, polyamide-imides, and polybenzimidazoles.

More recently, we have seen other new materials arrive, including several exotic materials which are mostly development products today and are very expensive. They will undoubtedly find their first uses, if any, in the aerospace industry, which can afford to pay for them. The growth of all plastics, in general, and engineering plastics, in particular, has more than kept pace with the general progress in technology. On a volume basis, U.S. plastics production in 1979 surpassed U.S. steel production, heralding the arrival of the plastics age in the United States. Plastics are still evolving, changing, and improving as the 21st century approaches.

Numerous plastics and fibers are produced from synthetic polymers: containers from polypropylene, coating materials form PVC, packaging film from polyethylene, experimental apparatus from Teflon, organic glasses from poly(methyl methacrylate), stockings from nylon fiber—there are simply too many to mention them all. The reason why plastics materials are popular is that they may offer such advantages as transparency, self-lubrication, light weight, flexibility, economy in fabricating and decorating. Properties of plastics can be modified through the use of fillers, reinforcing agents, and chemical additives. Plastics have thus found many engineering applications, such as mechanical units under stress, low-friction components, heat- and chemical-resistant units, electrical parts, high-light-transmission applications, housing, building construction functions, and many others. Although it is true that in these applications plastics have been used in a proper manner according to our needs, other ways of utilizing both natural and synthetic polymers may still remain. To investigate these further possibilities, active research has been initiated in a field called *specialty polymers*. This field relates to synthesis of new polymers with high additional value and specific functions.

Many of the synthetic plastics materials have found established uses in a number of important areas of engineering involving mechanical, electrical, telecommunication, aerospace, chemical, biochemical, and biomedical applications. There is, however, no single satisfactory definition of engineering plastics. According to one definition, "engineering plastics are those which possess physical properties enabling them to perform for prolonged use in structural applications, over a wide temperature range, under mechanical stress and in difficult chemical and physical environments." In the most general sense, however, all polymers are engineering materials, in that they offer specific properties which we judge quantitatively in the design of end-use applications.

For the purpose of this discussion, we will classify polymers into three broad groups: addition polymers, condensation polymers, and special polymers. By convention, polymers whose main chains consist entirely of C—C bonds are *addition polymers*, whereas those in which hetero atoms (e.g., O, N, S, Si) are present in the polymer backbone are considered to be *condensation polymers*. Grouped as *special polymers* are those products which have special properties, such as temperature and fire resistance, photosensitivity, electrical conductivity, and piezoelectric properties, or which possess specific reactivities to serve as functional polymers.

Further classification of polymers in the groups of addition polymers and condensation polymers has been based on monomer composition, because this provides an orderly approach, whereas classification based on polymer uses, such as plastics, elastomers, fibers, coatings, etc. would result in too much overlap. For example, polyamides are used not only as synthetic fibers but also as thermoplastic molding compounds and polypropylene, which is used as a thermoplastic molding compound has also found uses as a fiber-forming material.

All vinyl polymers are addition polymers. To differentiate them, the homopolymers have been classified by the substituents attached to one carbon atom of the double bond. If the substituent is hydrogen, alkyl, aryl, or halogen, the homopolymers are listed under polyolefines. Olefin homopolymers with other substituents are described under polyvinyl compounds, except where the substituent is a nitrile, a carboxylic acid, or a carboxylic acid ester or amide. The monomers in the latter cases being derivatives of acrylic acid, the derived polymers are listed under acrylics. Under olefin copolymers are listed products which are produced by copolymerization of two or more monomers.

Condensation polymers are classified as polyesters, polyamides, polyurethanes, and ether polymers, based on the internal functional group being ester (—COO—), amide (—CONH—), urethane (—OCONH—), or ether (—O—). Another group of condensation polymers derived by condensation reactions with formaldehyde is described under formaldehyde resins. Polymers with special properties have been classified into three groups: heat-resistant polymers, silicones and other inorganic polymers, and functional polymers. Discussions in all cases are centered on important properties and main applications of polymers.

PART I: ADDITION POLYMERS

Addition polymers are produced in largest tonnages among industrial polymers. The most important monomers are ethylene, propylene, and butadiene. They are based on low-cost petrochemicals or natural gas and are produced by cracking or refining of crude oil. Polyethylene, polypropylene, poly(vinyl chloride), and polystyrene are the four major addition polymers and are by far the least-expensive industrial polymers on the market. In addition to these four products, a wide variety of other addition polymers are commercially available.

For addition polymers four types of polymerization processes are known: free-radical-initiated chain polymerization, anionic polymerization, cationic polymerization, and coordination polymerization (with Ziegler–Natta catalysts). By far the most extensively used process is the free-radical-initiated chain polymerization. However, the more recent development of stereoregular polymers using certain organometallic coordination compounds called Ziegler–Natta catalysts, which has added a new dimension to polymerization processes, is expected to play a more important role in coming years. The production of linear low-density polyethylene (LLDPE) is a good example. Ionic polymerization is used to a lesser extent. Thus, anionic polymerization is used mainly in the copolymerization of olefins, such as the production of styrene-butadiene elastomers, and cationic polymerization is used exclusively in the production of butyl rubber.

Different processes are used in industry for the manufacture of polymers by free-radical chain polymerization. Among them *homogeneous bulk polymerization* is economically the most attractive and yields products of higher purity and clarity. But it has problems associated with the heat of polymerization, increases in viscosity, and removal of unreacted monomer. This method is nevertheless used for the manufacture of PVC, polystyrene, and poly(methyl methacrylate). More common processes are *homogeneous solution polymerization* and *heterogeneous suspension polymerization*.

Solution polymerization is used for the manufacture of polyethylene, polypropylene, and polystyrene, but by far the most widely used process for polystyrene and PVC is suspension polymerization. In the latter process (also known as *bead, pearl,* or *granular polymerization* because of the form in which the final products may be obtained), the monomer is dispersed as droplets (0.01– 0.05 cm in diameter) in water by mechanical agitation. Various types of stabilizers, which include water-soluble organic polymers, electrolytes, and water-insoluble inorganic compounds, are added to prevent agglomeration of the monomer droplets. Each monomer droplet in the suspension constitutes a small bulk polymerization system and is transformed finally into a solid bead. Heat of polymerization is quickly dissipated by continuously stirring the suspension medium, which makes temperature control relatively easy.

POLYOLEFINS
Polyethylene

$$+CH_2—CH_2+_n$$

Monomer	Polymerization	Major uses
Ethylene	LDPE: Free-radical-initiated chain polymerization	LDPE: Film and sheet (55%), housewares and toys (16%), wire and cable coating (5%)
	HDPE: Ziegler-Natta or metal-oxide catalyzed chain polymerization	HDPE: Bottles (40%), housewares, containers, toys (35%), pipe and fittings (10%), film and sheet (5%)

Polyethylene is a thermoplastic material composed of polymers of ethylene. The two main types are low-density polyethylene (LDPE) and high-density polyethylene (HDPE).

LDPE is manufactured by polymerization of ethylene under high pressures (15,000–50,000 psi, i.e., 103–345 MPa) and elevated temperatures (200–350°C) in the presence of oxygen (0.03–0.1%). It is an amorphous and branched product and has a melting range from 107 to 120°C. HDPE is produced in a low-pressure process either by Ziegler–Natta catalysis or by metal-oxide catalysis (usually referred to as Phillips catalysis). HDPE is essentially linear, having much less branching than LDPE, and has a melting range from 130 to 138°C. Recently, the production of linear low-density polyethylene (LLDPE) has been announced. These new polymers can be considered as linear polyethylenes having a significant number of branches (pendant alkyl groups). The linearity imparts strength, the branches impart toughness.

Polyethylene is partially amorphous and partially crystalline. Linearity of polymer chains affords more efficient packing of molecules and hence a higher degree of crystallinity. On the other hand, side-chain branching reduces the degree of crystallinity. Increasing crystallinity increases density, stiffness, hardness, tensile strength, heat and chemical resistance, creep resistance, barrier properties, and opacity, but it reduces stress-crack resistance, permeability, and impact strength. Table 4.1 shows a comparison of the three types of polyethylene.

Polyethylene has excellent chemical resistance and is not attacked by acids, bases, or salts. (It is, however, attacked by strong oxidizing agents.) The other characteristics of polyethylene which have led to its widespread use are low

TABLE 4.1 Types of Polyethylene

Material	Chain structure	Density (g/cm^3)	Crystallinity	Process
LDPE	Branched	0.912–0.94	50%	High pressure
LLDPE	Linear/less branched	0.92–0.94	50%	Low pressure
HDPE	Linear	0.958	90%	Low pressure

cost, easy processability, excellent electrical insulation properties, toughness and flexibility even at low temperatures, freedom from odor and toxicity, reasonable clarity of thin films, and sufficiently low permeability to water vapor for many packaging, building, and agricultural applications.

Major markets for LDPE are in packaging and sheeting, whereas HDPE is used mainly in blow-molded products (milk bottles, household and cosmetic bottles, fuel tanks), and in pipe, wire, and cable applications. Ultrahigh-molecular-weight polyethylene materials (see later), which are the toughest of plastics, are doing an unusual job in the textile machinery field.

Chlorinated Polyethylene

Low chlorination of polyethylene, causing random substitution, reduces chain order and thereby also the crystallinity. The low chlorine products (22–26% chlorine) of polyethylene are softer, more rubber-like, and more compatible and soluble than the original polyethylene. However, much of the market of such materials has been taken up by chlorosulfonated polyethylene (Hypalon, Du Pont), produced by chlorination of polyethylene in the presence of sulfur dioxide, which introduces chlorosulfonyl groups in the chain. Chlorosulfonated LDPE containing about 27% chlorine and 1.5% sulfur has the highest elongation. Chlorosulfonated polyethylene rubbers, designated as CSM rubbers, have very good heat, ozone, and weathering resistance together with a good resistance to oils and a wide spectrum of chemicals. The bulk of the output is used for fabric coating, film sheeting, and pit liner systems in the construction industry, and as sheathing for nuclear power cables, for offshore oil rig cables, and in diesel electric locomotives.

Cross-Linked Polyethylene

Cross-linking polyethylene enhances its heat resistance (in terms of resistance to melt flow) since the network persists even about the crystalline melting point of the uncross-linked material. Cross-linked polyethylene thus finds application in the cable industry as a dielectric and as a sheathing material. Three main approaches used for cross-linking polyethylene are (1) radiation cross-linking, (2) peroxide cross-linking, and (3) vinyl silane cross-linking.

Radiation cross-linking (cf. Fig. 1.27) is most suitable for thin sections. The technique, however, requires expensive equipment and protective measures against radiation. Equipment requirements for peroxide curing are simpler, but the method requires close control. The peroxide molecules break up at elevated temperatures, producing free radicals which then abstract hydrogen from the polymer chain to produce a polymer-free radical. Two such radicals can combine and thus cross-link the two chains. It is important, however, that the peroxide be sufficiently stable thermally so that premature cross-linking does not take place during compounding and shaping operations. Dicumyl peroxide is often used for LDPE but more stable peroxides are necessary for HDPE. For cross-linking polyethylene in cable coverings, high curing temperatures, using high-pressure steam in a long curing tube set into the extrusion line, are normally employed.

Copolymers of ethylene with a small amount of vinyl acetate are often preferred for peroxide cross-linking because the latter promotes the cross-linking process. Large amounts of carbon black may be incorporated into polyethylene that is to be cross-linked. The carbon black is believed to take part in the cross-linking process, and the mechanical properties of the resulting product are superior to those of the unfilled material.

In the vinyl silane cross-linking process (*Sioplas process*) developed by Dow, an easily hydrolyzable trialkoxy vinyl silane, $CH_2{=}CHSi(OR)_3$, is grafted onto the polyethylene chain, the site activation having been achieved with the aid of a small amount of peroxide. The material is then extruded onto the wire. When exposed to hot water or low-pressure steam, the alkoxy groups hydrolyze and then condense to form a siloxane

$$(-\overset{|}{\underset{|}{Si}}-O-\overset{|}{\underset{|}{Si}}-)$$

cross-link. The cross-linking reaction is facilitated by the use of a cross-linking catalyst, which is typically an organotin compound. There are several variations of the silane cross-linking process. In one process, compounding, grafting, and extrusion onto wire are carried out in the same extruder.

Cross-linked LDPE foam has been produced by using either chemical cross-linking or radiation cross-linking. These materials have been used in the automotive industry for carpeting, boot mats, sound deadening, and pipe insulation, and as flotation media for oil-carrying and dredging hose.

Linear Low-Density Polyethylene (LLDPE)

Chemically, LLDPE can be described as linear polyethylene copolymers with alpha-olefin comonomers in the ethylene chain. They are produced primarily at low pressures and temperatures by the copolymerization of ethylene with various alpha-olefins such as butene, hexene, octene, etc., in the presence of suitable catalysts. Either gas-phase fluidized-bed reactors or liquid-phase solution-process reactors are used. (In contrast, LDPE is produced at very high pressures and temperatures either in autoclaves or tubular reactors.) Polymer properties such as molecular weight, molecular-weight distribution, crystallinity, and density are controlled through catalyst selection and control of reactor conditions. Among the LLDPE processes, the gas-phase process has shown the greatest flexibility to produce resins over the full commercial range.

The molecular structure of LLDPE differs significantly from that of LDPE: LDPE has a highly branched structure, but LLDPE has the linear molecular structure of HDPE, though it has less crystallinity and density than the latter (see Table 4.1).

The stress-crack resistance of LLDPE is considerably higher than that of LDPE with the same melt index and density. Similar comparisons can be made with regard to puncture resistance, tensile strength, tensile elongation, and low- and high-temperature toughness. Thus LLDPE allows the processor to make a stronger product at the same gauge or an equivalent product at a reduced gauge.

LLDPE is now replacing conventional LDPE in many applications because of the combination of favorable production economics and product performance characteristics. For many applications (blow molding, injection molding, rotational molding, etc.) existing equipment for processing LDPE can be used to process LLDPE. LLDPE film can be treated, printed, and sealed by using the same equipment used for LDPE. Heat-sealing may, however, require slightly higher temperatures.

LLDPE films provide superior puncture resistance, high tensile strength, high impact strength, and outstanding low-temperature properties. The resins can be drawn down to thicknesses below 0.5 mil without bubble breaks. Slot-cast films combine high clarity and gloss with toughness. LLDPE films are being increasingly used in food packaging for such markets as ice bags and retail merchandise bags, and as industrial liners and garment bags.

Good flex properties and environmental stress-crack resistance combined with good low-temperature impact strength and low warpage make LLDPE suitable for injection-molded parts for housewares, closures, and lids. Extruded pipe and tubing made from LLDPE exhibit good stress-crack resistance and good bursting strength. Blow-molded LLDPE parts such as toys, bottles, and drum liners provide high strength, flex life, and stress-crack resistance. Lightweight parts and faster blow-molding cycle times can be achieved. The combination of good high- and low-temperature properties, toughness, environmental stress-crack resistance, and good dielectric properties suit LLDPE for wire and cable insulation and jacketing applications.

A new class of linear polyethylene copolymers with densities ranging between 0.890 and 0.915 g/cm^3, known as *very low-density polyethylene* (VLDPE), was introduced commercially in late 1984 by Union Carbide. These resins are produced by copolymerization of ethylene and alpha-olefins in the presence of a catalyst.

VLDPE provides flexibility previously available only in lower-strength materials, such as ethylene-vinyl acetate (EVA) copolymer (see later) and plasticized PVC, together with the toughness and broader operating temperature range of LLDPE. Its unique combination of properties makes VLDPE suited for a wide range of applications.

Generally, it is expected that VLDPE will be widely used as an impact modifier. Tests suggest that it is suited as a blending resin for polypropylene and in HDPE films for improved tear strength.

On its own, VLDPE should find use in applications requiring impact strength, puncture resistance, and dart drop resistance combined with flexibility. The drawdown characteristics of VLDPE allow for very thin films to be formed without pinholing. Soft flexible films for disposable gloves, furniture films, and high-performance stretch and shrink film are potential markets.

High-Molecular-Weight High-Density Polyethylene

High-molecular-weight high-density polyethylene (HMW-HDPE) is defined as a linear homopolymer or copolymer with a weight-average molecular weight ($\overline{M}_w$) in the range of approximately 200,000 to 500,000. HMW-HDPE resins are manufactured using predominantly two basic catalyst systems: Ziegler-type catalysts and chromium oxide-based catalysts. These catalysts produce linear polymers which can be either homopolymers when higher-density products are required or copolymers with lower density. Typical comonomers used in the latter type of products are butene, hexene, and octenes.

HMW-HDPE resins have high viscosity because of their high molecular weight. This presents problems in processing and, consequently, these resins are normally produced with broad molecular-weight distribution (MWD).

The combination of high molecular weight and high density imparts the HMW-HDPE good stiffness characteristics together with above-average abrasion resistance and chemical resistance. Because of the relatively high melting temperature, it is imperative that HMW-HDPE resins be specially stabilized with antioxidant and processing stabilizers. HMW-HDPE products are normally manufactured by the extrusion process; injection molding is seldom used.

The principal applications of HMW-HDPE are in film, pressure pipe, large blow-molded articles, and extruded sheet. HMW-HDPE film now finds application in T-shirt grocery sacks (with 0.6- to 0.9-mil thick sacks capable of carrying 30 lb of produce), trash bags, industrial liners, and specialty roll stock. Sheets 20 to 100 mils thick and 18 to 20 ft wide are available that can be welded in situ for pond and tank liners. HMW-HDPE piping is used extensively in gas distribution, water collection and supply, irrigation pipe, industrial effluent discharge, and cable conduit. The availability of pipe materials with significantly higher hydrostatic design stress (800 psi compared with 630 psi of the original HDPE resins) has given added impetus to their use. Large-diameter HMW-HDPE piping has found increasing use in sewer relining. Large blow-molded articles, such as 55-gal shipping containers, are produced. Equipment is now available to blow mold very large containers, such as 200- and 500-gal capacity industrial trash receptacles and 250-gal vessels to transport hazardous chemicals.

Ultrahigh-Molecular-Weight Polyethylene

Ultrahigh-molecular-weight polyethylene (UHMWPE) is defined by ASTM as "polyethylene with molecular weight over three million (weight average)." The resin is made by a special Ziegler-type polymerization. Being chemically similar to HDPE, UHMWPE shows the typical polyethylene characteristics of chemical inertness, lubricity, and electrical resistance, while its very long substantially linear chains provide greater impact strength, abrasion resistance, toughness, and freedom from stress cracking. However, this very high molecular weight also makes it difficult to process the polymer by standard molding and extrusion techniques. Compression molding of sheets and ram extrusion of profiles are the normal manufacturing techniques.

Compression molding is by far the most commonly employed process. Forms produced by compression molding or specialty extrusion can be made into final form by machining, sintering, or forging. Standard woodworking techniques are employed for machining; sharp tools, low pressures, and good cooling are used. Forging can be accomplished by pressing a preform and billet and then forging to the final shape. Parts are also formed from compression-molded or skived sheets by heating them above 300°F (~150°C) and stamping them in typical metal-stamping equipment. Such UHMWPE items have better abrasion resistance than do steel or polyurethanes. They also have high impact strength even at very low temperatures, high resistance to cyclic fatigue and stress cracking, low coefficient friction, good corrosion and chemical resistance, good resistance to nuclear radiation, and resistance to boiling water.

Fillers such as graphite, talc, glass beads or fibers, mica, and powdered metals can be incorporated to improve stiffness or to reduce deformation and deflection under load. Resistance to abrasion and deformation can be increased by peroxide cross-linking, described earlier.

UHMWPE was first used in the textile machinery field picker blocks and throw sticks, for example. Wear strips, timing wheels, and gears made of the UHMW polymer are used in material handling, assembly, and packaging lines. Chemical resistance and lubricity of the polymer are important in its applications in chemical, food, beverage, mining, mineral processing, and paper industries. All sorts of self-unloading containers use UHMWPE liners to reduce wear, prevent sticking, and speed up the unloading cycles. The polymer provides slippery surfaces that facilitate unloading even when the product is wet or frozen.

The polymer finds applications in transportation, recreation, lumbering, and general manufacturing. Metal equipment parts in some cases are coated or replaced with UHMWPE parts to reduce wear and prevent corrosion. Sewage plants have used this polymer to replace cast-iron wear shoes and rails, bearings, and sprockets. There is even an effort to use UHMW polymer chain to replace metal chain, which is corroded by such environments.

Porous UHMW polymer is made by sintering to produce articles of varied porosity. It has found growing use for controlled-porosity battery separators. Patents have been issued recently on the production of ultrahigh strength, very lightweight fibers from the UHMW polymer by gel spinning.

Polypropylene

$$-\!\!\!-\!\!\left[\!\mathrm{CH_2\!-\!CH}\!\right]\!\!\!-\!\!\!$$
$$\underset{\mathrm{CH_3}}{\overset{|}{}}{}_n$$

Monomer	Polymerization	Major uses
Propylene	Ziegler–Natta catalyzed chain polymerization	Fiber products (30%), housewares and toys (15%), automotive parts (15%), appliance parts (5%)

Polypropylene (PP) is made almost entirely by low-pressure processes using Ziegler–Natta catalysts. Commercial polymers are usually about 90 to 95% isotactic, the other structures being atactic and syndiotactic (a rough measure of isotacticity is provided by the "isotactic index"—the percentage of polymer insoluble in heptane): the greater the degree of isotacticity the greater the crystallinity and hence the greater the softening point, stiffness, tensile strength, modulus, and hardness.

Although very similar to HDPE, PP has a lower density (0.90 g/cm^3) and a higher softening point, which enables it to withstand boiling water and many steam sterilizing operations. It has a higher brittle point and appears to be free from environmental stress-cracking problems, except with concentrated sulfuric acid, chromic acid, and aqua regia. However, because of the presence of tertiary carbon atoms occurring alternately on the chain backbone, PP is more susceptible to UV radiation and oxidation at elevated temperatures. Whereas PE crosslinks on oxidation, PP undergoes degradation to form lower-molecular-weight products. Substantial improvement can be made by the inclusion of antioxidants, and such additives are used in all commercial PP compounds. The electrical properties of PP are very similar to those of HDPE.

Because of its reasonable cost and good combination of the foregoing properties, PP has found many applications, ranging from fibers and filaments to films and extrusion coatings. A significant portion of the PP produced is used in moldings, which include luggage, stacking chairs, hospital sterilizable equipment, toilet cisterns, washing machine parts, and various auto parts, such as accelerator pedals, battery cases, dome lights, kick panels, and door frames.

Although commercial PP is a highly crystalline polymer, PP moldings are less opaque when unpigmented than are corresponding HDPE moldings, because the differences between amorphous and crystal densities are less with PP (0.85 and 0.94 g/cm^3, respectively) than with polyethylene (0.84 and 1.01 g/cm^3, respectively).

A particularly useful property of PP is the excellent resistance of thin sections to continued flexing. This has led to the production of one-piece moldings for boxes, cases, and accelerator pedals in which the hinge is an integral part of the molding.

Monoaxially oriented polypropylene film tapes have been widely used for carpet backing and for woven sacks (replacing those made from jute). Combining strength and lightness, oriented PP straps have gained rapid and widespread acceptance for packaging.

Nonoriented PP film, which is glass clear, is used mainly for textile packaging. However, biaxially oriented PP film is more important because of its greater clarity, impact strength, and barrier properties. Coated grades of this material are used for packaging potato crisps, for wrapping bread and biscuits, and for capacitor dielectrics. In these applications PP has largely replaced regenerated cellulose. (The high degree of clarity of biaxially oriented PP is caused by layering of the crystalline structures. Layering reduces the variations in refractive index across the thickness of the film, which thus reduces the amount of light scattering.)

Polypropylene, produced by an oriented extrusion process, has been uniquely successful as a fiber. Its excellent wear, inertness to water, and microorganisms, and its comparatively low cost have made it extensively used in functional applications, such as carpet backing, upholstery fabrics, and interior trim for automobiles.

Poly(Vinyl Chloride)

$$+CH_2-CH+_n$$
$$|$$
$$Cl$$

Monomer	Polymerization	Major uses
Vinyl chloride	Free-radical-initiated chain polymerization	Pipe and fittings (35%), film and sheet (15%), flooring materials (10%), wire and cable insulation (5%), automotive parts (5%), adhesives and coatings (5%)

Poly(vinyl chloride) (PVC) is produced by polymerization of vinyl chloride by free-radical mechanisms, mainly in suspension and emulsion, but bulk and solution processes are also employed to some extent. (The control of vinyl chloride monomer escaping into the atmosphere in the PVC production plant has become important because cases of angiosarcoma, a rare type of liver cancer, were found among workers exposed to the monomer. This led to setting of stringent standards by governments and modification of manufacturing processes by the producers to comply with the standards.)

At processing temperatures used in practice (150–200°C), sufficient degradation may take place to render the product useless, Evidence points to the fact that dehydrochlorination occurs at an early stage in the degradation process and produces polyene structures:

$$\text{ⱮⱮCH}_2\text{—CH—CH}_2\text{—CHⱮⱮ} \xrightarrow{\text{—HCl}} \text{ⱮⱮCH}=\text{CH—CH}=\text{CHⱮⱮ}$$
$$\phantom{\text{ⱮⱮCH}_2\text{—}}|\phantom{\text{CH—}}|$$
$$\phantom{\text{ⱮⱮCH}_2}\text{Cl}\phantom{\text{—CH—}}\text{Cl}$$

It is believed that the liberated hydrogen chloride can accelerate further decomposition and that oxygen also has an effect on the reaction. However, incorporation of certain materials known as *stabilizers* retards or moderates the degradation reaction so that useful processed materials can be obtained. Many stabilizers are also useful in improving the resistance of PVC to weathering, particularly against degradation by UV radiation.

Stabilizers

The most important class of stabilizers are the lead compounds which form lead chloride on reaction with the hydrogen chloride evolved during decomposition. Basic lead carbonate (white lead), which has a low weight cost, is more commonly used. A disadvantage of lead carbonate is that it may decompose with the evolution of carbon dioxide at higher processing temperatures and lead to a porous product. For this reason, tribasic lead sulfate, which gives PVC products with better electrical insulation properties than lead carbonate, is often used despite its somewhat higher weight cost. Other lead stabilizers are of much more specific applications. For example, dibasic lead phthalate, which is an excellent heat stabilizer, is used in heat-resistant insulation compounds (e.g., in 105°C wire), in high-fidelity gramophone records, in PVC coatings for steel, and in expanded PVC formulation.

The use of lead compounds as stabilizers has been subjected to regulation because of its toxicity. Generally, lead stabilizers are not allowed in food-packaging PVC materials, but in most countries they are allowed in PVC pipes for conveying drinking water, with reduction in the level of use of such stabilizers.

Today the compounds of cadmium, barium, calcium, and zinc have gained prominence as PVC stabilizers. A modern stabilizing system may contain a large number of components. A typical cadmium-barium "packaged" stabilizer may have the following composition: cadmium-barium phenate 2 to 3 parts, epoxidized oils 3 to 5 parts, trisnonyl phenyl phosphite 1 part, stearic acid 0.5 to 1 part, and zinc octoate 0.5 part by weight. For flooring compositions, calcium-barium, magnesium-barium, and copper-barium compounds are sometimes used in conjunction with pentaerythritol (which has the function of reducing color by chelating iron present in asbestos).

Another group of stabilizers are the organotin compounds. Development of materials with low toxicity, excellent stabilizing performance, and improving relative price situation has led to considerable growth in the organotin market during the last decade. Though the level of toxicity of butyltins is not sufficiently low for application in contact with foodstuffs, many of the octyltins, such as dioctyltin dilaurate and dioctyltin octylthioglycolate, meet stringent requirements for use in contact with foodstuffs. Further additions to the class of organotins include the estertins characterized by low toxicity, odor, and volatility, and the methyltins having higher efficiency per unit weight compared with the more common organotins.

Organotin compounds that are salts of alkyltin oxides with carboxylic acids (e.g., dioctyltin dilaurate) are usually called organotin carboxylates. Organotin compounds with at least one tin-sulfur bond (e.g., dioctyltin octylthioglycolates) are generally called organotin mercaptides. The latter are considered to be the most efficient and most universal heat stabilizers. The important products which are on the market have the following structures:

$$\begin{array}{cc} R_1 \diagdown \quad \diagup S-R_2 & \diagup S-R_2 \\ \quad Sn & R_1-Sn-S-R_2 \\ R_1 \diagup \quad \diagdown S-R_2 & \diagdown S-R_2 \end{array}$$

Where R_1 is H_3C-, $n-C_4H_9-$, $n-C_8H_{17}-$, $n-C_{12}H_{25}$ or alkyl$-O-CO-CH_2-CH_2-$, and R_2 is $-CH_2-CO-O-$alkyl, $-CH_2-CH_2-CO-O-$alkyl, $-CH_2-CH_2-O-CO-$alkyl or $-$alkyl.

One of the most important properties—not only of the sulfur-containing tin stabilizers but also of the whole group of organotin stabilizers—is absolute crystal clarity, which can be achieved by means of a proper formulation. Clarity is required for bottles, containers, all kinds of packaging films, corrugated sheets, light panels, and also for hose, profiles, swinging doors, and transparent top coats of floor or wall coverings made from plasticized PVC.

Plasticizers

In addition to resin and stabilizers, a PVC compound may contain ingredients such as plasticizers, extenders, lubricants, fillers, pigments, polymeric processing aids, and impact modifiers.

Plasticizers (see also Chapter 1) are essentially nonvolatile solvents for PVC. At the processing temperature of about 150°C, molecular mixing occurs in a short period of time to give products of greater flexibility. Phthalates prepared from alcohols with about eight carbon atoms are by far the most important class and constitute more than 70% of plasticizers used. For economic reasons, diisooctyl phthalate (DIOP), di-2-ethylhexyl phthalate (DEHP or DOP), and the phthalate ester of the C7-C9 oxo-alcohol, often known as dialphanyl phthalate (DAP) because of the ICI trade name 'Alphanol-79' for the C7-C9 alcohols, are used. DIOP has somewhat less odor, whereas DAP has the greatest heat stability. Dibutyl phthalate and diisobutyl phthalate are also efficient plasticizers and continue to be used in PVC (except in thin sheets) despite their high volatility and water extractability.

Phosphate plasticizers such as tritolyl phosphate (TTP) and trixylyl phosphate (TXP) are generally used where good flame resistance is required, such as in insulation and mine belting. These materials, however, are toxic and give products with poor low-temperature resistance, i.e., with a high cold flex temperature (typically, −5°C).

For applications where it is important to have a compound with good low-temperature resistance, aliphatic ester plasticizers are of great value. Dibutyl sebacate (DBS), dioctyl sebacate (DOS), and, more commonly, cheaper esters of similar effect derived from mixed acids produced by the petrochemical industry are used. These plasticizers give PVC products with a cold flex temperature of −42°C.

Esters based on allyl alcohol, such as diallyl phthalate and various polyunsaturated acrylates, have proved useful in improving adhesion of PVC to metal. They may be considered as polymerizable plasticizers. In PVC pastes they can be made to cross-link by the action of peroxides or perbenzoates when the paste is spread on to metal, giving a "cured coating" with a high degree of adhesion.

Extenders

In the formulation of PVC compounds it is not uncommon to replace some of the plasticizer with an extender, a material that is not in itself a plasticizer because of its very low compatibility but that can be used in conjunction with a true plasticizer. Commercial extenders are cheaper than plasticizers and can often be used to replace up to one-third of the plasticizer without seriously affecting the properties of the compound. Three commonly employed types of extenders are chlorinated paraffin waxes, chlorinated liquid paraffinic fractions, and oil extracts.

Lubricants

In plasticized PVC it is common practice to incorporate a lubricant whose main function is to prevent sticking of the compound to processing equipment. The material used should have limited compatibility such that it will sweat out during processing to form a film between the bulk of the compound and the metal surfaces of the processing equipment. The additives used for such a purpose are known as *external lubricants*. In the United States normal lead stearate is commonly used. This material melts during processing and lubricates like wax. Also used is dibasic lead stearate, which does not melt but lubricates like graphite and improves flow properties. In Britain, stearic acid is mostly used with transparent products, calcium stearate with nontransparent products.

An unplasticized PVC formulation usually contains at least one other lubricant, which is mainly intended to improve the flow of the melt, i.e., to reduce the apparent melt viscosity. Such materials are known as *internal lubricants*. Unlike external lubricants they are reasonably compatible with the polymer and are more like plasticizers in their behavior at processing temperatures, whereas at room temperature this effect is negligible. Among materials usually classified as internal lubricants are montan wax derivatives, glyceryl monostearate, and long-chain esters such as cetyl palmitate.

Fillers

Fillers are commonly employed in opaque PVC compounds to reduce cost and to improve electrical insulation properties, to improve heat deformation resistance of cables, to increase the hardness of a flooring compound, and to reduce tackiness of highly plasticized compounds. Various calcium carbonates (such as whiting, ground limestone, precipitated calcium carbonate) are used for general-purpose work, china clay is commonly employed for electrical insulation, and asbestos for flooring applications. Also employed occasionally are the silicas and silicates, talc, light magnesium carbonate, and barytes (barium sulfate).

Pigments

Many pigments are now available commercially for use with PVC. Pigment selection should be based on the pigment's ability to withstand process conditions, its effect on stabilizer and lubricant, and its effect on end-use properties, such as electrical insulation.

Impact Modifiers and Processing Aids

Unplasticized PVC presents some processing difficulties due to its high melt viscosity; in addition, the finished product is too brittle for some applications. To overcome these problems and to produce toughening, certain polymeric additives are usually added to the PVC. These materials, known as *impact modifiers*, are generally semicompatible and often somewhat rubbery in nature. Among the most important impact modifiers in use today are butadiene-acrylonitrile copolymers (nitrile rubber), acrylonitrile-butadiene-styrene (ABS) graft terpolymers, methacrylate-butadiene-styrene (MBS) terpolymers, chlorinated polyethylene, and some polyacrylates.

ABS materials are widely used as impact modifiers, but they cause opacity and have only moderate aging characteristics. Many grades also show severe *stress whitening*, a phenomenon advantageously employed in labeling tapes, such as Dymotape. MBS modifiers have been used such as where tough PVC materials of high clarity are desired (e.g., bottles and film). Chlorinated polyethylene has been widely used as an impact modifier where good aging properties are required.

A number of polymeric additives are also added to PVC as *processing aids*. They are more compatible with PVC and are included mainly to ensure more uniform flow and thus improve the surface finish. In chemical constitution they are similar to impact modifiers and include ABS, MBS, acrylate-methacrylate copolymers, and chlorinated polyethylene.

Typical formulations of several PVC compounds for different applications are given in Appendix 5.

Properties and Applications

PVC may reasonably be considered as the most versatile of plastics; its usage ranges from building construction to toys and footwear. PVC compounds are made in a wide range of formulations, which makes it difficult to make generalizations about their properties. Mechanical properties are considerably affected by the type and amount of plasticizer. Table 4.2 illustrates differences in some properties of three distinct types of compound. To a lesser extent, fillers also affect the physical properties.

Unplasticized PVC (UPVC) is a rigid material, whereas the plasticized material is tough, flexible, and even rubbery at high plasticizer loadings. Relatively high plasticizer loadings are necessary to achieve any significant improvement in impact strength. Thus incorporation of less than 20% plasticizer does not give compounds with impact strength higher than that of unplasticized grades. Lightly plasticized grades are therefore used when the ease of processing is more important than achieving good impact strength.

PVC is resistant to most aqueous solutions, including those of alkalis and dilute mineral acids. The polymer also has a good resistance to hydrocarbons. The only effective solvents appear to be those which are capable of some form of interaction with the polymer. These include cyclohexanone and tetrahydrofuran.

TABLE 4.2 Properties of Three Types of PVC Compounds

Property	Unplasticized PVC	PVC + DIOP (50 parts per 100 resin)	Vinyl chloride-vinyl acetate copolymer (sheet)
Specific gravity	1.4	1.31	1.35
Tensile strength			
lbf/in^2	8500	2700	7000
MPa	58	19	48
Elongation at break (%)	5	300	5
Vicat softening (°C)	80	Flexible at room temperature	70

At ordinary temperatures, PVC compounds are reasonably good electrical insulators over a wide range of frequencies, but above the glass transition temperature their value as an insulator is limited to low-frequency applications. The volume resistivity decreases as the amount of plasticizer increases.

PVC has the advantage over other thermoplastic polyolefins of built-in fire retardancy because of its 57% chlorine content.

Copolymers of vinyl chloride with vinyl acetate have lower softening points, easier processing, and better vacuum-forming characteristics than the homopolymer. They are soluble in ketones, esters, and certain chlorinated hydrocarbons, and have generally inferior long-term heat stability.

About 90% of the PVC produced is used in the form of homopolymers, the other 10% as copolymers and terpolymers. The largest application of homopolymer PVC compounds, particularly unplasticized grades, is for rigid pipes and fittings, most commonly as suspension homopolymers of high bulk density compounded as *powder blends*. In addition to the aforesaid properties, UPVC has an excellent resistance to weathering. Moreover, when the cost of installation is taken into account, the material frequently turns out to be cheaper. UPVC is therefore becoming used increasingly in place of traditional materials. Important uses include translucent roof sheathing with good flame-retarding properties, window frames, and piping that neither corrodes nor rots.

As a pipe material PVC is widely used in soil pipes and for drainage and above-ground applications. Piping with diameters of up to 60 cm is not uncommon.

UPVC is now being increasingly used as a wood replacement due to its more favorable economics, taking into account both initial cost and installation. Specific applications include bench-type seating at sports stadia, window fittings, wall-cladding, and fencing. UPVC bottles have better clarity, oil resistance, and barrier properties than those made from polyethylene. Compared with glass they are also lighter, less brittle, and possess greater design flexibility. These products have thus made extensive penetration into the packaging market for fruit juices and beverages, as well as bathroom toiletry. Sacks made entirely of PVC enable fertilizers and other products to be stored outdoors.

The largest applications of plasticized PVC are wire and cable insulation and as film and sheet. PVC is of great value as an insulator for direct-current and low-frequency alternating-current carriers. It has almost completely replaced rubber in wire insulation. PVC is widely used in cable sheathing where polyethylene is employed as the insulator.

Other major outlets of plasticized PVC include floor coverings, leathercloth, tubes and profiles, injection moldings, laminates, and paste processes.

When a thin layer of plasticized PVC is laminated to a metal sheet, the bond may be strong enough that the laminate can be punched, cut, or shaped without parting the two layers. A pattern may be printed or embossed on the plastic before such fabrication. Typewriter cases and appliance cabinets have been produced with such materials.

PVC leathercloth has been widely used for many years in upholstery and trim in car applications, house furnishings, and personal apparel. The large-scale replacement of leather by PVC initiated in the 1950s and 1960s was primarily due to the greater abrasion resistance, flex resistance, and washability of PVC. Ladies handbags are frequently made from PVC leathercloth. House furnishing applications include kitchen upholstery, printed sheets, and bathroom curtains. Washable wallpapers are obtained by treating paper with PVC compounds.

Special grades of PVC are used in metal-finishing applications, for example, in stacking chairs. Calendered plasticized PVC sheet is used in making plastic rainwear and baby pants by the high-frequency welding technique. The application of PVC in mine belting is still important in terms of the actual tonnage of material consumption. All-PVC shoes are useful as beachwear and standard footwear. PVC has also proved to be an excellent abrasion-resistant material for shoe soles. PVC adhesives, generally containing a polymerizable plasticizer, are useful in many industries.

The two main applications of vinyl chloride-vinyl acetate copolymers are phonograph records and vinyl floor tiles. The copolymers contain an average of about 13% vinyl acetate. They may be processed at lower temperatures than those used for the homopolymer. Phonograph records contain only a stabilizer, lubricant, pigment, and, possibly, an antistatic agent; there are no fillers. Preformed resin biscuits are normally molded in compression presses at about 130 to 140°C. The press is a flash mold that resembles a waffle iron. The faces of the mold may be nickel negatives of an original disc recording that have been made by electrodeposition.

Floor tiles contain about 30 to 40 parts plasticizer per 100 parts copolymer and about 400 parts filler (usually a mixture of asbestos and chalk). Processing involves mixing in an internal mixer at about 130°C, followed by calendering at 110 to 120°C.

Pastes

A PVC paste is obtained when the voids between the polymer particles in a powder are completely filled with plasticizer so that the particles are suspended in it. To ensure a stable paste, there is an upper limit and a lower limit to the order of particle size. PVC paste polymers have an average particle size of about 0.2 to 1.5 μm. The distribution of particle sizes also has significant influences on the flow and fluxing characteristics of the paste.

The main types of PVC pastes are plastisols, organosols, plastisols incorporating filler polymers (including the rigisols), plastigels, hot-melt compounds, and compounds for producing cellular products. Typical formulations of the first three types are shown in Table 4.3. The processing methods were described in Chapter 2.

Plastisols are of considerable importance commercially. They are converted into tough, rubbery products by heating at about 160°C (*gelation*). *Organosols* are characterized by the presence of a volatile organic diluent whose sole function is to reduce the paste viscosity. The diluent is removed after application and before gelling the paste.

Another method of reducing paste viscosity is to use a *filler polymer* to replace a part of the PVC paste polymer. The filler polymer particles are too large to make stable pastes by themselves, but in the presence of paste-polymer particles they remain in stable suspension. Being very much larger than paste-polymer particles and having a low plasticizer absorption, they take up large volumes in the paste and make more plasticizer available for particle lubrication, thus reducing paste viscosity. The use of filler polymers has increased considerably in recent years. Pastes prepared using filler polymers and only small quantities of plasticizer (approximately 20 parts per 100 parts of polymer) are termed *rigisols*.

Hot-melt PVC compounds are prepared by fluxing polymer with large quantities of plasticizers and extenders. They melt at elevated temperatures and become very fluid, so they may be poured. These compounds are extensively used for casting and prototype work.

Sigma-blade trough mixers are most commonly used for mixing PVC pastes. It is common practice to mix the dry ingredients initially with part of the plasticizer so that the shearing stresses are high enough to break down the aggregates. The remainder of the plasticizer is then added to dilute the product. The mix is preferably deaerated to remove air bubbles before final processing.

TABLE 4.3 Typical Formulations[a] of Three Types of PVC Pastes

Ingredient	Plastisol	Organosol	Plastigel
PVC paste polymer	100	100	100
Plasticizer (e.g., DOP)	80	30	80
Filler (e.g., china clay)	10	10	10
Stabilizer (e.g., white lead)	4	4	4
Naphtha	—	50	—
Aluminum stearate	—	—	4

[a]Parts by weight.

A larger proportion of PVC paste is used in the manufacture of leathercloth by a *spreading technique*. A layer of paste is smeared on the cloth by drawing the latter between a roller or endless belt and a doctor blade against which there is a rolling bank of paste. The paste is gelled by passing through a heated tunnel or under infrared heaters. Embossing operations may be carried out by using patterned rollers when the gelled paste is still hot. The leathercloth is then cooled and wound up. Where it is desired that the paste should enter the interstices of the cloth, a shear-thinning (pseudo-plastic) paste is employed. Conversely, where strike-through should be minimized, a dilatant paste (viscosity increases with shear rate) is employed.

Numerous methods exist for producing cellular products (see also Chapter 2) from PVC pastes. Closed-cell products can be made if a blowing agent such as azodiisobutyronitrile is incorporated into the paste. The paste is then heated in a mold to cause the blowing agent to decompose and the compound to gel. Since the mold is full, expansion does not take place at this stage. The unexpanded block is removed after thoroughly cooling the mold and is heated in an oven at about 100°C to produce uniform expansion. One method of producing a flexible, substantially open-cell product is to blend the paste with carbon dioxide (either as dry ice or under pressure). The mixture is heated to volatilize the carbon dioxide to produce a foam, which is then gelled at a higher temperature.

Poly (Vinylidene Chloride)

$$
+\!\!\begin{array}{c} \mathrm{Cl} \\ | \\ \mathrm{CH_2-C} \\ | \\ \mathrm{Cl} \end{array}\!\!\Big)_{\!n}
$$

Monomer	Polymerization	Major uses
Vinylidene chloride	Free-radical-initiated chain polymerization	Film and sheeting for food packaging

The polymer may be prepared readily by free-radical mechanisms in bulk, emulsion, and suspension; the latter technique is usually preferred on an industrial scale. Copolymers of vinylidene chloride with vinyl chloride, acrylates, and acrylonitrile are also produced.

Since the poly(vinylidene chloride) molecule has an extremely regular structure (and the question of tacticity does not arise), the polymer is capable of crystallization. Because of the resultant close packing and the presence of heavy chloride atoms the polymer has a high specific gravity (1.875) and a low per meability to vapors and gases. The chlorine present gives a self-extinguishing polymer. Vinylidene chloride-vinyl chloride copolymers are also self-extinguishing and possess very good resistance to a wide range of chemicals, including acids and alkalies. Because of a high degree of crystallization, even in the copolymers, high strengths are attained even though the products have relatively low molecular weights (~20,000–50,000).

Both poly(vinylidene chloride) and copolymers containing vinylidene chloride are used to produce flexible films and coatings. Flexible films are used extensively for food packaging because of their superior barrier resistance to water and oxygen. The coating resins are used for cellophane, polyethylene, paper, fabric, and container liner applications. Dow's trade name for a copolymer of vinylidene chloride (87%) and vinyl chloride (13%) is *Saran*. Biaxially stretched Saran film is a useful, though expensive, packaging material possessing exceptional clarity, brilliance, toughness, and impermeability to water and gases.

Vinylidene chloride-vinyl chloride copolymers are used in the manufacture of filaments. The filaments have high toughness, flexibility, durability, and chemical resistance. They find use in car upholstery, deck-chair fabrics, decorative radio grilles, doll hair, filter presses, and other applications. A flame-resisting fiber said to be a 50:50 vinylidene chloride-acrylonitrile copolymer is marketed by Courtaulds with the name Teklan.

Polytetrafluoroethylene and Other Fluoropolymers

$$\left[\begin{array}{c} \begin{array}{cc} F & F \\ | & | \\ C - C \\ | & | \\ F & F \end{array} \end{array}\right]_n$$

Monomer	Polymerization	Major uses
Tetrafluoroethylene, hexafluoropropylene, chlorotrifluoroethylene, vinyl fluoride, vinylidene fluoride, perfluoroalkyl vinyl ether	Free-radical-initiated chain polymerization	Coatings for chemical process equipment, cable insulation, electrical components, nonsticking surfaces for cookware

Polytetrafluoroethylene (PTFE) was discovered in 1947. Today PTFE probably accounts for at least 85% of the fluorinated polymers and, in spite of its high cost, has a great diversity of applications. It is produced by the free-radical chain polymerization of tetrafluoroethylene. With a linear molecular structure of repeating —CF_2—CF_2— units, PTFE is a highly crystalline polymer with a melting point of 327°C. Density is 2.13 to 2.19 g/cm^3. Commercially PTFE is made by two major processes—one leading to the so-called granular polymer, the second to a dispersion of polymers of much finer particle size and lower molecular weight.

Since the carbon-fluorine bond is very stable and since the only other bond present in PTFE is the stable C—C bond, the polymer has a high stability, even when heated above its melting point.

Because of its high crystallinity (>90%) and incapability of specific interaction, PTFE has exceptional chemical resistance and is *insoluble in all organic solvents*. (PTFE dissolves in certain fluorinated liquids such as perfluorinated kerosenes at temperatures approaching the melting point of the polymer.) At room temperature it is attacked only by alkali metals and, in some cases, by fluorine. Treatment with a solution of sodium metal in liquid ammonia sufficiently alters the surface of a PTFE sample to enable it to be cemented to other materials by using epoxide resin adhesives.

PTFE has good weathering resistance. The polymer is not wetted by water and has negligible water absorption (0.005%). The permeability of gases is low—the rate of transmission of water vapor is approximately half that of poly-(ethylene terephthalate) and low-density polyethylene. PTFE is however degraded by high-energy radiation. For example, the tensile strength of a given sample may be halved by exposure to a dosage of 70 Mrad.

PTFE is a tough, flexible material of moderate tensile strength (2500–3800 psi, i.e., 17–21 MPa) at 23°C. Temperature has a considerable effect on its properties. It remains ductile in compression at temperatures as low as 4 K (−269°C). The creep resistance is low in comparison to other engineering plastics. Thus, even at 20°C unfilled PTFE has a measurable creep with compression loads as low as 300 psi (2.1 MPa).

The coefficient of friction is unusually low and is lower than almost any other material. The values reported in the literature are usually in the range 0.02 to 0.10 for polymer to polymer and 0.09 to 0.12 for polymer to metal. The polymer has a high oxygen index and will not support combustion.

PTFE has outstanding insulation properties over a wide range of temperatures and frequencies. The volume resistivity exceeds 10^{20} ohm-m. The power factor (<0.003 at 60 Hz and <0.0003 at 10^6 Hz) is negligible in the temperature range −60 to +250°C. PTFE's low dielectric constant (2.1) is unaffected by frequency. The dielectric strength of the polymer is 16 to 20 kV/mm (short time on 2-mm thick sheet).

Processing

PTFE is commonly available in three forms: (a) granular polymers with average particle size of 300 to 600 μm; (b) dispersion polymers (obtained by coagulation of a dispersion) consisting of agglomerates with an average diameter of 450 μm, which are made up of primary particles 0.1 μm in diameter; and (c) dispersion (lattices) containing about 60% polymer in the form of particles with an average diameter of about 0.16 μm.

PTFE cannot be processed by the usual thermoplastics processing techniques because of its exceptionally high melt viscosity ($\sim 10^{10}$–10^{11} poises at 350°C). Granular polymers are processed by press and sinter methods used in powder metallurgy. In principle, these methods involve preforming the sieved powder by compressing in a mold at 1.0 to 3.5 tonf/in^2 (16 to 54 MPa) usually at room temperature or at 100°C, followed by sintering at a temperature above the melting point (typically at about 370°C), and then cooling. Free-sintering of the preform in an oven at about 380°C is also satisfactory. The sintering period depends on the thickness of the sample. For example, a 0.5-in. (1.25-cm)-thick sample will need sintering for 3.5 hr. Granular polymers may also be extruded, though at very low rates (2.5–16 cm/min), by screw and ram extruders. The extrudates are reasonably free of voids.

PTFE moldings and extrudates may be machined without difficulty. Continuous film may be obtained by peeling a pressure-sintered ring and welding it to a similar film by heat sealing under pressure at about 350°C.

A PTFE dispersion polymer leads to products with improved tensile strength and flex life. Preforms are made by mixing the polymer with 15 to 25% of a lubricant and extruded. This step is followed by lubricant removal and sintering. In a typical process a mixture of PTFE dispersion polymer (83 parts) and petroleum ether (17 parts) with a 100 to 120°C boiling range is compacted into a preform billet which is then extruded by a vertical ram extruder. The extrudate is heated in an oven at about 105°C to remove the lubricant and is then sintered at about 380°C. Because of the need to remove the lubricant, only thin sections can be produced by this process. Thin-walled tubes with excellent fatigue resistance can be produced, or wire can be coated with very thin coatings of PTFE.

Tapes also may be made by a similar process. However, in this case the lubricant used is a nonvolatile oil. The preform is extruded in the shape of a rod, which is then passed between a pair of calendered rolls at about 60 to 80°C. The unsintered tape finds an important application in pipe-thread sealing. If sintered tape is required, the calendered product is first degreased by passing through boiling trichloroethylene and then sintered by passing through a salt bath. The tape made in this way is superior to that obtained by machining granular polymer moldings.

PTFE dispersions may be used in a variety of ways, such as coating metal with PTFE to produce nonadhesive surfaces with a very low coefficient of friction. Glass-coated PTFE laminates may be produced by piling up layers of glass cloth impregnated with PTFE dispersions and pressing at about 330°C. Asbestos-PTFE laminates may be produced in a similar way. The dispersions can also be used for producing filled PTFE molding material. The process typically involves stirring fillers into the dispersion, coagulating with acetone, drying at 280 to 290°C, and disintegrating the resulting cake of material.

Applications

Its exceptional properties make PTFE highly useful, but because of its high volume cost, it is not generally used to produce large objects. In many cases it is possible to coat a metal object with a layer of PTFE to meet the particular requirement. Nonstick home cookware is perhaps the best-known example. PTFE is used for lining chutes and coating other metal objects where low coefficients of friction, chemical inertness, or nonadhesive characteristics are required.

Because of its exceptional chemical resistance over a wide temperature range, PTFE is used in a variety of seals, gaskets, packings, valve and pump parts, and laboratory equipment.

Its excellent electrical insulation properties and heat resistance lead to its use in high-temperature wire and cable insulation, molded electrical components, insulated transformers, hermetic seals for condensers, laminates for printed circuitry, and many other electrical applications.

Reinforced PTFE applications include bushings and seals in compressor hydraulic applications, automotive applications, and pipe liners. A variety of moldings are used in aircraft and missiles and also in other applications where use at elevated temperatures is required.

Copolymers of tetrafluoroethylene were developed in attempts to provide materials with the general properties of PTFE and the melt processability of the more conventional thermoplastics. Two such copolymers are tetrafluoroethylene-hexafluoropropylene (TFE-HFP) copolymers (Teflon FEP resins by Du Pont; FEP stands for fluorinated ethylene propylene) with a melting point of 290°C and tetrafluoroethylene-ethylene (ETFE) copolymers (Tefzel by Du Pont) with a melting point of 270°C. These products are melt processable. A number of other fluorine containing melt processable polymers have been introduced.

Polychlorotrifluoroethylene (PCTFE) was the first fluorinated polymer to be produced on an experimental scale and was used in the United States early in World War II. It was also used in the handling of corrosive materials, such as uranium hexafluoride, during the development of the atomic bomb.

PCTFE is a crystalline polymer with a melting point of 218°C and density of 2.13 g/cm^3. The polymer is inert to most reactive chemicals at room temperature. However, above 100°C a few solvents dissolve the polymer, and a few, especially chlorinated types, swell it. The polymer is melt processable, but processing is difficult because of its high melt viscosity and its tendency to degrade, resulting in deterioration of its properties.

PCTFE is marketed by Hoechst as Hostaflon C2 and in the United States by Minnesota Mining and Manufacturing as Kel-F and by Allied Chemical as Halon. The film is sold by Allied Corp. as Aclar.

PCTFE is used in chemical processing equipment and cryogenic and electrical applications. Major applications include wafer boats, gaskets, O-rings, seals, and electrical components. PCTFE has outstanding barrier properties to gases, the PCTFE film has the lowest water-vapor transmission of any transparent plastic film. It is used in pharmaceutical packaging and other applications for its vapor barrier properties, including electroluminescent lamps.

Other melt-processable thermoplastics include ethylene-chlorotrifluoroethylene (ECTFE) copolymer (melting point 240°C), polyvinylidene fluoride (PVDF) (melting point 170°C), and polyvinyl fluoride (PVF), which is commercially available only as film.

PVF is tough and flexible, has good abrasion and staining resistance, and has outstanding weathering resistance. It maintains useful properties over a temperature range of -70 to 110°C. PVF can be laminated to plywood, hardboard, vinyl, reinforced polyesters, metal foils, and galvanized steel. These laminates are used in aircraft interior panels, lighting panels, wall coverings, and a variety of building applications. PVF is also used as glazing in solar energy collectors. PVF film is marketed by Du Pont as Tedlar.

Perfluoroalkoxy (PFA) resins are the newest commercially available class of melt-processable fluoroplastics. Its general chemical structure is

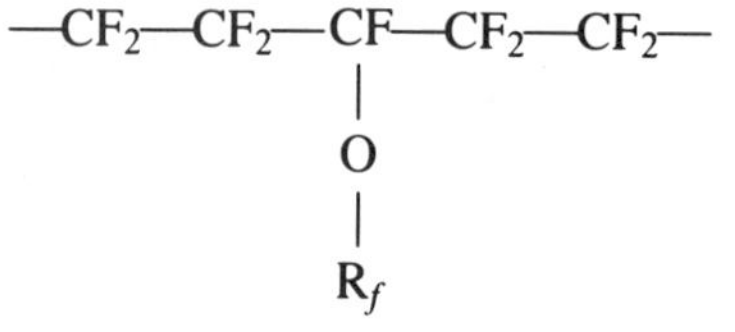

$$-CF_2-CF_2-CF-CF_2-CF_2-$$
$$|$$
$$O$$
$$|$$
$$R_f$$

where $R_f = -C_nF_{2n+1}$.

PFA resin has somewhat better mechanical properties than FEP above 150°C and can be used up to 260°C. In chemical resistance it is about equal to PTFE. PFA resin is sold by Du Pont under the Teflon trademark.

Polyisobutylene

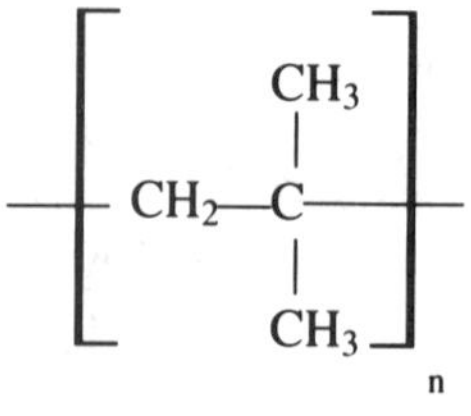

$$\left[\begin{array}{c} CH_3 \\ | \\ -CH_2-C- \\ | \\ CH_3 \end{array}\right]_n$$

Monomer	Polymerization	Major uses
Isobutylene	Cationic-initiated chain polymerization	Lubricating oils, sealants

High-molecular-weight polyisobutylene (PIB) is produced by cationic chain polymerization in methyl chloride solution at $-70°C$ using aluminum chloride as the catalyst. Such polymers are currently available from Esso (Vistanex) and BASF (Oppanol).

PIB finds a variety of uses. It is used as a motor oil additive to improve viscosity characteristics, as a blending agent for polyethylene to improve its impact strength and environmental stress-cracking resistance, as a base for chewing gum, as a tackifier for greases; it is also used in caulking compounds and tank linings. Because of its high cold flow, it has little use as a rubber in itself, but copolymers containing about 2% isoprene to introduce unsaturation for cross-linking are widely used (butyl rubber; see later).

Polystyrene

$$-\left[CH_2-\underset{\underset{\bigcirc}{|}}{CH}\right]_n-$$

Monomer	Polymerization	Major uses
Styrene	Free-radical-initiated chain polymerization	Packaging and containers (35%), housewares, toys and recreational equipment (25%), appliance parts (10%), disposable food containers (10%)

Polystyrene is made by bulk or suspension polymerization of styrene. Polystyrene is very low cost and is extensively used where price alone dictates. Its major characteristics include rigidity, transparency, high refractive index, no taste, odor, or toxicity, good electrical insulation characteristics, low water absorption, and ease of coloring and processing. Polystyrene has excellent organic acid, alkali, salts, and lower alcohol resistance. It is, however, attacked by hydrocarbons, esters, ketones, and essential oils. A more serious limitation of polystyrene in many applications is its brittleness. This limitation led to the development of rubber-modified polystyrenes (containing usually 5–15% rubber), the so-called *high impact polystyrenes* (HIPS). The most commonly used are styrene-butadiene rubber and *cis*-1,4-polybutadiene. The method of mixing the polystyrene and rubber has a profound effect on the properties of the product. Thus, much better results are obtained if the material is prepared by polymerization of styrene in the presence of the rubber rather than by simply blending the two polymers. The product of the former method contains not only polystyrene and straight rubber but also a graft copolymer in which polystyrene side chains are attached to the rubber.

Compared to straight or general-purpose polystyrenes, high-impact polystyrene materials have much greater toughness and impact strength, but clarity, softening point, and tensile strength are not as good. Expanded or foamed polystyrene (see Chapter 2), which has become very important as a thermal insulating material, has a low density, has a low weight cost, is less brittle, and can be made fire retarding.

End uses for all types of polystyrene are packaging, toys, and recreational products, housewares, bottles, lenses, novelties, electronic appliances, capacitor dielectrics, low-cost insulators, musical instrument reeds, light-duty industrial components, furniture, refrigeration, and building and construction uses (insulation). Packaging is by far the largest outlet: bottle caps, small jars and other injection-molded containers, blow-molded containers, toughened polystyrene liners (vacuum formed) for boxed goods, and oriented polystyrene film for foodstuffs are some of its uses. The second important outlet is refrigeration equipment, including door liners and inner liners (made from toughened polystyrene sheet), moldings for refrigerator furnishings, such as flip lids and trays, and expanded polystyrene for thermal insulation.

Expanded polystyrene products have widely increased the market for polystyrene resin (see the section on polystyrene foams in Chapter 2). With as light a weight as 2 lb/ft^3 (0.032 g/cm^3), the thermal conductivity of expanded polystyrene is very low, and its cushioning value is high. It is an ideal insulation and packaging material. Common applications include ice buckets, water coolers, wall panels, and general thermal insulation applications. Packaging uses of expanded polystyrene range from thermoformed egg boxes and individually designed shipping packages for delicate equipment, such as cameras and electronic equipment, to individual beads (which may be about 1 cm in diameter and up to 5 cm long) for use as a loose fill material in packages.

Polybutadiene (Butadiene Rubber)

$$+CH_2-CH=CH-CH_2+_n$$

Monomer	Polymerization	Major uses
Butadiene	Ziegler–Natta-catalyzed chain polymerization	Tires and tire products (90%)

Polybutadiene is made by solution polymerization of butadiene using Ziegler–Natta catalysts. Slight changes in catalyst composition produce drastic changes in the stereoregularity of the polymer. For example, polymers containing 97 to 98% of *trans*-1,4 structure can be produced by using Et_3Al/VCl_3 catalyst, those with 93 to 94% *cis*-1,4 structure by using $Et_2AlCl/CoCl_2$, and those with 90% 1,2-polybutadiene by using $Et_3Al/Ti(OBu)_4$. The stereochemical composition of polybutadiene is important if the product is to be used as a base polymer for further grafting. For example, a polybutadiene with 60% *trans*-1,4, 20% *cis*-1,4, and 20% 1,2 configuration is used in the manufacture of ABS resin.

Polybutadiene rubbers generally have a higher resilience than natural rubbers at room temperature, which is important in rubber applications. On the other hand, these rubbers have poor tear resistance, poor tack, and poor tensile strength. For this reason polybutadiene rubbers are usually used in conjunction with other materials for optimum combination of properties. For example, they are blended with natural rubber in the manufacture of truck tires and with styrene-butadiene rubber (SBR) in the manufacture of automobile tires.

Polybutadiene is also produced in low volume as specialty products. These include low-molecular-weight, liquid 1,2-polybutadienes (60–80% 1,2 content) used as potting compounds for transformers and submersible electric motors and pumps, liquid *trans*-1,4-polybutadienes used in protective coatings inside metal cans, and hydroxy-terminated polybutadiene liquid resins for use as a binder and in polyurethane and epoxy resin formulations.

Polyisoprene

$$\left[\begin{array}{c} CH_2-C=CH-CH_2 \\ | \\ CH_3 \end{array} \right]_n$$

Monomer	Polymerization	Major uses
Isoprene	Ziegler–Natta-catalyzed chain polymerization	Car tires (55%), mechanical goods, sporting goods, footwear, sealants, and caulking compounds

Polyisoprene is produced by solution polymerization using Ziegler-Natta catalysts. The *cis*-1,4-polyisoprene is a synthetic equivalent of natural rubber. However, the synthetic polyisoprenes have cis contents of only about 92 to 96%; consequently, these rubbers differ from natural rubber in several ways. The raw synthetic polyisoprene is softer than raw natural rubber (due to a reduced tendency for a stress-induced crystallization because of the lower cis content) and is therefore more difficult to mill. On the other hand, the unvulcanized synthetic material flows more readily; this feature makes it easier to injection mold. The synthetic product is somewhat more expensive than natural rubber.

Polyisoprene rubbers are used in the construction of automobile tire carcasses and inner liners and truck and bus tire treads. Other important applications are in mechanical goods, sporting goods, footwear, sealants, and caulking compounds.

Polychloroprene

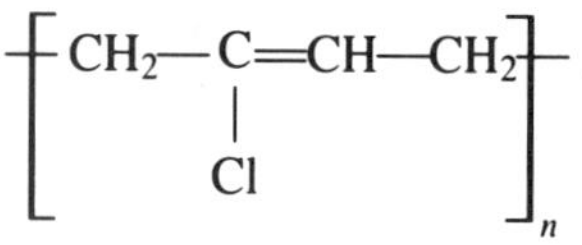

$$\left[CH_2 - \underset{\underset{Cl}{|}}{C} = CH - CH_2 \right]_n$$

Monomer	Polymerization	Major uses
Chloroprene (2-chlorobuta-1,3-diene)	Free-radical-initiated chain polymerization (mostly emulsion polymerization)	Conveyor belts, hose, seals and gaskets, wire and cable sheathing

The polychloroprenes were first marketed by Du Pont in 1931. Today these materials are among the leading special-purpose or non-tire rubbers and are well known under such commercial names as Neoprene (Du Pont), Baypren (Bayer), and Butachlor (Distagul).

A comparison of polychloroprene and natural rubber or polyisoprene molecular structures shows close similarities. However, while the methyl groups activates the double bond in the polyisoprene molecule, the chlorine atom has the opposite effect in polychloroprene. Thus polychloroprene is less prone to oxygen and ozone attack than natural rubber is. At the same time accelerated sulfur vulcanization is also not a feasible proposition, and alternative vulcanization or curing systems are necessary.

Vulcanization of polychloroprene rubbers is achieved with a combination of zinc and magnesium oxide and added accelerators and antioxidants. The vulcanizates are broadly similar to those of natural rubber in physical strength and elasticity. However, the polychloroprene vulcanizates show much better heat resistance and have a high order of oil and solvent resistance (though less resistant than those of nitrile rubber). Aliphatic solvents have little effect, although aromatic and chlorinated solvents cause some swelling. Because of chlorine content chloroprene rubber is generally self-extinguishing.

Because of their greater overall durability, chloroprene rubbers are used chiefly where a combination of deteriorating effects exists. Products commonly made of chloroprene rubber include conveyor belts, V-belts, diaphragms, hoses, seals, gaskets, and weather strips. Some important construction uses are highway joint seals, pipe gaskets, and bridge mounts and expansion joints. Latexes are used in gloves, balloons, foams, adhesives, and corrosion-resistant coatings.

OLEFIN COPOLYMERS
Styrene-Butadiene Rubber

$$\left[CH_2-CH=CH-CH_2-CH_2-CH \atop {\bigcirc} \right]_n$$

Monomers	Polymerization	Major uses
Styrene, butadiene	Free-radical-initiated chain polymerization (mostly emulsion polymerization)	Tires and tread (65%), mechanical goods (15%), latex (10%), automotive mechanical goods (5%)

In tonnage terms SBR is the world's most important rubber. Its market dominance is primarily due to three factors: low cost, good abrasion resistance, and a higher level of product uniformity than can be achieved with natural rubber.

Reinforcement of SBR with carbon black leads to vulcanizates which resemble those of natural rubber, and the two products are interchangeable in most applications. As with natural rubber, accelerated sulfur systems consisting of sulfur and an *activator* comprising a metal oxide (usually zinc oxide) and a fatty acid (commonly stearic acid) are used. A conventional curing system for SBR consists of 2.0 parts sulfur, 5.0 parts zinc oxide, 2.0 parts stearic acid, and 1.0 part N-t-butylbenzothiazole-2-sulfenide (TBBS) per 100 parts polymers.

The most important applications of SBR is in car tires and tire products, but there is also widespread use of the rubber in mechanical and industrial goods. SBR latexes, which are emulsions of styrene-butadiene copolymers (containing about 23–25% styrene), are used for the manufacture of foam rubber backing for carpets and for adhesive and molded foam applications.

Nitrile Rubber

$$\left[\!\!\begin{array}{c} CH_2\text{—}CH\text{—}CH_2\text{—}CH\!=\!CH\text{—}CH_2 \\ | \\ CN \end{array}\!\!\right]_n$$

Monomers	Polymerization	Major uses
Acrylonitrile, butadiene	Free-radical-initiated chain polymerization (mostly emulsion polymerization)	Gasoline hose, seals, gaskets, printing tools, adhesive, footwear

Nitrile rubber (acrylonitrile-butadiene copolymer) is a unique elastomer. The acrylonitrile content of the commercial elastomers ranges from 25 to 50% with 34% being a typical value. This non-hydrocarbon monomer imparts to the copolymer very good hydrocarbon oil and gasoline resistance. The oil resistance increases with increasing amounts of acrylonitrile in the copolymer. Nitrile rubber is also noted for its high strength and excellent resistance to abrasion, water, alcohols, and heat. Its drawbacks are poor dielectric properties and poor resistance to ozone.

Because of the diene component, nitrile rubbers can be vulcanized with sulfur. A conventional curing system consists of 2.5 parts sulfur, 5.0 parts zinc oxide, 2.0 parts stearic acid, and 0.6 parts N-t-butylbenzothiazole-2-sulfenamide (TBBS) per 100 parts polymer.

Nitrile rubbers (vulcanized) are used almost invariably because of their resistance to hydrocarbon oil and gasoline. They are, however, swollen by aromatic hydrocarbons and polar solvents such as chlorinated hydrocarbons, esters, and ketones.

Ethylene-Propylene Elastomer

$$\left[\!\!\begin{array}{c} CH_2\text{—}CH_2\text{—}CH_2\text{—}CH \\ | \\ CH_3 \end{array}\!\!\right]_n$$

Monomers	Polymerization	Major uses
Ethylene, propylene	Ziegler–Natta-catalyzed chain polymerization	Automotive parts, radiator and heater hoses, seals

Two types of ethylene-propylene elastomers are currently being produced: ethylene-propylene binary copolymers (EPM rubbers) and ethylene-propylene-diene ternary copolymers (EPDM rubbers). Because of their saturated structure, EPM rubbers cannot be vulcanized by using accelerated sulfur systems, and the less convenient vulcanization with free-radical generators (peroxide) is required. In contrast, EPDM rubbers are produced by polymerizing ethylene and propylene with a small amount (3–8%) of a diene monomer, which provides a cross-link site for accelerated vulcanization with sulfur. A typical vulcanization system for EPDM rubber consists of 1.5 parts sulfur, 5.0 parts zinc oxide, 1.0 part stearic acid, 1.5 parts 2-mercaptobenzothiazole (MBT), and 0.5 part tetramethylthiuram disulfide (TMTD) per 100 parts polymer.

The EPDM rubbers, though hydrocarbon, differ significantly from the diene hydrocarbon rubbers considered earlier in that the level of unsaturation in the former is much lower, giving rubbers much better heat, oxygen, and ozone resistance. Dienes commonly used in EPDM rubbers include dicyclopentadiene, ethylidene norbornene, and hexa-1,4-diene (Table 4.4). Therefore the double bonds in the polymer are either on a side chain or as part of a ring in the main

TABLE 4.4 Principal Diene Monomers Used in EPDM Manufacture.

Monomer	Predominant structure present in terpolymer
Dicyclopentadiene	
4-Ethylidenenorborn-2-ene	
Hexa-1,4-diene $CH_2{=}CH{-}CH_2{-}CH{=}CH{-}CH_3$	$-CH_2{-}CH-$ $\qquad\ \ \ CH_2{-}CH{=}CH{-}CH_3$

chain. Hence, should the double bond become broken, the main chain will remain substantially intact, which also accounts for the greater stability of the product.

The use of EPDM rubbers for the manufacture of automobile and truck tires has not been successful, mainly because of poor tire cord adhesion and poor compatibility with most other rubbers. However, EPDM rubbers have become widely accepted as a moderately heat-resisting material with good weathering, oxygen, and ozone resistance. They find extensive use in nontire automotive applications, including body and chassis parts, car bumpers, radiator and heater hoses, weatherstrips, seals, and mats. Other applications include wire and cable insulation, appliance parts, hoses, gaskets and seals, and coated fabrics.

These rubbers are now also being blended on a large scale with polyolefin plastics, particularly polypropylene, to produce an array of materials ranging from tough plastics at one end to the thermoplastic polyolefin rubbers (see later) at the other.

Butyl Rubber

$$\left[CH_2\!-\!\underset{\underset{CH_3}{|}}{\overset{\overset{CH_3}{|}}{C}}\!-\!CH_2\!-\!CH\!=\!\underset{\underset{CH_3}{|}}{C}\!-\!CH_2 \right]_n$$

Monomers	Polymerization	Major uses
Isobutylene, isoprene	Cationic chain polymerization of isobutylene with 0.5 to 2.5 mol % of isoprene	Tire inner tubes and inner liners of tubeless tires (70%), inflatable sporting goods

Although polyisobutylene described earlier is a nonrubbery polymer exhibiting high cold flow, the copolymer containing about 2% isoprene can be vulcanized with a powerful accelerated sulfur system to give rubbery polymers. Being almost saturated, they are broadly similar to the EPDM rubbers in many properties. The most outstanding property of butyl rubber is its very low air permeability, which has led to its extensive use in tire inner tubes and liners. A major disadvantage is its lack of compatibility with SBR, polybutadiene, and natural rubber.

Thermoplastic Elastomers

Monomers	Polymerization	Major uses
Butadiene, isoprene, styrene	Anionic block polymerization	Footwear, automotive parts, hot-melt adhesives

Conventional rubbers are vulcanized, that is, cross-linked by primary valence bonding. For this reason vulcanized rubbers cannot dissolve or melt unless the network structure is irreversibly destroyed. These products cannot therefore be reprocessed like thermoplastics. Hence, if a polymer could be developed which showed rubbery properties at normal service temperatures but could be reprocessed like thermoplastics, it would be of great interest. During the past two decades a few groups of materials have been developed that could be considered as being in this category. Designated as *thermoplastic elastomers*, they include (1) styrene-diene-styrene triblock copolymers; (2) thermoplastic polyester elastomers and thermoplastic polyurethane elastomers; and (3) thermoplastic polyolefin rubbers (polyolefin blends).

Styrene-Diene-Styrene Triblock Elastomers

The styrene-diene-styrene triblocks consist of a block of diene units joined at each end to a block of styrene units and are made by sequential anionic polymerization of styrene and a diene. In this way two important triblock copolymers have been produced—the styrene-butadiene-styrene (SBS) and styrene-isoprene-styrene (SIS) materials, developed by Shell. The commercial thermoplastic rubbers, Clarifex TR (Shell), are produced by joining styrene-butadiene or styrene-isoprene diblocks at the active ends by using a difunctional coupling agent. Similarly, copolymer molecules in the shape of a T, X, or star have been produced (e.g., Solprene by Phillips) by using coupling agents of higher functionality.

The outstanding behavior of these rubbers arises from the natural tendency of two polymer species to separate. However, this separation is restrained in these polymers since the blocks are covalently linked to each other. In a typical commercial SBS triblock copolymer with about 30% styrene content, the styrene blocks congregate into rigid, glassy domains which act effectively to link the butadiene segments into a network (Fig. 4.1) analogous to that of cross-linked rubber.

As the SBS elastomer is heated above the glass transition temperature (T_g) of the polystyrene, the glassy domains disappear and the polymer begins to flow

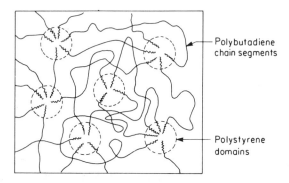

FIG. 4.1 Schematic representation of polystyrene domain structure in styrene-butadiene-styrene triblock copolymers. [After G. Holden, E. T. Bishop, and N. R. Legge, *J. Polymer Sci., Part C, 26*: 37 (1969).]

like a thermoplastic. However, when the molten material is cooled (below T_g), the domains reharden and the material once again exhibits properties similar to those of a cross-linked rubber.

Below T_g of polystyrene the glassy domains also fulfill another useful role by acting like a reinforcing particulate filler. It is also an apparent consequence of this role that SBS polymers behave like carbon-black-reinforced elastomers with respect to tensile strength.

The styrene-diene-styrene triblock copolymers are not used extensively in traditional rubber applications because they show a high level of creep. The block copolymers can, however, be blended with many conventional thermoplastics, such as polystyrene, polyethylene, and polypropylene, to obtain improved properties. A major area of use is in footwear, where blends of SBS and polystyrene have been used with remarkable success for crepe soles.

Other important uses are adhesives and coatings. A wide variety of resins, plasticizers, fillers, and other ingredients commonly used in adhesives and coatings can be used with styrene-diene-styrene triblock copolymers. With these ingredients properties such as tack, stiffness, softening temperatures, and cohesive strength can be varied over a wide range. With aliphatic resin additives the block copolymers are used for permanently tacky pressure-sensitive adhesives, and in conjunction with aromatic resins they are used for contact adhesives. The copolymers can be compounded into these adhesives by solution or hot-melt techniques.

The block copolymers are also used in a wide variety of sealants, including construction, industrial, and consumer-grade products. They are unique in that they can be formulated to produce a clear, water white product. Other applications include bookbinding and product assembly adhesives and chemical milling coatings.

Thermoplastic Polyester Elastomers

Because of the relatively low T_g of the short polystyrene blocks, the styrene-diene-styrene triblock elastomers have very limited heat resistance. One way to overcome this problem is to use a block copolymer in which one of the blocks is capable of crystallization and has a melting temperature well above room temperature. This approach coupled with polyester technology has led to the development of thermoplastic polyester elastomers (Hytrel by Du Pont and Arnitel by Akzo). A typical such polymer consists of relatively long sequences of tetramethylene terephthalate (which segregate into rigid domains of high melting point) and softer segments of polyether (see the section on polyesters for more details).

Being polar polymers, these rubbers have good oil and gasoline resistance. They have a wider service temperature range than many general-purpose rubbers, and they also exhibit a high resilience, good flex fatigue resistance, and mechanical abuse resistance. These rubbers have therefore become widely accepted in such applications as seals, belting, water hose, etc.

Thermoplastic Polyurethane Elastomers

Closely related to the polyether-ester thermoplastic elastomers are thermoplastic polyurethane elastomers, which consist of polyurethane or urethane terminated polyurea "hard" blocks, with T_g above normal ambient temperature, separated by "soft" blocks of polyol, which in the mass are rubbery in nature (see the section of polyurethanes for more details). The main uses of thermoplastic rubbers (e.g., *Estane* by Goodrich) are for seals, bushes, convoluted bellows, and bearings.

One particular form of thermoplastic polyurethane elastomers is the elastic fiber known as Spandex. Several commercial materials of this type have been introduced, which include Lycra (Du Pont), Dorlastan (Bayer), Spanzelle (Courtaulds), and Vyrene (U.S. Rubber). Spandex fibers have higher modulus, tensile strength, and resistance to oxidation, and are able to produce finer deniers than natural rubber. They have enabled lighter-weight garments to be produced. Staple fiber blends of Spandex fiber with nonelastic fibers have also been introduced.

Thermoplastic Polyolefin Elastomers

Blends of EPDM rubbers with polypropylene in suitable ratios have been marketed as thermoplastic polyolefin rubbers. Their recoverable high elasticity is believed to be due to short propylene blocks in the EPDM rubber cocrystallizing with segments of the polypropylene molecules so that these crystalline domains act like cross-linking agents. Having good weathering properties, negligible toxicity hazards, and easy processability, these rubbers have received rapid acceptance for use in a large variety of nontire automotive applications, such as bumper covers, headlight frames, radiator grilles, door gaskets, and other auto parts. They have also found use in cable insulation.

Ionic Elastomers

Ionic elastomers have been obtained using sulfonated EPDM. In one case an EPDM terpolymer consisting of 55% ethylene units, 40% propylene units, and 5% ethylidene norbornene units is sulfonated to introduce about 1 mol % sulfonate groups (appended to some of the unsaturated groups of the EPDM). The sulfonic acid group is then neutralized with zinc acetate to form the zinc salt. The ionized sulfonic groups create ionic cross-links in the intermolecular structure (Fig. 4.2), giving properties normally associated with a cross-linked elastomer. However, being a thermoplastic material, it can be processed in conventional molding machines. This rubber, however, has a very high melt viscosity, which must be reduced by using a polar flow promoter, such as zinc stearate, at levels of 9.5 to 19%.

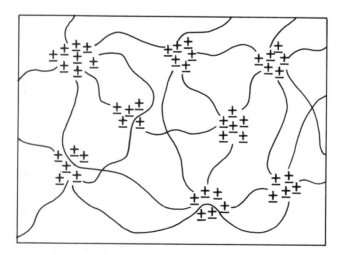

FIG. 4.2 Schematic representation of domain structure in ionic elastomers.

Fluoroelastomers

Monomers	Polymerization	Major uses
Vinylidene fluoride (CH_2=CF_2), chlorotrifluoroethylene (CF_2=$CFCl$), tetrafluoroethylene (CF_2=CF_2), hexafluoropropylene (CF_3CF=CF_2), perfluoromethyl vinyl ether (CF_2=$CFOCF_3$)	Free-radical-initiated chain polymerization	Aerospace industry (20%), industrial equipment, wire and cable jacketing, and other insulation applications

The fluoroelastomers are a general family of fluorinated olefin copolymers. To be rubbery, the copolymer must have a flexible backbone and be sufficiently irregular in structure to be noncrystalline. A number of important fluororubbers are based on vinylidene fluoride (CH_2=CF_2). Several common products are listed in Table 4.5. The most important of these are the copolymers of vinylidene fluoride and hexafluoropropylene (VF$_2$-HFP), as typified by the Du Pont product Viton A. The terpolymer of these two monomers together with tetrafluoroethylene (VF$_2$-HFP-TFE) is also of importance (e.g., Du Pont product Viton B). This terpolymer is the best among oil-resistant rubbers in its resistance to heat aging, although its actual strengths are lower than for some other rubbers. The copolymers of vinylidene fluoride and chlorotrifluoroethylene (VF$_2$-CTFE) are notable for their superior resistance to oxidizing acids such as fuming nitric acid.

Fluoroelastomers with no C—H groups will be expected to exhibit a higher thermal stability. Du Pont thus developed a terpolymer of tetrafluoroethylene, perfluoro(methyl vinyl ether) and, in small amounts, a cure site monomer of undisclosed composition. This product, marketed as Kalrez, has excellent air-oxidation resistance up to 315°C and exhibits extremely low swelling in a wide range of solvents, which is unmatched by any other commercial fluoroelastomer. Table 4.6 lists the presently used commercial elastomers with their main properties and applications.

Styrene-Acrylonitrile Copolymer

$$\left[CH_2\text{-}CH\text{-}CH_2\text{-}CH \atop \quad\ CN \qquad \bigcirc \right]_n$$

Monomers	Polymerization	Major uses
Acrylonitrile, styrene	Free-radical-initiated chain polymerization	Components of domestic appliances, electrical equipment and car equipment, picnic ware, housewares

Because of the polar nature of the acrylonitrile molecule, styrene-acrylonitrile (SAN) copolymers have better resistance to hydrocarbons, oils, and greases than polystyrene. These copolymers have a higher softening point, a much better resistance to stress cracking and crazing, and a higher impact strength than the homopolymer polystyrene, yet they retain the transparency of the latter. The toughness and chemical resistance of the copolymer increases with the acrylonitrile content but so do the difficulty in molding and the yellowness of the resin. Commercially available SAN copolymers have 20 to 30% acrylonitrile content. They are produced by emulsion, suspension, or continuous polymerization.

Due to their rigidity, transparency, and thermal stability, SAN resins have found applications for dials, knobs, and covers for domestic appliances, electrical equipment, car equipment, dishwasher-safe housewares, such as refrigerator meat and vegetable drawers, blender bowls, vacuum cleaner parts, humidifier parts, plus other industrial and domestic applications with requirements more stringent than can be met by polystyrene.

SAN resins are also reinforced with glass to make dashboard components and battery cases. Over 35% of the total SAN production is used in the manufacture of ABS blends.

Acrylonitrile-Butadiene-Styrene Terpolymer

Monomers	Polymerization	Major uses
Acrylonitrile, butadiene styrene	Free-radical-initiated chain polymerization	Pipe and fittings (30%), automotive and appliance (15%), telephones and business machine housings

A range of materials popularly referred to as ABS polymers first became available in the early 1950s. They are formed basically from three different

TABLE 4.5 Commercial Fluoroelastomers

Composition	Trade name (manufacturer)	Remarks
Vinylidene fluoride-hexafluoro-propylene copolymer	Viton A (Du Pont) Fluorel (MMM)	60–85% VF. Largest tonnage production among fluororubbers
Vinylidene fluoride-hexafluoro-propylene-tetrafluoroethylene terpolymer	Viton B (Du Pont) Daiel G-501 (Daikin Kogyo)	Superior resistance to heat, chemical, and solvent
Vinylidene fluoride-chlorotri-fluoroethylene copolymer	Kel-F 3700 (MMM) Kel-F 5500 (MMM)	Superior resistance to oxidiz-ing acids
Vinylidene fluoride-1-hydro-pentafluoropropylene - tetrafluoroethylene terpolymer	Tecnoflon T (Montecatini)	Superior resistance to oil, chemical, and solvent
Tetrafluoroethylene - perfluoro-(methyl vinyl ether) + cure site monomer terpolymer	Kalrez (Du Pont)	Excellent air oxidation re-sistance to 315°C
Tetrafluoroethylene - propylene + cure site monomer[a] terpolymer	Aflas (Asahi Glass)	Cross-linked by peroxides. Resistant to inorganic acids and bases. Cheaper alternative to Kalrez

[a]Suggested as triallyl cyanurate.

TABLE 4.6 Commercial Elastomer Products

Type	Properties	Major uses
Natural rubber	Excellent properties of vulcanizates under conditions not demanding high levels of heat, oil, and chemical resistance	Tires, bushings, couplings, seals, footwear and belting; second place in global tonnage
Styrene-butadiene (SBR)	Reinforcement with carbon black leads to vulcanizates which resemble those of natural rubber; more effectively stabilized by antioxidants than natural rubber	Tires, tire products, footwear, wire and cable covering, adhesives; highest global tonnage
Polybutadiene	Higher resilience than similar natural rubber compounds, good low-temperature behavior and adhesion to metals, but poor tear resistance, poor tack, and poor tensile strength	Blends with natural rubber and SBR; manufacture of high-impact polystyrene
Polyisoprene	Similar to natural rubber, but excellent flow characteristics during molding	Tires, belting, footwear, flooring

Butyl	Outstanding air-retention property, but low resiliency and poor resistance to oils and fuels	Tire inner tubes, inner liners, seals, coated fabrics
Ethylene-propylene	Outstanding resistance to oxygen and ozone, poor fatigue resistance, poor tire-cord adhesion	Nontire automotive parts, radiator and heater hoses, wire and cable insulation
Nitrile	Excellent resistance to oils and solvents, poor low-temperature flexibility and poor resistance to weathering	Hoses, seals, gaskets, footwear
Chloroprene	High order of oil and solvent resistance (but less than nitrile rubber), good resistance to most chemicals, oxygen and ozone, good heat resistance, high strength but difficult to process	Mechanical automotive goods, conveyor belts, diaphragms, hose, seals and gaskets
Silicone	Outstanding electrical and high-temperature properties, retention of elasticity at low temperature, poor tear and abrasion resistance, relatively high price	Gaskets and sealing rings for jet engines, ducting, sealing strips, vibration dampers and insulation equipment in aircraft, cable insulation in naval craft, potting and encapsulation

TABLE 4.6 Continued

Polyurethane	High tensile strength, tear, abrasion and oil resistance, relatively high price	Oil seals, shoe soles and heels, forklift truck tires, diaphragms, fiber coatings resistant to dry cleaning, variety of mechanical goods
Chlorosulfonated polyethylene	Very good heat, ozone, and weathering resistance, good resistance to oil and a wide range of chemicals, high elasticity, good abrasion resistance	Wire and cable coating, chemical plant hose, fabric coating, film sheeting, footwear, pond liners
Polysulfide	Excellent oil, solvent, and water resistance, high impermeability to gases, low strength, unpleasant odor (particularly during processing)	Adhesive, sealants, binders, hose
Epichlorohydrin	Low air permeability, low resilience, excellent ozone resistance, good heat resistance, flame resistance, and weathering resistance	Seals, gaskets, wire and cable coating
Fluoroelastomers	Outstanding heat resistance, superior oil, chemical, and solvent resistance, highest-priced elastomer	Aerospace applications, high quality seals, and gaskets

monomers: acrylonitrile, butadiene, and styrene. Acrylonitrile contributes chemical resistance, heat resistance, and high strength; butadiene contributes toughness, impact strength, and low-temperature property retention; styrene contributes rigidity, surface appearance (gloss), and processability. Not only may the ratios of the monomers be varied, but the way in which they can be assembled into the final polymer can also be the subject of considerable variations. The range of possible ABS-type polymers is therefore very large.

The two most important ways of producing ABS polymers are (1) blends of styrene-acrylonitrile copolymers with butadiene-acrylonitrile rubber, and (2) interpolymers of polybutadiene with styrene and acrylonitrile, which is now the most important type. A typical blend would consist of 70 parts styrene-acrylonitrile (70:30) copolymer and 40 parts butadiene-acrylonitrile (65:35) rubber.

Interpolymers are produced by copolymerizing styrene and acrylonitrile in the presence of polybutadiene rubber (latex) by using batch or continuous emulsion polymerization. The resultant materials are a mixture of polybutadiene, SAN copolymer, and polybutadiene grafted with styrene and acrylonitrile. The mixture is made up of three phases: a continuous matrix of SAN, a dispersed phase of polybutadiene, and a boundary layer of SAN graft.

ABS polymers are processable by all techniques commonly used with thermoplastics. They are slightly hygroscopic and should be dried 2 to 4 hr at 180 to 200°F (82 to 93°C) just prior to processing. A dehumidifying circulating air-hopper dryer is recommended. ABS can be hot stamped, painted, printed, vacuum metallized, electroplated, and embossed. Common fabrication techniques are applicable, including sawing, drilling, punching, riveting, bonding, and incorporating metal inserts and threaded and nonthreaded fasteners. The machining characteristics of ABS are similar to those of nonferrous metals.

ABS materials are superior to the ordinary styrene products and are commonly described as tough, hard, and rigid. This combination is unusual for thermoplastics. Moreover, the molded specimens generally have a very good surface finish, and this property is particularly marked with the interpolymer type ABS polymers. Light weight and the ability to economically achieve a one-step finished appearance part have contributed to large-volume applications of ABS.

Adequate chemical resistance is present in the ABS materials for ordinary applications. They are affected little by water, alkalis, weak acids, and inorganic salts. Alcohol and hydrocarbon may affect the surfaces. ABS has poor resistance to outdoor UV light; significant changes in appearance and mechanical properties will result after exposure. Protective coatings can be applied to improve resistance to UV light.

ABS materials are employed in thousands of applications, such as household appliances, business machine and camera housings, telephone handsets, electrical hand tools (such as drill housings), handles, knobs, cams, bearings, wheels, gears, pump impellers, automotive trim and hardware, bathtubs, refrigerator liners, pipe and fittings, shower heads, and sporting goods. Business machines, consumer electronics, and telecommunications applications represent the fastest-growing areas for ABS. Painted, electroplated, and vacuum-metallized parts are used throughout the automotive, business machine, and electronics markets.

Multilayered laminates with an ABS outer layer can be produced by coextrusion. In this process two or three different polymers may be combined into a multilayered film or sheet. Adhesion is enhanced by cooling the extruded laminate directly from the melt rather than in a separate operation after the components of the sheet have been formed and cooled separately. In one process flows from individual extruders are combined in a *flow block* and then conveyed to a single manifold die. All the polymer streams should have approximately the same viscosity so that laminar flow can be maintained.

Multilayered films and sheets have the advantage that a chemically resistant sheet can be combined with a good barrier to oxygen and water diffusion, or a decorative glossy sheet can be placed over a tough, strong material. One commercial example is an ABS–high-impact-polystyrene sheet, which can be thermoformed (see Chapter 2) to make the inside door and food compartment of a refrigerator. Another example is a four-layered sheet comprising ABS, polyethylene, polystyrene, and rubber-modified polystyrene for butter and margarine packages. However, not all combinations adhere equally well, so there are limits to the design of such structures.

In recent years there has been an increased demand for a variety of special ABS grades. These include high-temperature-resistant grades (for automotive instrument panels, power tool housings), fire-retardant grades (for appliance housings, business machines, television cabinets), electroplating grades (for automotive grilles and exterior decorative trim), high-gloss, low-gloss, and matte-finish grades (for molding and extrusion applications), clear ABS grades (using methyl methacrylate as the fourth monomer), and structural foam grades (for molded parts with high strength-to-weight ratio). The structural foam grades are available for general-purpose and flame-retardant applications. The cellular structure can be produced by injecting nitrogen gas into the melt just prior to entering the mold or by using chemical blowing agents in the resin (see also the section on foaming processes in Chapter 2).

Ethylene-Methacrylic Acid Copolymers (Ionomers)

$$\left[CH_2-CH_2-CH_2-\underset{\underset{COO^-}{|}}{\overset{\overset{CH_3}{|}}{C}} \right]_n$$

Monomers	Polymerization	Major uses
Ethylene, methacrylic acid	Free-radical-initiated chain polymerization	Packaging film, golf ball covers, automotive parts, footwear

Ionomers

Ionomer is a generic name for polymers containing interchain ionic bonding. Introduced in 1964 by Du Pont, they have the characteristics of both thermoplastics and thermosetting materials and are derived by copolymerizing ethylene with a small amount (1–10% in the basic patent) of an unsaturated acid, such as methacrylic acid, using the high-pressure process. The carboxyl groups in the copolymer are then neutralized by monovalent and divalent cations, resulting in some form of ionic cross-links (see Fig. 4.2) which are stable at normal ambient temperatures but which reversibly break down on heating. These materials thus possess the advantage of cross-linking, such as enhanced toughness and stiffness, at ambient temperatures, but they behave as linear polymers at elevated temperatures, so they may be processed and even reprocessed without undue difficulty.

Copolymerization used in making ionomers has had the effect of depressing crystallinity, although not completely eliminating it, so the materials are also transparent. Ionomers also have excellent oil and grease resistance, excellent resistance to stress cracking, and a higher water vapor permeability than does polyethylene.

The principal uses of ionomers are for film lamination and coextrusion for composite food packaging. The ionomer resin provides an outer layer with good sealability and significantly greater puncture resistance than an LDPE film. Sporting goods utilize the high-impact toughness of ionomers. Most major golf ball manufacturers use covers of durable ionomer. Such covers are virtually cut proof in normal use and retain a greater resiliency over a wider temperature range; they are superior to synthetic *trans*-polyisoprene in these respects.

Automotive uses (bumper pads and bumper guards) are based on impact toughness and paintability. In footwear applications, resilience and flex toughness of ionomers are advantages in box toes, counters, and shoe soles. Ski boot and ice skate manufacturers produce light weight outer shells of ionomers. Sheet and foamed sheet products include carpet mats, furniture tops, ski lift seat pads, boat bumpers, and wrestling mats.

Ionomers should be differentiated from polyelectrolytes and ion-exchange resins, which also contain ionic groups. Polyelectrolytes show ionic dissociation in water and are used, among other things, as thickening agents. Common examples are sodium polyacrylate, ammonium polymethacrylate (both anionic polyacrylate) and poly-(*N*-butyl-4-vinyl-pyridinium bromide), a cationic polyelectrolyte. Ion-exchange resins used in water softening, in chromatography, and for various industrial purposes, are cross-linked polymers containing ionic groups. Polyelectrolytes and ion-exchange resins are, in general, intractable materials and not processable on conventional plastics machinery. In ionomers, however, the amount of ionic bonding is limited to yield useful and tractable plastics. Using this principle, manufacturers can produce rubbers which undergo ionic cross-linking to give the effect of vulcanization as they cool on emergence from an extruder or in the mold of an injection-molding machine (see the section on thermoplastic polyolefin rubbers).

ACRYLICS

Acrylic polymers may be considered structurally as derivatives of acrylic acid and its homologues. The family of acrylics includes a range of commercial polymers based on acrylic acid, methacrylic acid, esters of acrylic acid and of methacrylic acid, acrylonitrile, acrylamide, and copolymers of these compounds. By far the best known applications of acrylics are acrylic fibers and acrylonitrile copolymers such as NBR, SAN, and ABS.

Polyacrylonitrile

$$\left[\begin{array}{c} CH_2-CH \\ | \\ CN \end{array} \right]_n$$

Monomer	Polymerization	Major uses
Acrylonitrile	Free-radical-initiated chain polymerization	Fibers in apparel (70%) and house furnishings (30%)

Polyacrylonitrile and closely related copolymers have found wide use as fibers. The development of acrylic fibers started in the early 1930s in Germany. In the United States they were first produced commercially about 1950 by Du Pont (Orlon) and Monsanto (Acrilan).

In polyacrylonitrile appreciable electrostatic forces occur between the dipoles of adjacent nitrile groups on the same polymer molecule. This restricts the bond rotation and leads to a stiff, rodlike structure of the polymer chain. As a result, polyacrylonitrile has a very high crystalline melting point (317°C) and is soluble in only a few solvents, such as dimethylformamide and dimethylacetamide, and in concentrated aqueous solutions of inorganic salts, such as calcium thiocyanate, sodium perchlorate, and zinc chloride. Polyacrylonitrile cannot be melt processed because its decomposition temperature is close to the melting point. Fibers are therefore spun from solution by either wet or dry spinning (see Chapter 2).

Fibers prepared from straight polyacrylonitrile are difficult to dye. To improve dyeability, manufacturers invariably add to commercial fibers minor amounts of one or two comonomers, such as methyl acrylate, methyl methacrylate, vinyl acetate, and 2-vinyl-pyridine. Small amounts of ionic monomers (sodium styrene sulfonate) are often included for better dyeability. Modacrylic fibers are composed of 35 to 85% acrylonitrile and contain comonomers, such as vinyl chloride, to improve fire retardancy.

Acrylic fibers are more durable than cotton, and they are the best alternative to wool for sweaters. A major portion of the acrylic fibers produced are used in apparel (primarily hosiery). Other uses include pile fabrics (for simulated fur), craft yarns, blankets, draperies, carpets, and rugs.

Polyacrylates

$$\left[\begin{array}{c} CH_2{-}CH \\ | \\ COOR \end{array} \right]_n$$

Monomers	Polymerization	Major uses
Acrylic acid esters	Free-radical-initiated chain polymerization	Fiber modification, coatings, adhesives, paints

The properties of acrylic ester polymers depend largely on the type of alcohol from which the acrylic acid ester is prepared. Solubility in oils and hydrocarbons increases as the length of the side chain increases. The lowest member of the series, poly(methyl acrylate), has poor low-temperature properties and is water sensitive. It is therefore restricted to such applications as textile sizes and leather finishes. Poly(ethyl acrylate) is used in fiber modification and in coatings; and poly(butyl acrylate) and poly(2-ethylhexyl acrylate) are used in the formulation of paints and adhesives.

The original acrylate rubbers first introduced in 1948 by B. F. Goodrich and marketed as Hycar 4021 were a copolymer of ethyl acrylate with about 5% of 2-chloroethyl vinyl ether ($CH_2{=}CH{-}O{-}CH_2{-}CH_2Cl$) acting as a cure site monomer. Such polymers are vulcanized through the chlorine atoms by amines (such as triethylenetetramine), which are, however, not easy to handle. Therefore, 2-chloroethyl vinyl ether has now been replaced with other cure site monomers, such as vinyl and allyl chloroacetates; the increased reactivity of the chlorine in these monomers permits vulcanization with ammonium benzoate (which decomposes on heating to produce ammonia, the actual cross-linking agent) rather than amines.

Acrylate rubbers have good oil resistance. In heat resistance they are superior to most rubbers, exceptions being the fluororubbers, the silicones, and the fluorosilicones. It is these properties which account for the major use of acrylate rubbers, i.e., in oil seals for automobiles. They are, however, inferior in low-temperature properties.

A few acrylate rubbers (such as Hycar 2121X38) are based on butyl acrylate. These materials are generally copolymers of butyl acrylate and acrylonitrile ($\sim 10\%$) and may be vulcanized with amines. They have improved low-temperature flexibility compared to ethyl acrylate copolymers but swell more in aromatic oils.

Polymethacrylates

$$\left[\begin{array}{c} CH_3 \\ | \\ CH_2 - C - \\ | \\ COOR \end{array} \right]_n$$

A large number of alkyl methacrylates, which may be considered as esters of poly(methacrylic acid), have been prepared. By far the most important of these polymers is poly(methyl methacrylate), which is an established major plastics material. As with other linear polymers, the mechanical and thermal properties of polymethacrylates are largely determined by the intermolecular attraction, spatial symmetry, and chain stiffness.

As the size of the ester alkyl group increases in a series of poly(n-alkyl methacrylate)s, the polymer molecules become spaced further apart and the intermolecular attraction is reduced. Thus, as the length of the side chain increases, the softening point decreases, and the polymers become rubbery at progressively lower temperatures (Fig. 4.3). However, when the number of carbon atoms in the side chain exceeds 12, the polymers become less rubbery, and the softening point, brittle point, and other properties related to the glass transition temperature rise with an increase in chain length (Tables 4.7 and 4.8). As with the polyolefins, this effect is due to side-chain crystallization.

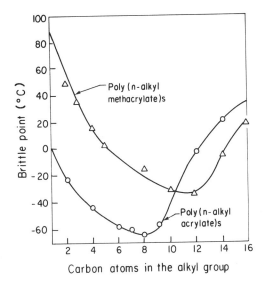

Carbon atoms in the alkyl group

FIG. 4.3 Brittle points of poly(*n*-alkyl acrylate)s and poly(*n*-alkyl methacrylate)s. [After C. E. Rehberg and C. H. Fisher, *Ind. Eng. Chem.*, *40*: 1431 (1948).]

Poly(alkyl methacrylate)s in which the alkyl group is branched have higher softening points (see Table 4.7) and are harder than their unbranched isomers. This effect is not simply due to the better packing possible with the branched isomers. The lumpy branched structures impede rotation about the carbon-carbon bond on the main chain, thus contributing to stiffness of the molecule and consequently a higher transition temperature. Similarly, since the α-methyl group in polymethacrylates reduces chain flexibility, the lower polymethacrylates have higher softening points than the corresponding polyacrylates do.

Poly(methyl methacrylate) (PMMA) is by far the predominant polymethacrylate used in rigid applications because it has crystalclear transparency, excellent weatherability (better than most other plastics), and a useful combination of stiffness, density, and moderate toughness. The glass transition temperature of the polymer is 105°C (221°F), and the heat deflection temperatures range from 75 to 100°C (167 to 212°F). The mechanical properties of PMMA can be further improved by orientation of heat-cast sheets.

PMMA is widely used for signs, glazing, lighting, fixtures, sanitary wares, solar panels, and automotive tail and stoplight lenses. The low index of refraction (1.49) and high degree of uniformity make PMMA an excellent lens material for optical applications.

Methyl methacrylate has been copolymerized with a wide variety of other monomers, such as acrylates, acrylonitrile, styrene, and butadiene. Copolymerization with styrene gives a material with improved melt-flow characteristics.

TABLE 4.7 Vicat Softening Points of Polymethacrylates Derived from Monomers of Type $CH_2{=}C(CH_3)COOR$

R—	Softening point (°C)	R	Softening point (°C)
CH_3-	119	$(CH_3)_3C-$	104
CH_3-CH_2-	81	$(CH_3)_2CH-CH_2-CH_2-$	46
$CH_3-CH_2-CH_2-$	55	$(CH_3)_3C-CH_2-$	115
$CH_3-CH_2-CH_2-CH_2-$	30	$(CH_3)_3C-CH-$ $\quad\quad\quad\quad\ \ CH_3$	119
$CH_3-CH_2-CH_2-CH_2-CH_2-$	a		
$(CH_3)_2CH-$	88		
$(CH_3)_2CH-CH_2-$	67		

[a]Too rubbery for testing.

TABLE 4.8 Glass Transition
Temperatures of Polymethacrylates

Ester group	T_g (°C)
Methyl	105
Ethyl	65
n-Butyl	20
n-Decyl	-70
n-Hexadecyl	-9

Copolymerization with either butadiene or acrylonitrile, or blending PMMA with SBR, improves impact resistance. Butadiene-methyl methacrylate copolymer has been used in paper and board finishes.

Higher n-alkyl methacrylate polymers have commercial applications. Poly(n-butyl-), poly(n-octyl-) and poly(n-nonyl methacrylate)s are used as leather finishes; poly(lauryl methacrylate) is used to depress the pour point and improve the viscosity-temperature property of lubricating oils.

Mention may also be made here of the 2-hydroxyethyl ester of methacrylic acid, which is the monomer used for soft contact lenses. Copolymerization with ethylene glycol dimethacrylate produces a hydrophilic network polymer (a *hydrogel*). Hydrogel polymers are brittle and glassy when dry but become soft and plastic on swelling in water.

Terpolymers based on methyl methacrylate, butadiene, and styrene (MBS) are being increasingly used as tough transparent plastics and as additives for PVC.

Polyacrylamide

$$\left[CH_2-CH \atop \underset{CONH_2}{|} \right]_n$$

Monomer	Polymerization	Major uses
Acrylamide	Free-radical-initiated chain polymerization	Flocculant, adhesives, paper treatment, water treatment, coatings

Polyacrylamide exhibits strong hydrogen bonding and water solubility. Most of the interest in this polymer is associated with this property. Polymerization of acrylamide monomer is usually conducted in an aqueous solution, using free-radical initiators and transfer agents.

Copolymerization with other water-soluble monomers is also carried out in a similar manner. Cationic polyacrylamides are obtained by copolymerizing with ionic monomers such as dimethylaminoethyl methacrylate, dialkyldimethylammonium chloride, and vinylbenzyltrimethylammonium chloride. These impart a positive charge to the molecule. Anionic character can be imparted by copolymerizing with monomers such as acrylic acid, methacrylic acid, 2-acrylamido-2-methyl-propanesulfonic acid, and sodium styrene sulfonate. Partial hydrolysis of polyacrylamide, which converts some of the amide groups to carboxylate ion, also results in anionic polyacrylamides.

Polyacrylamides have several properties which lead to a multitude of uses in diverse industries. Table 4.9 lists the main functions of the polymers and their uses in various industries.

Polyacrylamides are used as primary flocculants or coagulant aids in water clarification and mining application. They are effective for clarification of raw river water. The capacity of water clarifiers can be increased when the polymer is used as a secondary coagulant in conjunction with lime and ferric chloride. Polyacrylamides, and especially cationic polyacrylamides, are used for condi-

TABLE 4.9 Applications of Polyacrylamide

Function	Application	Industry
Flocculation	Water clarification	General
	Waste removal	Sewage
	Solids recovery	Mining
	Retention aid	Paper
	Drainage aid	Paper
Adhesion	Dry strength	Paper
	Wallboard cementing	Construction
Rheology control	Waterflooding	Petroleum
	Viscous drag reduction	Petroleum
		Fire fighting
		Irrigation pumping

tioning municipal and industrial sludges for dewatering by porous and empty sand beds, vacuum filters, centrifuges, and other mechanical devices.

Certain anionic polyacrylamides are approved by the U.S. Environmental Protection Agency for clarification of potable water. Polymer treatment also allows filters to operate at higher hydraulic rates. The function of clarification is not explained by a simple mechanism. The long-chain linear polymer apparently functions to encompass a number of individual fine particles of the dispersed material in water, attaching itself to the particles at various sites by chemical bonds, electrostatic attraction, or other attractive forces. Relatively stable aggregates are thus produced, which may be removed by filtration, settling, or other convenient means.

Polyacrylamides are useful in the paper industry as processing aids, in compounding and formulating, and as filler-retention aids. Polyacrylamides and copolymers of acrylamide and acrylic acid are used to increase the dry strength of paper.

Polyacrylamides are used as flooding aids in secondary oil recovery from the producing oil. Water, being of low viscosity, tends to finger ahead of the more viscous oil. However, addition of as little as 0.05% polyacrylamide to the waterflood reduces oil bypass and gives significantly higher oil to water ratios at the producing wellhead. Greatly increased yields of oil result from adding polymer to waterflooding.

Solutions containing polyacrylamide are very slippery and can be used for water-based lubrication. Small amounts of polymer, when added to an aqueous solution, can significantly reduce the friction in pipes, thereby increasing the throughput or reducing the power consumption.

Other applications include additives in coatings and adhesives and binders for pigments.

Poly(Acrylic Acid) and Poly(Methacrylic Acid)

$$\left[\begin{array}{c} CH_2-CH \\ | \\ COOH \end{array} \right]_n \quad \text{and} \quad \left[\begin{array}{c} CH_3 \\ | \\ CH_2-C \\ | \\ COOH \end{array} \right]_n$$

Monomer	Polymerization	Major uses
Acrylic acid and methacrylic acid	Free-radical-initiated chain polymerization	Sodium and ammonium-salts as polyelectrolytes, thickening agents

Poly(acrylic acid) and poly(methacrylic acid) may be prepared by direct polymerization of the appropriate monomer, namely, acrylic acid or methacrylic acid, by conventional free-radical techniques, with potassium persulfate used as the initiator and water as the solvent (in which the polymers are soluble); or if a solid polymer is required, a solvent such as benzene, in which the polymer is insoluble, can be used, with benzoyl peroxide as a suitable initiator.

The multitude of applications of poly(acrylic acid) and poly(methacrylic acid) is reminiscent of the fable of the man who blew on his hands to warm them and blew on his porridge to cool it. They are used as adhesives and release agents, as flocculants and dispersants, as thickeners and fluidizers, as reaction inhibitors and promoters, as permanent coatings and removable coatings, etc. Such uses are the direct result of their varied physical properties and their reactivity. Many of the applications thus depend on the ability of these polymers to form complexes and to bond to substrates. Monovalent metal and ammonium salts of these polymers are generally soluble in water. These materials behave as anionic polyelectrolytes and are used for a variety of purposes, such as thickening agents, particularly for rubber latex.

Acrylic Adhesives

Acrylic adhesives are essentially acrylic monomers which achieve excellent bonding upon polymerization. Typical examples are cyanoacrylates and ethylene glycol dimethacrylates. Cyanoacrylates are obtained by depolymerization of a condensation polymer derived from a malonic acid derivative and formaldehyde.

$$
\begin{array}{c}
CN \\
| \\
CH_2 + CH_2O \\
| \\
COOR
\end{array}
\xrightarrow{-H_2O}
\begin{bmatrix}
CN \\
| \\
C-CH_2 \\
| \\
COOR
\end{bmatrix}_n
\xrightarrow{\Delta}
\begin{array}{c}
CN \\
| \\
C=CH_2 \\
| \\
COOR
\end{array}
$$

Cyanoacrylates are marketed as contact adhesives. Often popularly known as *superglue*, they have found numerous applications. In dry air and in the presence of polymerization inhibitors, methyl- and ethyl-2-cyanoacrylates have a storage life of many months. As with many acrylic monomers, air can inhibit or severely retard polymerization of cyanoacrylates. These monomers are, however, prone to anionic polymerization, and even a very weak base such as water can bring about rapid polymerization.

$$
\begin{array}{c}
\text{CN} \\
| \\
\text{C}{=}\text{CH}_2 + \text{H}_2\text{O} \longrightarrow \\
| \\
\text{COOR}
\end{array}
\left[
\begin{array}{c}
\text{CN} \\
| \\
\text{C}{-}\text{CH}_2 \\
| \\
\text{COOR}
\end{array}
\right]_n
$$

In practice, a trace of moisture occurring on a substrate is adequate to cause polymerization of the cyanoacrylate monomer to provide strong bonding within a few seconds of closing the joint and excluding air. Cyanoacrylate adhesives are particularly valuable because of their speed of action, which obviates the need for complex jigs and fixtures. The amount of monomer applied should be minimal to obtain a strong joint. Larger amounts only reduce the strength. Notable uses of cyanoacrylates include surgical glue and dental sealants; morticians use them to seal eyes and lips.

Dimethacrylates, such as tetramethylene glycol dimethacrylate, are used as *anaerobic adhesives*. Air inhibition of polymerization of acrylic monomers is used to advantage in this application because the monomers are supplied along with a curing system (comprising a peroxide and an amine) as part of a one-part pack. When this adhesive is placed between mild steel surfaces, air inhibition is prevented since the air is excluded, and polymerization can take place. Though the metal on the surface acts as a polymerization promoter, it may be necessary to use a primer such as cobalt naphthenate to expedite the polymerization. The anaerobic adhesives are widely used for sealing nuts and bolts and for miscellaneous engineering purposes.

Dimethacrylates form highly cross-linked and, therefore, brittle polymers. To overcome brittleness, manufacturers often blend dimethacrylates with polyurethanes or other polymers such as low-molecular-weight vinyl-terminated butadiene-acrylonitrile copolymers and chlorosulfonated polyethylene. The modified dimethacrylate systems provide tough adhesives with excellent properties. These can be formulated as two-component adhesives, the catalyst component being added just prior to use or applied separately to the surface to be bonded. One-component systems also have been formulated which can be conveniently cured by ultraviolet radiation.

VINYL POLYMERS

If the R substituent in an olefin monomer ($CH_2{=}CHR$) is either hydrogen, alkyl, aryl, or halogen, the corresponding polymer in the present discussion is grouped under polyolefins. If the R is a cyanide group or a carboxylic group or its ester or amide, the substance is an *acrylic* polymer. *Vinyl* polymers include those polyolefins in which the R substituent in the olefin monomer is bonded to the unsaturated carbon through an oxygen atom (vinyl esters, vinyl ethers) or a nitrogen atom (vinyl pyrrolidone, vinyl carbazole).

Vinyl polymers constitute an important segment of the plastics industry. Depending on the specific physical and chemical properties, these polymers find use in paints and adhesives, in treatments for paper and textiles, and in special applications. The commonly used commercial vinyl polymers are described next.

Poly(Vinyl Acetate)

$$\left[CH_2 - CH \atop \underset{\underset{O}{\|}}{OCCH_3} \right]_n$$

Monomer	Polymerization	Major uses
Vinyl acetate	Free-radical-initiated chain polymerization (mainly emulsion polymerization)	Emulsion paints, adhesives, sizing

Since poly(vinyl acetate) is usually used in an emulsion form, it is manufactured primarily by free-radical-initiated emulsion polymerization. The polymer is too soft and shows excessive cold flow, which precludes its use in molded plastics. The reason is that the glass transition temperature of 28°C is either slightly above or (at various times) below the ambient temperatures.

Vinyl acetate polymers are extensively used in emulsion paints, as adhesives for textiles, paper, and wood, as a sizing material, and as a "permanent starch." A number of commercial grades are available which differ in molecular weight and in the nature of comonomers (e.g., acrylate, maleate, fumarate) which are often used. Two vinyl acetate copolymers of particular interest to the plastics industry are ethylene-vinyl acetate and vinyl chloride-vinyl acetate copolymers.

Ethylene-vinyl acetate (EVA) copolymers with a vinyl acetate content of 10 to 15 mol % are similar in flexibility to plasticized PVC and are compatible with inert fillers. Both filled and unfilled EVA copolymers have good low-temperature flexibility and toughness. The absence of leachable plasticizers in these products is a clear advantage over plasticized PVC in certain applications.

The EVA copolymers are slightly less flexible than normal rubber compounds but have the advantage of simpler processing since no vulcanization is necessary. The materials have thus been largely used in injection molding in place of plasticized PVC or vulcanized rubber. Typical applications include turntable mats, base pads for small items of office equipment, buttons, car door protection strips, and for other parts where a soft product of good appearance is required.

A substantial use of EVA copolymers is as wax additives and additives for hot-melt coatings and adhesives. Cellular cross-linked EVA copolymers are used in shoe parts.

EVA copolymers with only a small vinyl acetate content ($\sim$3 mol %) are best considered as a modification of low-density polyethylene. These copolymers have less crystallinity and greater flexibility, softness, and, in case of film, surface gloss.

Poly(Vinyl Alcohol)

$$-\left[\begin{array}{c} CH_2 - CH \\ | \\ OH \end{array}\right]_n$$

Manufacture	Major uses
Alcoholysis of poly(vinyl acetate)	Paper sizing, textile sizing, cosmetics

Poly(vinyl alcohol) (PVA) is produced by alcoholysis of poly(vinyl acetate), because vinyl alcohol monomer does not exist in the free state.

$$\sim\!\!CH_2-CH\!\!\sim + CH_3OH \longrightarrow \sim\!\!CH_2-CH\!\!\sim + CH_3COCH_3$$

with pendant $OCCH_3$ (C double bond O) on left becoming OH on right, and the byproduct CH_3COCH_3 having $\overset{\parallel}{O}$

(The term "hydrolysis" is often used incorrectly to describe this process.) Either acid or base catalysts may be employed for alcoholysis. Alkaline catalysts such as sodium hydroxide or sodium methoxide give more rapid alcoholysis. The degree of alcoholysis, and hence the residual acetate content, is controlled by varying the catalyst concentration.

The presence of hydroxyl groups attached to the main chain renders the polymer hydrophilic. PVA therefore dissolves in water to a greater or lesser extent according to the degree of hydrolysis. Polymers with a degree of hydrolysis in the range of 87 to 89% readily dissolve in cold water. Solubility decreases with an increase in the degree of hydrolysis, and fully hydrolyzed polymers are water soluble only at higher temperatures (>85°C). This apparently anomalous behavior is due to the higher degree of crystallinity and the greater extent of hydrogen bonding in the completely hydrolyzed polymers.

Commercial PVA is available in a number of grades which differ in molecular weight and degree of hydrolysis. The polymer finds a variety of uses. It functions as a nonionic surface active agent and is used in suspension polymerization as a protective colloid. It also serves as a binder and thickener and is widely used in adhesives, paper coatings, paper sizing, textile sizing, ceramics, and cosmetics.

Completely hydrolyzed grades of PVA find use in quick-setting, water-resistant adhesives. Combinations of fully hydrolyzed PVA and starch are used as a quick-setting adhesive for paper converting. Borated PVA, commonly called "tackified," are combined with clay and used in adhesive applications requiring a high degree of wet tack. They are used extensively to glue two or more plies of paper together to form a variety of shapes such as tubes, cans, and cores.

Since PVA film has little tendency to adhere to other plastics, it can be used to prevent sticking to mold. Films cast from aqueous solution of PVA are used as release agents in the manufacture of reinforced plastics.

Partially hydrolyzed grades have been developed for making tubular blown film (similar to that with polyethylene) for packages for bleaches, bath salts, insecticides, and disinfectants. Use of water-soluble PVA film for packaging preweighed quantities of such materials permits their addition to aqueous systems without breaking the package or removing the contents, thereby saving time and reducing material losses. Film made from PVA may be used for hospital laundry bags that are added directly to the washing machine.

A process has been developed in Japan for producing fibers from poly(vinyl alcohol). The polymer is wet spun from a warm aqueous solution into a concentrated aqueous solution of sodium sulfate containing sulfuric acid and formaldehyde, which insolubilizes the alcohol by formation of *formal* groups (see below). These fibers are generally known as *vinal* or *vinylon* fibers.

Poly(Vinyl Acetals)

Poly(vinyl acetals) are produced by treating poly(vinyl alcohol) with aldehydes.

(They may also be made directly from poly(vinyl acetate) without separating the alcohol.) Since the reaction with aldehyde involves a pair of neighboring hydroxyl groups on the polymer chain and the reaction occurs at random, some hydroxyl groups become isolated and remain unreacted. A poly(vinyl acetal) molecule will thus contain acetal groups and residual hydroxyl groups. In addition, there will be residual acetate groups due to incomplete hydrolysis of poly(vinyl acetate) to the poly(vinyl alcohol) used in the acetalization reaction. The relative proportions of these three types of groups may have a significant effect on specific properties of the polymer.

When the aldehyde in this reaction is formaldehyde, the product is poly(vinyl formal). This polymer is, however, made directly from poly(vinyl acetate) and formaldehyde without separating the alcohol. The product with low hydroxyl (5–6%) and acetate (9.5–13%) content (the balance being *formal*) is used in wire enamel and in structural adhesives (e.g., Redux). In both applications the polymer is used in conjunction with phenolic resins and is heat cured.

When the aldehyde in the acetalization reaction is butyraldehyde, i.e., R = $CH_3CH_2CH_2$—, the product is poly(vinyl butyral). Sulfuric acid is the catalyst in this reaction. Poly(vinyl butyral) is characterized by high adhesion to glass, toughness, light stability, clarity, and moisture insensitivity. It is therefore extensively used as an adhesive interlayer between glass plates in the manufacture of laminated safety glass and bullet-proof composition.

Poly(Vinyl Cinnamate)

Poly(vinyl cinnamate) is conveniently made by the Schotten-Baumann reaction using poly(vinyl alcohol) in sodium or potassium hydroxide solution and cinnamoyl chloride in methyl ethyl ketone. The product is, in effect, a copolymer of vinyl alcohol and vinyl cinnamate, as shown. The polymer has the ability to cross-link on exposure to light, which has led to its important applications in photography, lithography, and related fields as a *photoresist* (see also Chapter 5).

Poly(Vinyl Ethers)

$$\left[\begin{array}{c} CH_2-CH \\ | \\ OR \end{array} \right]_n$$

Commercial uses have developed for several poly(vinyl ethers) in which R is methyl, ethyl, and isobutyl. The vinyl alkyl ether monomers are produced from acetylene and the corresponding alcohols, and the polymerization is usually conducted by cationic initiation using Friedel-Craft-type catalysts.

Poly(vinyl methyl ether) is a water-soluble viscous liquid which has found application in the adhesive and rubber industries. One particular application has been as a heat sensitizer in the manufacture of rubber-latex dipped goods.

Ethyl and butyl derivatives have found use as adhesives. *Pressure-sensitive adhesive tapes* made from poly(vinyl ethyl ether) incorporating antioxidants are said to have twice the shelf life of similar tapes made from natural rubber. Copolymers of vinyl isobutyl ether with vinyl chloride, vinyl acetate, and ethyl acrylate are also produced.

Poly(Vinyl Pyrrolidone)

$$\left[CH_2-CH \right]_n$$

Poly(vinyl pyrrolidone) is produced by free-radical-initiated chain polymerization of N-vinyl pyrrolidone. Polymerization is usually carried out in aqueous solution to produce a solution containing 30% polymer. The material is marketed in this form or spray dried to give a fine powder.

Poly(vinyl pyrrolidone) is a water-soluble polymer. Its main value is due to its ability to form loose addition compounds with many substances. It is thus used in cosmetics. The polymer has found several applications in textile treatment because of its affinity for dyestuffs. In an emergency it is used as a blood plasma substitute. Also, about 7% polymer added to whole blood allows it to be frozen, stored at liquid nitrogen temperatures for years, and thawed out without destroying blood cells.

Poly(Vinyl Carbazole)

Poly(vinyl carbazole) is produced by polymerization of vinyl carbazole using free-radical initiation or Ziegler-Natta catalysis.

Poly(vinyl carbazole) has a high softening point, excellent electrical insulating properties, and good photoconductivity, which has led to its application in xerography.

PART II: CONDENSATION POLYMERS

According to the original classification of Carothers, condensation polymers are formed from bi- or polyfunctional monomers by reactions which involve elimination of some smaller molecule. A condensation polymer, according to this definition, is one in which the repeating unit lacks certain atoms which were present in the monomer(s) from which the polymer was formed.

With the development of polymer science and synthesis of newer polymers, this definition of condensation polymer was found to be inadequate. For example, in polyurethanes, which are classified as condensation polymers, the repeat unit has the same net composition as the two monomers—that is, a diol and a diisocyanate, which react without the elimination of any small molecule. Similarly the polymers produced by the ring-opening polymerization of cyclic monomers, such as cyclic ethers and amides, are generally classified as condensation polymers based on the presence of functional groups, such as the ether and amide linkages, in the polymer chains, even though the polymerization occurs without elimination of any small molecule.

To overcome such problems, an alternative definition has been introduced. According to this definition, polymers whose main chains consist entirely of C—C bonds are classified as *addition polymers*, whereas those in which heteroatoms (O, N, S, Si) are present in the polymer backbone are considered to be condensation polymers. A polymer which satisfies both the original definition (of Carothers) and the alternative definition or either of them, is classified as a condensation polymer. Phenol-formaldehyde condensation polymers, for example, satisfy the first definition but not the second.

Condensation polymers described in Part II are classified as polyesters, polyamides, formaldehyde resins, polyurethanes, and ether polymers.

POLYESTERS

Polyesters were historically the first synthetic condensation polymers studied by Carothers in his pioneering work in the early 1930s. Commercial polyesters were manufactured by polycondensation reactions, the methods commonly used being melt polymerization of diacid and diol, ester interchange of diester and diol, and interfacial polymerization (Schotten-Baumann reaction) of diacid chloride and diol. In a polycondensation reaction a by-product is generated which has to be removed as the reaction progresses.

Thermoplastic saturated polyesters are widely used in synthetic fibers and also in films and molding applications. The production of polyester fibers accounts for nearly 30% of the total amount of synthetic fibers. Unsaturated polyesters are mainly used in glass-fiber reinforced plastic products.

Poly(Ethylene Terephthalate)

$$\left[\begin{array}{c} \underset{O}{\overset{}{C}} - \bigcirc\!\!\!\!\!\bigcirc - \underset{O}{\overset{}{C}} - O - CH_2 - CH_2 - O \end{array}\right]_n$$

Monomer	Polymerization	Major uses
Dimethyl terephthalate or terephthalic acid, ethylene glycol	Bulk polycondensation	Apparel (61%), home furnishings (18%), tire cord (10%)

Whinfield and Dixon, in England, developed polyethylene terephthalate fibers (Dacron, Terylene). The first Dacron polyester plant went into operation in 1953. *Ester interchange* (also known as *ester exchange* or *alcoholysis*) was once the preferred method for making polyethylene terephthalate (PET) because dimethyl terephthalate can be readily purified to the high quality necessary for the production of the polymer. The process is carried out in two steps. Dimethyl terephthalate is heated at 150°C with an excess of ethylene glycol (mole ratio 1:2.1–2.2) in the presence of catalysts such as antimony trioxide and cobaltous acetate to give a mixture consisting of dihydroxyethyl terephthalate and small amounts of higher oligomers. Further heating to 270°C under vacuum in the presence of a catalyst produces the final polymer. In this second stage an ester interchange reaction occurs in which the dihydroxyethyl terephthalate serves as both ester and alcohol, and successive interchanges result in the formation of a polyester.

In recent years methods have been developed to produce terephthalic acid with satisfactory purity, and direct polycondensation reaction with ethylene glycol is now the preferred route to this polymer.

PET is widely used in synthetic fibers designed to simulate wool, cotton, or rayon, depending on the processing conditions. They have good wash-and-wear properties and resistance to wrinkling. In the production of fiber the molten polymer is extruded through spinnerets and rapidly cooled in air. The filaments thus formed are, however, largely amorphous and weak. They are therefore drawn at a temperature (80°C) above T_g and finally heated at 190°C under tension, whereby maximum molecular orientation, crystallinity, and dimensional stability are achieved. The melting point of highly crystalline PET is 271°C.

Crystalline PET has good resistance to water and dilute mineral acids but is degraded by concentrated nitric and sulfuric acids. It is soluble at normal temperatures only in proton donors which are capable of interaction with the ester group, such as chlorinated and fluorinated acetic acids, phenols, and anhydrous hydrofluoric acid.

PET is also used in film form (Melinex, Mylar) and as a molding material. The manufacture of PET film closely resembles the manufacture of fiber. The film is produced by quenching extruded sheet to the amorphous state and then reheating and stretching the sheet approximately threefold in the axial and transverse directions at 80 to 100°C. To stabilize the biaxially oriented film, it is annealed under restraint at 180 to 210°C. This operation increases the crystallinity of PET film and reduces its tendency to shrink on heating. The strength of PET in its oriented form is outstanding.

The principal uses of biaxially oriented PET film are in capacitors, in slot liners for motors, and for magnetic tape. Although a polar polymer, its electrical insulation properties at room temperature are good (even at high frequencies) because at room temperature, which is well below T_g (69°C), dipole orientation is severely restricted. The high strength and dimensional stability of the polyester film have also led to its use for x-ray and photographic film and to a number of graphic art and drafting applications. The film is also used in food packaging, including boil-in-bag food pouches. Metallized polyester films have many uses as a decorative material.

Because of its rather high glass transition temperature, only a limited amount of crystallization can occur during cooling after injection molding of PET. The idea of molding PET was thus for many years not a technical proposition. Toward the end of the 1970s Du Pont introduced Rynite, which is a PET nucleated with an ionomer, containing a plasticizer and only available in glass-fiber-filled form (at 30, 45, and 55% fill levels). The material is very rigid, exceeding that of polysulfone, is less water sensitive than an unfilled polymer, and has a high heat-deflection temperature (227°C at 264 psi).

In the late 1970s the benefits of biaxial stretching PET were extended from film to bottle manufacture. Producing carbonated beverages PET bottles by blow molding has gained prominence (particularly in the United States) because PET has low permeability to carbon dioxide. The process has been extended, particularly in Europe, to produce bottles for other purposes, such as fruit juice concentrates and sauces, wide-necked jars for coffee, and other materials. Because of its excellent thermal stability, PET is also used as material for microwave and conventional ovens.

Poly(Butylene Terephthalate)

Monomers	Polymerization	Major uses
Dimethyl terephthalate or terephthalic acid, butanediol	Bulk polycondensation	Machine parts, electrical applications, small appliances

Poly(butylene terephthalate), often abbreviated to PBT or PBTP, is manufactured by condensation polymerization of dimethyl terephthalate and butane-1,4-diol in the presence of tetrabutyl titanate. The polymer is also known as poly(tetramethylene terephthalate), PTMT in short. Some trade names for this engineering thermoplastic are Tenite PTMT (Eastman Kodak), Valox (General Electric), Celanex (Celanese) in America and Arnite PBTP (Akzo), Ultradur (BASF), Pocan (Bayer), and Crastin (Ciba-Geigy) in Europe.

Because of the longer sequence of methylene groups in the repeating unit poly(butylene terephthalate) chains are both more flexible and less polar than poly(ethylene terephthalate). This leads to lower values for melting point (about 224°C) and glass transition temperature (22–43°C). The low glass transition temperature facilitates rapid crystallization when cooling in the mold, and this allows short injection-molding cycles and high injection speeds.

PBT finds use as an engineering material due to its dimensional stability, particularly in water, and its resistance to hydrocarbon oils without showing stress cracking. PBT also has high mechanical strength and excellent electrical properties but a relatively low heat-deflection temperature 130°F (54°C) at 264 psi (1.8 MPa). The low water absorption of PBT—less than 0.1% after 24-hr immersion—is outstanding. Both dimensional stability and electrical properties are retained under conditions of high humidity. The lubricity of the resin results in outstanding wear resistance.

As with PET, there is particular interest in glass-filled grades of PBT. The glass has a profound effect on such properties as tensile strength, flexural modulus, and impact strength, as can be seen from the values of these properties for unfilled and 30% glass-filled PBT: 8200 vs. 17,000 psi (56 vs. 117 MPa), 340,000 vs. 1.1 to 1.2 × 10^6 psi (2350 vs. 7580 to 8270 MPa) and 0.8 to 1.0 vs. 1.3 to 1.6 ft-lbf/in^2 (Izod), respectively. Reinforcing with glass fiber also results in an increase in heat-deflection temperature to over 400°F (204°C) at 264 psi (1.8 MPa).

Typical applications of PBT include pump housings, impellers, bearing bushings, gear wheels, automotive exterior and under-the-hood parts, and electrical parts such as connectors and fuse cases.

Poly(Dihydroxymethylcyclohexyl Terephthalate)

In 1958, Eastman Kodak introduced a more hydrophobic polyester fiber under the trade name Kodel. The raw material for this polyester is dimethyl terephthalate. Reduction leads to 1,4-cyclohexylene glycol, which is used with dimethyl terephthalate in the polycondensation (ester exchange) reaction.

Eastman Kodak also introduced in 1972 a copolyester based on 1,4-cyclohexylene glycol and a mixture of terephthalic and isophthalic acids. The product is now sold as Kodar PETG. Being irregular in structure, the polymer is amorphous and gives products of brilliant clarity. In spite of the presence of the heterocyclic ring, the deflection temperature under load is as low as that of the poly(butylene terephthalate)s, and the polymer can be thermoformed at draw ratios as high as 4:1 without blushing or embrittlement. Because of its good melt strength and low molding shrinkage, the material performs well in extrusion blow molding and in injection molding. The primary use for the copolyester is extrusion into film and sheeting for packaging.

Ethylene glycol-modified polyesters of the Kodel type are used in blow-molding applications to produce bottles for packaging liquid detergents, shampoos, and similar products. One such product is Kodar PETG 6703 in which one acid (terephthalic acid) is reacted with a mixture of glycols (ethylene glycol and 1,4-cyclohexylene glycol). A related glass-reinforced grade (Ektar PCTG) has also been offered.

The principle of formation of segmented or block copolymers (see the section on thermoplastic elastomers) has also been applied to polyesters, with the "hard" segment formed from butanediol and terephthalic acid, and the "soft" segment provided by a hydroxyl-terminated polyether [polytetramethylene ether glycol (PTMEG)] with molecular weight 600 to 3000. In a typical preparation, dimethyl terephthalate is transesterified with a mixture of PTMEG and a 50% excess of butane-1,4-diol in the presence of an ester exchange catalyst. The stoichiometry is such that relatively long sequences of tetramethylene terephthalate (TMT) are produced which, unlike the polyether segments, are crystalline and have a high melting point. Since the sequences of TMT segregate into rigid domains, they are referred to as "hard" segments, and the softer polyether terephthalate (PE/T) segments are referred to as "soft" segments (Fig. 4.4).

Du Pont markets this polyester elastomer under the trade name Hytrel. These elastomers are available in a range of stiffnesses. The harder grades have up to 84% TMT segments and a melting point of 214°C, and the softest grades contain as little as 33% TMT units and have a melting point of 163°C.

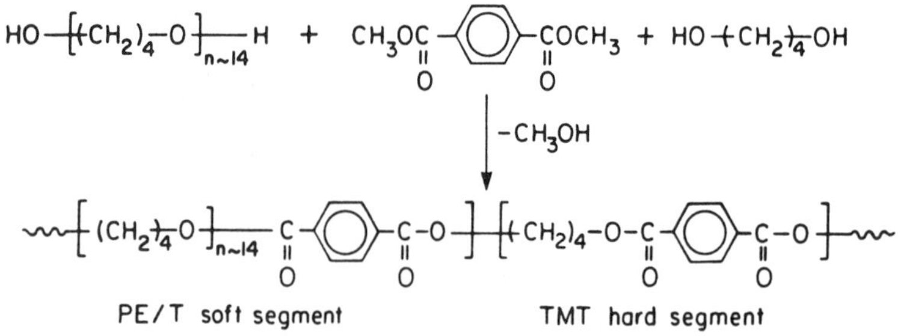

FIG. 4.4 Formation of polyester-type segmented or block copolymer.

Processing of these thermoplastic rubbers is quite straightforward. The high crystallization rates of the hard segments facilitate injection molding, while the low viscosity at low shear rates facilitates a low shear process, such as rotational molding.

These materials are superior to conventional rubbers in a number of properties (see the section on thermoplastic rubbers). Consequently, in spite of their relatively high price they have become widely accepted as engineering rubbers in many applications.

Unsaturated Polyesters

Monomers	Polymerization	Major uses
Phthalic anhydride, maleic anhydride, fumaric acid, isophthalic acid, ethylene glycol, propylene glycol, diethylene glycol, styrene	Bulk polycondensation followed by free-radical-initiated chain polymerization	Construction, automotive applications, marine applications

Unsaturated polyester laminating resins are viscous materials of a low degree of polymerization (i.e., oligomers) with molecular weights of 1500 to 3000. They are produced by condensing a glycol with both an unsaturated dicarboxylic acid (maleic acid) and a saturated carboxylic acid (phthalic or isophthalic acid). The viscous polyesters are dissolved in styrene monomer (30–50% concentration) to reduce the viscosity. Addition of glass fibers and curing with peroxide initiators produces a cross-linked polymer (solid) consisting of the original polyester oligomers, which are now interconnected with polystyrene chains (Fig. 4.5). The unsaturated acid residues in the initial polyester oligomer provide a site for cross-linking in this curing step, while the saturated acid reduces the brittleness of the final cross-linked product by reducing the frequency of cross-links.

In practice, the peroxide curing system is blended into the resin before applying the resin to the reinforcement, which is usually glass fiber (E type), as preform, cloth, mat, or rovings, but sisal or more conventional fabrics may also be used. The curing system may be so varied (in both composition and quality) that curing times may range from a few minutes to several hours, and the cure may be arranged to proceed either at ambient or elevated temperatures. The two most important peroxy materials used for room temperature curing of polyester resins are methyl ethyl ketone peroxide (MEKP) and cyclohexanone peroxide. These are used in conjunction with a cobalt compound such as a naphthenate,

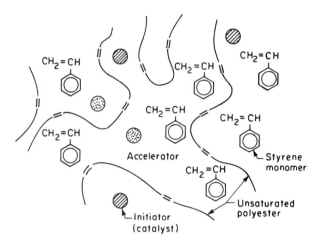

(a)

(b)

FIG. 4.5 Curing of unsaturated polyesters. (a) Species in polyester resin ready for laminating. (b) Structures present in cured polyester resin. Cross-linking takes place via an addition copolymerization reaction. The value of $n \sim 2$ to 3 on average in general-purpose resins.

octoate, or other organic-solvent-soluble soap. The peroxides are referred to as *catalysts* (though, strictly speaking, these are polymerization initiators) and the cobalt compound is referred to as an *accelerator*.

In room-temperature curing it is obviously necessary to add the resin to the reinforcement as soon as possible after the curing system has been blended and before gelation can occur. Benzoyl peroxide is most commonly used for elevated-temperature curing. It is generally supplied as a paste ($\sim$50%) in a liquid such as dimethyl phthalate to reduce explosion hazards and to facilitate mixing.

Since the cross-linking of polyester-styrene system occurs by a free-radical chain-reaction mechanism across the double bonds in the polyesters, with styrene providing the cross-links, the curing reaction does not give rise to volatile by-products (unlike phenolic and amino resins) and it is thus possible to cure without applying pressure. This fact as well as that room temperature cures are also possible makes unsaturated polyesters most useful in the manufacture of large structures such as boats and car bodies.

Unsaturated polyesters find applications mainly in two ways: polyester-glass-fiber laminates and polyester molding compositions.

Polyester-Glass-Fiber Laminates (GRP, FRP)

Methods of producing FRP laminates with polyesters have been described in Chapter 2. The major process today is the hand layup technique in which the resin is brushed or rolled into the glass mat (or cloth) by hand (see Fig. 2.41). Since unsaturated polyesters are susceptible to polymerization inhibition by air, surfaces of the hand layup laminates may remain undercured, soft, and, in some cases, tacky if freely exposed to air during the curing. A common way of avoiding this difficulty is to blend a small amount of paraffin wax (or other incompatible material) in with the resin. This blooms out on the surface and forms a protective layer over the resin during cure.

For mass production purposes, matched metal molding techniques involving higher temperatures and pressures are employed (see Fig. 2.42). A number of intermediate techniques also exist involving vacuum bag, pressure bag, pultrusion, and filament winding (see Figs. 2.43–2.45).

Glass fibers are the preferred form of reinforcement for polyester resins. Glass fibers are available in a number of forms, such as glass cloth, chopped strands, mats, or rovings (see p. 198). Some typical properties of polyester-glass laminates with different forms of glass reinforcements are given in Table 4.10. It may be seen that laminates can have very high tensile strengths.

Being relatively cheaper, polyesters are preferred to epoxide and furan resins for general-purpose laminates. Polyesters thus account for no less than 95% of the low-pressure laminates produced. The largest single outlet is in sheeting for

TABLE 4.10 Typical Properties of Polyester-Glass Laminates

Property	Mat laminate (hand layup)	Mat laminate (press formed)	Fine square woven cloth laminate	Rod from rovings
Specific gravity	1.4–1.5	1.5–1.8	2.0	2.19
Tensile strength				
10^3 lbf/in.^{2}	8–17	18–25	30–45	150
MPa	55–117	124–173	210–310	1,030
Flexural strength				
10^3 lbf/in.^{2}	10–20	20–27	40–55	155
MPa	69–138	138–190	267–380	1,100
Flexural modulus				
10^5 lbf/in.^{2}	5	6	10–20	66
MPa	3440	4150	6890–1380	45,500
Dielectric constant (10^6 Hz)	3.2–4.5	3.2–4.5	3.6–4.2	—
Power factor (10^6 Hz)	0.02–0.08	0.02–0.08	0.02–0.05	—
Water absorption (%)	0.2–0.8	0.2–0.8	0.2–0.8	—

roofing and building insulation. For the greatest transparency of the laminate the refractive indices of glass-cured resins and binder should be identical.

The second major outlet is in land transport. Polyester-glass laminates are used in the building of sports car bodies, translucent roofing panel in lorries, and in public transport vehicles. In such applications the ability to construct large polyester-glass moldings without complicated equipment is used to advantage. Polyester resins in conjunction with glass cloth or mat are widely used in the manufacture of boat hulls up to 153 ft ($\sim$46 m) in length. Such hulls are competitive in price and are easier to maintain and to repair.

The high strength-to-weight ratio of polyester-glass laminates—particularly when glass is used—microwave transparency, and corrosion resistance of the laminates have led to their use in air transport applications as in aircraft radomes, ducting, spinners, and other parts. Land, sea, and air transport applications account for nearly half the polyester resin produced. Other applications include such diverse items as chemical storage vessels, chemical plant components, swimming pools, stacking chairs, trays, and sports equipment.

Polyester Molding Compositions

Four types of polyester molding compounds may be recognized: (1) dough-molding compound (DMC), (2) sheet-molding compound (SMC), (3) alkyd-molding compositions, sometimes referred to as *polyester alkyds*, (4) diallyl phthalate (DAP) and diallyl isophthalate (DAIP) compounds.

Dough-molding compounds of puttylike consistency are prepared by blending resins, catalyst, powdered mineral filler, reinforcing fiber (chopped strand), pigment, and lubricant in a dough mixer, usually of the Z-blade type. Formulations for three typical DMC grades are given in Table 4.11.

The tendency of thick sections of DMC structural parts to crack has been overcome by using low-profile polyester resins (or low-shrink resins). These are prepared by making a blend of a thermoplastic (e.g., acrylic) polymer-styrene system with a polyester-styrene system. Moldings of this blend cured at elevated temperatures exhibit negligible shrinkage and minimal warpage and have very smooth surfaces, to which paint may be applied with very little pretreatment. A wide spectrum of properties may be obtained by varying the ratios of thermoplastics, polyester, and styrene in the blend. Among the thermoplastics quoted in the literature for such blending are poly(methyl methacrylate), polystyrene, PVC, and polyethylene. High-gloss DMCs using low-shrink resins have found uses in kitchen appliances such as toaster end plates, steam iron bases, and casings for electric heaters.

Since the manufacture of DMC involves intensive shear that causes extensive damage to fibers, DMC moldings have less strength than GRP laminates. This problem is largely avoided with the sheet-molding compounds (SMC). In the SMC process, unsaturated polyester resin, curing systems, filler thickening

TABLE 4.11 Typical Formulations of DMC Grades

Ingredients	Low-cost general-purpose	High-grade mechanical	High-grade electrical
Polyester resin	100	100	100
E glass (1/4-in. length)	20	—	90
E glass (1/2-in. length)	—	85	—
Sisal	40	—	—
Calcium carbonate	240	150	—
Benzoyl peroxide	1	1	1
Pigment	2	2	2
Calcium stearate	2	2	2

agents, and lubricant are blended together and coated onto two polyethylene films. Chopped-glass rovings are supplied between the resin layers, which are then sandwiched together and compacted as shown in Figure 4.6. Thickening occurs by the reaction of free carboxyl end groups with magnesium oxide. This converts the soft, sticky mass to a handleable sheet, which takes usually a day or two. A typical formulation consists of 30% chopped-glass fiber, 30% ground limestone, and resin. For molding, the sheet may be easily cut to the appropriate

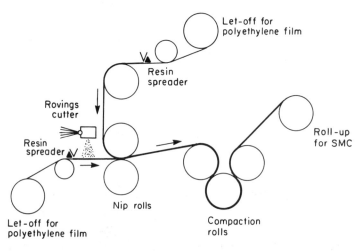

FIG. 4.6 Outline of machine for preparing sheet-molding compounds.

weight and shape and placed between the halves of the heated mold. The main applications of SMCs are in car parts, baths, and doors.

The polyester alkyd molding compositions are also based on a polyester resin similar to those used for laminating. (The term "alkyd" is derived from *al*cohol and a*cid*.) They are prepared by blending the resin with cellulose pulp, mineral filler, pigments, lubricants, and peroxide curing systems on hot mills to the desired flow properties. The mix is then removed, cooled, crushed, and ground.

Diallyl phthalate (DAP) is a diester of phthalic acid and allyl alcohol and contains two double bonds.

$$\text{(benzene ring)} \begin{array}{c} \overset{O}{\underset{\|}{C}}-O-CH_2-CH=CH_2 \\ \underset{\underset{O}{\|}}{C}-O-CH_2-CH=CH_2 \end{array}$$

On heating with a peroxide, DAP therefore polymerizes and eventually cross-links, forming an insoluble network polymer. However, it is possible to heat the DAP monomer under carefully controlled conditions, to give a soluble and stable partial polymer in the form of a white powder. The powder may then be blended with peroxide catalysts, fillers, and other ingredients to form a molding powder in the same manner as polyester alkyds. Similar products can be obtained from diallyl isophthalate (DAIP).

Both DAP and DAIP alkyd moldings are superior to the phenolics in their tracking resistance and in their availability in a wide range of colors; however, they tend to show a higher shrinkage on cure. The DAIP materials are more expensive than DAP but have better heat resistance. They are supposed to be able to withstand temperatures as high as 220°C for long periods. The polyester alkyd resins are cheaper than the DAP alkyd resins but are mechanically weaker and do not maintain their electrical properties as well under severe humid conditions. Some pertinent properties of the polyester molding compositions are compared in Table 4.12 along with those of a GP phenolic composition. The alkyd molding compositions are used almost entirely in electrical applications where the cheaper phenolic and amino resins are not suitable.

Aromatic Polyesters

Monomers	Polymerization	Major uses
p-Hydroxy benzoic acid, bisphenol A, diphenyl isophthalate	Bulk polycondensation	High-temperature engineering thermoplastics, plasma coatings, abradable seals

TABLE 4.12 Properties of Thermosetting Polyester Moldings

Property	Phenolic (GP)	DMC (GP)	Polyester alkyd	DAP alkyd	DAIP alkyd
Molding temperature (°C)	150–170	140–160	140–165	150–165	150–165
Cure time (cup flow test) (sec)	60–70	25–40	20–30	60–90	60–90
Shrinkage (cm/cm)	0.007	0.004	0.009	0.009	0.006
Specific gravity	1.3	2.0	1.7	1.6	1.8
Impact strength (ft-lb)	0.12–0.2	2.0–4.0	0.13–0.18	0.12–0.18	0.09–0.13
Volume resistivity (ohm-m)	$10^{12}-10^{14}$	10^{16}	10^{16}	10^{16}	10^{16}
Dielectric constant (10^6 Hz)	4.5–5.5	5.6–6.0	4.5–5.0	3.5–5.0	4.0–6.0
Dielectric strength (90°C) (kV/cm)	39–97	78–117	94–135	117–156	117–156
Power factor (10^6 Hz)	0.03–0.05	0.01–0.03	0.02–0.04	0.02–0.04	0.04–0.06
Water absorption (mg)	45–65	15–30	40–70	5–15	10–20

Recently several polyesters with high aromatic content have been introduced mainly because of their outstanding heat resistance. In the 1960s the Carborundum Company introduced the homopolymer of *p*-hydroxybenzoic acid under the trade name Ekonol. It is used in plasma coating. This wholly aromatic homopolyester is produced in practice by the self-ester exchange of the phenyl ester of *p*-hydroxybenzoic acid.

$$HO-\hspace{-4pt}\langle \bigcirc \rangle\hspace{-4pt}-\underset{\underset{O}{\|}}{C}-O-C_6H_5 \xrightarrow{-C_6H_5OH} \left[O-\hspace{-4pt}\langle \bigcirc \rangle\hspace{-4pt}-\underset{\underset{O}{\|}}{C} \right]_n$$

The homopolyester (mol. wt. 8,000–12,000) is insoluble in dilute acids and bases and all solvents up to their boiling points. It melts about 500°C and is difficult to fabricate. It can be shaped only by hammering (like a metal), by impact molding and by pressure sintering (420°C at 35 MPa). The difficulty in fabrication has severely limited the wider application of these polymers.

The homopolyester is available as a finely divided powder in several grades, based on particle size. The average particle size ranges from 35 to 80 μm. The material can be blended with various powdered metals, such as bronze, aluminum, and nickel-chrome, and is used in flame-spray compounds. Plasma-sprayed coatings are thermally stable, self-lubricating, and wear and corrosion resistant. Applications include abradable seals for jet aircraft engine parts.

The polymer can also be blended up to 25% with PTFE. Such blends have good temperature and wear resistance and are self-lubricating. Applications include seals, bearings, and rotors.

Copolymeric aromatic polyesters, though possessing a somewhat lower level of heat resistance are easier to fabricate than are the wholly aromatic polymers; they also possess many properties that make them of interest as high-temperature engineering materials. These materials, called *polyarylates*, are copolyesters from bisphenols and dicarboxylic acids. One such material is a copolyester of terephthalic acid, isophthalic acid, and bisphenol A in the ratio of 1:1:2.

The use of two isomeric acids leads to an irregular chain which inhibits crystallization. This allows the polymer to be processed at much lower temperatures than would be possible with a crystalline homopolymer. Nevertheless the high aromatic content of these polyesters ensures a high T_g ($\sim$190°C). The polymer is self-extinguishing with a limiting oxygen index of 34 and a self-ignition temperature of 545°C. The heat-deflection temperature under load (1.8 MPa) is about 175°C.

Among other distinctive properties of the polyarylate are its good optical properties (luminous light transmission 84 to 88% with 1 to 2% haze, refractive index 1.61), high impact strength between that of polycarbonate and polysulfone, exceptionally high level of recovery after deformation (important in applications such as clips and snap fasteners), good toughness at both elevated and low temperatures with very little notch sensitivity, and high abrasion resistance which is superior to that of polycarbonates. Polyarylate's weatherability and flammability (high oxygen index, low flame spread) are inherent and are achieved without additives. The weatherability properties therefore do not deteriorate significantly with time. (Tests show that over 5000 hr of accelerated weathering results in virtually no change in performance with respect to luminous light transmittance, haze, gloss, yellowness, and impact.) Having no flame-retardant additives, the combustion products of polyarylate are only carbon dioxide, carbon monoxide, and water, with no formation of toxic gas.

Several companies have marketed polyarylates under the trade names: U-polymer (Unitika of Japan), Arylef (Solvay of Belgium), Ardel (Union Carbide), and Arylon (Du Pont). These are noncrystallizing copolymers of mixed phthalic acids with a bisphenol and have repeat units of the type shown above. They are melt processable with T_g and heat distortion temperatures in the range 150–200°C and have similar mechanical properties to polycarbonate and polyethersulfones (see later). The polymers are useful for electrical and mechanical components that require good heat resistance and for lighting fixtures and consumer goods that operate at elevated temperatures, such as microwave ovens and hair dryers. Potential uses of the somewhat cheaper Arylon type polyesters include exterior car parts, such as body panels and bumpers.

A newer approach to obtaining polymers with good melt processability coupled with high softening point has led to another type of aromatic co-polyester, the so-called *liquid crystalline polymers* (LCPs). In these polymers, marketed under the trade name Vectra (Celanese), the retention of liquid crystalline order in the melt gives lower melt viscosities than would otherwise be achieved. Heat distortion temperatures are also in the high range of 180–240°C. LCPs may thus herald a new era of readily molded engineering and electrical parts for high temperature use. Typical properties of aromatic polyesters cited above are shown in Table 4.13.

TABLE 4.13 Properties of Unfilled Aromatic Polyesters

Property	Ekonol	Arylef or Ardel	Arylon	Vectra (range for various grades)
Tensile strength				
10^3 lbf/in^2	11	10	10	20–35
MPa	74	70	70	140–240
	(flexural)	(yield)		
Elongation at break (%)	—	50	25	1.6–7
Tensile modulus				
10^5 lbf/in^2	—	3.0	2.9	14–58
GPa	—	2.1	2.0	10–40
Flexural modulus				
10^5 lbf/in^2	10	2.9	3.0	14–51
GPa	7.1	2.0	2.1	10–35
Heat distortion temperature (°C)	>550	175	155	180–240
Impact strength, notched Izod				
ft-lb/in	—	2.8–4.7	5.4	1–10
J/m	—	150–250	288	53–530
Limiting oxygen index (%)	—	34	26	35–50

Wholly Aromatic Copolyester

A high-performance, wholly aromatic copolyester suitable for injection molding was commercially introduced in late 1984 by Dartco Mfg. under the trade name Xydar. Xydar injection-molding resins are based on terephthalic acid, *p,p'*-dihydroxybiphenyl, and *p*-hydroxybenzoic acid.

Polymers of this class contain long, relatively rigid chains which are thought to undergo parallel ordering in the melt, resulting in low melt viscosity and good injection-molding characteristics, although at relatively high melt temperatures—750 to 806°F (400 to 430°C). The melt solidifies to form tightly packed fibrous chains in the molded parts, which give rise to exceptional physical properties. The tensile modulus of the molded unfilled resin is 2.4 × 10^6 psi (16,500 MPa) at room temperature and 1.2 × 10^6 psi (8300 MPa) at 575°F (300°C). Tensile strength is about 20,000 psi (138 MPa), compressive strength is 6000 psi (41 MPa), and elongation is approximately 5%. Mechanical properties are claimed to improve at subzero temperatures.

The wholly aromatic copolyester is reported to have outstanding thermal oxidative stability, with a decomposition temperature in air of 1040°F (560°C) and

1053°F (567°C) in a nitrogen atmosphere. The resin is inherently flame retardant and does not sustain combustion. Its oxygen index is 42, and smoke generation is extremely low.

The resin is extremely inert, resists attack by virtually all chemicals, including acids, solvents, boiling water, and hydrocarbons. It is attacked by concentrated, boiling caustic but is unaffected by 30 days of immersion in 10% sodium hydroxide solution at 127°F (53°C). It withstands a high level of UV radiation and is transparent to microwaves.

The wholly aromatic copolyester for injection molding is available in filled and unfilled grades. It can be molded into thin-wall components at high speeds. The high melt flow also enables it to be molded into heavy-wall parts. No mold release is required because of the inherent lubricity and nonstick properties. No post-curing is necessary because the material is completely thermoplastic in nature. The material is expected to have many applications because of its moldability and its resistance to high temperatures, fire, and chemicals.

Polycarbonates

Monomers	Polymerization	Major uses
Bisphenol A, phosgene	Interfacial polycondensation, solution polycondensation, transesterification	Glazing (37%), electrical and electronics (15%), appliances (15%), compact discs

The major process for polycarbonate manufacture include (1) transesterification of bisphenol A with diphenyl carbonate:

(2) solution phosgenation in the presence of an acid acceptor such as pyridine:

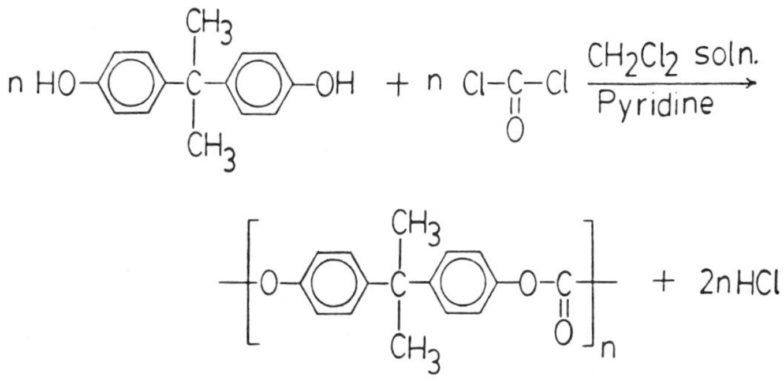

and (3) interfacial phosgenation in which the basic reaction is the same as in solution phosgenation, but it occurs at the interface of an aqueous phase and an organic phase. Here the acid acceptor is aqueous sodium hydroxide, which dissolves the bisphenol A and a monohydric phenol used for molecular-weight control (without which very high-molecular-weight polymers of little commercial value will be obtained), and the organic phase is a solvent for phosgene and the polymer formed. A mixture of methylene chloride and chlorobenzene is a suitable solvent. The interfacial polycondensation method is the most important process at present for the production of polycarbonate. Interestingly, polycarbonate represents the first commercial application of interfacial polycondensation. Fire-retardant grades of polycarbonates are produced by using tetrabromobisphenol A as comonomer.

Polycarbonate resin is easily processed by all thermoplastic-molding methods. Although it is most often injection molded or extruded into flat sheets, other options include blow molding, profile extrusion, and structural foam molding. Polycarbonate sheet can be readily thermoformed. The resin should be dried to less than 0.02% moisture before processing to prevent hydrolytic degradation at the high temperatures necessary for processing.

The chemical resistance of polyester materials is generally limited due to the comparative ease of hydrolysis of the ester groups, but the bisphenol A polycarbonates are somewhat more resistant. This resistance may be attributed to the shielding of the carbonate group by the hydrophobic benzene rings on either side. The resin thus shows resistance to dilute mineral acids; however, it has poor resistance to alkali and to aromatic and chlorinated hydrocarbons.

Polycarbonates have an unusual combination of high impact strength (12 to 16 ft-lbf per inch notch for 1/2-in. × 1/8-in. bar), heat-distortion temperature (132°C), transparency, very good electrical insulation characteristics, virtually self-extinguishing nature, and physiological inertness. As an illustration of the toughness of polycarbonate resins, it is claimed that an 1/8-in.-thick molded disc will stop a .22 caliber bullet, causing denting but not cracking. In creep resistance, polycarbonates are markedly superior to acetal and polyamide thermoplastics.

Because of a small dipole polarization effect, the dielectric constant of polycarbonates (e.g., 3.0 at 10^3 Hz) is somewhat higher than that for PTFE and the polyolefins (2.1 to 2.5 at 10^3 Hz). The dielectric constant is also almost unaffected by frequency changes up to 10^6 Hz and temperature changes over the normal range of operations. (Note that for satisfactory performance electrical insulating materials should have a *low* dielectric constant and *low* dissipation factor but *high* dielectric strength. For dielectrics used in capacitors, however, a high dielectric constant is desirable.) At low frequencies (60 Hz) and in the ordinary temperature range (20–100°C), the power factor of polycarbonates (~0.0009) is remarkably low for a polar polymer. It increases, however, at higher frequencies, reaching a value of 0.010 at 10^6 Hz. The polycarbonates have a high volume resistivity (2.1 × 10^{20} ohm-cm at 23°C) and a high dielectric strength (400 kV/in., 1/8-in. sample). Because of the low water absorption, these properties are affected little by humidity. Polycarbonates, however, do have a poor resistance to tracking.

Although the electrical properties of polycarbonates are not as impressive as those observed with polyethylene, they are adequate for many purposes. These properties, coupled with the high impact strength, heat and flame resistance, transparency, and toughness have led to the extensive use of these resins in electronics and electrical engineering, which remains the largest single field of their application. Polycarbonate is the only material that can provide such a combination of properties, at least at a reasonable cost.

Polycarbonate covers for time switches, batteries, and relays utilize the good electrical insulation characteristics in conjunction with transparency, toughness, and flame resistance of the polymer. Its combination of properties also accounts for its wide use in making coil formers. Many other electrical and electronic applications include moldings for computers, calculating machines, and magnetic disc pack housing, contact strips, switch plates, and starter enclosures for fluorescent lamps. Polycarbonate films of high molecular weight are used in the manufacture of capacitors.

Traditional applications of polycarbonate in the medical market, such as filter housings, tubing connectors, and surgical staplers, have relied on the material's unique combination of strength, purity, transparency, and ability to stand all sterilization methods (steam, ethylene oxide gas, and gamma radiation). Recently developed polycarbonate-based blends and copolymers have further extended the material's usefulness to medical applications.

Recent years have seen a continuing growth of the market for polycarbonate glazing and light transmission units. Applications here include lenses and protective domes as well as glazing. The toughness and transparency of polycarbonates have led to many successful glazing applications of the polymer, such as bus shelters, telephone kiosks, gymnasium windows, lamp housings for street lighting, traffic lights, and automobiles, strip-lighting covers at ground level, safety goggles, riot-squad helmets, armor, and machine guards. The limited scratch and weathering resistance of the polycarbonates is a serious drawback in these applications, and much effort is being directed at overcoming these problems. One approach is to coat the polycarbonate sheet with a glasslike composition by using a suitable priming material (e.g., *Margard*, marketed by the General Electric Company) to ensure good adhesion between coating and the base plastic.

Polycarbonates modified with ABS (acrylonitrile and styrene grafted onto polybutadiene) and MBS (methyl methacrylate and styrene grafted onto polybutadiene) resins have been available for many years. Usually used to the extent of 2 to 9%, the styrene-based terpolymers are claimed to reduce the notch sensitivity of the polycarbonate and to improve its resistance to environmental stress cracking while retaining for some grades the high impact strength of the unmodified polycarbonate. These materials find use in the electrical industry, in the automotive industry (instrument panels and glove compartment flaps), and for household appliances (coffee machine housings, hair drier housings, and steam handles). Elastomer modified polycarbonates have been used for automobile front ends and bumpers (e.g., 1982 Ford Sierra).

Polycarbonate is presently used for making compact audio discs, which are based on digital recording and playback technology and can store millions of bits of information in the form of minute pits in an area only a few inches in diameter. It is the presence or absence of the pits which is read by the laser. Each 'track' comprising a spiral of these pits is laid in polycarbonate which is backed with reflective aluminum and coated with a protective acrylic layer. The processability of the polycarbonate, or any other material used as the substrate, is crucial in the manufacture of all optical discs. Bayer's polycarbonate grade Makrolon CD-2000 has been specially developed to fit such requirements.

POLYAMIDES

The early development of polyamides started with the work of W. H. Carothers and his colleagues, who, in 1935, first synthesized nylon-6,6—a polyamide of hexamethylene diamine and adipic acid—after extensive and classical researches into condensation polymerization. Commercial production of nylon-6,6 and its conversion into fibers was started by the Du Pont Company in 1939. In a parallel development in Germany, Schlack developed polyamides by ring-opening polymerization of cyclic lactams, and nylon-6 derived from caprolactam was introduced in 1939. Today nylon-6,6 and nylon-6 account for nearly all of the polyamides produced for fiber applications.

Nylon-6,6 and nylon-6 are also used for plastics applications. Besides these two polyamides very many other aliphatic polyamides have been prepared in the laboratory, and a few of them (nylon-11, nylon-12, and nylon-6,10 in particular) have become of specialized interest as plastics materials. However, only about 10% of the nylons produced are used for plastics production. Virtually all of the rest goes for the production of fibers where the market is shared, roughly equally, between nylon-6 and nylon-6,6. (Nylon is the trade name for the polyamides from unsubstituted, nonbranched aliphatic monomers. A polyamide made from either an amino acid or a lactam is called nylon-x, where x is the number of carbon atoms in the repeating unit. A nylon made from a diamine and a dibasic acid is designated by two numbers, in which the first represents the number of carbons in the diamine chain and the second the number of carbons in the dibasic acid.)

For a variety of technical reasons the development of aromatic polyamides was much slower in comparison. Commercially introduced in 1961, the aromatic polyamides have expanded the maximum temperature well above 200°C. High-tenacity, high-modulus polyamide fibers (aramid fibers) have provided new levels of properties ideally suited for tire reinforcement. More recently there has been considerable interest in some new aromatic glassy polymers, in thermoplastic polyamide elastomers, and in a variety of other novel materials.

Aliphatic Polyamides

$$\left[NH(CH_2)_6NH-\underset{\underset{O}{\|}}{C}-(CH_2)_4-\underset{\underset{O}{\|}}{C} \right]_n$$

Nylon-6,6

$$\left[NH(CH_2)_5-\underset{\underset{O}{\|}}{C} \right]_n$$

Nylon-6

Monomers	Polymerization	Major uses
Adipic acid, hexamethylenediamine, caprolactam	Bulk polycondensation	Home furnishings, apparel, tire cord

Aliphatic polyamides are produced commercially by condensation of diamines with dibasic acids, by self-condensation of an amino acid, or by ring-opening polymerization of a lactam. To obtain polymers of high molecular weight, it is important that there should be stoichiometric equivalence of amine and acid groups of the monomers. For amino acids and lactams the stoichiometric balance is ensured by the use of pure monomers; for diamines and dibasic acids this is readily obtained by the preliminary formation of a 1:1 ammonium salt, often referred to as a *nylon salt*. Small quantities of monofunctional compounds are often used to control the molecular weight.

The nylon-6,6 salt (melting point 190–191°C) is prepared by reacting hexamethylenediamine and adipic acid in boiling methanol, the comparatively insoluble salt precipitating out. A 60% aqueous slurry of the salt together with a trace of acetic acid to limit the molecular weight to the desired level (9000–15,000) is heated under a nitrogen blanket at about 220°C in a closed autoclave under a pressure of about 20 atmospheres (atm). The polymerization proceeds to approximately 80 to 90% without removal of by-product water. The autoclave temperature is then raised to 270 to 300°C, and the steam is continuously driven off to drive the polymerization to completion.

$$n\text{H}_2\text{N(CH}_2)_6\text{NH}_2 + n\text{HO}_2\text{C(CH}_2)_4\text{CO}_2\text{H} \longrightarrow n \begin{bmatrix} ^-\text{O}_2\text{C(CH}_2)_4\text{CO}_2^- \\ ^+\text{H}_3\text{N(CH}_2)_6\text{NH}_3^+ \end{bmatrix}$$

$$\left[\text{NH---(CH}_2)_6\text{---NH---}\underset{\underset{\text{O}}{\|}}{\text{C}}\text{---(CH}_2)_4\text{---}\underset{\underset{\text{O}}{\|}}{\text{C}} \right]_n \text{---OH} + (2n - 1)\text{H}_2\text{O}$$

The later stages of the polymerization reaction constitute a *melt polycondensation*, since the reaction temperature is above the melting point of the polyamide. The molten polymer is extruded by nitrogen pressure on to a water-cooled casting wheel to form a ribbon which is subsequently disintegrated. In a continuous process for the production of nylon-6,6 similar reaction conditions are used, but the reaction mixture moves slowly through various zones of a reactor.

Nylon-6,10 is prepared from the salt (melting point 170°C) of hexamethyl-enediamine and sebacic acid by a similar technique. Nylon-6,9 uses azelaic acid. Decane-1,10-dicarboxylic acid is used for nylon-6,12.

In a typical batch process for the production of nylon-6 by ring-opening po-lymerization, a mixture of caprolactam, water (5–10% by weight), which acts as a catalyst, and a molecular-weight regulator [e.g., acetic acid (~0.1%)] is heated in a reactor under a nitrogen blanket at 250°C for about 12 hr, a pressure of about 15 atm being maintained by venting off steam. The product consists of high-molecular-weight polymer (about 90%) and low-molecular-weight material (about 10%), which is mainly monomer. To obtain the best physical properties, the low-molecular-weight materials may be removed by leaching and/or by vacuum distillation. In the continuous process, similar reaction conditions are used. In one process a mixture of molten caprolactam, water, and acetic acid is fed continuously to a reactor operating at about 260°C. Residence time is 18 to 20 hr.

A newer technique for the preparation of nylon-6 is the *polymerization cast-ing* of caprolactam in situ in the mold. In this process rapid formation of poly-mer is achieved by anionic polymerization, initiated by strong bases such as metal amides, metal hydrides, and alkali metals. However, the anionic polymer-ization of lactams by strong bases alone is relatively slow because it is associ-ated with an *induction period* due to a slow step in the initiation sequence leading to an *N*-acyl lactam which participates in the propagation reaction. The induction period may be eliminated by adding along with the strong base a pre-formed *N*-acyl lactam or related compound at the start of the reaction.

A typical system for polymerization casting of caprolactam thus uses as a catalyst 0.1 to 1 mol % *N*-acetyl caprolactam and 0.15 to 0.50 mol % of the sodium salt of caprolactam. The reaction temperature is initially about 150°C, but during polymerization it rises to about 200°C. The technique is especially applicable to the production of large, complex shapes that could not be made by the more conventional plastics processing techniques. Important advantages of the process are the low heats and low pressures involved. Although the poly-merization process is exothermic, the relatively low heat of polymerization of caprolactam, coupled with its low melting point, makes the process easy to con-trol and simplifies the heat transfer problem generally associated with the pro-duction of massive parts. Moldings of cast nylon-6 up to one ton are claimed to have been produced by these techniques.

Nylon parts made by polymerization casting of caprolactam exhibit higher molecular weights and a highly crystalline structure and are, therefore, slightly harder and stiffer than conventionally molded nylon-6.

Applications for cast nylon-6 include huge gears (e.g., a 150-kg nylon gear for driving a large steel drum drier) and bearings, gasoline and fuel tanks, buck-ets, building shutters, and various components for paper production machinery and mining and construction equipment. Later development has centered on adding reinforcing materials to the monomer before polymerization to produce parts with higher heat distortion temperature, impact strength and tensile strength.

Nylon-12 is produced by the ring-opening polymerization of laurolactam (dodecyl lactam) such as by heating the lactam at about 300°C in the presence of aqueous phosphoric acid. Unlike the polymerization of caprolactam, the polymerization of dodecyl lactam does not involve an equilibrium reaction. Hence, an almost quantitative yield of nylon-12 polymer is obtained by the reaction, and the removal of low-molecular-weight material is unnecessary.

Nylon-11 is produced by the condensation polymerization of ω-aminoundecanoic acid at 200 to 220°C with continuous removal of water. The latter stages of the reaction are conducted under reduced pressure to drive the polymerization to completion.

Nylon copolymers can be obtained by heating a blend of two or more different nylons above the melting point so that amide interchange occurs. Initially, block copolymers are formed, but prolonged reaction leads to random copolymers. For example, a blend of nylon-6,6 and nylon-6,10 heated for 2 hr gives a random copolymer (nylon-6,6—nylon-6,10) which is identical with a copolymer prepared directly from the mixed monomers. Other copolymers of this type are available commercially.

Properties

Aliphatic polyamides are linear polymers containing polar —CONH— groups spaced at regular intervals by aliphatic chain segments. The principal structural difference between the various types of nylon is in the length of aliphatic chain segments separating the adjacent amide groups. The polar amide groups give rise to high interchain attraction in the crystalline zones, and the aliphatic segments impart a measure of chain flexibility in the amorphous zones. This combination of properties yields polymers which are tough above their glass transition temperatures. The high intermolecular attraction also accounts for high melting points of nylons, which are usually more than 200°C. The melting point, however, decreases (which facilitates processing) as the length of the aliphatic segment in the chain increases, as indicated in Table 4.14.

Because of the high cohesive energy density and their crystalline state, the nylons are resistant to most solvents. They have exceptionally good resistance to hydrocarbons and are affected little by esters, alkyl halides, and glycols. There are only a few solvents for the nylons, of which the most common are formic acid, glacial acetic acid, phenols, and cresols. Alcohols generally have some swelling action and may dissolve some copolymers (e.g., nylon-6,6, nylon-6,10, nylon-6). Nylons have very good resistance to alkalis at room temperature.

TABLE 4.14   Melt Temperatures of
Aliphatic Polyamides

Polyamide	T_m (°C)
Nylon-6,6	265
Nylon-6,8	240
Nylon-6,10	225
Nylon-6,12	212
Nylon-6	230
Nylon-7	223
Nylon-11	188
Nylon-12	180

Mineral acids attack nylons, but the rate of attack depends on the nature and concentration of acids and the type of nylon. Nitric acid is generally active at all concentrations.

Because of the presence of amide groups, the nylons absorb water. Figure 4.7 shows how the equilibrium water absorption of different nylons varies with humidity at room temperature, and Figure 4.8 shows how the rate of moisture absorption of nylon-6,6 is affected by the environmental conditions. Since dimensional changes may occur as a result of water absorption, this effect should be considered when dimensional accuracy is required in a specific ap-

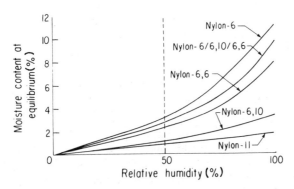

FIG. 4.7   Effect of relative humidity on the equilibrium moisture absorption of the nylons.

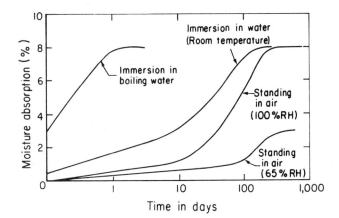

FIG. 4.8 Effect of environmental conditions on rate of moisture absorption of nylon-6,6 (1/8-in.-thick specimens).

plication. Manufacturers commonly supply data on the dimensional changes of their products with ambient humidity.

The various types of nylon have generally similar physical properties, being characterized by high toughness, impact strength, and flexibility (Table 4.15). Mechanical properties of nylons are affected significantly by the amount of crystallization in the test piece, ambient temperature (Fig. 4.9), and humidity (Fig. 4.10), and it is necessary to control these factors carefully in the determination of comparative properties. Moisture has a profound plasticizing influence on the modulus. For example, the Young's modulus values for nylon-6,6 and nylon-6 decreases by about 40% with the absorption of 2% moisture.

Nylons have extremely good abrasion resistance. This property can be further enhanced by addition of external lubricants and by providing a highly crystalline hard surface to the bearings. The surface crystallinity can be developed by the use of hot injection molds and by annealing in a nonoxidizing fluid at an elevated temperature (e.g., 150–200°C for nylon-6,6).

The coefficient of friction of nylon-6,6 is lower than mild steel but is higher than the acetal resins. The frictional heat buildup, which determines the upper working limits for bearing applications, is related to the coefficient of friction under working conditions. The upper working limits measured by the maximum LS value (the product of load L in psi on the projected bearing area and the peripheral speed S in ft/min) are 500 to 1000 for continuous operation of unlubricated nylon-6,6. For intermittent operation initially oiled nylon bearings can be used at LS values of 8000. Higher LS values can be employed with continuously lubricated bearings.

TABLE 4.15 Comparative Properties[a] of Typical Commercial Grades of Nylon

Property	6,6	6	6,10	11	12	6,6/6,10/6 (40:30:30)
Specific gravity	1.14	1.13	1.09	1.04	1.02	1.09
Tensile stress at yield						
10^3 lbf/in.^{2}	11.5	11.0	8.5	5.5	6.6	—
MPa	80	76	55	38	45	—
Elongation at break (%)	80–100	100–200	100–150	300	200	300
Tension modulus						
10^5 lbf/in.^{2}	4.3	4	3	2	2	2
10^2 MPa	30	28	21	14	14	14
Impact strength						
ft-lbf/1/2-in. notch	1.0–1.5	1.5–3.0	1.6–2.0	1.8	1.9	—
Rockwell hardness	R118	R112	R111	R108	R107	R83

Heat distortion temperature (264 lbf/in.^{2}) (°C)	75	60	55	55	51	30
Coefficient of linear expansion, 10^{-5} cm/cm-°C	10	9.5	15	15	12	—
Volume resistivity, ohm-m (dry)	$>10^{17}$	$>10^{17}$	$>10^{17}$	—	—	—
ohm-m (50% RH)	10^{15}	—	10^{16}	—	—	10^{15}
Dielectric constant (10^3 Hz dry)	3.6–6.0	3.6–6.0	3.6–6.0	—	—	—
Power factor (10^3 Hz dry)	0.04	0.02–0.06	0.02	—	—	—
Dielectric strength (kV/cm) (25°C, 50% RH)	>100	>100	>100	—	—	—

^{a}ASTM tests for mechanical and thermal properties.

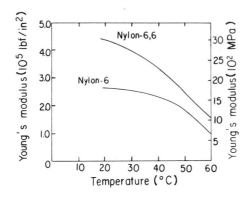

(a)

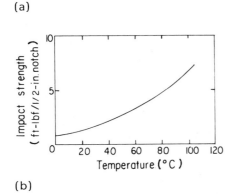

(b)

FIG. 4.9 Effect of temperature on (a) Young's modulus of nylon-6,6 and nylon-6 and (b) impact strength of nylon-6,6.

The electrical insulation properties of the nylons are reasonably good at room temperature, under conditions of low humidity, and at low frequencies. Because of the presence of polar amide groups, they are not good insulators for high-frequency work, and since they absorb water, the electrical insulation properties deteriorate as the humidity increases (see Fig. 4.11).

The properties of nylons are considerably affected by the amount of crystallization and by the size of morphological structures, such as spherulites, which, in turn, are generally influenced by the processing conditions. Thus, a molding of nylon-6, slowly cooled and subsequently annealed, may be 50 to 60% crystalline, whereas a rapidly cooled thin-walled molding may be only 10% crystalline.

Slowly cooled melts may form bigger spherulites, but rapidly cooled polymers may form only fine aggregates. Consequently, in an injection molding the

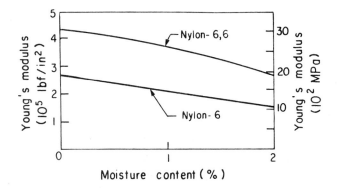

(a)

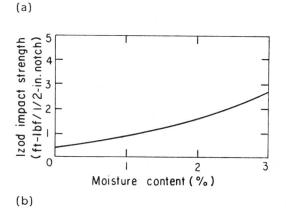

(b)

FIG. 4.10 Effect of moisture content on (a) Young's modulus of nylon-6,6 and nylon-6 and (b) impact strength of nylon-6,6.

morphological form of rapidly cooled surface layers may be quite different from that of the more slowly cooled centers. The use of nucleating agents (e.g., about 0.1% of a fine silica) can give smaller spherulites and thus a more uniform structure in an injection molding. Such a product may have greater tensile strength, hardness, and abrasion resistance at the cost of some reduction in impact strength and elongation at break: the higher the degree of crystallinity the less the water absorption, and hence the less will be the effect of humidity on the properties of the polymer.

Nylon molding materials are available in a number of grades which may differ in molecular weight and/or in the nature of additives which may be present. The various types of additives used in nylon can be grouped as heat stabilizers, light stabilizers, lubricants, plasticizers, pigments, nucleating agents, flame retarders, and reinforcing fillers.

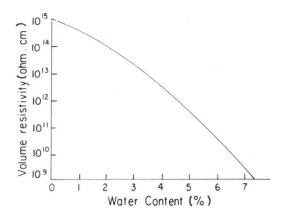

FIG. 4.11 Effect of moisture content on the volume resistivity of nylon-6,6.

Heat stabilizers include copper salts, phosphoric acid esters, mercaptoben-zothiazole, mercaptobenzimidazole, and phenyl-β-naphthyl-amine. Among light stabilizers are carbon black and various phenolic materials. Self-lubricating grades of nylon which are of value in some gear and bearing applications incorporate lubricants such as molybdenum disulfide (0.2%) and graphite (1%).

Plasticizers may be added to nylon to lower the melting point and to improve toughness and flexibility particularly at low temperatures. A plasticizer used commercially is a blend of o- and p-toluene ethyl sulfonamide.

Substances used as nucleating agents include silica and phosphorous compounds. Nucleating agents are used to control the size of morphological structures of the molding.

There have been substantial efforts to improve the flame resistance of nylons. Various halogen compounds (synergized by zinc oxide or zinc borate) and phosphorus compounds have been used (see the section on flame retardation in Chapter 1). They are, however, dark in color.

Glass-reinforced nylons have become available in recent years. Two main types of glass fillers used are glass fibers and glass beads. From 20 to 40% glass is used. Compared to unfilled nylons, glass-fiber reinforcement leads to a substantial increase in tensile strength (160 vs. 80 MPa), flexural modulus (8000 vs. 3000 MPa), hardness, creep resistance (at least three times as great), and heat-distortion temperature under load (245 vs. 75°C under 264 psi), and to a significant reduction in coefficient of expansion (2.8×10^{-5} vs. 9.9×10^{-5} cm/cm-°C).

The glass-fiber-filled types can be obtained in two ways. One route involves passing continuous lengths of glass fiber (as rovings) through a polymer melt or solution to produce glass-reinforced nylon strand that is chopped into pellets.

Another route involves blending a mixture of resin and glass fibers about 1/4 in. (0.6 cm) long in an extruder. Usually E-grade glass with a diameter of about 0.001 cm treated with a coupling agent, such as a silane, to improve the resin-glass bond is used.

Nylons filled with 4.0% of glass spheres have a compressive strength about eightfold higher than unfilled grades, besides showing good improvement in tensile strength, modulus, and heat-distortion temperature. Having low melt viscosity, glass-bead-filled nylons are easier to process than the glass-fiber-filled varieties. They are also more isotropic in their mechanical properties and show minimum warpage. Glass fillers, both fibers and beads, tend to improve self-extinguishing characteristics of nylons.

Applications

The most important application of nylons is as fibers, which account for nearly 90% of the world production of all nylons. Virtually all of the rest is used for plastics applications. Because of their high cost, they have not become general-purpose materials, such as polyethylene and polystyrene, which are available at about one-third the price of nylons. Nylons have nevertheless found steadily increasing application as plastics materials for specialty purposes where the combination of toughness, rigidity, abrasion resistance, reasonable heat resistance, and gasoline resistance is important.

The largest plastics applications of nylons have been in mechanical engineering—nylon-6, nylon-6,6, nylon-6,10, nylon-11, and nylon-12 being mainly used. These applications include gears, cams, bushes, bearings, and valve seats. Nylon moving parts have the advantage that they may often be operated without lubrication, and they may often be molded in one piece, they are smooth running. Among the aforesaid nylons, nylon-11 and nylon-12 have the lowest water absorption and are easy to process, but there is some loss in mechanical properties. For the best mechanical properties, nylon-6,6 would be considered, but this material is also the most difficult to process and has high water absorption. Nylon-6 is easier to process but has even higher water absorption (see Fig. 4.7).

Other applications include sterilizable nylon moldings in medicine and pharmacy, nylon hair combs, and nylon film for packaging foodstuffs and pharmaceutical products. The value of nylon in these later applications is due to its low odor transmission and the boil-in-the bag feature.

Besides film, other extruded applications of nylons are as monofilaments, which have found applications in surgical sutures, brush tufting, wigs, sports equipment, braiding, outdoor upholstery, and angling.

Production of moldings by polymerization casting of caprolactam and the ability to produce large objects in this way have widened the use of nylon plastics in engineering and other applications. The process gives comparatively stress-free moldings having a reasonably consistent morphological structure with a 45 to 50% crystallinity, which is higher than melt-processed materials, and thus leads to higher tensile strength, modulus, hardness, and resistance to creep. Products made by polymerization casting include main drive gears for use in the textile and papermaking industries, conveyor buckets used in the mining industry, liners for coal-washing equipment, and propellers for small marine craft.

Glass-filled nylons form the most important group of glass-filled varieties of thermoplastics. Glass-reinforced nylon plastics have high rigidity, excellent creep resistance, low coefficient of friction, high heat-deflection temperature, good low-frequency electrical insulation properties, and they are nonmagnetic in nature. Therefore they have replaced metals in many applications.

Nylons reinforced with glass fibers are thus widely used in domestic appliances, in housings and casings, in car components, including radiator parts, and in the telecommunication field for relay coil formers and tag blocks. Glass-bead-filled nylons have found use in bobbins. Carbon-fiber reinforcement has been used with nylon-6 and nylon-6–nylon-12 mixtures. These materials have found use in the aerospace field and in tennis rackets. A more recent development is the appearance of supertough nylon plastics, which are blends of nylon-6,6 with other resins, such as an ionomer resin used in the initial grades or a modified ethylene-propylene-diene terpolymer rubber (EPDM rubber) used in more recent materials.

Interest has been aroused by the appearance of novel elastomeric polyamides. The products introduced by Huls under the designation XR3808 and X4006 may be considered as the polyether-amide analogue of the polyether-ester thermoplastic elastomers introduced in the 1970s by Du Pont as Hytrel (see Fig. 4.4). The polyether-amide is a block copolymer prepared by the condensation of polytetramethylene ether glycol (i.e., polytetrahydrofuran) with laurin lactam and decane-1,10-dicarboxylic acid. The elastomeric polyamide XR3808 is reported to have a specific gravity of 1.02, a yield stress of 24 MPa, a modulus of elasticity of 300 MPa, and an elongation at break of 360%.

Aromatic Polyamides

In recent years there has been considerable interest in aromatic polyamide fibers, better known as aramid fibers. These fibers have been defined as ''a long chain synthetic polyamide in which at least 85% of the amide linkages are attached directly to two aromatic rings.'' The first significant material of this type was introduced in 1961 by Du Pont as Nomex. It is poly(*m*-phenyleneisophthalamide), prepared from *m*-phenylenediamine and isophthaloyl chloride by interfacial polycondensation.

The fiber may be spun from a solution of the polymer in dimethylformamide containing lithium chloride. In 1973, Du Pont commenced production of another aromatic polyamide fiber, a poly(*p*-phenylene terephthalamide) marketed as Kevlar. It is produced by the reaction of *p*-phenylenediamine with terephthaloyl chloride in a mixture of hexamethylphosphoramide and *N*-methyl pyrrolidone (2:1) at −10°C.

Kevlar fibers are as strong as steel but have one-fifth the weight. Kevlar is thus ideally suited as tire cord materials and for ballistic vests. The fibers have a high T_g (>300°C) and can be heated without decomposition to temperatures exceeding 500°C. The dimensional stability of Kevlar is outstanding: It shows essentially no creep or shrinkage as high as 200°C. In view of the high melting temperatures of the aromatic polyamides and their poor solubility in conventional solvents, special techniques are required to produce the fibers. For example, Kevlar is wet spun from a solution in concentrated sulfuric acid.

Similar fiber-forming materials have been made available by Monsanto. Thus the product marketed as PABH-T X-500 is made by reacting *p*-aminobenzhydrazide with terephthaloyl chloride.

Polymers have also been prepared from cyclic amines such as piperazine and bis(p-aminocyclohexyl)methane. The latter amine is condensed with decanedioic acid to produce the silklike fiber Qiana (Du Pont).

$$H_2N-\!\!\bigcirc\!\!-CH_2-\!\!\bigcirc\!\!-NH_2 \;+\; HOC(CH_2)_{10}COH \longrightarrow$$
$$\underset{O \qquad\qquad O}{}$$

$$\left[\!\!-HN-\!\!\bigcirc\!\!-CH_2-\!\!\bigcirc\!\!-NHC(CH_2)_{10}C-\!\!\right]_n$$
$$\underset{O \qquad\qquad O}{}$$

Qiana fibers have a high glass transition temperature (135°C, as compared to 90°C for nylon-6,6), which assures that the polymer will remain in the glassy state during fabric laundering and resist wrinkles and creases.

Synthetic fibers range in properties from low-modulus, high-elongation fibers like Lycra (see the section on polyurethanes) to high-modulus high-tenacity fibers such as Kevlar. A breakthrough in fiber strength and stiffness has been achieved with Kevlar. Another high-performance fiber in commercial application is graphite. The use of these new fibers has resulted in the development of superior composite materials, generally referred to as fiber-reinforced plastics or FRPs (see Chapters 2 and 3), which have shown promise as metal-replacement materials by virtue of their low density, high specific strength (strength/density), and high specific modulus (modulus/density).

Today a host of these FRP products are commercially available as tennis rackets, golf clubs shafts, skis, ship masts, and fishing rods, which are filament wound with graphite and Kevlar fibers. Significant quantities of graphite composites and graphite/Kevlar hybrid composites are used in Boeing 757 and 767 planes, which make possible dramatic weight saving. Boron, alumina, and silicon carbide fibers are also high-performance fibers but they are too expensive for large-scale commercial applications.

Partially aromatic, melt processable, polyamides are produced as random copolymers, which do not crystallize and are therefore transparent, but are sill capable of high-temperature use because of their high T_g values. Several commercial polymers of this type that have glass-like clarity, high softening point, and oil and solvent resistance have been developed. For example, Trogamid T (Dynamit Nobel) contains repeat units of

$$\left[\!\!-\underset{O}{\overset{}{C}}-\!\!\bigodot\!\!-\underset{O}{\overset{}{C}}-NH-CH_2-CH_2-\underset{CH_3}{\overset{}{C}H}-CH_2-\underset{CH_3}{\overset{CH_3}{\overset{}{C}}}-CH_2-NH-\!\!\right]$$

and of the 2,4,4-trimethyl isomer, and has a T_g of about 150°C. Grilamid TR (Emser) with a T_g of about 160°C is a copolymer with units of

and of

A crystalline, partially aromatic polyamide, poly-*m*-xylylene-adipamide, (also known as MXD-6) with repeat units of

is available as a heat resistant engineering plastic (e.g., Ixef by Solvay) generally similar in properties to nylon-6,6, having a T_m of 243°C but with reduced water absorption, greater stiffness, and a T_g of about 90°C. Although its heat distortion temperature is only 96°C, with 30% glass filling this is increased to about 270°C. Typical properties of some of these polyamides, along with those of nylon-6,6 for comparison, are shown in Table 4.16.

Copolymers containing amide and imide units, the polyamideimides, are described in the following section on polyimides.

Polyimides

TABLE 4.16 Properties of Polyamides

Property	Nylon-6,6	Ixef (high-impact grade)	Trogamid T	Nomex (fiber)	Kevlar
Tensile strength					
10^3 lbf/in^2	9.4	25.7	8.7	97.2	435
MPa	65	177	60	670	3000
Tensile modulus					
10^5 lbf/in^2	4.6	19.4	4.4	25.5	194
GPa	3.2	13.4	3.0	17.6	134
Elongation at break (%)	100	2.7	132	22	2.6
Flexural modulus					
10^5 lbf/in^2	4.8	14.8	—	—	—
GPa	3.3	10.2	—	—	—
Impact strength, notched Izod					
ft-lb/in	1.3	3.0	—	—	—
J/m	69	159	—	—	—

The polyimides have the characteristic functional group

$$-N\begin{matrix}\diagup CO-\\ \diagdown CO-\end{matrix}$$

and are thus closely related to amides. The branched nature of the imide functional group enables production of polymers having predominantly ring structures in the backbone and hence high softening points. Many of the structures exhibit such a high level of thermal stability that they have become important for application at much higher service temperatures than had been hitherto achieved with polymers.

The use of tetracarboxylic acid anhydride instead of the dicarboxylic acids used in the manufacture of polyamides yields polyimides. The general method of preparation of the original polyimides by the polymerization of pyromellitic dianhydride and aromatic diamine is shown in Figure 4.12a. A number of diamines have been investigated, and it has been found that certain aromatic amines, which include *m*-phenylenediamine, benzidine, and di-(4-aminophenyl)ether, give polymers with a high degree of oxidative and thermal stability.

The aromatic amine di-(4-aminophenyl)ether is employed in the manufacture of polyimide film, designated as Kapton (Du Pont). Other commercial materials of this type introduced by Du Pont in the early 1960s included a coating resin

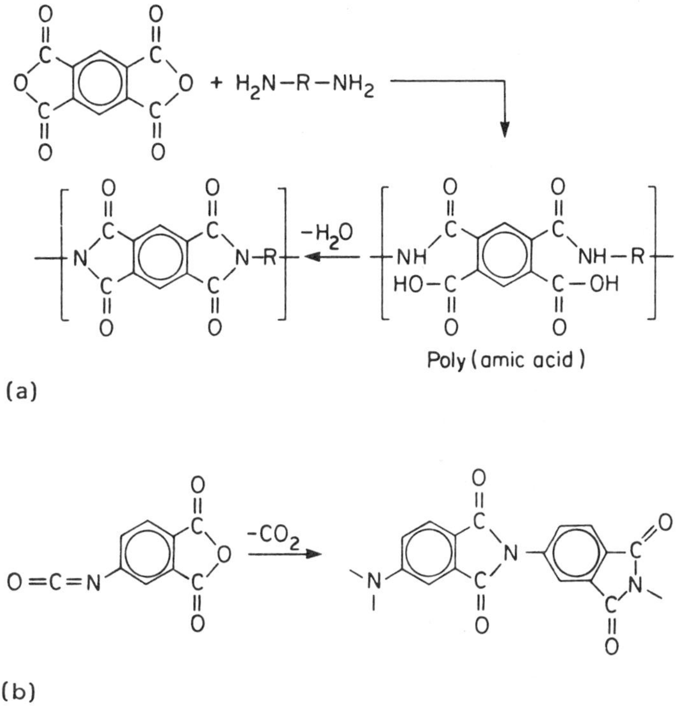

(a)

(b)

FIG. 4.12 (a) Synthesis of polyimides by polycondensation. (b) Self-condensation of isocyanate of trimellitic acid.

(Pyre ML) and a machinable block form (Vespel). In spite of their high price these materials have found established uses because of their exceptional heat resistance and good retention of properties at high temperatures.

Since the polyimides are insoluble and infusible, they are manufactured in two stages. The first stage involves an amidation reaction carried out in a polar solvent (such as dimethylformamide and dimethylacetamide) to produce an intermediate poly(amic acid) which is still soluble and fusible. The poly(amic acid) is shaped into the desired physical form of the final product (e.g., film, fiber, coating, laminate) and then the second stage of the reaction is carried out. In the second stage the poly(amic acid) is cyclized in the solid state to the polyimide by heating at moderately high temperatures above 150°C. A different approach, avoiding the intermediate poly(amic acid) step, has been pioneered by Upjohn. The Upjohn process involves the self-condensation of the isocyanate of trimellitic acid, and the reaction product is carbon dioxide (Fig. 4.12b).

Polypyromellitimides (Fig. 4.12a) have many outstanding properties: flame resistance, excellent electrical properties, outstanding abrasion resistance, exceptional heat resistance, and excellent resistance to oxidative degradation, most chemicals (except strong bases), and high-energy radiation. After 1000 hours of exposure to air at 300°C the polymers retained 90% of their tensile strength, and after 1500 hours exposure to a radiation of about 10 rad at 175°C, they retained form stability, although they became brittle.

The first commercial applications of polypyromellitimides were as wire enamels, as insulating varnishes, as coating for glass cloth (Pyre ML, Du Pont), and as film (Kapton, Du Pont). A fabricated solid grade was marketed as Vespel (Du Pont). Laminates were produced by impregnation of glass and carbon fiber, with the polyimide precursor followed by pressing and curing at about 200°C and further curing at temperatures of up to 350°C. Such laminates could be used continuously at temperatures up to 250°C and intermittently to 400°C. The laminates have thus found important application in the aircraft industry, particularly in connection with supersonic aircraft.

At the present time the applications of the polyimides include compressor seals in jet engines, sleeves, bearings, pressure discs, sliding and guide rolls, and friction elements in data processing equipment, valve shafts in shutoff valves, and parts in soldering and welding equipment.

Polyimides have also found a number of specialist applications. Polyimide foams (Skybond by Monsanto) have been used for sound deadening of jet engines. Polyimide fibers have been produced by Upjohn and by Rhone-Poulenc (Kermel). A particular drawback of the polyimides is that they have limited resistance to hydrolysis and may crack in water or steam at temperatures above 100°C. Consequently, polyimides have now encountered competition from polyetheretherketones (PEEK), which are not only superior in this regard but are also easier to mold.

Modified Polyimides

The application potential of polyimides is quite limited because, being infusible, they cannot be molded by conventional thermoplastics techniques. In trying to overcome this limitation, scientists, in the early 1970s, developed commercially modified polyimides, which are more tractable materials than polyimides but still possessing significant heat resistance. The important groups of such modified polyimides are the polyamide-imides (e.g., Torlon by Amoco Chemicals), the polybismaleinimides (e.g., Kinel by Rhone-Poulenc), the polyester-imides (e.g., Icdal Ti40 by Dynamit Nobel), and the polyether-imides (e.g, Ultem by General Electric).

If trimellitic anhydride is used instead of pyromellitic dianhydride in the reaction shown in Figure 4.12a, then polyamide-imide is formed (see Fig. 4.13a). Other possible routes to this type of product involve the reaction of trimellitic

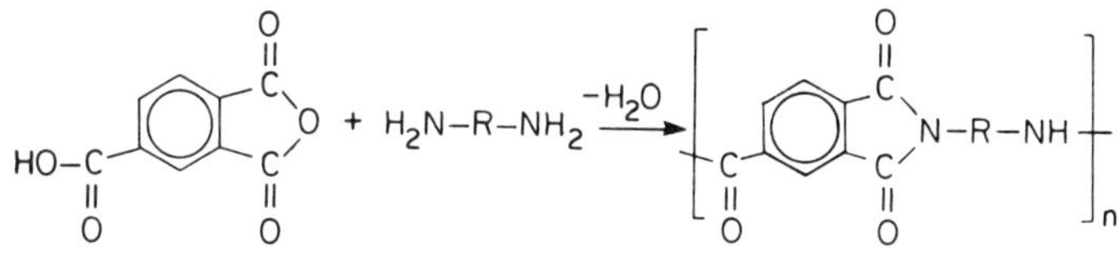

(a)

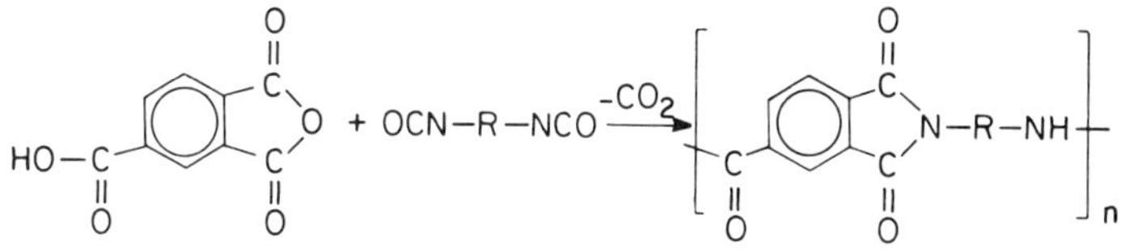

(b)

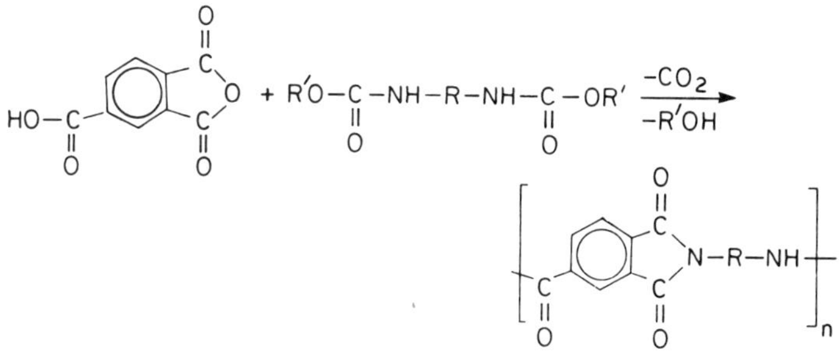

(c)

FIG. 4.13 Synthesis of polyamide-imides from trimellitic anhydride and (a) diamine, (b) diisocyanate, and (c) diurethane.

anhydride with diisocyanates (Fig. 4.13b) or diurethanes (Fib. 4.13c). Closely related is the Upjohn process for polyimide by self-condensation of the isocyanate of trimellitic acid, as illustrated in Figure 4.12b, although the product in this case is a true polyimide rather than a polyamide-imide.

Polyamide-imides may also be produced by reacting together pyromellitic dianhydride, a diamine, and a diacid chloride. Alternatively, it may be produced in a two-stage process in which a diacid chloride is reacted with an excess of diamine to produce a low-molecular-weight polyamide with amine end groups which may then be chain extended by reaction with pyromellitic dianhydride to produce imide linkages.

The Torlon materials produced by Amoco Chemicals are polyamide-imides of the type shown in Figure 4.13a. Torlon has high strength, stiffness, and creep resistance, shows good performance at moderately high temperatures, and has excellent resistance to radiation. The polymers are unaffected by all types of hydrocarbons (including chlorinated and fluorinated products), aldehydes, ketones, ethers, esters, and dilute acids, but resistance to alkalis is poor.

Torlon has been marketed both as a compression-molding grade and as an injection-molding grade. The compression-molding grade, Torlon 2000, can accept high proportions of filler without seriously affecting many of its properties. For compression molding, the molding compound is preheated at 280°C before it is molded at 340°C at pressures of 4350 psi (30 MPa); the mold is cooled to 260°C before removal. For injection molding, the melt at temperatures of about 355°C is injected into a mold kept at about 230°C. To obtain high-quality moldings, prolonged annealing cycles are recommended.

Uses of polyamide-imides include pumps, valves, refrigeration plant accessories, and electronic components. The polymers have low coefficient of friction, e.g., 0.2 (to steel), which is further reduced to as little as 0.02 to 0.08 by blending with graphite and Teflon. In solution form in N-methyl-2-pyrrolidone, Torlon has been used as a wire enamel, as a decorative finish for kitchen equipment, and as an adhesive and laminating resin in spacecraft.

The polyimides and polyamide-imides are produced by condensation reactions which give off volatile low-molecular-weight by-products. The polybismaleinimides may be produced by rearrangement polymerization with no formation of by-products. The starting materials in this case are the bismaleimides, which are synthesized by the reaction of maleic anhydride with diamines (Fig. 4.14).

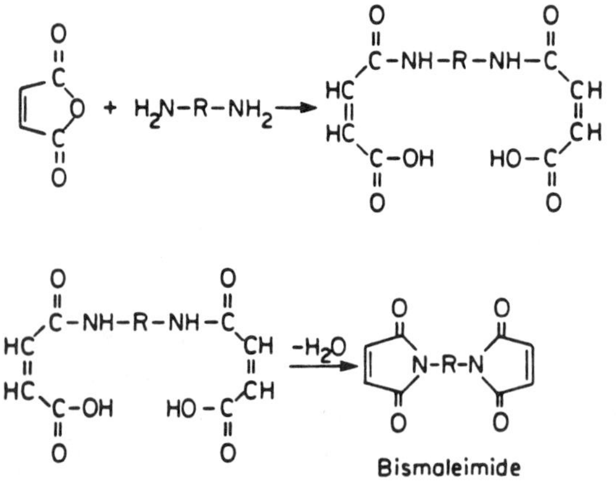

FIG. 4.14 Synthesis of bismaleimides by the reaction of maleic anhydride with diamines.

The bismaleimides can be reacted with a variety of bifunctional compounds to form polymers by rearrangement reactions. These include amines, mercaptans, and aldoximes (Fig. 4.15). If the reaction is carried out with a deficiency of the bifunctional compound, the polymer will have terminal double bonds to serve as a cure site for the formation of a cross-linked polymer via a double-bond polymerization mechanism during molding. The cross-linking in this case occurs without the formation of any volatile by-products.

The Kinel materials produced by Rhone-Poulenc are polybismaleinimides of the type shown in Figure 4.15. These materials having chain-end double bonds, as explained previously, can be processed like conventional thermosetting plastics. The properties of the cured polymers are broadly similar to the polyimides and polyamide-imides. Molding temperatures are usually from 200 to 260°C. Post-curing at 250°C for about 8 hr is necessary to obtain the optimum mechanical properties.

Polybismaleinimides are used for making laminates with glass- and carbon-fiber fabrics, for making printed circuit boards, and for filament winding. Filled grades of polybismaleinimides are available with a variety of fillers such as asbestos, glass fiber, carbon fiber, graphite, Teflon, and molybdenum sulfide. They find application in aircraft, spacecraft, and rocket and weapons technology. Specific uses include fabrication of rings, gear wheels, friction bearings, cam discs, and brake equipment.

The polyester-imides constitute a class of modified polyimide. These are typified by the structure shown in Figure 4.16. Polyether-imides form yet another class of modified polyimide. Ultem, introduced by General Electric in 1982, is

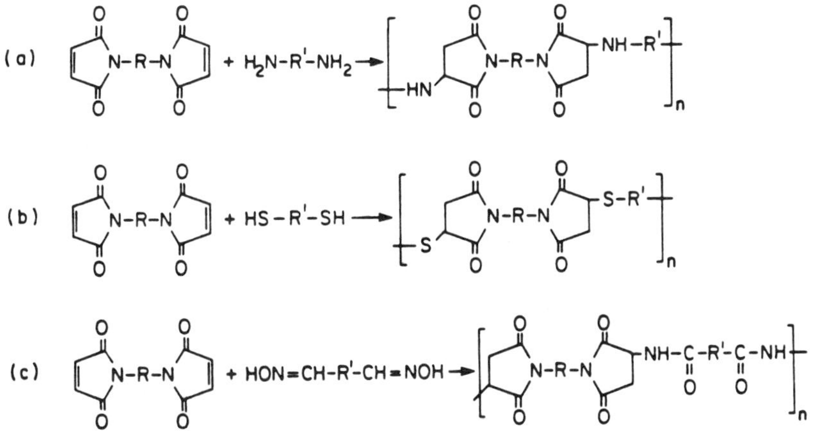

FIG. 4.15 Formation of polymers by reaction of bismaleimides with (a) amines, (b) mercaptans, and (c) aldoximes.

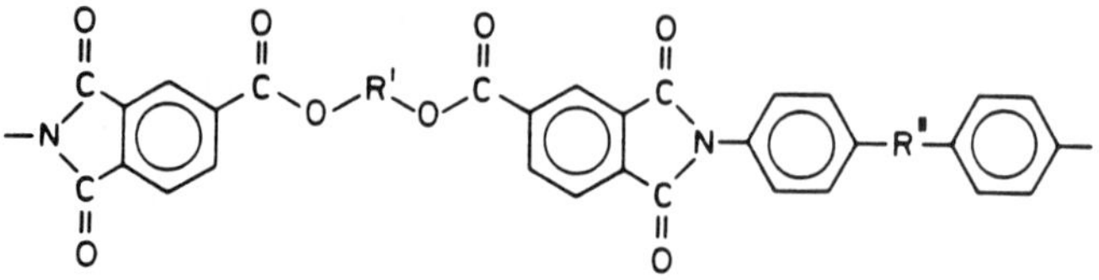

FIG. 4.16 Typical structure of polyester-imides.

a polyether-imide. It was designed to compete with heat- and flame-resistory, high-performance engineering polymers, polysulfones, and polyphenylene sulfide. Some typical properties of Ultem 1000 are specific gravity 1.27, tensile yield strength 105 MPa, flexural modulus 3300 MPa, hardness Rockwell M109, Vicat softening point 219°C, heat-distortion temperature (1.82 MPa) 200°C, and limiting oxygen index 47. Specific applications include circuit breaker housings and microwave oven stirrer shafts.

Typical properties of unfilled polyimides are compared in Table 4.17.

FORMALDEHYDE RESINS

The phenol-formaldehyde and urea-formaldehyde resins are the most widely used thermoset polymers. The phenolic resins were the first truly synthetic polymers to be produced commercially. Both phenolic and urea resins are used in the highly cross-linked final form (C-stage), which is obtained by a stepwise polymerization process. Lower-molecular-weight prepolymers are used as precursors (A-stage resins), and the final form and shape are generated under heat and pressure. In this process water is generated in the form of steam because of the high

TABLE 4.17 Properties of Unfilled Polyimides

Property	Vespel (ICI)	Torlon (Amoco)	Kinel (Rhone-Poulenc)	Ultem (General Electric)
Tensile strength (MPa)				
25°C	90	186	~40	100
150°C	67	105	—	—
260°C	58	52	~25	—
Flexural modulus (GPa)				
25°C	3.5	4.6	3.8	3.3
150°C	2.7	3.6	—	2.5
260°C	2.3	3.0	2.8	—
Heat distortion temperature (°C)	357	282	—	200
Limiting oxygen index (%)	35	42	—	47

processing temperatures. Fillers are usually added to reduce resin content and to improve physical properties. The preferred form of processing is compression molding.

The phenolics and urea resins are high-volume thermosets which owe their existence to the relatively low cost of the starting materials and their superior thermal and chemical resistance. Today these resins are widely used in molding applications, in surface coatings and adhesives, as laminating resins, casting resins, binders and impregnants, and in numerous other applications. However, as with all products based on formaldehyde, there is concern about the toxicity of these resins during processing and about the residual traces of formaldehyde in the finished product.

Phenol-Formaldehyde Resins

Monomers	Polymerization	Major uses
Phenol, formaldehyde	Base- or acid-catalyzed stepwise polycondensation	Plywood adhesives (34%), glass-fiber insulation (19%), molding compound (8%)

The reaction of phenol with formaldehyde involves a condensation reaction which, under appropriate conditions, leads to a cross-linked or network polymer structure. Since the cross-linked polymer is insoluble and infusible, it is necessary for commercial applications to produce first a tractable and fusible low-molecular-weight prepolymer which may, when desired, be transformed into the cross-linked polymer. The initial phenol-formaldehyde products (prepolymers) may be of two types: resols and novolacs.

Resols

Resols are produced by reacting a phenol with a molar excess of formaldehyde (commonly about 1:1.5 to 2) by using a basic catalyst (ammonia or sodium hydroxide). This procedure corresponds to Baekeland's original technique. Typically, reaction is carried out batchwise in a resin kettle equipped with stirrer and jacketed for heating and cooling. The resin kettle is also fitted with a condenser such that either reflux or distillation may take place as required. A mixture of phenol, formalin, and ammonia (1 to 3% on the weight of phenol) is heated under reflux at about 100°C for 0.25 to 1 hr, and then the water formed is removed by distillation, usually under reduced pressure to prevent heat hardening of the resin.

Two classes of resins are generally distinguished. Resols prepared with ammonia as catalysts are spirit-soluble resins having good electrical insulation properties. Water-soluble resols are prepared with caustic soda as catalyst. In aqueous solutions (with a solids content of about 70%) these are used mainly for mechanical grade paper and cloth laminates and in decorative laminates.

The reaction of phenol and formaldehyde in alkaline conditions results in the formation of o- and p-methylol phenols. These are more reactive towards formaldehyde than the original phenol and undergo rapid substitution with the formation of di- and trimethylol derivatives. The methylol phenols obtained are relatively stable in an alkaline medium but can undergo self-condensation to form dinuclear and polynuclear phenols (of low molecular weight) in which the phenolic nuclei are bridged by methylene groups. Thus in the base-catalyzed condensation of phenol and formaldehyde, there is a tendency for polynuclear phenols, as well as mono-, di-, and trimethylol phenols to be formed. Liquid resols have an average of less than two phenolic nuclei per molecule, and a solid resol may have only three or four. Because of the presence of methylol groups, the resol has some degree of water tolerance. However, for the same reason, the shelf life of resols is limited.

Resols are generally neutralized or made slightly acidic before cure (cross-linking) is carried out. Network polymers are then obtained simply by heating, which results in cross-linking via the uncondensed methylol groups or by more complex mechanisms (see Fig. 4.17). Above 160°C it is believed that quinone methide groups, as depicted on the bottom of Figure 4.17, are formed by condensation of the ether linkages with the phenolic hydroxyl groups. These quinone methide structures can be cross-linked by cycloaddition and can undergo other chemical reactions. It is likely that this formation of quinone methide and other related structures is responsible for the dark color of phenolic compression moldings made at higher temperatures. Note that cast phenol-formaldehyde

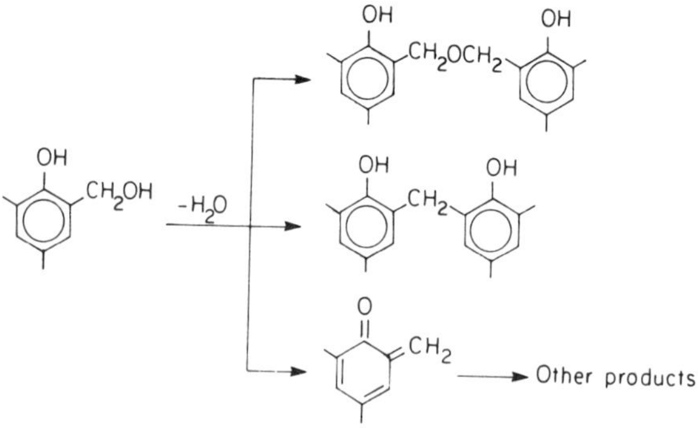

FIG. 4.17 Curing mechanisms for resols.

resins, which are cured at much lower temperatures, are water white in color. If they are heated to about 180°C, they darken considerably.

Novolac

The resols we have described are sometimes referred to as *one-stage resins*, since cross-linked products may be made from the initial reaction mixture only by adjusting the pH. The resol process is also known as the *one-stage* process. On the other hand, the novolacs are sometimes referred to as *two-stage* resins because, in this case, it is necessary to add, as we will show, some agent to enable formation of cross-linked products.

Novolac resins are normally prepared by the reaction of a molar excess of phenol with formaldehyde (commonly about 1.25:1) under acidic conditions. The reaction is commonly carried out batchwise in a resin kettle of the type used for resol manufacture. Typically, a mixture of phenol, formalin, and acid is heated under reflux at about 100°C. The acid is usually either hydrochloric acid (0.1 to 0.3% on the weight of phenol) or oxalic acid (0.5 to 2%). Under acidic conditions the formation of methylol phenols is rather slow, and the condensation reaction thus takes approximately 2 to 4 hr. When the resin reaches the requisite degree of condensation, it becomes hydrophobic, and the mixture appears turbid. Water is then distilled off until a cooled sample of the residual resin shows a melting point of 65 to 75°C. The resin is then discharged and cooled to give a hard, brittle solid (novolac).

Unlike resols, the distillation of water for novolac is normally carried out without using a vacuum. Therefore the temperature of the resin increases as the water is removed and the reaction proceeds, the temperature reaching as high as 160°C at the end. At these temperatures the resin is less viscous and more easily stirred.

The mechanism of phenol-formaldehyde reaction under acidic conditions is different from that under basic conditions described previously. In the presence of acid the products *o*- and *p*-methylol phenols, which are formed initially, react rapidly with free phenol to form dihydroxy diphenyl methanes (Fig. 4.18). The latter undergo slow reaction with formaldehyde and phenolic species, forming polynuclear phenols by further methylolation and methylol link formation. Reactions of this type continue until all the formaldehyde has been used up. The final product thus consists of a complex mixture of polynuclear phenols linked by *o*- and *p*-methylene groups. The average molecular weight of the final product (novolac) is governed by the initial molar ratio of phenol and formaldehyde. A typical value of average molecular weight is 600, which corresponds to about six phenolic nuclei per chain. The number of nuclei in individual chains is usually 2 to 13.

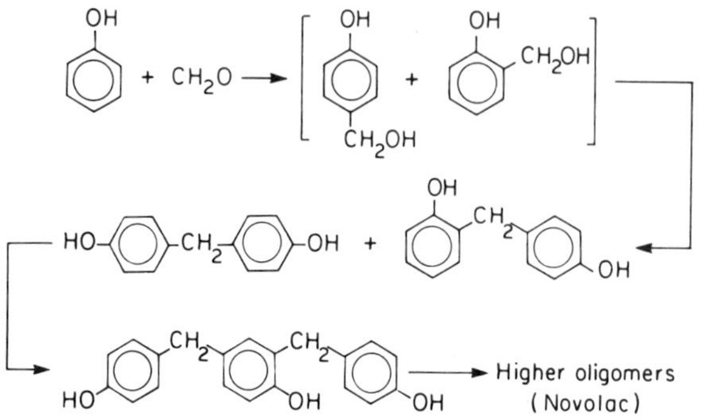

FIG. 4.18 Formation of novolac in an acid-catalyzed reaction of phenol and formaldehyde.

A significant feature of novolacs is that they represent completed reactions and as such have no ability to continue increasing in average molecular weight. Thus there is no danger of gelation (cross-linking) during novolac production. Resols, however, contain reactive methylol groups and so are capable of cross-linking on heating. To convert novolacs into network polymers, the addition of a cross-linking agent (hardener) is necessary. Hexamethylenetetramine (also known as hexa or hexamine) is invariably used as the hardener. The mechanism of the curing process is complex.

Because of the exothermic reaction on curing and the accompanying shrinkage, it is necessary to incorporate inert fillers to reduce resin content. Fillers also serve to reduce cost and may give additional benefits, such as improving the shock resistance. Commonly used fillers are wood flour, cotton flock, textile shreds, mica, and asbestos. Wood flour, a fine sawdust preferably from soft woods, is the most commonly used filler. Good adhesion occurs between the resin and the wood flour, and some chemical bonding may also occur. Wood flour reduces exotherm and shrinkage, improves the impact strength of the moldings, and is cheap. For better impact strength cotton fabric or chopped fabric may be incorporated. Asbestos may be used for improved heat and chemical resistance, and iron-free mica powder may be used for superior electrical insulation resistance characteristics.

Other ingredients which may be incorporated into a phenolic molding powder include *accelerators* (e.g., lime or magnesium oxide) to promote the curing reaction, *lubricants* (e.g., stearic acid and metal stearates) to prevent sticking to molds, *plasticizers* (e.g., naphthalene, furfural, and dibutyl phthalate) to improve flow properties during cure, and *pigments* or *dyes* (e.g., nigrosine) to color the product. Some typical formulations of phenolic molding powders are given in Table 4.18.

TABLE 4.18 Typical Formulations[a] of Phenolic Molding Grades

Ingredient	General-purpose grade	Medium shock-resisting grade	High-shock-resisting grade	Electrical grade
Novolac resin	100	100	100	100
Hexa	12.5	12.5	17	14
Magnesium oxide	3	2	2	2
Magnesium stearate	2	2	3.3	2
Nigrosine dye	4	3	3	3
Wood flour	100	—	—	—
Cotton flock	—	110	—	—
Textile shreds	—	—	150	—
Asbestos	—	—	—	40
Mica	—	—	—	120

[a]Parts by weight.

The bulk of phenol-formaldehyde molding compositions is traditionally processed on compression- and transfer-molding machines with a very small amount being extruded. More recently, the injection-molding process as modified for thermosetting plastics is being increasingly used, though it is still on a smaller scale than the traditional processes. Since the phenolic resins cure with evolution of volatiles, compression molding is performed using molding pressures of 1 to 2 ton/in.^{2} (15 to 30 MPa) at 155 to 170°C. Phenolic molding compositions may be preheated by high frequency or other methods. Preheating reduces cure time, shrinkage, and required molding pressures. It also enhances the ease of flow, with consequent reduction of mold wear and danger of damage to inserts. Molding shrinkage of general-purpose grades is about 0.005 to 0.08 in./in. Highly loaded mineral-filled grades exhibit lower shrinkage.

Properties and Applications

Since the polymer in phenolic moldings is highly cross-linked and interlocked, the moldings are hard, infusible, and insoluble. The chemical resistance of the moldings depends on the type of resin and filler used. General-purpose PF grades are readily attacked by aqueous sodium hydroxide, but cresol- and xylenol-based resins are more resistant. Phenolic moldings are resistant to acids except formic acid, 50% sulfuric acid, and oxidizing acids. The resins are ordinarily stable up to 200°C.

The mechanical properties of phenolic moldings are strongly dependent on the type of filler used (Table 4.19). Being polar, the electrical insulation properties of phenolics are not outstanding but are generally adequate. A disadvantage of phenolics as compared to aminoplasts and alkyds is their poor tracking resistance under high humidity, but this problem is not serious, as will be evident from the wide use of phenolics for electrical insulation applications.

Perhaps the most well-known applications of PF molding compositions are in domestic plugs and switches. However, in these applications PF has now been largely replaced by urea-formaldehyde plastics because of their better antitracking property and wider range of color possibility. (Because of the dark color of the phenolic resins molded above 160°C, the range of pigments available is limited to relatively darker colors—blacks, browns, deep blues, greens, reds, and oranges.) Nevertheless, phenolics continue to be used as insulators in many applications because their properties have proved quite adequate. There are also many applications of phenolics where high electrical insulation properties are not as important, and their heat resistance, adequate shock resistance, and low cost are important features: for example, knobs, handles, telephones, and instrument cases. In some of these applications phenolics have been replaced by urea-formaldehyde, melamine-formaldehyde, alkyd, or newer thermoplastics because of the need for brighter colors and tougher products.

TABLE 4.19 Properties of Phenolic Moldings[a]

Property	General-purpose grade	Medium shock-resisting grade	High shock-resisting grade	Electrical grade
Specific gravity	1.35	1.37	1.40	1.85
Shrinkage (cm/cm)	0.006	0.005	0.002	0.002
Tensile strength				
lbf/in.^{2}	8000	7000	6500	8500
MPa	55	48	45	58
Impact strength				
ft-lbf	0.16	0.29	0.8–1.4	0.14
J	0.22	0.39	1.08–1.9	0.18
Dielectric constant				
at 800 Hz	6.0–10.0	5.5–5.7	6.0–10.0	4.0–6.0
at 10^6 Hz	4.5–5.5	—	—	4.3–5.4

TABLE 4.19 Continued

Dielectric strength (20°C)				
V/mil	150–300	200–275	150–250	275–350
kV/cm	58–116	78–106	58–97	106–135
Power factor				
at 800 Hz	0.1–0.4	0.1–0.35	0.1–0.5	0.03–0.05
at 10^6 Hz	0.03–0.05	—	—	0.01–0.02
Volume resistivity (ohm-m)	10^{12}–10^{14}	10^{12}–10^{14}	10^{11}–10^{13}	10^{13}–10^{16}
Water absorption (24 hr, 23°C)				
mg	45–65	30–50	50–100	2–6

[a]Testing according to BS 2782

In general, phenolics have better heat and moisture resistance than urea-formaldehyde moldings. Heat-resistant phenolics are used in handles and knobs of cookware, welding tongs, electric iron parts, and in the automobile industry for fuse box covers, distributor heads, and other applications where good electrical insulation together with good heat resistance is required. Bottle caps and closures continue to be made in large quantities from phenolics. The development of machines for injection molding of thermosetting plastics and availability of fast-curing grades of phenolics have stimulated the use of PF for many small applications in spite of competition from other plastics.

Among the large range of laminated plastics available today, the phenolics were the first to achieve commercial significance, and they are still of considerable importance. In these applications one-stage resins (resols) are used, since they have sufficient methylol groups to enable curing without the need of a curing agent. Caustic soda is commonly used as the catalyst for the manufacture of resols for mechanical and decorative laminates. However, it is not used in electrical laminates because it adversely affects the electrical insulation properties. For electrical-grade resols ammonia is the usual catalyst, and the resins are usually dissolved in industrial methylated spirits. The use of cresylic acid (*m*-cresol content 50–55%) in place of phenol yields laminating resins of better electrical properties.

In the manufacture of laminates for electrical insulation, paper (which is the best dielectric) is normally used as the base reinforcement. Phenolic paper laminates are extensively used for high-voltage insulation applications.

Besides their good insulation properties, phenolic laminates also possess good strength, high rigidity, and machinability. Sheet, tubular, and molded laminates are employed. Phenolic laminates with cotton fabric reinforcement are used to manufacture gear wheels which are quiet running but must be used at lower working stresses than steel. Phenolic-cotton or phenolic-asbestos laminates have been used as bearings for steel rolling mills to sustain bearing loads as high as 3000 psi (21 MPa). Because of the advent of cheaper thermoplastics, cast phenolic resins (resols) are no longer an important class of plastics materials.

Urea-Formaldehyde Resins

Monomers	Polymerization	Major uses
Urea, formaldehyde	Stepwise polycondensation	Particle-board binder resin (60%), paper and textile treatment (10%), molding compounds (9%), coatings (7%)

Aminoresins or *aminoplastics* cover a range of resinous polymers produced by reaction of amines or amides with aldehydes. Two such polymers of current commercial importance in the field of plastics are the urea-formaldehyde and melamine-formaldehyde resins. Formaldehyde reacts with the amino groups to form aminomethylol derivatives which undergo further condensation to form resinous products. In contrast to phenolic resins, products derived from urea and melamine are colorless.

Urea and formaldehyde resins are usually prepared by a two-stage reaction. In the first stage, urea and formaldehyde (mole ratio in the range 1:1.3 to 1:1.5) are reacted under mildly alkaline (pH 8) conditions, leading to the production of monomethylol urea (Fig. 4.19 I) and dimethylol urea (Fig. 4.19 II). If the product of the first stage, which in practice usually also contains unreacted urea and formaldehyde, is subjected to acid conditions at elevated temperatures (stage 2), the solution increases in viscosity and sets to an insoluble and irreversible gel. The gel eventually converts with evolution of water and formaldehyde to a hard, colorless, transparent, insoluble, and infusible mass having a network molecular structure.

The precise mechanisms involved during the second stage are not fully understood. It does appear that in the initial period of the second stage methylol ureas condense with each other by reaction of a —CH_2OH group of one mol-

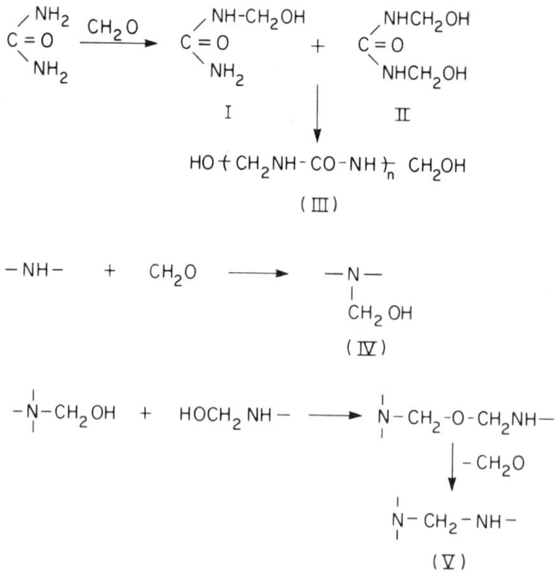

FIG. 4.19 Reactions in the formation of urea-formaldehyde resins.

ecule with an —NH$_2$ of another molecule, leading to linear polymers of the form shown in Figure 4.19 III. These polymers are relatively less soluble in aqueous media and tend to form amorphous white precipitates on cooling to room temperature. More soluble resins are formed on continuation of heating. This probably involves the formation of pendant methylol groups (Fig. 4.19 IV) by reactions of the —NH— groups with free formaldehyde. These methylol groups and the methylol groups on the chain ends of the initial reaction product can then react with each other to produce ether linkages, or with amine groups to give methylene linkages (Fig. 4.19 III). The ether linkages may also break down on heating to methylene linkages with the evolution of formaldehyde (Fig. 4.19V). An idealized network structure of the final cross-linked product is shown in Figure 1.14.

Molding Powder

The urea-formaldehyde (UF) molding powder will contain a number of ingredients. Most commonly these include resin, filler, pigment, accelerator, stabilizer, lubricant, and plasticizer.

Bleached wood pulp is employed as a filler for the widest range of bright colors and in slightly translucent moldings. Wood flour, which is much cheaper, may also be used.

A wide variety of pigments is now used in UF molding compositions. Their principal requirements are that they should be stable to processing conditions and be unaffected by service conditions of the molding.

To obtain a sufficiently high rate of cure at molding temperatures, it is usual to add about 0.2 to 2.0% of an accelerator (hardener)—a latent acid catalyst which decomposes at molding temperatures to yield an acidic body that will accelerate the rate of cure. Many such materials have been described, the most prominent of them being ammonium sulfamate, ammonium phenoxyacetate, trimethyl phosphate, and ethylene sulfite. A stabilizer such as hexamine is often incorporated into the molding powder to improve its shelf life.

Metal stearates, such as zinc, magnesium, or aluminum stearates are commonly used as lubricants at about 1% concentration. Plasticizers (e.g., monocresyl glycidyl ether) are used in special grades of molding powders. They enable more highly condensed resins to be used in the molding powder and thus reduce curing shrinkage while maintaining good flow properties.

In a typical manufacturing process, the freshly prepared UF first-stage reaction product is mixed with the filler (usually with a filler-resin dry weight ratio of 1:2) and other ingredients except pigment in a trough mixer at about 60°C for about 2 hr. Thorough impregnation of the filler with the resin solution and further condensation of the resin takes place during this process. Next, the wet mix is in a turbine or rotary drier for about 2 hr at 100°C or about 1 hr in a countercurrent of air at 120 to 130°C. The drying process reduces the water content from about 40% to about 6% and also causes further condensation of the resin.

After it is removed from the drier, the product is ground in a hammer mill and then in a ball mill for 6 to 9 hr. The pigments are added during the ball-milling process, which ensures a good dispersion of the pigment and gives a fine powder that will produce moldings of excellent finish. The powder, however, has a high bulk factor and needs densification to avoid problems of air and gas trappings during molding. There are several methods of densification. In one method, the heated powder is formed into strips by passing through the nip of a two-roll mill. The strips are then powdered into tiny flat flakes in a hammer mill. Other processes involve agglomeration of the powder by heating in an internal mixer at about 100°C or by treatment with water or steam and subsequent drying. More recently, continuous compounders, such as the Buss Ko-Kneader, have been used.

Processing

Urea-formaldehyde molding powders have a limited storage life. They should therefore be stored in a cool place and should be used, wherever possible, within a few months of manufacture. Conventional compression and transfer molding are commonly used for UF materials, the former being by far the most important process in terms of tonnage handled. Compression molding pressures usually range from 1 to 4 tons/in.^{2} (15 to 60 MPa), the higher pressures being used for deep-draw articles. Molding temperatures from 125 to 160°C are employed. The cure time necessary depends on the mold temperature and on the thickness of the molding. The cure time for a 1/8 in. thick molding is typically about 55 sec at 145°C. Bottle caps (less than 1/8 in. thick) and similar items, however, are molded industrially with much shorter cure times (~10 to 20 sec) at the higher end of the molding temperature range. For transfer molding of UF molding powders, pressures of 4 to 10 ton/in.^{2} (60 to 150 MPa), calculated on the area of the transfer pot, are generally recommended.

Special injection grades of UF molding powder have been developed for injection-molding applications which call for molding materials with good flow characteristics between 70 and 100°C and that are unaffected by long residence time in the barrel but capable of almost instant cure in the mold cavity at a higher temperature. Although the transition from compression molding to injection molding has been extensive for phenolics, the same cannot be said for UF materials, because they are more difficult to mold, possibly because the UF are more brittle than a phenolic resin and so are less able to withstand the stress peaks caused by filler orientation during molding. A combination of compression and injection processes has therefore been developed in which a screw preplasticizing unit delivers preheated and softened material directly to a compression-mold cavity.

Properties and Applications

The wide color range possible with UF molding powders has been an important reason for the widespread use of the material. These moldings have a number of other desirable features: low cost, good electrical insulation properties, and resistance to continuous heat up to a temperature of 70°C. Some typical values of physical properties of UF molding compositions are given in Table 4.20. They do not impart taste and odor to foodstuffs and beverages with which they come in contact and are resistant to detergents and dry-cleaning solvents.

The foregoing properties account for major uses of UF in two applications, namely, bottle caps and electrical fittings. It is also used for colored toilet seats, vacuum flasks, cups and jugs, hair drier housings, toys, knobs, meat trays, switches, lamp shades, and ceiling light bowls. In the latter applications it is important to ensure adequate ventilation to prevent overheating and consequent cracking of the molded articles.

However, only about 3% of UF resins are used for molding powders. The bulk (about 85%) of the resins are used as adhesives in the particle-board, plywood,

TABLE 4.20 Properties of Urea-Formaldehyde and Melamine-Formaldehyde Moldings[a]

Property	Urea-formaldehyde (α-cellulose filled)	Melamine-formaldehyde (cellulose filled)
Specific gravity	1.5–1.6	1.5–1.55
Tensile strength		
10^3 lbf/in.^{2}	7.5–11.5	8–12
MPa	52–80	55–83
Impact strength (ft-lbf)	0.20–0.35	0.15–0.24
Dielectric strength (90°C) V/0.001 in.	120–200	160–240
Volume resistivity (ohm-m)	10^{13}–10^{15}	10^9–10^{10}
Water absorption (mg)		
24 hr at 20°C	50–130	10–50
30 min at 100°C	180–460	40–110

[a]Testing according to BS 2782

and furniture industries. Resins for these applications are commonly available with UF molar ratios ranging from 1:1.4 to 1:2.2.

To prepare a suitable resin for adhesive applications, urea is dissolved in formalin (initially neutralized to pH 7.5) to give the desired UF molar ratio. After boiling under reflux for about 15 minutes to give dimethylol urea and other low- molecular products, the resin is acidified, conveniently with formic acid, to pH 4, and reacted for a further period of 5 to 20 min. The resulting water-soluble resin with approximately 50% solids content is stabilized by neutralizing to a pH 7.5 with alkali. For use as an aqueous solution, as is normally the case, the resin is then partially dehydrated by vacuum distillation to give a 70% solids content.

Phosphoric acid, or more commonly ammonium chloride, is used as a hardener for UF resin adhesives. Ammonium chloride reacts with formaldehyde to produce hexamine and hydrochloric acid, and the latter catalyzes the curing of the resin. In the manufacture of plywood a resin (with UF molar ratio typically 1:1.8) mixed with hardener is applied to wood veneers, which are then plied together and pressed at 95 to 110°C under a pressure of 200 to 800 psi (1.38 to 5.52 MPa). The UF resin-bonded plywood is suitable for indoor applications but is generally unsuitable for outdoor use. For outdoor applications phenol-formaldehyde, resorcinol-formaldehyde, or melamine-formaldehyde resins are more suitable.

Large quantities of UF resin are used in general wood assembly work. For joining pieces of wood the resin-hardener solution is usually applied to the surfaces to be joined and then clamped under pressure while hardening occurs. Alternatively, the resin may be applied to one surface and the hardener to the other, allowing them to come into contact in situ. This method serves to eliminate pot-life problems of the resin-hardener mixture. Gap-filling resins are produced by incorporating into UF resins plasticizers, such as furfuryl alcohol, and fillers to minimize shrinkage and consequent cracking and crazing.

In the manufacture of wood chipboard, which now represents one of the largest applications of UF resins, wood chips are mixed with about 10% of a resin-hardener solution and pressed in a multidaylight press at 150°C for about 8 min. Since some formaldehyde is released during the opening of the press, it is necessary to use a resin with a low formaldehyde content. Because it has no grain, a wood chipboard is nearly isotropic in its behavior and so does not warp or crack. However, the water resistance of chipboard is poor.

Melamine-Formaldehyde Resins

Monomers	Polymerization	Major uses
Melamine (trimerization of cyanamide), formaldehyde	Stepwise polycondensation	Dinnerware, table tops, coatings

Reaction of melamine (2,4,6-triamino-1,3,5-triazine) with neutralized formalin at about 80 to 100°C leads to the production of a mixture of water-soluble methylol melamines. The methylol content of the mixture depends on the initial ratio of formaldehyde to melamine and on the reaction conditions. Methylol melamines possessing up to six methylol groups per molecule are formed (Fig. 4.20).

On further heating, the methylol melamines undergo condensation reactions, and a point is reached where hydrophobic resin separates out. The rate of resinification depends on pH. The rate is minimum at about pH 10.0 to 10.5 and increases considerably both at lower and higher pH. The mechanism of resinification and cross-linking is similar to that observed for urea-formaldehyde (Fig. 4.19) and involves methylol-amine and methylol-methylol condensations.

$$\text{NH·CH}_2\text{OH} + \text{H}_2\text{N} \longrightarrow \text{NH·CH}_2\text{·NH} + \text{H}_2\text{O}$$

$$\text{NH·CH}_2\text{OH} + \text{HO·CH}_2\text{NH} \longrightarrow \text{NH·CH}_2\text{·O·CH}_2\text{·NH} + \text{H}_2\text{O}$$

$$\text{NH·CH}_2\text{·O·CH}_2\text{·NH} \longrightarrow \text{NH·CH}_2\text{·NH} + \text{CH}_2\text{O}$$

In industrial practice, resinification is carried out to a point close to the hydrophobe point. This liquid resin is either applied to the substrate or dried and converted into molding powder before proceeding with the final cure.

In a typical process a jacketed resin kettle fitted with stirrer and reflux condenser is charged with 240 parts of 40% w/v formalin (pH adjusted to 8.0 to 8.5 using a sodium carbonate solution) and 126 parts of melamine (to give a melamine-formaldehyde ratio of 1:3), and the temperature is raised to 85°C. The melamine forms methylol derivatives and goes into solution. This water-soluble A-stage resin may be used for treatment of paper, leather, and fabrics to impart crease resistance, stiffness, shrinkage control, water repellency, and fire retardance. It may be spray dried to give a more stable, water-soluble product.

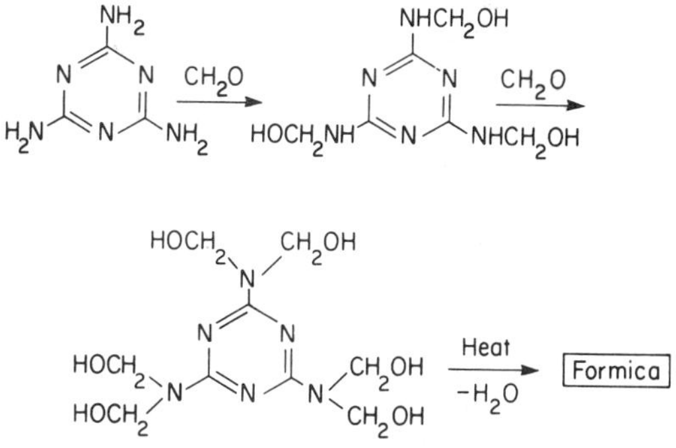

FIG. 4.20 Reactions in the synthesis of formica.

For laminating and other purposes the initial product is subjected to further condensation reactions at about 85°C with continuous stirring for more than 30 min. The hydrophilicity of the resin, as shown by its water tolerance, decreases with increasing condensation. The reaction is usually continued until a stage is reached when addition of 3 cm^3 of water will cause 1 cm^3 of resin to become turbid. The condensation reactions may be carried out at higher temperatures and lower pH values to achieve this stage more rapidly.

In aqueous solutions the hydrophobic resins have a shelf life of just a few days. The resin may be diluted with methylated spirit to about 50% solids content and pH adjusted to 9.0 to 9.5 to achieve greater stability. The addition of about 0.1% borax (calculated on the weight of the solids content) as an aqueous solution is useful in obtaining this pH and maintaining it for several months. The stabilized resin is stored preferably at 20 to 35°C, because too low a storage temperature will cause precipitation and too high a temperature will cause gelation.

Melamine-formaldehyde molding powders are generally prepared by methods similar to those used for UF molding powders. In a typical process an aqueous syrup of MF resin with melamine-formaldehyde ratio of 1:2 is compounded with fillers, pigments, lubricants, stabilizers, and accelerators in a dough-type mixer. The product is then dried and ball-milled by processes similar to those used for UF molding powders. Alpha-cellulose is used as a filler for the more common decorative molding powders. Industrial-grade MF materials use fillers such as asbestos, silica, and glass fiber. These fillers are incorporated by dry blending methods. The use of glass fiber gives moldings of higher mechanical strength, improved dimensional stability, and higher heat resistance than other fillers. The mineral-filled MF moldings have superior electrical insulation and heat resistance and may be used when phenolics and UF compositions are unsuitable.

MF moldings are superior to UF products in lower water absorption (see Table 4.20), greater resistance to staining by aqueous solutions such as fruit juices and beverages, better retention of electrical properties in damp conditions, better heat resistance, and greater hardness. Compared with the phenolic resins, MF resins have better color range, track resistance, and scratch resistance. MF resins are, however, more expensive than general-purpose UF and PF resins.

MF compositions are easily molded in conventional compression- and transfer-molding equipment. Molding temperatures from 145 to 165°C and molding pressures 2 to 4 tonf/in.^{2} (30 to 60 MPa) are usually employed. In transfer/molding pressures of 5 to 10 tonf/in.^{2} (75 to 150 MPa) are used. The cure time for an 1/8-in.-thick molding is typically 2 1/2 min at 150°C.

Largely because of their wide color range, surface hardness, and stain resistance, MF resins are used as molding compositions for a variety of mechanical parts or household goods and as laminating resins for tops for counters, cabinets, and tables. The mineral-filled molding powders are used in electrical application and knobs and handles for kitchen utensils. An interesting application of MF resins in compression molding involves decorative foils made by impregnating a printed or decorated grade of paper with resin and then drying. The foil may be applied to a compression molding shortly before the cure is complete, and the resin in the foil may be cured in that position to produce a bonding.

In a typical process of laminating paper layers to make materials useful in electrical applications as well as decorative laminates (best known as *formica*), kraft paper, about the weight used in shopping bags, is run through a solution of melamine-formaldehyde prepolymer. Drying out water or driving off the solvent leaves an impregnated sheet that can be handled easily, since the brittle polymer does not leave the surface sticky. As many as a dozen or more layers are piled up. For decorative purposes a printed rag or decorated cloth paper is put on top and covered with a translucent paper layer. The entire assembly is heated between smooth plates in a high-pressure press to carry out the thermosetting (curing) reaction that binds the sheets together into a strong, solvent-resistant, heat-resistant, and scratch-resistant surfacing material. The laminate, which is only about 1.5-mm thick, can be glued to a plywood base for use in tabletops, countertops, and the like.

POLYURETHANES

A *urethane* linkage (—NHCOO—) is formed by the reaction of an isocyanate (—NCO) and an alcohol: RNCO + R′OH→RNHCOOR′. By the same reaction, polyhydroxy materials will react with polyisocyanates to yield polyurethanes. The development of polyurethanes can be traced to the work of German chemists attempting to circumvent the Du Pont patents on nylon-6,6. O. Bayer and his team (1937) succeeded in producing fiber-forming polymer by reacting aliphatic diisocyanates and aliphatic diols (glycols). Subsequent work resulted in the production of other useful products by using polymeric hydroxyl-containing compounds such as polyesters to give rubbers, foams, adhesives, and coatings. Today polyurethanes are by far the most versatile group of polymers, because products ranging from soft thermoplastic elastomers to hard thermoset rigid foams are readily produced from liquid monomers.

The basic building blocks for polyurethanes are polyisocyanates and macroglycols, also called *polyols*. The commonly used polyisocyanates are tolylenediisocyanate (TDI), diphenylmethane diisocyanate or methylenediphenyl isocyanate (MDI), and polymeric methylenediphenyl isocyanate (PMDI) mixtures manufactured by phosgenating aromatic polyamines derived from the acid-catalyzed condensation of aniline and formaldehyde. MDI and PMDI are produced by the same reaction, and separation of MDI is achieved by distillation. The synthetic routes in the manufacture of commercial polyisocyanates are summarized in Figure 4.21. A number of specialty aliphatic polyisocyanates have been introduced recently in attempts to produce a light-stable polyurethane coating. Triisocyanate made by reacting hexamethylene diisocyanate with water (Fig. 4.21c) is reported to impart good light stability and weather resistance in polyurethane coatings and is probably the most widely used aliphatic polyisocyanate.

The macroglycols used in the manufacture of polyurethanes are either polyether or polyester based. Polyether diols are low-molecular-weight polymers prepared by ring-opening polymerization of olefin oxides (see also the section on polyethers), and commonly used polyester polyols are polyadipates. A polyol produced by ring-opening polymerization of caprolactone, initiated with low-molecular-weight glycols, is also used. The reactions are summarized in Figure 4.22.

Isocyanates are highly reactive materials and enter into a number of reactions with groups or molecules containing active hydrogen, such as water, amine, amide, and also urethane. Isocyanates are also toxic and care should be exercised in their use. Their main effect is on the respiratory system.

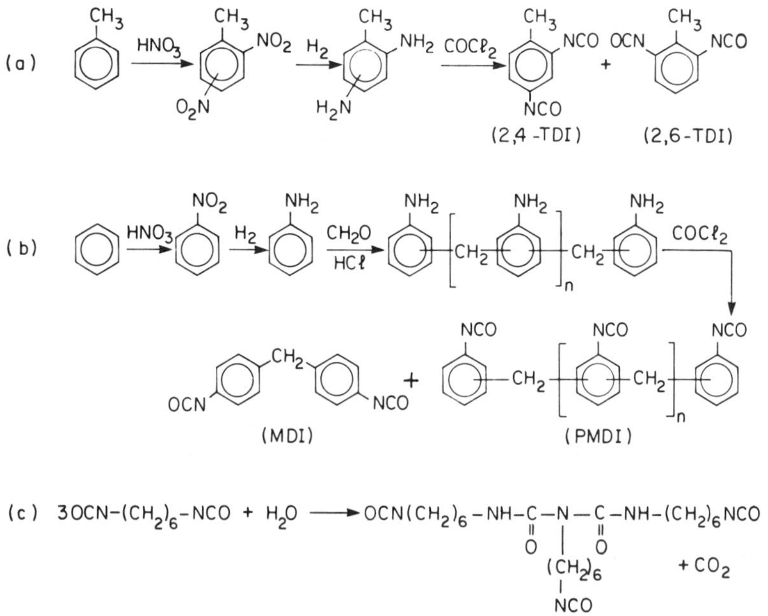

FIG. 4.21 Reactions used in the manufacture of commercial isocyanates. (a) TDI. (b) PMDI and MDI. (c) Aliphatic triisocyanate.

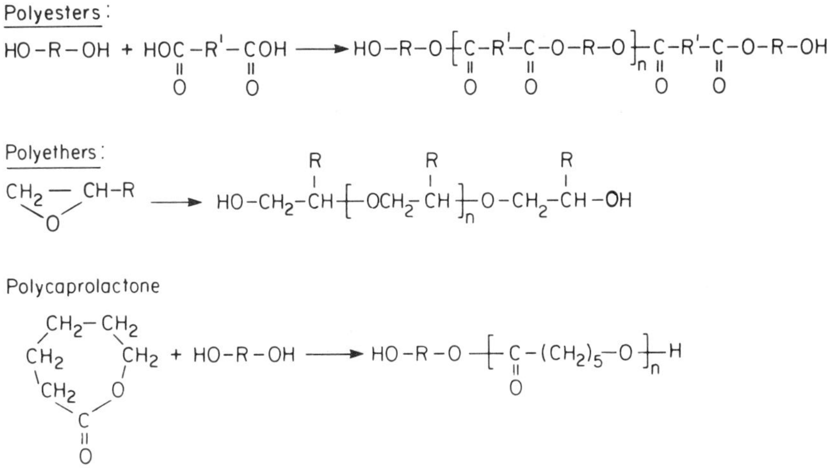

Polyesters:

$$HO-R-OH + HOC-R'-COH \longrightarrow HO-R-O\left[C-R'-C-O-R-O\right]_n C-R'-C-O-R-OH$$

with carbonyl groups shown as C=O below HOC, COH, and the repeating/terminal C groups.

Polyethers:

$$CH_2 \overset{\displaystyle R}{-} CH-R \longrightarrow HO-CH_2-\overset{\displaystyle R}{CH}\left[OCH_2 \overset{\displaystyle R}{CH}\right]_n O-CH_2-\overset{\displaystyle R}{CH}-OH$$

Polycaprolactone

$$\begin{array}{c} CH_2-CH_2 \\ CH_2 \quad\quad CH_2 \\ CH_2 \quad\quad O \\ C \\ \| \\ O \end{array} + HO-R-OH \longrightarrow HO-R-O\left[\overset{\|}{\underset{O}{C}}-(CH_2)_5-O\right]_n H$$

FIG. 4.22 Reactions used in the manufacture of macroglycols.

As previously mentioned, the initial research on polyurethanes was directed towards the preparation of fiber-forming polymers. Many diisocyanates and glycols were used, and the properties of the polymers were compared. From the consideration of properties and the commercial availability of reactants with the desired purity, 6,4-polyurethane produced by the reaction of hexamethylene diisocyanate and butane-1,4-diol was chosen for commercial production in Germany during World War II. The linear polyurethanes used to make fibers (known as Perlon U) can also be used as thermoplastics (marketed as Durethan U by Farbenfabriken Bayer) and may be processed by injection-molding and extrusion techniques.

The greatest difference in properties between 6,4-polyurethane and nylon-6,6, both for fibers and molding compositions, is in water absorption: the 6,4-polyurethane molding absorbs only about one-sixth the moisture of nylon-6,6 under comparable conditions. Thus it has better dimensional stability and a good retention of electrical insulation properties in conditions of high humidity. However, the thermoplastic polyurethanes are costlier than nylon-6,6 and nylon-6. For engineering applications requiring low water absorption, nylon-11, acetal resin, and, in certain instances, polycarbonates are cheaper and usually as good. Polyurethane crystalline fibers and the corresponding thermoplastics are thus no longer of importance. However, elastic polyurethane fibers, commonly known as *spandex* fibers, are significant. They are considered in a later section.

Major polyurethane products today include cellular materials such as water-blown flexible foams or fluorocarbon-blown rigid foams, elastomers, coatings, and elastic fibers, which are described subsequently. Closely related to polyure-thanes is an isocyanate-based product called *isocyanurate foam.*

Polyurethane Rubbers and Spandex Fibers

By careful formulations it is possible to produce polyurethane rubbers with a number of desirable properties. The rubbers can be thermoplastic (linear) or thermoset (slightly cross-linked) products.

Cross-Linked Polyurethane Rubbers

The starting point in the preparation of this type of rubber, typified by Vulkollan rubbers, is a polyester prepared by reacting a glycol such as ethylene or propy-lene glycol with adipic acid. The glycol is in excess so that the polyester formed has hydroxyl end groups. This polyester macroglycol is then reacted with an ex-cess of a diisocyanate such as 1,5-naphthalene diisocyanate or MDI (Fig. 4.21). The molar excess of diisocyanate is about 30%, so the number of polyesters joined together is only about 2 to 3, and the resulting prepolymer has isocyanate end groups (see Fig. 4.23a). The prepolymer can be chain extended with water, glycols, or amines which link up prepolymer chains by reacting with terminal isocyanate groups (see Fig. 4.23b). (The water reaction liberates carbon diox-ide, so it must be avoided in the production of elastomers, but it is important in the manufacture of foams.) The urea and urethane linkages formed in the chain extension reactions also provide sites for branching and cross-linking, since these groups can react with free isocyanate or terminal isocyanate groups to form *biuret* and *allophanate* linkages, respectively (see Fig. 4.23c). Biuret links, however, predominate since the urea group reacts faster than the urethane groups. The degree of cross-linking can to some extent be controlled by adjust-ing the amount of excess isocyanate, whereas more highly cross-linked struc-tures may be produced by the use of a triol in the initial polyester.

Vulkollan-type rubbers suffer from the disadvantage that the prepolymers are unstable and must be used within a day or two of their production. Moreover, these rubbers cannot be processed with conventional rubber machiners, so the products are usually made by a casting process. Attempts were then made to develop other polyurethane rubbers which could be processed by conventional techniques.

One approach was to react the diisocyanate with a slight excess of polyester so that the prepolymer produced has terminal hydroxyl groups. The prepolymers are rubberlike gums and can be compounded with other ingredients on two-roll mills. Final curing can be done by the addition of a diisocyanate or, preferably, a latent diisocyanate, i.e., a substance which produces an active diisocyanate under the conditions of molding. Polyurethane rubbers of this class are exem-plified by Chemigum SL (Goodyear), Desmophen A (Bayer), and Daltoflex 1 (ICI), which used polyester-amide for the manufacture of prepolymer.

(a) Prepolymer formation

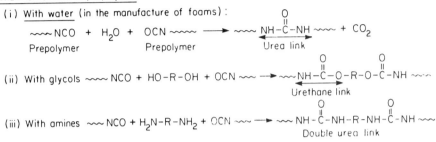

Di-isocyanate   Glycol

Urethane prepolymer

(b) Chain extension of prepolymer

  (i) With water (in the manufacture of foams):

     $\sim\sim\sim$NCO + H_2O + OCN$\sim\sim\sim$ $\longrightarrow$ $\sim\sim\sim$NH–C–NH$\sim\sim\sim$ + CO_2

     Prepolymer   Prepolymer   Urea link

  (ii) With glycols $\sim\sim\sim$NCO + HO–R–OH + OCN$\sim\sim$ $\longrightarrow$ $\sim\sim\sim$NH–C–O–R–O–C–NH$\sim\sim\sim$

                                               Urethane link

  (iii) With amines $\sim\sim\sim$NCO + H_2N–R–NH_2 + OCN$\sim\sim$ $\longrightarrow$ $\sim\sim\sim$NH–C–NH–R–NH–C–NH$\sim\sim\sim$

                                             Double urea link

(c) Cross-linking of chain-extended polyurethane

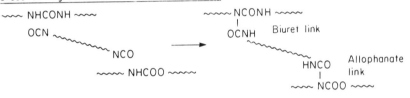

FIG. 4.23 Equations for preparation, chain extension, and curing of polyurethanes.

Another approach has been adopted by Du Pont with the product Adiprene C, a polyurethane rubber with unsaturated groups that allow vulcanization with sulfur.

Polyurethane rubbers, in general, and the Vulkollan-type rubbers, in particular, possess certain outstanding properties. They usually have higher tensile strengths than other rubbers and possess excellent tear and abrasion resistance. The urethane rubbers show excellent resistance to ozone and oxygen (in contrast to diene rubbers) and to aliphatic hydrocarbons. However, they swell in aromatic hydrocarbons and undergo hydrolytic decomposition with acids, alkalis, and prolonged action of water and steam.

Though urethane rubbers are more costly than most other rubbers, they are utilized in applications requiring superior toughness and resistance to tear, abrasion, ozone, fungus, aliphatic hydrocarbons, and dilute acids and bases. In addition, they excel in low-temperature impact and flexibility. Urethane rubbers have found increasing use for forklift tires, shoe soles and heels, oil seals, diaphragms, chute linings, and a variety of mechanical applications in which high elasticity is not an important prerequisite.

Emphasis on reaction injection-molding (RIM) technology in the automotive industry to produce automotive exterior parts has created a large potential for thermoset polyurethane elastomers. *Reaction injection molding*, originally known as reaction casting, is a rapid, one-step process to produce thermoset polyurethane products from liquid monomers. In this process liquid monomers are mixed under high pressure prior to injection into the mold. The polymerization occurs in the mold. Commercial RIM polyurethane products are produced from MDI, macroglycols, and glycol or diamine extenders. The products have the rigidity of plastics and the resiliency of rubber.

Recent advances include short, glass-fiber reinforced, high-modulus (flexural modulus greater than 300,000 psi, i.e., 2070 MPa) polyurethane elastomers produced by the reinforced RIM process. These reinforced high-modulus polyurethane elastomers are considered for automotive door panels, trunk lids, and fender applications.

Though originally developed for the automotive industry for the production of car bumpers, the RIM process has found its greatest success in the shoe industry, where semiflexible polyurethane foams have proved to be good soling materials.

Thermoplastic Polyurethane Rubbers

The reactions of polyols, diisocyanates, and glycols, as described, do tend to produce block copolymers in which hard blocks with glass transition temperatures well above normal ambient temperature are separated by soft rubbery blocks. These polymers thus resemble the SBS triblock elastomers and, more closely, the polyether-ester thermoplastic elastomers of the Hytrel-type described earlier.

In a typical process of manufacturing thermoplastic polyurethane elastomers, a prepolymer is first produced by reacting a polyol, such as a linear polyester with terminal hydroxyl groups, or a hydroxyl-terminated polyether, of molecular weights in the range 800 to 2500, with an excess of diisocyanate (usually of the MDI type) to give a mixture of isocyanate-terminated polyol prepolymer and free (unreacted) diisocyanate. This mixture is then reacted with a chain extender such as 1,4-butanediol to give a polymer with long polyurethane segments whose block length depends on the extent of excess isocyanate and the corresponding stoichiometric glycol. The overall reaction is shown in Figure 4.24a. Provided that R (in free diisocyanate) and R' (in glycol) are small and regular, the polyurethane segments will show high intersegment attraction (such as hydrogen bonding) and may be able to crystallize, thereby forming hard segments. In such polymers hard segments with T_g well above normal ambient temperature are separated by polyol soft segments, which in the mass are rubbery in nature. Hard and soft segments alternate along the polymer chain. This structure closely

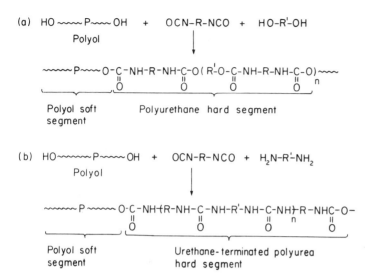

FIG. 4.24 Reactions for the manufacture of polyurethane block copolymers.

resembles that of polyester-polyether elastomers (Fig. 4.4). Similar reactions occur when an amine is used instead of a glycol as a chain extender (see Fig. 4.24b). The polymer in this case has polyurea hard segments separated by polyol soft segments.

The polymers produced by these reactions are mainly thermoplastic in nature. Though it is possible that an excess of isothiocyanate may react with urethane groups in the chain to produce allophanate cross-links (see Fig. 4.23c), these cross-links do not destroy the thermoplastic nature of the polymer because of their thermal lability, breaking down on heating and reforming on cooling. However, where amines have been used as chain extenders, urea groups are produced (Fig. 4.24b), which, on reaction with excess isocyanate, may give the more stable biuret cross-links (see Fig. 23c).

Many of the commercial materials designated as thermoplastic polyurethanes are in reality slightly cross-linked. This cross-linking may be increased permanently by a post-curing reaction after shaping. The polyurethane product Estane (Goodrich) may, however, be regarded as truly thermoplastic. The thermoplastic rubbers have properties similar to those of Vulkollan-type cast polyurethane rubbers, but they have higher values for compression set.

Thermoplastic polyurethane elastomers can be molded and extruded to produce flexible articles. Applications include wire insulation, hose, tracks for all-terrain vehicles, solid tires, roller skate wheels, seals, bushings, convoluted bellows, bearings, and small gears for high-load applications. In the automobile industry thermoplastic polyurethanes are used primarily for exterior parts. Their ability to be painted with flexible polyurethane-based paints without pretreatment is valuable.

Spandex Fibers

One particular form of thermoplastic polyurethane elastomers is the elastic fiber known as Spandex. The first commercial material of this type was introduced by Du Pont in 1958 (Lycra). It is a relatively high-priced elastomeric fiber made on the principle of segmented copolymers. Here again the soft block is a polyglycol, and the hard block is formed from MDI (4,4'-diisocyanatodiphenyl methane) and hydrazine. The reactions are shown in Figure 4.25. Note that this fiber-forming polymer contains urethane and semicarbazide linkages in the chain. The product is soluble in amide solvent, and the fiber is produced by dry spinning from a solution. Major end uses of Lycra are in apparel (swimsuits and foundation garments).

Subsequently several other similar materials have been introduced, including Dorlastan (Bayer), Spanzelle (Courtaulds), and Vyrene (U.S. Rubber).

The polyol component with terminal hydroxyl groups used in the production of the foregoing materials may be either a polyether glycol or a polyester glycol (see Fig. 4.22). For example, Du Pont uses polytetrahydrofuran (a polyether glycol) for Lycra. U.S. Rubber originally used a polyester of molecular weight of about 2000, obtained by condensation of adipic acid with a mixture of ethylene glycol and propylene glycol, and a polyether-based mixture was used for Vyrene 2, introduced in 1967. These polyols are reacted with an excess of diisocyanate to yield an isocyanate-terminated prepolymer which is then chain extended by an amine such as hydrazine (NH_2NH_2) or ethylenediamine (see Fig. 4.25). Fibers are usually spun from solution in dimethylformamide.

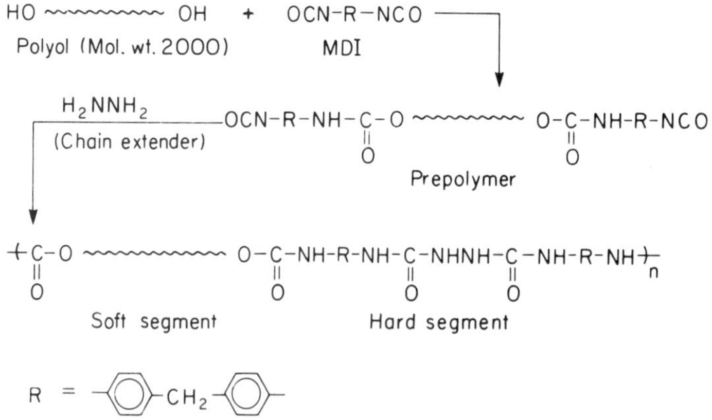

FIG. 4.25 Reactions for the manufacture of segmented elastomeric fiber (Lycra).

Possessing higher modulus, tensile strength, resistance to oxidation, and ability to be produced at finer deniers, spandex fibers have made severe inroads into the natural rubber latex thread market. Major end uses are in apparel. Staple fiber blends with nonelastic fibers have also been introduced.

Flexible Polyurethane Foam

Although polyurethane rubbers are specialty products, polyurethane foams are well known and widely used materials. About half of the weight of plastics in modern cars is accounted for by such foams.

Flexible urethane foam is made in low densities of 1 to 12 lb/ft^3 (0.016 to 0.019 g/cm^3), intermediate densities of 1.2 to 3 lb/ft^3 (0.019 to 0.048 g/cm^3), high resilience (HR) foams of 1.8 to 3 lb/ft^3 (0.029 to 0.048 g/cm^3), and semiflexible foams of 6 to 12 lb/ft^3 (0.096 to 0.192 g/cm^3). Filled foams of densities as high as 45 lb/ft^3 (0.72 g/cm^3) have also been made.

In many respects the chemistry of flexible urethane foam manufacture is similar to that of the Vulkollan-type rubbers except that gas evolution reactions are allowed to occur concurrently with chain extension and cross-linking (see Fig. 4.23). Most flexible foams are made from 80/20 TDI, which refers to the ratio of the isomeric 2,4-tolylenediisocyanate to 2,6-tolylenediisocyanate. Isocyanates for HR foams are about 80% 80/20 TDI and 20% PMDI, and those for semiflexible foams are usually 100% PMDI.

Polyols usually are polyether based, since these are cheaper, give greater ease of foam processing, and provide more hydrolysis resistance than polyesters. Polyether polyols can be diols, such as polypropylene glycols with molecular weight of about 2000, or triols with molecular weight of 3000 to 6000. The most common type of the latter is triol adduct of ethylene oxide and propylene oxide with glycerol with a molecular weight of 3000. For HR foams polyether triols with molecular weight of 4500 to 6000, made by reacting ethylene oxide with polypropylene oxide-based triols, are used.

The reaction of isocyanate and water that evolves carbon dioxide (see Fig. 4.23b) is utilized for foaming in the production of flexible foams. The density of the product, which depends on the amount of gas evolved, can be reduced by increasing the isocyanate content of the reaction mixture and by correspondingly increasing the amount of water to react with the excess isocyanate. For greater softness and lower density some fluorocarbons (F-11) may be added in addition to water.

The processes for producing flexible polyurethane foams have been described in Chapter 2.

Applications

The largest-volume use of flexible foam is furniture and bedding, which account for nearly 47% of the flexible polyurethane foam market. Almost all furniture cushioning is polyurethane. Automobile seating is either made from flexible slabstock or poured directly into frames, so-called deep seating, using HR chemical formulations. Semiflexible polyurethane foams find use in crash pads, arm and head rests, and door panels.

Much flexible foam is used in carpet underlay. About 43% is virgin foam, and 57% is scrap rebonded with urethane adhesives under heat and pressure.

Thermal interlining can be made by flame bonding thin-sliced, low-density (1 to 1.5 lb/ft^3) polyester-based foam to fabric. The flame melts about one-third of the foam thickness, and the molten surface adheres to the fabric.

Flexible foams also find use in packaging applications. Die-cut flexible foam is used to package costly goods such as delicate instruments, optical products, and pharmaceuticals. Semiflexible foam lining is used for cart interiors to protect auto and machine parts during transportation.

Rigid and Semirigid Polyurethane Foams

Most rigid foam is made from polymeric isocyanate (PMDI) and difunctional polyether polyols. PMDI of functionality 2.7 (average number of isocyanate groups per molecule) is used in insulation foam manufacture. Functionalities greater than 2 contribute rigidity through cross-linking. Higher-functionality, low-molecular-weight polyols are sometimes added because they contribute rigidity by cross-linking and short chain length. Such polyols are made by reaction of propylene oxide with sucrose, pentaerythritol, or sorbitol.

Rigid insulation foams are usually hydrocarbon blown to produce a closed-cell foam with excellent insulation properties. High-density foams are water blown where structural and screw-holding strength is needed, and halocarbons are used as blowing agents when high-quality, decorative surfaces are required. A compromise between these two aims can be achieved by the addition of water to halocarbon-blown formulations.

Formulation, processing methods, properties, and applications of rigid and semirigid polyurethane foams have been described in Chapter 2. Major uses of these foams are in building and construction (56%), transportation (12%), furniture, and packaging. Low-density, rigid foam is the most efficient thermal insulation commercially available and is extensively used in building construction. In transportation, urethane insulation is used in rail cars, containers, truck trailer bodies, and in ships for transporting liquefied natural gas. This requires thinner insulation and so yields more cargo space. Rigid foam is also used to give flotation in barge compartments.

High-density rigid foam, 5 to 15 lb/ft^3 (0.08 to 0.24 g/cm^3), is used for furniture items such as TV and stereo cabinets, chair shells, frames, arms, and legs, cabinet drawer fronts and doors, and mirror and picture frames. RIM-molded, integral-skin, high-density foams with core densities of 10 to 20 lb/ft^3

(0.16 to 0.32 g/cm^3) and skin densities of 55 to 65 lb/ft^3 (0.88 to 1.04 g/cm^3) are used in electronic, instrument, and computer cabinets.

Industrial uses of rigid foams include commercial refrigeration facilities as well as tanks and pipelines for cryogenic transport and storage.

Polyisocyanurates

Though closed-cell rigid polyurethane foams are excellent thermal insulators, they suffer from the drawback of unsatisfactory fire resistance even in the presence of phosphorus- and halogen-based fire retardants. In this context, polyisocyanurates, which are also based on isocyanates, have shown considerable promise. Isocyanurate has greater flame resistance than urethane. Although rigid polyurethane is specified for use temperatures up to 200°F (93°C), rigid polyisocyanurate foams, often called *trimer foams*, withstand use temperatures to 300°F (149°C). Physical properties and insulation efficiency are similar for both types.

The underlying reaction for polyisocyanurate formation is the trimerization of an isocyanate under the influence of specific catalysts (Fig. 4.26a). The most commonly used isocyanate is a polymeric isocyanate (PMDI) prepared by reacting phosgene with formaldehyde-aniline condensates, as shown in Figure 4.21b. PMDIs are less reactive than monomeric diisocyanate but are also less volatile. The polyisocyanurate produced from this material will be of the type shown in Figure 4.26b. Some of the catalysts used for the polytrimerization reactions are alkali-metal phenolates, alcoholates and carboxylates, and compounds containing *o*-(dimethylaminomethyl)-phenol groups.

To produce foams, fluorocarbons such as trichlorofluoromethanes are used as the sole blowing agents. Polyisocyanurate foams may be prepared by using standard polyurethane foaming equipment and a two-component system, with isocyanate and fluorocarbon forming one component and the activator or activator mixture forming the second component. Because of the high cross-link density of polyisocyanurates, the resultant foam tends to be brittle. Consequently, there has been a move toward making polyisocyanurate-polyurethane combinations. For example, the isocyanate trimerization reaction has been carried out with isocyanate end-capped TDI-based prepolymers to make isocyanurate-containing polyurethane foams. Isocyanate trimerization in the presence of polyols of molecular weights less than 300 has also been employed to produce foams by both one-shot and prepolymer methods.

Polyurethane Coatings

Polyurethane systems are also formulated for surface-coating applications. A wide range of such products has become available in recent years. These include simple solutions of finished polymer (linear polyurethanes), one-component

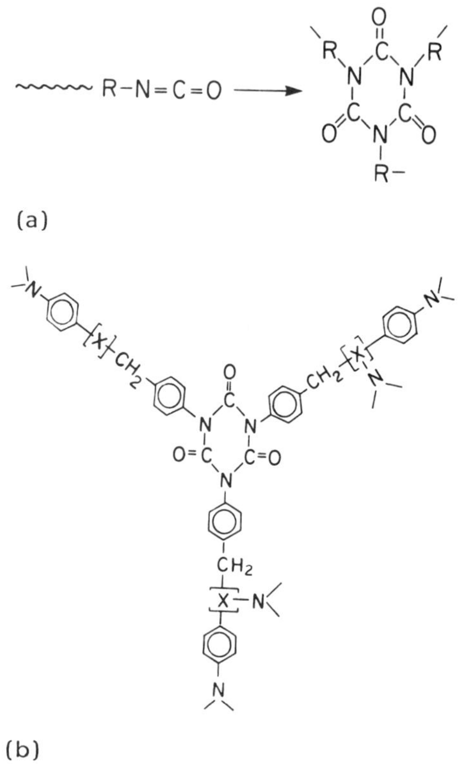

(a)

(b)

FIG. 4.26 (a) Trimerization of an isocyanate. (b) Structure of polyisocyanurate produced from polymeric MDIs.

systems containing blocked isocyanates, two-component systems based on polyester and isocyanate or polyether and isocyanate, and a variety of prepolymer and adduct systems.

Coatings based on TDI and MDI gradually discolor upon exposure to light and oxygen. In contrast, aliphatic diisocyanates such as methylene dicyclohexyl isocyanate, hexamethylene diisocyanate derivatives, and isophorone diisocyanate all produce yellowing resistant, clear, or color-stable pigmented coatings. Of the polyols used, half are polyester type and half are polyether type. The coatings can vary considerably in hardness and flexibility, depending on formulation.

Polyurethane coatings are used wherever applications require toughness, abrasion resistance, skin flexibility, fast curing, good adhesion, and chemical resistance. Uses include metal finishes in chemical plants, wood finishes for

TABLE 4.21 Blocked Isocyanates for One-Component System

Blocking agent	Unblocking temperature	
	°C	°F
Phenol	160	320
m-Nitrophenol	130	266
Acetone oxime	180	356
Diethyl malonate	130–140	266–284
Caprolactam	160	320
Hydrogen cyanide	120–130	248–266

boats and sports equipment, finishes for rubber goods, and rain-erosion resistant coatings for aircraft.

The one-component coating systems require blocking of the isocyanate groups to prevent polymerization in the container. Typical blocking agents are listed in Table 4.21.

Generation of the blocking agent upon heating to cause polymerization is a disadvantage of blocked one-component systems. This problem can be overcome by using a masked aliphatic diisocyanate system, as shown in Figure 4.27. The cyclic bisurea derivative used in such a system is stable in the polyol or in water emulsion formulated with the polyol at ordinary temperatures. Upon heating, ring opening occurs, generating the diisocyanate, which reacts instantaneously with the macroglycol to form a polyurethane coating.

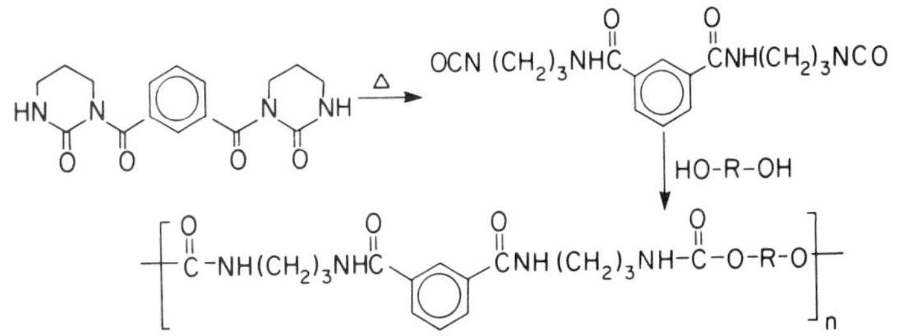

FIG. 4.27 Reactions of macrocyclic ureas used as masked diisocyanates.

Polyurethane adhesives involving both polyols and isocyanates are used. These materials have found major uses in the boot and shoe industry.

ETHER POLYMERS

For the purposes of this chapter ether polymers or polyethers are defined as polymers which contain recurring ether groupings

$$-\overset{|}{\underset{|}{C}}-O-\overset{|}{\underset{|}{C}}-$$

in their backbone structure. Polyethers are obtained from three different classes of monomers, namely, carbonyl compounds, cyclic ethers, and phenols. They are manufactured by a variety of polymerization processes, such as ionic polymerization (polyacetal), ring-opening polymerization (polyethylene oxide, polypropylene oxide, and epoxy resins), oxidative coupling (polyphenylene oxide), and polycondensation (polysulfone).

Polyacetal and polyphenylene oxide are widely used as engineering thermoplastics, and epoxy resins are used in adhesive and coating applications. The main uses of poly(ethylene oxide) and poly(propylene oxide) are as macroglycols in the production of polyurethanes. Polysulfone is one of the high-temperature-resistant engineering plastics.

Polyacetal

$$-\!\!\left[\!-OCH_2\!-\!\right]_n\!\!-$$

Monomer	Polymerization	Major uses
Formaldehyde, trioxane	Cationic or anionic chain polymerization	Appliances, plumbing and hardware, transportation

Polyoxymethylene (polyacetal) is the polymer of formaldehyde and is obtained by polymerization of aqueous formaldehyde or ring-opening polymerization of trioxane (cyclic trimer of formaldehyde, melting point 60 to 62°C), the latter being the preferred method. The polymerization of trioxane is conducted in bulk with cationic initiators. In contrast, highly purified formaldehyde is polymerized in solution using either cationic or anionic initiators.

Polyacetal strongly resembles polyethylene in structure, both polymers being linear with a flexible chain backbone. Since the structures of both the polymers are regular and the question of tacticity does not arise, both polymers are capable of a high degree of crystallization. However, the acetal polymer molecules have a shorter backbone ($-C-O-$) bond and so pack more closely together than polyethylene molecules. The acetal polymer is thus harder and has a higher melting point (175°C for the homopolymer).

Being crystalline and incapable of specific interaction with liquids, acetal homopolymer resins have outstanding resistance to organic solvents. No effective solvent has yet been found for temperatures below 70°C (126°F). Above this temperature, solution occurs in a few solvents such as chlorophenols. Swelling occurs with solvents of similar solubility parameter to that of the polymer [δ = 11.1 $(cal/cm^3)^{1/2}$ = 22.6 $MPa^{1/2}$]. The resistance of polyacetal to inorganic reagents is not outstanding, however. Strong acids, strong alkalis, and oxidizing agents cause a deterioration in mechanical properties.

The ceiling temperature for the acetal polymer is 127°C. Above this temperature the thermodynamics indicate that depolymerization will take place. Thus it is absolutely vital to stabilize the polyacetal resin sufficiently for melt processing at temperatures above 200°C. Stabilization is accomplished by capping the thermolabile hydroxyl end groups of the macromolecule by etherification or esterification, or by copolymerizing with small concentrations of ethylene oxide. These expedients retard the initiation or propagation steps of chain reactions that could cause the polymer to "unzip" to monomer (formaldehyde). End-group capping is more conveniently achieved by esterification using acetic anhydride.

$$HOCH_2 \underset{n}{\underbrace{-[-OCH_2-]}} O-CH_2OH \xrightarrow{Ac_2O}$$

$$\underset{O}{\overset{\|}{CH_3C}}-O-CH_2 \underset{n}{\underbrace{-[-OCH_2-]}} O-CH_2-O-\underset{O}{\overset{\|}{C}}CH_3$$

If formaldehyde is copolymerized with a second monomer, which is a cyclic ether such as ethylene oxide and 1,3-dioxolane, end-group capping is not necessary. The copolymerization results in occasional incorporation of molecules containing two successive methylene groups, whereby the tendency of the molecules to "unzip" is markedly reduced. This principle is made use of in the commercial products marketed as Celcon (Celanese), Hostaform (Farbwerke Hoechst), and Duracon (Polyplastic).

Degradation of polyacetals may also occur by oxidative attack at random along the chain leading to chain scission and subsequent depolymerization ("unzipping"). Oxidative chain scission is reduced by the use of antioxidants (see Chapter 1); in recent formulations hindered phenols seem to be preferred. For example, 2,2'-methylene-bis(4-methyl-6-*t*-butylphenol) is reported to be used in Celcon (Celanese) and 4,4'-butylidene bis(3-methyl-6-*t*-butylphenol) in Delrin (Du Pont). Acid-catalyzed cleavage of the acetal linkage can also cause initial chain scission. To reduce this acid acceptors are believed to be used in commercial practice. Epoxides, nitrogen-containing compounds, and basic salts are all quoted in the patent literature.

Polyacetal is obtained as a linear polymer about 80% crystalline with an average molecular weight of 30,000 to 50,000. Comparative values for some properties of typical commercial products are given in Table 4.22. The principal

TABLE 4.22 Typical Values for Some Properties of Acetal Homopolymers and Copolymers

Property	Acetal homopolymer	Acetal copolymer
Specific gravity	1.425	1.410
Crystalline melting point (°C)	175	163
Tensile strength (23°C)		
lbf/in.^{2}	10,000	8,500
MPA	70	58
Flexural modulus (23°C)		
lbf/in.^{2}	410,000	360,000
MPa	2,800	2,500
Deflection temperature (°C)		
at 264 lbf/in.^{2}	100	110
at 66 lbf/in.^{2}	170	158
Elongation at break (23°C) (%)	15–75	23–35
Impact strength (23°C)		
ft-lbf/in. notch	1.4–2.3	1.1
Hardness, Rockwell M	94	80
Coefficient of friction	0.1–0.3	0.2
Water absorption (%)		
24-hr immersion	0.4	0.22
50% RH equilibrium	0.2	0.16
Continuous immersion equilibrium	0.9	0.8

features of acetal polymers which render them useful as engineering thermoplastics are high stiffness, mechanical strength over a wide temperature range, high fatigue endurance, resistance to creep, and good appearance. Although similar to nylons in many respects, acetal polymers are superior to them in fatigue resistance, creep resistance, stiffness, and water resistance (24-hr water absorption at saturation 0.22% for acetal copolymer vs. 8.9% for nylon-6,6). The nylons (except under dry conditions) are superior to acetal polymers in impact toughness. Various tests indicate that the acetal polymers are superior to most other plastics and die cast aluminum.

The electrical properties of the acetal polymers may be described as good but not outstanding. They would thus be considered in applications where impact toughness and rigidity are required in addition to good electrical insulation characteristics.

The end-group capped acetal homopolymer and the trioxane-based copolymers are generally similar in properties. The copolymer has greater thermal stability, easier moldability, better hydrolytic stability at elevated temperatures, and much better alkali resistance than the homopolymer. The homopolymer, on the other hand, has slightly better mechanical properties, e.g., higher flexural modulus, higher tensile strength, and greater surface hardness.

Acetal polymers and copolymers are engineering materials and are competitive with a number of plastics materials, nylon in particular, and with metals. Acetal resins are being used to replace metals because of such desirable properties as low weight (sp. gr. 1.41 to 1.42), corrosion resistance, resistance to fatigue, low coefficient of friction, and ease of fabrication. The resins may be processed without difficulty on conventional injection-molding, blow-molding, and extrusion equipment. The acetal resins are used widely in the molding of telephone components, radios, small appliances, links in conveyor belts, molded sprockets and chains, pump housings, pump impellers, carburetor bodies, blower wheels, cams, fan blades, check valves, and plumbing components such as valve stems and shower heads.

Because of their light weight, low coefficient of friction, absence of slipstick behavior, and ability to be molded into intricate shapes in one piece, acetal resins find use as bearings. The lowest coefficient of friction and wear of acetal resin are obtained against steel. With other metals, aluminum in particular, high friction and greater wear are known to occur. The use of acetal to acetal in bearings is not desirable because of the tendency to build up heat except with very light loads. Where a nonmetallic material has to be used, acetal and nylon in conjunction are found to give better results compared to either on its own.

Though counted as one of the engineering plastics, acetal resins with their comparatively high cost cannot, however, be considered as general-purpose thermoplastics in line with polyethylene, polypropylene, PVC, and polystyrene.

Poly(Ethylene Oxide)

$$+OCH_2CH_2+$$

Monomers	Polymerization	Major Uses
Ethylene oxide	Ring-opening polymerization	Molecular weights 200 to 600—surfactants, humectants, lubricants
		Molecular weights >600—pharmaceutical and cosmetic bases, lubricants, mold release agents
		Molecular weights 10^5 to 5×10^6—water-soluble packaging films and capsules

Poly(ethylene oxide) of low molecular weight, i.e., below about 3000, are generally prepared by passing ethylene oxide into ethylene glycol at 120 to 150°C (248 to 302°F) and about 3 atm pressure (304 kPa) by using an alkaline catalyst such as sodium hydroxide. Polymerization takes place by an anionic mechanism.

$$CH_2-CH_2 \xrightarrow{\text{NaOH}} HO+CH_2-CH_2-O+_n CH_2CH_2O^-Na^+$$
$$\diagdown O \diagup$$

$$HO+CH_2-CH_2-O+_n CH_2CH_2O^-Na^+ + H_2O \rightleftharpoons$$

$$HO+CH_2-CH_2-O+_n CH_2CH_2OH + NaOH$$

The polymers produced by these methods are thus terminated mainly by hydroxyl groups and are often referred to as *polyethylene glycols* (PEGs). Depending on the chain length, PEGs range in physical form at room temperature from water white viscous liquids (mol. wt. 200 to 700), through waxy semisolids (mol. wt. 1000 to 2000), to hard, waxlike solids (mol. wt. 3000 to 20,000 and above). All are completely soluble in water, bland, nonirritating, and very low in toxicity; they possess good stability, good lubricity, and wide compatibility.

Since PEGs form a homologous series of polymers, many of their properties vary in a continuous manner with molecular weight. The freezing temperature, which is less than −10°C for PEG of molecular weight 300, rises first rapidly with molecular weight through the low-molecular-weight grades, then increases more gradually through the solids while approaching 66°C, the true crystalline melting point for very high-molecular-weight poly(ethylene oxide) resins (see p. 513). Other examples of such continuous variations are the increase in viscosity, flash points, and fire points with an increase in molecular weight and a slower increase in specific gravity. In a reverse relationship, hygroscopicity decreases as molecular weight increases, as does solubility in water.

In 1958, commercial poly(ethylene oxide)s of very high molecular weight became available from Union Carbide (trademark Polyox), which is now the largest of the worldwide suppliers of these resins. Since then, two Japanese companies, Meisei Chemical Works, Ltd. and Seitetsu Kagaku Company, Ltd., have begun producing poly(ethylene oxide) under the trademarks Alkox and PEO, respectively. Details of techniques used to manufacture have not been disclosed, but the essential feature is the use of heterogeneous catalyst systems, which are mainly of two types, namely, alkaline earth compounds (e.g., oxides and carbonates of calcium, barium, and strontium) and organometallic compounds (e.g., aluminum and zinc alkyls and alkoxides, usually with cocatalysts).

Commercial poly(ethylene oxide) resins supplied in the molecular weight range 1×10^5 to 5×10^6 are dry, free-flowing, white powders soluble in an unusually broad range of solvents. The resins are soluble in water at temperatures up to 98°C and also in a number of organic solvents, which include chlorinated hydrocarbons such as carbon tetrachloride and methylene chloride, aromatic hydrocarbons such as benzene and toluene, ketones such as acetone and methyl ethyl ketone, and alcohols such as methanol and isopropanol.

Applications

Polyethylene Glycol

The unusual combination of properties of PEGs has enabled them to find a very wide range of commercial uses as cosmetic creams, lotions, and dressings; textile sizes; paper-coating lubricants; pharmaceutical salves, ointments, and suppositories; softeners and modifiers; metal-working lubricants; detergent modifiers; and wood impregnants. In addition, the chemical derivatives of PEGs, such as the mono- and diesters of fatty acids, are widely used as emulsifiers and lubricants.

The PEGs themselves show little surface activity, but when converted to mono- and diesters by reaction with fatty acids, they form a series of widely useful nonionic surfactants. The required balance of hydrophilic-hydrophobic character can be achieved by suitable combination of the molecular weight of the PEG and the nature of the fatty acid. For large-volume items a second production route of direct addition of ethylene oxide to the fatty acids is often preferred. End uses for the fatty acid esters are largely as textile lubricants and softeners and as emulsifiers in food products, cosmetics, and pharmaceuticals.

The PEGs have found a variety of uses in pharmaceutical products. Their water solubility, blandness, good solvent action for many medicaments, pleasant and nongreasy feel on the skin, and tolerance of body fluids are the reasons why they are frequently the products of choice. Blends of liquid and solid grades are often selected because of their desirable petrolatumlike consistency. An especially important example of pharmaceutical application of PEGs is as bases for suppositories, where the various molecular grades can be blended to provide any desired melting point, degree of stability, and rate of release of medication. The fatty acid esters of the PEGs are often used in pharmaceuticals as emulsifiers and suspending agents because of their nonionic nature, blandness, and desirable surface activity. The solid PEGs also find use as lubricants and binders in the manufacture of medicinal tablets.

The PEGs, providing they contain not over 0.2% ethylene and diethylene glycols, are permitted as food additives. The PEG fatty acid esters are especially useful emulsifying agents in food products.

For many of the same reasons which account for their use in pharmaceuticals, the PEGs and their fatty acid esters find many applications in cosmetics and toiletries. Their moisturizing, softening, and skin-smoothing characteristics are especially useful. Typical examples of applications are shaving creams, vanishing creams, toothpastes, powders, shampoos, hair rinses, suntan lotions, pomades and dressings, deodorants, stick perfumes, rouge, mascara, and so on. The nonionic surface-active PEG fatty acid esters find use in a variety of detergents and cleaning compositions.

Liquid PEGs, solid PEGs, or their solutions are often used in a variety of ink preparations, such as thixotropic inks for ballpoint pens, water-based stencil inks, steam-set printing inks, and stamp-pad inks.

There are numerous other industrial uses for the PEGs in which they serve primarily as processing aids and do not remain as integral components of the products. The PEGs add green strength and good formability to various ceramic compositions to be stamped, extruded, or molded. They can be burned out cleanly during subsequent firing operations. In electroplating baths small amounts of PEGs improve smoothness and grain uniformity of the deposited coatings. Solid PEGs are effective lubricants in paper-coating compositions and promote better gloss and smoothness in calendering operations. PEGs, and more particularly their fatty acid esters, are quite widely used as emulsifiers, lubricants, and softeners in textile processing. PEGs and their esters find use as components of metal corrosion inhibitors in oil wells where corrosive brines are present. Fatty acid esters of PEGs are useful demulsifiers for crude oil-water separation.

The water solubility, nonvolatility, blandness, and good lubricating abilities are some of the reasons for the use of PEGs in metalworking operations. Metalworking lubricants for all but the most severe forming operations are made with the PEGs or with their esters or other derivatives.

Poly(Ethylene Oxide)

Since their commercialization in 1958, the reported and established applications for high-molecular-weight poly(ethylene oxide) resins have been numerous and diversified. Table 4.23 summarizes in alphabetical order the main applications of these resins in various industries. In addition to their water solubility and blandness, the main functions and effects of poly(ethylene oxide) resins which lead to these diverse applications are lubrication, flocculation, thickening, adhesion, hydrodynamic drag reduction, and formation of association complexes. The resins are relatively nontoxic and have a very low level of biodegradability (low BOD). Extensive testing has indicated that poly(ethylene oxide) resins with molecular weights from 1×10^5 to 1×10^7 have a very low level of oral toxicity and are not readily absorbed from the gastrointestinal tract. The resins are relatively nonirritating to the skin and have a low sensitizing potential.

Poly(ethylene oxide) resins can be formed into various shapes by using conventional thermoplastic processing techniques. Commercially, thermoplastic processing of these resins has been, however, limited almost exclusively to the manufacture of film and sheeting. Generally, the medium-molecular weight resins (4×10^5 to 6×10^5) possess melt rheology best suited to thermoplastic processing. The films are produced by calendering or blown-film extrusion techniques. Usually produced in thicknesses from 1 to 3 mils, the films have very good mechanical properties combined with complete water solubility.

Poly(ethylene oxide) films have been used to produce seed tapes, which consist of seeds sandwiched between two narrow strips of film sealed at the edges. When the seed tape is planted, water from the soil dissolves the water-soluble film within a day or two, releasing the seed for germination. Because the seeds are properly spaced along the tape, the process virtually eliminates the need for thinning of crops.

Films for poly(ethylene oxide) are also used to manufacture water-soluble packages for preweighed quantities of fertilizers, pesticides, insecticides, detergents, dyestuffs, and the like. The packages dissolve quickly in water, releasing the contents. They eliminate the need for weighing and offer protection to the user from toxic or hazardous substances.

Poly(ethylene oxide) forms a water-insoluble association complex with poly(acrylic acid). This is the basis of microencapsulation of nonaqueous printing inks. Dry, free-flowing powders obtained by this process can be used to produce "carbonless" carbon papers. When pressure is applied to the paper coated with the microencapsulated ink, the capsule wall ruptures and the ink is released.

TABLE 4.23　Applications of Poly(Ethylene Oxide) Resins

Industry	Applications
Agriculture	Water-soluble seed tapes
	Water-soluble packages for agricultural chemicals
	Hydrogels as soil amendments to increase water retention
	Soil stabilization using association complexes with poly(acrylic acid)
	Drift control agent for sprays
Ceramics and glass	Binders for ceramics
	Size for staple glass-fiber yarns
Chemical	Dispersant and stabilizer in aqueous suspension polymerization
Electrical	Water-soluble, fugitive binder for microporous battery and fuel-cell electrodes
Metals and mining	Flocculant for removal of silicas and clays in hydrometallurgical processes
	Flocculant for clarification of effluent streams from coal-washing plants
Paper	Filler retention and drainage aid in the manufacture of paper
	Flocculant for clarification of effluent water
Personal-care products	Lubricant and toothpaste
	Thickener in preparation of shaving stick
	Opthalmic solution for wetting, cleaning, cushioning, and lubricating contact lenses
	Adhesion and cushioning ingredient in denture fixatives
	Hydrogels as adsorptive pads for catamenial devices and disposable diapers

TABLE 4.23 Continued

Industry	Applications
Petroleum	Thickener for bentonite drilling muds
	Thickener for secondary oil-recovery fluids in waterflooding process
Pharmaceutical	Water-soluble coating for tablets
	Suspending agent to inhibit settling of ceramic lotion
Printing	Microencapsulation of inks
Soap and detergent	Emollient and thickener for detergent bars and liquids
Textile	Additive to improve dyeability and anti-static properties of polyolefin, polyester, and polyamide fibers

Various water-immiscible oils and solids can be microencapsulated in the poly(ethylene oxide)-polycarboxylic acid association complex by taking advantage of the pH dependence of the complex formation. The material to be encapsulated is emulsified with a suitable nonionic surfactant in an aqueous solution of poly(ethylene oxide). The emulsion is acidified to form the association complex at the interface, producing a coating around the water-immiscible phase and precipitation of the microcapsules, which are then dried to produce free-flowing powders.

The formation of a water-insoluble association complex of poly(ethylene oxide) and poly(acrylic acid) is also the basis for a soil stabilization process to prevent erosion of soil on hillsides and river banks.

A variety of different types of adhesives can be produced by forming association complexes of poly(ethylene oxide) with tannin or phenolic resins. Examples include wood glue, water-soluble quickset adhesive, and pressure-sensitive adhesives.

High-molecular-weight poly(ethylene oxide) resins are effective flocculants for many types of clays, coal suspensions, and colloidal silica, and so find application as process aids in mining and hydrometallurgy.

The turbulent flow of water through pipes and hoses or over surfaces causes the effect known as *hydrodynamic drag*. High-molecular-weight poly(ethylene oxide) resins are most effective in reducing the hydrodynamic drag and thus find use in fire fighting, where small concentrations of these resins (50 to 100 ppm) reduce the pressure loss in fire hoses and make it possible to deliver as much as 60% more water through a standard 2.5-in. (6.35-cm)-diameter fire hose. The Union Carbide product UCAR Rapid Water Additive, which contains high-molecular-weight poly(ethylene oxide) as the active ingredient, is a currently available hydrodynamic-drag-reducing additive for this application.

The ability of poly(ethylene oxide) to reduce hydrodynamic drag has also led to its use in fluid-jet systems used for cutting soft goods, such as textiles, rubber, foam, cardboard, etc. In these systems specially designed nozzles produce a very-small-diameter water jet at a pressure of 30,000 to 60,000 psi (200 to 400 MPa). Although a plain water disperses significantly as it leaves the nozzle, with poly(ethylene oxide) addition the stream becomes more cohesive and maintains its very small diameter up to 4 in. (10 cm) from the nozzle.

Chemical or irradiation cross-linking of poly(ethylene oxide) resins yields *hydrogels*, which are not water soluble but water absorptive, capable of absorbing 25 to 100 times their own weight of water. These hydrogels are reportedly useful in the manufacture of absorptive pads for catamenial devices and disposable diapers. The water absorbed by the hydrogels is also readily desorbed by drying the hydrogel. This characteristic is the basis for the use of these hydrogels as so-called *soil amendments*. When mixed with ordinary soil in a concentration of about 0.001 to 5.0 wt % of the soil, these hydrogels will reduce the rate of moisture loss due to evaporation but will still release water to the plants and thus eliminate the need for frequent watering of the soil.

Poly(Propylene Oxide)

$$-\left[OCH_2-\underset{\underset{CH_3}{|}}{CH} \right]_n-$$

Monomer	Polymerization	Major uses
Propylene oxide	Base-catalyzed ring-opening polymerization	Polyols for polyurethane foams, surfactants, lubricants, cosmetic bases

Propylene oxide is polymerized by methods similar to those described in the preceding section for poly(ethylene oxide). Like the latter, low- and high-molecular-weight polymers are of commercial interest.

Poly(propylene oxide)s of low molecular weight (i.e., from 500 to 3500), often referred to as *polypropylene glycols* (PPGs), are important commercial materials mainly because of their extensive use in the production of polyurethane foams (see Chapters 2 and 4). PPGs are less hydrophilic and lower in cost and may be prepared by polymerizing propylene oxide in the presence of propylene glycol as an initiator and sodium hydroxide as a catalyst at about 160°C. The polymers have the general structure

$$HO-\overset{\overset{\displaystyle CH_3}{|}}{CH}-CH_2-O-\left[CH_2-\overset{\overset{\displaystyle CH_3}{|}}{CH}-O\right]_n-CH_2-\overset{\overset{\displaystyle CH_3}{|}}{CH}-OH$$

The end hydroxy groups of the polymer are secondary groups and are ordinarily rather unreactive in the urethane reaction. Initially, this limitation was overcome by the preparation of isocyanate-terminated prepolymer and by the use of block copolymers with ethylene oxide. The latter products are known as *tipped polyols* and are terminated with primary hydroxy groups of enhanced activity.

$$HO-[CH_2-CH_2-O-]_x\left[CH_2-\overset{\overset{\displaystyle CH_3}{|}}{CH}-O\right]_y-[CH_2-CH_2-O-]_z CH_2-CH_2-OH$$

(Note that straight PEG is not satisfactory for polyurethane foam production due to its water sensitivity and tendency to crystallize.)

The latter advent of more powerful catalysts, however, made it possible for straight PPG to be used in the preparation of flexible polyurethane foams (see one-shot processes in Chapter 2) without recourse to the foregoing procedures. Also, today the bulk of the polyethers used are triols rather than diols, since these lead to slightly cross-linked flexible foams with improved load-bearing characteristics. The polyether triols are produced by polymerizing propylene oxide by using 1,1,1-trimethylolpropane, 1,2,6-hexane triol, or glycerol as the initiator. The use of, for example, trimethylolpropane leads to the following polyether triol.

$$CH_3-CH_2-C\begin{cases}CH_2-O-[CH_2-CH(CH_3)-O-]_x CH_2-CH(CH_3)-OH\\ CH_2-O-[CH_2-CH(CH_3)-O-]_y CH_2-CH(CH_3)-OH\\ CH_2-O-[CH_2-CH(CH_3)-O-]_z CH_2-CH(CH_3)-OH\end{cases}$$

For flexible polyurethane foams, polyether triols of molecular weights from 3000 to 3500 are normally used because they give the best balance of properties. For the production of rigid foams, polyether triols of lower molecular weight (about 500) are used to increase the degree of cross-linking. Alternatively, polyether polyols of higher functionality, such as produced by polymerizing propylene oxide with pentaerythritol or sorbitol, may be used.

Copolymerization of ethylene oxide and propylene oxide yields quite valuable functional fluids of various sorts. The random copolymers of ethylene and propylene oxides of relatively low molecular weights are water soluble when the proportion of ethylene oxide is at least 40 to 50% by weight. The block copolymers consist of sequences or "blocks" of all-oxypropylene or all-oxyethylene groups, as shown.

$$CH_2-CH_2 \longrightarrow HO\text{-}[CH_2-CH_2-O]_x H \xrightarrow{\quad CH_2-CH-CH_3 \quad}$$

$$HO\text{-}[CH_2-CH-O]_y\text{-}[CH_2-CH_2-O]_x\text{-}[CH_2-CH-O]_z H$$
$$\qquad\quad CH_3 \qquad\qquad\qquad\qquad\qquad CH_3$$

Properties vary considerably, depending on the lengths and arrangements of these blocks. The block copolymers comprise unique and valuable surface-active agents. They can act as breakers for water-in-oil emulsions, as defoamers, and as wetting and dispersing agents.

Epoxy Resins

$$CH_2-CH-CH_2\text{-}[O-C_6H_4-\underset{CH_3}{\overset{CH_3}{C}}-C_6H_4-O-CH_2-\underset{OH}{CH}-CH_2]_n\text{-}O-C_6H_4-\underset{CH_3}{\overset{CH_3}{C}}-C_6H_4-O-CH_2-CH-CH_2$$

Monomer	Polymerization	Major uses
Bisphenol A, epichlorohydrin	Condensation and ring-opening polymerization	Surface coating (44%), laminates and composites (18%), moldings (9%), flooring (6%), adhesives (5%)

Epoxide or epoxy resins contain the epoxide group, also called the epoxy, oxirane, or ethoxyline group, which is a three-membered oxide ring

$$-C\text{---}C-$$

(The simplest compound in which the epoxy group is found is ethylene oxide.) In the uncured stage epoxies are polymers with a low degree of polymerization. They are most often used as thermosetting resins which cross-link to form a three-dimensional nonmelting matrix. A curing agent (hardener) is generally used to achieve the cross-linking. In room-temperature curing the hardener is generally an amine such as diethylene triamine or triethylenetetramine. For elevated temperature curing a number of different curing agents could be utilized, including aromatic amines and acid anhydrides.

Epoxy resins were first synthesized by P. Schlack of I. G. Farben in 1934. The important curing reactions with polyfunctional amines and anhydrides were discovered by P. Castan in Switzerland in 1943. At the Swiss Industries Fair in 1946, CIBA A. G. of Basle first demonstrated the use of an epoxy resin adhesive, Araldite Type I, to bond light alloys. Parallel with this European activity, the paint company Devoe and Raynolds in the U.S.A. had been working with the Shell Chemical Corporation to develop epoxy resins suitable for the surface-coating industry. This led to a long series of patents (first patent to Greenlee 1943) which covered methods of preparing epoxy resins of higher molecular weight and important ways of modifying or combining them with other materials to form surface coatings.

Much has happened since the early days of epoxy resin technology in the late 1940s and early 1950s, when CIBA and Shell first began to manufacture and market these new and versatile products. Many more resin manufacturers have begun to produce epoxies. Many newer types of epoxy resins and curing agents have been developed, and an extensive literature covering all aspects of their science and technology has been built up.

Epoxy resins first developed commercially and still completely dominating the worldwide markets are those based on 2,2-bis(4'-hydroxyphenyl) propane, more commonly known as bisphenol A, and 1-chloro-2,3-epoxy-propane, also known as epichlorohydrin. It can be seen from the general formula of these resins that the molecular species concerned is a linear polyether with terminal glycidyl ether group

$$-O-CH_2-\overset{\displaystyle O}{\overset{\diagup\ \diagdown}{CH}-CH_2}$$

and secondary hydroxyl groups occurring at regular intervals along the length of the macromolecule. The number of repeating units (n) depends essentially on the molar ratio of bisphenol A and epichlorohydrin. When $n = 0$, the product is diglycidyl ether and the molecular weight is 340. When $n = 10$, the molecular weight is about 3000. Commercial liquid epoxy resins based on bisphenol A and epichlorohydrin have average molecular weights from 340 to 400, and it is therefore obvious that these materials are composed largely of diglycidyl ether. The liquid resin is thus often referred to as DGEBA (diclycidyl ether of bisphenol A).

DGEBA may be reacted with additional quantities of bisphenol A in an advancement reaction. This advancement produces higher-molecular-weight solid resins possessing a higher melting point (>90°C). Advancement generally increases flexibility, improves salt fog corrosion resistance, and increases hydroxyl content, which can be utilized later for cross-linking. Possessing generally low functionality (number of epoxy groups), their major use is in coatings. They provide outstanding adhesion and good salt fog corrosion resistance.

Commercial solid epoxy resins seldom have average molecular weights exceeding 4000, which corresponds to an average value of n of about 13. Resins with molecular weights above 4000 are of limited use since their high viscosity and low solubility make subsequent processing difficult.

Resin Preparation

The molecular weights of epoxy resins depend on the molar ratio of epichlorohydrin and bisphenol A used in their preparation (see Table 4.24). In a typical process for the production of liquid epoxy resins, epichlorohydrin and bisphenol A in the molar ratio of 10:1 are added to a stainless steel kettle fitted with a powerful anchor stirrer. The water content of the mixture is reduced to below 2% by heating the mixture until the epichlorohydrin–water distils off. After condensation, the epichlorohydrin layer is returned to the kettle, the water being discarded. When the necessary water content of the reaction mixture is reached, the reaction is started by the slow addition of sodium hydroxide (2 mole per mole of bisphenol A) in the form of 40% aqueous solution, the temperature being maintained at about 100°C. The water content of the reaction mixture should

TABLE 4.24 Effect of Reactant Ratio on Molecular Weight of Epoxy Resins

Molar ratio epichlorohydrin/ bisphenol A	Molecular weight	Epoxide equivalent[a]	Softening point (°C)
10:1	370	192	9
2:1	451	314	43
1.4:1	791	592	84
1.33:1	802	730	90
1.25:1	1133	862	100
1.2:1	1420	1176	112

[a]This is the weight of resin (in grams) containing one epoxide equivalent. For a pure diglycidyl ether (mol. wt. 340) with two epoxy groups per molecule, epoxide equivalent = 340/2 = 170.

be maintained between 0.3 and 2% by weight throughout the reaction if high yields (90 to 95%) are to be obtained. (No reaction occurs under anhydrous conditions, and undesired by-products are formed if the water content is greater than 2%.) Besides distilling off the water as an azeotrope with the epichlorohydrin, the water content can also be partly controlled by the rate of addition of the caustic soda solution.

When all the alkali has been added, which may take 2 to 3 hr, the excess epichlorohydrin is recovered by distillation at reduced pressure. A solvent such as toluene or methyl isobutyl ketone is then added to the cooled reaction product to dissolve the resin and leave the salt formed in the reaction.

The resin solution is washed with hot water, and after filtration and further washing, the solvent is removed by distillation and the resin is dried by heating under vacuum.

Though the pure diglycidyl ether of bisphenol A is a solid (m.p. 43°C), the commercial grades of the resin which contain a proportion of high-molecular-weight materials are supercooled liquids with viscosities of about 100 to 140 poise at room temperature.

The high-molecular-weight epoxy resins usually manufactured have values of n in the general formula ranging from 2 to 12. These resins are synthesized by allowing epichlorohydrin and bisphenol A to interact in the presence of excess sodium hydroxide. In the *taffy process* usually employed, a mixture of bisphenol A and epichlorohydrin (the molar ratio of the reactants used depends on the resin molecular weight required; see Table 4.24) is heated to 100°C and aqueous sodium hydroxide (NaOH–epichlorohydrin molar ratio 1.3:1) is added slowly with vigorous stirring, the reaction being completed in 1 to 2 hr. A white puttylike taffy (which is an emulsion of about 30% water in resin and also contains salt and sodium hydroxide) rises to the top of the reaction mixture. The lower layer of brine is removed; the resinous layer is coagulated and washed with hot water until free from alkali and salt. The resin is then dried by heating and stirring at 150°C under reduced pressure, poured into cooling pans, and subsequently crushed and bagged.

To avoid the difficulty of washing highly viscous materials, higher-molecular-weight epoxy resins may also be prepared by a two-stage process (advancement reaction, described earlier). This process consists of fusing a resin of lower molecular weight with more bisphenol A at about 190°C. The residual base content of the resin is usually sufficient to catalyze the second-stage reaction between the phenolic hydroxyl group and the epoxy group. Typical data of some commercial glycidyl ether resins are given in Table 4.25.

Solid resins have been prepared with narrow molecular-weight distributions. These resins melt sharply to give low-viscosity liquids, which enables incorporation of larger amounts of fillers with a consequent reduction in cost and coefficient of expansion. These resins are therefore useful in casting operations.

TABLE 4.25 Typical Properties of Some Commercial Glycidyl Ether Resins

Resin	Average molecular weight	Epoxide equivalent	Melting point (°C)
A	350–400	175–210	—
B	450	225–290	—
C	700	300–375	40–50
D	950	450–525	64–76
E	1,400	870–1025	95–105
F	2,900	1,650–2,050	125–132
G	3,800	2,400–4,000	145–155

Curing

To convert the epoxy resins into cross-linked structures, it is necessary to add a curing agent. Most of the curing agents in common use can be classified into three groups: *tertiary amines, polyfunctional amines*, and *acid anhydrides*.

Examples of tertiary amines used as curing agents for epoxy resins include

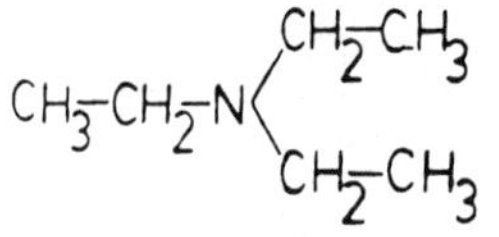

Triethylamine (TEA)

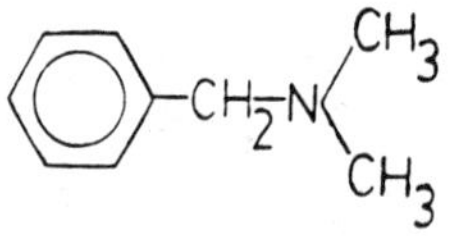

Bezyldimethylamine (BDA)

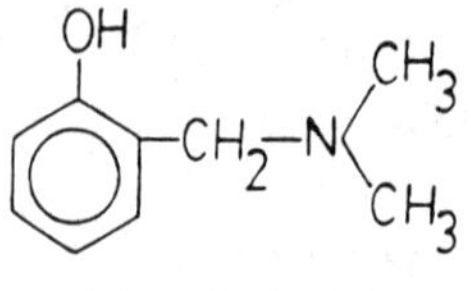

Dimethylaminomethylphenol (DMAMP)

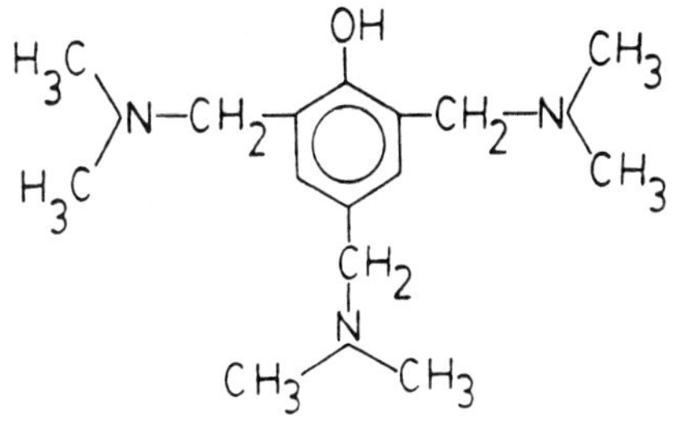

Tri(dimethylaminomethyl) phenol (TDMAMP)

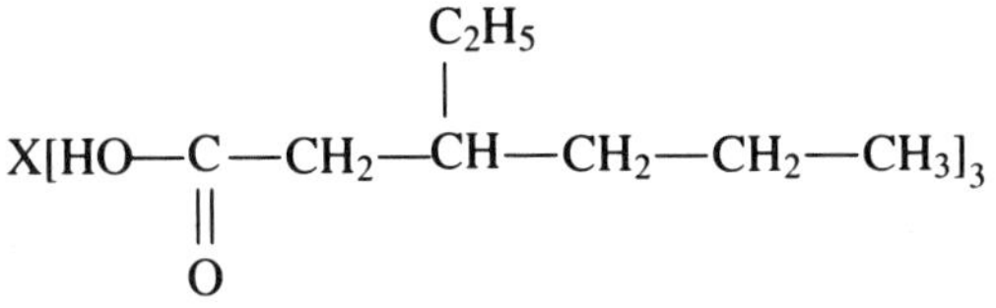

$$X[HO-\underset{\underset{O}{\|}}{C}-CH_2-\underset{\underset{C_2H_5}{|}}{CH}-CH_2-CH_2-CH_3]_3$$

Tri-2-ethylhexoate salt of TDMAMP

where X = TDMAMP

Tertiary amines are commonly referred to as catalytic curing agents since they induce the direct linkage of epoxy groups to one another. The reaction mechanism is believed to be as follows:

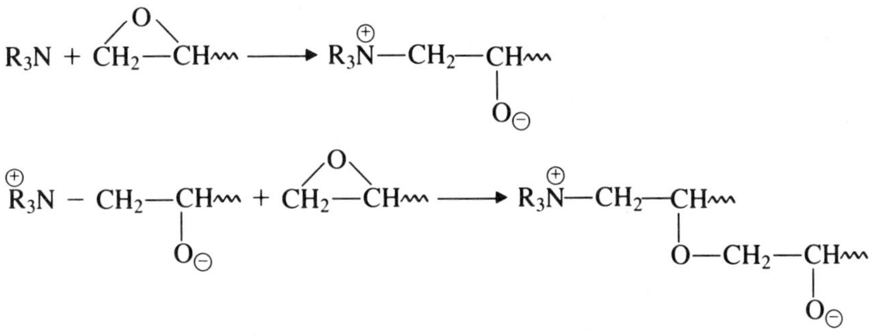

Since the reaction may occur at both ends of the diglycidyl ether molecule, a cross-linked structure will be built up. The overall reaction may, however, be more complicated because the epoxy group, particularly when catalyzed, also reacts with hydroxyl groups. Such groups may be present in the higher-molecular-weight homologues of the diglycidyl ether of bisphenol A, or they may be formed as epoxy rings are opened during cure.

In contrast to tertiary amine hardeners, which, as shown, cross-link epoxide resins by a catalytic mechanism, polyfunctional primary and secondary amines act as reactive hardeners and cross-link epoxy resins by bridging across epoxy molecules.

An amine molecule with two active hydrogen atoms can link across two epoxy molecules, as shown:

$$\text{wCH—CH}_2 + H \quad \overset{R}{\underset{}{N}} \quad H + CH_2\text{—CHw} \longrightarrow \text{wCH—CH}_2\text{—N—CH}_2\text{—CHw}$$

With polyfunctional aliphatic and aromatic amines having three or more active hydrogen atoms in amine groups, this type of reaction results in a network polymer (see Fig. 1.15). Generally speaking, aliphatic amines provide fast cures and are effective at room temperature, whereas aromatic amines are somewhat less reactive but give products with higher heat-distortion temperatures. Polyfunctional amines are widely used in adhesive, casting, and laminating applications.

Diethylenetriamine (DETA), $H_2N\text{—}CH_2\text{—}CH_2\text{—}NH\text{—}CH_2\text{—}CH_2\text{—}$ NH_2, and triethylenetetramine (TETA), $H_2N\text{—}CH_2\text{—}CH_2\text{—}NH\text{—}CH_2\text{—}$ $CH_2\text{—}NH\text{—}CH_2\text{—}CH_2\text{—}NH_2$, are highly reactive primary aliphatic amines with five and six active hydrogen atoms available for cross-linking, respectively. Both materials will cure (harden) liquid epoxy resins at room temperature and produce highly exothermic reactions. For example, with DETA the exothermic temperature may reach as high as 250°C in 200-g batches. With this amine used in the stoichiometric quantity of 9 to 10 pts phr (parts per hundred parts resin), the room temperature pot life is less than an hour. The actual time, however, depends on the ambient temperature and the size of the batch. With TETA, 12 to 13 pts phr are required.

Dimethylaminopropylamine (DMAPA), $(CH_3)_2N\text{—}CH_2\text{—}CH_2\text{—}CH_2\text{—}$ NH_2, and diethylaminopropylamine (DEAPA), $(C_2H_5)_2N\text{—}CH_2\text{—}CH_2\text{—}$ $CH_2\text{—}NH_2$, are slightly less reactive and give a pot life of about 140 min (for a 500-g batch) and are sometimes preferred. Note that both DMAPA and DEAPA have less than three hydrogen atoms necessary for cross-linking by reaction with epoxy groups. Other examples of amine curing agents with less than three hydrogen atoms are diethanolamine (DEA), $NH(CH_2CH_2OH)_2$, and piperidine (see later). These curing agents operate by means of two-part reaction. Firstly, the active hydrogen atoms of the primary and secondary amine groups are utilized in the manner already described. Thereafter, the resulting tertiary amines, being sufficiently reactive to initiate polymerization of epoxy groups, function as catalytic curing agents, as described previously.

As a class, the amines usually suffer from the disadvantage that they are pungent, toxic, and skin sensitizers. To reduce toxicity, the polyfunctional amines are often used in the form of *adducts*. A number of such modified amines have been introduced commercially. For example, reaction of the amine with a mono-

or polyfunctional glycidyl material will give a higher-molecular-weight product with less volatility.

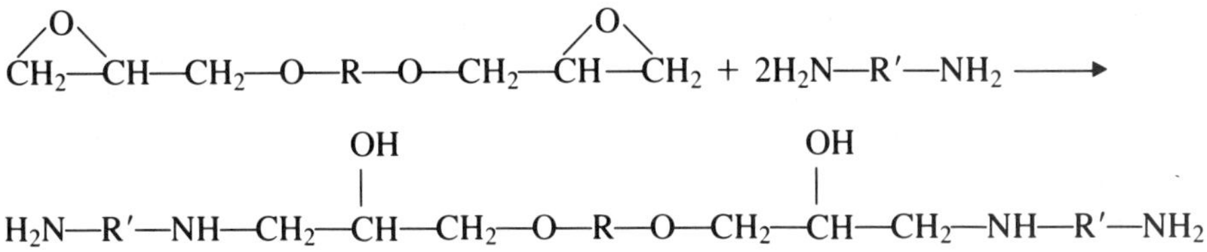

$$CH_2\!-\!CH\!-\!CH_2\!-\!O\!-\!R\!-\!O\!-\!CH_2\!-\!CH\!-\!CH_2 + 2H_2N\!-\!R'\!-\!NH_2 \longrightarrow$$

$$H_2N\!-\!R'\!-\!NH\!-\!CH_2\!-\!\underset{\underset{OH}{|}}{CH}\!-\!CH_2\!-\!O\!-\!R\!-\!O\!-\!CH_2\!-\!\underset{\underset{OH}{|}}{CH}\!-\!CH_2\!-\!NH\!-\!R'\!-\!NH_2$$

(I)

An advantage of such modified amines is that because of higher molecular weight larger quantities are required for curing, and this helps to reduce errors in metering the hardener. These hardeners are also extremely active, and the pot life for a 500-g batch may be as little as 10 min.

The glycidyl adducts are, however, skin irritants, being akin to the parent amines in this respect. Substitution of the hydroxyethyl group and its alkyl and aryl derivatives at the nitrogen atom is effective in reducing the skin sensitization effects of primary aliphatic amines. Both ethylene and propylene oxides have thus been used in the preparation of adducts from a variety of amines, including ethylene diamine and diethylenetriamine.

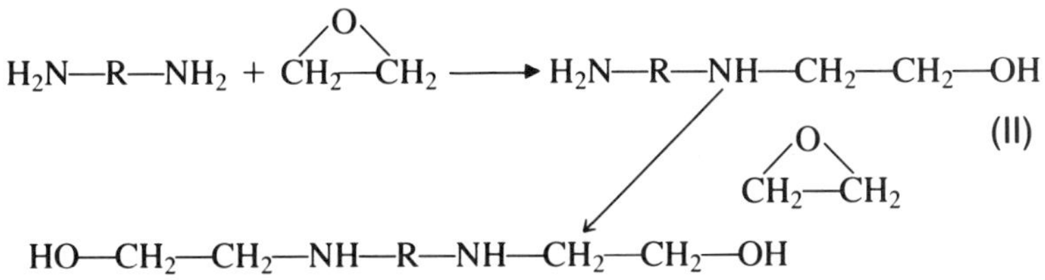

$$H_2N\!-\!R\!-\!NH_2 + CH_2\!-\!CH_2 \longrightarrow H_2N\!-\!R\!-\!NH\!-\!CH_2\!-\!CH_2\!-\!OH$$

(II)

$$HO\!-\!CH_2\!-\!CH_2\!-\!NH\!-\!R\!-\!NH\!-\!CH_2\!-\!CH_2\!-\!OH$$

(III)

Such adducts from diethylenetriamine appears free of skin sensitizing effects. A hardener consisting of a blend of the reaction products II and III is a low-viscosity liquid giving a pot life (for a 500-g batch) of 16 to 18 min at room temperature.

Modification of primary amines with acrylonitrile results in hardeners with reduced activity.

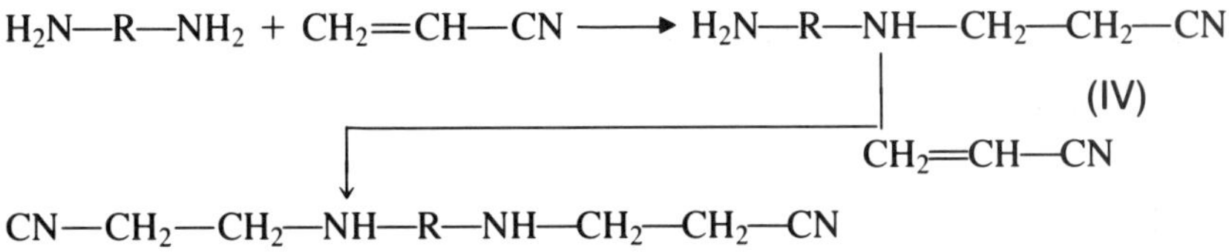

$$H_2N\!-\!R\!-\!NH_2 + CH_2\!\!=\!\!CH\!-\!CN \longrightarrow H_2N\!-\!R\!-\!NH\!-\!CH_2\!-\!CH_2\!-\!CN$$

(IV)

$$CH_2\!\!=\!\!CH\!-\!CN$$

$$CN\!-\!CH_2\!-\!CH_2\!-\!NH\!-\!R\!-\!NH\!-\!CH_2\!-\!CH_2\!-\!CN$$

(V)

Commercial hardeners are mixtures of the addition compounds IV and V. Since accelerating hydroxy groups are not present in IV and V (in contrast to I, II, and III), these hardeners have reduced activity and so give longer pot lives.

A number of aromatic amines also function as epoxy hardeners. Since they introduce the rigid benzene ring structure into the cross-linked network, the resulting products have significantly higher heat-distortion temperatures than are obtainable with the aliphatic amines (see Table 4.26). For example, metaphenylenediamine (MPDA)

gives cured resins with a heat distortion temperature of 150°C and very good chemical resistance. The hardener finds use in the manufacture of chemical resistance laminates.

Diaminodiphenylmethane (DADPM)

and diaminodiphenyl sulfone (DADPS)

used in conjunction with an accelerator, provide even higher heat-distortion temperatures but at some expense to chemical resistance.

Piperidine, a cyclic aliphatic amine,

$$\begin{array}{ccc} H_2C\!\!-\!\!CH_2 & & \\ \diagup & \diagdown & \\ H_2C & & NH \\ \diagdown & \diagup & \\ H_2C\!\!-\!\!CH_2 & & \end{array}$$

TABLE 4.26 Some Characteristics of Amine Hardeners for Use in Low-Molecular-Weight Bisphenol A-Based Epoxy Resins

Hardener[a]	Parts used per 100 parts resin	Pot life (500-g batch)	Typical cure schedule	Max HDT of cured resin (°C)	Applications
DETA	10–11	20 min	room temp	110	General purpose
DEAPA	7	140 min	room temp	97	General purpose
DETA-glycidyl adduct	25	10 min	room temp	75	Adhesives laminating, fast cure
DETA-ethylene oxide adduct	20	16 min	room temp	92	—
MPDA	14–15	6 hr	4–6 hr at 150°C	150	Laminates, chemical resistance
DADPM	28.5	—	4–6 hr at 165°C	160	Laminates
DADPS	30	—	8 hr at 160°C	175	Laminates
Piperidine	5–7	8 hr	3 hr at 100°C	75	General purpose
Triethylamine	10	7 hr	room temp	—	Adhesives
BDA	15	75 min	room temp	—	Adhesives
TDMAMP	6	30 min	room temp	64	Adhesives, coatings
2-Ethyl hexoate salt of TDMAMP	10–14	3–6 hr	—	—	Encapsulation

[a]For chemical formulas see text.

has been in use as an epoxy hardener since the early patents of Castan. Although a skin irritant, this hardener is still used for casting larger masses than are possible with primary aliphatic amines. Note that because it has only one active hydrogen atom in the amine group, piperidine operates by a two-part reaction in curing. Typical amine hardeners and their characteristics are summarized in Table 4.26.

Cyclic acid anhydrides are widely employed as curing agents for epoxy resins. Compared with amines they are less skin sensitive and generally give lower exotherms in curing reaction. Some acid curing agents provide cured resins with very high heat-distortion temperatures and with generally good mechanical, electrical, and chemical properties. The acid-cured resins, however, show less alkali resistance than amine-cured resins because of the susceptibility of ester groups to hydrolysis. Anhydride curing agents find use in most of the important applications of epoxy resins, particularly in casting and laminates.

In practice, acid anhydrides are preferred to acids since the latter are generally less soluble in the resin and also release more water on cure, leading to foaming of the product. Care must be taken, however, during storage since the anhydrides in general are somewhat hygroscopic. Examples of some anhydrides which are used are shown in Figure 4.28.

The mechanism of hardening by anhydride is complex. In general, however, two types of reactions occur: (1) opening of the anhydride ring with the formation of carboxy groups, and (2) opening of the epoxy ring. The most important reactions which may occur are shown in what follows, with phthalic anhydride as an example.

(1) The first stage of the interaction between an acid anhydride and an epoxy resin is believed to be the opening of the anhydride ring by (a) water (traces of which may be present in the system) or (b) hydroxy groups (which may be present as pendant groups in the original resin or may result from reaction (2a).

$$\tag{1}$$

(2) The epoxy ring may then be opened by reaction with (a) carboxylic groups formed by reactions (1a, b) or (b) hydroxyl groups [see (1)].

(a)   [benzene ring]–C(=O)–OH + CH₂–CH~~ (epoxide) ⟶ [benzene ring]–C(=O)–O–CH₂–CH(OH)~~

~~CH₂–CH–CH₂~~   ~~CH₂–CH–CH₂~~

(b)   ~~CH₂–CH(OH)–CH₂~~ + CH₂–CH~~ (epoxide) →(cat) ~~CH₂–CH–CH₂~~ with O–CH₂–CH(OH)~~

$$(2)$$

The reaction between an epoxy resin and an anhydride is rather sluggish. In commercial practice, the curing is accelerated by the use of organic bases to catalyze the reaction. These are usually tertiary amines such as α-methylbenzyl-dimethylamine and butylamine. The tertiary amine appears to react preferentially with the anhydride to generate a carboxy anion ($-COO^-$). This anion then opens an epoxy ring to give an alkoxide ion ($\geq$C$-O^-$), which forms another carboxy anion from a second anhydride molecule, and so on.

Phthalic anhydride is the cheapest anhydride curing agent, but it has the disadvantage of being rather difficult to mix with the resin. Liquid anhydrides (e.g., dodecenylsuccinic anhydride and nadic methyl anhydride), low-melting anhydrides (e.g., hexahydrophthalic anhydride), and eutectic mixtures are more easily incorporated into the resin. Since maleic anhydride produces brittle products, it is seldom used by itself and is used as a secondary hardener in admixture with other anhydrides. Dodecenylsuccinic anhydride imparts flexibility into the casting, whereas chlorendic anhydride confers flame resistance. Anhydride-cured resins generally have better thermal stability. Pyromellitic dianhydride with higher functionality produces tightly cross-linked products of high heat-distortion temperatures. Heat-distortion temperatures as high as 290°C have been quoted. Table 4.27 summarizes the characteristics of some of the anhydride hardeners.

In addition to the amine and anhydride hardeners, many other curing agents have been made available. Among them are the so-called fatty polyamides. These polymers are of low molecular weight (2000 to 5000) and are prepared by treating dimer acid (which is a complex mixture consisting of 60 to 75% dimerized fatty acids together with lesser amounts of trimerized acids and higher polymers) with stoichiometric excess of ethylenediamine or diethylenetriamine so that the resultant amides have free amine groups. Fatty polyamides are used to cure epoxy resins where a more flexible product is required, particularly in adhesive and coating applications. An advantage of the system is that roughly similar quantities of hardener and resin are required and, because it is not critical, metering can be done visually without the need of measuring aids. They thus form the basis of some domestic adhesive systems. Also used with epoxy resins for adhesives is dicyanodiamide, $H_2N \cdot C(=NH)NH \cdot CN$. It is insoluble in com-

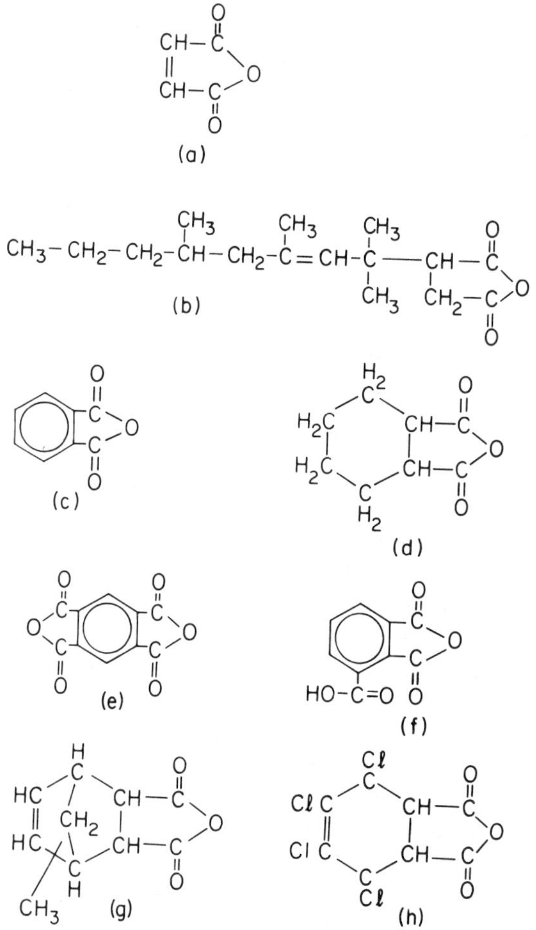

FIG. 4.28 Anhydride curing agents. (a) Maleic anhydride. (b) Dodecenylsuccinic anhydride. (c) Phthalic anhydride. (d) Hexahydrophthalic anhydride. (e) Pyromellitic anhydride. (f) Trimellitic anhydride. (g) Nadic methyl anhydride. (h) Chlorendic anhydride.

mon resins at room temperature but is soluble at elevated temperatures and thus forms the basis of a one-pack system.

Other Epoxies

Resins containing the glycidyl ether group

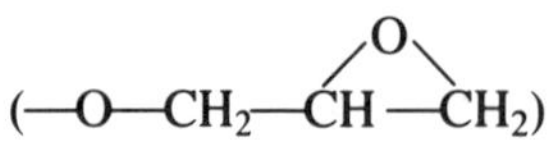

$$(-O-CH_2-CH-CH_2)$$

result from the reaction of epichlorohydrin and hydroxy compounds. Although bisphenol A is the most commonly used hydroxyl compound, a few glycidyl ether resins based on other hydroxy compounds are also commercially available, including novolac epoxies, polyglycol epoxies, and halogenated epoxies.

A typical commercial novolac epoxy resin (VI) produced by epoxidation of phenolic hydroxyl groups of novolac (see the section on phenol-formaldehyde resins) by treatment with epichlorohydrin has an

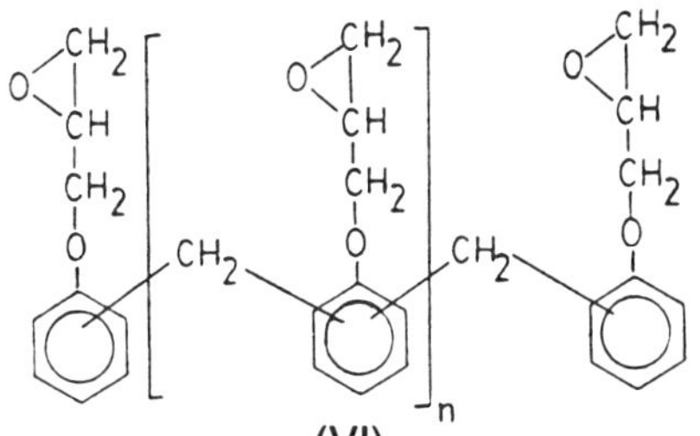

(VI)

average molecular weight of 650 and contains about 3.6 epoxy groups per molecule. Because of higher functionality, the novolac epoxy resins give, on curing, more tightly cross-linked products than the bisphenol A-based resins. This results in greater thermal stability, higher heat-deflection temperatures, and improved chemical resistance. Their main applications have been in high-temperature adhesives, heat-resistant structural laminates, "electrical" laminates resistant to solder baths, and chemical-resistant filament-wound pipes. The use of novolac epoxies has been limited, however, by their high viscosity and consequent handling difficulties.

Polymeric glycols such as polypropylene glycol may be epoxidized through the terminal hydroxy groups to give diglycidyl ethers (VII).

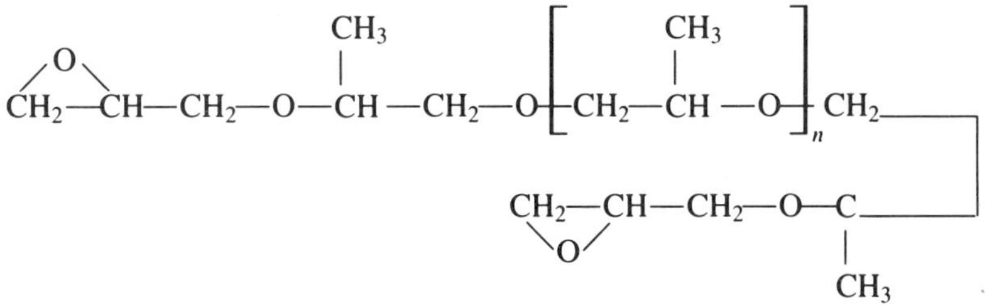

(VII)

TABLE 4.27 Properties of Some Anhydride Hardeners Used in Low-Molecular-Weight Bisphenol A-Based Epoxy Resins

Hardener (anhydride)	Parts used per 100 parts resin	Typical cure schedule	Max HDT of cured resin (°C)	Application
Phthalic	35–45	24 hr at 120°C	110	Casting
Hexahydrophthalic (+ accelerator)	80	24 hr at 120°C	130	Casting
Pyromellitic	26	20 hr at 220°C	290	High HDT
Nadic methyl	80	16 hr at 120°C	202	High HDT
Dodecenylsuccinic (+ accelerator)		2 hr at 100°C + 2 hr at 150°C	38	Flexibilizing
Chlorendic	100	24 hr at 180°C	180	Flame retarding

In commercial products n usually varies from 1 to 6. Alone, these resins give soft products of low strength on curing. So they are normally used in blends with bisphenol A- or novolac-based resins. Added to the extent of 10 to 30%, they improve resilience without too large a loss in strength and are used in such applications as adhesives and encapsulations.

Epoxies containing halogens have flame-retardant properties and may be prepared from halogenated hydroxy compounds. Halogenated epoxies are available based on tetrabromophenol A and tetrachlorobisphenol A (VIII).

(VIII)

The brominated resins are more effective than the chlorinated resins and have become more predominant commercially. The ability of the resins to retard or extinguish burning is due to the evolution of hydrogen halide at elevated temperatures. Brominated epoxy resins are generally blended with other epoxy resins to impart flame retardance in such applications as laminates and adhesives.

Nonglycidyl ether epoxy resins are usually prepared by treating unsaturated compounds with peracetic acid.

Two types of nonglycidyl ether epoxy resins are commercially available: cyclic aliphatic epoxies and acyclic aliphatic epoxies.

Cyclic aliphatic epoxy resins were first introduced in the United States. Some typical examples of commercial materials are 3,4-epoxy-6-methylcyclohexylmethyl-3,4-epoxy-6-methylcyclohexane carboxylate (Unox epoxide 201, liquid) (IX), vinylcyclohexene dioxide (Unox epoxide 206, liquid) (X), and dicyclopentadiene dioxide (Unox epoxide 207, solid) (XI).

(IX)   (X)   (XI)

Generally, acid anhydrides are the preferred curing agents since amines are less effective. A hydroxy compound, such as ethylene glycol, is often added as initiator.

Because of their more compact structure, cycloaliphatic resins produce greater density of cross-links in the cured products than bisphenol A-based glycidyl resins. This generally leads to higher heat-distortion temperatures and to increased brittleness. The products also are clearly superior in arc resistance and are track resistant. Thus although bisphenol A-based epoxies decompose in the presence of a high-temperature arc to produce carbon which leads to tracking and insulation failure, cycloaliphatic epoxies oxidize to volatile products which do not cause tracking. This has led to such applications as heavy-duty electrical castings and laminates, tension insulators, rocket motor cases, and transformer encapsulation.

Acyclic aliphatic resins differ from cyclic aliphatic resins in that the basic structure of the molecules in the former is a long chain, whereas the latter, as shown, contains ring structures. Two types of acyclic aliphatic epoxies are commercially available, namely, epoxidized diene polymers and epoxidized oils.

Typical of the epoxidized diene polymers are the products produced by treatment of polybutadiene with peracetic acid. Epoxidized diene polymers are not very reactive toward amines but may be cross-linked with acid hardeners. Cured resins have substantially higher heat-distortion temperatures (typically, 250°C) than do the conventional amine-cured diglycidyl ether resins.

Epoxidized oils are obtained by treatment of drying and semidrying oils (unsaturated), such as linseed and soybean oils, with peracetic acid. Epoxidized oils find use primarily as plasticizers—stabilizers for PVC.

Applications

Epoxy resins have found a wide range of applications and a steady rate of growth over the years mainly because of their versatility. Properties of the cured products can be tailored by proper selection of resin, modifier, cross-linking agent, and the curing schedule. The main attributes of properly cured epoxy systems are outstanding adhesion to a wide variety of substrates, including metals and concrete; ability to cure over a wide temperature range; very low shrinkage on cure; excellent resistance to chemicals and corrosion; excellent electrical insulation properties; and high tensile, compressive, and flexural strengths. In general, the toughness, adhesion, chemical resistance, and corrosion resistance of epoxies suit them for protective coating applications. It is not surprising that about 50% of epoxy resins are used in protective coating applications.

Two types of epoxy coatings are formulated: those cured at ambient temperature and those that are heat cured. The first type uses amine hardening systems, fatty acid polyamides, and polymer-captans as curing agents. Very high cure rates may be achieved by using mercaptans. Heat-cured types use acid anhydrides and polycarboxylic acids as well as formaldehyde resins as curing agents.

Typical coating applications for phenol-formaldehyde resin-modified epoxies include food and beverage can coatings, drum and tank liners, internal coatings for pipes, wire coating, and impregnation varnishes. Urea-formaldehyde resin-modified epoxies offer better color range and are used as appliance primers, can linings, and coatings for hospital and laboratory furniture.

Environmental concerns have prompted major developmental trends in epoxy coating systems. Thus epoxy coating systems have been developed to meet the high-temperature and hot-acid environments to which SO_2 scrubbers and related equipment are subjected. Ambient curable epoxy systems have been developed to provide resistance to concentrated inorganic acids.

In the development of maintenance and marine coatings, the emphasis has been on the development of low-solvent, or solvent-free, coatings to satisfy EPA volatile organic content standards. Thus liquid epoxy systems based on polyamidamines have been developed. Epoxies have been formulated as powdered coatings, thus completely eliminating solvents.

Pipe coatings still represent a major market for epoxies. High-molecular-weight powdered formulations are used in this application.

The important maintenance coating area, particularly for pipe and tank coatings, is served by epoxy systems cured with polyamine or polyamidamines. Two-component, air-dried, solventless systems used in maintenance coatings provide tough, durable, nonporous surfaces with good resistance to water, acids, alkalies, organic solvents, and corrosion.

Emulsifiable epoxies of varying molecular weights, water-dilutable modified epoxies, and dispersions of standard resins represent promising developments for coating applications. Can coatings are an important waterborne resin application. Two-component waterborne systems are also finding use as architectural coatings.

For electrical and electronic applications epoxy formulations are available with low or high viscosity, unfilled or filled, slow or fast curing at low or high temperatures. Potting, encapsulation, and casting of transistors, integrated circuits, switches, coils, and insulators are a few electrical applications of epoxies. With their adhesion to glass, electrical properties, and flexural strength, epoxies provide high-quality printed circuit boards. Epoxies have been successfully used in Europe for outdoor insulators, switchgear, and transformers for many years. In these heavy electrical applications, the advantage of cycloaliphatic epoxies over porcelain has been demonstrated.

In a relatively new development, epoxy photopolymers have been used as solder masks and photoresists in printed circuit board fabrications.

Adhesion properties of epoxies, complete reactivity with no volatiles during cure, and minimal shrinkage make the materials outstanding for adhesives, particularly in structural applications. The most commonly known adhesive applications involve the two-component liquids or pastes, which cure at room or elevated temperatures. A novel, latent curing system, which gives more than one year pot life at room temperature, has increased the use of epoxies for specialty adhesives and sealants, and for vinyl plastisols. The one-pack system provides fast cure when heated—for example 5 min at 100°C.

Epoxies are used in fiber-reinforced composites, providing high strength-to-weight ratios and good thermal and electrical properties. Filament-wound epoxy composites are used for rocket motor casings, pressure vessels, and tanks. Glass-fiber reinforced epoxy pipes are used in the oil, gas, mining, and chemical industries.

Sand-filled epoxies are used in industrial flooring. Resistance to a wide variety of chemicals and solvents and adhesion to concrete are key properties responsible for this use. Epoxies are also used in patching concrete highways. Decorative flooring and exposed aggregate systems make use of epoxies because of their low curing shrinkage, and the good bonding of glass, marble, and quartz chip by the epoxy matrix.

Poly(Phenylene Oxide)

Monomer	Polymerization	Major uses
2,6-Dimethylphenol	Condensation polymerization by oxidative coupling	Automotive, appliances, business machine cases, electrical components

Poly(2,6-dimethyl-1,4-phenylene oxide), commonly called poly(phenylene oxide) or PPO, was introduced commercially in 1964. PPO is manufactured by oxidation of 2,6-dimethyl phenol in solution using cuprous chloride and pyridine as catalyst. The monomer is obtained by the alkylation of phenol with methanol. End-group stabilization with acetic anhydride improves the oxidation resistance of PPO.

PPO is counted as one of the "engineering plastics." The rigid structure of the polymer molecules leads to a material with a high T_g of 208°C. It is characterized by high tensile strength, stiffness, impact strength, and creep resistance, and low coefficient of thermal expansion. These properties are maintained over a broad temperature range (-45 to 120°C). One particular feature of PPO is its exceptional dimensional stability among the so-called engineering plastics. The polymer is self-extinguishing.

PPO has excellent resistance to most aqueous reagents and is unaffected by acids, alkalis, and detergents. The polymer has outstanding hydrolytic stability and has one of the lowest water absorption rates among the engineering thermoplastics. PPO is soluble in aromatic hydrocarbons and chlorinate solvents. Several aliphatic hydrocarbons cause environmental stress cracking.

PPO has low molding shrinkage. The polymer is used for the injection molding of such items as pump components, domestic appliance and business machines, and electrical parts such as connectors and terminal blocks.

The high price of PPO has greatly restricted its application and has led to the introduction of the related and cheaper thermoplastic materials in 1966 under the trade name Noryl by General Electric. If PPO ($T_g = 208°C$) is blended with polystyrene ($T_g \sim 90°C$) in equal quantities, a transparent polymer is obtained with a single T_g of about 150°C, which apparently indicates a molecular level of mixing. Noryl thermoplastics may be considered as being derived from such polystyrene-PPO blends. Since the electrical properties of the two polymers are very similar, the blends also have similar electrical characteristics. In addition to Noryl blends produced by General Electric, grafts of styrene onto PPO are also now available (Xyron by Asahi-Dow).

CELLULOSIC POLYMERS

Cellulose is a carbohydrate with molecular formula $(C_6H_{10}O_5)_n$, where n is a few thousand. Complete hydrolysis of cellulose by boiling with concentrated hydrochloric acid yields D-glucose, $C_6H_{12}C_6$, in 95 to 96% yield. Cellulose can thus be considered chemically as a polyanhydroglucose. Careful chemical analysis of stepwise degradation products of cellulose reveals that the anhydroglucose units making up a cellulose molecule are joined together by beta-glucosidic (1,4) linkages. Thus since the linkage between the anhydroglucose units occurs at the reducing carbon atom, cellulose is nonreducing, unlike glucose which is a strongly reducing sugar. The structure of cellulose is

The regularity of the cellulose chain and extensive hydrogen bonding between hydroxyl groups of adjacent chains cause cellulose to be a tightly packed crystalline material which is insoluble and infusible. As a result, cellulose cannot be processed in the melt or in solution. However, cellulose derivatives in which there is less hydrogen bonding are processable. The most common means of preparing processable cellulose derivatives are esterification and etherification of the hydroxyl groups. In the following, the more important commercial cellulosic polymers are described.

Regenerated Cellulose

As mentioned, many derivatives of cellulose are soluble, though cellulose itself is insoluble. A solution of a cellulose derivative can be processed (usually by extrusion) to produce the desired shape (commonly fiber or film) and then treated to remove the modifying groups to reform or regenerate unmodified cellulose. Such material is known as *regenerated cellulose*.

Modern methods of producing regenerated cellulose can be traced to the discovery in 1892 by Cross, Bevan, and Beadle that cellulose can be rendered soluble by xanthate formation by treatment with sodium hydroxide and carbon disulfide and regenerated by acidification of the xanthate solution. This process is known as the *viscose process*. The reactions can be indicated schematically as

$$
\text{R—OH} \xrightarrow{\text{NaOH}} \text{R—ONa} \xrightarrow{\text{CS}_2} \text{R—O—}\overset{\overset{\text{S}}{\|}}{\text{C}}\text{—S—Na} \xrightarrow{\text{H}^+} \text{R—OH}
$$

| Cellulose | Alkali cellulose | Cellulose xanthate | Regenerated cellulose |

The viscose process is used for the production of textile fibers, known as *viscose rayon*, and transparent packaging film, known as *cellophane* (the name is coined from *cell*ulose and dia*phane*, which is French for "transparent").

A suitably aged solution of cellulose xanthate, known as *viscose*, is fed through spinnerets with many small holes (in the production of fiber), through a slot die (in the production of film), or through a ring die (in the production of continuous tube used as sausage casing) into a bath containing 10 to 15% sulfuric acid and 10 to 20% sodium sulfate at 35 to 40°C, which coagulates and completely hydrolyzes the viscose. Cellulose is thus regenerated in the desired shape and form.

It is possible to carry out a drawing operation on the fiber as it passes through the coagulating bath. The stretching (50 to 150%) produces crystalline orientation in the fiber. The product, known as *high-tenacity rayon*, has high strength and low elongation and is used in such applications as tire cord and conveyor belting.

For the production of cellophane, the regenerated cellulose film is washed, bleached, plasticized with ethylene glycol or glycerol, and then dried; sometimes a coating of pyroxylin (cellulose nitrate solution) containing dibutyl phthalate as plasticizer is applied to give heat sealability and lower moisture permeability. Cellophane has been extensively and successfully used as a wrapping material, particularly in the food and tobacco industries. However, the advent of polypropylene in the early 1960s has produced a serious competitor to this material.

Cellulose Nitrate

Cellulose nitrate or nitrocellulose (as it is often erroneously called) is the doyen of cellulose ester polymers. It is prepared by direct nitration with nitric and sulfuric acid mixtures at about 30 to 40°C for 20 to 60 min. Complete substitution at all three hydroxyl groups on the repeating anhydroglucose unit will give cellulose trinitrate containing 14.14% nitrogen:

$$[C_6H_7O_2(OH)_3]_n \xrightarrow{\text{HNO}_3/\text{H}_2\text{SO}_4} [C_6H_7O_2(ONO_2)_3]_n + 3nH_2O$$

This material is explosive and is not made commercially, but esters with lower degrees of nitration are of importance. The degree of nitration may be regulated by the choice of reaction conditions. Industrial nitrocelluloses have a degree of substitution somewhere between 1.9 and 2.7 and are generally characterized for their various uses by their nitrogen content, usually about 11% for plastics, 12% for lacquer and cement base, and 13% for explosives.

The largest use of cellulose nitrate is as a base for lacquers and cements. Butyl acetate is used as a solvent. Plasticizers such as dibutyl phthalate and tritolyl phosphate are necessary to give films of acceptable flexibility and adhesion.

For use as plastic in bulk form, cellulose nitrate is plasticized with camphor. The product is known as *celluloid*. In a typical process alcohol-wet cellulose nitrate is kneaded at about 40°C with camphor (about 30%) to form a viscous plastic mass. Pigments or dyes may be added at this stage. The dough is then heated at about 80°C on milling rolls until the alcohol content is reduced to about 15%. The milled product is calendered into sheets about 1/2-in. (1.25-cm) thick. A number of sheets are laid up in a press and consolidated into a block. The block is sliced into sheets of thickness 0.005 to 1 in. (0.012 to 2.5 cm), which are then allowed to season for several days at about 50°C so that the volatile content is reduced to about 2%. Celluloid sheet and block may be machined with little difficulty if care is taken to avoid overheating.

The high inflammability and relatively poor chemical resistance of celluloid severely restrict its use in industrial applications. Consequently the material is used because of its desirable characteristics, which include rigidity, dimensional stability, low water absorption, reasonable toughness, after-shrinkage around inserts, and ability of forming highly attractive colored sheeting. Today the principal outlets of celluloid are knife handles, table-tennis balls, and spectacle frames. Celluloid is marketed as Xylonite (BX Plastics Ltd.) in England.

Cellulose Acetate

The acetylation of cellulose is usually carried out with acetic anhydride in the presence of sulfuric acid as catalyst. It is not practicable to stop acetylation short of the essentially completely esterified triacetate. Products of lower acetyl content are thus produced by partial hydrolysis of the triacetate to remove some of the acetyl groups:

$$[C_6H_7O_2(OH)_3]_n \xrightarrow{\text{HOAc, Ac}_2\text{O}} [C_6H_7O_2(OAc)_3]_n \xrightarrow[\text{Heat}]{\text{H}_2\text{O}}$$

Cellulose   Triacetate
(44.8%acetyl)

$$[C_6H_7O_2(OAc)_2OH]_n$$

Diacetate
(34.9%acetyl)

Cellulose triacetate is often known as *primary cellulose acetate*, and partially hydrolyzed material is called *secondary cellulose acetate*. Many physical and chemical properties of cellulose acetylation products are strongly dependent on the degree of esterification, which is measured by the acetyl content (i.e., the weight of acetyl radical (CH_3CO—) in the material) or acetic acid yield (i.e., the weight of acetic acid produced by complete hydrolysis of the ester). The commercial products can be broadly distinguished as cellulose acetate (37–40% acetyl), high-acetyl cellulose acetate (40–42% acetyl), and cellulose triacetate (43.7–44.8% acetyl).

Cellulose acetate containing 37 to 40% acetyl is usually preferred for use in general-purpose injection-molding compounds. Cellulose acetate, however, decomposes below its softening point, and it is necessary to add a plasticizer (e.g., dimethyl phthalate or triphenyl phosphate), usually 25 to 35%, to obtain a moldable material. The use of cellulose acetate for molding and extrusion is becoming small owing largely to the competition of polystyrene and other polyolefins. At the present time the major outlets of cellulose acetate are in the fancy goods trade as toothbrushes, combs, hair slides, etc.

Cellulose acetate with a slightly higher degree of esterification (38.7–40.1% acetyl) is usually preferred for the preparation of fibers, films, and lacquers because of the greater water resistance. A significant application of cellulose acetate film has been found in sea-water desalination by reverse osmosis.

High-acetyl cellulose acetate (40–42% acetyl) has found occasional use in injection-molding compounds where greater dimensional stability is required. However, processing is more difficult.

Cellulose triacetate (43.7–44.8% acetyl) finds little use in molding compositions because its very high softening temperature is not greatly reduced by plasticizers. It is therefore processed in solution. A mixture of methylene chloride and methanol is the commonly used solvent. The sheeting and fibers are made from cellulose triacetate by casting or by extruding a viscous solution and evaporating the solvent. The sheeting and film are grainless, have good gauge uniformity, and good optical clarity. The products have good dimensional stability and are highly resistant to water, grease, oils, and most common solvents such as alcohol and acetone. They also have good heat resistance and high dielectric constant.

Sheeting and films of cellulose triacetate are used in the production of visual aids, graphic arts, greeting cards, photographic albums, and protective folders. Cellulose triacetate is extensively used for photographic, x-ray, and cinematographic films. In these applications cellulose triacetate has displaced celluloid mainly because the triacetate does not have the great inflammability of celluloid.

Other Cellulose Esters

Homologues of acetic acid have been employed to make other cellulose esters. Of these, cellulose propionate, cellulose acetate-propionate, and cellulose acetate-butyrate are produced on a commercial scale. They are produced in a manner similar to that described previously for cellulose acetate. The propionate and butyrate esters are made by substituting propionic acid and propionic anhydride or butyric acid and butyric anhydride for some of the acetic acid and acetic anhydride.

Cellulose acetate-butyrate (CAB) has several advantages in properties over cellulose acetate: lower moisture absorption, greater solubility and compatibility with plasticizer, higher impact strength, and excellent dimensional stability. CAB used in plastics has about 13% acetyl and 37% butyryl content. It is an excellent injection-molding material (Tenite Butyrate by Kodak, Cellidor B by Bayer). Principal end products of CAB have been for tabulator keys, automobile parts, toys, pen and pencil barrels, steering wheels, and tool handles. In the United States CAB has been used for telephone housings, and extruded CAB piping has been used for conveying water, oil, and natural gas. CAB sheet is readily vacuum formed and is especially useful for laminating with thin-gauge aluminum foil. It also serves particularly well for vacuum metallizing.

Cellulose propionate (Forticel by Celanese) is very similar in both cost and properties to CAB. It has been used for similar purposes as CAB. Cellulose acetate-propionate (Tenite Propionate by Kodak) is similar to cellulose propionate. It finds wide use in blister packages and formed containers, safety goggles, motor covers, metallized flash cubes, brush handles, steering wheels, face shields, displays, and lighting fixtures.

Cellulose Ethers

Of cellulose ethers only ethyl cellulose has found application as a molding material. Methyl cellulose, hydroxyethyl cellulose, and sodium carboxymethyl cellulose are useful water-soluble polymers. The first step in the manufacture of each of these materials is the preparation of alkali cellulose (soda cellulose) by treating cellulose with concentrated sodium hydroxide. Ethyl cellulose is made by reacting alkali cellulose with ethyl chloride.

$$[[R(OH)_3]_n + NaOH, H_2O] + Cl\!-\!CH_2CH_3 \xrightarrow[\text{6-12 hr}]{90-150°C}$$
$$\text{where } R = C_6H_7O_2$$

$$[R(OCH_2CH_3)_m(OH)_{3-m}]_n$$
$$m = 2.4 - 2.5$$

Ethyl cellulose is produced in pellet form for molding and extrusion and in sheet form for fabrication. It has good processability, is tough, and is moderately flexible; its outstanding feature is its toughness at low temperatures. The principal uses of ethyl cellulose moldings are thus in those applications where good impact strength at low temperatures are required, such as refrigerator bases and flip lids and ice-crusher parts. Ethyl cellulose is often employed in the form of hot melt for strippable coatings used for protection of metal parts against corrosion and marring during shipment and storage. A recent development is the use of ethyl cellulose gel lacquers for permanent coatings.

Methyl cellulose is prepared by reacting alkali cellulose with methyl chloride at 50 to 100°C (cf. ethyl cellulose). With a degree of substitution of 1.6 to 1.8, the resultant ether is soluble in cold water but not in hot water. It is used as a thickening agent and emulsifier in cosmetics, pharmaceuticals, ceramics, as a paper size, and in leather-tanning operations. Hydroxyethyl cellulose, produced by reacting alkali cellulose with ethylene oxide, is employed for similar purposes.

Reaction of alkali cellulose with sodium salt of chloroacetic acid yields sodium carboxymethyl cellulose (SCMC),

$$[[R(OH)_3]_n + NaOH, H_2O] + Cl\!-\!CH_2\!-\!CO\!-\!ONa \xrightarrow[\text{2-4 hr}]{40-50°C}$$

$$[R(OCH_2CO\!-\!ONa)_m(OH)_{3-m}]_n$$

where R = $C_6H_7O_2$ and m = 0.65–1.4. SCMC appears to be physiologically inert and is very widely used. Its principal application is as a soil-suspending agent in synthetic detergents, but it is also used as a sizing and finishing agent for textiles, as a surface active agent, and as a viscosity modifier in emulsion and suspensions. Purified grades of SCMC are used in ice cream to provide a smooth texture and in a number of cosmetic and pharmaceutical products. SCMC is also the basis of a well-known proprietary wallpaper adhesive.

Poly(Phenylene Sulfide)

Monomers	Polymerization	Major uses
p-Dichlorobenzene, sodium sulfide	Polycondensation	Electrical components, mechanical parts

Poly(phenylene sulfide) (PPS) is the thio analogue of poly(phenylene oxide) (PPO). The first commercial grades were introduced by Phillips Petroleum in 1968 under the trade name Ryton. Other manufacturers also have introduced PPS (e.g., Tedur by Bayer). The commercial process involves the reaction of *p*-dichlorobenzene with sodium sulfide in a polar solvent.

PPS is an engineering plastic. The thermoplastic grades of PPS are outstanding in heat resistance, flame resistance, chemical resistance, and electrical insulation characteristics. The linear polymers are highly crystalline with melting point in the range 285 to 295°C and T_g of 193 to 204°C. The material is soluble only above 200°C in aromatic and chlorinated aromatic solvents. It has the ability to cross-link by air-oxidation at elevated temperatures, thereby providing an irreversible cure. Thermogravimetric analysis shows no weight loss below 500°C in air but demonstrates complete decomposition by 700°C. It is found to retain its properties after four months at 233°C (450°F) in air.

Significant increases in mechanical properties can be achieved with glass-fiber reinforcement. In the unfilled form the tensile strength of the material is 64 to 77 MPa at 21°C, 33 MPa at 204°C, and the flexural modulus is 4200 MPa at 21°C. The corresponding values for PPS-glass fiber (60:40) composites are 150, 33, and 15,500 MPa. Although rigidity and tensile strength are similar to other engineering plastics, PPS does not possess the toughness of amorphous materials, such as the polycarbonates and the polysulfones (described later), and are somewhat brittle. On the other hand, PPS does show a high level of resistance to environmental stress cracking.

Being one of the most expensive commercial moldable thermoplastics, the use of PPS is heavily dependent on its particular combination of properties. Good electrical insulation characteristics, including good arcing and arc-tracking resistance, has led to PPS replacing some of the older thermosets in electrical parts. These include connectors, terminal blocks, relay components, brush holders, and switch components. PPS is used in chemical process plants for gear pumps. More recently, it has found application in the automotive sector, in such specific uses as carburetor parts, ignition plates, flow control valves for heating systems, and exhaust-gas return valves to control pollution. The material also finds uses in sterilizable medical, dental, and general laboratory equipment, cooking appliances, and hair dryer components.

Injection-molded products include high-temperature lamp holders and reflectors, pump parts, valves, and, especially when filled for example with PTFE or graphite, bearings. Processing temperatures are 300–350°C with mold temperatures of up to 200°C. PPS is also used for encapsulation of electronic components and as a high temperature surface coating material.

Polysulfone

Polysulfones are a family of engineering thermoplastics with excellent high-temperature properties. The simplest aromatic polysulfone, poly(p-phenylene sulfone) does not show thermoplastic behavior, melting with decomposition

above 500°C. Hence, to obtain a material capable of injection molding in conventional machines, the polymer chain is made more flexible by incorporating ether links into the backbone. The structures and glass transition temperatures of several commercial polysulfones are listed in Table 4.28. The polymers have different degrees of spacing between the p-phenylene groups and thus have a spectrum of glass transition temperatures which determine the heat-distortion temperature (or deflection temperature under load), since the materials are all amorphous.

The first commercial polysulfone (Table 4.28I) was introduced in 1965 by Union Carbide as Bakelite Polysulfone. This material, now renamed Udel, has a continuous-use temperature of 150°C and a maximum-use temperature of 170°C, and it can be fabricated easily by injection molding in conventional machines. In 1967, Minnesota Mining and Manufacturing (3M) introduced Astrel 360 (Table 4.28II), an especially high-performance thermoplastic, which re-

TABLE 4.28 Commercial Polysulfones

Type of structure	Tg (°C)	Trade name
I $\left[\text{(structure: phenyl—C(CH}_3)_2\text{—phenyl—O—phenyl—SO}_2\text{—phenyl—O)}\right]$	190	Udel (Union Carbide)
II $\left[\text{(a) phenyl—phenyl—SO}_2\right]\left[\text{(b) phenyl—O—phenyl—SO}_2\right]$ (a) Predominates	285	Astrel (3M Corp.)
III $\left[\text{phenyl—O—phenyl—SO}_2\right]$	230	Victrex (ICI)
IV $\left[\text{(a) phenyl—phenyl—SO}_2\right]\left[\text{(b) phenyl—O—phenyl—SO}_2\right]$ (b) Predominates	250	Polyether- Sulfone 720 P (ICI)
V $\left[\text{phenyl—SO}_2\text{—phenyl—O—phenyl—phenyl—O}\right]$	–	Radel (Union Carbide)

quires specialized equipment with extra heating and pressure capabilities for processing. ICI's polyether sulfones, introduced in 1972—Victrex (Table 4.28III) and polyethersulfone 720P (Table 4.28IV)—are intermediate in performance and processing. In the late 1970s, Union Carbide introduced Radel (Table 4.28V), which has a higher level of toughness. Note that all of the commercial materials mentioned in Table 4.28 may be described as polysulfones, polyarylsulfones, polyether sulfones, or polyaryl ether sulfones.

In principle, there are two main routes to the preparation of polysulfones: (1) polysulfonylation and (2) polyetherification.

Polysulfonylation reactions are of the following general types:

$$\text{H—Ar—H} + \text{ClSO}_2\text{—Ar}'\text{—SO}_2\text{Cl} \xrightarrow[\text{catalyst}]{\text{Friedel-Crafts}}$$

$$\text{—}(\text{Ar—SO}_2\text{—Ar}'\text{—SO}_2)_n\text{—} + \text{HCl}$$

$$\text{H—Ar—SO}_2\text{Cl} \longrightarrow \text{—}(\text{Ar—SO}_2)_n\text{—} + \text{HCl}$$

The Ar and/or Ar' group(s) will contain an ether oxygen, and if Ar = Ar', then basically identical products may be obtained by the two routes.

In the *polyetherification* route the condensation reaction proceeds by reactions of types

$$HO-Ar-OH + Cl-Ar'-Cl \xrightarrow{NaOH} -(O-Ar-O-Ar')_n + NaCl$$

$$HO-Ar-Cl \xrightarrow{NaOH} -(O-Ar)_n + NaCl$$

The Ar and/or Ar' group(s) will contain sulfone groups, and if Ar = Ar', then identical products may be obtained by the two routes.

Polyetherification processes form the basis of current commercial polysulfone production methods. For example, the Udel-type polymer (Union Carbide) is prepared by reacting 4,4'-dichlorodiphenylsulfone with an alkali salt of

bisphenol A. The polycondensation is conducted in highly polar solvents, such as dimethylsulfoxide or sulfolane.

Properties

In spite of their linear and regular structure the commercial polysulfones are amorphous. This property might be attributed to the high degree of chain stiffness of polymer molecules which makes crystallization difficult. Because of their high in-chain aromaticity and consequent high chain stiffness, the polymers have high values of T_g (see Table 4.28), which means that the processing temperatures must be above 300°C.

Commercial polymers generally resist aqueous acids and alkalis but are attacked by concentrated sulfuric acid. Being highly polar, the polymer is not dissolved by aliphatic hydrocarbons but dissolves in dimethyl formamide and dimethyl acetamide.

In addition to the high heat-deformation resistance, the polymers also exhibit a high degree of chemical stability. This has been ascribed to an enhanced bond strength arising from the high degree of resonance in the structure. The polymers are thus capable of absorbing a high degree of thermal and ionizing radiation without cross-linking.

The principal features of commercial polysulfones are their rigidity, transparency, self-extinguishing characteristics, exceptional resistance to creep, especially at somewhat elevated temperatures, and good high-temperature resistance. The use temperatures of the major engineering thermoplastics are com-

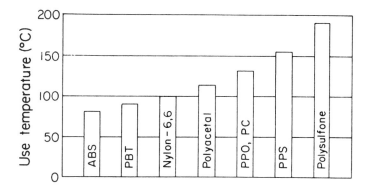

FIG. 4.29   Use temperatures of major engineering thermoplastics.

pared in Figure 4.29. Polysulfones are among the higher-priced engineering thermoplastics and so are only considered when polycarbonates or other cheaper polymers are unsuitable. In brief, polysulfones are more heat resistant and have greater resistance to creep, whereas polycarbonates have a somewhat higher Izod and tensile impact strength besides being less expensive.

In many fields of use polysulfones have replaced or are replacing metals, ceramics, and thermosetting plastics, rather than other thermoplastics. Since commercial polysulfones can be injection molded into complex shapes, they avoid costly machining and finishing operations. Polysulfones can also be extruded into film and foil. The latter is of interest for flexible printed circuitry because of its high-temperature performance.

Polysulfones have found widespread use where good dimensional stability at elevated temperatures is required and fabrication is done by injection molding. Some products made from polysulfones are electrical components, connectors, coil bobbins, relays, and appliances operating at high temperatures (e.g., hair driers, fan heaters, microwave ovens, lamp housings and bases). The materials are transparent (though often slightly yellow), have low flammability (limiting oxygen index typically 38), and burn with little smoke production. Typical properties of some of the commercial polysulfones are shown in Table 4.29.

Polyether Ether Ketone

The chemistry and technology of aromatic polyether ketones may be considered as an extension to those of the polysulfones. The two polymer classes have strong structural similarities, and there are strong parallels in preparative methods. Preparations reported for aromatic polyether ketones are analogous to the polysulfonylation and polyetherification reactions with the polysulfones. Several

TABLE 4.29 Properties of Aromatic Polysulfones

Property	Udel (Union Carbide)	Victrex (ICI)	Astrel (3M)
Tensile strength			
$\quad 10^3$ lbf/in^2	9.3	12.2	13.0
$\quad$MPa	64	84	90
Tensile modulus			
$\quad 10^5$ lbf/in^2	3.6	3.5	3.8
$\quad$GPa	2.5	2.4	2.6
Elongation at break (%)	50–100	40–80	10
Flexural modulus			
$\quad 10^5$ lbf/in^2	3.0	3.8	4.0
$\quad$GPa	2.1	2.6	2.8
Heat distortion temperature (°C)	174	203	274
Impact strength, notched Izod			
$\quad$ft-lb/in	—	1.9	—
$\quad$J/m	—	100	—
Limiting oxygen index (%)	—	38	—

aromatic polyether ketones have been prepared. The polyether ether ketone (PEEK)

was test marketed in 1978 by ICI.

PEEK is a high-temperature-resistant thermoplastic suitable for wire coating, injection molding, film, and advanced structural composite fabrication. The wholly aromatic structure of PEEK contributes to its high-temperature performance. The polymer exhibits very low water absorption, very good resistance to water at 125°C (under which conditions other heat-resisting materials, such as aromatic polyamides, are liable to fail), and is resistant to attack over a wide pH range, from 60% sulfuric acid to 40% sodium hydroxide at elevated temperatures, although attack can occur with some concentrated acids. PEEK has outstanding resistance to both abrasion and dynamic fatigue. It has low flammability with a limiting oxygen index of 35% and generates an exceptionally low level of smoke. Other specific features are excellent resistance to gamma radiation and good resistance to environmental stress cracking.

TABLE 4.30 Properties of Polyetheretherketone

Property	Unfilled	30% Glass fiber filled
Tensile strength		
10^3 lbf/in^2	13.3	22.8
MPa	92	157
Elongation at break (%)	4.9 (yield)	2.2
Flexural modulus		
10^5 lbf/in^2	5.4	14.9
GPa	3.7	10.3
Impact strength, notched Izod		
ft-lb/in	1.6	1.8
J/m	83	96
Heat distortion temperature (°C)	140	315
Limiting oxygen index (%)	35	—

PEEK has superior heat resistance to poly(phenylene sulfide) and is also markedly tougher (and markedly more expensive). PEEK is melt processable and may be injection molded and extruded on conventional equipment.

Typical applications of PEEK include coating and insulation for high-performance wiring, particularly for the aerospace and computer industries, military equipment, nuclear plant applications, oil wells, and compressor parts.

Composite prepregs with carbon fibers have been developed for structural aircraft components. Typical properties of PEEK are shown in Table 4.30.

Polybenzimidazole

Monomers	Polymerization	Major uses
Tetraaminobiphenyl, terephthalic acid	Polycondensation	Fiber

Polybenzimidazole (PBI) is the most well-known commercial example of aromatic heterocycles used as high-temperature polymers. The synthesis of PBI is carried out as follows (see also Fig. 1.22). The tetraaminobiphenyl required for the synthesis of PBI is obtained from 3,3'-dichloro-4,4'-diaminodiphenyl (a dye intermediate) and ammonia. Many other tetraamines and dicarboxylic acids have been condensed to PBI polymeric systems.

The high thermal stability of PBI (use temperature about 400°C compared to about 300°C for polyimides) combined with good stability makes it an outstanding candidate for high-temperature application despite its relatively high cost. Fibers have been wet spun from dimethylacetamide solution, and a deep gold woven cloth has been made from this fiber by Celanese. The cloth is said to be more comfortable than cotton (due to high moisture retention) and has greater flame resistance than Nomex (oxygen index of 28% for PBI compared to 17% for Nomex). The U.S. Air Force has tested flight suits of PBI and found them superior to other materials. Other applications of PBI are in drogue parachutes and lines for military aircraft as well as ablative heat shields. The PBI fibers have also shown promise as reverse osmosis membranes and in graphitization to high-strength, high-modulus fibers for use in composites. The development of ultra-fine PBI fibers for use in battery separator and fuel cell applications has been undertaken by Celanese.

A new technique of simple precipitation has been used to process PBI polymers into films. High-strength molecular-composite films have been produced with tensile strength in the region of 20,000 psi (137 MPa). The PBI polymer has also been fabricated as a foam. The material provides a low-weight, high-strength, thermally stable, machinable insulation, much needed in the aerospace industry. PBIs exhibit good adhesion as films when cast from solution onto glass plates. This property leads to their use in glass composites, laminates, and filament-wound structures.

SILICONES AND OTHER INORGANIC POLYMERS

The well-known thermal stability of minerals and glasses, many of which are themselves polymeric, has led to intensive research into synthetic inorganic and semi-inorganic polymers. Numerous such polymers have been synthesized, but only a few have found industrial acceptance, due to the difficulties encountered in processing them. These polymers can be classified into the following main groups: (1) polymers containing main-chain silicon atoms, (2) polymetallosiloxanes, (3) polymetalloxanes, (4) phosphorus-containing polymers, (5) boron-containing polymers, (6) sulfur-containing polymers.

Among the polymers containing main-chain silicon atoms, most important commercially are the silicones, which are described in a later section. Several other polymers which contain silicon atoms in the main chain have been studied in recent years. These include silicon sulfides (Fig. 4.30a) and polymers containing silicon and nitrogen in the main chain (Fig. 4.30b). Unfortunately, neither of these materials show hydrolytic stability. Of greater interest and potential value are the polymetallosiloxanes which have been investigated by Andrianov and co-workers in the Soviet Union. Several types of these polymers, which may more appropriately be classed as polyorganosiloxymetalloxanes, are shown in Fig. 4.30 c–e. Where the metal is more electropositive than silicon, the metal-oxygen bond will be more polar than the silicon-oxygen bond, and the material will be expected to show greater thermal stability. Andrianov has produced thermally stable polyorganosiloxymetalloxanes which are soluble in organic solvents and may therefore be cast on films and lacquers, or they may be used in laminating resins.

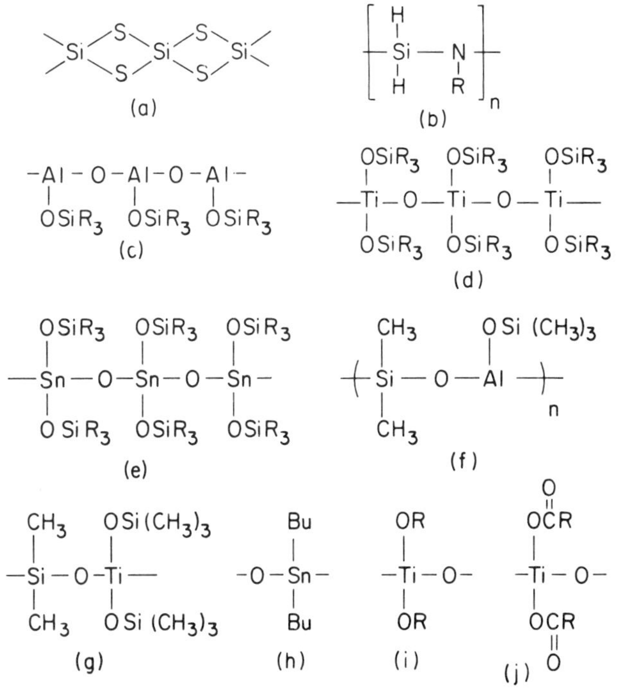

FIG. 4.30 Inorganic copolymers. (a) Silicon sulfides. (b) Polymer containing silicon and nitrogen in the main chain. (c) Polyorganosiloxyaluminoxanes. (d) Polyorganosiloxytitanoxanes. (e) Polyorganosiloxystannoxanes. (f) & (g) Polymetallosiloxanes containing silicon atom and a metal atom in the main chain. (h) Dibutyltin oxide. (i) Polyalkoxytitanoxanes. (j) Polymeric titanate ester.

A different class of polymetallosiloxanes have been produced in America. These polymers contain both the metal atom and the silicon atom incorporated into the main chain (e.g., Fig. 4.30f and g). The metallosiloxanes are liable to hydrolysis, but the rate of hydrolysis depends on the metal incorporated in the polymer chain. For example, the relative rates of hydrolysis of tin, aluminum, and titanium derivatives of low molecular weight are 2220:27:1.

Polymetalloxanes are akin to the silicones but contain, for example, tin, germanium, and titanium instead of silicon. Of the polyorganostannoxanes, dibutyltin oxide (Fig. 4.30h) finds use as a stabilizer for PVC and as a silicone cross-linking agent. Titanium does not form a stable bond with carbon. However, organotitanium polymers involving a Ti—O bond, such as polyalkoxytitanoxanes (Fig. 4.30i) and polymeric titanate esters (Fig. 4.30j), have been prepared. Butyl titanate polymers are used in surface coatings.

Among phosphorus-based polymers, several phosphonitrilic polymers have proved to be of some interest (see later). Though many polymers containing boron in the main chain have been prepared, most of them have either low molecular weight or intractable cross-linked structures. Polyboron nitride is a heteropolymer of boron and nitrogen. It is obtained readily from borax and ammonium chloride.

$$Na_2B_4O_7 + 2NH_4Cl \xrightarrow{1000°C} 2 \text{---}\!\!\left[B\!\!=\!\!N \right]\!\!\text{---} + B_2O_3 + 2NaCl + 4H_2O$$

Polyboron nitride melts at 3000°C under pressure. Two structural modifications are known: a hexagonal layer structure similar to graphite, and a tetrahedral structure isomorphic and isoelectronic with diamond. The former is used in insulators, and the latter is used as an abrasive.

Sulfur-containing polymers generally lack hydrolysis stability or, alternatively, have a tendency to revert to simple forms.

Although some inorganic polymers have found limited use as laminating resins, surface-coating resins, and wire enamels, the fruits of research to date on inorganic polymers have been rather disappointing. Polymers of high thermal stability have often lacked hydrolytic stability and in other cases have been brittle or intractable.

Silicones

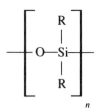

Monomers	Polymerization	Major uses
Chlorosilanes	Polycondensation	Elastomer, sealants, and fluids

Silicones are by far the most important inorganic polymers and are based on silicon, an element abundantly available on our planet. The silicone polymers are available in a number of forms, such as fluids, greases, rubbers, and resins. Because of their general thermal stability, water repellency, antiadhesive characteristics, and constancy of properties over a wide temperature range, silicones have found many and diverse applications. [The structure used as the basis of the nomenclature of the silicon compounds is silane SiH_4, corresponding to methane CH_4. Alkyl-, aryl-, alkoxy-, and halogen-substituted silanes are referred to by prefixing "silane" by the specific group present. For example, $(CH_3)_2SiH_2$ is dimethyl silane, and CH_3SiCl_3 is trichloromethylsilane. Polymers in which the main chain consists of repeating —Si—O— units together with predominantly organic side groups are referred to as *polyorganosiloxanes* or, more loosely, as *silicones*.]

The basis of modern silicone chemistry was laid by F. S. Kipping at the University College, Nottingham during the period 1899–1944. The commercial production of a broad variety of products from a few basic monomers followed the development of an economically attractive direct process for chlorosilanes, discovered by E. G. Rochow in 1945 at the G. E. Research Laboratories. The process involves reaction of alkyl or aryl halides with elementary silicon in the presence of a catalyst, e.g., copper for methyl- and silver for phenyl-chlorosilanes. The basic chemistry can be described as

$$SiO_2 + C \longrightarrow Si + 2CO$$

$$Si + RX \longrightarrow R_nSiX_{4-n} \qquad (n = 0 - 4)$$

In the alkylation of silicon with methyl chloride, mono-, di-, and trimethyl chlorosilanes are formed. The reaction products must then be fractionated. Because the dimethyl derivative is bifunctional, it produces linear methylsilicone polymers on hydrolysis.

Since monomethyl trichlorosilane has a functionality of 3, the hydrolysis leads to the formation of a highly cross-linked gel.

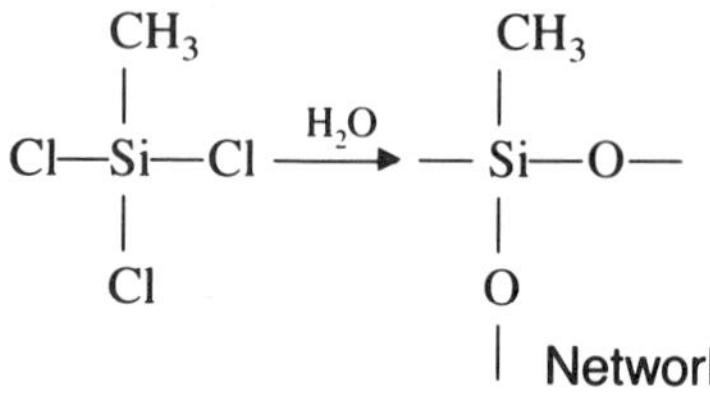

$$
\begin{array}{ccc}
& CH_3 & & CH_3 \\
& | & & | \\
Cl\!-\!Si\!-\!Cl & \xrightarrow{\;H_2O\;} & -\!Si\!-\!O\!- \\
& | & & | \\
& Cl & & O
\end{array}
$$

| Network polymer

Since the trimethyl monochlorosilane is monofunctional, it forms only a disiloxane.

$$
(CH_3)_3SiCl \xrightarrow{\;H_2O\;} (CH_3)_3Si\!-\!O\!-\!Si(CH_3)_3
$$

Products of different molecular-weight ranges and degrees of cross-linking are obtained from mixtures of these chlorosilanes in different ratios. In characterizing commercial branched and network structures, the CH_3/Si ratio (or, generally, R/Si ratio) is thus a useful parameter. For example, the preceding three idealized products have CH_3/Si ratios of 2:1, 1:1, and 3:1, respectively. A product with a CH_3/Si ratio of 1.5:1 will thus be expected to have a moderate degree of cross-linking.

Many different silicon products are available today. The major applications are listed in Table 4.31.

Silicone Fluids

The silicone fluids form a range of colorless liquids with viscosities from 1 to 1,000,000 centistokes (cs). The conversion of chlorosilane intermediates into polymer is accomplished by hydrolysis with water, which is followed by spontaneous condensation. In practice, the process involves three important stages:

TABLE 4.31 Major Applications of Silicones

Mold-release agents	Greases and waxes
Water repellants	Cosmetics
Antifoaming agents	Insulation
Glass-sizing agents	Dielectric encapsulation
Heat-exchange fluids	Caulking agents (RTV)
Hydraulic fluids	Gaskets and seals
Surfactants	Laminates
Coupling agents	Biomedical devices

(1) hydrolysis, condensation, and neutralization (of the HCl evolved on hydrolysis); (2) catalytic equilibration; and (3) devolatilization.

The product after the first stage consists of an approximately equal mixture of cyclic compounds, mainly the tetramer, and linear polymer. To achieve a more linear polymer and also to stabilize the viscosity, it is common practice to equilibrate the products of hydrolysis by heating with a catalyst such as dilute sulfuric acid. For fluids of viscosities below 1000 cs, this equilibrium reaction is carried out for hours at 100 to 150°C. After addition of water, the oil is separated from the aqueous acid layer and neutralized. To produce nonvolatile silicone fluids, volatile low-molecular products are removed by using a vacuum still. Commercial nonvolatile fluids have a weight loss of less than 0.5% after 24 hr at 150°C.

Dimethylsilicone fluids find a wide variety of applications mainly because of their water repellency, lubrication and antistick properties, low surface tension, a high order of thermal stability, and a fair constancy of physical properties over a wide range of temperature (-70 to 200°C). As a class the silicone fluids have no color or odor, have low volatility, and are nontoxic. The fluids have reasonable chemical resistance but are attacked by concentrated mineral acids and alkalis. They are soluble in aliphatic, aromatic, and chlorinated hydrocarbons.

A well-known application of the dimethylsilicone fluids is as a polish additive. The value of the silicone fluid in this application is due to its ability to lubricate, without softening, the microcrystalline wax plates.

Dilute solutions or emulsions containing 0.5 to 1% of a silicone fluid are extensively used as a release agent for rubber molding. However, their use has been restricted with thermoplastics because of the tendency of the fluids to cause stress cracking in polymers.

Silicone fluids are used in shock absorbers, hydraulic fluids, dashpots, and in other damping systems in high-temperature operations.

The silicones have established their value as water-repellent finishes for a range of natural and synthetic textiles. Techniques have been developed which result in the pickup of 1 to 3% of silicone on the cloth. Leather also may be made water repellent by treatment with solutions or emulsions of silicone fluids. These solutions are also used for paper treatment.

Silicone fluids and greases are useful as lubricants for high-temperature operations for applications depending on rolling friction. Greases may be made by blending silicone with an inert filler such as fine silicas, carbon black, or a metallic soap. The silicone/silica greases are used as electrical greases for such applications as aircraft and car ignition systems. Silicone greases have also found uses in the laboratory for lubricating stopcocks and for high-vacuum work.

Silicone fluids are used extensively as anitfoams, although concentration needed is normally only a few parts per million. The fluids have also found a number of uses in medicine. Barrier creams based on silicone fluids are particularly useful against cutting oils used in metal machinery processes.

High-molecular-weight dimethylsilicone fluids are used as stationary phase for columns in vapor-phase chromatographic apparatus.

Surfactants based on block copolymers of dimethylsilicone with poly(ethylene oxide) are unique in regulating the cell size in polyurethane foams. One route to such polymers used the reaction between a polysiloxane and an allyl ether of poly(ethylene oxide),

$$
\begin{array}{cc}
CH_3 & CH_3 \\
| & | \\
CH_3\text{-}(Si\text{-}O\text{-})_m\text{-}Si\text{-}H & + \quad CH_2\text{=}CH\text{-}CH_2\text{-}(OCH_2CH_2\text{-})_n\text{-}OH \xrightarrow{Pt} \\
| & | \\
CH_3 & CH_3
\end{array}
$$

$$
\begin{array}{cc}
CH_3 & CH_3 \\
| & | \\
CH_3\text{-}(Si\text{-}O\text{-})_m\text{-}Si\text{-}CH_2CH_2CH_2\text{-}(OCH_2CH_2\text{-})_n\text{-}OH \\
| & | \\
CH_3 & CH_3
\end{array}
$$

where $m = 2$–5 and $n = 3$–20. Increasing the silicone content makes the surfactant more lipophilic, whereas a higher poly(ethylene oxide) content makes it more hydrophilic.

Silicone Resins

Silicone resins are manufactured batchwise by hydrolysis of a blend of chlorosilanes. For the final product to be cross-linked, a certain amount of trichlorosilane must be incorporated into the blend. (In commercial practice, R/Si ratios are typically in the range 1.2:1 to 1.6:1.) The cross-linking of the resin is, of course, not carried out until it is in situ in the finished product. The cross-linking takes place by heating the resin at elevated temperatures with a catalyst, several of which are described in the literature (e.g., triethanolamine and metal octoates).

The resins have good heat resistance but are mechanically much weaker than cross-linked organic plastics. The resins are highly water repellent and are good electrical insulators particularly at elevated temperatures and under damp conditions. The properties are reasonably constant over a fair range of temperature and frequency.

Methyl phenyl silicone resins are used in the manufacture of heat-resistant glass-cloth laminates particularly for electrical applications. These are generally superior to PF and MF glass-cloth laminates. The dielectric strength of silicone-bonded glass-cloth laminates is 100 to 120 kV/cm compared to 60 to 80 kV/cm for both PF and MF laminates. The insulation resistance (dry) of the former (500,000 Ω) is significantly greater than those for the PF and MF laminates

(10,000 and 20,000 Ω, respectively). The corresponding values after water immersion are 10,000, 10, and 10 Ω.

Silicone laminates are used principally in electrical applications such as printed circuit boards, transformers, and slot wedges in electric motors, particularly class H motors. Compression-molding powders based on silicone resins are available and have been used in the molding of switch parts, brush ring holders, and other electrical applications that need to withstand high temperatures.

Silicone Rubbers

Dimethylsilicone rubbers consist of very-high-molecular-weight ($\sim 0.5 \times 10^6$) linear polymers which are cross-linked after fabrication. To obtain high-molecular-weight silicone polymers, the cyclic tetramer (octamethylcyclotetrasiloxane) is equilibrated with a trace of alkaline catalyst for several hours at 150 to 200°C, the molecular weight being controlled by careful addition of monofunctional siloxane. The product is a viscous gum with no elastic properties.

Before fabrication it is necessary to compound the gum with fillers, curing agent, and other special additives on a two-roll mill or in an internal mixer (cf. Rubber Compounding, Chapter 2). Unfilled polymers have negligible strength, whereas reinforced silicone rubbers may have strengths up to 2000 psi (14 MPa).

Silica fillers are generally used with silicone rubbers. These materials with particle sizes in the range 0.003 to 0.03 μm are prepared by combustion of silicon tetrachloride (fume silicas), by precipitation, or as an aerogel.

Benzoyl peroxide, 2,4-dichlorobenzoyl peroxide, and *t*-butyl perbenzoate in quantities of 0.5 to 3% are used as curing agents for the dimethylsilicones. These materials are stable in the compounds at room temperature for several months but will start to cure at about 70°C. The curing (cross-linking) is believed to take place by the sequence of reactions shown in Figure 4.31.

Dimethyl silicone rubbers show a high compression set. (For example, normal cured compounds have a compression set of 20 to 50% after 24 hr at 150°C.) Substantially reduced compression set values may be obtained by using a polymer containing small amounts of methylvinylsiloxane. Rubbers containing vinyl groups can be cross-linked by weaker peroxide catalysts. Where there is a high vinyl content (4 to 5% molar), it is also possible to vulcanize with sulfur.

Room-temperature vulcanizing silicone rubbers (RTV rubbers) are low-molecular-weight liquid silicones with reactive end groups and loaded with reinforcing fillers. Several types are available on the market. One type, which is typical of two-pack RTVs, cures by reaction of silanol end groups with silicate esters in the presence of a catalyst such as tin octoate or dibutyltin dilaurate (Fig. 4.32). In another type, which is typical of a one-pack system, the terminal silicon atoms carry acetoxy end groups which undergo hydrolysis with water.

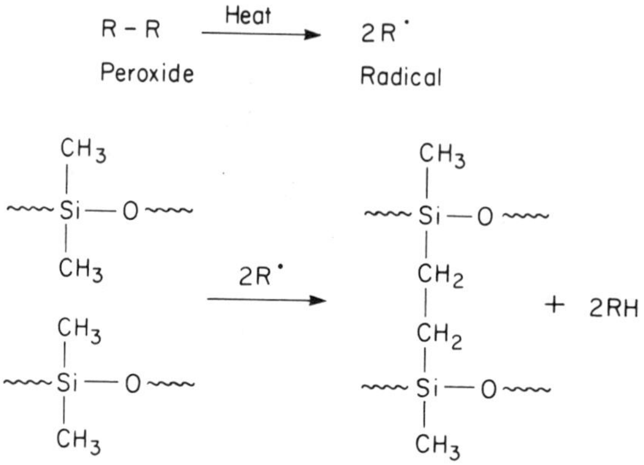

FIG. 4.31 Peroxide curing of silicone rubbers.

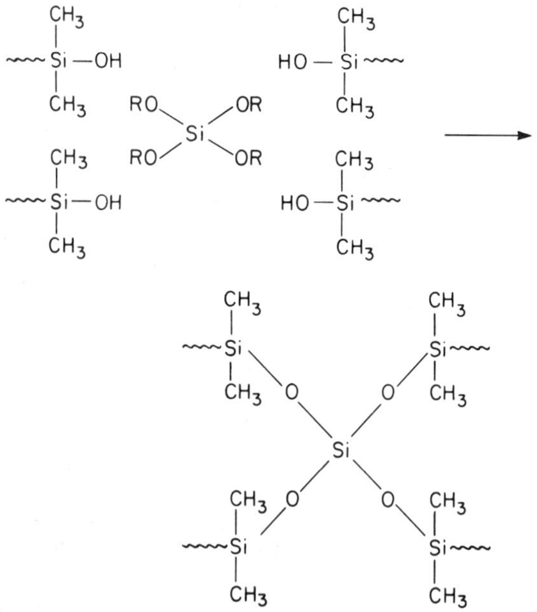

FIG. 4.32 Curing of RTV rubbers by reaction of silanol end groups with silicate esters in the presence of a catalyst such as tin octoate or dibutyltin dilaurate.

Water in the atmosphere may be sufficient for such hydrolysis, which is then followed by a condensation leading to a network structure.

RTV rubbers have proved of considerable value as they provide a method for producing rubbery products with the simplest equipment. These rubbers find use in the building industry for caulking and in the electrical industry for encapsulation.

Nontacky self-adhesive rubbers (fusible rubbers) are obtained if small amounts of boron (~1 boron atom per 300 silicon atoms) are incorporated into the polymer chain. They may be obtained by condensing dialkylpolysiloxanes end-blocked with silanol groups with boric acid or by reacting ethoxyl end-blocked polymers with boron triacetate.

Bouncing putty is somewhat similar in that the Si—O—B bond occurs occasionally along the chain. It is based on a polydimethylsiloxane polymer modified with boric acid, additives, fillers, and plasticizers to give a material that shows a high elastic rebound when small pieces are dropped on a hard surface but flows like a viscous fluid on storage or slow application of pressure.

The applications of the rubbers stem from their important properties, which include thermal stability, good electrical insulation properties, nonstick properties, physiological inertness, and retention of elasticity at low temperatures. The temperature range of general-purpose material is approximately -50 to $+250°C$, and the range may be extended with special rubbers. Silicone rubbers are, however, used only as special-purpose materials because of their high cost and inferior mechanical properties at room temperature as compared to conventional rubbers (e.g., natural rubber and SBR).

Modern passenger and military aircraft each use about 500 kg of silicone rubber. This is to be found in gaskets and sealing rings for jet engines, vibration dampers, ducting, sealing strips, and electrical insulators. Silicone cable insulation is also used extensively in naval craft since the insulation is not destroyed in the event of fire but forms an insulating layer of silica.

The rubbers find use in diverse other applications which include electric iron gaskets, domestic refrigerators, antibiotic container closures, and for nonadhesive rubber-covered rollers for handling such materials as confectionary and adhesive tape.

Due to their relative inertness, new applications have emerged in the biomedical field. A silicone rubber ball is used in combination with a fluorocarbon seal to replace defective human heart valves. Silicone rubber has had many applications in reconstructive surgery on or near the surface of the body. Prosthetic devices are very successfully used in all parts of the body.

The cold-curing silicone rubbers are of value in potting and encapsulation.

In recent years there has been considerable interest in a new form of silicone materials, namely, the *liquid silicone rubbers*. These may be considered as a development from the RTV silicone rubbers but have a better pot life and improved physical properties, including heat stability (in the cured state) similar to that of conventional silicone elastomers. Liquid silicone rubbers range from a flow consistency to a paste consistency and are usually supplied as a two-pack system, which requires simple blending before use. The materials cure rapidly above 110°C. In injection molding of small parts at high temperatures (200 to 250°C), cure times may be as small as a few seconds. One example of application is in baby bottle nipples, which, although more expensive, have a much longer working life. Liquid silicone rubbers have also been used in some extruded applications. Vulcanization of the extruded material may be carried out by using infrared heaters or circulated hot air. The process has been applied to wire coating, ignition cables, optical fibers, various tapes, and braided glass-fiber sleeving, as well as for covering delicate products.

Polyphosphazenes

$$\left[\begin{array}{c} OCH_2CF_2CHF_2 \\ | \\ -N{=}P{-} \\ | \\ OCH_2CF_2CHF_2 \end{array} \right]_n$$

Monomers	Polymerization	Major uses
Phosphorus penta-chloride, ammonium chloride, fluorinated alcohols	Polycondensation followed by nucleophilic replacement of chloro-groups	Aerospace, military, oil exploration applications

Polyphosphazenes containing nitrogen and phosphorus have been synthesized by replacing the chlorine atoms on the backbone chain of polymeric phosphonitrilic chloride (dichlorophosphazene) by alkoxy or fluoroalkoxy groups. These derivative polymers do not exhibit the hydrolytic instability of the parent polymer. The general synthesis scheme is

$$PCl_5 + NH_4Cl \longrightarrow \left[\begin{array}{c} Cl \\ | \\ N{=}P \\ | \\ Cl \end{array} \right]_n \xrightarrow{RONa} \left[\begin{array}{c} OR \\ | \\ N{=}P \\ | \\ OR \end{array} \right]_n$$

Industrial Polymers

With mixtures of alkoxy substituents with longer alkyl chains, crystallization can be avoided to produce an amorphous rubber. The product is referred to as phosphonitrilic fluoroelastomer, a "semiorganic" rubber. The rubber can be cross-linked with free-radical initiators or by radiation. A commercial rubber (PNF by Firestone Tire and Rubber Co.) is based on alkoxides of trifluoroethyl alcohol or heptafluoroisobutyl alcohol.

The polyphosphazene rubbers have excellent resistance to oils and chemicals (except alcohols and ketones), good dynamic properties, good abrasion resistance, and a broad range of use temperatures ($-65°$ to $+117°C$). Disadvantages are water resistance that is only fairly good.

Polythiazyl

Monomers	Polymerization	Major uses
S_4N_4	Ring-opening polymerization	Semiconductive polymers

The solid four-member ring reaction product S_2N_2 obtained by pyrolysis of gaseous S_4N_4 under vacuum is polymerized at room temperature by ring-opening polymerization to give the linear chain polythiazyl

$$S_4N_4 \xrightarrow[\text{1 torr}]{300°C/Ag} \begin{matrix} S-N \\ | \quad | \\ N-S \end{matrix} \longrightarrow \left(S=N \right)_n$$

The product is a brasslike solid material that behaves like a metal or alloy but is lighter and more flexible. Polythiazyl has electrical conductivity (3700 reciprocal ohm-cm, or siemens/cm) at room temperature and superconductivity at 0.3 K. Some doped polymers are photoconductive.

POLYBLENDS [23–25]

The concept of physically blending two or more existing polymers to obtain new products (polymer blends) or for problem solving is now attracting widespread interest and commercial utilization. Polymer blends are often referred to by the contraction "polyblends" and sometimes as "alloys" to borrow a term from metallurgy. The two terms "blend" and "alloy" are considered synonymous. The need for improved balance of properties and the potential ability of polymer blends to satisfy this need have converged in the development of polymer blending as a major area for vigorous growth in the past several years. Properties of plastics that have most often been improved by polymer blending include processability, tensile strength, ductility, impact strength, abrasion resistance, heat deflection temperature, low-temperature flexibility, flame retardance, and environmental stress-crack resistance.

The first commercial blend of two dissimilar polymers was Noryl, a blend of poly(phenylene oxide) and polystyrene, introduced by General Electric in the 1960s. Since that time, the market for polyblends has grown at a rapid rate, and a large number of different blends have been introduced. Blending technology offers many advantages. Gaps in thermoplastic technology can be filled more easily by combining polymers that are already available than by development of new products. For example, the time normally required to develop a new blend ready for the market is 3–5 years compared to 8–10 years required for new plastic materials. This reduces the cost of development of a new product. Cost is also reduced if an expensive polymer can be diluted by a less expensive one.

Blending provides a convenient way of combining the mechanical, physical, or thermal properties of more than one material. One example is the blend of ABS and PVC. The ABS contributes high heat distortion temperature, toughness, and easy moldability, while the PVC imparts weatherability and flame retardance, besides reducing the cost of the blend. Applications of the blend include automotive interior trim, luggage shells, and canoes.

Important disadvantages of a basic polymer (for example, difficulties of fabrication) may be overcome by blending with one or more other polymers. In some cases, synergistic improvement in properties, exceeding the value for either polymer alone, is achieved by blending, though such cases are relatively rare. A good example is the improved Izod notched impact strength shown when ABS and polycarbonate (PC) are blended. At subzero temperatures, the blend has better notched impact strength than either of its component polymers.

The variation of properties with the polymer/polymer ratio in a polyblend is shown schematically in Fig. 4.33. In the case of complete miscibility, we would expect properties to follow a simple monotonic function, more or less proportional to the ratio of the two polymer components in the blend (Fig. 4.33a). This is particularly useful for processors because it permits them to inventory a few commodity polymers and simply blend them in different proportions to meet the specific requirements of each product they manufacture. However, for two polymers to be completely miscible, optimum requirements are: similar polarity, low molecular weight, and hydrogen-bonding or other strong intermolecular attraction.

Most polymer pairs do not meet the above requirements for complete theoretical miscibility. The free energy of mixing is positive, and they tend to separate into two phases. If they are slightly immiscible, each phase will be a solid solution of minor polymer in major polymer, and the phases will separate into submicroscopic domains with the polymer present in major amount forming the continuous matrix phase and contributing most toward its properties. Plots of

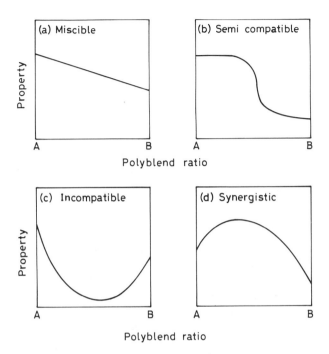

FIG. 4.33 Properties versus polymer/polymer ratio in a polyblend.

properties versus the ratio of the two polymers in the blend will be S-shaped showing an intermediate transition region where there is a phase inversion from one continuous phase to the other (Fig. 4.33b). Most commercial polyblends are of this type, with the major polymer forming the continuous phase and retaining most of its useful properties, while the minor polymer forms small discrete domains, which contribute synergistically to certain specific properties.

When the polymer components in a blend are less miscible, phase separation will form larger domains with weaker interfacial bonding between them. The interfaces will therefore fail under stress and properties of polyblends are thus likely to be poorer than for either of the polymers in the blend. U-shaped property curves (Fig. 4.33c) thus provide a strong indication of immiscibility. In most cases they also signify practical incompatibility, and hence lack of practical utility.

A fourth type of curve for properties versus polyblend ratio representing synergistic behavior (Fig. 4.33d) has been obtained in a few cases of polymer blending. This is synergistic improvement of properties, beyond what would be expected from simple monotonic proportionality, and sometimes far exceeding the value for either polymer alone. Discovery of practical synergistic polyblends is difficult and not predictable with the present state of understanding.

Compatibilization of Polymers

Blends made from incompatible polymers are usually weak. These poor blends are a result of high interfacial tension and poor adhesion between the two phases. For many years, blending of polymers was unsuccessful due to the fact that many polymers were incompatible. Lately new polymer blends have been successfully made by the use of compatibilizing agents. This has yielded polymers of unique properties not attainable from either of the polymer components of the blends. Sometimes these blends are preferred over completely miscible blends since they may combine important characteristics of both polymers. Their performance greatly depends on the size and morphology of the dispersed phase.

Polymers can be compatibilized in a number of ways. Polymers that act as mutual solvents can be used as compatibilizing agent, e.g., polycaprolactone in a blend of polycarbonate and poly(styrene-co-acrylonitrile). Polymers can also be compatibilized by the use of functionalized polymers. Thus polymers can be post-reacted with a reactive monomer to form functional groups that react with a second polymer to form grafts in a blend, e.g., EPDM-g-maleic anhydride/nylon-6,6 blends are produced by grafting maleic anhydride onto EPDM, followed by blending with nylon-6,6. The grafted maleic anhydride reacts with the terminal $-NH_2$ groups of the nylon to form the following copolymer:

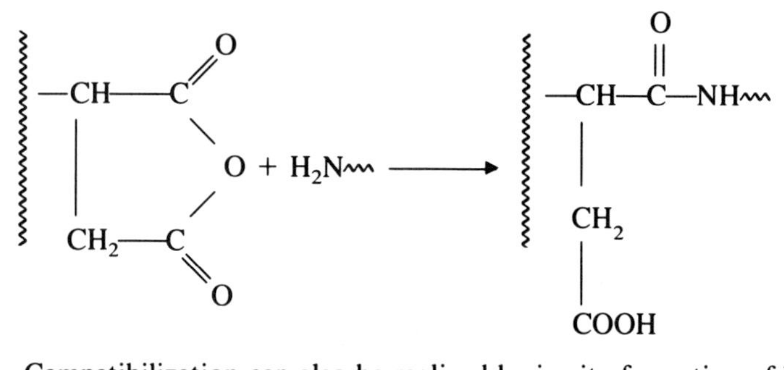

Compatibilization can also be realized by in situ formation of block or graft copolymers. For example, blends of EPDM and PMMA are obtained by the reactive extrusion of the homopolymers and a peroxide in a twin screw extruder. Block or graft copolymers can also be added separately to produce blends with controlled particle sizes. This type of compatibilizing agent is called an interfacial agent since it acts at the boundary between two immiscible polymers.

An interfacial compatibilizing agent functions as surfactant, somewhat the same as a low molecular weight surfactant used in emulsion polymerization. It is usually a block or graft copolymer and is located at the interface between two polymers (Fig. 4.34). It reduces interfacial tension, alleviates gross segregation, and promotes adhesion. The effective concentrations are from 0.1% to 5% depending on the desired particle size and the effectiveness of the compatibilizing agent.

The block or graft copolymer used as a compatibilizing agent must have the ability to separate into their respective phases and not be miscible as a whole molecule in one phase. The best choice of block or graft copolymers would be to choose polymers identical to the blended polymers, e.g., poly(A-b-B) or poly(A-g-B) copolymers for blending poly(A) and poly(B). Another alternative is to use poly(C-*b*-D) or poly(C-*g*-D) where C and D are compatible with A and B, respectively or where C and D adhere to A and B, respectively, but are not miscible with them. Examples of such compatibilized systems that have been studied include EPDM/PMMA blends compatibilized with EPDM-g-MMA, polypropylene/polyethylene with EPM or EPDM, polystyrene/nylon-6 with polystyrene/nylon-6 block copolymer, and poly(styrene-co-acrylonitrile)/poly (styrene-co-butadiene) with butadiene rubber/PMMA block copolymer.

Reactive compatibilizing agents of the type A-C can also compatibilize an A/B blend as long as C can chemically react with B. Recently studied systems include polyethylene (PE)/nylon-6 blends compatibilized with carboxyl functional PE, polypropylene (PP)/poly(ethylene-terephthalate) with PP-*g*-acrylic acid, nylon-6,6/EPDM with poly(styrene-co-maleic anhydride), and nylon-6/PP with PP-*g*-maleic anhydride. For the last cited system, for example, maleic anhydride was grafted onto PP and the latter was then blended with nylon-6 in

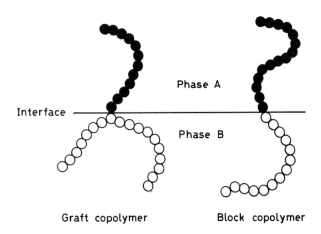

FIG. 4.34 Ideal location of block and graft copolymers at the interface between polymer phases A and B in a polyblend.

order to compatibilize an immiscible blend of nylon-6/PP by A-C method through anhydride-amine reaction.

Industrial Polyblends

The automotive sector is an important area for polymer blends. Alloys of poly-carbonate (PC) and poly(butylene terephthalate) (PBT), which combine the high impact strength of PC with the resistance of gasoline and oil and processing ease of PBT, are being increasingly used in car bumpers. General Electric's PC/PBT alloy is used for bumpers in several cars made in the United States and Europe. Bayer's PC/PBT blends, called Makrolon PR, are used for several automobile impact-resistant parts. Makrolon EC900 is also used for crash helmets.

An important requirement for plastics materials in the automotive industry is paintability alongside metal components in high temperature areas. Noryl GTX series of General Electric has been developed to meet this requirement. The blend consisting of polyphenylene oxide (PPO) in a nylon matrix combines the heat and dimensional stability and the low water absorption of PPO with the flow and semicrystalline properties of nylons. Its impact strength, however, is not as high as PC/PBT blends. Several grades of Noryl GTX are available, some of these being glass-filled. The unfilled grades are used for automobile exterior components such as fenders, spoilers and wheeltrims, whereas the glass-reinforced grades are used in under-the-bonnet applications, such as cooling fans, radiator end caps, and impeller housings, where in addition to high heat resistance extra stiffness is required. Bayer's Bayblend PC/ABS alloys are used in dashboard housings, interior mirror surrounds, and steering column covers. The high stability of the alloy allows pigmentation in pastel colors.

Some alloys of PPO have been developed specifically for office equipment and housings. Noryl AS alloys, which are blends of PPO/PS and special fillers, are expected to replace metals in such applications as structural chassis for type-writers and printers.

Monsanto Chemical Co. produce injection molded Nylon/ABS alloys (Triax) that are characterized by good toughness, high chemical resistance, and easy moldability. Applications of these alloys include industrial power tools, lawn and garden equipment housings and handles, gears pump impellers, and car fascia.

Bayer and BASF have developed a PC/ASA blend where ASA is the rubber variant of ABS in which the polybutadiene phase is completely replaced by a polyacrylate rubber. Compared to PC/ABS alloys, the PC/ASA blend has greater resistance to weathering, which permits outdoor applications and an ex-tra heat stability that reduces the risk of yellowing in processing. BASF's alloy, Terblend S, is recommended for outdoor applications as it has six times the UV resistance of PC/ABS and an even higher resistance when pigmented.

A number of polyphenylene ether (PPE) blends have been introduced by Borg-Warner, Hulls, and BASF. These blends have the advantage that they can be made flame retardant without halogens. Borg-Warner offers many grades of PPE/high impact polystyrene (HIPS) blend under the trade name Prevex. The materials are characterized by high impact resistance and high heat distortion temperatures. Applications are for office equipment, business machines, and telecommunication equipment. Luranyl PPE/HIPS blends of BASF include general purpose types, reinforced grades, impact modified types, and flame-proof blends. BASF also has a PPE/nylon blend tradenamed Ultranyl, which is aimed at a similar market.

Blends have also been used to a limited extent in packaging. Du Pont's Solar is a nylon/polyethylene blend that can be blow-molded into bottles and can contain volatile hydrocarbons found in household cleaners and agricultural chemicals. Further developments in such barrier blends are expected.

INTERPENETRATING POLYMER NETWORKS

An interpenetrating polymer network (IPN) has been defined [26] as "an intimate combination of two polymers both in network form, at least one of which is synthesized or cross-linked in the presence of the other." There are no induced covalent bonds between the two polymers, i.e., monomer A reacts only with other molecules of monomer A. A schematic representation of an ideal IPN (*full IPN*) in which both the polymers are cross-linked is shown in Fig. 4.35a. If one of the two polymers is in network form (cross-linked) and the other a linear polymer, i.e., not cross-linked, the product is called a *semi-IPN* (Fig. 4.35b). An IPN is distinguished from simple polymer blends and block or graft copolymers in two ways: it swells but does not dissolve in solvents, and creep and flow are suppressed [26].

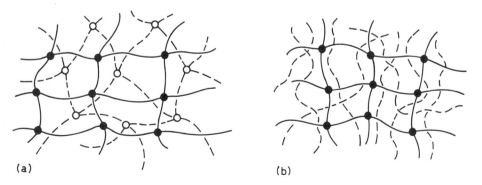

(a)   (b)

FIG. 4.35 Schematic representation of (a) a full IPN and (b) a semi-IPN.

IPNs represent a third mechanism, in addition to mechanical blending and copolymerization, by which different polymers can be intimately combined. IPNs can be synthesized either by sequential polymerization or simultaneous polymerization of two monomers. In a sequential synthesis, monomer I is polymerized using an initiator and a cross-linking agent to form network I. Network I is then swollen in monomer II containing cross-linking agent and initiator to form network II by polymerization *in situ*. Such polymers are called *sequential interpenetrating networks* (SIPN). On the other hand, polymer networks synthesized by mixing monomers or linear polymers (prepolymers) of the monomers together with their respective cross-linking agents and catalysts in melt, solution, or dispersion, followed, usually immediately, by simultaneous polymerization by noninterfering modes, are called *simultaneous interpenetrating networks* (SIN). In the latter process, the individual monomers are polymerized by chain or stepwise polymerization, while reaction between the polymers is usually prevented due to different modes of polymerization.

The advantage of SINs is the ease of processing; for example, polyurethane-epoxy or unsaturated polyester SINs are used in reaction injection molding (RIM) applications. An A-component consists of the polyurethane polyol, while liquid MDI blended with liquid epoxy or unsaturated polyester forms the B-component. The two components may be mixed together for about 30 seconds and elastomeric sheets cast in heated aluminum molds. The modulus of such SIN-RIM part is enhanced by modifying the polyurethane with glassy epoxy or unsaturated polyester, unlike conventional reinforced RIM (RRIM), which utilizes glass fibers to increase the modulus (glass has the disadvantage of increasing the viscosity and wearing down machine pumps by abrasion).

Interpenetrating polymerization is the only way of combining cross-linked polymers. This technique can be used to combine two or more polymers into a mixture in which phase separation is not as extensive as it would be otherwise. Normal blending or mixing, it may be noted, result in a multiphase morphology unless the polymers are completely compatible. However, complete compatibility, which is almost impossible, is not necessary for complete phase mixing by interpenetrating polymerization, since permanent entanglements produced by interpenetration prevent phase separation. With moderate compatibility, intermediate and complex phase behavior results. Thus IPNs with dispersed phase domains have been reported ranging from a few micrometers (the largest) to a few hundred nanometers (intermediate) to those without a resolvable domain structure (complete mixing). With highly incompatible polymers, on the other hand, the thermodynamics of phase separation is so powerful that it occurs before it can be prevented by cross-linking.

Combining polymeric networks in different compositions often results in IPNs with controlled different morphologies showing synergistic behavior. For example, a glassy polymer (T_g above room temperature) combined with an elastomeric polymer (T_g below room temperature) gives a reinforced rubber, if the elastomeric phase is predominant and continuous, or a high impact plastic, if the glassy phase is continuous. More complete phase mixing enhances mechanical properties due to increased physical cross-link density. IPNs have thus been synthesized in various network compositions with optimum bulk properties such as tensile strength, impact strength, and thermal resistance.

Industrial IPNs

Commercial IPNs have been developed to combine useful properties of two or more polymer systems. For example, high levels of silicone have been combined with the thermoplastic elastomer (TPE) based on Shell's Kraton styrene-ethylene/butadiene-styrene TPE and Monsanto's Santoprene olefin TPE. These IPN TPEs are said to provide the biocompatibility and release properties of silicone with tear and tensile strength up to five times greater than medical-grade silicone. Thermal and electronic properties and elastic recovery are also improved. IPN TPEs offer physical and thermal properties of thermoset rubber, the processability of a thermoplastic, and a wider hardness range than available to other TPEs.

Rimplast materials from Petrarch are silicone/thermoplastic IPNs that combine the warpage and wear resistance properties of silicone with nylon, thermoplastic polyurethane, or styrene/butadiene block copolymers. The combination of properties these IPNs possess make them suitable for high quality, high tolerance gear and bearing applications that can benefit from the addition of internal lubrication and isotropic shrinkage of silicone to the fatigue endurance and chemical resistance of crystalline resin, while processability still remains the same.

Interpenetrating polymer/metal networks are now on the market. One of these, produced by Princeton Polymer Laboratories, imparts conductive properties to the IPN. Low melting alloys such as zinc and tin are melt-blended with a resin, forming a fibrous network within the resin matrix. The process of metal addition eliminates the problem of breaking up of fiber or flake additives and has very little effect on processability.

REFERENCES

1. A. Renfrew and P. Morgan (eds.), *Polythene—The Technology and Uses of Ethylene Polymers*, 2nd ed., Iliffe, London (1960).
2. T. O. J. Kresser, *Propylene*, Van Nostrand Reinhold, New York (1960).
3. P. D. Ritchie (ed.), *Vinyl and Allied Polymers*, Vol. 1. *Aliphatic Polyolefins and Polydienes: Fluoro-Olefin Polymers*, Iliffe, London (1968).
4. H. A. Sarvetnick, *Polyvinyl Chloride*, Van Nostrand Reinhold, New York (1969).

5. J. H. L. Henson and A. Whelan (eds.), *Developments in PVC Technology*, Applied Science Publishers, London (1973).
6. A. Whelan and J. L. Craft (eds.), *Developments in PVC Production and Processing* I, Applied Science Publishers, London (1977).
7. L. I. Nass, *Encyclopedia of PVC*, Marcel Dekker, New York, Vol. 1 (1976), Vol. 2 (1977).
8. W. E. Nelson, *Nylon Plastics Technology*, Newnes-Butterworths, London (1976).
9. M. I. Kohan, *Nylon Plastics*, John Wiley, New York (1973).
10. L. N. Phillips and D. B. V. Parker, *Polyurethanes—Chemistry, Technology and Properties*, Iliffe, London (1964).
11. A. Whelan and J. A. Brydson (eds.), *Developments with Thermosetting Plastics*, Applied Science Publishers, London (1974).
12. W. G. Potter, *Epoxide Resins*, Iliffe, London (1970).
13. C. A. May and Y. Tanaka (eds.), *Epoxy Resins: Chemistry and Technology*, Marcel Dekker, New York (1973).
14. P. E. Cassidy, *Thermally Stable Polymers*, Marcel Dekker, New York (1980).
15. N. H. Ray, *Inorganic Polymers*, Academic Press, London (1978).
16. H. Ulrich, *Introduction to Industrial Polymers*, Hanser Publishers, Munchen (1982).
17. J. A. Brydson, *Plastics Materials*, 4th ed., Butterworth Scientific, London (1982).
18. W. E. Driver, *Plastics Chemistry and Technology*, Van Nostrand Reinhold, New York (1979).
19. J. H. Dubois and F. W. John, *Plastics*, Van Nostrand Reinhold, New York (1981).
20. K. J. Saunders, *Organic Polymer Chemistry*, Chapman and Hall, London (1973).
21. H. F. Mark and N. G. Gaylord, *Encyclopedia of Polymer Science and Technology*, Interscience, New York, Vols. 1–16 (1964–1972).
22. *Modern Plastics Encyclopedia 1985–1986*, McGraw-Hill, New York (1985).
23. R. B. Seymour and H. F. Mark (eds.), *Applications of Polymers*, Plenum Press, New York (1988).
24. D. R. Paul and Seymour Newman (eds.), *Polymer Blends*, Vol. 1, Academic Press, New York (1978).
25. J. Baker, *Recent Developments in Polymer Technology*, RAPRA Technology Ltd., Shawbury, England (1987).
26. L. H. Sperling, *Interpenetrating Polymer Networks and Related Materials*, Plenum Press, New York (1981).

5

Polymers in Special Uses

INTRODUCTION

There are a number of polymeric materials that distinguish themselves from others by virtue of their limited use, high prices, or very specific application or properties. The expression *specialty polymer* is, however, slightly ambiguous to use for such materials as the definition covers any polymeric material that does not have high volume use. There are thus some materials that were originally developed as specialties have now become high volume commodities, while a number of materials developed some years ago still fall into the specialty category. Some examples of the latter are polytetrafluoroethylene, polydimethyl siloxane, poly(vinylidene fluoride), and engineering materials such as poly-(phenylene oxide), poly(phenylene sulfide), polyether sulfone, polyether ether ketone, and polyetherimide.

In this chapter, attention is focused on a number of polymers that are either themselves characterized by special properties or are modified for special uses. These include high-temperature and fire-resistant polymers, electroactive polymers, polymer electrolytes, liquid crystal polymers, polymers in photoresist applications, ionic polymers, and polymers as reagent carriers and catalyst supports.

HIGH-TEMPERATURE AND FIRE-RESISTANT POLYMERS [1–3]

Compared to traditional materials, especially metals, organic polymers show high sensitivity to temperature. Most importantly, they exhibit very low softening points, which is attributed to the intrinsic flexibility of their molecular chains. Thus, whereas most metals do not soften appreciably below their melting points, which may be 1000°C or higher, many polymers commonly used as plastics such as polyethylene, polystyrene, and poly(vinyl chloride), soften sufficiently by about 100°C to be of no use in any load-bearing applications. The poor thermal resistance of common polymers has greatly restricted some of their application potential. In two particular application areas, namely electrical and transport applications, this restriction has long been particularly evident.

Owing to their unique electric insulation properties, polymers are widely used in electrical products. However, many electrical components are required to operate at high temperatures, for example, electric motors and some domestic appliances.

Another characteristic property of polymers, namely their high specific stiffness and strength (which are due to their low density, especially when used in fiber-reinforced composite materials), has led to the use of polymers in transport applications, especially in aerospace industries, where weight saving is of vital importance and materials cost is secondary. However, here again many applications also demand high temperature resistance.

More recently, road vehicle manufacturers in their attempt to save weight (and hence reduce fuel consumption) have been replacing heavy metal components with light plastic ones. The ease of molding plastics into intricately shaped parts has also been used to advantage for fabricating many under-the-bonnet products. Here again, resistance to elevated temperatures is often also needed, necessitating use of high-temperature resistant polymers.

Although electrical and transport applications have perhaps provided the biggest demand for thermally resistant specialty polymers, such polymers are also being sought increasingly for use in more mundane consumer goods, especially appliances where exposure to elevated temperature can occur, such as hair dryers, toasters, and microwave ovens.

Beside low softening points, a further restriction to the use of common polymers at high temperature is their susceptibility to atmospheric oxidation when used at elevated temperatures. This oxidation usually leads to polymer degradation by chain scission by way of a free-radical chain reaction. In the absence of atmospheric oxygen, however, polymers do not undergo degradation until they are at much higher temperatures, usually in excess of 300°C. While this purely thermal degradation resistance is not relevant to most polymer applications since polymers are usually used in air, it is to be noted that purely thermal degradation is the first stage in the burning of polymers. Thus, polymers with good thermal stability and a higher T_m should also offer good fire resistance. Moreover, since the softening point is usually just below the T_g for an amorphous polymer or between T_g and T_m for a partially crystalline polymers, polymers having a high T_g and/or T_m will also have high softening points.

Temperature-Resistant Polymers

To measure the thermal stability of polymers, one must define the thermal stress in terms of both time and temperature. An increase in either of these factors shortens the expected lifetime. In general terms, for a polymer to be considered thermally stable, it must retain its physical properties at 250°C for extended periods, or up to 1000°C for a very short time (seconds). As compared to this, some of the more common engineering thermoplastics such as ABS, polyacetal, polycarbonates, and the molding grade nylons have their upper limit of use temperatures (stable physical properties) at only 80 to 120°C.

For a polymer to be thermally stable in its practical applications it must have, in addition to high melting (softening) temperature, resistance to oxidative degradation at elevated temperatures, resistance to other (nonoxidative) thermolytic processes, and stability to radiation and chemical reagents. The principal ways to improve the thermal stability of a polymer are to increase crystallinity, introduce cross-linking, increase inherent stiffness of the polymer chain, and remove thermoxidative weak links. Although cross-linking of oligomers is certainly useful and does make a real change in properties (see Chapter 1), crystallinity development has limited application for very high temperatures, since higher crystallinity results in lower solubility and more rigorous processing conditions. Chain stiffening or elimination of weak links is a more fruitful approach.

The weakest bond in a polymer chain determines the overall thermal stability of the polymer molecule. The aliphatic carbon-carbon bond has a relatively low bond energy (see Table 5.1). Oxidation of alkylene groups is also observed during prolonged heating in air. Thus the weak links to be avoided are mostly alkylene, alicyclic, unsaturated, and nonaromatic hydrocarbon. On the other hand, the functions proven to be desirable are aromatic (benzenoid or heterocyclic) ether, sulfone, and some carboxylic acid derivatives (amide, imide, etc.). Aromatic rings in the polymer chain also give intrinsically stiff backbone.

It follows from this reasoning that aromatic polymers will have greater thermal stability. For example, poly(*p*-phenylene) was synthesized by Marvel by stereospecific 1,4-cyclopolymerization of cyclohexadiene, followed by dehydrogenation:

TABLE 5.1 Bond Energies of Common Organic and
and Inorganic Polymers

Bond	Bond energy	
	(kcal/mol)	(kJ/mol)
$C_{al}-C_{al}$	83	347
$C_{ar}-C_{ar}$	98	410
$C_{al}-H$	97	405
$C_{ar}-H$	102	426
C—F	116	485
B—N	105	439
Si—O	106	443

The polymer was found to be infusible and insoluble. Another example of aromatic polymer is poly(p-xylene), which is usually vacuum deposited as thin films on substrates. In one application the monomer is di-p-xylylene formed by the pyrolysis of p-xylylene at 950°C in the presence of steam. When this monomer is heated at about 550°C at a reduced pressure, a diradical results in the vapor phase, which, when deposited on a surface below 70°C, polymerizes instantaneously and forms a thin, adherent coating.

Metals or other substrates can be coated in this way. A major application has been the production of miniature capacitors that have the polymer as a dielectric.

Commercial polymers based on the principle of synthesis of polyaromatic compounds include the previously discussed commercial polymers—aromatic polyamides, polyimides, poly(phenylene oxide), polysulfone, and polybenzimidazole (see Chapter 4).

Another approach to achieve thermal stability is to synthesize *ladder polymers*, so called because of their ladderlike structure (⊏⊐⊏⊐). For example, pyrolysis of polyacrylonitrile gives a ladder polymer of high thermal stability:

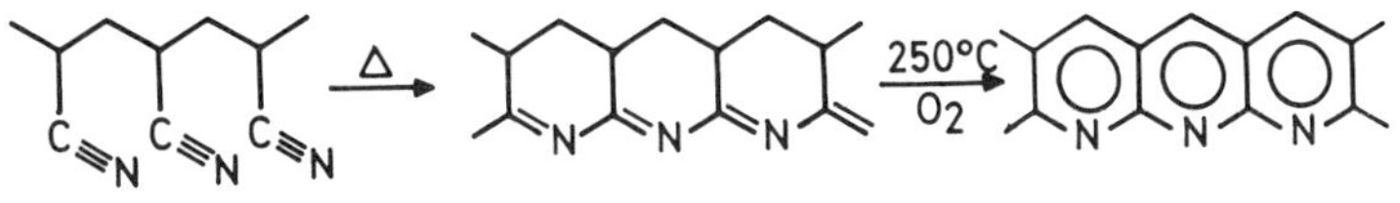

The product (black orlon) is so stable that it can be held directly in a flame in the form of woven cloth and not be changed physically or chemically. Further heating of black orlon to 1400 to 1800°C and simultaneous stretching produces graphite HT. If the heating and stretching is conducted at 2400 to 2500°C, the high modulus graphite HM is obtained. Other carbonizable polymers that produce carbon fibers on heating include poly(vinyl acetate), poly(vinyl chloride), poly(vinylidene chloride), and cellulose. Thermosets, such as phenolic resins, generally produce nongraphitizing or glassy carbon.

Ladder polymers are also produced by polycondensation reactions of tetrafunctional monomers. If a tetrafunctional monomer is reacted with a bifunctional monomer, as in the formation of polyimides, the derived polymer is referred to as a partial ladder or *stepladder* polymer.

If two tetrafunctional monomers are used, as in the formation of polyquinoxaline from an aromatic tetramine and an aromatic tetraketone, the resulting polymer is a ladder polymer:

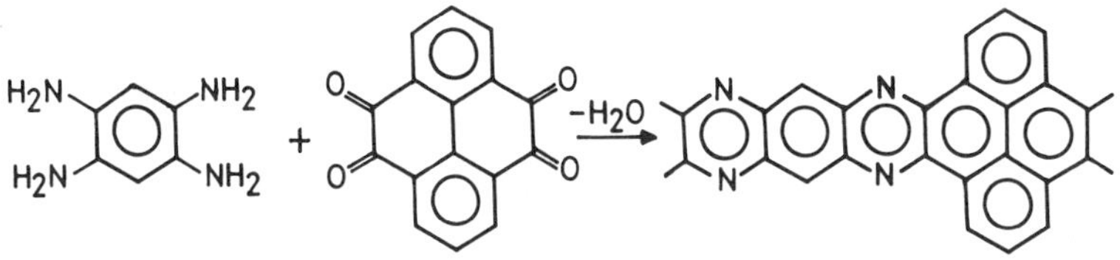

Other tetraketones have also been used in the preparation of polyquinoxalines (PQs). The PQs have proven to be one of the better high-temperature polymers with respect to both stability and potential application. The PQs are also one of the three most highly developed systems—the others being benzimidazoles and oxadiazoles. The interest in the PQs increased considerably after the development in 1967 of the soluble phenylated PQ (PPQ). PQs are stable to 550°C and are used for high-temperature composites and adhesives.

Fire-Resistant Polymers

To understand the burning of polymers better, the combustion region may be divided into five zones, as shown in Fig. 5.1. The first zone is defined by the polymer layer where pyrolysis takes place due to heat produced in the flame,

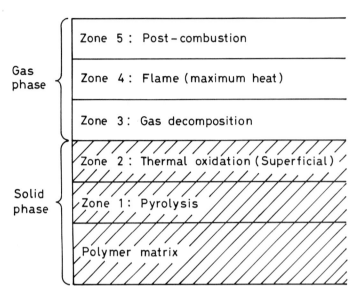

FIG. 5.1 Schematic representation of the combustion region of a burning polymer.

but very little oxidation takes place. Thermal oxidation takes place in the second superficial zone. In the third zone, which is a gaseous one, low-molecular-weight products formed in these two zones are mixed with heated air and are decomposed or oxidized by oxygen or by radicals coming out of the flame. The fourth zone begins where the concentration of degradation products is sufficient for the flame to grow, emitting light radiation. This is where most of the thermal energy is released, and the temperature reaches its maximum value. The fifth zone, also called the post-combustion zone, is where oxidation reactions terminate and combustion products are found in maximum concentration. The thermal energy produced in this zone and in the fourth zone, together with the light radiation, are responsible for the degradation that occurs in the inner part of the polymer not touched by the flame.

It is evident from the above discussion that the burning of polymers may be considered to consist of three principal stages, namely, (1) thermal (nonoxidative) decomposition of the polymer, (2) evolution and transport of volatiles from the burning surface to the flame, and (3) high temperature (600–900°C) oxidation of the volatiles in the flame itself. Interruption of the processes occurring at any of the above three stages will reduce polymer flammability. Thus, polymers with good thermal stability that are, therefore, resistant to thermal decomposition also offer good fire resistance.

Although, because many factors are involved, no single test can provide an adequate picture of the fire hazards of polymers, the limiting oxygen index

(LOI) is the single most widely used test of polymer flammability. The test (ASTM D-2863) is made in a very simple way. A small rod of a plastic material is burned in a Pyrex tube in the presence of oxygen and nitrogen (see Fig. 3.60). The tubing is connected to flowmeters and the minimum amount of oxygen, in an oxygen-nitrogen mixture, required to just sustain burning is determined. The limiting oxygen index is defined by

$$\text{LOI} = \frac{[O_2]}{[O_2] + [N_2]} \times 100$$

Thus a smaller LOI value indicates a greater flammability. LOI values for some common and not so common polymers are shown in Table 5.2.

The polymers listed in Table 5.2 have been classified into non-self-extinguishing and self-extinguishing types depending on their LOI values. It is seen that aromatic polymers have low flammability and are self-extinguishing, which reflects their much greater thermal stability compared with aliphatic polymers. It is because of this property that aromatic polymers are often specified in fire-hazardous electrical insulation applications or for fire-resistant textiles. In contrast, most of the common polymers, with the exception of PVC, are non-self-extinguishing and flammable.

The presence of halogen in a polymer produces the most dramatic increase in LOI and decrease in flammability. Thus, rigid (unplasticized) PVC, with a LOI value of 42, has the lowest flammability of any common and many not so common polymers. This advantage is, however, offset by the large amount of smoke and toxic gas (hydrochloric acid) produced when PVC and other chloropolymers

TABLE 5.2 Limiting Oxygen Indices (LOI) of Polymers[a]

Non-self extinguishing polymer	LOI	Self-extinguishing polymer	LOI
Polyoxymethylene	16	Polycarbonate	27
Polyethylene	17	Polyarylate	34
Polypropylene	18	Polyethersulfone	38
Polystyrene	18	Polyether ether ketone	40
Poly(methyl methacrylate)	18	Poly(vinyl chloride), rigid	42
Polyisoprene (natural rubber)	18	Polyamide-imide	43
ABS	18	Poly(phenylene sulfide)	44
Nylon-6,6	24	Polyether-imide	47
Poly(ethylene terephthalate)	25	Polypyromellitimide	50
Polychloroprene	26	Poly(vinylidene chloride)	60
		Polytetrafluoroethylene	95

[a]The LOI value for self-extinguishing behavior is often taken as 27, not 21 (which is the volume % concentration of oxygen in air), to correct for a lack of convective heating in the LOI test.

burn. Moreover, plasticization of PVC markedly reduces the LOI; for example, the value decreases to 25 when 50 phr dioctylphthalate is used as a plasticizer. Fluoropolymers also have low flammability. The fully fluorinated polymer polytetrafluoroethylene, for example, burns only in 95% oxygen under the test conditions of LOI. The burning, however, produces a highly toxic and corrosive gas hydrogen fluoride.

ELECTROACTIVE POLYMERS [4–17]

Not many years ago, plastics were best known for their (electrical and thermal) insulating properties and great strides were made in electronics, in electrical component design, and electrical equipment as the dielectric strength of plastics was improved. The field of thermal insulation was advanced greatly as ways were found to produce plastic fibers and foams that could give an order of magnitude better insulation than ever before. And yet today researchers are looking for ways to make plastics into better conductors of electricity and heat. The reasons are related to the fact that plastics have certain other special properties besides their insulating abilities, and to make full use of the former in many newer applications, the plastics should be electrically and/or thermally conducting.

Three of the special properties of plastics referred to above are flexibility, low density, and the ability, through appropriate processing, to form intricate shapes and patterns. Die-cast aluminum, zinc, and their alloys are typical materials that are frequently used to make conductive housings. However, structures made from these metals and alloys are not flexible and often weigh more than desired. Also their costs have risen markedly. As computers and sophisticated electronics devices began to move out of their shielded rooms into offices, stores, and homes, it has become highly desirable to take advantage of the light weight, low cost, and aesthetics that could be gained by substituting plastics for metal housings for the instruments. Conductive plastics have therefore been increasingly used to provide flexible, lightweight, and moldable parts having good static bleed-off and electromagnetic interference (EMI) shielding properties. A variety of uses of such materials are encountered, ranging from compliant gasketing to rigid housings for business machines.

Conductive polymers fall into two distinct categories: *filled polymers*, which are used for a wide range of anti-static and static-discharging applications, and *intrinsically conductive polymers*, which contain no metals but conduct electricity when chemically modified with dopants.

A number of other polymeric solids have also been the subject of much interest because of their special properties, such as polymers with high photoconductive efficiencies, polymers having *nonlinear optical* properties, and polymers with *piezoelectric*, *pyroelectric* and *ferroelectric* properties. Many of these polymeric materials offer significant potential advantages over the traditional materials used for the same application, and in some cases applications not possible by other means have been achieved.

Filled Polymers

Polymers can be made to coduct electricity relatively easily by compounding them with high loadings of conductive materials. The plastic matrix in such a case functions as a forming, containing a protective medium in which the conductive particles provide the electrical properties. Apart from the inherent properties of the fillers, parameters such as concentration, particle form (sphere, flake, fiber), size, distribution, and orientation are deciding factors that influence the properties of filled polymers.

When choosing a filler, the following requirements merit consideration: (1) the filler has high conductivity to avoid excess weight; (2) it does not impair the physical/mechanical properties of the plastic; (3) it is easily dispersed in the plastic; (4) it does not cause wear on forming tools (injection molding, extrusion); (5) it has favorable cost picture; and (6) it produces good surface structure of the finished product. The commonly used electrically conducting fillers are carbon, aluminum, and steel. The most common metallic conductor, copper, is not used because it oxidizes within the plastic and impairs its physical properties.

Typical dimensions for common fillers are:

Aluminum flakes: length 1–1.4 mm, thickness 25–40 μm
Carbon black: spherical, diameter 25–50 μm
Graphite fiber: length mm-cm, thickness 8 μm
Steel fiber: length 3–5 mm, diameter 2–22 μm

All of them are to be found in a number of qualities with varying prices and conduction properties.

Depending on particle form and orientation, there is a certain critical volume concentration at which the resistance decreases, i.e., the conductivity increases, drastically. This is shown in Fig. 5.2. At the *critical concentration*, the filler can form a continuous phase through the matrix in the form of microscopic conductive channels. As shown in Fig. 5.2, the specific resistance of filled polymers also depends on the inherent conductivity and particle form of the filler besides its concentration. The critical concentration can be reduced to low levels by using conductive particles that are fibrous in shape. The reduction in critical volume loading is proportional to the magnitude of the fiber's aspect ratio (length/diameter).

It has been shown that even extremely small concentrations of additives can make plastics conductive if they are in the form of conductive fibers with length

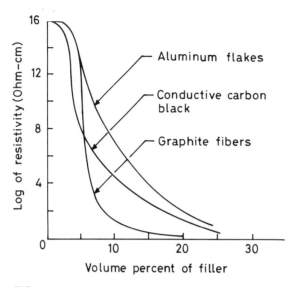

FIG. 5.2 Composite resistivity as a function of filler volume loading and filler type [4].

to diameter (*L/D*) ratio of 100 or more. The striking difference between the use of chunky fragments and fibrous materials in their effect on conductivity can be seen from the diagram in Fig. 5.3.

If one loads a plastic with 5 percent by volume of 6-mil size in a random distribution (Fig. 5.3a), there will be, on the average, a 6-mil gap to the nearest sphere. Heat or electrons flowing through such a matrix would thus cross alternate paths of about equal lengths in the two media. The picture, however, becomes quite different if the same 5 percent of material is dispersed as 1-mil diameter and 100-mil long fibers. Even if the fibers are all arranged parallel to

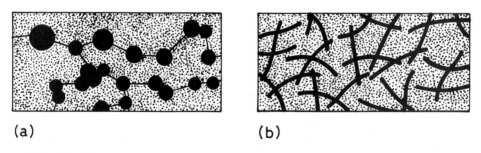

(a)   (b)

FIG. 5.3 Comparison of typical flow path through composites using the same volume percentage of material as spheres and fibers (*L/D* = 100). (a) Chunky particles—long path through plastic; (b) fibers—short gaps in plastic.

one another and at an even spacing, they would be only 3 mils apart. It is thus inevitable that at random orientations they will touch one or more of their nearest neighbors, as shown in Fig. 5.3b, thereby providing an almost continuous path through the composite along the conductive fibers. Fibers will therefore be more effective in lowering the electrical resistivity of plastic and increasing the thermal conductivity than chunky fragments.

The above concept led to the creation of *metalloplastics*—a family of conductive plastics in which the conductivity is great enough to make the material resemble metal both electrically and thermally, as opposed to the so-called semiconductor levels achieved by adding carbon to a plastic. To serve as a practical engineering material, metalloplastics should have electrical resistivities of less than 1 ohm-cm and thermal conductivities of at least a factor of ten higher than those of normal plastics.

The importance of high thermal conductivity in plastics is being recognized as the automobile business tries hard to use more and more plastic parts to reduce the weight of the automobile. Also the industry is required to make parts very quickly (taking only seconds or a fraction of a minute at most to make each one) since they must make millions of parts a year. With conventional plastics having poor thermal conductivity, the problem of getting heat into the part to form it and then getting the heat out again results in cycle times of the order of minutes, not seconds. This implies an increase of one to two orders of magnitude in the number of tools necessary to make the same number of parts. So, for the auto industry, an order of magnitude increase in thermal conductivity could be a very significant advance.

Metalloplastics with electrical resistivities as low as 0.01 ohm-cm and up to 100-fold higher thermal conductivity than ordinary plastics have been developed by the addition of a few percent of metal and/or metallized glass fibers (L/D of the order of 100/1) to plastics. Use of such materials can have very significant effects on molding cycle times, uniform heating and cooling rates, and heat transfer rates in the final product.

Though fibers are more effective than chunky fragments in lowering the electrical resistivity and increasing the thermal conductivity, the effects of low concentrations of fibers on these two properties are quite different. Whereas, thermal conductivity is proportional to the concentration of conductive fillers and is increased even by such low concentrations of the fibers that one fiber does not touch its neighbors, the electrical resistivity is not significantly modified until an almost continuous path is available through the conductive fibers. There can thus be a region of low concentrations and short fiber lengths in which one has no significant electrical conductivity and gets significant improvements in thermal conductivity. Plastics can thus be developed that are improved in thermal conductivity but can be used for electrical insulation or resistive heating. Suitably filled polymers are thus used to drain off heat in pressure switches as well as in polymeric tapes intended for self-regulating, resistive heating of water pipes, railroad switches, etc. In the latter case, use is made of the shrinkage of plastics at decreasing temperatures, which increases conductivity continuously.

Filled conductive polymers are 10 to 12 orders of magnitude more conductive than unfilled polymers but are still several orders of magnitude lower than copper. Carbon-black filled polymers are the most common. Fillers other than carbon black include finely divided metal flakes and fibers, metallized glass fibers, and metallized inorganics such as mica.

Filled conductive polymers used for packaging include polycarbonate, polyolefin, and styrenics incorporating fillers such as carbon, aluminum, and steel flakes and fibers. A polycarbonate/ABS blend (Bayblend ME) introduced by Bayer is 4% aluminum filled and suited to many screening functions.

The electrically conductive polymers have their greatest use in EMI-shielded casings and protective housings of electronic and telecommunications equipment, where they have rapidly substituted metal because of their low weight and easy workability. Steel fiber is mostly used in filled conductive polymers intended for EMI-shielding applications, which will be discussed below.

Conductive rubber has been used in a number of applications. In most cases, silicone rubber is used because of its greater resistance to temperature, UV light, oxygen, corrosive gases, chemicals, and solvents as compared to organic rubber. Conductive rubber is used in tires to leak off static electricity. Ethylene-propylene conductive rubber is used as a covering in cables to reduce high voltage problems. In all these cases, carbon black is generally used as the filler, one reason being that it also acts as a reinforcing filler, increasing the strength and tear resistance of the product. In the electronics industry, conductive rubber finds use as a connector for liquid crystal displays in electronic digital watches.

EMI Shielding

Electromagnetic interference (EMI) is the random, uncontrolled, broad-range frequency radiation emitted from many natural and man-made sources. The recent rapid growth in man-made sources such as computers, telecommunication, and other business machinery has led to the legislation in the United States and Europe covering the levels of electromagnetic radiation emitted by electronic devices. The EMI-shield on most computers and electronic digital equipment on the American market has to conform to Federal Communications Commission (FCC) standards within the frequency range 30–1000 MHz. Similar regulations exist in Germany and are being implemented in other countries.

Metal or carbon particles in plastic reflect and absorb electromagnetic radiation. When dimensioning a shielding material, the absorption of nonreflected radiation must be made sufficient. An electromgnetic wave is absorbed according to the following correlation [4]:

$$A = 3.34 \, t\sigma\mu\sqrt{F} \tag{1}$$

where

A = absorption expressed in decibels (dB)

t = material thickness in one thousandths of an inch (mils)

σ = conductivity relative to copper

μ = relative magnetic permeability

F = frequency in MHz

Experimentally, EMI shielding is determined by the attenuation of a high frequency signal transmitted through the test sample. The attenuation or shielding effect α expressed in decibels is calculated from the equation

$$\alpha \, (dB) = 20 \, \log_{10}(E_b/E_a) \tag{2}$$

where E_b is the field intensity (V/m) without shielding and E_a is the field intensity (V/m) with shielding.

Data for some homogeneous metallic materials are given in Table 5.3. Aluminum and steel are the commonly used metallic fillers, and of these, steel provides the most effective EMI-shielding because of its high relative magnetic permeability.

Steel fibers give very good shielding properties at considerably low percentages. At only 1 percent of steel fiber in plastic, a shielding of 45–60 dB at 1 GHz is obtained. The conductivity of the filled plastic then is approximately 1 $(ohm-cm)^{-1}$. One great advantage of such low percentages permitted by steel fibers is that the other properties of the plastics are left almost unaffected.

In carbon, the conductivity varies from 10^{-2} $(ohm-cm)^{-1}$ for amorphous carbon to approximately 300 $(ohm-cm)^{-1}$ in the longitudinal direction for PAN-based high modulus carbon fibers. Apart from relatively low conductivity, carbon has the same magnetic permeability as aluminum, i.e., approximately 1. In order to obtain a given damping, carbon-based fillers have to be added in higher concentrations in comparison with metallic fillers such as steel. However, special carbon black grades with microporous structure and increased conductivity can now be found that allow the construction of a conductive network at relatively low concentrations.

A comparison between the shielding properties of a 3 mm thick nylon-6,6 with 40% by weight aluminum flakes and carbon black, respectively, is shown in Table 5.4. Note that while the absorption in both cases decreases with increasing frequency (in contrast to what happens in homogeneous metallic materials), according to formula (1) and Table 5.3, the difference between them is greatest at low frequencies (30 MHz), and they tend to converge towards about 40 dB at high frequencies (2 GHz).

TABLE 5.3 Conductivity, Permeability, and Absorption Characteristics of Major Metals [4]

Metal	Relative[a] conductivity	Relative[a] permeability (100 kHz)	Absorption loss (dB/mil)		
			100 Hz	10 kHz	1 MHz
Copper	1.00	1	0.03	0.33	3.33
Silver	1.05	1	0.03	0.34	3.40
Aluminum	0.61	1	0.03	0.26	2.60
Zinc	0.29	1	0.02	0.17	1.70
Brass	0.26	1	0.02	0.17	1.70
Nickel	0.20	1	0.01	0.15	1.49
Iron	0.17	1,000	0.44	4.36	43.60
Steel (SAE 1045)	0.10	1,000	0.33	3.32	33.20
Stainless steel	0.02	1,000	0.15	1.47	14.70
Mu-metal	0.03	80,000	1.63	16.30	163.00
Permalloy	0.03	80,000	1.63	16.30	163.00

[a]Relative to copper

Although a plastic with conductive fillers does not provide the same degree of shielding as metal, a great number of filled plastic materials have been introduced commercially, which in 1–3 mm thicknesses give shielding effects of 30–40 dB. As can be seen from the relationship given in Table 5.5 between shielding effect expressed in dB and in percent, respectively, these dB numbers correspond to 99.9 to 99.99 percent damping of the interference signal, which is considered satisfactory.

TABLE 5.4 EMI-Shielding for Loaded Nylon-6,6 at Various Frequencies [4]

Frequency	EMI-shielding (dB) of 3-mm nylon-6,6	
	Aluminum flake-filled[a]	Carbon-black filled
30 MHz	95	63
60 MHz	67	62
120 MHz	69	50
250 MHz	49	41
500 MHz	49	39
1.0 GHz	40	37
2.0 GHz	42	39

[a]40% by weight

TABLE 5.5 Shielding Effect Expressed in Decibels, Percentage of Leakage, and Classification of Shielding Effect

Shielding effect $(\alpha)^a$, dB	EMI leakage (%)	Value judgement (technical)
0–10	10^2–10^1	Insignificant
10–30	10^1–10^{-1}	Minimal
30–60	10^{-1}–10^{-4}	Average
60–90	10^{-4}–10^{-7}	Above average
90–120	10^{-7}–10^{-10}	Maximum (surpassing present technology)

^{a}Defined by equation (2)

There are a range of conductive polymers on the market that are based on metal fillers such as aluminum flake, brass fibers, stainless steel fibers, graphite-coated fibers, and metal-coated graphite fibers. However, the most cost effective conductive filler is carbon black. Mention should also be made of other more exclusive fillers, such as fiber glass with a metallized surface (aluminum) and silver-coated glass beads.

Where color is not critical, carbon black filled conductive polymers provide the most cost effective way of producing EMI shielding. A range of carbon black filled conductive polymers are produced by Cabot Plastics under the trade name Cabelec. One application is in telephone microcomponents.

In addition to being invariably black, another disadvantage of carbon-black-filled polymers is that they do not have the impact resistance of commonly used materials. One solution to these problems has been found by the use of sandwich molding. The process, originally developed by ICI in the 1960s, involves the production of a housing with an inner core of conductive plastics surrounded by an outer skin of conventional engineering plastic. The process thus gives the designer the full aesthetic freedom of design and also gives a good in-molded surface finish.

A system for sandwich molding developed by Aron Kasei uses fiber-reinforced ABS as the outer skin and a PLS conductor (see below) as the inner core. Applications include keyboard housings, printer housings, CRT enclosures, and medical equipment housings.

Conductor PLS, offered by Aron Kasei, is a range of brass or aluminum-filled thermoplastics such as ABS, polypropylene, PBT, and polycarbonate. Housing for personal computers made of PLS is claimed to cost less than housing made of plastics coated with conductive paint.

Ube Industries offer a brass fiber-filled nylon-6 that is reported to have excellent EMI-shielding properties, high resistance to abrasion, high conductivity, and easy malleability.

Mobay has introduced a 40% aluminum flake-filled, flame-retardant grade polycarbonate/ABS blend that is molded into internal cover configurations for EMI shielding.

LNP Corporation offers conductive plastics with fillers such as aluminum flakes, nickel-coated fibers, stainless steel fibers, and carbon fibers in a range of engineering polymers including nylon-6,6, polycarbonate, ABS, PPS, and PEEK. The company's 40% flake-filled polycarbonate is used for microprocessor covers in mainframe computers and nickel coated fiber-filled PEEK for avionic enclosures.

A 10% stainless steel fiber-filled ABS, produced by Mitech Incorporated under the tradename Magnex DC is used to shield electronic components in Xerox's Model 2100 laser printer. Mitech also produces a flame retardant stainless steel fiber-filled ABS for large interior covers.

Lacana Mining Company has developed nickel-coated mica fillers (Suzerite E. Micon) that can withstand compounding, extrusion, and injection molding with no special treatment. The coating is done by a hydrometallurgical process patented by Sheritt Gordon Company. Wilson-Fiberfil also offers nickel-coated mica that at 45% loadings in polyolefins provides 40 dB shielding and can be used for small-part moldings.

Wilson-Fiberfil offers 1% steel fiber-filled polycarbonate (Electrapil R-5147) for EMI-shielding, which has the advantage that the other properties of the plastic are left mostly unaffected during both tooling and use.

Conductive Coating

Some of the more commonly used methods to produce EMI shielding consist of coating the plastic with a conductive layer by post-molding or in-molding processes. The coating methods mostly used are conductive paints, zinc arc spraying, vacuum metallizing, electroplating, and foil application.

The simplest method of coating is the use of conventional coating systems such as brush coating or spray gun. Nickel, copper, silver, or graphite can be coated in this way onto vinyl polymers, acrylics, and polyurethane. This method is the least expensive but it suffers from the disadvantage that the resin has to be tailored to the substrate to avoid cracking.

Nickel coating is the most popular. A film thickness of 50–70 microns can achieve attenuation levels of 30–60 dB. A new acrylic-based nickel coating, MDT1001, from Mitsubishi Rayon Company is claimed to be 100 times more effective as an EMI shield than normal carbon-based coating; it dries at room temperature and does not require multiple coatings.

For solvent-sensitive plastics where paint systems are unsuitable, zinc arc spraying is sometimes used. In this method, the metal is melted by an electric arc and sprayed as droplets by compressed air onto the part to be coated. Consequently, only materials with very high melting points can be used. The method gives a hard, dense coating of good conductivity, but it requires expensive equipment.

In vacuum metallizing, which also requires a major capital investment in equipment, a metallic film, usually aluminum, is deposited onto a plastic by condensation from vapor under high vacuum. By rotating parts during application, a controlled film thickness can be obtained producing high levels of attenuation with layers as thin as 4–6 microns.

Electroplating with copper, nickel, or chromium using conventional techniques provides good conductivity and durability. It, however, requires paint finishing. Electroless plating, which differs from electroplating in that no external source is required to initiate and control the process, has proved to be an effective EMI-shielding process for components that may be conductive in some areas and nonconductive in others, such as computer housing.

In a selective electroless plating, a conductive film of copper or nickel is deposited onto a specific part of the molding as opposed to overall plating by conventional techniques that require paint finishing. Attenuation levels of 50 dB can be obtained at layers of 1–3 microns. In one electroless plating technique, which reportedly achieves attenuation levels of 70 dB in layers of 1–2 microns thick, the molded part is immersed in an aqueous solution of copper salt, reducing agent, and initiator to obtain copper deposition as the primary shielding layer. Nickel is then added to form an outer layer that provides corrosion resistance and improved impact strength.

In the process of metal coating by ion plating or cathode sputtering, which is claimed to produce better adhesion of metal to the substrate compared to vacuum metallizing, positively charged gas ions are discharged between two electrodes and bombarded onto a metal target. As a result, the metal vaporizes and condenses onto the molding. A major drawback of the process is the cost of ion-plating equipment.

A relatively new approach to EMI shielding is in-mold coating. In a process developed by Dai Nippon Toryo and Tokai Kogyo, tradenamed NTS, the mold surface is coated with a metallic material tradenamed Metrafilm by electrostatic spraying. The coating and compound then bond during injection producing excellent adhesion. Since the metallic powder is encapsulated in a resin and does not contain any solvents, the process can be used for resins that are sensitive to solvents such as polystyrene.

Signature Materials

Developments during the last decade have demonstrated that the operative conditions of weapons carriers can be drastically influenced (*stealth technology*) by the so-called *signature adaptation*, which provides reduced probability of detection by radar, infrared, visual, or acoustical reconnaissance. The active materials are applied mostly as top layers, such as camouflage painting, radar layers, IR-layers, and acoustical damping layers, and are often used in combination. Thickness and material properties are adjusted to give minimal reflection against incident electromagnetic radiation. During radar reconnaissance, this radiation falls in the frequency range of 3–30 GHz. In other cases, it may be a question of damping the natural radiation from the weapons carrier in the IR range of 2–20 μm.

In order to obtain low radar reflection with very thin layers, the following correlations should be fulfilled:

$$\epsilon' = \mu'$$
$$\epsilon'' = \mu'' \tag{3}$$

where ϵ' and ϵ'' are, respectively, the real part and the imaginary part of the permittivity ϵ, which in alternating fields is generalized to a complex quantity:

$$\epsilon = \epsilon' - j\epsilon'' \tag{4}$$

The imaginary part ϵ'' is a measure for the dielectric losses of the material. Similarly, μ' and μ'' are defined in relation to the permeability μ:

$$\mu = \mu' - j\mu'' \tag{5}$$

The correlations (3) above, result in the layer having the same wave impedance as air (120 π), i.e., there is no reflection from the outer surface. The radiation passing through the layer is reflected back by the metallic base. If the dielectric and magnetic losses, represented by ϵ'' and μ'' respectively, are large, the greater part of the radiation can be absorbed by the layer even if it is thin.

For several decades, efforts have been made to produce matereials with these properties. Since the end of the sixties, great interest has been shown towards magnetic absorption materials based on ferrite powder dispersed in organic binder agents. Ferrite represents a group of ceramic materials with the principal formula $(M^{II}O)(F_2O_3)$, where M^{II} is a divalent metal, e.g., Fe, Co, Ni, Mn, Zn, or Cd.

The magnetic losses of the ferrite produce a high resonance peak at a certain frequency, above which they diminish linearly to the wavelength. The phenomenon can be used to obtain thin, absorbent, wide-band layers. No reflection occurs at the outer surface when the impedance at the surface air/ferrite becomes equal to in-impedance of the layer. The in-impedance is, in effect, a sum impedance because the greatest part of the incident radiation is reflected several times between the interfaces of the layer.

The development in the ferrite area is intensive and is aimed at increasing the medium-frequency in the GHz-range with preserved band-width. Much of the development in the area is covered in secrecy.

Mention should also be made of signature adaptation that makes use of *negative interference* to reduce the probability of detection. In this case, the signature protective layer is adjusted so that negative interference occurs between the radar radiation that is reflected respectively from the outer surface and from the underlying metal, thus causing a black-out of the net reflection from the surface. The reflections in this case need to be adjusted so that 50% is reflected at the outer layer and 50% at the metal surface and so that the phase shift between these reflected radiations becomes half a wavelength at the outer surface.

Interference layers in the form of filled polymers have been used on warships. A simple system is polyester filled with TiO_2. It, however, provides protection only within a very narrow frequency range.

The structural composition and design of the signature layer are as important as the inherent properties of its material components in developing a good signature defense. For example, considerably higher efficiency is obtained if the damping properties of the layer increases from the surface inwards, as when gradient materials and so-called elimine materials are used. The development in this area is also aimed at increasing integration of the absorption properties of the supporting structure, such as by using suitable composite materials and sandwich-type construction elements.

Inherently Conductive Polymers

Polymers used as plastic materials have numerous technological applications because of their low cost, lightness, and flexibility. The discovery of electrical and optical activities in one class of polymers has resulted in the development of one of today's exciting areas of academic and industrial research, not only because of these polymers' theoretically interesting properties, but also because of their technologically promising future.

Interest in electrically conducting polymers began in the mid-1960s with the suggestion of Little [8] in his theoretical studies that some organic materials could become superconductors. The first example that has been extensively studied is tetrathiofulvalene-tetracyanoquinodimethane (TTF-TCNQ) and its derivatives. The first covalent polymer exhibiting the electronic properties of a metal was polymeric sulfur nitride, $(SN)_x$, which attracted much attention in the mid- to late 1970s. In 1977, the first covalent organic polymer, polyacetylene $(CH)_x$, that could be doped through the semiconducting to the metallic range, was reported.

Polyacetylene has been investigated much more extensively than any other conducting polymer and has served as a prototype for the synthesis and study of

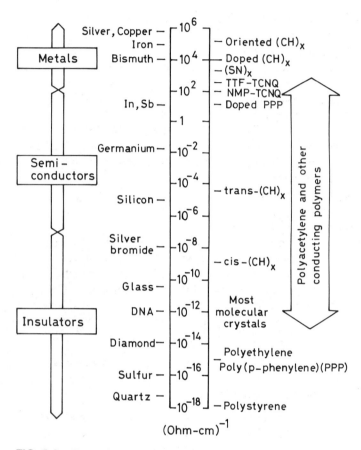

FIG. 5.4 Chart of conductivities for selected materials.

a large number of other conjugated, dopable organic polymers. Of these, the greatest research efforts are devoted to polypyrrole, poly(phenylene sulfide), poly(*p*-phenylene), polythiophene, and polyaniline.

In a nondoped state, the basic polymers have low conductivity (Fig. 5.4). The two polyacetylene conformations *cis* and *trans* are in the semiconductor range; poly(*p*-phenylene) (PPP) is a good insulator.

The synthesis of the polyacetylene powder has been known since the late 1950s, when Natta used transition metal derivatives that have since become known as Ziegler-Natta catalysts. The characterization of this powder was difficult until Shirakawa and coworkers [9] succeeded in synthesizing lustrous, silvery, polycrystalline films of polyacetylene and in developing techniques for controlling the content of *cis* and *trans* isomers:

    cis   trans

Coordinative polymerization of acetylene using the Ziegler-Natta catalytic system, $Ti(OBu)_4$-$AlEt_3$, is the most powerful way of forming polyacetylene powders, gels, or films. Polyacetylene films may be prepared by simply wetting the inside walls of a glass vessel with a toluene solution of $AlEt_3$ and $Ti(OBu)_4$ and then immediately admitting acetylene gas at any pressure up to 1 atm. A cohesive film grows over a period of time with thickness varying from 10^{-5} cm to 0.5 cm, depending on the pressure of acetylene and the temperature used. The film is formed completely as a *cis* isomer if a polymerization temperature of $-78°C$ is used and as the *trans* isomer if a temperature of $150°C$ is used (with decane solvent), while with room temperature polymerization, the film is approximately 60% *cis* and 40% *trans* isomer.

The *cis* isomer may be converted to the *trans* isomer, which is the thermodynamically stable form, by heating at $150°C$ for 30 min to 1 hr. The *cis-trans* content is conveniently determined from the relative intensities of bands in the IR spectra. The morphology of all polyacetylenes made with various Ziegler-Natta catalysts is fibrillar. The conductivity of films of polyacetylene depends on the *cis-trans* content, varying from 10^{-5} (ohm-cm)$^{-1}$ for the *trans* material to 10^{-9} (ohm-cm)$^{-1}$ for the *cis* isomer. Doping increases the conductivity of both *cis*- and *trans*-polyacetylene drastically.

Doping

Because electron mobility is almost nonexistent in most organic polymers, a mechanism for the creation of that mobility is needed and is generally accomplished by the doping process (oxidizing or reducing). In practice, doping occurs by exposing the polymer, most often in the form of thin film or powder, to the doping agent in gas phase or in liquid phase. The doping can also be done electrochemically by using the polymer as an electrode. The principal methods of doping are as follows [10, 11].

1. *Chemical doping.* The doping is accomplished by exposing the polymer to the vapor of a dopant such as I_2, AsF_3, or H_2SO_4, or by immersing the polymer films in a dopant solution, such as sodium naphthalide in tetrahydrofuran, I_2 in hexane, and so on. The amount of dopant incorporated in the material (doping level) depends on the vapor pressure of the dopant or its concentration, the doping time, and the temperature.

2. *Electrochemical doping.* Polymers an be doped electronically in an electrochemical cell, such as by immersing the material as an electrode in an organic electrolyte solution (LiClO$_4$ in tetrahydrofuran or propylene carbonate, with lithium as a counter electrode) or in aqueous electrolytes (PbSO$_4$/PbO$_2$ in sulfuric acid solutions, with lead as a counter electrode). Figure 5.5 shows, schematically, electrochemical doping in comparison with chemical doping. During electrochemical doping, the electrons are driven by an outside source of current. The lack of electrons and then the surplus of electrons which occurs, is balanced electrostatically by the dissociated ions in the electrolyte, which diffuse into the polymer. The end result is a polymer that is *p*-doped and *n*-doped, respectively.

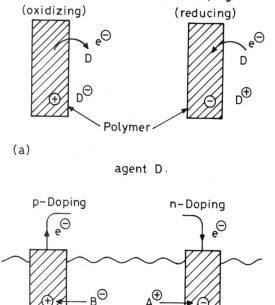

FIG. 5.5 Schematic comparison between two kinds of doping. For electrochemical doping an outside power source is required (not drawn). (a) Direct charge transfer with doping agent D. (b) Electrochemical charge transfer.

3. *Ion implantation.* Polymers, when exposed to proper ion beams, can become conducting due to the implantation of ions in the polymer's lattice, which form covalent bonds with the material. The doping level depends on the energy of the ion beam.

4. *Photochemical doping.* This is accomplished by treating the polymer with a dopant species that is initially inert towards the materials, followed by irradiation. For example, diphenyliodonium hexafluoride arsenate in methylene chloride or triarylsulfonium salts, followed by ultraviolet irradiation.

The academic research in the area of conductive polymers was mainly directed towards polyacetylene, while industry, gave priority to more stable basic materials such as poly(p-phenylene), polypyrrole, poly(phenylene sulfide), and polythiophene. Today, the academic research is extensive even in those areas, as well as a number of other conjugating polymers. A large number of dopants have been used to dope polyacetylene and other conjugated polymers.

Electron-attracting dopants, such as Br_2, I_2, AsF_5, H_2SO_4 and $HClO_4$ may oxidize polyacetylene to produce p-type conductivity. As shown in Fig. 5.6, the conductivity increases continuously with the degree of doping to a saturation level. Some of the powerful dopants have been used to oxidize other conjugated polymers. Some examples of p-type dopants and their measured room temperature conductivities are given in Table 5.6.

Electron-donating dopants may reduce polyacetylene or other conjugated polymers giving rise to n-type conductivity. Doping occurs when the polymer is immersed in a tetrahydrofuran solution of radical anion/alkalide where the alkalide components can be Li, Na, K, Cs, or Rb, and the radical anion can be

TABLE 5.6 Examples of Oxidation/p-Type Doping [11]

Doped polymer	Conductivity, $(ohm\text{-}cm)^{-1}$
$[CH(I_3^-)_{0.10}]_n{}^a$	5.5×10^2
$[CH(AsF_5)_{0.10}]_n{}^a$	1.2×10^3
$[CH(SbF_5)_{0.006}]_n{}^a$	5.0×10^1
$[CH(H_3O^+ HSO_4^-)_{0.10}]_n{}^a$	1.2×10^3
$CH(FeCl_3)_{0.09}]_n{}^a$	8.5×10^2
$[CH(MoCl_5)_{0.10}]_n{}^a$	2.0×10^2
$[CH(ClO_4)_{0.065}]_n{}^b$	5.0×10^2
$[C_6H_4 (AsF_5)_{0.26}]_n{}^a$	1.5×10^2
$[C_6H_4S(AsF_5)_x]_n{}^a$	2.5×10^1
Polypyrrole/$(C_6H_5)_2IAsF_6{}^c$	3.0×10^{-3}
Polyaniline/60% protonation	5.0×10^0

[a]Chemical doping
[b]Electrochemical doping
[c]Photochemical doping

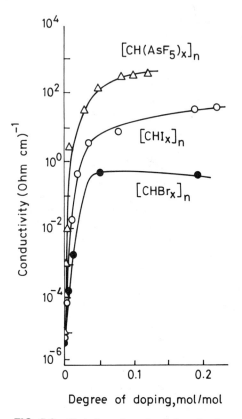

FIG. 5.6 Variation of conductivity of polyacetylene on degree of doping [12].

naphthalene, anthracene, or benzophenone. A doping reaction such as the following occurs (Nph = naphthalene):

$$(CH)_x + 0.10x(Na^+Nph^-) \longrightarrow [Na^{0.1}(CH)]_x + 0.10x Nph \qquad (6)$$

When the polymer is allowed to contact a K/Na alloy at room temperature, n-doping also occurs. In n-doped polymers, the polymer chain is considered as a polycarbanion associated with the corresponding M^+ metal ion. A few examples are given in Table 5.7.

Polyacetylene, because it represents the simplest of the conjugated organic polymers, has received much attention as a fundamental material, although other polymers such as polyphenylene, polypyrrole, polyquinoline, polyquinoxyline, polyphenylene sulfide, and polythiophene are also being increasingly studied. Poly-p-phenylene (PPP) was the first example of a nonacetylene hydrocarbon

TABLE 5.7 Examples of Reduction/n-Type Doping [11]

Doped polymer	Conductivity, (ohm-cm)$^{-1}$
$[Li_{0.30}(CH)]_n$[a]	2.0×10^2
$Na_{0.21}(CH)]_n$[a]	2.5×10^1
$[K_{0.16}(CH)]_n$[a]	5.0×10^1
$[(n\text{-}Bu_4N)_{0.03}(CH)]_n$[b]	2.5×10^1
Polyquinoline/Li[b]	2.0×10^1
Polyquinoline/Na[a]	5.6×10^{-1}
$(CH)_n/LiAlH_4$[a]	6.0×10^0

[a]Chemical doping
[b]Electrochemical doping

polymer having an intrinsic conductivity of less than 10^{-14} (ohm-cm)$^{-1}$, but, as with polyacetylene, PPP can be doped with AsF_5 to give conductivities of around 10^2 (ohm-cm)$^{-1}$. It may also be doped with alkali metals to provide highly conductive n-type materials.

The main advantage of poly-p-phenylene is that, due to its nonacetylenic composition, it has a much higher thermal stability (450°C in air and 550°C in inert atmosphere). Potential applications of poly-p-phenylene are similar to those envisaged for polyacetylene, such as Schottky barriers in photocells. (A Schottky barrier is a metal semiconductor contact that has rectifying characteristics similar to a p-n junction.)

Polypyrroles have prompted considerable research because they are the only conducting polymers that can be produced directly in the doped state. When solutions of pyrrole and tetraethyl ammonium tetrafluoroborate ($Et_4N^+BF_4^-$) in acetonitrile is electrolyzed, a blue-black conducting polymer is produced at the anode. The polymer can then be removed from the electrode and conductivities of up to 100 (ohm-cm)$^{-1}$ have been reported.

One of the principal advantages of polypyrrole over other doped polymers is its excellent thermal stability in air. It is thermally stable up to 250°C, showing little degradation of its conducting properties below this temperature. Above 250°C, the rate of weight loss increases with temperature.

Poly(phenylene sulfide) (PPS) is unique amongst polymers utilized for electrical conduction in that it is available commercially as an engineering thermoplastic (Ryton by Philips Petroleum) and is finding use as a molding component for the encapsulation of electronic components. Unfilled PPS is a white polymer with a melting point of 288°C and is intrinsically insulating, having a conductivity of less than 10^{-16} (ohm-cm)$^{-1}$. Doping with AsF_5 leads to an increase in conductivity to around 1–10 (ohm-cm)$^{-1}$, although there is also a deterioration in mechanical properties and a color change to dark blue.

Liquid-phase doping of polymer solutions is being studied by Allied Corporation (United States). PPS has been doped and dissolved simultaneously by exposing PPS particles suspended in AsF_3 to the doping agent AsF_5. Films of doped PPS with good mechanical properties and conductivities up to 200 (ohm-cm)$^{-1}$ can be cast from such solutions of AsF_5-doped PPS in AsF_3. The resulting films show improved strength and flexibility compared to those obtained by standard doping of solid PPS with AsF_5.

Applications

The interest on the part of industry in inherently conductive polymers is great because of the possibilities that lie in the workable qualities and light weights of polymer materials. The greatest problem with polyacetylene and many of the other conductive polymers is their instability when exposed to oxygen and normal atmospheric moisture. In addition, they are insoluble in typical solvents, difficult to mold or extrudate, and hard to mass produce. Consequently, no significant commercial uses have yet emerged. Presently, the only well-known commercial application is the incorporation of an intrinsically conductive polymer in the electrostatic grounding brush of Xerox's 1075 copiers.

One of the most promising areas of application of inherently conducting polymers is as electrode materials in rechargeable batteries. Conducting polymers are promising charge storage materials because of their high charge carrier concentrations.

A polymeric electrode is doped electrochemically when the battery is charged, and is undoped at discharge (see Fig. 5.5). The fundamentally appealing aspect of polymer electrodes is that charge and discharge are based on purely physical and reversible processes. There are no chemical processes to consume material.

A battery cell where both the electrodes consist of dopable polymer is shown in Fig. 5.7. The electrolyte in this case consists of $Li^+ClO_4^-$ dissolved in an inert organic solvent, usually tetrahydrofuran or propylene carbonate.

It is also possible for only one electrode to consist of dopable polymer and for the other be of metal, usually lithium (Fig. 5.8), which is preferred because of its low weight and high thermodynamic electrode potential.

The comparable data for two conventional and two polymeric batteries, which are given in Table 5.8, show that, in energy absorption per unit weight, the polymeric batteries are not superior to the lead-acid battery. What is most significant, however, is the fact that the power density is calculated here to be higher by an order of magnitude. This reflects, apart from the higher voltage, the considerably higher current density in polymer batteries that is attributed to the fact that the ions in the electrolyte are able to wander quickly in and out in the electrode as it is doped and undoped. This is facilitated by the porous nature

CHARGE

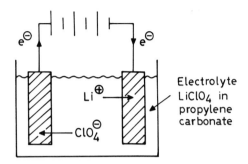

Electrode reactions at charge:

Anode: $(CH)_x + xy\,(ClO_4^-) \longrightarrow [(CH)^{+y}(ClO_4^-)_y]_x + xye^-$

(p-type doping)

Cathode: $(CH)_x + xy\,(Li^+) + xye^- \longrightarrow [Li_y^+(CH)^{-y}]_x$

(n-type doping)

DISCHARGE

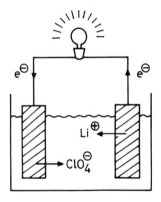

Electrode reactions at discharge are as above but in reverse.

FIG. 5.7 Secondary battery with PAc electrodes. Here PAc is given the chemical symbol $(CH)_x$.

of polyacetylene film consisting of very thin fibers with a diameter of about 200 A (2×10^{-8} m), which only occupies one-third of the volume of the film and gives the material an exceptionally large effective surface (40–50 m^2/g).

Though the rate of diffusion for a doping agent in polyacetylene itself is very low ($D = 10^{-15}$ cm^2/s for Li in polyacetylene), it is compensated for by the fiber

CHARGE

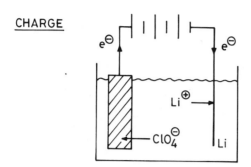

Electrode reactions at charge :

Anode : $(CH)_x + xy\,ClO_4^- \longrightarrow \left[(CH)^{+y}(ClO_4^-)_y\right]_x + xy\,e^-$

Cathode: $xy\,Li^+ + xy\,e^- \longrightarrow xy\,Li$

DISCHARGE

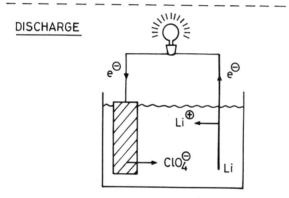

Electrode reactions at discharge are as above but in reverse.

FIG. 5.8 Electrochemical oxidation and reduction reactions in polyacetylene/lithium batteries.

structure, which reduces the diffusion length when the material is impregnated with a liquid electrolyte.

Though the charging and discharging of polymer electrodes are based on purely physical and reversible processes, in practice, however, disturbing side reactions occur that give reduced efficiency. This happens especially at complete discharge-recharge. Therefore, only about one-third of the total charge should be utilized. (However, this also applies to lead-acid batteries where the normal

TABLE 5.8 Comparison of Conventional and Polymeric Batteries [4]

Battery type	Initial voltage (V)	Energy density (Wh/kg)	Power density (kW/kg)
Lead-acid battery	2	30	0.4
Zinc-bromine	1.8	65	0.14
$(CH)_x/(CH)_x$ [a]	3.7	30	5
$(C_6H_4)_x/(C_6H_4)_x$ [b]	4.4	40	5

[a]Polyacetylene electrodes
[b]Poly(p-phenylene) electrodes

operating range is between 40 and 80% of full charge.) Better cell construction and more stable electrolytes need to be developed in order to increase efficiency and useful life of polymeric batteries.

In order for electrically operated vehicles to become generally accepted, batteries with higher energy- and power-density in combination with lower weight and volume are required. Polymeric batteries show promise of making this a reality.

An all-plastic battery may have many advantages. For example, a car battery made of polyacetylene could weigh only one-tenth of that of a conventional lead-acid battery. Moreover, batteries with plastic electrodes could be fabricated into odd shapes, such as a flat disc that could be slotted into a car door. Prototype batteries have been made using polyacetylene and poly-p-phenylene electrodes, but a number of technical problems, such as long-term mechanical integrity need to be solved.

A derivative of polyacetylene that is more stable in air and soluble in organic solvents when doped with iodine, is polytrimethyl-silylacetylene (PTMSA). Though the derivative has a conductivity only one ten-millionth that of doped polyacetylene, it can be used as precursor to a wide range of other conductive derivatives. One such derivative is polyfluoroacetylene (polyacetylene in which all of the hydrogen atoms are replaced by fluorine), which according to calculations, would be electrically conductive without the need for dopants.

Several potential technological applications for conductive polymers have been suggested, but according to market analysis there will not be a demand for any large volumes of conductive polymers before the turn of the century. Crucial difficulties that have to be mastered are how to produce doped polymers in desired shapes with good mechanical properties as well as obtaining long-term stability. Research has extended toward solving these problems, and progress has recently been made in that direction. If satisfactory solutions to the problems

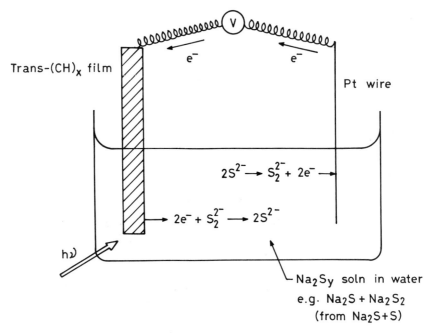

FIG. 5.9   Photoelectrochemical photovoltaic cell using polyacetylene as an electrode.

of intractability and instability in ambient environment are found, the use of conducting polymers in many applications will become possible. Common suggestions [10] are uses in photoelectrochemical solar cells (Fig. 5.9), electrochromic display devices, and electronic components as *p-n* junctions and Schottky barriers.

Photoconductive Polymers

The enhanced flow of current under the influence of an applied electric field that occurs when a semiconductor is exposed to visible light or other electromagnetic radiation is known as photoconduction. Poly (*N*-vinyl carbazole) (I) and various other vinyl derivatives of polynuclear aromatic compounds such as poly(2-vinyl carbazole) (II) and poly(vinyl pyrene) (III) have high photoconductive efficiencies. The excellent photoconductivities of these polymers are believed to be due to their helical conformation with successive aromatic side chains lying parallel to each other in a stack along which electron transfer takes place relatively easily.

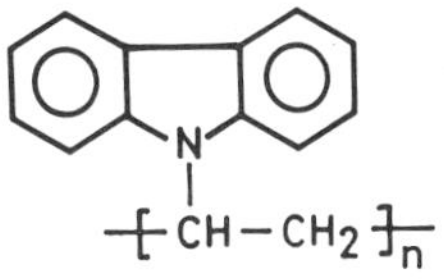

(I)

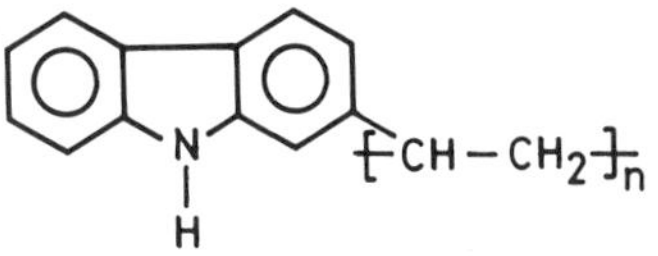

(II)

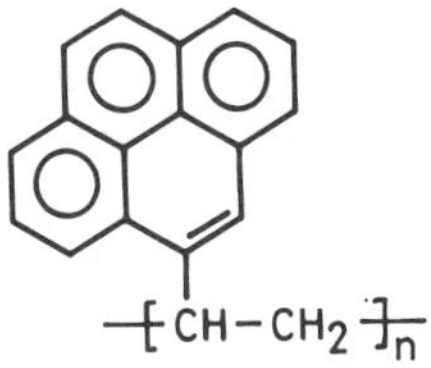

(III)

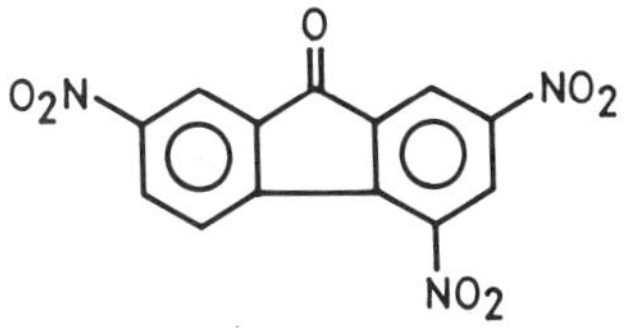

(IV)

Poly(*N*-vinyl carbazole) absorbs light in the 360 nm region to undergo electronic excitation and ionization in an electric field. The photogeneration efficiency of the polymer can be greatly enhanced by the addition of an equimolar amount of 2,4,7-trinitrofluorenone (TNF) (IV), which shifts the absorption of poly(*N*-vinyl carbazole) into the visible range by the formation of a charge transfer state, rendering it photoconductive at 550 nm. While the polymer alone is a hole conductor, the addition TNF creates electron carriers and the conduction mechanism actually becomes electron dominated.

Photoconduction forms the basis of electroreprography. In this photocopying process, or xergraphy as it is sometimes known, a photoconductive material is coated onto a metal drum and uniformly charged (sensitized) in darkness by a corona discharge. The drum is then exposed to the bright image of the object to be copied whereupon the illuminated areas of the photoconductive material become conductive and dissipate their charge to the metal drum (earthed). The photoconductive material in the dark areas is still charged and so is able to attract a positively charged black resin-coated toner powder forming a latent image of the object. The latent image is then transferred to a piece of negatively charged paper which is heated to fuse and fix the resin, thereby making the image permanent. Early photoconductive materials were based on selenium compounds such as As_2Se_3. Such a material is difficult to handle, because it needs to be applied by vacuum sublimation and it is rather brittle. Poly(*N*-vinyl carbazole)-based materials are now replacing the selenium-based compounds.

Much work has been reported on the development of carbazole derivatives and also noncarbazole photoconductive polymers. An example of the latter group is poly(bis(2-naphthoxy)phosphazene), which is intrinsically an insulator with a very low photosensitivity towards both UV and visible radiation, but when doped with trinitrofluorenone in a 1:1 molar ratio, it is a strong photo-conductor.

Polymers in Nonlinear Optics

Materials that have nonlinear optical properties are of considerable interest because of their potential for optical switching and waveguiding applications in telecommunication. Until recently, nearly all of the nonlinear optical materials commonly in use were inorganic solids. However, there is currently a growing interest in organic and polymeric solids because of their exceptionally large nonlinear, second-order optical properties and the larger variety of asymmetrical crystal structures available. Possible applications of these materials include amplifiers, frequency doublers, waveguides, Q-switches, (used for building up power output in lasers) and fitters.

Nonlinear second-order optical properties include *second harmonic generation* and the *linear electro-optic effect*. Second harmonic generation is the ability of a material to double the frequency of light passing through it, and the linear electro-optic effect is a change in the refractive index of a material under the application of a low-frequency electric field. For these effects to occur, the material must not have a center of symmetry, while for maximum second harmonic generation, a crystal should possess propagation directions where the crystal birefringence cancels the natural dispersion, leading to the condition of equal refractive indices at the fundamental and second harmonic frequencies (*phase matching*).

Phase-matched second harmonic generation in single crystal polymers was first observed in 2-methyl-4-nitroaniline substituted diacetylene polymers. Subsequently, a number of other diacetylene structures have been synthesized that exhibit orders of magnitude greater phase-matched second harmonic generation than lithium iodate.

Besides single crystals, disubstituted diacetylene polymer thin films can be obtained for waveguide applications by evaporation, solidification, or deposition of the diacetylene monomers by the Langmuir-Blodgett approach, which allows film thicknesses to be controlled at the molecular level. These diacetylene films can be patterned using selective polymerization techniques such as developed for UV, X-ray, and electron beam lithography.

Piezo- and Pyroelectric Polymers

Piezoelectricity is defined as an electric polarization that occurs in certain crystalline materials at mechanical deformation. The polarization is proportionate to the deformation and the polarity changes with change in deformation. In reverse, electric polarization produces mechanical deformation in piezoelectric crystals. The piezoelectric materials also possess pyroelectric properties, i.e., electric polarization is generated at temperature change.

If a polymer has a larger dipole moment in its molecule than can be aligned to form a polar crystal, there is every likelihood it will exhibit strong piezo- and pyroelectricity. However, though many polar polymers such as poly(vinyl chloride), polyacrylonitrile, and nylon-11 have relatively large dipole moments, they exhibit only weak piezoelectric activities because the dipoles cannot be aligned very well by an electric field.

A major advance was made in 1969 when a strong piezoelectric effect was discovered in poly(vinylidene fluoride) (PVDF). The effect is much greater than for other polymers. In 1971, the pyroelectric properties of PVDF were also first reported, and as a consequence, considerable research and development has continued during the last two decades.

Poly(vinylidene fluoride) is a semicrystalline thermoplastic polymer consisting of $-\!\!-\!\!(\!-CH_2CF_2\!-\!)\!-\!$ units in long chains. The crystallinity is normally about 55%, and the polymer forms lamellar crystals typically 10 nm thick by 100 nm long that are dispersed in an amorphous phase that has a glass transition temperature of about 40°C. The molecule, which may extend through several crystalline and amorphous regions, can have three morphological forms known as α, β, and γ-PVDF.

The preferred form for piezoelectric activity is the β-PVDF, which has all-*trans* conformation (Fig. 5.10) with dipoles essentially normal to the molecular axis. The α-PVDF, on the other hand, has *trans-gauche-trans-gauche* ($tg^+ tg^-$) conformation (Fig. 5.10) having components of the dipole moment both parallel and perpendicular to the chain axis. When packed in the respective unit cells, the dipoles of the chain interact in form β giving a large dipole moment, but are antiparallel in form α. In order to obtain strongly piezoelectric (pyroelectric) PVDF, the nonactive α-form must therefore be converted to the active β-form. This may be done by special treatment during the extrusion process.

In order to impart stronger piezoelectric properties to PVDF, a number of processing steps are undertaken. The material is uniaxially stretched at 150°C and annealed and then subjected to *poling* (a term used to describe methods of inducing polarization), either thermally or by a corona discharge procedure.

In the case of *thermal poling*, electrodes are evaporated on to the PVDF and a field of 0.4–1.4 MV/cm is applied at temperatures of 90–100°C for 1 hr. Cooling with the field maintained stabilizes the polar orientation of the crystallites and produces permanent polarization (poling).

For *corona poling*, which is a faster method used industrially, the nonmetallized or half-metallized PVDF film is subjected to a corona discharge from a needle electrode a few centimeters away. The resulting charge build-up on the

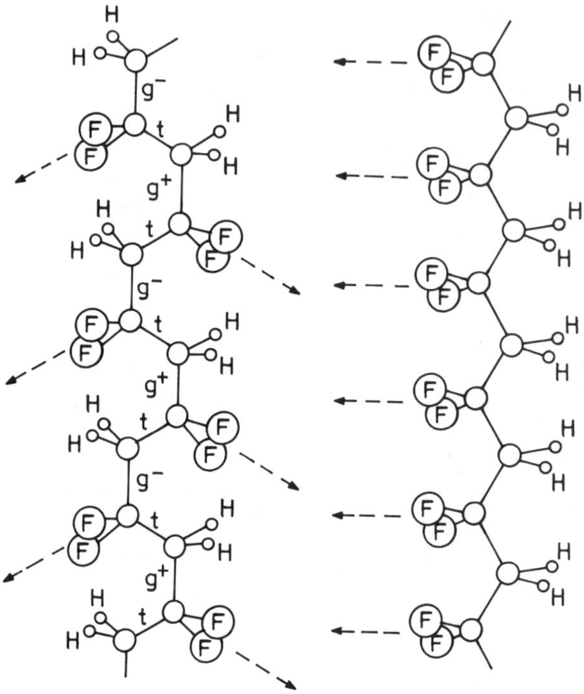

$$tg^+ tg^-$$
(a)

All-trans
(b)

FIG. 5.10 Schematic presentation of the two most common crystalline chain conformations in PVF_2: (a) $tg^+ tg^-$ (α form) and (b) all-*trans* (β form). The arrows indicate projections of the $-CF_2$ dipole directions on planes defined by the carbon backbone. The $tg^+ tg^-$ conformation, where t = trans, g= gauche, and g^- = gauche minus, has components of the dipole moment both parallel and perpendicular to the chain axis. The all-*trans* conformation has all dipoles essentially normal to the molecular axis.

film generates a field in the sample that is sufficient to align the polar chains, even at room temperature. A very high piezoelectric activity can be obtained by a combination of stretching and corona poling.

The parallel dipolar alignment in PVDF film treated as above leads to a residual polarization, which is the basis for several proposed mechanisms that explain the piezoelectric and pyroelectric behavior of the material. Although various observations support the suggestion that PVDF is also ferroelectric, its Curie point has not been detected. (Most ferroelectrics show a phase transition at a temperature known as the *Curie point* above which they become paraelectric.)

PVDF film, as produced from the melt, is largely in the nonpolar α-form, the β phase only being obtained after subsequent processing operations, as described above. If, however, vinylidene fluoride is copolymerized with as little as 7% by weight of trifluoroethylene, a copolymer is formed with crystallites completely in the β-form. This obviates the need for stretching after synthesis and the copolymer can be processed by way of conventional routes, such as injection molding. Moreover, unlike PVDF, copolymers of vinylidene fluoride and trifluoroethylene have been shown to demonstrate the ferroelectric to paraelectric transition. For a copolymer with a composition of 55% vinylidene fluoride and 45% trifluoroethylene, a phase transition is observed near 70°C, and with 90% vinylidene difluoride, a phase transition at 130°C.

Applications

Technological applications exist for both the piezo- and pyroelectric phenomena, but piezoelectric applications are the most common. Piezoelectric materials are of practical importance because they permit conversion of mechanical energy into electrical energy and vice-versa. Compared to inorganic piezoelectric material, the great advantages of piezoelectric polymers lie in their workability, mechanical flexibility, high shock resistance, low cost, low density, chemical stability, and availability in large sizes.

A large number of applications have been proposed for piezoelectric polymers. The types of applications can be grouped into five major categories: sonar hydrophones, ultrasonic transducers, audio-frequency transducers, pyroelectric sensors, and electromechanical devices. The principal polymers of interest in these applications are PVDF and copolymers of vinylidene fluoride and trifluoroethylene.

The greatest use of PVDF is as a source of pressure. The fact that PVDF has an acoustic impedance that is close to that of water or the human body makes it especially suitable as a source of pressure in underwater technology and medicine. PVDF also has an extreme band width, i.e., a large frequency range.

With its low acoustic impedance, extreme band width, high piezoelectric coefficient, and low density (only one-quarter the density of ceramic materials), PVDF is ideally suited as a transducer for broad band underwater receivers in lightweight hydrophones. The softness and flexibility of PVDF give it a compliance 30 times greater than ceramic. PVDF can thus be utilized in a hydrophone structure using various device configurations, such as compliant tubes, rolled cylinders, discs, and planar stacks of laminated material.

A 100 element 360° scanning sonar transducer utilizing PVDF has been developed by Marconi to provide a 360° view of the acoustic scene on a radar type display. Hydrophones for submarine reconnaisance are of great interest to the military. A comparison between PVDF and a ceramic material PZT-4 commonly used in sources of pressure is shown in Table 5.9. The so-called *g-d* product, which is considered to be the standard of suitability for hydrophone applications, is 2.5 times greater for PVDF as compared to the corresponding value for PZT-4. PVDF is thus superior for acoustical recording in water. Basic constructions of hydrophones with various sensitivity and pressure resistance are shown in Fig. 5.11.

There is much interest in the application of PVDF in medical imaging because of its close acoustic impedance match with body tissues. Monolithic silicon-PVDF devices have been produced in which a sheet of PVDF is bonded to a silicon wafer containing an array of metal oxide semiconductor field effect transistor (MOSFET) amplifiers arranged in such a way that when an acoustic wave is detected, the electrical signal resulting from the piezoelectric action in the PVDF appears directly on the gate of an MOS transistor. The device is therefore known as a piezoelectric oxide semiconductor field effect transistor (POSFET).

PVDF is generally an ideal material for transducers operating at frequencies above 0.5 MHz in hydrophones and pulse echo probes for medical and nonmedical testing. A 64-element linear array transducer has been produced that, operating at 5 MHz, offers a wide-bandwidth pulse response, sharp ultrasonic field distribution, and a high energy conversion efficiency.

While PVDF has piezoelectric properties similar to those of piezoelectric ceramic materials, the pyroelectric coefficient is too low to be useful except for specialized applications where the main requirements may be speed and resolution, and sensitivity is a minor consideration. PVDF, for example, has been used to form an array of 50 detectors for the energy profile determination of various wavelength laser beams. PVDF has also been employed in a pyroelectric vidicon

TABLE 5.9 Comparison of the Properties of PVDF-Film (Kureha 19) with Typical Values of PZT-4 Ceramic for Hydrophone Applications [4]

Property	Units	PVDF polymer	PZT-4 ceramic
Piezoelectric constant— field/stress	10^{-3} Vm/N	174	11.1
Piezoelectric constant— charge density/stress	10^{-12} m/V	20	123
g-d product	10^{-12} m²/N	3.48	1.37
Piezoelectric coupling factor	—	0.102	0.334
Relative dielectric constant	—	13	1300
Elastic compliance	10^{-12} m²/N	330	10.9
Density	10^{3} kg/m³	1.78	7.5

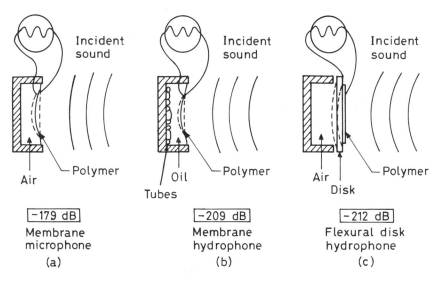

FIG. 5.11 Cross-sections of cylindrical piezoelectric polymer receivers. (a) Membrane microphone formed by fixing a taut film over an air-filled cylindrical chamber. (b) Membrane hydrophone with an oil-filled chamber containing plastic compliant tubes to provide compliance while withstanding hydrostatic pressure. (c) Flexural disk hydrophone backed by air. The disk both excites the polymer film and provides strength against hydrostatic pressure.

to obtain a high resolution picture, because its low thermal conductivity reduces thermal spreading.

Pyroelectric detectors made with piezo films can be used in a number of specialized applications. A pyroelectric detector shown in Fig. 5.12, consists of a thin semitransparent electrode placed on the front of the piezo film and a highly reflective rear electrode. A blacking layer added to the front electrode results in far more efficient broadband absorption of incident radiation. The incident radiation passes into the piezo film, which absorbs strongly in the wavelength range of 8–11 μm. Pyroelectric detectors made with piezo film are used in intrusion detection systems and energy management systems. Pyroelectric detectors can also be used for humidity monitoring and certain gas detection, a good example being carbon monoxide.

Much interest has been shown in the use of PVDF for elements in telephone handsets. Plessey Telecommunications have produced bimorph microphone units using PVDF. It has also been employed in high fidelity tweeters and headphone transducers for audio applications.

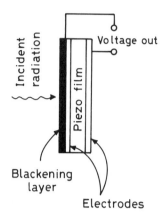

FIG. 5.12 A pyroelectric detector made with piezo film.

POLYMERIC ELECTROLYTES [17–20]

The organic solvents currently used the most, tetrahydrofuran and propylene carbonate, have clear limitations in regards to stability. The potential at which p-doping (indiffusion of negative ions) of polyacetylene electrode occurs is close to the anodic limit for propylene carbonate and higher than the limit for tetrahydrofuran. Thus one of the major problems with polymeric batteries is the problem of slow electrochemical breakdown of organic solvents at actual doping potentials.

An alternative to employing liquid electrolytes is to use a solid electrolyte to produce an all solid battery. This may have several advantages—an absence of electrolyte leakage or gassing, the likelihood of an extremely long shelf-life, and the capability of operating over a wide temperature range.

Polymeric materials, because of their low weight and ease of fabrication, have been investigated as electrolytes for lithium batteries. Early investigations included ion-exchange membranes, such as polystyrene sulfonic acid. However, it was found that in the dry state, the conductivity of these materials was extremely low (10^{-12}–10^{-15} ohm^{-1}-cm^{-1}), and the addition of aprotic solvents, such as propylene carbonate, did not appreciably loosen the ionic clusters formed within these membranes.

Polymer gels based on polymers such as poly(vinylidene fluoride), polyacrylonitrile, and aprotic solvents containing added alkali metal salts, gave appreciable room-temperature conductivity. However, solvent volatility and voltage stability of the electrolyte were serious problems.

A new perspective in polymer electrolytes was obtained in 1978 when Armand [17] suggested the use of poly(ethylene oxide)-alkali metal salt complexes for alkali metal rechargeable batteries. Poly(ethylene oxide) (PEO) and poly-(propylene oxide) (PPO) form ion complexes with, for instance, NaI, $LiClO_4$, $LiCF_3SO_3$, and others. Perhaps the most important advantage of such polymer electrolytes is the ability of the complex to form a good interface with solid electrodes, thereby permitting faster kinetics at the ion transfer between electrode and electrolyte.

Fig. 5.13 shows the structure of the PEO-salt complexes proposed by Armand. The regular PEO helix is filled by metal ions (M^+) solvated by ether oxygens, and the counterions (X^-) are expelled from the strands of the parallel helices. (Pure PEO has a helical structure in the crystalline state with seven CH_2CH_2O units in two turns of the helix.)

Initial measurements carried out on PEO-alkali metal salt complexes indicated that the observed conductivities were mostly ionic with little contribution

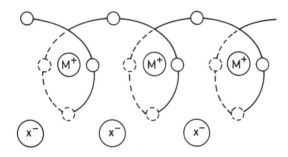

Low-temperature form

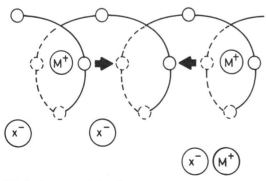

High-temperature form

FIG. 5.13 Proposed helical structure of PEO-salt complexes.

from electrons. It should be noted that the ideal electrolyte for lithium rechargeable batteries is a purely ionic conductor and, furthermore, should only conduct lithium ions. Contributions to the conductivity from electrons reduces the battery performance and causes self-discharge on storage. Salts with large bulky anions are used in order to reduce ion mobility, since contributions to the conductivity from anions produces a concentration gradient that adds an additional component to the resistance of the electrolyte.

Though the conductivity of the PEO-alkali metal complexes (10^{-3} ohm^{-1}-cm^{-1} at 140°C) is fairly low in comparison with inorganic solid electrolytes such as β-alumina(Na), $RbCu_{16}I_7Cl_{13}$ and $RbAg_4I_5$ at the same temperature, this can be compensated for by the facile production of thin films typically 25–500 μm thick. A cell may thus consist of a lithium or lithium-based foil as anode, an alkali metal salt-PEO complex, such as $(PEO)_9$ $LiCF_3SO_3$ (25–50 μm), as the electrolyte, and a composite cathode (50–75 μm) containing a vanadium oxide (V_6O_{13}) as the active ingredient. The vanadium oxide is one of a number of "insertion" compounds that permits the physical insertion of lithium ions reversibly into their structure and thus allows recharging of the cell.

While the advantages of PEO-alkali metal complexes are their good electrochemical stability as well as faster kinetics at the ion transfer between electrode and electrolyte, the disadvantage is the low conductivity of the polymeric electrolyte at normal operating temperatures. Thus, in order to obtain high-power densities, it is necessary to work at relatively high temperatures ($\sim$120°C). Although this temperature of operation may be suitable for certain application, for example, vehicle traction, quite clearly it would be unsuitable for other applications, such as consumer products.

Recently, a new polymeric electrolyte consisting of polyester-substituted polyphosphazene [18] has been developed. This polymer, which is designated MEEP, forms complexes with a large number of metallic salts; the complexes having a higher conductivity at room temperature than earlier polymer electrolytes. For example, MEEP-LiF_3SO_3 has a conductivity of 10^{-4} ohm^{-1}-cm^{-1}, which is sufficient for battery use, since the polymeric electrolyte can be shaped as films between the electrodes. This holds promise of very light batteries with potentially very great energy density.

While it is possible that future polymeric batteries will be all-polymeric solid-state batteries, it is predicted, however, that the most promising solid state batteries will combine polymeric electrolytes with nonpolymeric electrode materials such as TiS_2, V_6O_{13}, Li, or LiAl, the specific capacity of which surpasses that of polyacetylene electrodes.

As a great number of combination possibilities now exist for electrode as well as electrolyte materials, the probability of developing rechargeable batteries with considerably increased performance may be considered to be high. For potential use in electric cars and other electrically operated vehicles, which may become an environmental requirement in densely populated areas, such developments in polymeric batteries are eagerly awaited.

Drastic improvement in rechargeable batteries is also of great interest militarily. One example is submarine batteries. Compared to lead-acid batteries, the polymeric batteries enjoy potential advantages in this area: faster charge, smaller volume, and great freedom in shaping the batteries to available space. Though the energy density (KWh/kg) of polymeric batteries are, in current estimation, comparable to that of the lead-acid battery, charging of the batteries, which is the most critical routine operation of a diesel-powered submarine and requires about two hours to complete, can be reduced drastically with the use of polymeric batteries.

LIQUID CRYSTAL POLYMERS [21, 22]

Ordinarily, a crystalline solid melts sharply at a single, well-defined temperature to produce a liquid phase that is amorphous or isotropic. A liquid crystal material, on the other hand, is characterized by the appearance of an intermediate state (*mesomorphic phase* or *mesophase*) between the crystalline solid and isotropic liquid phase. A substance in this intermediate state possesses an ordered structure with some resemblance to a crystalline solid, exhibits strongly anisotropic behavior in some of its properties, and yet exhibits a certain degree of fluidity, which in some cases may be comparable to that of an ordinary liquid.

Some polymers are able to form a highly oriented crystalline polymer network when subjected to shear in the melt phase. Such polymers are named liquid crystal polymers (LCPs). The network formed by the orientation in the melt phase gives the polymer a woodlike appearance at fracture surfaces and also gives it self-reinforcing properties that are at least as good as those of conventional fiber-reinforced thermoplastics. Excellent physical properties, low combustibility and smoke generation, resistance to chemicals and solvents, and stability towards radiation are some of the important properties of LCPs.

The liquid crystal chains of LCPs do not reorient when they solidify from the melt, which ensures that molded parts can be produced with minimum warpage and distortion. The dimensional stability of the molded parts is further enhanced by the property of low coefficient of thermal expansion of LCPs.

There are several other factors that must be taken into consideration when processing LCPs compared to other engineering thermoplastics. Due to the anisotropic nature of LCP formulations, the physical properties of an injection-molded part depend on the direction and degree of molecular and filler orientation. This presents problems for the mold designer. Because orientation is imparted through shear as the melt flows, the flow pattern in the cavity must be carefully considered. Proper gate design is also important, as LCPs exhibit little die swell and incorrect gating may produce jetting.

High shear sensitivities of LCPs also necessitate careful streamlining of dies and flow channels in extrusion processes. As LCPs set up almost immediately upon leaving the die, products must be extruded to final dimensions at the die exit.

In vacuum forming of LCPs, good detail definition can be obtained and deep draws are possible. Moreover, as the material can be demolded while it is very hot, little cooling is needed. Unfilled LCPs are, however, not suitable for thermoforming. This is because they are so highly oriented in the machine direction that they do not draw sufficiently along the axis. However, using fillers to partly disrupt orientation makes thermoforming much easier. Thus, the LCP formulations Vectra A540 (40% mineral filled) and A130 (30% glass filled) may be processed by thermoforming.

The first LCP to be launched commercially was Dartco's Xydar, introduced in late 1984. Xydar injection-molding resins are aromatic polyesters based on terephthalic acid, parahydroxy benzoic acid, and p,p'-dihydroxybiphenyl. Xydar has a high melting point (close to 400°C), which necessitates certain modifications to processing equipment. It also has a high melt viscosity making it difficult to mold in small sections. An advantage of Xydar, however, is that it offers high retention of mechanical properties even at temperatures in excess of 300°C.

A second range of LCPs have been introduced by Celanese under the trade name Vectra. Vectra is an aromatic polyester condensate. Its main advantages are its fast cycling and ease of processing, the melt temperature for both injection and extrusion grades being typically about 285°C. The material is noncorrosive to molds and clean running, while its exceptionally low melt viscosity ensures easy filling of complex, small-part geometries. No modifications are required to processing equipment. However, as with filled or reinforced polymers, tools should be hardened to minimize the effects of abrasion.

Applications

One of the first applications of LCPs was a range of dual-ovenable cookware of Xydar made by Tupperware, a subsidiary of Dartco Manufacturing.

The resistance of LCPs to chemicals and solvents and good performance in hostile environments have led to their several specialized applications. Vectra materials have been used commercially for the molding of formic acid separation tower packings as they are proving to be more efficient and longer lasting than conventional ceramic packing materials. Parts made of Vectra have also been used in the fuel system components of aircraft and automotives.

Most future opportunities for LCPs are believed to lie in the electrical/electronic and telecommunications industry. LCPs are finding use as replacement for epoxy and phenoloc resins in electrical and electronic components, printed circuit boards, and in fiber optics. In these applications, the high mechanical properties, low coefficient of thermal expansion, inherent nonflammability, good barrier properties and ease of processing of LCPS (Vectra in particular) are important. In fiber optics applications, LCPs can be extruded into a variety of shapes and sizes using conventional extrusion equipment found in typical fiber optic wire and cable production plants.

The ability of LCPs to withstand contact with solder, fluxes, and chlorinating agents is important in their use for printed circuit boards. The ability of LCPs to be molded to extremely thin walled parts at very fast cycle times with high temperature resistance and dimensional stability has assumed much significance in relation to vapor-phase soldering applications. In several electronic applications, such as connectors and capacitors, LCPs are also being used in preference to high-performance materials, such as poly(phenylene sulfide), because of their ease of processing, greater impact resistance, and overall lower part cost.

POLYMERS IN PHOTORESIST APPLICATIONS [23–28]

Polymers and polymeric systems that can undergo imagewise light-induced reactions are of great technological importance. Photoresists are polymers or polymeric systems (polymer binders containing dispersed or dissolved photoactive compounds) which, applied as a surface coating to an underlying substrate, undergo some type of reaction upon exposure to visible or ultraviolet radiation such that the differences in solubility of the exposed and unexposed areas can be exploited to obtain selective removal of either the exposed or unexposed area of the film (a process known as development). Thus, by providing suitable radiation exposure of the polymer-coated surface through a pattern or mask, such as a negative, an image can be generated on the substrate. The image can then be transferred permanently to the substrate by etching, deposition, etc. Since such images are raised above the substrate on which the polymer has been coated (see Fig. 5.14), they are called relief images. In order for the image to be transferred to the substrate, the polymer must "resist" the chemical activity of the particular etchant, which can be either a wet (solvent) or dry (plasma) process. Thus, these polymers are called *photoresists*.

There are a number of types of photoresist that function by a variety of mechanisms, such as:

1. Cross-linking of a linear polymer backbone by the light-induced decomposition of a photosensitizer to generate active species (e.g., cyclized rubber with azide type photosensitizer).

2. Cross-linking of a polymer containing photosensitive groups withn its own structure [e.g., poly(vinyl cinnamate)].

3. Polymerization of a monomeric material to yield a reduced solubility polymer (e.g., methylene bis-acrylamide monomer and benzoin photoinitiator in polyamide as the carrier polymer).

4. Enhancement of solubility by a photoinduced molecular rearrangement (e.g., naphthaquinone diazide in a novolac binder resin).

5. Enhancement of solubility by reduction in molecular weight by bond scission (e.g., poly(methyl methacrylate) deep-UV resist).

6. Depolymerization by high energy irradiation to allow a kind of ablative imaging (e.g., by irradiation with laser or an electron beam).

Figure 5.14 shows schematically the positive and negative modes of lithographic processes with a photoresist film. When exposure of the photoresist film to radiation produces an image that is less soluble in the developer, a negative image results. Negative images are generally produced by photochemical reac-

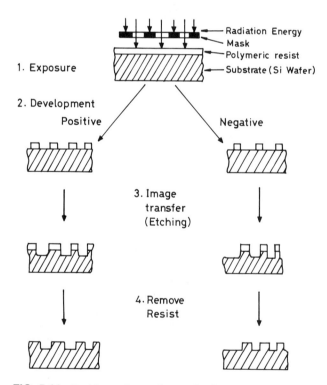

FIG. 5.14 Positive and negative modes for a photoresist film.

tions that lead to crosslinking and network formation in the polymer film in the exposed areas. Such network polymers are insoluble in the developer solvent. A second means of generating a negative image is to produce a photoinduced change which makes the exposed areas less soluble in a selective developer.

A *positive* image results when the areas of the film exposed to radiation are more soluble than the unexposed areas in the developer (Fig. 5.14). There are two classes of chemical reactions that can lead to enhanced solubility of the exposed areas. The first is a photochemical change that converts the exposed area to a different polarity as compared to the unexposed areas. A selective dissolution of the exposed areas can then be obtained with the proper choice of a developer. The second chemical change that can occur is a backbone cleavage of the resist polymer such that the exposed area of the polymer film is degraded to a low molecular weight (or all the way to monomer) that has a dissolution rate significantly higher than the rate for the unexposed higher-molecular-weight area in the developer.

There are two important aspects of the performance of a photoresist: imaging (lithography) and image transfer (resistance). An image must be formed in the resist polymer and then the image must be transferred permanently into the substrate or to other materials by etching, deposition, etc. The term "photolithography" is also used to refer to both the processes. Traditionally, photolithography has had great commercial impact on the printing industry, photomachining of fine parts, and so on.

More recently, the revolutions in the field of electronic have been made possible by the photopolymers (microresists) used to delineate the tiny features that make up modern integrated circuits. The microlithographic applications of photopolymers have been the driving force behind most of the technological and scientific activity in the area over the last decade. A plot of the decrease in device feature size (the smallest feature that must be resolved in the photoresist to produce the device) versus the year that device was first manufactured, shown in Fig. 5.15, indicates how fast this technology has progressed. However, most of the devices that have been manufactured using microlithography have employed one of two resist systems. Negative working resists based on cyclized rubbers and bisazides were the original mainstays of this industry. Although these resists are still in use today, novolac/naphthaquinone-diazide-based positive photoresists are used for most features less than 3 μm in size.

The drive to smaller feature sizes in microlithography has brought in its wake a trend toward shorter wavelengths for the exposure radiation. To a very rough approximation, the resolution limit of an optical exposure source is given by 2λ, where λ is the wavelength. As the wavelength decreases, the potential resolution thus becomes smaller, i.e., better. This, however, requires resists that are sensitive at these shorter wavelengths and also exposure sources that are capable of sufficient output at these wavelengths.

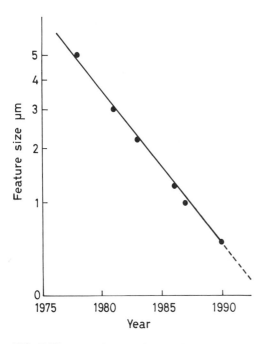

FIG. 5.15   Plot of device feature size versus approximate year of initial production.

Almost all commercial exposure sources use super-high-intensity Hg or Hg/Xe lamps with high-intensity outputs at 436, 405, 365, 330, and 313 nm. The first three wavelengths are normally grouped and called the *near-UV*. The next two wavelengths are called the *mid-UV*. Radiation from these lamps and other sources with wavelengths below 280 nm comprises the *deep-UV*. With the development of excimer lasers high-output deep-UV sources have become readily available.

The majority of current microlithography for semiconductor processing is carried out using radiation in the near-UV wavelength range. It is generally agreed, however, that near-UV lithography is becoming resolution-limited by its relatively long wavelength. The submicron resolution that is required for increasing sophistication and degree of integration is more readily achievable by going to shorter wavelengths, and there is currently much interest in using not only deep-UV but also X-rays, ion beams, and electron beams for exposing resists. Each of these technologies has specific resist requirements, and various polymeric resist systems have been developed and are under development.

The other change in resist processing involves the use of multilayer schemes in place of a simple photoresist coating. Microimaging using multilayer materials and technology has revolutionized the electronics industry. Many silicon chips used in calculators and computers are produced in some variation of the following sequence of operations:

A silicon wafer that has one surface oxidized to a controlled depth is coated (on the oxide surface) with a photoresist, such as poly(vinyl cinnamate), to produce a thin and uniform coating several micrometers thick when dry. Exposure to UV light through a mask insolubilizes part of the polymer. The uncross-linked polymer is washed off with solvents. The bare substrate parts that thus reappear are etched through the oxide layer down to the silicon layer by a fluoride solution in water or by a plasma that contains reactive ions.

Operations are then performed to alter the chemical composition of the etched regions. The operations may include ion implantation to introduce dopants that make semiconductors of the diffused-base transistor type and depositing a layer of aluminum to act as a conductor or a layer of other materials to act as insulators. After removing all of the remaining polymer by solvents, plasmas, or baking, the wafer is recoated and a new pattern is imposed and processed. The sequence may be repeated as needed to produce integrated circuits with many such layers and with amazing complexity. The minimum line widths on chips produced by photolithography are usually in the 5-μm range. However, to produce large-scale integrated circuit (LSI) devices, minimum line widths must be decreased to pack more circuits on a chip.

Negative Photoresists

Most negative-working resist systems are based on the fact that the dissolution rate of a polymer decreases as the molecular weight increases, and the polymer ultimately becomes insoluble in any solvent as a cross-linked network is formed.

Probably the earliest recorded photoresist system was the process of Niepce around 1826. He found that bitumen (a complex hydrocarbon polymer mixture with some unsaturation present) derived from Judea asphalt became insoluble in a mixture of lavender oil and mineral spirit after prolonged exposure to light (several hours). He used this phenomenon to make images on silvered glass, the first photographs. A more widely used system took advantage of the photosensitization reaction of the dichromate/poly(vinyl alcohol) or dichromate/gelatin system. The mechanism of these dichromate resists is obscure, but it is thought that the formation of the chromate ion from the dichromate ion is an important initial step:

$$Cr_2O_7^{2-} \xrightarrow{h\nu} CrO_3 + CrO_4^{2-} \tag{7}$$

This followed by the transition of chromium from Cr(VI) to a low state of oxidation, and an interaction process with the hydrophilic polymer to form a chromium/polymer complex in which binding between the polymer and photoreleased chromium compound involves either primary forces or physical forces of absorption. The resulting complex causes a solubility decrease in the aqueous system used for development.

A search for novel compounds to supersede dichromates resulted in the production of the diazo compounds. The diazo resist systems are based not on crosslinking but on photofunctional change. A diazonium salt and a binder resin are coated from an aqueous solution. Upon irradiation, the diazonium salt decomposes, giving nitrogen, and a neutral product is formed, presumably via a radical reaction. While the diazonium salt enhances the binder solubility in an aqueous developer, the neutral compound formed reduces the solubility. However, like the dichromate resists, the diazo resist systems also suffer from limited stability and limited exposure wavelength sensitivity and are very slow photographically.

All of the above processes were displaced for resist applications by better-characterized negative photoresists using photodimerization (cinnamate type resists), photocrosslinking (bisazide resists), and photoinitiated polymerization (free-radical or cationic) to yield insoluble polymer networks.

The simplest photodimerizable resist is poly(vinyl cinnamate), made by esterification of poly(vinyl alcohol) with cinnmoyl chloride. Upon exposure, the cinnamate groups can dimerize to yield truxillate or truxinate. If the two cinnamate groups involved in this reaction are from two chains, the cyclic product represents a cross-link (Fig. 5.16).

There are many cinnamate resins, including derivatives of poly(vinyl alcohol), cellulose, starch, and epoxy resins. It is the epoxy resins that have found the most applications in lithographic materials.

As cinnamate resins are water-insoluble, a suitable solvent, such as trichloroethylene, is used to achieve the solubility differential for image development. An emulsion of the solvent dispersed in an aqueous phase of gum arabic and phosphoric acid is generally used. The light-exposed regions of the coating are rendered insoluble due to cross-linking and form the printing image.

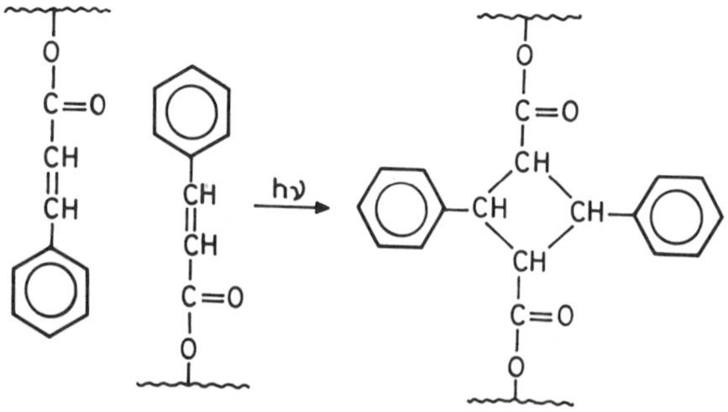

FIG. 5.16 Photodimerization of poly(vinyl cinnamate).

Poly(vinyl cinnamate) itself is only a weakly absorbent above 320 nm. Its photoresponse is generally of the order of a tenth of that of dichromated colloids, but the rate can be accelerated by the use of photosensitizers, such as nitroamines (increase of the order of 100 times), quinones (increase of the order of 200 times for specific ones), and aromatic amino ketones (increase of the order of 300 times for specific ones). A commonly used aromatic ketone is 4,4'-bis(dimethylamino)-benzophenone, also known as Michler's ketone.

The Kodak photoresist based upon the poly(vinyl cinnamate) system is particularly suitable for printed circuit manufacture. It also finds application in some invert halftone photogravure processes and photolithographic plates. The property of superior adhesion to metal of cinnamic esters of particular epoxy resins has resulted in the preferred use of these resins for platemaking. Very few of the cinnamate-type resists have, however, been used in micro applications.

A negative-working resist system that has seen extensive use in microlithography and continues to be used where high resolution is not required, is that based on the photochemistry of bisazide as the photoactive compound in a cyclized or partially cyclized polyisoprene rubber as the binder resin. Since poly(*cis*-isoprene) has a low glass-transition temperature and is too soft for use in lithography, partial cyclization is performed to yield a polymer with good adhesion and film forming properties. The cyclized units increase the T_g of the rubber while reducing the solution viscosity so that high-solids coatings are possible.

The azide-type photoactive compounds used with the above type of resist have good thermal stability and decompose efficiently upon UV irradiation from mercury lamps to give intermediates known as nitrenes. The nitrenes being highly reactive undergo several reactions, e.g., (1) insertion reactions with carbon-hydrogen bonds to yield secondary amines, (2) addition reactions with carbon-carbon double bonds to produce aziridines, and (3) abstraction reactions that generate radicals. These chemistries are shown in Fig. 5.17. Since a bifunctional azide is used, two polymer chains can be linked and eventually a cross-linked network is formed. The overall effect of the nitrene formation is thus to produce a cross-linked, reduced solubility matrix.

Resists that use bisazides are mostly near UV (436 nm) resists. Because of solubility limitations, it has been difficult to make a bisazide resist that is sensitive beyond 436 nm.

The drawback of the bisazide-based cyclized rubber resist becomes manifest during development. A good solvent must be used to remove the unexposed high polymer. However, this solvent also penetrates and swells the images that are to be left behind. Though with large (>10 μm) spaces between images, the image returns to its original position, with small (<3μm) spaces, two neighboring images may swell, touch, and coalesce to leave a polymeric bridge between the two images. Such a bridge will finally translate into a short in the electronic circuit.

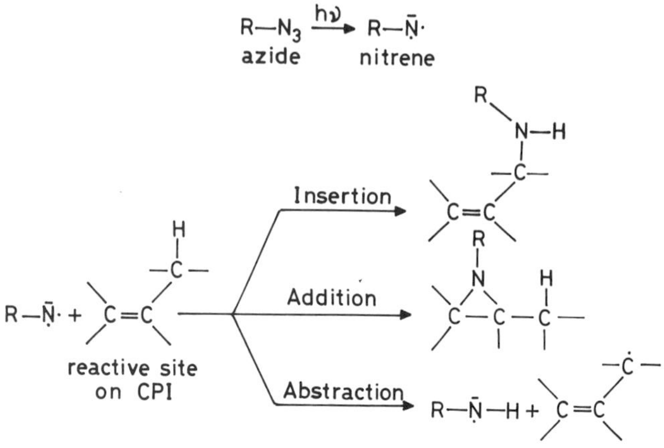

$$R-N_3 \xrightarrow{h\nu} R-\ddot{N}\cdot$$
$$\text{azide} \qquad \text{nitrene}$$

FIG. 5.17 Reaction of nitrene, formed by photodecomposition of azide, with cyclized polyisoprene (CPI) polymer.

Because of their very high reactivities, bisazide compounds have also been used with polymers other than the cyclized rubbbers in attempts to obtain higher resolution. Thus resists that use bisazides along with aqueous base-soluble polymeric binders, such as novolac resins and acyclic acid copolymers, have been shown to produce superior resolution properties with little swelling during development. The bisazides have been tuned for deep-UV absorbance in order to get the highest possible resolution. A deep-UV negative resist with less swell has been formulated from a short-wavelength bisazide, 3,3'-diazidodiphenyl-sulfone (V) and poly(4-hydroxy styrene) (VI).

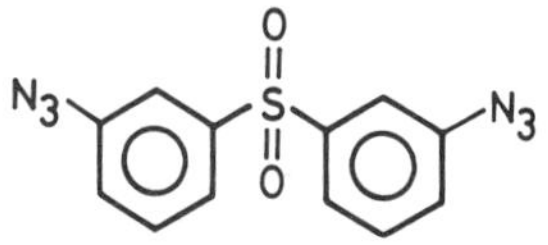

(V)

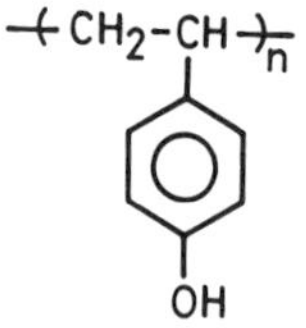

(VI)

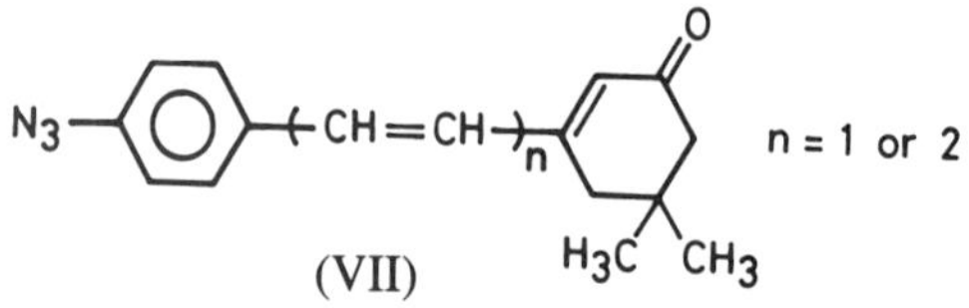

(VII)   n = 1 or 2

Another approach for general lithography has used a monoazide (VII) with an aqueous base-soluble polymer, poly(4-hydroxy styrene). The monoazide can be tuned readily to have absorbance maxima at or near the mercury lamp emission lines, and the image network does not swell strongly during development. The wavelength sensitivities of this type of negative resist are determined entirely by the azide. Thus, it has been shown that novolac resins can be sensitized to all common exposure sources, including the deep-UV, simply by altering the azide structure.

Many negative-working resist systems based on photoinitiated free-radical polymerization have been investigated. Most of these systems use a multifunctional acrylate or methacrylate monomer along with a photolabile polymerization initiator. These systems have, however, been little used in microlithography because they are difficult to spin-coat and are sensitive to the amount of oxygen in the coatings.

The oxygen sensitivity of photoinitiated polymerization may be avoided by using cationic rather than free-radical polymer systems. There has been considerable activity to develop negative working resists based on photolabile acid generators. Many acid generators are possible, but the most widely used have been sulfonium and iodonium salts that decompose to form a strong acid during exposure. Subsequent heating of the coating can cause several types of acid-catalyzed reactions including polymerization.

A novel microlithographic application for photoinitiated polymerization involves the polymerization of a monomer and the locking-in of a plasma-sensitive host polymer so that plasma techniques can be used to carry out all-dry development, thus avoiding the problems of swelling and resolution limitation associated with standard resists. Some plasma-developable resists are described later in this section.

Positive Photoresists

Positive photoresists have been increasingly important because of their higher resolution capability and better thermal stability. They become more soluble in exposed areas by increasing their acidity or by undergoing bond scission and degradation of polymer chains.

Near-UV Application

At near-UV wavelengths it is not possible to enhance solubility by photoinduced bond scission because the light is not energetic enough, and consequently an alternative mechanism is used. Positive resists with sensitivity in the near-UV usually comprise a large photoactive molecule that is insoluble in basic solvents and also sufficiently bulky to inhibit dissolution of a base soluble polymer. On exposure, the photoactive compound breaks down to form a base-soluble product and the exposed region can then be washed out using a basic solvent. The base soluble polymeric binder is typically a novolac resin of relatively low molecular weight.

The most commonly used photoresists for resolving features of 5 μm and below are based on the photolysis of diazonaphthoquinone sulfonate esters (DNS), which are used as the photo-sensitive compound in a novolac binder resin. Upon exposure at near-UV wavelengths, the photosensitive compound undergoes a Wolf rearrangement. Nitrogen is given off in this reaction producing a ketene intermediate (Fig. 5.18a), which reacts with water in the film to form an indene carboxylic acid (Fig. 5.18b). Under normal ambient working conditions, enough water is present in the novolac binder to ensure the formation of the carboxylic acid. However, in the absence of water, the ketene intermediate has been shown to react with the phenolic hydroxyl groups on the novolac to form pendant ester linkages (Fig. 5.18c).

The unexposed DNS in the masked areas acts as a dissolution inhibitor for the novolac binder, presumably due to its hydrophobic nature, while the indene carboxylic acid that is formed from the photolysis reaction in the exposed areas is a dissolution accelerator for the novolac. The rate of development in an aqueous base is much higher for the exposed areas, which promotes good developer discrimination between the exposed and unexposed areas leading to high-contrast, high-resolution images. It has been proposed, however, that this simplistic picture of hydrophobic-to-hydrophilic switch may not represent all of the factors that enable the high discrimination and high contrast obtainable in this type of resists. The enhancement in dissolution rate may also be a result of the gaseous nitrogen that is formed during the DNS photolysis (see Fig. 5.18). It is hypothesized that the nitrogen evolution causes microvoids and stresses in the novolac film and that this promotes the diffusion of the developer into the polymer film leading to an increase in the dissolution rate.

Novolacs produced by cresol-formaldehyde condensation have been the polymers of choice for the DNS photochemistry. These novolac resins are usually synthesized from commercial cresol mixtures, which contain about 60% m-cresol, 30% p-cresol and 10% of various other aromatic phenols. These polymers are soluble in spin-coating solvents, show good film-forming and coating properties, provide excellent adhesion to most substrates, and exhibit enough balance in the needed binder properties to be used almost exclusively in all commercial DNS containing positive photoresist formulations.

With the increase in demands on photoresist formulations due to smaller imagery and new image transfer steps, such as ion implantation and plasma etching, novolacs are, however, found to fall short in two areas. First, a specific chemical composition and molecular size is difficult to reproduce because the exact composition of the cresol starting materials can vary from lot to lot and because the condensation polymerization of the cresol with formaldehyde is difficult to control, which results in a relatively low-molecular-weight polymer with M_n around 1000 and with a broad dispersity (about 20–40). These variations can lead to irreproducible lithographic performance of the high-resolution

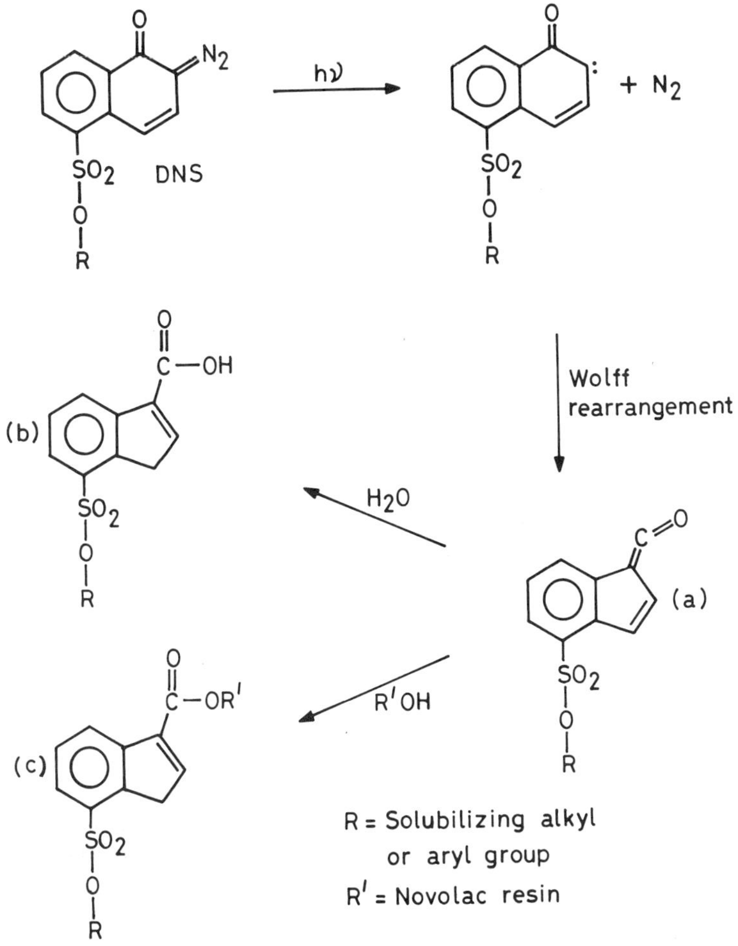

FIG. 5.18 Photolysis of diazonaphthoquinone sulfonate esters (DNS) followed by Wolff rearrangement to a ketene intermediate (a), which reacts with water to form an indene carboxylic acid (b) or with phenolic hydroxyl groups (in absence of water) to form ester linkages (c).

resists. Second, novolacs have low glass transition temperatures (T_g), generally in the range of 20–120°C depending on molecular weight, which often lead to unacceptable image distortions when the image is subjected to plasma etching or ion implantation process steps.

For several years, many of the commercial positive photoresists have contained multifunctional, principally trifunctional, DNS derivatives, some

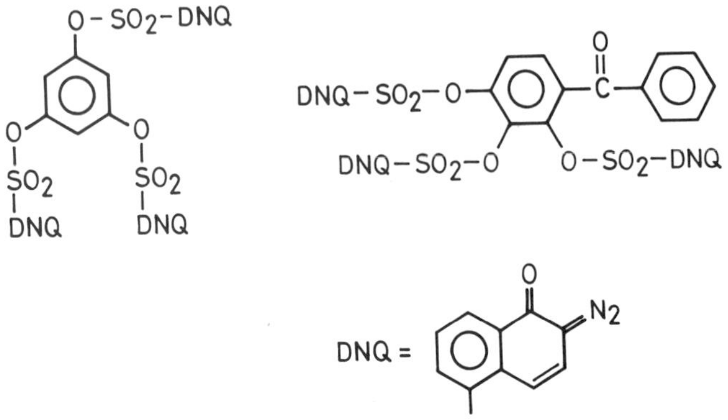

FIG. 5.19 Typical examples of multifunctional diazonaphthoquinone sulfonate esters used in photoresists.

examples of which are shown in Fig. 5.19. They contribute to enhanced image stability during the image transfer step in addition to higher contrast and faster development in the microlithographic process.

In normal resist processing, after the development step, the resist image is baked at about 130°C for 30 minutes to "harden" the novolac image. At this temperature, the DNS in the unexposed areas decomposes to the ketene, and a large portion of it reacts with the phenolic hydroxyl groups on the novolac to form ester linkages (see Fig. 5.18c). Since the multifunctional DNS forms a cross-linked network in the image area, it leads to a considerable improvement of the performance of these images to subsequent image transfer steps, such as plasma etching and ion implantation, by preventing the flow or distortion of the image at high temperature.

The multifunctional photosensitive compounds described above suffer from the disadvantage of low solubility. At the heart of almost every one of the proprietary formulations utilizing these compounds, there is thus a specific compound or a combination of compounds and isomers that possess the requisite solubility in the limited number of environmentally permissible spin-coating solvents that are available at present.

As stated earlier, the developer used to dissolve or develop out the exposed regions of the DNS-based positive photoresist coatings is an aqueous base. There are two general types of these aqueous-base developers. The first type, the metal-ion containing developers, is based on metallic hydroxides. The need to eliminate metal ion contaminants in semiconductor processing has led to the development of a second category of metal-ion free developers. The most important members in this category are quaternary alkylammonium hydroxides, such as tetramethylammonium hydroxide, tetraethylammonium hydroxide, etc., used in aqueous solutions.

The DNS/novolac-based positive photoresist can also be used to obtain negative images by an image reversal process involving post-exposure treatment of the photoresist film with a reactive amine. This amine-treated film is heated and then flood-exposed. Development with a conventional aqueous base at this stage produces a high resolution negative image of the original positive image. This process is shown in Fig. 5.20. An amine (e.g., monazoline, imidazole, triethanolamine) reacts with the indene carboxylic acid in the originally exposed areas to form an ammonium salt that decomposes on heat treatment to indene. Subsequent flood exposure gives indene carboxylic acid in the previously unexposed areas that are then washed out in the aqueous base developer. The hydrocarbon indene from the original exposure and subsequent process protects the novolac just as the original DNS would do.

Mid- and Deep-UV Photoresists

Though conventional positive resists have performed well at 436, 405, and 365 nm wavelength (common outputs of commercial exposure sources), the potential resolution gain has been small. This has given a push toward shorter wavelengths for achieving higher resolution.

There are two main problems in trying to use a conventional positive photoresist in the mid- and deep-UV regions. First, the quinone diazide yields a photoproduct that absorbs at these wavelengths. Second, the novolac resin absorbs strongly at wavelengths below 295 nm. (Neither problem exists, however, for near-UV exposures, since both the novolac and the photoproduct are essentially transparent in the near-UV.) There have been many attempts, therefore, to make a short-wavelength quinone diazide resist suitable for lithographic processes.

It is observed that naphthaquinone photoactive compounds, which have the sulfonyl group at the 4-position rather than at the 5-position, have suitable absorption in the mid-UV and, moreover, yield a photoproduct with less absorption in the mid-UV.

A DNS-based positive photoresist could be used for deep-UV exposures if the binder resin in the photoresist itself did not absorb. In this context, it is interesting to note that novolacs made from the condensation of formaldehyde with pure *p*-cresol (instead of commercial cresol mixtures) are found to have very transparent windows at about 250 nm. A DNS/*p*-cresol novolac-based photoresist thus gave a positive image after deep-UV exposure and development with aqueous base. A copolymer of styrene and maleimide was also used in place of novolac as a binder for DNS-based deep-UV positive resists.

The photochemistry of *o*-nitrobenzyl esters (Fig. 5.21) has been used as the basis for another dissolution inhibitor approach to making deep-UV resists. In

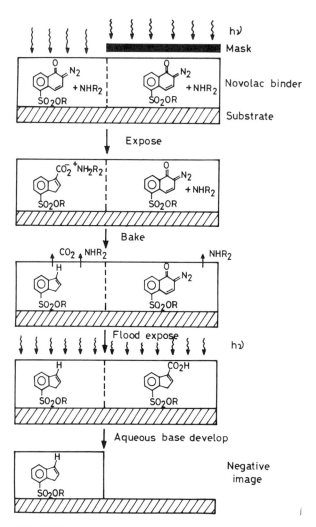

FIG. 5.20 Schematic of image reversal process of DNS-novolac positive photoresist.

this case, the polymer binders are base-soluble copolymers containing methacrylic acid and the dissolution inhibitor is an *o*-nitrobenzyl ester of a large organic molecule such as cholic acid. The ester cleaves, upon exposure, to form products (Fig. 5.21) that no longer interfere with the resin dissolution, and thus a positive relief image is obtained after development.

A widely studied resist for the deep-UV is poly(methyl methacrylate) (PMMA). The polymer undergoes chain scission upon exposure around 230 nm. The exposed areas are thus lower in molecular weight than the unexposed areas and a positive relief image results on development, with an appropriate organic solvent such as chlorobenzene.

While PMMA is capable of very high resolution, it has several major drawbacks as a deep-UV resist—a very low extinction coefficient, a very narrow waveband for absorbance, and low quantum efficiency for scission. PMMA also suffers from a relatively poor resistance to the plasma etching treatments that are necessary for producing high-resolution images. One way to improve the etch resistance is to blend in a more resistant polymer. Thus, the addition of 10% of an aromatic polymer to PMMA was found to cause a four-fold decrease in the etch rate of the PMMA in a fluorine plasma without any significant loss of photographic speed.

A related polymer, poly(methyl isopropenyl ketone), PMIPK (VIII) have better spectral response and speed (five times faster than PMMA) in the deep-UV, as would be expected in view of the more efficient photochemistries of ketones. Moreover, chemical sensitization of PMIPK with substituted benzoic acids gives speeds that are more than 25 times faster than that of PMMA, depending on the wavelength of exposure. PMIPK does not, however, have improved plasma resistance over PMMA. The etch resistance is improved by blending in a more resistant polymer. For example, PMIPK had a five-fold decrease in etch rate in a chloride plasma using 10% of an added aromatic.

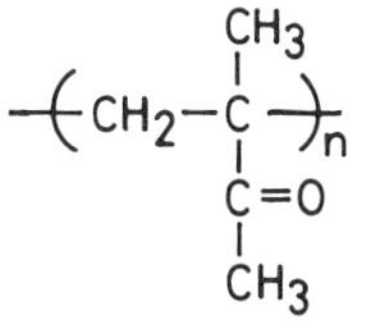

(VIII)

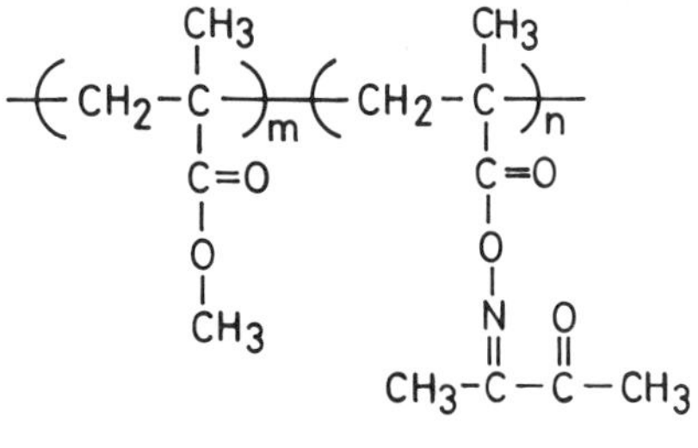

(IX)

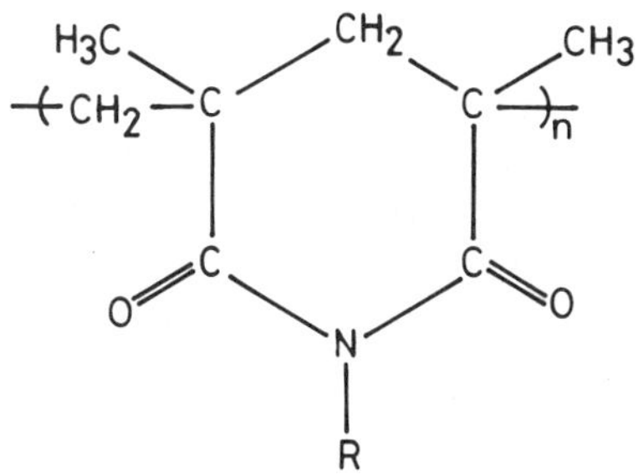

R = H, alkyl

(X)

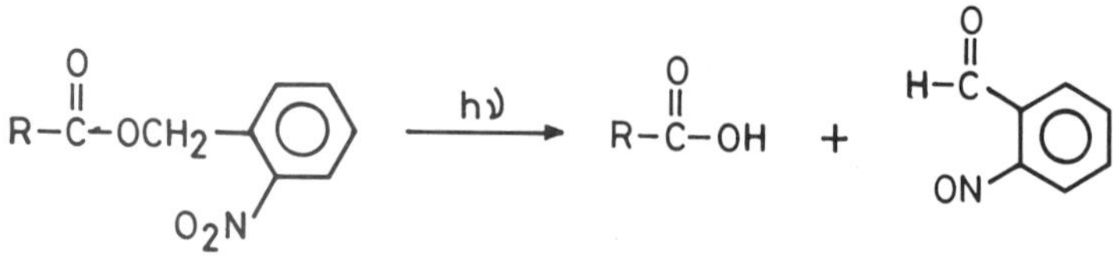

FIG. 5.21 Photochemistry of *o*-nitrobenzyl ester.

A copolymer (IX) of methyl methacrylate and 3-oximino-2-butanone methacrylate provides a broader absorption band in the deep UV than PMMA and is relatively thermally stable. The addition of 37% of the oximino monomer gave a 50-fold increase in speed over PMMA itself. Further improvement in speed was obtained by chemical sensitization using *p-t*-butylbenzoic acid.

Deep-UV resists are used in two principal modes: as surface resist in single layer mode, or as the thick planarizing layer in portable conformable mask (PCM) bilayer mode. In the single layer mode, the use of deep-UV resists provides improved resolution, while in the PCM mode, a deep-UV resist is used in combination with a conventional resist to make use of the wavelength selectivity of the two types of resists. The PCM process (Fig. 5.22) first exposes and develops the conventional novolac-based thin top resist using a 436 nm masked exposure. A second full exposure is then made in the deep-UV, which is masked by the novolac-based resist image of the 436 nm exposure. The unmasked (exposed) positive-working deep-UV resist is developed to give a thick, high-resolution image for subsequent processing.

PMMA is used as the planarizing layer in most PCM systems since speed is not a problem because of the blanket exposure, and since the poor fabrication resistance of PMMA is also less of a problem due to the thin novolac cap remaining in place. Copolymers of methyl methacrylate and analogues of PMMA have been worked on in search of improved planarizing deep-UV resists. For example, a terpolymer of methyl methacrylate, methacrylic acid, and methacrylonitrile has provided a planarizing deep-UV resist with a two-fold speed increase and a 40°C increase in T_g.

A further improvement in the physical performance of the planarizing layer has been realized with poly(dimethyl glutarimide), PMGI (X). It should be noted that the polymer (X) is actually a copolymer with the relative amounts of −H and −alkyl being chosen to provide the best compromise of solubility for coating and development. The copolymer PMGI behaves as a scissioning resist, very similar to PMMA, but it absorbs at longer wavelengths, 250 nm rather than 220 nm, and has the additional advantage of being developable with the same aqueous base that is used for the conventional 436 nm resist rather than requiring an organic developer.

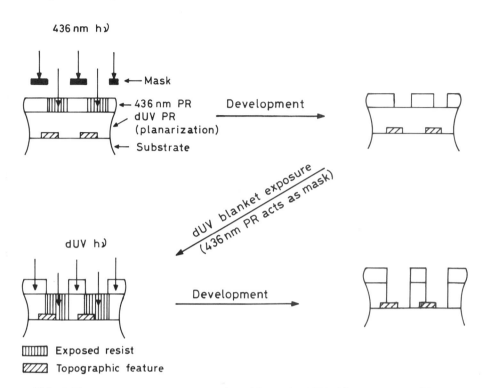

FIG. 5.22 Schematic of portable conformable mask (PCM) bilayer process. (PR = photoresist; dUV = deep-UV).

The use of a photolabile acid generator to deprotect imagewise a functional group or depolymerize a polymer is the best-documented example of photochemical change systems that are known to proceed by a chain reaction mechanism. A new term, *chemical amplification*, has been coined to describe any of several such systems. Many photolabile acid generators are possible, but the most widely used have been iodonium and sulfonium salts, which have their natural sensitivity between 200 and 300 nm but can be substituted and sensitized out into the visible.

Polymers with acid-labile protecting group such as *t*-butyl carbonate and *t*-butyl esters have been doped with iodonium and sulfonium salts, which produce acid on radiation exposure in resist matrix. Imagewise exposure of these resist formulations to generate acid and then heating produces a relief image (Fig. 5.23a). Either positive or negative images may be developed depending on the solvent chosen.

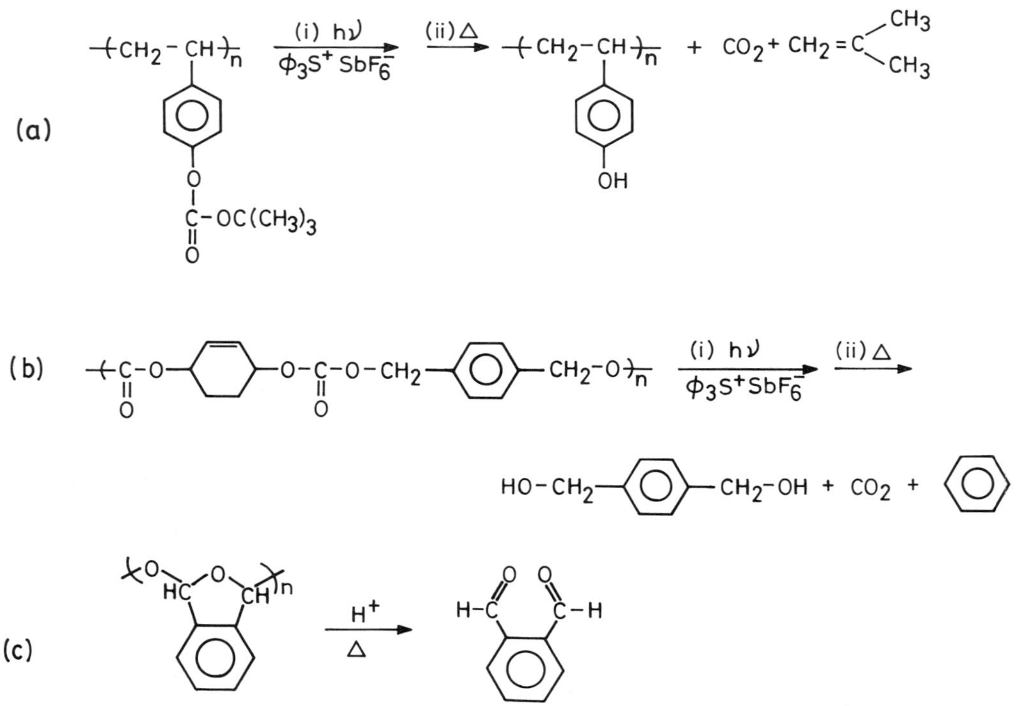

FIG. 5.23 Photochemistry of poly(4-t-butoxycarbonyloxy styrene) doped with triphenylsulfonium hexafluoroantimonate (a), a dry developable polycarbonate (b), and self-developing polyphthalaldehyde (c).

TABLE 5.10 Excimer Laser Outputs

Complex	Wavelength (nm)
Ar-F	193
KrCl	222
KrF	248
XeBr	282
XeCl	308
XeF	351

The design and synthesis of polymers with carbonate units in the backbone that allow a facile degradation of the polymer by the chemical amplification scheme with photoacid have been reported. The exposed areas of such resists can be developed by a solvent giving high-resolution positive images. On the other hand, polymers containing precursors to more volatile fragments such as copolycarbonate (Fig. 5.23b) may be dry developed by thermal treatment alone. Thus, after exposure of the copolycarbonate in the presence of a photoacid and thermal treatment at 60°C for 2 min, complete image cleanout was observed.

Polymers capable of facile depolymerization upon irradiation may be suitable as "self-developing" positive resists. One polymer where depolymerization has been successful is polyphthalaldehyde (Fig. 5.23c). End-capped polyphthalaldehyde is stable to a temperature of 150°C, although its ceiling temperature is −40°C. When polyphthalaldehyde is irradiated with a high-powered excimer laser, it is blown away as monomer during the exposure. This type of *ablative imaging* is possible because of the high-energy output of the laser. Moreover, in the presence of an onium salt, the depolymerization can be catalyzed imagewise.

The excimer laser mentioned above filled a long-standing need for a workable exposure source for deep-UV lithography. The excimer laser can deliver large amounts of light at several deep-UV wavelengths, depending on the mixture in the laser gas chamber. Several of the possible outputs of this tunable tool according to the excited complex that emits light are listed in Table 5.10.

The extremely high-energy output of the laser also allows the potential of the above kind of ablative imaging with other polymers. For example, when PMMA is used with an ArF (193 nm) laser, the exposure produces a large amount of scissioning in the top surface. The fragments are blown out of the exposed hole by the kinetic energy of the photopulse, while continued pulses "drill" through the resist. Generally, the energy output is so great that almost any polymer can be ablated.

Electron Beam Resists

With the demand for greater and greater device integration, as in the technology of random access memories where the number of devices to be accommodated on a piece of silicon has increased dramatically with no concomitant increase in the device area, the geometry of the features on a chip has continued to fall and routine production is now in the 1–2 μm range for some advanced devices. Imaging such small geometries call for new lithographic processes. One of the most important of these new processes uses an electron beam to irradiate the resist and is known as *electron beam lithography*.

An electron beam is of much shorter wavelength than the radiation used in standard microlithography and, therefore, provides greater resolution possibilities. For resist exposure a beam of electrons can be used whose position is controlled by a computer-driven beam deflector, thus obviating the need for a mask. Both positive- and negative-working resists are used for electron beam lithography.

Since an electron beam having shorter wavelength and higher energy can cause covalent bond scission relatively easily, polymer degradation by exposure to the beam forms the basis for a positive photoresist system. Poly(methyl methacrylate) and its analogues, such as poly(hexafluorobutyl methacrylate), have been widely studied as positive-working electron resists. Exposure to an electron beam causes extensive main chain scission in these polymers, but depolymerization back to monomer does not occur. Moreover, PMMA is not very sensitive and so other materials have been employed. Polyolefin sulfones, for example, are extremely sensitive positive-working electron resists, undergoing chain scission back to the olefin and SO_2.

Polymers containing reactive cross-linking groups such as epoxy or allyl are commonly used for negative-working electron beam resists. A typical example is a copolymer of glycidyl methacrylate and ethyl acrylate. Copolymers of glycidyl methacrylate and 3-chlorostyrene have been shown to have high sensitivity, good adhesion, and plasma resistance. These materials are commercially available.

Plasma-Developable Photoresists

One important area of resist research in the last few years has involved attempts to develop plasma-developable resist systems. The aim of plasma developable resists is to use nonsolvent, all dry development methods to avoid the problems of swelling and consequent resolution limitation associated with conventional resists. Much of the semiconductor fabrication process now utilizes plasma techniques as they are capable of providing high resolution images. An important consideration in this is that the plasma-developable resist images should stand up well to the plasma etching treatments.

Plasma resistance of organic polymers can increase significantly due to the presence of aromatic and heteroatomic groups. So if a resist could be made such that after exposure, the exposed area contained more of these groups than the unexposed area a differential plasma etch rate would be achieved leading to selective development (Fig. 5.24).

A process known as *photolocking* based on the above method was developed at Bell Laboratories. The process used a resist consisting of a plasma degradable acrylic polymer such as poly(dichloropropylacrylates) intimately mixed with a readily polymerizable volatile monomer such as *N*-vinyl carbazole or acenaphthylene. Upon exposure to UV radiation through a mask, the monomer only in the exposed areas polymerizes and becomes locked in the plasma sensitive host polymer. After exposure, the monomer in the unexposed regions is removed by application of vacuum leaving the plasma-sensitive poly(dichloropropylacrylate), while in the exposed regions the photolocked polymers of

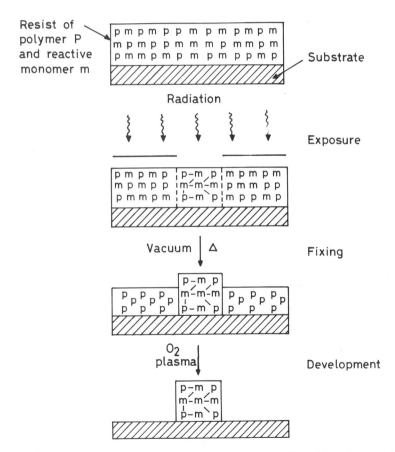

FIG. 5.24 Schematic of photolocking process used to bind etch-resistant molecules imagewise in a polymer film.

N-vinyl carbazole or acenaphthylene provides plasma resistance, thus facilitating selective development. This type of photolocking system is particularly suitable for X-ray lithography.

In photolocking systems, *N*-vinyl carbazole, which is rather toxic, was subsequently replaced by a number of silicon containing monomers. A novel feature of such silicon-based monomers (XI) is that, on oxygen plasma treatment for final resist removal, they deposit a thin layer of silicon dioxide, which is itself a useful protective dielectric layer.

$$CH_2 = CHCOO(CH_2)_n - \underset{\underset{CH_3}{|}}{\overset{\overset{CH_3}{|}}{Si}} - O - \underset{\underset{CH_3}{|}}{\overset{\overset{CH_3}{|}}{Si}} - (CH_2)_n - OOCCH = CH_2$$

(XI)

PHOTORESIST APPLICATIONS FOR PRINTING

Photoresists find applications in several of the printing processes. The major areas of interest include printing plates, photoengraving, silkscreen printing, printed circuits, collotype, and proofing systems.

Printing Plates

Photopolymerization by actinic radiation is used extensively in the preparation of printing plates. The kind of plates may be categorized into three groups: (1) relief or raised image; (2) planographic, photolithography; and (3) gravure or intaglio-photogravure. Photopolymerization, however, finds use primarily in (1) and (2).

Relief or Raised-Image Plates

The traditional copper, zinc, and magnesium plates used by the bulk of letterpress printers and platemakers are now being replaced by photopolymer platemaking, because the latter is a more rapid and simpler process, and permits faster processing. Photopolymer plates seem to have more favorable ink transference with less impression and squash than metal plates in general, and permit a contrast reduction between the main shadow and low middle tones, particularly on monochrome work. Photopolymer plates can be made accurate to the negative. The principle of relief plate making is depicted in Fig. 5.25. Press life for photopolymer plate systems is satisfactory, generally up to about half a million impressions, though current systems claim longer.

Photosetting in conjunction with photopolymer plates has a vast potential in three chief areas of printing, namely, newspaper production, flatbed cylinder letterpress, and rotary letterpress commercial printing.

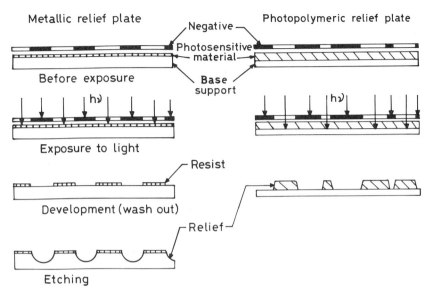

FIG. 5.25 Principle of relief plate making.

Photopolymer relief plates fall into two principal categories: liquid types and solid types. Presensitized liquid and dry raised-image or relief plates are used for letterpress, dry offset, and flexography.

The first liquid photopolymer system was the *letterflex* plate. The process relies on the cross-linking reaction of a polyurethane possessing allylic unsaturation and a polythiol, i.e.,

$$\sim\!\!\sim\!\!R\!\!-\!\!CH_2\!\!-\!\!CH\!\!=\!\!CH_2 + HS\!\!-\!\!R'\!\!\sim\!\!\sim$$

$$\downarrow \begin{array}{l} \text{Actinic} \\ \text{radiation} \\ + \text{ Benzophenone} \end{array}$$

$$\sim\!\!\sim\!\!R\!\!-\!\!CH_2\!\!-\!\!CH_2\!\!-\!\!CH_2\!\!-\!\!S\!\!-\!\!R'\!\!\sim\!\!\sim$$

The liquid photosensitive resin is employed as a top coating on an aluminum support. A negative is then used and exposed to actinic radiation from a Xenon lamp. The unexposed areas are removed with a solution of surface-active agents using an ultrasonic washing technique.

A similar product to letterflex is the Asahi photopolymer liquid resin plate system. The resin in this case is an unsaturated polyester that is dispersed on a transparent film and backed up with a film of polyester-based material. Liquid plates have chiefly found application in newspaper printing.

Solid photopolymer plates for printing were pioneered by Dupont with the introduction of *Dycril* in 1957. The photosensitive layer in Dycril is a 0.3–1.0 mm thick binder composed of a water-soluble cellulose derivative such as cellulose acetate succinate, a monomer of the divinyl type such as triethylene glycol diacrylate, and a photo-initiator such as 2-ethylanthraquinone and a thermal stabilizer (inhibitor) of the *p*-methoxyphenol type. The mixture is applied on an aluminum support or steel sheet with an adhesive sheet and antihalation sheet between the pair. Exposure to actinic radiation from a carbon or mercury arc lamp is performed through a negative placed over the photosensitive layer. Post-exposure treatment involves washing with a 0.2–0.5% caustic soda solution that removes the unexposed area and leaves a relief on the support. Dycril is fairly cheap, has a fairly high resolving power, and is durable.

BASF introduced *Nyloprint*, a solid type of photosensitizer compound coated plate, around 1967. The solid photopolymerizing material is a combination of an alcohol-soluble polyamide (such as a polyamide copolymer of hexamethylene diammonium adipate and ε-caprolactam), a vinyl monomer (such as bis-acrylamide, hexamethylene bis-acrylamide, triethylene glycol diacrylate, etc.), a photo-initiator such as benzoin methyl ether, and a inhibitor (such as hydroquinone) and maleic anhydride to function as a compatibilizer for polyamide with vinyl monomers. Post-exposure treatment of Nyloprint, however, involves washing with alcohol.

Dyna-flex, a solid photosensitive plate marketed by Dynaflex, has a photosensitive emulsion on an aluminum backing or carrier plate and is thought to be based on poly(vinyl alcohol) and dichromate chemistry.

Photolithography, Planographic Plates

These are lithographic plates and find application on offset presses. The photosensitive materials used include diazo monomers, diazo polymers, and diazo compounds, in conjunction with additive film formers such as epoxy resins, phenolics, poly(vinyl acetals), polyamides, azide compounds, unsaturated chalcone, cinnamate, stilbene, and vinyl polymers and dichromated colloids. Three kinds of metallic photolithographic plates are used. These are surface plates, deep-etch plates, and bimetallic or trimetallic plates.

Surface plates are either negative- or positive-working and direct use is made of the coating material or the resultant compounds of photochemical change as the image bearer. The other two types of plates may be regarded as "reversal" positive working processes in which the coating acts as a hydrophilic stencil, and the image is in essence composed of the plate surface backed up by the lacquer application. Deep-etch plates and trimetallic plates require a positive transparency for their preparation, whereas bimetallic plates are generally processed from photographic negatives. A bimetallic plate has an oleophilic metal, such as copper or zinc, in the image area and another one, such as chromium or iron, in the nonimage areas. For trimetallic plates, backing occurs from a third metal for support. Etching is done with ferric nitrate solution, which dissolves the copper but does not attack the metal underneath. The photoresist is removed subsequent to etching, the plate is then inked, and desensitization of the nonimage areas is carried out.

Commercial presensitized plates are formulated on two chief groups of photoresponsive compounds, namely, (1) diazo resins and diazo oxides, and (2) cinnamate resins, some of which have been previously described.

Photogravure

In *photogravure*, ink is transferred from recesses in a metal plate or cylinder to the surface to be printed on. In contrast to the halftone processes (see below), depth of color variations are produced by varying the depth of the recesses and, consequently, the ink quantity transferred to the printed surface.

Gelatin or poly(vinyl alcohol)/dichromate photoresist may be used to sensitize the paper on which a mesh-like pattern or hardened gelatin enclosing small squares of unhardened gelatin (corresponding to the cells that will finally constitute the printing areas) is created by exposure underneath a screen made of two sets of transparent parallel lines arranged orthogonally. A positive transparency is then used to expose the paper so that every square is hardened to a depth dependent on the incident actinic radiation intensity. The paper is then transferred to a copper plate or cylinder with the metal against the gelatin. The backing paper is then removed and the image is developed by washing with water to remove the unhardened gelatin. A photoresist image remains, consisting of squares of varying depths partitioned by walls of completely hardened gelatin. Ferric chloride is used as the etching agent to prevent gas production, as this might dislodge the resist. The greatest depth of etch occurs in the least hardened areas and, conversely, the least depth in the most hardened areas, thus producing shadow and highlight regions. The resist remaining is removed prior to printing.

Photoengraving

Photoengraving is the process by which blocks for illustrations are made. The illustrative material uses either line drawings with lines of differing thickness representing various tones or photographs reproduced as halftone blocks. For halftone work, the continuous tones of the original photograph are transformed to a uniform pattern of varying sized dots by either a crossline screen placed a short distance in front of the negative or a contact screen placed in contact with the negative prior to exposure.

Clean copper, zinc, or magnesium plates are used, with the fresh photosensitive coating applied over the surface such that on drying it is about 1–4 μm thick. The plate is put under a negative in a vacuum printing frame and exposed to actinic radiation from a carbon arc lamp. After exposure, the plate is washed in running water and the unexposed film areas removed with cotton-wool. The plate is electrolytically or chemically etched after coating the reverse side of the plate with an acid-resistant lacquer such as an alcoholic shellac solution. Zinc or magnesium plates are chemically etched with dilute nitric acid, whereas copper plates are etched with ferric chloride solution. Electrolytic etching is often carried out with copper plates, where they are rendered as the anode in a cell with an iron cathode. A solution of ammonium and sodium chloride is used as an electrolyte.

In both types of etching, it is necessary to prevent undercutting of the lines and dots by lateral etching. In the chemical etching process, this is achieved by taking the plate out of the etch bath from time to time, drying it, and brushing powdered "dragon's blood" resin onto the sides of the dots and then heating it to fuse the resin and then recommence the etch process. In the "Dow-etch" process, lateral etching is prevented by emulsifying diethylbenzene into the etch solution. The solution is sprayed onto the plate and the hydrocarbon covers the sides of the dots. In electrolytic etching, the current density may be monitored to control any lateral etching.

Silkscreen Printing

Stencils for screen printing may be made independently of the screen (indirect photostencil) or directly on the screen itself (direct photostencil). By impregnating the screen with dichromated gelatin and exposing it to actinic radiation under a line or halftone positive, stencils may be made to reproduce fine and complicated subjects. The image is developed by washing away the gelatin from the unexposed areas with warm water. Photosensitizers used for these emulsions are either alkali metal and ammonium dichromates or diazo compounds. While gelatin has been mostly used as binder for the photosensitizer, combinations of poly(vinyl alcohol) and poly(vinyl acetate) are among those currently used, due to their superiority in heat resistance compared to gelatin.

Photostencils are called indirect when processing is performed before the film is transferred to the mesh. There are many variations in their processing techniques. A typical base film would be a polyester coated with a presensitized gelatin emulsion. Exposure in this case is made to visible or near-UV light through a positive in intimate contact with the photostencil. The emulsion from the unexposed areas is washed away with warm water, and the stencil is then mounted onto the screen. The base support is peeled off after drying. The indirect stencil system has the advantages of handling flexibility and high resolution without ink spread, as the stencil film adheres to the screen's underside thus ensuring intimate contact with the printing stock. Print runs up to about 5000 are feasible.

In the direct process of photostencil preparation, the photosensitive liquid emulsion is coated onto both sides of the mesh, two or more coats being applied to embed the mesh in the photoemulsion. After drying with warm air or at ambient temperature, the whole screen is exposed to actinic radiation through a positive transparency kept in contact with the screen in a vacuum frame. Warm water is used to wash out the unexposed areas of the stencil and the screen is then dried. The method provides durable, long-run stencils. However, the stencils thus made have the disadvantage of limitation in quality because of the fact that the emulsion does not readily form a flat surface across the individual mesh filaments on drying, but tends to follow the contour of the mesh, which results in ink spread due to inadequate substrate contact. As emulsions often cause the stencil to contract and shrink along the edge of the printed image during drying, a problem of sawtoothing is often encountered with direct photostencil.

A combination of the two previous systems retaining some of the best features of both of them is used in the direct/indirect method of making photostencil.

Printed Circuits

The use of printed circuits reduces significantly the time necessary to wire electronic apparatus. The design of the circuit ensures that contacts between the components may be provided by a metal coating on an insulating base that is covered with a thin copper layer. Poly(vinyl alcohol)/dichromate or any usable photosensitive system is applied to the copper. The resist is hardened, and the copper regions not forming part of the circuit are removed by etching with aqueous ferric chloride or peroxodisulfate solution.

Collotype and Proofing Systems

Collotype is similar to photolithography as the image areas are made by photocuring dichromated gelatin, with the exception that it has a nonplanar printing surface that is covered with microscopic reticulations. The use of halftones is dispensed with in collotype printing, and the process approaches continuous tone production more closely than any of the other printing methods.

With the advent of proofing systems, photoresist technology allows the graphic arts industry to approximate final full-color print. Photoresist coatings on polyester clear film in the three primary colors, yellow, magenta and cyan, and also in black, can be contacted with color-separation negatives. After exposure, these plastic-supported resists are developed in a manner similar to offset plates and, on superimposition on each other in close registration, afford a good approximation to the final full-color work.

In the Dupont Cromalin TM system, each photopolymer layer is exposed through a color-separation negative, and the superimposed photopolymer layers are dusted with various toners. As the exposed areas are photoconverted to non-tackiness and the unexposed areas remain tacky, the latter accepts the powdered pigment toner. Superimposition of the exposed separations occurs in the transfer system.

IONIC POLYMERS [29–35]

Ionic polymers are polymers containing chemically bound ions within their structure. These are specialist materials, some of which have limited, though important, commercial applications. Those containing few ions are melt-processable thermoplastics called *ionomers*; those containing many ions are either water-soluble polymers called *polyelectrolytes* or cross-linked polymers containing ionic groups called *ion-exchange resins*. Polyelectrolytes and ion-exchange resins are, in general, intractable materials and not processable on conventional plastics machinery.

Ionic polymers may be classified according to the type of the bound ion, its position within the structure, and the amount of bound ions present along a given length of polymer chain. They may be further classified according to the nature of the counterion or the nature of the supporting polymeric backbone.

The bound ion is usually either an anion such as sulfonate $(-SO_3^-)$, phosphonate $(-PO_3^{2-})$ or carboxylate $(-CO_2^-)$ or a cation such as quaternary $(-NR_3^+)$ or amine $(-NH_3^+)$. But both types can occur together and the polymer is then said to be *ampholytic*. In each case, the polyion may be strong or weak according to the degree to which it ionizes. For example, polysulfonates and quaternary type polyions are strong, polycarboxylates and amine type polyions are weak, while polyphosphonates are intermediate in nature.

The bound ion can be pendant to the polymer's covalent backbone, as in a polysulfonate (XII) or it can be integral or enchained, as in an *ionene* (XIII).

$$
\begin{array}{cc}
\text{SO}_3^- & \text{R} \\
\text{(XII)} & \text{(XIII)}
\end{array}
$$

Almost any conventional polymer can be modified to form an ionic polymer of the pendant type. However, ionic polymers of the integral type, like the ionenes, are specialized structures whose backbones can exist only in the ionic form.

Ionic polymers contain counterions that neutralize the charges on the bound ions. The counterions may be grouped into three types: (1) univalent, (2) di- or trivalent and (c) polymeric. Polymers with polymeric counterions are often called polysalts, polyelectrolyte complexes, polyion complexes, simplexes, or coacervates.

The methods used for synthesis of ionic polymers can be divided mainly into three types: (1) direct synthesis, (2) post-functionalization of a standard pre-formed polymer, and (3) post-functionalization of a special pre-formed polymer.

Direct synthesis is used for polyelectrolytes such as carboxy polymers, e.g., poly(acrylic acid) and poly(methacrylic acid), which are probably the most common ionic polymers of all. Post-functionalization of pre-formed polymers is a frequently used synthetic method, provided that the polymer is sufficiently reactive. Common examples are the sulfonation of polystyrene and the grafting of thioglycollic acid on to polybutadiene using free radicals.

Special post-functionalizable copolymers have also been used to derive acid ionomers by hydrolysis, thus avoiding the difficulties of copolymerizing ionic and nonionic monomers. To this end there are many examples where carboxylic acid polymers are formed by hydrolyzing copolymers containing acrylate esters, acrylonitrile, or maleic anhydride. As described later, a sulfonic acid ionomer, *Nafion*, is formed by hydrolysis of tetrafluoroethylene copolymerized with a sulfonyl fluoride.

All of the aforementioned ionomers are anionic polymers. Cationic polymers are less common, though equally important. Pendant cations are usually of the quaternary ammonium type and made by a multi-step post-functionalization in which a precursor containing labile chloride is reacted with a tertiary amine:

$$\text{CH}_2\text{Cl} \xrightarrow{\text{N(CH}_3)_3} \text{CH}_2\overset{+}{\text{N}}(\text{CH}_3)_3 \quad + \text{Cl}^- \quad\quad (8)$$

Sometimes, the inverse of this is also done when a pre-formed polyamine is quaternized by reacting with an alkyl halide:

$$\text{SO}_2\text{NH(CH}_2)_3\text{N(CH}_3)_2 \xrightarrow{\text{CH}_3\text{I}} \text{SO}_2\text{NH(CH}_2)_3\text{N}^+(\text{CH}_3)_3 \quad + \text{I}^- \quad (9)$$

Physical Properties and Applications

Synthetic ionic polymers have three distinctive properties that dictate their usage: (a) ionic cross-linking, (b) ion-exchange capability, and (c) hydrophilicity. Major applications for organic ionic polymers in relation to their ion content and dominant property are summarized in Table 5.11.

Ionic Cross-Linking

Since the ions in ionic polymers are held by chemical bonds within a low dielectric medium consisting of a covalent polymer backbone material with which they are incompatible, the polymer backbone is forced into conformations that

TABLE 5.11 Applications of Organic Ionic Polymers in Relation to the Amount of Ion Present

Ion content (equiv/kg)	Designation	Dominant property	Application	Example
<0.5	Ionomer	Ionic cross-linking	Thermoplastic	Ethylene-acrylic acid copolymer
1–1.5	Ionomer	Ion-exchange	Membranes (electrodialysis, etc.)	PTFE copolymer (Nafion, Flemion)
1–1.5	Ionomer	Hydrophilicity	Membranes (reverse osmosis)	Sulfonated polyethersulfones
>4	Polyelectrolyte	Water solubility	Thickeners, dispersants, flocculants and sizes	Polyacrylic acid or copolymer
>4	Polyelectrolyte (cross-linked)	Ion-exchange	Ion-exchange resins	Sulfonated polystyrene, aminated polystyrene
14	Polyelectrolyte	Ionic cross-linking	Dental cements	Polyacrylic acid

allow the ions to associate with each other. Because these ionic associations involve ions from different chains they behave as cross-links, but because they are thermally labile they reversibly break down on heating. Ionomers therefore behave as cross-linked, yet melt-processable, thermoplastic materials, or if the backbone is elastomeric, as thermoplastic rubbers. It should be noted that it is with the slightly ionic polymers, the ionomers, where the effect of ion aggregation is exploited to produce melt-processable, specialist thermoplastic materials. With highly ionic polymers, the polyelectrolytes, the ionic cross-linking is so extreme that the polymers decompose on melting or are too viscous for use as thermoplastics.

Two types of ionic aggregates are found in ionomers, called *multiplets* and *clusters*, which coexist in equilibrium within the matrix of covalent backbone material. The multiplets, very small and very numerous, are associations of ion-pairs up to eight in number that are purely ionic and are devoid of trapped segments of the covalent backbone, while the clusters are large associations of these multiplets between which are trapped segments of the backbone. Evidence for this theory of microphase separation into multiplets and clusters is provided by many types of physical measurement, the important techniques being small-angle X-ray scattering, and infrared and Raman spectroscopy.

Ion-Exchange

Ionic polymers contain two types of ions, namely bound ions, which are part of the structure, and the counterions, which are free. In a medium in which the ionic polymer is insoluble, the counterions are exchanged for similar ions from the surrounding medium and an equilibrium is established, the kinetics of the process being dependent on factors such as the physical form of the insoluble polyion, its porosity, and surface area. The fundamental fact in this exchange is that the counterions have free movement into and out of the polymer, while ions of the same type of charge as the bound ion do not. This barrier to ion movement is known as *Donnan exclusion*.

The property of ion-exchange has important consequences and three different types of application depend on it. They are (1) ion-exchange resins, (2) ion-exchange membranes, and (3) heterogeneous catalysis.

Ion-exchange resins are insoluble beads of an ionic polymer. Their best-known use is in deionization or demineralization of tap water to distilled water. In this process, the water is contacted with two types of resin, one a polyacid and the other a polybase, so that the net effect is to exchange M^+ and X^- from the tap water for H^+ and OH^-, i.e., H_2O, from the polyion beads:

$$\text{(P)}SO_3^-H^+ + M^+X^- \rightleftharpoons \text{(P)}SO_3^-M^+ + H^+X^- \tag{10}$$

$$\text{(P)}NR_3^+OH^- + H^+X^- \rightleftharpoons \text{(P)}NR_3^+X^- + H_2O \tag{11}$$

To be effective, ion-exchange resins should have high ionic content. They are therefore polyelectrolytes and, since polyelectrolytes are inherently water-soluble, they are chemically cross-linked during manufacture to make them insoluble. When immersed in water they undergo swelling, which facilitates rapid ion-exchange. The degree of swelling is, however, inversely related to the density of the cross-links.

A polyion in the form of a thin membrane is used as *ion-exchange membrane* in another application of the ion-exchange phenomenon. When exposed to an electrolyte, an ion-exchange membrane will allow counterions to pass through it, but will act as a barrier to the complementary ion, and is therefore said to be *permselective*. Thus a polyanionic membrane will allow passage of cations and a polycationic membrane that of anions, so that under the influence of an electric current, continuous fluxes of cations and anions, respectively, can be set up across these membranes. This principle is exploited in *electrodialysis* and in chlor-alkali cells as described later.

The dialyzer used for desalination of water by electrodialysis (Fig. 5.26) is an electric cell divided into a series of sub-cells that are separated by alternate polyanion and polycation membranes. While the ions move under the influence of the applied electric potential, the complimentary permselectivities of the ion-exchange membranes insure that salt is removed from alternate sub-cells and

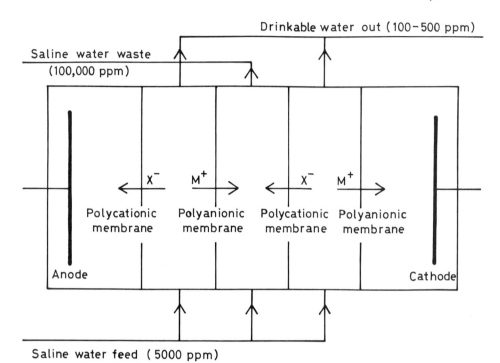

FIG. 5.26 Desalination of water by electrodialysis.

concentrated in the others. Desalination dialyzers used in practice have up to 100 such sub-cells. Other applications of dialyzers include concentration of brine and the desalting of cheese whey. (It is important not to confuse electrodialysis with ordinary dialysis, which is not an ion-exchange process but is a separation based on differences in size and hence diffusivity between large and small molecules through a membrane, which, generally speaking, is not ionic.)

A related application of ion-exchange membrane is in ion-selective electrode. Thus when an electrode with an ion-exchange membrane encasing is immersed in an ionic medium, it will develop a potential proportional to the activity of the ion to which the membrane is permselective.

A high ion content is not necessary for an ion-exchange membrane, since its function is not governed by its exchange capacity. Though a high ion content can be advantageous as it decreases the electrical resistance, it can in fact be counterproductive if the membrane is so highly swollen by water that its permselectivity is reduced. The ion content of an ion-exchange membrane is thus often intermediate between that of an ionomer and an ion-exchange resin. While chemical cross-linking is used for ion-exchange resins to limit this swelling or to prevent the resin from dissolving, this method is generally not desirable for

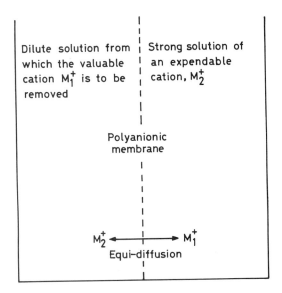

FIG. 5.27 Ion-exchange dialysis.

membranes, because cross-linking interferes with the process of membrane fabrication.

Ion-exchange dialysis (or *Donnan dialysis*) like ordinary dialysis is a diffusion-controlled separation process, but unlike the latter it involves ion-exchange and so needs an ionic membrane (though without an applied current). If a polyanionic membrane is used, as shown in Fig. 5.27, the cations diffuse each way across the membrane but the anions cannot so that, in effect, cations are separated as they swap over, but the amount of electrolyte on each side of the membrane remains constant. This type of dialysis, using continuous cells, is being developed as a means of stripping and concentrating radioactive ions from dilute solutions of radioactive wastes.

Hydrophilicity

Ionic polymers are hydrophilic. Those of moderate ion content swell on contact with water and those of high ion content dissolve, unless they are cross-linked. With moderately ionic polymers, which swell without dissolving, the hydrophilicity has the advantage of making the structure more permeable to ions for ion-exchange. The hydrophilicity of moderately ionic polymers leads to another type of membrane application, that of *reverse osmosis*.

The method of reverse osmosis involves application of pressure to the surface of a saline solution, thus forcing pure water to pass from the solution through a semipermeable membrane that does not permit passage of ions. Since the natural

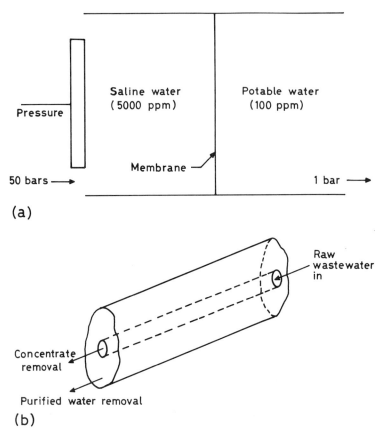

FIG. 5.28 (a) Schematic representation of reverse osmosis process. (b) Tubular configuration system for wastewater treatment by reverse osmosis.

osmotic pressure tends to force the water from the region of low ionic concentration to that of the high, to achieve a flow in the opposite direction, as in reverse osmosis, a pressure has to be applied exceeding the osmotic pressure.

In the process used for desalination, a saline, brackish water or sea water is forced at very high pressure through a hydrophilic membrane resulting in water of drinkable quality (Fig. 5.28). Reverse osmosis is also utilized for recovering waste water for paper mill operations, pollution control, industrial water treatment, chemical separations, and food processing.

The compartments indicated in Fig. 5.28a are a schematic representation of reverse osmosis process. In practice, the reverse osmosis process is conducted in a tubular configuration system (Fig. 5.28b). Raw waste water flows under high pressure (greater than osmotic pressure) through an inner tube made of a semipermeable membrane material and designed for high pressure operation. Purified water is removed from the outer tube that is at atmospheric pressure and is made of ordinary material.

Although the membrane used for reverse osmosis may be ionic and ion-exchange can occur, there is, however, no overall transmission of salt because of Donnan exclusion (and unlike electrodialysis no electric current is applied). Thus, while an essential requirement of a membrane for reverse osmosis is hydrophilicity, it need not be ionic. In fact, the most successful reverse osmosis membranes developed are made of nonionic cellulose acetate. Certain ionic polymers, such as the sufonated polyaromatics, has been used because of their greater chemical stability and resistance to biological degradation.

Several mechanisms have been proposed to explain reverse osmosis. According to the preferential sorption-capillary flow mechanism of Sourirajan [33], reverse osmosis separation is the combined result of an interfacial phenomenon and fluid transport under pressure through capillary pores. Figure 5.29a is a conceptual model of this mechanism for recovery of fresh water from aqueous salt solutions. The surface of the membrane in contact with the solution has a preferential sorption for water and/or preferential repulsion for the solute, while a continuous removal of the preferentially sorbed interfacial water, which is of a monomolecular nature, is effected by flow under pressure through the membrane capillaries. According to this model, the critical pore diameter for a maximum separation and permeability is equal to twice the thickness of the preferentially sorbed interfacial layer (Fig. 5.29b).

From an industrial standpoint, the two basic parameters characterizing reverse osmosis systems for a given separation are: (1) rejection factor, which involves the choice of the appropriate chemical nature of the film surface, and (2) water flux, which depends on methods for preparing films containing the largest numbers of pores of the required size. This approach is the basis of the successful development of the Sourirajan-Loeb type of porous cellulose acetate membranes for desalination and other applications.

To achieve high flux in reverse osmosis, the membrane should have a high surface area and be very thin (0.02–1.0 µm), but to withstand the high applied pressure it also needs to be very strong. To meet these requirements, which are in conflict, the semipermeable membrane is used as a skin mounted on a support, which is another membrane that is very porous, like a filter paper, and that is nonselective, and very thick (>100 µm). In some such bilayers each layer is of the same polymer and the membrane is then called *integral* or *asymmetric*, but in others the layers are of different polymers and the membrane is called a *composite*. Composite membranes have the advantage of each layer being individually optimized—the skin for its permeability and the support for its strength—but they are harder to make than integral membranes.

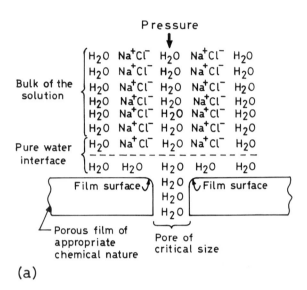

(a)

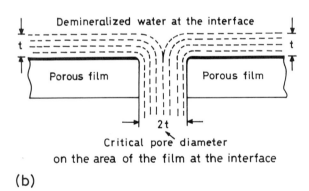

(b)

FIG. 5.29 (a) Schematic representation of preferential sorption capillary flow mechanism. (b) Critical pore diameter for maximum separation and permeability [33].

Polyelectrolytes, except those which are covalently cross-linked, are soluble in water, and their applications are based on this property. Basically, these applications depend on the polyelectrolyte altering the fluid properties of an aqueous medium, or modifying the behavior of particles in aqueous slurries or colloidal suspensions.

Polyelectrolytes raise the viscosity of aqueous solutions thus acting as *thickeners*, and the magnitude of the effect increases with the polymer's molecular weight. Naturally occurring gums and acidic polysaccharides have been traditionally used as thickening agents in food stuffs and pharmaceutical products, but, more recently, synthetic polyelectrolytes have been used in these roles. Much use is also made of polyelectrolytes to modify the characteristics of latex paints and similar proprietary fluids.

Polyelectrolytes can also stabilize particles in aqueous suspension and so act as *dispersants*. In this reaction, the hydrophobic backbones of polyelectrolytes are absorbed onto the surface of the particles by van der Waals attraction, while their ions form a hydrophilic surface and interacts with water. In suspension or bead polymerization, for example, a hydrophobic vinyl monomer is dispersed in water by agitation to form droplets that are stabilized by a polyelectrolyte. Each stabilized droplet of the monomer acts as a bulk polymerization system and on polymerization gets the shape of a bead.

Polyelectrolytes, depending on their ionic charge, can interact with colloidal particles and neutralize the stabilizing hydrophilic charges, thus acting as flocculating agents. They have been used in this way to coagulate slurries and industrial wastes.

There are other applications of polyelectrolytes that depend on their behavior in water in various ways. They are thus used as *sizes* in the textile industry and in paper manufacturing, and as additives to drilling muds and to soil for conditioning purpose.

A very different kind of application of polyelectrolytes is their use as dental cements. Here a divalent cation is added to an aqueous solution of a polyanion to form a highly cross-linked precipitate of great strength.

Ionomers

Polyethylene Ionomers

Ionomers based on polyethylene were described earlier (Chapter 4). They are mainly copolymers with pendant carboxylate groups in which the polyethylene backbone is the major component (>90%). Ethylene is directly copolymerized with methacrylic acid in a continuous process developed from the high-pressure method used to make low-density polyethylene by free-radical initiation. The comonomers are mixed in appropriate proportions, allowing for the greater reactivity of the methacrylic acid, and introduced with a peroxide initiator to a reactor at a pressure of about 30,000 psi (207 MPa) and a temperature of 250–280°C. Because methacrylic acid reacts disproportionately rapidly and has to be replenished frequently, the conversion to polymer is kept low per pass, and at 15–20% conversion the residual monomer is removed, then recycled with fresh monomer. This ensures that the methacrylic acid units, and hence the ionic cross-links formed later, are randomly distributed and uniform in composition.

The distinctive properties of the ionomers become manifest only when the carboxylic acid groups are neutralized. The neutralization is a kind of post-treatment that is performed by melting the polyacid in a mill at 150°C and adding a base or basic salt in powder form or in solution. In these neutralizations, the melt starts as a soft, fluid, opaque mass of polyacid and then becomes stiff, rubbery, and transparent as the ionomer is formed.

The loss in fluidity, or gain in melt strength, is a significant factor determining the usefulness of ionomers. That the effect is due to the formation of strong ionic cross-links is clearly shown by a comparison of the melt viscosity of an ionomer with its acid precursor (Fig. 5.30). The melt viscosity of a copolymer of ethylene and methacrylic acid containing, for example, 2 mole per cent of acid is increased by only about 50% compared to an otherwise equivalent polyethylene homopolymer, but when neutralized, and therefore ionized, the melt viscosity increases twenty-fold. (The slight increase with the acid is attributed to the weak hydrogen-bonded cross-links.) The melt-strength of an ionomer is such that its molten film can be drawn over the sharp edges of a nail without it being punctured or torn.

Though, as just explained, the viscosity of an ionomer at low shear is very much greater than its acid precursor, at high shear, however, the two are nearly the same (Fig. 5.31), which is attributed to breaking down of the physical bonding of the cross-links. The high shear sensitivity of melt viscosity is an important

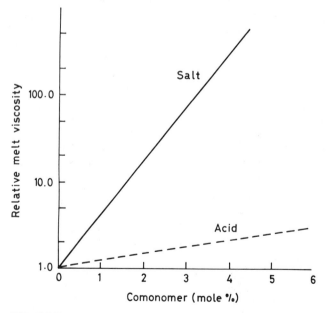

FIG. 5.30 Melt viscosity of poly(ethylene-co-methacrylic acid) and its sodium salt relative to polyethylene [29].

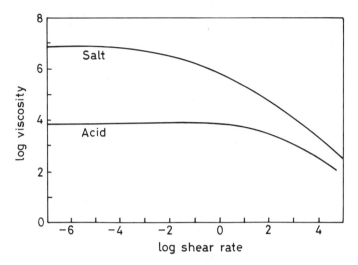

FIG. 5.31 Melt viscosities (160°C) at various shear rates of poly(ethylene-co-methacrylic acid) with 3.5 mole% comonomer and its sodium salt [29].

characteristic of ionomers and is a useful feature in a number of melt-fabrication processes. Thus the high melt-strength at low shear makes ionomers useful for extrusion or blow-molding, or for any process where the melt is partially supported.

A striking property of the ethylene ionomers is that they are transparent, unlike their acid precursors or polyethylene itself, which are not. The haze in polyethylene is due to the fact that the polymer is partially crystalline, and minute crystallites within it agglomerate into spherulites, which are of a size to scatter light. Ionomers also contain crystallites, and the crystallite formation is, in fact, helped by the ions, the domains of which serve to nucleate them; but the crystallites are unable to agglomerate because of the high viscosity of their surroundings. Thus, microcrystallinity is enhanced by the ions, but macrocrystallinity, which causes haze, is inhibited. The transparency of ionomers is useful in packaging applications.

Some physical properties of polyethylene ionomers are compared with those of polyethylene and the acid copolymer, poly(ethylene-co-methacrylic acid) in Table 5.12. Ionomer is generally tougher and, as shown in Table 5.12, relative to the acid copolymer, its tensile strength is increased by 27–53% and its stiffness is nearly tripled.

Polyethylene ionomers are described as flexible and tough with good impact toughness at low temperature, and with good resistance to grease and solvents, to stress-cracking, and to abrasion. They are blow-molded into films, sheets,

TABLE 5.12 Comparative Physical Properties of Polyethylene Ionomers and Their Acid Precursor

| | | Copolymer[a] | Ionomer | |
Property	PE	$(-CO_2 H)$	Na^+	Zn^{2+}
Appearance	Hazy	Hazy	Transparent	Transparent
Melt index (g/10 min)[b]		5.8	0.03	0.09
Yield strength				
(10^3 lbf/in^2)	1.20	0.88	1.91	1.93
(MPa)	8.3	6.1	13.2	13.3
Elongation (%)	600	553	330	313
Tensile strength				
(10^b lbf/in^2)	1.8	3.4	5.2	4.3
(MPa)	12.4	23.4	35.9	29.6
Stiffness (relative)		1.0	2.8	3.0

[a]Poly(ethylene-co-methacrylic acid) with 3.5 mole% methacrylic acid comonomer.
[b]ASTM-D-1238-57T

bottles, and blister packs, and injection-molded into various objects. Their resistance to grease and solvents has made them useful in meat packaging. They are very effective as external surface coatings on glass bottles to contain breakages.

Elastomeric Ionomers

A number of ionic polymers exist that have a recognized elastomer as the covalent backbone and have a small ionic content, so they may be called elastomeric ionomers. The ions provide at least a part of the cross-links in these polymers. Those elastic ionomers that are cross-linked exclusively by their ions have, however, the useful feature of being thermoplastic.

Carboxylated polybutadiene ionomers, which are close relatives of the polyethylene ionomers described above, have an essentially polybutadiene backbone that contains some acrylonitrile and styrene to adjust its flexibility and toughness, and, in addition, up to 6% by weight of acrylic or methacrylic acid. Like the polyethylene ionomers, they are usually made by direct copolymerization with the carboxylic acid monomer using, however, emulsion methods. Typically the monomers are slurried in water with sodium dodecylbenzene sulfonate as the emulsifier and potassium persulfate as the free-radical initiator. The tendency of the carboxylic acid monomer to dissolve in the aqueous phase instead of remaining in the butadiene-rich phase is suppressed by making the aqueous phase acidic so that the monomer remains in the nonionized form.

The carboxylated polybutadienes, when neutralized, undergo ionic cross-linking, producing the effect of vulcanization. The neutralization can be done by

TABLE 5.13 Vulcanization of Carboxylated Polybutadienes[a]

Form	Tensile strength		Elongation (5)
	10^3 lbf/in.^{2}	MPa	
Acid copolymer	0.1	0.7	1600
Sodium vulcanizate	1.7	11.7	900
Zinc vulcanizate	6.0	41.4	400

[a]Poly(butadiene-co-methacrylic acid) with 6.7 mole% methacrylic acid

treating with aqueous sodium hydroxide then heating, or by heating directly with zinc oxide. The ionic cross-link formed with the sodium ion is of moderate strength at room temperature and dissociates at 100°C. With zinc, the ionic cross-link formed is much stronger, although the polymer is capable of substantial flow at higher temperatures.

Some properties of sodium and zinc vulcanizates are compared with those of the acid precursor in Table 5.13. High tensile strength is a characteristic of ionic vulcanizates. As a comparison, a standard polybutadiene elastomer vulcanized with sulfur gives a strength of 1.9–5.8 MPa (275–841 lbf/in.^{2}), whereas an equivalent copolymer containing 1.5 mole% methacrylic acid, and vulcanized with magnesium oxide, gives 29.0 MPa (4.2 × 10^3 lbf/in.^{2}). They also respond differently to fillers such as carbon black—sulfur vulcanizates are reinforced, while ionic vulcanizates are weakened.

Carboxylated polybutadienes have not been used much as thermoplastic elastomers, principally because they have poor compression set, high stress relaxation, and poor performance at higher temperatures. Carboxylated polybutadienes, supplied as latices, are used mainly in dipping and coating processes, applications, which often do not involve ionized carboxylate, at least directly. The applications include adhesives, paper coating, glove-dipping, carpet-backing, binding nonwoven fabrics, and also shrink-proofing woollen garments where the carboxyl groups are believed to react with pendant amino groups in the proteins of the wool. A wide range of carboxylated latices are available as well-developed items of commerce; some of these are styrene-butadiene rubbers (SBRs) and others acrylontrile-butadiene rubbers (NBRs).

Mixed vulcanizations, such as with zinc oxide and sulfur or zinc oxide and peroxide, are used for carboxylated NBRs in order to combine the advantages of the ionic method with the conventional methods. The mixed vulcanizates have high tensile strength and notable resistance to abrasion, oil, and fuel. They are used in applications such as industrial rollers and wheels and shoe soles. Dry NBRs are available in fewer grades than latices. Examples are Krynac by Doverstrand and Hycar by Goodrich.

Elastomeric ionomers have also been developed from ethylene-propylene-diene ternary copolymers known as EPDM rubbers. The diene is commonly ethylidene norbornene. Du Pont made carboxylated ionomers by free-radical grafting of maleic anhydride (0.5–5%) onto the diene moiety of the polymer and neutralized the product with rosin salt. The ethylidene norbornene can also be sulfonated, thus:

$$\tag{12}$$

Ionomer properties of sulfonated EPDM are said to develop only when the acids are neutralized. The best properties are given by zinc salts, particularly when they are plasticized by zinc stearate. (See the section on thermoplastic elastomers in Chapter 4 for properties of ionic elastomers.)

Ionomers Based on Polytetrafluoroethylene

Polymers with a polytetrafluoroethylene (PTFE) backbone and pendant perfluorosulfonate or perfluorocarboxylate groups have become commercially important materials, although they are expensive. The sulfonates were introduced as Nafion by Du Pont in the early 1970s and the carboxylates as Flemion by Asahi Glass in 1978. They are made by free-radical copolymerization of tetrafluoroethylene and perfluorovinyl monomers giving precursor copolymers, (XIV) and (XV), which can be post-functionalized by hydrolysis to generate sulfonic and carboxylic acid groups. The perfluorovinyl monomers themselves are made from tetrafluorethylene by multi-step synthesis using hexafluoropropylene oxide.

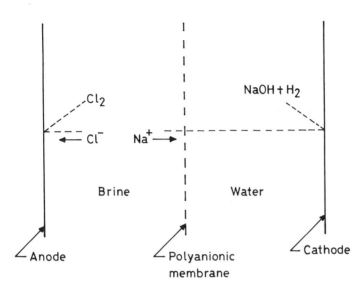

FIG. 5.32 Chlor-alkali cell.

Although originally developed for use as a membrane in fuel cells, Nafion has found more useful applications in various other electrolytic separation processes. In these applications, use is made of its cation-exchange properties and its ability to survive in extremely aggressive chemical environments. An outstanding example is the use of Nafion as a membrane in the chlor-alkali cell where it is gradually replacing the traditional diaphragm and mercury cells. In the chlor-alkali process a cell is partitioned by the polyanionic membrane into an anode compartment, to which brine is added, and a cathode compartment to which water is added (Fig. 5.32). On application of electric potential, Na^+ passes from the brine around the anode and through the membrane to the water around the cathode where it forms sodium hydroxide. The membrane keeping the brine and water separate allows transfer of Na^+ ions by ion-exchange and acts as a barrier to Cl^- and OH^-.

The Nafion membranes have ionic contents of 0.66–0.91 Eq/kg. A bilayer of these two extremes of composition is made and also a bilayer with a perfluorocarboxylate (Flemion) polymer. The bilayered membranes are made by melt processing the precursor copolymer, which then, in membrane form, is hydrolyzed.

Perfluorocarboxylates, the Flemions, were introduced with the idea that a carboxylate at a given ion content would be less hydrophilic than a sulfonate and so would be of help in balancing the ionic content of an ion-exchange membrane with its permselectivity and hydrophilicity as required. The bilayered Nafions also resulted from similar thinking and attempt to combine certain advantages of the two types.

The efficiency with which power is consumed in the electrolysis and the highest concentration of uncontaminated sodium hydroxide that can be produced determine the membrane performance in the chlor-alkali process. Rapid progress in membrane performance has been made in recent years, particularly since the introduction of carboxylate membranes, leading to an efficiency level of 95% and a concentration of 33%.

Bilayer carboxylate membranes can be produced by surface modification of Nafion-type membranes. In a process used by Asahi Chemical the sulfonate on a Nafion-type surface is reduced to sulfinic and sulfenic acids, then oxidized to a carboxylate layer of 2–10 μm thickness:

$$\text{\textasciitilde\textasciitilde OCF}_2\text{CF}_2\text{SO}_2\text{Cl} \xrightarrow{\text{Reduction}} \text{\textasciitilde\textasciitilde OCF}_2\text{CF}_2\text{SOOH}$$

$$\downarrow \begin{array}{l} -\text{SO}_2 \\ \text{and} \\ \text{oxidation} \end{array}$$

$$\text{\textasciitilde\textasciitilde OCF}_2\text{COOH} \tag{13}$$

Tokoyama Soda makes Neosepta F membranes by treating a Nafion-type precursor with an alcohol and then oxidizing the surface in air.

Nafions in their acid forms are *super-acidic*, i.e., stronger than sulfuric acid and so have high catalytic power. Nafion powders in acid form have therefore been used as catalysts in many types of reactions, such as esterification and Friedel-Crafts reactions at low temperatures.

Ionomers Based on Polysulfones

Ionomers of commercial polysulfones, principally sulfonates, are being developed as membranes, particularly for purifying water by reverse osmosis. In this application, they are superior to conventional membrane materials because they are resistant to oxidation by the chlorine used in water treatment, to harsh chemical cleaning operations, to biological fouling, and to compaction under high operating pressures. The traditional polymer, cellulose acetate, is less strong, while Nafion, though very inert chemically, is too costly and difficult to fabricate in a form suitable for reverse osmosis. The sulfones, having completely amorphous backbones and solubility, can be easily cast into membranes from solvents. The optimum ion content is about one Eq/kg, which provides a balance of properties between a high flux of the permeant (water) and a low leakage of the rejected species (the dissolved salts).

Commercial polysulfones (see Table 4.27) such as Udel (Union Carbide) and Victrex (ICI) are sulfonated in a post-functionalization step to make ionomers.

When dissolved in dichloroethane and treated with a complex of sulfur trioxide and triethyl phosphate, Udel (XVI) becomes monosulfonated on the rings marked "R" (as the rings connected to the sulfone group are deactivated by the sulfone group).

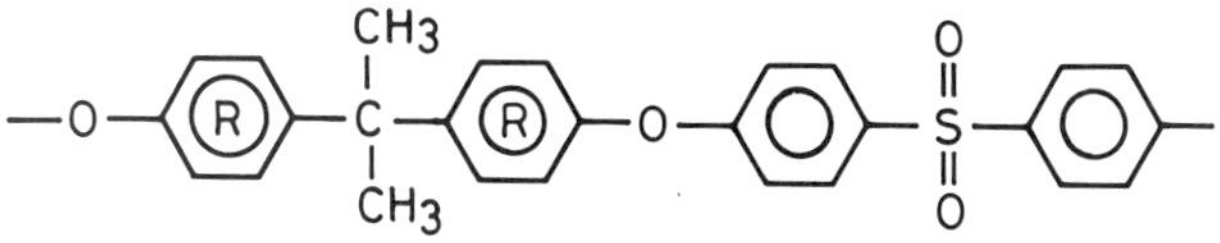

(XVI)

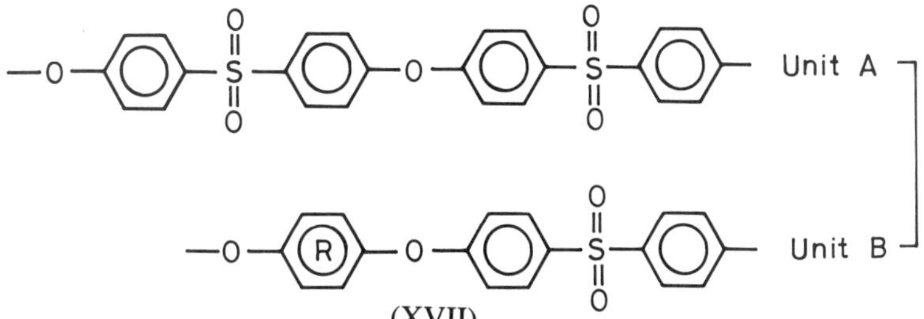

(XVII)

Victrex is a random copolymer of two units (XVII) in which the ring marked "R" of unit B becomes monosulfonated by the simple process of dissolving it in sulfuric acid. The other aromatic rings and the whole of unit A are inert. The extent of the reaction is predetermined by the proportion of unit B in the copolymer. The strength of the ionic associations in sulfonated Victrex is indicated by the increase in T_g, which occurs linearly with the content of unit B despite the already high value of 230°C for the parent polymer.

Polyelectrolytes

Ion-Exchangers

The most important class of ion-exchangers are the organic ion-exchange resins. Their framework, the so-called *matrix*, consists of an irregular, macromolecular, three-dimensional network of hydrocarbon chains. The matrix carries ionic groups such as $-SO_3^-$, $-COO^-$, $-PO_3^{2-}$, and $-AsO_3^{2-}$ in *cation exchangers* and $-NH_3^+$, $\rangle NH_2^+$, $\rangle N^+\langle$, and $-\rangle S^+$ in *anion exchangers*. Ion-exchange resins thus are cross-linked polyelectrolytes. The first completely synthetic ion-exchange resins were prepared by B. A. Adams and E. C. Holmes in England in 1935. Today more than 100 synthetic ion-exchange resins are

TABLE 5.14 Trade Names and Manufacturers of Ion-exchange Resins

Trade name	Manufacturer
Amberlite	Rohm and Haas Co., Philadelphia, PA
De-Acidite	The Permutit Co. Ltd., London, England
Dowex	Dow Chemical Co., Midland, MI
Duolite	Chemical Process Co., Redwood City, CA
Imac	Industrieele Mij. Activit N. V., Amsterdam, Netherlands
Ionac[a]	Ionac Co. Ltd., New York, NY
Lewatit	Farbenfabriken Bayer, Leverkusen, Germany
Nalcite[b]	National Aluminate Corp., Chicago, IL
Permutit	Permutit Co., New York, NY
Wofatit	VEB Farbenfabrik Wolfen, Wolfen, Kr. Bitterfeld, Germany
Zeo-Karb	The Permutit Co. Ltd., London, England

[a]Product of Permutit Co., marketed by Ionac Co.
[b]Product of Dow Chemical Co., marketed by National Aluminate Corp.

marketed throughout the world by various companies. Some of the largest manufacturing companies and the trade names of their resins are listed in Table 5.14.

Ion-exchange resins are supplied as insoluble, water-swellable beads that have either a dense internal structure (*gel-type*) or a porous, multi-channelled one (*macroporous-type* or *macroreticular*). The gel-type PS resin was the first to be introduced (1947) and the macroporous-type came later (1959).

Ion-exchangers are broadly classified as *cation exchangers* and *anion-exchangers*. Carriers of exchangeable cations are called cation exchangers; they have acidic functional groups bound to the resin matrix. Carriers of exchangeable anions are called anion exchangers; they have basic functional groups bound to the resin matrix (Table 5.15). Chelating ion-exchange resins contain chemically bound chelating functional groups, which sorb metal ions by chelation.

Polystyrene (PS) cross-linked with divinyl benzene (DVB) is the matrix on which most of the commercial ion-exchange resins are based; the ionic groups are introduced by post-functionalization of the cross-linked polymer matrix. By varying the DVB content, the degree of cross-linking can be adjusted in a simple and reproducible manner. The *nominal DVB content*, which refers to mole% of DVB in the polymerization mixture, is used to indicate the degree of cross-linking. General-purpose ion-exchangers contain between 8 and 12 mole% DVB, the most common being 8 mole%. For special purposes, resins with as little as 0.25% DVB and as much as 25% DVB have been prepared. Resins with low DVB content swell strongly and are soft and gelatinous, while those with very high DVB content swell very little and are tough and mechanically more stable.

TABLE 5.15 Types of Ion-Exchange Resins

Active group	Structure
Cation-exchange resins	
Sulfonic acid	
Carboxylic acid	
Phosphonic acid	
Anion-exchange resins	
Quaternary ammonium salt	
Secondary amine	
Tertiary amine	

The styrene-DVB copolymer beads are prepared by suspension (pearl) poly-merization technique. The monomers are mixed and a polymerization catalyst such as benzoyl peroxide is added. The mixture is then dispersed into small droplets in a thoroughly agitated aqueous solution that is kept at a temperature required for polymerization (usually 85–100°C). A suspension stabilizer (gela-tin, polyvinyl alcohol, sodium oleate, magnesium silicate, etc.) in the aqueous phase prevents agglomeration of the droplets. The size of the droplets depends chiefly on the stablizer, the viscosity of the solution, and the agitation, and it can be varied within wide limits. As polymerization takes place, the droplets are transformed into polymer beads. For most purposes, a bead size of 0.1–0.5 mm is preferred, but beads from 1 μm—2 mm in diameter can be prepared without much difficulty.

The above method gives beads of gel-type polymer matrix. Porous beads can be made by incorporating a component (for example, styrene homopolymer) that is soluble in the monomer mixture. After polymerization this component is re-moved from the matrix with, for example, toluene, thus leaving pores within the structure in the final product.

Highly porous, so-called *macroreticular* beads can be prepared by a variation of the conventional pearl polymerization technique. An organic solvent is added in which the mixture of monomers is soluble, but the polymer, when it is formed, is insoluble. Thus, as polymerization progresses, the solvent is squeezed out by the growing polymer regions. In this way, one can obtain spherical beads with large pores (several hundred angstrom units), which guarantee access to the in-terior of the beads even when nonpolar solvents are used.

Styrene-DVB copolymer beads are sulfonated to produce the most widely used strong-acid-type cation-exchange resins. To make them, the copolymer precursor beads are dispersed in about 10 times their weight of concentrated sulfuric acid and heated slowly to 150°C. The sulfonic acid group is normally introduced into the para position. Though the reaction is very simple in princi-ple, it involves delicate operations and close control of parameters in order to achieve beads of suitable structure and durability. A fully mono-sulfonated, polystyrene has a theoretical ion content of 5.1 equivalents/kg (dry) but many commercial resins have about 4.4–5.2 eq/kg. Amberlite IR-120, Dowex-50, Nalcite HCR, Permutit Q, Duolite C-20 and C-25, and Lewatit S-100 are resins of this type.

Weak-acid, cation exchange resins are prepared by copolymerization of an organic acid or acid anhydride and a cross-linking agent. As a rule, acrylic or methacrylic acid is used in combination with divinyl benzene, ethylene dimethacrylate, or similar compounds with at least two vinyl groups. The pearl polymerization technique described above can be used if esters instead of the water-soluble acids are polymerized. The esters are hydrolyzed after polymer-ization. The final products have ionic contents of 9–10 Eq/kg (dry). Resins of

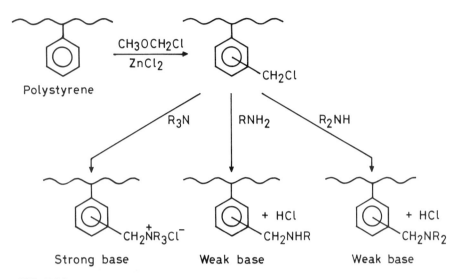

FIG. 5.33 Reaction scheme for the preparation of weak-base and strong-base resins starting with polystyrene.

this type are Amberlite IRC-50, Duolite CS-101, Permutit H-70, and Wofatit CP-300.

Strong-base anion-exchange resins, also very common, are made (Fig. 5.33) by chloromethylating the styrene-DVB copolymer, then aminating the product with a tertiary alkyl amine. The quarternization of chloromethylated resins with tertiary alkyl amines takes place smoothly and quantitatively. The two most common strong-base resins are made with trimethylamine and contain the quarternary amine groups $-N(CH_3)_3$ and $-N(CH_3)_2CH_2OH$. The first type of resin is classed as a Type I strong-base anion exchanger and the second type as a Type II exchanger. Dowex-1, Amberlite IRA-400, Permutit S-1, Nalcite SBR, Duolite A-42, and De Acidite FF are Type I resins. Dowex-2, Amberlite IRA-410, Permutit S-2, Nalcite SA-R, and Duolite A-40 are Type II resins. The resins Amberlite IRA-401 and 411 differ from the standard types 400 and 410 only by having a lower DVB content. The ion contents of strong-base resins are 3–4 Eq/ kg (dry).

Type I resins have better thermal and oxidative stability and maintain the integrity of the quarternary groups over a long period of time. Type II resins, when used in the hydroxide form, are limited to a maximum temperature of approximately 40°C and should not be used under oxidizing conditions. Type II resins are often used because of lower operation costs. They regenerate somewhat more easily and have higher operating capacities than the Type I products. The resins have a useful industrial life of 3–5 years.

Weak-base resins are made by treating a chloromethylated intermediate with primary or secondary alkyl amines (Fig. 5.33). The treatment with secondary amines leads to *monofunctional* weak-base resins having tertiary amino groups. By treatment with primary alkyl amines, however, *polyfunctional* weak-base resins having secondary and tertiary amino groups are obtained, the latter being formed by the reaction of the primary amine with two chloromethyl groups:

$$\cdots -CH_2Cl \qquad\qquad \cdots -CH_2$$
$$+ \ H_2NR \longrightarrow \qquad\qquad \Big\rangle \ NRH^+Cl^- + HCl$$
$$\cdots -CH_2Cl \qquad\qquad \cdots -CH_2 \qquad\qquad\qquad (14)$$

Additional cross-linking results if the chloromethyl groups belong to different chains. (Occasionally, polyamines such as tetramethylenepentamine are used; they can react in a similar way with two or more chloromethyl groups.) The resins Amberlite IR-45, Dowex-3, Nalcite WBR, and Duolite A-14 are polyfunctional weak-base anion exchangers. De-Acidite G is a monofunctional resin with tertiary amino groups.

Deionization by ion-exchange is usually confined to relatively dilue solutions, i.e., concentrations less than 0.03N (1500 ppm $CaCO_3$). Since ion-exchange resins have finite capacities, economics are most favorable for dilute solutions, although concentrations up to 0.1N (5000 ppm $CaCO_3$ can be treated quite satisfactorily.

Cation exchange resins having strongly acidic sulfonic acid groups are the major cation exchangers used for deionization purposes. When dilute solutions are passed through beds of such resins, either in acid form or sodium salt form, cation exchange takes place according to the following reactions (shown for $MgCl_2$):

$$(1/2)MgCl_2 + \text{(P)}SO_3^-H^+ \rightleftharpoons \text{(P)}SO_3^-(Mg^{2+})/2 + HCl \qquad (15)$$

or

$$(1/2)MgCl_2 + \text{(P)}SO_3^-Na^+ \rightleftharpoons \text{(P)}SO_3^-(Mg^{2+})/2 + NaCl \qquad (16)$$

The above reactions are reversible, and the exhausted resin can be regenerated with moderate concentrations of strong acids or NaCl. The relative affinities for the resin of various cations have an important influence on the efficiency of exchange. In general, the order of ease of replacement of common cations is $Li^+ > H^+ > Na^+ > K^+ > Mg^{2+} > Ca^{2+} > Al^{3+} > Fe^{3+}$.

Strongly basic anion-exchange resins can act as acid neutralizers or can split salts in the same manner as the sulfonic acid cation exchangers. The following are typical reactions:

$$\text{(P)}NR_4^+OH^- + H^+X^- \longrightarrow \text{(P)}NR_4^+X^- + H_2O \qquad (17)$$

or

$$\textcircled{P}NR_4^+OH^- + M^+X^- \longrightarrow \textcircled{P}NR_4^+X^- + M^+OH^- \tag{18}$$

The resins show marked affinity relationships depending on ion size and valence. The order of affinity for common anions is: $SO_4^{2-} > Cl^- > HCO_3^- > F^- >> HSiO_3^-$. Reaction (18) can be reversed by regenerating with moderate concentrations of sodium hydroxide solutions. However, since the reactions are not as easily reversed, regeneration levels of 150–200% of the stoichiometric requirements are frequently employed.

In addition to the commonly employed sulfuric acid cation exchanger, resins based on the carboxylic acid groups are sometimes employed under special conditions. These resins are effective for exchange reactions in neutral and alkaline solutions. For example, the effluent from a salt-splitting reaction, Eq. (18), can be treated with a carboxylic exchanger according to the reaction:

$$\textcircled{P}COOH + MOH \longrightarrow \textcircled{P}COOM + H_2O \tag{19}$$

The reaction is reversed very effectively by the hydrogen ion, and regeneration can be accomplished readily at efficiencies approaching 100%.

The role of weak-base anion-exchange resins in deionization is often confined to acid neutralization, as shown by the following equation:

$$RNH_3OH + HX \longrightarrow RNH_3X + H_2O \tag{20}$$

The reaction can be reversed by addition of some alkaline reagent such as NaOH, Na_2CO_3, NH_3, and $NaHCO_3$. The regeneration is accomplished readily at efficiencies approaching 100%. The ease with which the hydroxyl ion can be replaced by other anions is $SO_4^{2-} > Br^- > F^- > CH_3COO^- >> HCO_3^-$.

The weak-acid and weak-base resins do not totally dimineralize the water like their strong counterparts. However, the water they produce, containing 100–200 ppm of dissolved solids, is adequate for many purposes. The regeneration of the "weak" resins is also accomplished more readily than the "strong" resins.

A relatively recent development is a thermally regenerable ampholytic resin containing both weak acid and a weak-base functionality within the one bead (Sirotherm by ICI), which absorbs significant quantities of salt at ambient temperatures and releases salt on heating to 70–90°C. Thus, when the resin is fully exchanged or exhausted, it can be rinsed in hot water and reused. The adsorption step involves the transfer of protons from carboxylic acid groups to amino groups to form the cation and anion exchange sites:

$$\textcircled{P}\Big\langle {}^{COOH}_{NR_2} + Na^+ + Cl^- \rightleftharpoons \textcircled{P}\Big\langle {}^{COO^- \; Na^+}_{\overset{+}{NR_2}H \; Cl^-} \tag{21}$$

The equilibrium is temperature sensitive, with both types of groups showing weaker electrolyte behavior on heating. The large increase in the ionization of water that occurs on heating, about 30-fold from 25°C–85°C, releases additional protons and hydroxyl ions, which suppress the ionization of the weak electrolyte resins. The hot water can thus be looked on as providing the acidic and basic regenerants that usually have to be added separately.

Operating data indicate that Sirotherm TR-20 resins can produce waters of salinities as low as 50–100 ppm dissolved salts, the economic upper range of salinities in feed water being restricted to 2000–3000 ppm. The resins are expected to find application in the dimineralization of mildly brackish surface and underground waters for industrial and municipal use and as a roughing stage in the production of high-quality boiler feed water.

Some polystyrene resins (cross-linked with DVB) are specially modified to have chelating functional groups bound to the matrix so as to make them selective towards certain ions. Such resins with iminodiacetic acid groups are marketed under the trade names Dowex A-1 (Dow Chemical) and Chelex 100 (Bio-Rad Laboratories). The complex (XVIII) formation constants with metal ions of the chelating resin are so large that the resin adsorbs metal ions equivalent to the iminodiacetic acid groups (used in sodium salt form), i.e., the efficiency of metal ion adsorption is near 100%. A particular metal ion can be removed by controlling the pH of aqueous solution. For example, at pH 2, mercury and copper ions are preferentially adsorbed, while zinc, cobalt, and cadmium ions are little adsorbed.

(XVIII)   (XIX)

Polystyrene resins that have aminophosphonate chelating groups are highly selective towards calcium ion. Such a resin, e.g., Duolite ES467 (Rohm and

Haas), when added to a strong brine solution (25%) contaminated with 10 mg/liter of calcium ion will selectively remove the calcium ion, forming calcium amino-phosphonate complex (XIX), until only 0.02 mg/liter remains. This process is particularly useful for purifying the brine used in the chlor-alkali cell described earlier.

Many other specific resins have been prepared. To give only a few examples, resins with hydroxamic acid groups (XX) are specific for Fe^{3+} ions and those with mercapto groups (XXI) prefer Hg^{2+} ions. Resins containing *chlorophyll* and *haemin* derivatives or similar compounds form extremely strong chelates with ions such as Fe^{3+}. In fact, the chelated counterions are held so strongly that they can hardly be displaced. Another undesired consequence of such strong association is that the mobility of the counterion in the resin is greatly reduced. Hence for any application one should choose the resin carefully, seeking a reasonable compromise between selectivity, ease of regeneration, and rate of ion exchange.

~~~CH — CH₂~~~
         |
         C=O
         |
         HNOH

(XX)

~~~CH — CH₂~~~
 |
⬡
 |
SH   (XXI)

Prior to the development of polystyrene resins, phenol-formaldehyde (P-F) condensates were used as matrices, but they have now been replaced. A few weak-base types still exist (e.g., Duolite ES562 of Rohm and Haas), which are made by adding an amine during polycondensation. These P-F condensates are used for enzyme fixation.

Applications

Applications of ion-exchange resins are extremely varied, ranging from water-softening to purification of chemicals and therapeutic applications. An extremely useful industrial development of the ion-exchange technique is the production of demineralized water rivalling that of distilled water in purity.

Most ion-exchange reactions for industrial applications are done with columns of resin in which the ions in solution (say B) are depleted, proceeding from top to the bottom of the column, by exchange with ions (say A) in the resin. With the progress of ion exchange the resin bed shows an exhaused portion at the top, an ion-exchange zone in the middle and a regenerated portion at the bottom (Fig. 5.34). As the ion-exchange zone moves down through the resin column, the B ions eventually reach the outlet, at which time 'breakthrough' occurs.

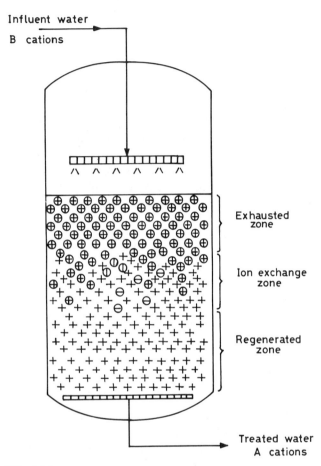

FIG. 5.34 Ion-exchange column in service. +, resin containing A cations; ⊕, resin containing B cations.

In most ion-exchange installations, the vertical column is the most commonly employed unit. The system may consist of a single column with one type of resin (*single column system*), two or more columns containing a variety of cation and/ or anion exchange resins (*multiple-bed system*), or a single column containing a mixed bed of two (or more) resins (*monobed* or *mixed-bed* system).

Much of the appeal of the ion-exchange process stems from the simplicity of the single-column system. Water softening by ion-exchange is the most widely used example of this system (Fig. 5.35). The hardness ions, calcium and magnesium, are exchanged for sodium as the hard water flows down through a column of cation exchange resin used in the sodium form [see Eq. (16)].

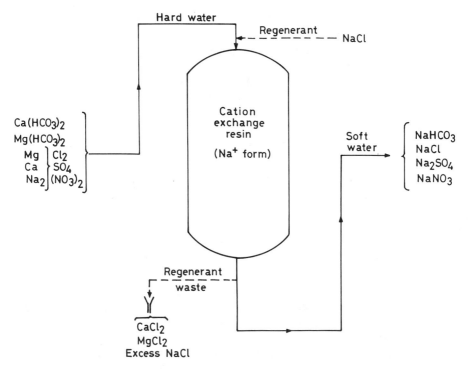

FIG. 5.35   Water softening by ion-exchange in single column system.

Conversion of the exhausted resin back to the sodium form [i.e., reverse of Eq. (16)] is accomplished in a *regeneration* step by contacting the column with an excess of sodium chloride solution. Water softeners range in size from small household units (e.g., 20 cm in diameter and 60 cm deep) to large industrial units (e.g., 360 cm in diameter and 150 cm deep).

It will be noted that in reactions (15) and (17) the overall result of ion-exchange is the complete elimination of dissolved salts with the formation of an amount of water equivalent to the amount of electrolyte removed. Deionization requires contacting the water with both cation and anion exchange resins, which can be done in a multiple-bed system.

Two-bed deionization of water is widely practiced. Such systems generally use a strong-acid cation exchange followed by either a weak-base or a strong-base anion exchanger. Reactions similar to Eqs. (15), (17), or (20) occur. Three beds or more are used either to achieve operating economy or a greater degree of deionization. For example, a three-bed system of strong acid cation exchanger/weak-base anion exchanger/strong-base anion exchanger offers economy in regeneration. In service, the weak-base resin removes the mineral acids coming from the cation exchanger and the strong-base resin removes principally carbonic and silicic acids.

For installations requiring very high effluent water quality, *monobed* or *mixed-bed* deionization has been used widely, yielding a demineralized water in one operation. The standard mixed-bed is a special case of a single column. It usually consists of an intimate mixture of a strong-acid cation exchange resin and a strong-base anion exchange resin in the hydrogen and hydroxide forms, respectively. When water is passed through a fixed bed of such a mixture, the ions in solution are alternately exposed to numerous contacts with cation and anion exchange sites, the effect being similar to that obtained with a very large number of multiple beds, with the result that almost complete deionization occurs. The process follows the reaction:

$$MX + \begin{Bmatrix} \text{\textcircled{P}}SO_3^- H^+ \\ \text{\textcircled{P}}NR_3^+ OH^- \end{Bmatrix} \longrightarrow H_2O + \begin{Bmatrix} \text{\textcircled{P}}SO_3^- M^+ \\ \text{\textcircled{P}}NR_3^+ X^- \end{Bmatrix} \qquad (22)$$

Following exhaustion of a mixed bed and prior to regeneration, the resins in the bed are separated by applying *backwash* at a flow rate sufficiently high to fluidize the bed. By virtue of the fact that anion exchange resins have a lower density than cation exchange resins, hydraulic separation into two layers occurs. The regenerant caustic soda then contacts the upper layer of anion resin and the acid regenerant flows through the lower layer of cation resin. Then following a rinse with relatively pure water, the resins are air-mixed prior to the next ion-exchange run.

Ion-exchange methods are established for treating various effluents arising from metal finishing processes such as plating and anodizing. The use of strong-base resins for decolorizing sugar liquors is widely practiced. The coloring bodies are organic anions that are sorbed by weakly cross-linked strong-base gel resins.

Even wines are sometimes treated by column cation exchange. Potassium hydrogen tartrate, which causes an unpleasant precipitate in wines, is converted to the more soluble sodium salt by treatment with polystyrene sulfonic acid resin in the sodium (Na^+) form.

Ion-exchange resins are used for metal recovery from low-grade ores and dilute leach liquors. One of the best examples is the recovery of uranium.

Ion-exchange chromatography is well known for separating mixtures of ions in solution. Possibly the best-known organic analytical ion-exchange application is the chromatographic separation and isolation of amino acids. Commercially, the most significant application is the recovery of antibiotics such as strepto−

mycin and neomycin. The fermentation broth containing the impure antibiotic is treated with a polyacrylic weak-acid resin on which the antibiotic is sorbed to the exclusion of other organic impurities. The product is recovered by elution with dilute mineral acid.

Polycarboxylates

Polyacrylate-type homopolymers are polyelectrolytes and are the most ionic of the organic polymers. They dissolve in water giving aqueous solutions with unusual and useful physical properties. They are generally made by free-radical polymerization in aqueous solution. Very-high-molecular-weight (e.g., 4 × 10^6) polymers can be obtained that give very viscous solutions. Polyacrylic (XXII), polymethacrylic (XXIII), and polyitaconic (XXIV) acids are the three main types having theoretical capacities of 13.9, 11.6, and 15.4 Eq/kg, respectively. Aqueous solutions or dry powders of these materials are commercially available.

$$\begin{array}{ccc}
\text{H} & \text{CH}_3 & \text{CH}_2\text{COOH} \\
| & | & | \\
\sim\!\!\sim\!\!\text{CH}_2\text{–C}\!\!\sim\!\!\sim & \sim\!\!\sim\text{CH}_2\text{–C}\!\!\sim\!\!\sim & \sim\!\!\sim\text{CH}_2\text{–C}\!\!\sim\!\!\sim \\
| & | & | \\
\text{COOH} & \text{COOH} & \text{COOH} \\
\text{(XXII)} & \text{(XXIII)} & \text{(XXIV)}
\end{array}$$

Versicols (Allied Colloids) and Texigels (scott Bader) are homopolymers of acrylic or methacrylic acids or their copolymers with acrylamide. They are used as stabilizers, and protective colloids and thickeners for aqueous dispersions, binders, and flocculants. Carbopols (B. F. Goodrich) are different grades of polyacrylic acid of varied molelcular weight having excellent suspending, thickening, and gel-forming properties. Carbosets (B. F. Goodrich) are acrylic copolymers and have a similiar carboxyic content but generally can be dissolved in alkaline solutions. They are used mainly in coating applications. In some applications, they are covalently cross-linked with epoxides and so on, but some applications use ionic cross-links made with zinc ions.
  A notable application of polyacrylic acid is for cements in dentistry. These are made by mixing an aqueous solution of the polymer with zinc oxide when the zinc salt precipitates as a highly cross-linked gel that rapidly sets to a hard mass under oral conditions. In a variation of this reaction, the zinc oxide is replaced with a tooth-colored glass powder that releases Al^{3+} and Ca^{2+} ions. These cements, called ASPA (aluminosilicate polyacrylic acid) or glass ionomer, set very rapidly, bond well to tooth enamel, and are compatible with living tissue.

Integral Polyelectrolytes

Polyelectrolytes having bound ions integrated in the polymer backbone are called *ionenes*. Some ionenes have been studied for their bacteriostatic and bactericidal activity. Ionenes with segments of polypropylene oxide in the backbone have been evaluated as thermolastic elastomers.

Polyethylenimine (PEI) is an integral polyelectrolyte that is available commercially, e.g., Polymin (BASF). It is formed by the ring-opening polymerization [reaction (23)] of ethyleneimine (aziridine). The resulting polyamine has about 50% of the expected secondary-amine functionality and about 50% primary and tertiary due to branching:

$$
\underset{\underset{NH}{\diagdown\diagup}}{CH_2{-}CH_2} \xrightarrow{H^+} {+}CH_2{-}CH_2{-}NH{+}_x {+} CH_2{-}CH_2{-}\underset{\underset{\underset{NH_2}{|}}{\underset{\underset{CH_2}{|}}{CH_2}}}{N}{-}CH_2{-}CH_2{-}NH{+}_y \qquad (23)
$$

PEI has typical polyelectrolyte properties; it is a highly viscous hygroscopic liquid, completely miscible with water and lower alcohols, insoluble in benzene, and reactive toward cellulose. PEI is mainly used as a size, flocculating agent, or protective colloid, notably in the paper and textile industries, because of its ability to bind to cellulosic fibers.

Membranes based on PEI were introduced for use in reverse osmosis to desalinate water. These membranes, known as NS100 and NS101, are made by forming a PEI skin on polysulfone support and insolubilizing it by treatment with toluene di-isocyanate or phthaloyl dichloride to produce a polyurea or polyamide (Fig. 5.36).

POLYMER SUPPORTS IN ORGANIC SYNTHESIS [36–38]

The field of organic chemistry has seen the most extensive use of polymeric materials as aids in effecting chemical transformation and product isolations. Insoluble polymer supports have been widely used as "handles" to facilitate the synthesis of polypeptides, polynucleotides, and even polysaccharides by repetitive sequential-type organic synthesis. More recently, polymer-bound reagents suitable for nonsequential single-step reactions have been utilized by many research groups.

In many synthetic organic chemical procedures, a chemical reagent is used, which after reaction gives a by-product. This by-product can sometimes be

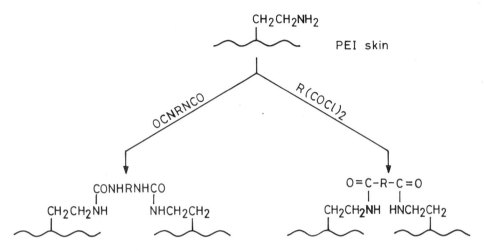

FIG. 5.36 Insolubilization of polyethylenimine by treatment with toluene diisocyanate or phthaloyl dichloride.

difficult to separate from the desired product of the reaction. If the chemical reagent can be bound to an insoluble polymer carrier and successfully used in organic syntehsis, then the by-product of the reagent, after reaction, will remain attached to the insoluble polymer and can be separated from the desired product of reaction by simple filtration. One of the advantages of such polymeric reagents, in common with solid-phase methods, is the simplification of separation procedures and the possibility of regeneration and reuse of reagents.

Polymeric reagents have been used to carry out many reactions. The reagents can be classified broadly as polymeric oxidizing agents, polymeric reducing agents, and polymeric acetylating agents. The advantages of organic synthesis on insoluble polymer supports have been put to most effective use in polypeptide synthesis (*Merrifield synthesis*) described later.

Polymeric Oxidizing Reagents

The preparation of polymer-supported peroxyacids and their use in epoxidation of olefins are two of the more thoroughly investigated areas of polymer-supported reagents chemistry. The most useful polymer-supported peroxyacids to-date are cross-linked polystyrenes containing aromatic peroxyacid residues [(4) in Fig. 5.37]. These can be prepared from the corresponding polymer acid (1), acid chloride (2), or aldehyde (3) (Fig. 5.37). High yields (97%) of the peracid can be obtained by treating the carboxylic acid resins (1) with 85% H_2O_2 in methanesulfonic acid or with 70% H_2O_2 in p-toluenesulfonic acid/THF, the acid

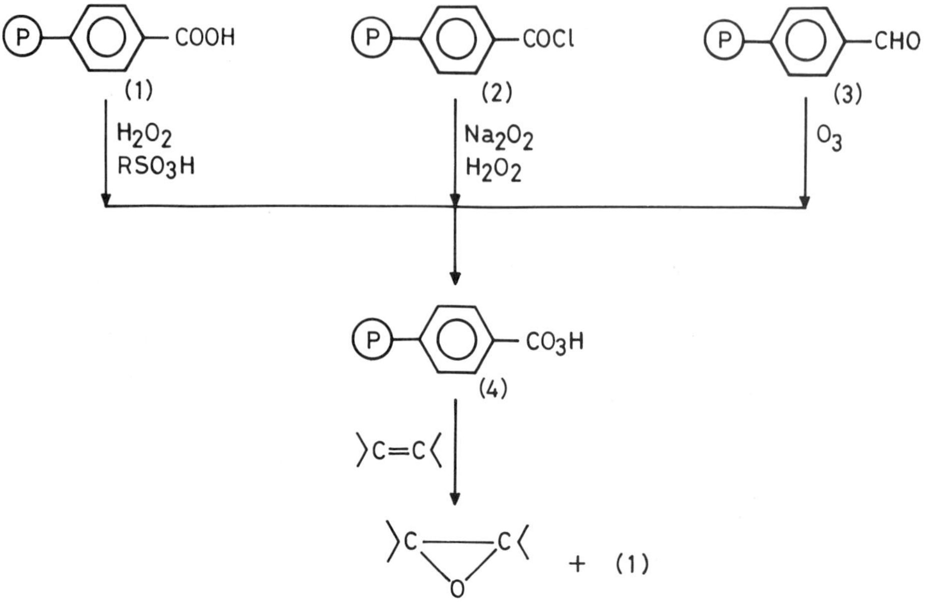

FIG. 5.37 Preparation of polymer-supported aromatic peroxyacid and its use in epoxidation of substituted alkenes.

chloride resin (2) with Na_2O_2, 70% H_2O_2, and THF, or the aldehyde resin (3) with ozone or with 85% H_2O_2 in methanesulfonic acid. Activities as high as 3.5–4.0 mmol/g have been obtained, and the resins can be stored in a refrigerator for several months without much loss of activity.

Tri- and di-substituted alkenes are generally epoxidized in good yield (>50%) by the above polymer-supported peroxyacids, and sulfides, including penicillins and deacetoxycephalosporins, are efficiently oxidized to sulfoxides and/or sulfones. Penicillin G, for example, has been oxidized in 91% yield by passing a solution in acetone down a column of the resin. When the peroxyacid resins epoxidize or oxidize the substrate, the carboxyl group is reformed (see Fig. 5.37), and so the spent resin can be recycled.

Two polymer-supported chromium-containing reagents have been used to oxidize alcohols. One is the polymer-supported analogue of pyridinium chlorochromate (XXV) and is prepared by treating cross-linked poly(4-vinyl pyridine) (XXVI) with chromium trioxide in hydrochloric acid. The second reagent is the polymer-supported chromic acid, prepared by treating the chloride form of a macroporous resin (XXVII) with aqueous chromium trioxide. The reagent is believed to contain acid chromate anions, and it oxidizes primary and secondary alcohols to aldehydes and ketones, respectively, in excellent yields.

The Moffat oxidation (carbodiimide, DMSO, H^+ catalyst) is a useful mild method for oxidizing alcohols:

$$R_2CHOH + R'\!\!-\!\!N\!\!=\!\!C\!\!=\!\!N\!\!-\!\!R' + (CH_3)_2S\!\!=\!\!O \xrightarrow{H^+}$$
$$R_2C\!\!=\!\!O + R'\!\!-\!\!NH\!\!-\!\!CO\!\!-\!\!NH\!\!-\!\!R' + (CH_3)_2S$$

$$(24)$$

However, the separation of the product from the substituted urea formed as a by-product often proves to be difficult. A solution to the problem is to use the polymer-supported carbodiimide (XXVIII), which can be prepared with an activity of about 2 mmol/g from chloromethylated cross-linked polystyrene, as shown in Fig. 5.38.

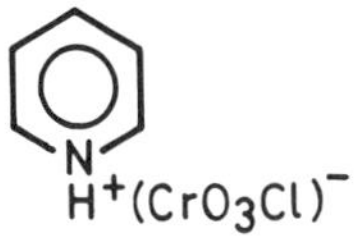

(XXV) 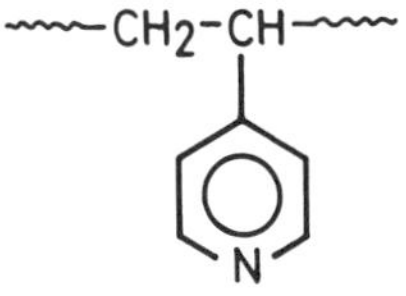 (XXVI) 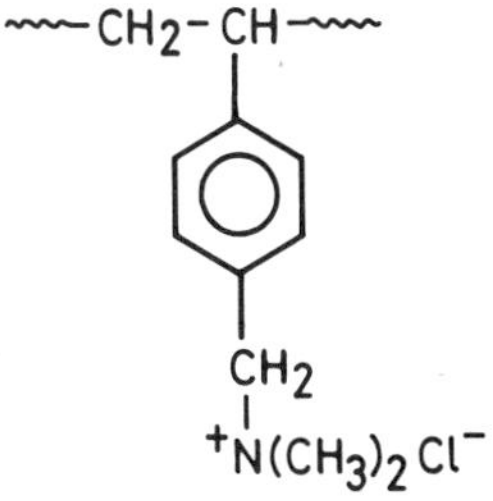 (XXVII)

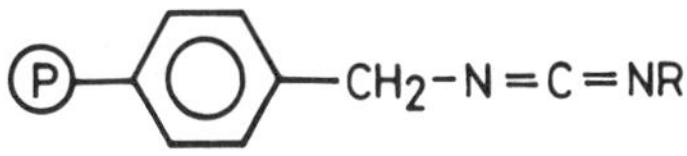

(XXVIII)

More types of polymeric structures have been functionalized to contain a carbodiimide group than any other functional group. The resulting polymeric carbodiimides have been used in peptide synthesis and for preparation of carboxylic anhydrides [reaction (25)], as well as in Moffat oxidation, previously described. During these reactions, the carbodiimides are converted into the corresponding ureas as the end-product. The urea by-product can be reconverted to the carbodiimide (see Fig. 5.38).

$$\underset{\underset{O}{\|}}{RC}\!\!-\!\!OH + R'\!\!-\!\!N\!\!=\!\!C\!\!=\!\!N\!\!-\!\!R' \longrightarrow \underset{\underset{O}{\|}}{RC}\!\!-\!\!O\!\!-\!\!\underset{\underset{O}{\|}}{C}R + R'\!\!-\!\!NH\!\!-\!\!\underset{\underset{O}{\|}}{C}\!\!-\!\!NH\!\!-\!R'$$

$$(25)$$

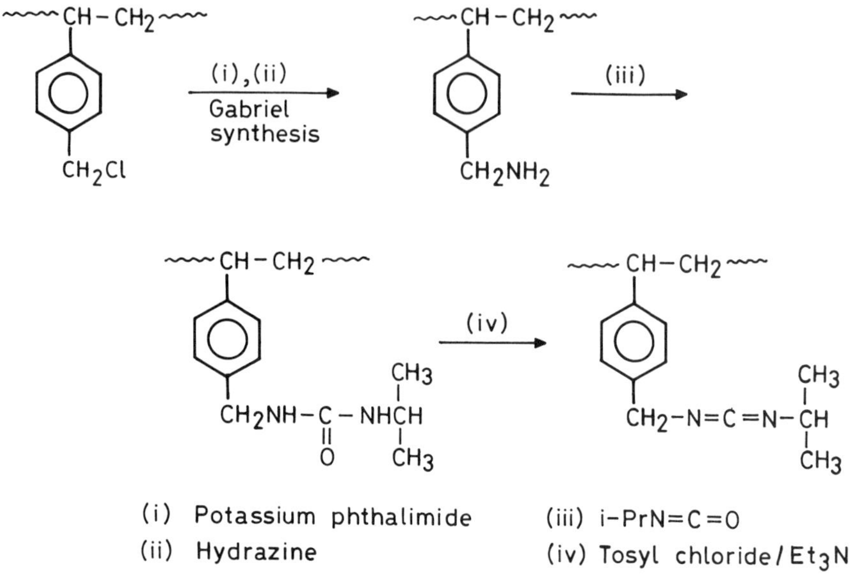

(i) Potassium phthalimide   (iii) i-PrN=C=O
(ii) Hydrazine   (iv) Tosyl chloride/Et₃N

FIG. 5.38 Scheme of preparation of polystyrene-supported carbodiimide.

Polymeric Reducing Agents

Far fewer polymer-supported reducing agents than oxidizing agents have been described. These have included reports of the various reductions brought about by polymer-supported metal hydrides. The most extensively studied among these reagents is the polymer-supported tin hydride. This reagent, which has an activity of 2.0–2.5 mmol of active hydride per gram, can be prepared from a cross-linked macroporous polystyrene via the reactions outlined in Fig. 5.39. An advantage of the reagent over nonpolymeric tin hydrides is that it is odorless and nontoxic. Aldehydes and ketones can be reduced to alcohols with generally good yields using polystyrene-supported tin hydride. The same reagent has also been used to replace iodo and bromo groups by hydrogen in both aliphatic and aromatic halides in excellent yields.

Polymeric Acetylating Agents

Numerous polymer-supported reagents have been used to acylate amines and alcohols. The earliest and best-known example of polymeric acetylating agents is poly(4-hydroxy-3-nitrostyrene) (XXIX). By attaching benzoic or acetic acid to this reagent, active polymeric esters (XXX) are obtained, which can then be used in acylation of amines:

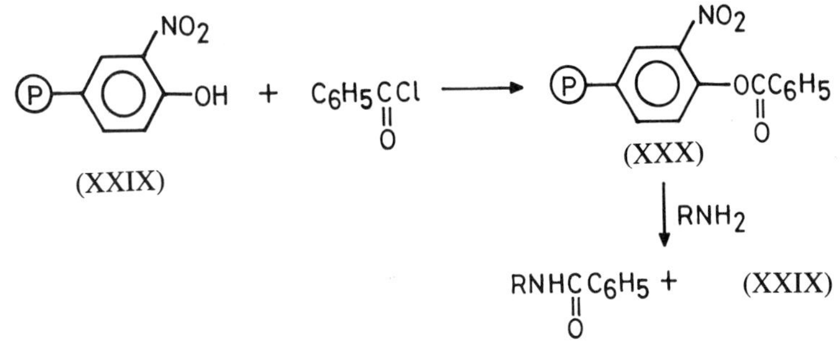

This approach gave rise to a general method for peptide synthesis using a multistep procedure (Fig. 5.40) in which each aminoacid residue is first bound to the nitrophenol polymer as an active ester that is then treated with a soluble aminoacid or peptide having a free amino group. The product of each step in the synthesis is thus obtained in solution and so may be purified if necessary. By employing excess polymeric reagent, high yields of product are obtained. As an example of a multistep synthesis by this method, bradykinin (a nonapeptide) was synthesized in an overall yield of 39%.

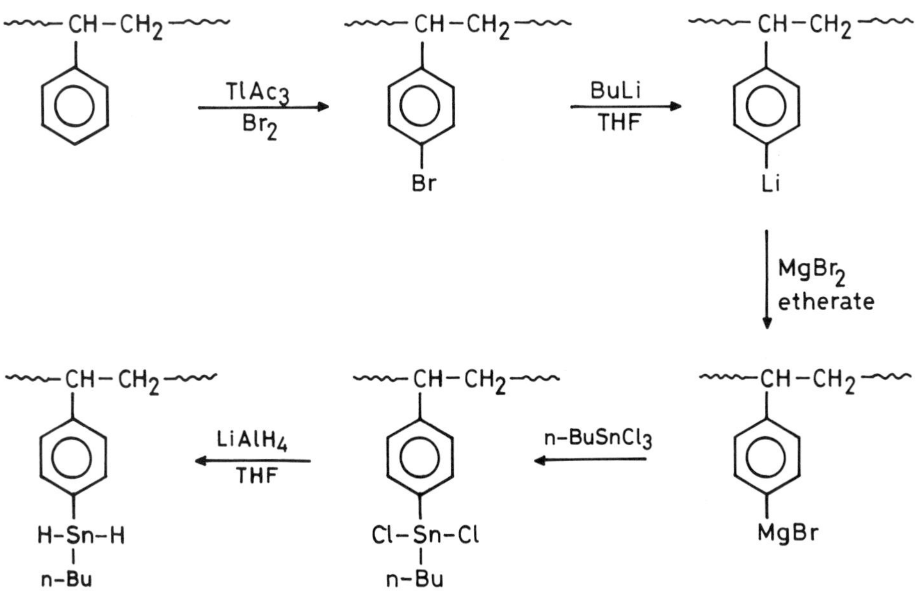

FIG. 5.39 Scheme of preparation of polystyrene-supported tin hydride.

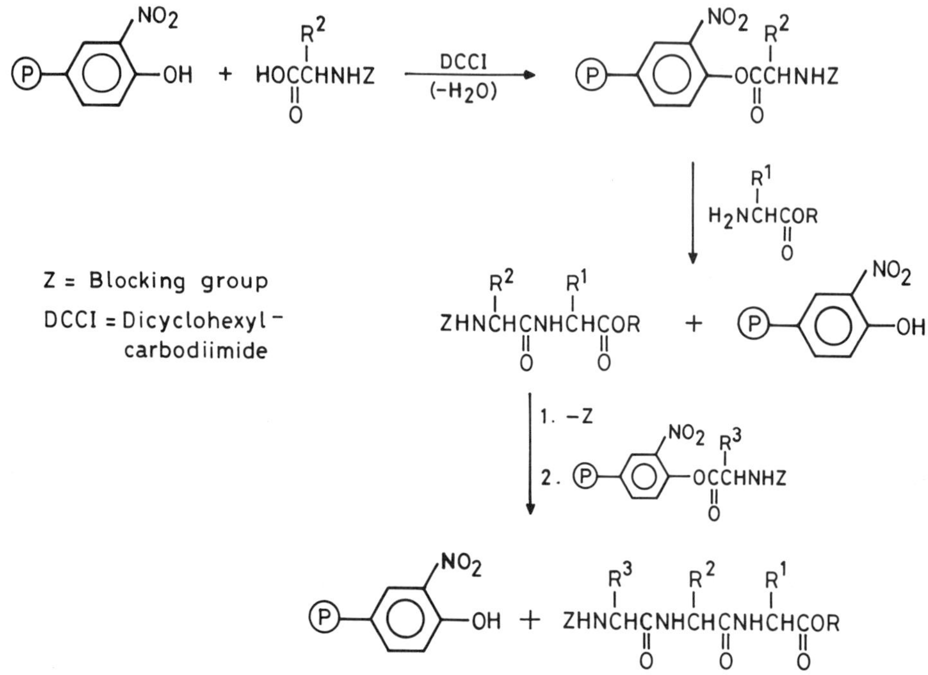

FIG. 5.40 Multi-step synthesis of polypeptide using nitrophenol polymer.

Besides the nitrophenol polymer (XXIX) mentioned above, numerous other polymer-supported reagents have been synthesized and used to acylate amines and alcohols. Polymer-bound activated phenols and N-hydroxysuccinimides constitute two important classes of these reagents.

In the above method of peptide synthesis based on polymeric acetylating agent, the peptide product is the only nonpolymer-supported product. The method is the reverse of that developed by Merrifield (see below) in which the peptide is the only polymer-supported product.

Merrifield Synthesis [38]

Peptides consisting of 30–50 amino acid residues play important structural and functional roles in biochemistry, pharmacology, and neurobiology. A rapid, efficient, and reliable method for the chemical synthesis of peptides is therefore of utmost importance. A large number of techniques have been developed since the beginning of the twentieth century for the chemical assembly of amino acids to form peptides. They can be subdivided into solution and solid-phase methods. Recognizing the many difficulties associated with the solution method, in 1959

R. B. Merrifield of Rockfeller University devised the solid-phase approach for peptide synthesis. While early reports demonstrated the feasibility of the method for both manual and automatic peptide synthesis, subsequent efforts from numerous laboratories refined the particulars for peptides and also applied the technique to other oligomeric systems, particularly DNA.

Solid-phase synthesis is based on the covalent attachment (anchoring) of the growing peptide chain to an insoluble polymeric support, so that unreacted reagents remaining in solution can be removed by filtration and washing without losses. The peptide residue anchored to the insoluble resin is then extended by a series of addition cycles, with excess soluble agents being added to drive the reactions to completion. Because of the speed and simplicity of the repetitive steps, the solid-phase procedure is amenable to automation. After the required chain sequence has been established, the peptide is cleaved from the support under conditions that do not affect sensitive residues in the sequence.

A general scheme for stepwise solid-phase synthesis of linear polypeptide is shown in Fig. 5.41. An insoluble polymer support is functionalized with group X, e.g., chloromethyl or aminomethyl. Building blocks are the amino acids with temporary protection of the side chains. (Most useful N^{α} and side-chain protected amino acid derivatives are commercially available.) A bifunctional spacer (handle) is used to covalently attach the terminal amino acid, AA^1 to the polymeric support. One end of the handle reacts with the polymer-bound functional group X and the other with the free carboxyl of the protected amino acid derivative. After selective removal of the temporary protecting group to release the α-amino group (deprotection step), the second partially protected amino acid, AA^2, is introduced and a coupling step carried out in which a water molecule is eliminated to form the peptide bond. This deprotection-coupling cycle is repeated until all the amino acids required are introduced. Finally, the peptide is cleaved from the polymeric support and the protecting groups are removed.

The choice and optimization of the protection-scheme chemistry is a key factor that influences the successful outcome of any synthetic effort. At least two levels of protecting group stability are required. Thus, the permanent protecting groups that are used to prevent branching or other reactions on the side chains of the amino acids must be stable enough to withstand repeated quantitative removal of the temporary *N*-amino protecting group. On the other hand, the permanent groups must be such that conditions can be found to remove them with minimal side reactions that affect the desired product. The necessary stability is often achieved by kinetic "fine-tuning," which relies on reaction rate differences whenever the same chemical mechanism (usually acidolysis) is involved in the removal of both classes of protecting groups.

The Merrifield synthesis, which represents one of the first applications of a macromolecule as a solid support in organic synthesis, has served to revolutionize the field of peptide fabrication. Condensing specific sequences of amino

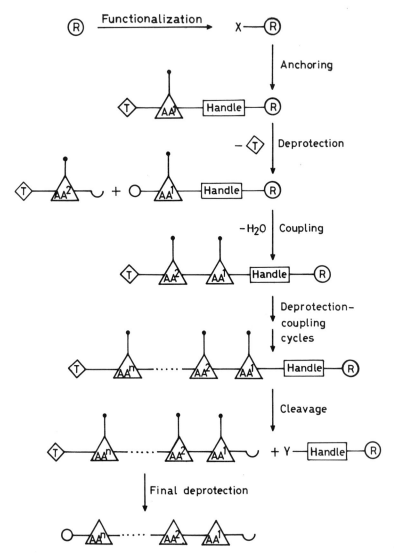

FIG. 5.41 Stepwise solid-phase synthesis of linear polypeptides. ®: insoluble polymer support; AA^1, AA^2, ... AA^n: amino acids; T: temporary protection; •: permanent protection; ⎯⎯⎯ : free carboxyl; ∘ = free amino group [38].

acids onto a growing peptide, which is attached to an insoluble polymer, has changed peptide synthesis from one of tedious, multifaceted separations to a procedure that has been successfully automated. In recognition of the impact and importance of this solid-phase method, Merrifield was honored with the 1984 Nobel prize in chemistry.

POLYMER-SUPPORTED CATALYSTS [39–41]

Attempts have long been made to use polymers as catalysts for chemical reactions. Enzymes, the natural polymeric catalysts, are characterized by high specificity and high catalytic activity, which are properties that have been the goal of catalyst development. The first advances in this direction were made with hydrolytically active polymers. It was observed that polymeric acids and polymeric bases, such as those that occur in ion-exchange resins, can catalyze hydrolytic reactions. Since this discovery, many soluble and insoluble polymeric catalysts have been prepared by attaching conventional catalytic species to polymeric carriers.

A common advantage in using a polymer-bound catalyst is found in the ease of separation at any state of the reaction, thus offering the possibility of arresting further progress of the reaction. Also, an immobilized catalyst is sometimes more stable to atmospheric conditions and does not cause undesirable side reactions, as is the case with homogeneous catalysts. In industrial applications, they provide for the adoption of continuous processes in place of batch operations and, in many cases, the catalyst can be regenerated and reused many times, thus lowering the cost of operation.

In recent years, the discovery and utility of homogeneous transition metal catalysts for hydrogenation, isomerization, oligomerization, hydroformylation, etc., have prompted efforts to effectively "heterogenize" them by anchoring them to polymers. This has the advantage that the valuable catalyst is easily removed from the reaction mixture and is available for reuse.

Catalysis by Acidic and Basic Ion-Exchange Resins

Catalysis by ion-exchange resins represents perhaps one of the earliest examples of the use of a polymer-supported species. Virtually all of the organic transformations that are catalyzed by homogeneous acid or base have also been carried out in the presence of the corresponding ion-exchange resin. The major groups of reactions are: esterification, hydration, hydrolysis, alkylation, acylation, condensation, cyclization, isomerization, and polymerization. A considerable amount of literature exists describing particular applications. Some examples are given in Table 5.16.

TABLE 5.16 Reactions Catalyzed by Polymeric Acid-Base Catalysts

Reaction type	Reactants	Resin form	References
Acid-catalyzed reactions			
Ester hydrolysis	General esters, mainly aliphatic	Amberlite IR-100(H$^+$)	Davies, C. W. and Thomas, G. G., *J. Chem. Soc.*, 1607 (1952).
Esterification	Fatty acids including higher ones, and alcohols or olefins	Ⓟ-SO$_3$H	George, F. V. and Virgil, I. S., *J. Org. Chem.*, 36: 2548 (1971)
Glycoside synthesis from carbohydrates and alcohol	Glucose and methanol, benzaldehyde and glycerol	Dowex-SO$_3$H	Mowery, Jr. D. F., *J. Org. Chem.* 26: 3484 (1961)
Phenol alkylation, with alkenes	Phenol and isobutene	Ⓟ-SO$_3$H	Loev, B. and Massengale, J. T., *J. Org. Chem.*, 22: 988 (1957)
Acylation of phenols	Pyrogallol, acetic anhydride	Amberlite IR-120(H$^+$)	Price, P. and Israelstam, S. S., *J. Org. Chem.*, 29: 2800 (1964)
Hydrolysis of amides	General amides	Amberlite IR-100(H$^+$)	Collins, R. F., *Chem. Ind.*, 736 (1957)

Reaction	Substrate	Resin/Catalyst	Reference
Hydration of isoolefins	Propene and other olefins	(P)-SO₃H	Neier, W. and Woellner, J., *Chem. Technol.*, 95 (1973)
Inversion of sucrose	Sucrose	P-SO₃H	Bodamer, G. and Kunin, R., *Ind. Eng. Chem.*, 43: 1082 (1951)
Base-catalyzed reactions			
Aldehydes or ketones, condensation with active methylene group	(1)Acetone, cyclopentadiene (2)Furfural and aliphatic aldehyde	Dowex 1-X10 (P)-NH₃⁺OH⁻	McCain, G. H., *J. Org. Chem.*, 23: 632 (1958) Mastagli, A., Floch, A., and Durr, G., *Comp. Rend.*, 235: 1402 (1952)
Knoevenagel reaction	Aldehyde, cyanoacetic acid	IR-4B or Dowex-3 (acetate form)	Hein, R. W., Astle, M. J., and Shetton, J. R., 26: 4874 (1961)
Hydration of nitriles to anides	Nicotinonitrile	IRA-400(OH⁻)	Bobbitt, J. M. and Doolittle, R. E., *J. Org. Chem.*, 29: 2298 (1964)
O-Alkylation of carboxylate ions	RCOOH + R'-X →RCOOR'	IRA-904	Cainelli, G. and Manescalchi F., *Synthesis*, 723 (1975)

Polymeric Esterolytic Catalysts

Nucleophiles, such as imidazole (XXXI), pyridine, and hydroxylamine, are known to be powerful catalysts for ester hydrolyses. An extensive study of polymeric analogs of these compounds has shown that they too act as effective esterolytic catalysts.

As catalysts, polymeric imidazoles are the synthetic prototype of esterolytic enzymes, because, like the enzymes, they have binding centers for the substrate as well as for the catalytic groups. In the case of polyvinylimidazole, PVIm (XXXII), for example, the partially protonated groups (depending on the pH of the medium) can act as binding sites, provided that the substrate has a negative charge. Thus at pH 7.5, when 5% of the basic groups in PVIm are protonated, it acts as a highly effective catalyst for the solvolysis of substrates carrying negative charge, e.g., *p*-nitrophenyl hydrogen phthalate (XXXIII).

Polymer-Supported Phase-Transfer Catalysts

When two reagents reside in essentially separate phases, such as when one reagent is an inorganic salt, as an aqueous solution, and the other an organic molecule in the liquid state or dissolved in an organic solvent, then reaction between these is generally very slow, and in some cases undetectable. A convenient solution to this problem involves the use of so-called phase-transfer catalysts. These are *onium* salts, usually quaternary ammonium or phosphonium species, crown ethers, cryptands, or linear polyethers.

One disadvantage of the conventional phase-transfer catalysis is that it involves the addition of catalysts that must be removed from the reaction mixture at some later stage. Moreover, some of the catalysts that are particularly useful, such as crown ethers, cryptands and chiral onium salts are rather expensive and certainly worth some effort expended in their recovery. Attaching these species to a cross-linked polymer support, with all the attendant advantages of retention, isolation, and recovery, is therefore a highly attractive proposition.

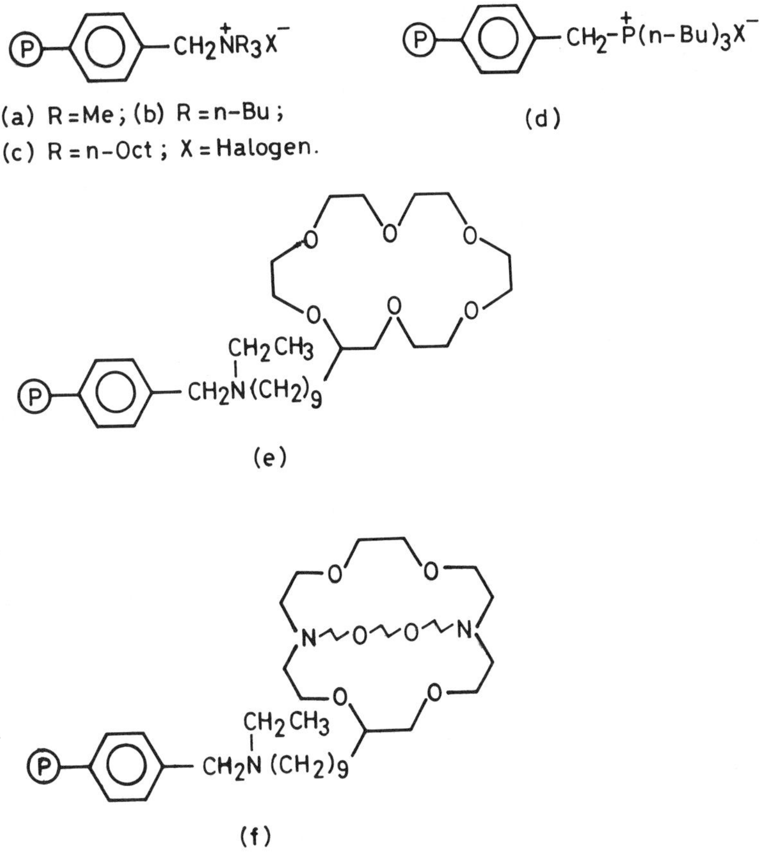

(a) R = Me; (b) R = n-Bu;
(c) R = n-Oct; X = Halogen.

(d)

(e)

(f)

FIG. 5.42 Polymer-supported phase-transfer catalysts with different catalytic functional groups: (a, b, and c) quaternary ammonium salts; (d) phosphonium salts; (e) crown ethers; (f) cryptands.

The use of phase-transfer catalysts that are bound to polymeric supports has been reported. The catalytic functional groups anchored to the polymer were (1) quaternary ammonium salts (Fig. 5.42 a, b, and c), (2) phosphonium salts (Fig. 5.42d), (3) crown ethers (Fig. 5.42e), and (4) cryptands (Fig. 5.42f). Chloromethylated, 2–4% cross-linked polystyrene resins (gel type) were mostly used as the support polymers and the catalyst groups were anchored either by the reaction with the corresponding amine or phosphine. Spacer arms were made for linking the crown ether and cryptand (Fig. 5.42e and f).

Crown polyethers have also been synthesized by polymerization of the corresponding vinyl monomer, e.g., 4'-vinyl monobenzo-15-crown-5 (Fig. 5.43). The polymeric 15C5 crown ethers were obtained as amorphous solids with an average molecular weight of 11,600 and softening at 122–128°C. The polymeric crown ethers, just like the monomeric crown ethers, have the ability to bind alkali metal (particularly potassium) ions and hence have been used for extraction of potassium.

If a mechanism similar to one for unbound catalysts operates in polymer-supported systems, then it is necessary to imagine the phase boundary existing at the catalytic sites as shown below:

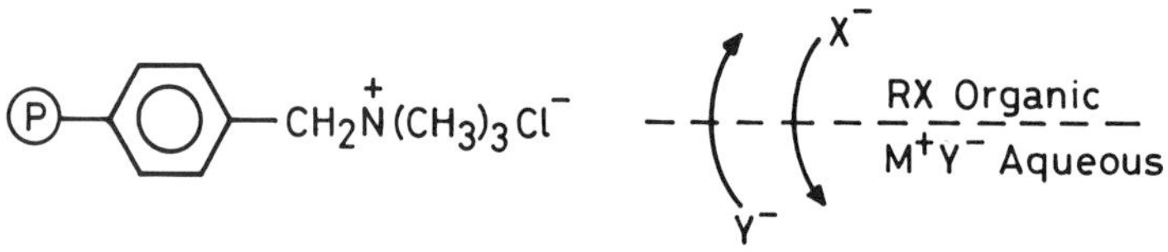

The transportation of anions back and forth across the boundary can then occur via local backbone and side-chain motion of the catalyst support. According to this model, increasing the length, and hence the flexibility, of the linkage between the catalyst and the polymeric backbone by introducing a "spacer arm" would improve the efficiency of catalysis, and this has been verified in many instances.

Polymer-Supported Transition Metal Catalysts

Numerous soluble transition metal complexes have been described that are useful catalysts for hydrogenation, isomerization, hydroformylation, etc. In contrast to heterogeneous catalysts that are widely used in industry, the homogeneous catalysts have greater accessibility to the reactants and often have better reproducibility in the catalyzed reactions. Even with these advantages, homogeneous catalysts have not been preferred because their recovery by separation from reactants and products is difficult and results in a considerable loss of the expensive catalyst (metal). In addition, in many cases the material is not reusable. To circumvent these problems, an intermediate system has been developed that involves the preparation of a homogeneous catalyst in a heterogeneous form, which is best describd as a *heterogenized-homogeneous* catalyst.

Heterogenization of a homogeneous catalyst might be done by attaching homogeneous catalysts to organic polymers (or inorganic supports) in a way such that the ligand sphere of the metal is essentially unchanged, and the attached complex is bathed by solvent and reactants. Thus the resulting catalyst could function mechanistically as if it were in solution, but it would operate as a separate immobile phase. To emphasize this dual behavior, the term *hybrid-phase catalyst* has been applied to this type of catalyst.

Preparation of Catalysts

Most of the monomeric transition metal catalysts contain triphenylphosphine ligands (XXXIV), and the polymer-supported analogs of these catalysts have generally been prepared by replacing one or more of these ligands with phosphine residues of phosphinated polystyrene resin (XXXV). Thus, when a suspension of the resin in toluene is stirred with homogeneous hydrogenation catalyst (Ph$_3$P)$_3$RhCl (Wilkinson's catalyst), the polymer-supported hydrogenation catalyst (XXXVI) is obtained.

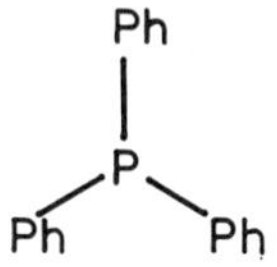

(XXXIV) 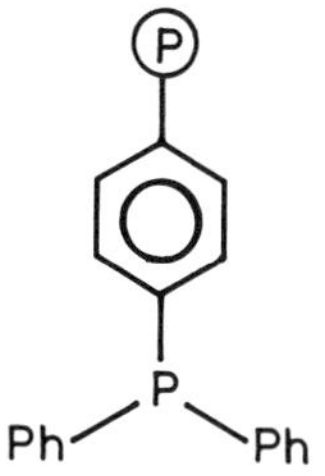 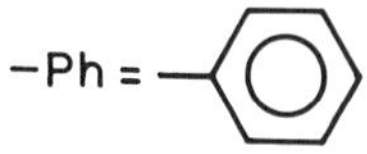 (XXXV)

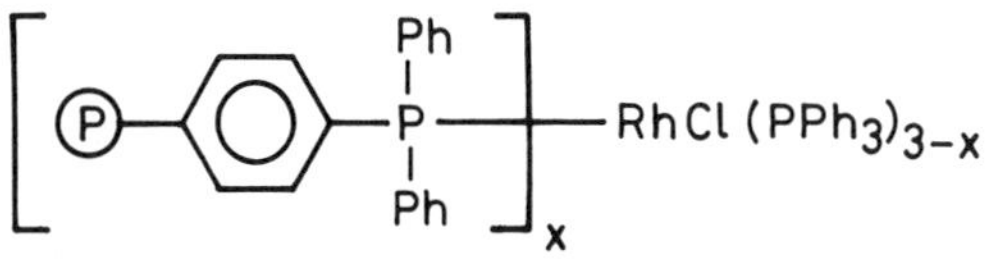

(XXXVI)

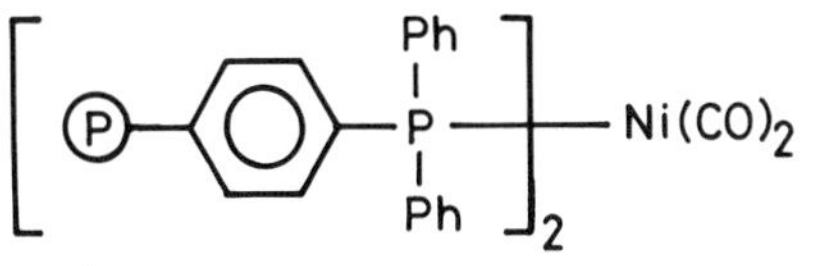

(XXXVII)

Reaction of the resin (XXXV) with homogeneous cyclooligomerization catalyst (Ph$_3$P)$_2$Ni(CO)$_2$ gives the polymer-supported cyclooligomerization catalyst (XXXVII). The extremely active hydroformylation catalyst (PPh$_3$)$_3$RhH(CO) was similarly anchored to give a bound catalyst (XXXVIII).

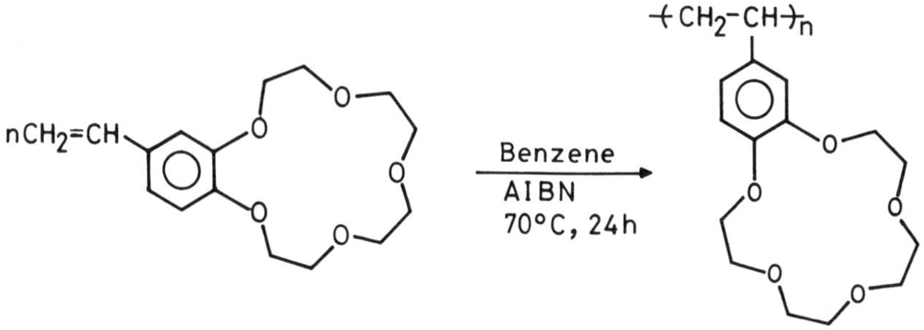

FIG. 5.43 Synthesis of crown polyethers by polymerization of the corresponding vinyl monomer (4'-vinylmonobenzo-15-crown-5).

Polyenes can be partially hydrogenated to their corresponding monoenes with remarkably high selectivity using the homogeneous catalyst $(PPh_3)_2RuCl_2(CO)_2$ (Fahey's catalyst). This catalyst was anchored to the phosphinated resin (XXXV) to give a bound catalyst (XXXIX).

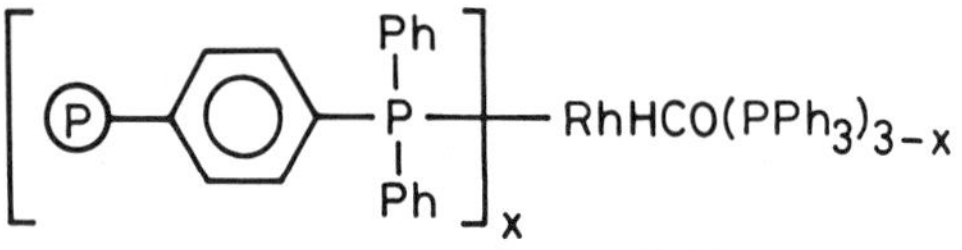

(XXXVIII)

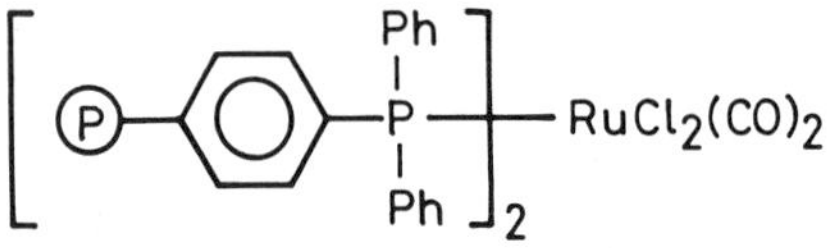

(XXXIX)

One of the advantages of incorporating metals into polymeric supports is exemplified by a set of titanocene complexes. Titanocene dichloride, when reduced by a variety of reducing agents such as sodium naphthalide, alkyllithium compounds, alkylaluminium compounds, or magnesium, produces reactive titanocene species having considerable catalytic utility in the hydrogenation of olefins and in the reduction of nitrogen to give ammonia. However, soluble titanocene species are known to rapidly form dimers or higher oligomers that have no catalytic activity. But attachment of titanocene to a 20% cross-linked styrene-divinyl benzene copolymer (see Fig. 5.44) produced an insoluble catalyst that is

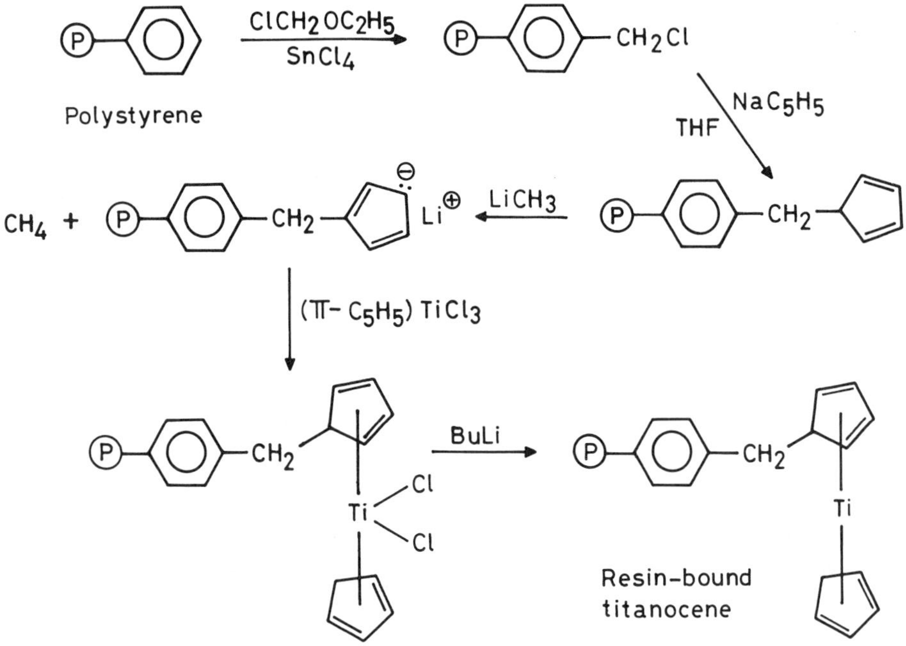

FIG. 5.44 Scheme of reactions for the preparaton of polystyrene-bound titanocene.

60 times more active than the homogeneous catalysts. This is an excellent example of the ability of a polymer to stabilize reactive coordinatively unsaturated metallic sites.

Types of Reactions

Tris(triphenyl phosphine) chlororhodium (Wilkinson's catalyst) has been the most widely studied homogeneous hydrogenation catalyst. Analogous polymeric catalysts, such as (XXXVI), have been prepared for use in hydrogenation. A variety of alkenes, cycloalkenes, dienes, and alkynes have been reduced by such polymer-bound catalysts, indicating the wide scope of their use.

Polymer supports may also offer other advantages. The polymer matrix can operate to increase the selectivity of a catalyst to molecules of different sizes (*size selectivity*). As an example, the rate of hydrogenation with a polymeric catalyst (XXXVI) has been found to be 6.25 times greater for cyclohexene than that for cyclooctene. This substrate selectivity has been attributed to the size dependence of the substrate entering the pores of the resin beads and indicates that the reaction mainly takes place inside the resin beads.

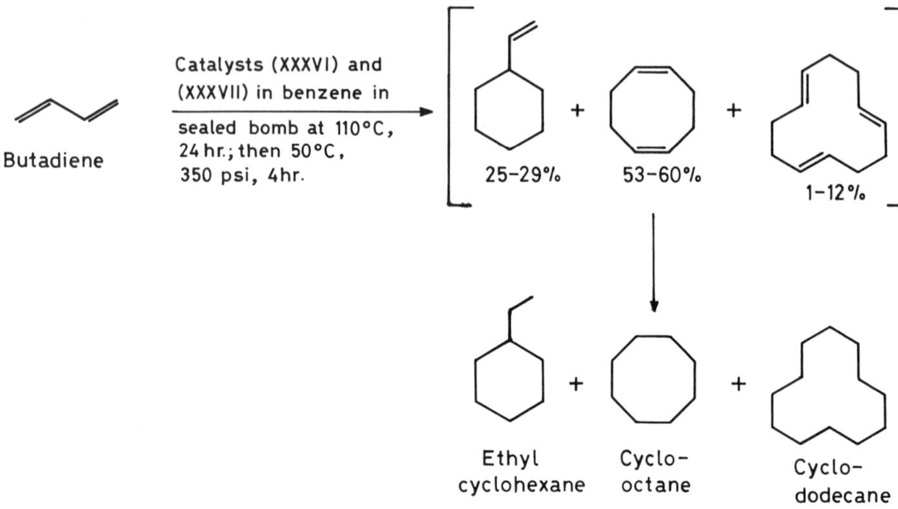

FIG. 5.45 Cyclooligomerization of butadiene followed by quantitative hydrogenation of cyclooligomers is effected with polymer-supported catalysts in the same reaction vessel.

Polymers may also exert a *polar selectivity* in catalytic reactions. The rate of cyclohexene reduction on catalyst (XXXVI), for example, was found to increase by a factor of 2.4 when the solvent is changed from benzene to 1:1 benzene-ethanol. This is attributed to the formation of a polar gradient in the pores, which favors a higher concentration of the olefin within the resin when ethanol is used.

One of the most interesting developments in the use of polymer-supported catalysts is their application to sequential multistep catalytic reactions. While the addition of more than one homogeneous catalysts and each catalysts' particular ligand to the same reaction vessel is often not permissible due to possible interactions, two or more catalysts may, however, be used in the same reaction vessel if they are anchored to the same resin or to different resin supports. Several successful examples of this approach have been documented.

Sequential butadiene cyclooligomerization and the hydrogenation of those cyclooligomers were carried out in the same reaction vessel using a supported nickel carbonyl cyclooligomerization catalyst (XXXVII) and a supported hydrogenation catalyst (XXXVI), either on the same polymer or attached to different sets of resin beads (see Fig. 5.45).

Polymer-supported catalysts have been used extensively for hydroformylation reactions. For example, catalyst (XXXVIII) readily catalyzed the hydroformylation of 1-pentene to high yields of hexan-1-al and 2-formylpentane:

$$\text{CH}_2\text{=CH-CH}_2\text{-CH}_3 \xrightarrow[\text{6.9 MPa, 62°C}]{\substack{\text{Catalyst}\\ \text{H}_2 : \text{CO} = 1:1}} \text{CH}_3\text{CH}_2\text{CH}_2\text{CH}_2\text{CHO} \underset{\sim 68\,\%}{} + \underset{\sim 15\,\%}{\overset{\text{CHO}}{\text{CH}_3\text{CH}_2\text{CH(CH}_3)}}$$

The polymeric catalyst could be repeatedly recycled and even exposed to water without drastic loss of activity.

Often the polymer-supported catalysts are somewhat less active than their soluble counterparts, probably because the substrates need to diffuse into the resins. Polymer-supported catalysts are not always less active, however. For example, the supported titanocene catalyst (see Fig. 5.44) is about 60 times more active as a hydrogenation catalyst than the nonsupported analog, probably because the active groups are well separated (*site isolation*) on a 20% cross-linked resin, and dimerization leading to inactive products occurs less readily than in solution.

Among the reactions that can be catalyzed by polymer-bound transition metal complexes (or the corresponding low-molecular weight homogeneous catalysts) are oxidative polymerization of phenols, acetoxylation, hydrogen-deuterium exchange in alcohols, acetylene cyclotrimerization and tetramerization, oligomerization and polymerization, ketone synthesis, and olefin metathesis.

REFERENCES

1. P. E. Cassidy, *Thermally Stable Polymers*, Marcel Dekker, New York (1980).
2. J. P. Critchley, High-temperature resistant heterocyclic-aromatic polymers. Synthesis, properties and applications, *Angew. Makromol. Chem.*, *109/110*: 41 (1982).
3. M. Wolf, Developments in applications technology for polyaromatics, *Kunststoffe*, 77(6): 23 (June, 1987).
4. A. Wirsen, *Electroactive Polymer Materials*, Technomic, Lancaster, Pennsylvania (1987).
5. R. B. Seymour (ed.), *Conductive Polymers*, Plenum Press, New York (1981).
6. J. M. Margolis (ed.), *Conductive Polymers and Plastics*, Chapman and Hall, New York (1989).
7. J. Matthan, A. Uusimaki, H. Trovela, and S. Leppavuori, Past and future impact of electroactive polymers on the electronics sector, *Makromol. Chem.*, *Makromol. Symp.*, 22:161 (1988).
8. W. A. Little, Possibility of synthesizing an organic superconductor, *Phys. Rev.*, *134*: 1416 (15 June 1964).
9. T. Ito, H. Shirakawa, and S. Ikeda, Simultaneous polymerization and formation of polyacetylene film on the surface of concentrated soluble Ziegler-type catalyst solution, *J. Polym. Sci.*, *12*: 11 (1974).
10. W. D. Gill, T. C. Clarke, and G. B. Streat, Electrically conducting polymers, *Appl. Phys. Commun.*, 2(4): 211 (1982).

11. M. Aldissi, *Inherently Conducting Polymers*, Noyes Data Corp., Park Ridge, New Jersey (1989).

12. C. K. Chiang, S. C. Gau, C. R. Fincher, Jr., Y. W. Park, A. G. Macdiarmid, and A. J. Heeger, Polyacetylene (CH)$_x$: *n*-type and *p*-type doping and compensation, *Appl. Phys. Lett.* *33*(1): 18 (1 July 1978).

13. J. M. Bowden and S. R. Turner (eds.), Electronics and photonic applications of polymers, *Advances in Chemistry Series*, vol. 218, American Chemical Society, Washington, D.C. (1988).

14. A. F. Garito, C. C. Teng, K. Y. Wong, and O. Zammani Khamiri, Molecular optics: nonlinear optical processes in organic and polymer crystals, *Mol. Cryst. Liq. Cryst.*, *106*(3–4): 219 (1984).

15. P. N. Prasad and D. R. Ulrich (eds.), *Nonlinear Optical and Electroactive Polymers*, Plenum, New York (1988).

16. T. T. Wang, J. M. Herbert, and A. Glass (eds.), *The Applications of Ferroelectric Polymers*, Blackie, Glasgow (1988).

17. M. B. Armand, J. M. Chabagno, and M. Duclot, Abstract 6, Second International Conference on Solid Electrolytes, St. Andrews (20–22 Sept 1978).

18. P. M. Blonsky, D. F. Shriver, P. Austin, and H. R. Allcock, Polyphosphazene solid electrolytes, *J. Am. Chem. Soc.*, *106*: 6854 (1984).

19. K. M. Abraham, M. Alamgir, and R. K. Reynolds, Polyphosphazene-poly(olefinic oxide) mixed polymer electrolytes, *J. Electrochem. Soc.*, *136*(12): 3576 (1989).

20. H. Wu, H. Shy, and H. Ko, Application of chemically synthesized polypyrrole for a rechargeable lithium battery, *J. Power Sources*, *27*(1): 59 (1989).

21. A. Blumstein (ed.), *Polymeric Liquid Crystals*, Proceedings of the Second Symposium at the American Chemical Society Meeting, Washington D.C. (28–31 Aug 1983).

22. A. V. Volokhina, Yu. K. Godovskii, and G. I. Kudryavtsev, *Liquid Crystal Polymers*, Khimiya, Moscow (1988).

23. W. S. De Forest, *Photoresist: Materials and Processes*, McGraw-Hill, New York (1975).

24. L. F. Thompson and R. E. Kerwin, Polymer resist systems for photo and electron lithography, *Ann. Rev. Mat. Sci.*, 6:267 (1976).

25. C. G. Roffery, *Photopolymerization of Surface Coatings*, John Wiley, New York (1982).

26. N. S. Allen (ed.), *Photopolymerization and Photoimaging Science and Technology*, Elsevier, Barking, Essex, England (1989).

27. E. Reichmanis and L. F. Thompson, Polymer materials for microlithography, *Chem. Rev.*, *89*(6): 1273 (1989).

28. M. Hatzakis, K. J. Stewart, J. M. Shaw, and S. A. Rishton, New high resolution and high sensitivity deep UV, X-ray and electron beam resists, *Microelect. Eng. 11* (1–4): 487 (1990).

29. L. Holliday (ed.), *Ionic Polymers*, Halsted Press, New York (1975).

30. A. Eisenberg and M. King, *Ion Containing Polymers*, Academic Press, New York (1977).

31. R. Longworth, *Developments in Ionic Polymers I.*, Applied Science Publishers, London (1983).

32. D. S. Flett (ed.), *Ion-Exchange Membranes*, Ellis Horwood (1983).
33. S. Sourirajan and J. P. Agarwal, Reverse osmosis, *Ind. Eng. Chem.*, *61*: 62 (1969).
34. R. Kunin, *Ion-Exchange Resins*, 2nd ed., R. E. Krieger, Melbourne, Florida (1972).
35. F. Osawa, *Polyelectrolytes*, Marcel Dekker, New York (1970).
36. P. Hodge and D. C. Sherrington (eds.), *Polymer Supported Reactions in Organic Synthesis*, John Wiley, Chichester, England (1980).
37. N. K. Mathur, C. K. Narang and R. E. Williams, *Polymers as Aids in Organic Chemistry*, Academic Press, New York (1980).
38. C. Birr, *Aspects of the Merrifield Peptide Synthesis*, Springer Verlag, New York (1978).
39. W. T. Ford (ed.), *Polymeric Reagents and Catalysts*, ACS Symposium Series No. 308, American Chemical Society, Washington D.C. (1986).
40. Y. Chauvin, D. Commereuc and F. Dewans, Polymer supported catalysts, *Prog. Polym. Sci.*, *5*: 95 (1977).
41. J. Kahovec and F. Svec (eds.), *Polymer Supported Organic Reagents and Catalysts*, Elsevier, Amsterdam (1988).

6

Trends in Polymer Applications

INTRODUCTION

While the plastics industry has witnessed a spectacular growth over the last four decades, the acceleration in consumption rates of plastics has taken place in several phases since World War II. Much of the use of plastics just after the war was as a cheap substitute for traditional materials (in other cases, the materials was used for its novelty value), and in many instances the result was detrimental to the industry. It required several years of painstaking work by the technical service departments of major plastics manufacturers before confidence was regained in the use of plastics. Even today the public image of plastics is not entirely positive, and the important role of plastics in contributing to raising the standard of living and quality of life is not fully appreciated.

In some areas of applications, plastics materials have been long established. A well-known example is in the electrical industries where the excellent combination of properties such as insulation characteristics, toughness, durability, and, where desired, flame-retardation capacity have led to increasing acceptance of plastics for plugs, sockets, wire and cable insulation, and for housing electrical and electronic equipment. This has lead to the increased use of the more general-purpose plastics. Plastics are also finding use in sophisticated techniques. For example, the photoconductive behavior of poly(vinyl carbazole) is made use of in photo-copying equipment and in the preparation of holographs, while the notable piezoelectric and pyro-electric properties of poly(vinylidene fluoride) are utilized in transducers, loudspeakers, and detectors.

As the plastics industry has matured, it has been realized that it is better to emphasize those applications where plastics are preferable to traditional materials. In the building industry, this approach has led to a number of uses of plastics that include piping, guttering and conduit, flooring, insulation, wall cladding, damp course layers, and window frames. In the domestic and commercial furniture industry, which forms another important outlet for plastics, uses include stacking chairs, armchair body shells, foam upholstery, cupboard drawers, and decorative laminates.

Plastics have found wide acceptance as packaging materials. Substituting plastics bottles for glass containers have been particularly appreciated in the bathroom where breakage of glass has often been the cause of many serious accidents. Small containers for medicines are also widely made from plastics. The ability of many plastics to withstand corrosive chemicals and their lighter weight compared with glass, which reduces the energy required for transportation, have been of benefit to the chemical and related industries. The wide use of plastics film for wrapping, bags, and sacks is well known. However, the quantity of plastics used in this area results in a large volume of waste plastics film, which is left around as an all-too-durable eyesore and a litter problem and has stimulated research into biodegradable plastics.

For many years the main uses of plastics in the automotive industry were associated with car electrical equipment, such as batteries, plugs, switches, and flex and distributor caps, as well as, with interior body trim including light fittings and seating upholstery. Subsequently there has been increased use in under-the-bonnet (under-the-hood) applications, such as radiator fans, drain plugs, gasoline tubing, and coolant water reservoirs. In recent years, the need for lighter cars for greater fuel economy and the emphasis on increased occupant safety have led to a substantially increased use of plastics materials for bumpers, radiator grills, and fascia assemblies. As a result, the automotive industry is now a major user of plastics, with the weight of plastics being used per car increasing every year.

Plastics are also finding increased use in the construction of vehicles for both water and air transport. Their light weight, noncorrosive nature, antimagnetic characteristics, ease of maintenance, and the economy in manufacture of glass-fiber reinforced plastics have led to their wide use for making boats and hulls. Plastics are particularly useful in aircraft, mainly due to their low density.

Plastics have been widely accepted in industrial equipment development as materials for many critical components and for accessories where such features as toughness, abrasion resistance, corrosion resistance, electrical insulation capability, nonstick properties, and transparency are of importance. For example, PVC, polyacetal, and PTFE are used for pipes, pumps, and valves because of their excellent corrosion resistance. Nylons are used for such diverse applications as mine conveyor belts and main drive gears for knitting machines and paper making equipment because of their excellent abrasion resistance, toughness, and low coefficient of friction.

In agriculture and horticulture, the most widely known use of plastics is in film form. But plastics also find use for automotive watering equipment, water piping containers, and potato chitting trays, to name but a few uses.

In the medical sector, plastics uses range from medical packaging and disposables (e.g., bandages and injection syringes) to nontoxic sterilizable items (such as tubing and catheters) to spare-part surgery (such as hip joints and heart valves). A wide-range of polymeric materials are found in these applications though medical grades of PVC and polystyrene are the most commonly used. Engineering resins, however, are finding an increasing number of uses in this area.

While natural and synthetic fibers have remained the major materials for clothing, plastics have also found increasing use in this area. Footwear has provided a major outlet for plastics as they are used not only in soles and uppers, but increasingly in an all-plastics molded shoe that is in high demand in underdeveloped regions of the world. In rainwear, plastics and rubbers continue to be used for waterproof lining and in the manufacture of all-plastics packable raincoat. In cold-weather apparel, polyurethane foam has found increasing use as an insulation layer and for giving ''body'' to clothing.

The photographic industry was one of the earliest users of plastics, which is for photographic film. There has also been widespread use of plastics in darkroom equipment. More recently, plastics have found increasing acceptance in cameras, both inexpensive and expensive, as it is recognized that well-designed camera bodies made from properly chosen plastics materials are more able to withstand rough usage and resist denting than metal camera bodies. In the audio field, while the use of plastics for tapes and discs is well established, plastics are now almost standard for the housings of reproduction equipment. Sports goods manufacturers make increasing use of glass- and even carbon-fiber reinforced plastics for such diverse articles as fishing rods, racquets, canoes, and so on.

While the above paragraphs indicate some of the major uses of plastics in a number of areas, the more recent trends in plastics applications in these areas form the subject of subsequent discussion in this chapter.

POLYMERS IN PACKAGING

The packaging industry is the most important user of plastics both in Europe and the United States, and it is expected to increase with an annual growth rate of 3–5%. Major innovations for this decade are expected to be retortable plastics as a replacement for the can, hot-fillable rivals to glass, and high-barrier-coated or laminated bottles for carbonated drinks.

Retorting

Retorting is the most widely used method for processing food. In this method, the raw or partially cooked food material is placed in a container with sufficient liquid to fill it. Prior to sealing the container, it is heated under pressure to at least 121°C to kill all bacteria.

American Can's Omni retortable plastic can marks a major development in food packaging. The five layer can, currently being used by Hormel, Del Monte, and Campbell Soup Company, consists of polyolefin structural layers, adhesive layers incorporating desiccants, and ethylene vinyl alcohol (EVOH) polymer as the barrier layer.

A newer development in this area is the production of retortable, multi-layer, blow-molded containers that incorporate polycarbonate as the tough outer layer, with EVOH or poly(vinylidene chloride) (PVDC) as the barrier, and polypropylene (PP) as the food contact layer. The advantages of polycarbonate are its light weight, improved heat stability, and good optics compared to the polyolefins that are normally used.

Asceptic Packaging

For asceptic packaging, both the food product and the container are sterilized separately and then brought together in a sterile environment. Both low-acid foods, like yogurt and soup, and high-acid foods, such a juices and drinks, are packaged by this method.

Combibloc cartons for Bowater PK and Tetra-Brik from Tetra Pak are used for packaging high-acid pure fruit juices and give a shelf life of six months. These are multi-layer structures with a paperboard base and polyolefin and aluminum foil for barrier layers. The cartons are also used in the range of adult nutritional products for hospitals. Tetra Pak's range of asceptic packaging for pasteurized products includes Tetra-Tals, Tetra-Top, and Tetra-King, all of which feature improved opening and closing facilities.

The Connofast thermoform-label-fill-seal system of Continental Can Company can be used to produce a microwaveable asceptic container for low-acid products. The container is a coextruded tray or cup combined with a foil-free lid. LDPE provides the food contact surface while the barrier layer is either PVDC or EVOH.

Hot-Filling

In the hot-filling technique, which is being used as a cheap alternative to asceptic packaging, food is cooked or pasteurized by heating at a high enough temperature to kill the bacteria and then placed into the container at a temperature above 80°C to kill all the bacteria in the container. With this method flexible materials can be used to replace metal cans at a reduced cost.

In the Q-box system introduced by Mead packaging for hot-filling high-acid juices, a web made of LDPE, paperboard, ionomer, foil, polyester, and LDPE is used. Products are pasteurized at 90°C before filling. Also used for high-acid products, is an oriented polyester bottle from Monsanto Company.

Hot-fill pouches are used for both low- and high-acid applications. Heinz Company's form-fill seal Pouch Pack, which is claimed to give a shelf life of one year, has a web of 3-mil LLDPE, ethylene-acrylic acid copolymer, 35-gauge foil, ethylene-acrylic acid copolymer, and biaxially oriented nylon, from inside outward. The Pouch Pack system from Du Pont offers pouches of sizes 4 oz to 1.5 L to fill a range of high- and low-acid food applications at various temperatures, including fruit toppings at 80°C.

Controlled-Atmosphere Packaging

By the turn of the century, controlled-atmosphere packaging (CAP) is expected to consume more film and sheet than all other types of plastic-based barrier structures combined. Most CAP used until recently PVDC and EVOH polymer-based oxygen barrier films to package red meat and baked goods. These films still allow the food to continue to respire after being packed, thus altering the balance of CAP gases introduced at the time of packing. However, selectively permeable films can now be used to control the amount of carbon dioxide, oxygen, and moisture in contact with the produce after it has been packed. In this way, CAP can more than double the shelf life of certain fruits and vegetables.

One such film that has been successfully used to package tomatoes is DRG's Ventaflex selectively permeable film. It is a polyester and HDPE strip-laminated film that has a laminating adhesive applied in diagonal strips. The film allows the carbon dioxide produced by the fruit to pass through the HDPE layer into the atmosphere while air penetration is limited by the outer polyester film. Besides tomatoes, other products commercially packaged in Ventaflex film include cheese, coffee beans, and bean-sprouts.

High-Barrier Films

Resins with high-barrier properties are being developed by a number of companies in response to a big increase in the demand for high-barrier films. Selar PA (Du Pont) amorphous nylon resins offer a good oxygen and flavor/odor barrier, high clarity, good mechanical properties at elevated temperatures, and good processability. The films are used for meat wrap, snack food bags, and cereal box liners.

Selar PT amorphous poly(ethylene terephthalate) (PET) blends combine the inherent barrier properties of PET with the ability to be processed on conventional equipment without orientation or crystallization steps. Besides food packaging grades, other industrial grades of Selar PT are available that rival PP or HDPE for paint pails and other containers that require a hydrocarbon or solvent barrier.

Dow's Saran wrap, which is a copolymer of vinylidene chloride and vinyl chloride, is a useful packaging material that possesses exceptional clarity, toughness, and impermeability to water and gases. Also available from Dow is a range of Saran resin and films with improved barrier properties. For example, Saran HB monolayer lamination films are claimed to exhibit an oxygen barrier 10 times greater than the conventional Saran wrap films.

In the snack food area, metallized plastics are being increasingly used. Metallized oriented PP is particularly popular for this application.

Plastic Bottles

Plastic bottles for carbonated beverages are mostly made of PET as it offers high tensile strength and good barrier properties. Blow-molded PET bottles account for 33% of the total carbonated soft drink market. For packaging beer, which is oxygen sensitive, PET bottles are coated with PVDC, while the base cups that support the bottle are usually made of HDPE.

However, since PET barrier properties decrease with a decrease in size of the bottle, smaller size poses a problem. Two possible ways to solve this problem are to employ coated PET or to use coextrusion technology. As an example of the latter case, PET has been successfully injected with nylon to form a PET-nylon-PET construction, with nylon improving the barrier properties.

In addition to carbonated drinks, PET bottles are also being used in other markets, which include toiletries, cosmetic ranges, spirit miniatures, and PET aerosols for cosmetic and pharmaceutical applications.

Biaxially stretched oriented PVC (OPVC) pressure bottles have been used as an alternative to PET for packaging some carbonated drinks. In this application, the level of carbon dioxide inside the bottle and the temperature at which the bottles are kept are important. Producing OPVC by extrusion stretch blow molding results in greater clarity, increases impact resistance, and reduces cost by up to 14% compared to ordinary PVC. Fruit, squashes, and edible oils are commonly packed in PVC bottles.

American Can pioneered the concept of multilayer squeezable plastic bottles with the introduction of its Gamma bottle. Another such product is Continental Plastic Containers' Lamicon bottle. The bottles are being used for food products such as ketchup, jelly, mayonnaise, and barbecue sauce. In a new development, polycarbonate has been used by Holmicon Plast as the outer layer, rather than polyolefin, which is normally used. Polycarbonate gives the bottle a more shiny appearance.

Chemical Containers

Household and industrial chemical containers are increasingly being made of plastics. HDPE has almost completely replaced glass for packaging household liquids. However, for solvents and flammable liquids, a plastic is required with higher barrier properties than conventional HDPE.

Three methods can be employed to obtain higher barrier properties. These are fluorination, coextrusion, and laminar blends. Fluorination of HDPE bottles is used most commonly for packaging solvent-based products and agricultural chemicals in particular. One process consists of treating the surface of HDPE bottles with a mixture of fluorine and nitrogen gas to produce a fluorocarbon barrier layer. The process developed by Air Products Limited, creates an internal barrier layer during the blow-molding process, while Bettix Limited's fluorination process, Fluoro-seal, is post-molded and particularly suited to short runs. The fluorination process suffers from the disadvantage that it is potentially hazardous.

Coextrusion to form an effective barrier layer has been used by Plysu Containers Limited and Reed Plastic Containers, both producing tough solvent-resistant containers. The five-layer construction used by Plysu has HDPE both on the inside and outside, and a central barrier layer with adhesive layers on either side to allow lamination to HDPE. The Reedpac range of blow-molded containers developed by Reed Plastic Containers have polyethylene, polypropylene, polycarbonate, or PET combined with barrier resins such as polyamide, PVDC, and EVOH polymer.

The laminated resin blend technology using Du Pont's barrier resins, trade-named Selar, has been employed to produce barrier bottles. Blowmocan Limited has used the Selar RB polyamide barrier resin in a one-step process that blends with polyethylene to form overlapping discontinuous plates within the wall system. The container can be used for many cleaning chemicals and agricultural products.

Metal Box has developed a copolymer polypropylene container, Polycan, which has an anti-static additive and is suitable for water-based paints and coating products. Another development is the first bag-in-box paints from Ashwell Paints. A smooth flow is obtained in this system by the use of an open bung and screw top rather than a tap.

Dual Ovenables

Dual ovenable containers must be able to perform under freezable conditions as well as in both microwave and conventional ovens. Materials commonly used for dual ovenables are temperature-resistant plastics such as crystallized PET (CPET), thermoset polyesters, and polycarbonate. CPET, the dominant material, has a temperature stability up to 232°C.

Attention has been paid to the possible use of multilayer coextrusion of engineering resins for dual ovenable containers, as multilayer packaging could increase shelf stability and thus eliminate the need for freezing or refrigeration. One multilayer combination already in commercial use is a sheet of Lexan polycarbonate sandwiched between layers of Ultem polyetherimide. Thermoformed trays of this material are reported to have better break resistance at lower temperatures and higher stiffness at temperatures up to 232°C, as compared to CPET. They are also inert, compatible with polyester lidding materials, and have lower oxygen permeability than PET materials.

Most plastics used in dual ovenables are two to six times more expensive than the aluminum containers they replace in frozen food applications. Less expensive plastic containers are made of molded fiber and polyester and paper laminates. However, due to the popularity of the microwave oven, in many cases microwaveable-only containers, instead of dual-ovenables, are used. For example, Campbell uses a polypropylene microwaveable-only container for its Soup du Jour, thereby reducing the material costs considerably.

With the increase in the market of dual-ovenable and microwaveable trays there has been an increase in the demand for special lidding material, as the lid has to provide an oxygen barrier and a 100% hermetic seal to prevent food spoilage, besides being easy to remove. Grades of Du Pont's Mylar OL polyester lidding can be used to seal CPET, thermoset polyesters, and paper-board trays.

Closures

Tamper-evident closures are used to prevent customers from opening containers before they buy them. Several tamper-evident features were originally introduced including neck bands, tamper-rings, and shrink wrapping. More recently, a number of new tamper-evident and child-resistant closures (CRC) have been introduced.

Clic-loc is a CRC available from United Closures and Plastics. A specialist version is provided with a thermoset inner liner for chemical applications.

A double tamper-evident cap sealer has been introduced by Paklink. It has a ring style enclosure incorporating an inner foil seal which, once heat sealed, provides a hermetic barrier suitable for products in the food, drug, and dairy industries.

A number of CRCs, including the Childgard and Safecap, are available from Johnson and Jorgensen. The Childgard is a one-piece jigger cap style CRC used for pharmaceuticals, while Safecap is a two-piece palm and turn style CRC used for a variety of pharmaceutical and household products.

POLYMERS IN BUILDING AND CONSTRUCTION

After the packaging industry, the building and construction industry is next in order of importance as a user of plastics. It has proved to be the fastest growing market for plastics use in the last ten years with increased sales in roofing, flooring, pipe and conduit, windows and doors, plumbing fixtures, and insulators. With its high potential for replacement of nonplastic materials, this market is expected to grow more rapidly, equalling packaging by the end of the century. One of the relatively new areas of building applications has been in sports surfaces, which are discussed in a later section on "Polymers in Sport."

Roofing

The most dramatic change in commercial roofing has been the recent shift from bitumen-based built-up roofing (BUR) to single-ply membrane roofing, particularly in the United States. Pioneered by Carlisle Corporation with its Sureseal ethylene propylene diene copolymer (EPDM) membrane, single-ply roofing became the fastest growing product in U.S. rubber industry history, presently accounting for over 50% of America's nonresidential flat roofing market. The ease of installation and durability of single-ply roofing together with its resistance to weathering, chemicals, and ozone make it a cost-effective roofing system. A distinctive feature of single-ply roofing is that it cures with weathering and increases its elasticity as it ages. Fire-retardant grades of EPDM are also being developed.

Although the material most used for single-ply membranes is EPDM, in competition with it are several other materials—Du Pont's Hypalon (a chlorosulfonated polyethylene), its Neoprene products, and modified PVC. It is believed that future growth in the roofing market will depend on products, such as Hypalon, that are easy to handle and have good ozone resistance, and on PVCs modified with polymeric plasticizers.

Window Frames

The use of PVC window frames is increasing at a steady pace. In Germany nearly 43% of all windows and 70% of all replacement windows are in PVC.

PVC formulations are generally ethylene vinyl acetate (EVA) polymer-based, chlorinated polyethylene (CPE)-based, or acrylic-based. The majority of new grades are acrylic-based and are claimed to exhibit greater impact strength than the other two types.

Grades of PVC are being tailored to meet different requirements of impact strength and weathering resistance for different climates. Solvay, for example, has developed an unmodified material in granular form that is suitable for use in European countries, an acrylic-modified powder for moderate and Mediterranean climates, lead-stabilized grades for extremely hot climates, and an unmodified grade with very high weathering resistance. Other compounds with high weathering resistance include Dow Chemical's Rovel, General Electric's Geloy, and Hul's Vestolit H1 powder.

The majority of window profiles, until fairly recently, were white. Coloring techniques were developed to produce colored window profiles, thus widening the market. Hoechst, for example, uses in-line bonding of acrylic foil to PVC profiles, allowing a wide range of colors and designs including wood-grain finishes. Both production and precision in window extrusion has been improved with the development of new screw designs, better dies, and microprocessor control of production parameters.

Pipes

Plastic pipe demand has been growing at a steady rate since 1967. In the United States its share of the total pipe market is expected to increase to more than 60% by the turn of the century. Factors contributing to the accelerating use of plastic pipes include their low cost relative to conventional materials, excellent corrosion resistance, ease of installation, and wider building acceptance.

The plastic pipe market is dominated by PVC pipes, which hold 80% of the market. Polyethylene, principally HDPE, is used for about 12% of pipe production. Following polyethylene are other higher priced, specialty pipes made from ABS terpolymer, chlorinated PVC, polypropylene, polybutylene, and fiber-reinforced epoxy and polyester.

Insulation

The major uses of insulation in the construction industry are in roofing, residential sheathing, and walls. In these applications, polymeric foams offer advantages over traditional insulation such as glass fiber, and these include higher insulating value per inch of thickness and lower costs. The use of polymeric foam for insulation increased markedly since the 1970s due to increased awareness of the need for energy conservation.

Foams are available as rigid sheets or slabs (which are used in the majority of roofing systems), as beads and granules (used in cavity wall insulation), and also as spray and pour-in applications. The market is dominated by polyurethane (PU) foams, in particular polyisocyanurate products, expanded polystyrene (PS), and extruded polystyrene.

Polystyrene foam (PS) holds much of the sheathing market. In masonry and brick walls, PS foams are mainly used because of their better moisture resistance. In cavity walls, loose fill PS is used, while exterior wall applications, which have no limitation on space, use low-cost expanded PS.

Polyurethane/polyisocyanurate products have higher insulation value and good flammability ratings and are expected to continue to be the leading products in plastic foam market as sheets and slabs.

A shift toward single-ply roofing as compared to built-up roof systems has an important influence on the type of foam being utilized. Thus the lower cost of expanded PS has promoted its use in preference to PU foam and extruded PS in single-ply applications. This is facilitated by the fact that the problems of damage to expanded PS foam from hot pitch when used in built-up roof systems are not encountered in single-ply systems.

Polymer Concrete

Polymers are added to concrete to form strong, durable, fast-setting, and corrosion-resistant materials. Polymer concretes were initially used to patch up concrete potholes. For such purposes methyl methacrylate is mixed with Portland cement. This mix cures to its full strength in two hours compared to the 28 hours required for Portland cement, thus reducing the time necessary for repair work.

Besides fast curing, polymer concretes offer several other advantages. Thus, as the polymer is virtually impermeable to moisture, there is very little cracking of the concrete caused by freezing and expansion of moisture within the cured mix. Polymer concretes also resist road salts and other agents that eat away at the reinforcing steel within the concrete. In Germany, these properties have led to the use of polymer concrete in water treatment and sewage plants.

Although polymer concretes do cost more initially, they can be put down in thinner layers than Portland cement concrete to give the same strength at lower volume, thus allowing a weight reduction of up to 50%. Polymer concrete overlays are normally ⅜–½ in. thick, while Portland concretes average 2½–3 in. thick. For bridge overlays, large stretches of epoxy and polyester resins are combined with Portland cement. These mixtures have a stress failure point of 10,000 psi (69 MPa) compared to 2,000–5,000 psi (13.8–34.5 MPa) for Portland cement.

A more recent application of polymer concretes is in pre-cast structural pieces including sinks, bathtubs, and manhole covers. An example is the Quazite polymer concrete of Quazite Corporation. It is a mixture of various mineral aggregates with selected binders including acrylics, epoxies, polyester, and glass. The material is claimed to possess the formability of glass fiber and the hypermeability of glass and to exhibit twice the flexural strength of graphite and dielectric properties superior to porcelain.

PLASTICS IN AUTOMOTIVE APPLICATIONS

The use of plastics in automotive applications is increasing at a significant rate. In Europe and the United States, plastics currently make up between 7% and 10% of the mass of an automobile. The original impetus for using plastics in automobiles stemmed from the desire to reduce overall weight in order to achieve higher mileage. However, the impetus is now divided more or less equally between weight reduction and other considerations such as performance, parts consolidation, and processing economics. For example, by replacing a part assembly involving a number of individually cast components by a single molded plastic part, much labor can be saved and, at the same time, weight is reduced and perhaps performance properties are improved.

Exterior Body Parts

The original application of plastics in automobiles, which was for radiator grills (in plated ABS), has been overtaken by the use of plastics body parts to meet styling and aerodynamic requirements, which now represents a major growth area for engineering plastics. New materials have been developed for exterior body parts to fulfill the functional properties of heat stability (140–160°C), high impact strength, toughness (no splintering), petrol resistance, and stiffness. The materials being used for body panels include Du Pont's Bexlay range, based on polyamide technology, which replaced reinforced reaction injection molded (RRIM) polyurethane (PU) for Pontiac's Fiero rear quarter panel; XBR alloy, a Celanese product, which is believed to be a thermoplastic polyester alloyed with an acrylate; another Celanese product, Duraloy 2000, which is a polymer blend based on PBT; and Borg Warner's Elemid, which is a nylon/ABS blend.

In 1985 General Electric launched two new resins for car exteriors—Noryl GTX, a polyphenylene oxide (PPO)/nylon blend for fenders and other panels, and Lomod polyether ester elastomers for bumper systems. Noryl GTX combines the heat and dimensional stability and the low water absorption of PPO with the chemical resistance of semicrystalline nylons.

An important requirement for plastic materials in automotive industry is paintability alongside metal components at high temperature. Noryl GTX series has the important requirement of paintability in high-temperature (150–160°C) ovens. Several grades of Noryl GTX are available, some of which are glass filled. The unfilled grades are for exterior components such as fenders, spoilers, and wheeltrims, while the glass-reinforced grades are used in under-the-bonnet applications such as cooling fans, radiation end caps, and impeller housings, where, in addition to heat resistance, extra stiffness is needed.

RRIM polyurethanes are widely used for body panels. Bayflex 150 (of Bayer) was chosen by Pontiac for its 1987 Fiero, a well-publicized all-plastic car. The car's panels hang from a metal space frame. Bayflex 150 is used on vertical panels while the bonnet and roof are made of sheet molded components. Bayflex 150 is said to offer a number of advantages (improved thermal stability, stiffness, impact resistance, and dimensional stability) over conventional ureas without their processing limitations.

One problem in using plastics for body panels is the requirement for stiffness under heat along with the need for a high-quality surface finish. To overcome this problem, ICI separated the two requirements and molded two compatible materials in sandwich form, using coupled glass reinforced polypropylene (PP) for the outer printable skin.

Ford Sierra featured the first all-plastic bumper in Europe, and the Ford Taurus and Sable featured the first of these bumpers in the United States. The bumper is molded from GE's Xenoy polycarbonate (PC)/polybutylene terephthalate (PBT) blend. The high impact strength of PC, combined with the resistance to gasoline and oil and the processing ease of PBT, is increasing the use of PC/PBT blends in car bumpers. Bayer's PC/PBT blends are called Makrolon PR. They are used for several automobile impact-resistant parts.

Foamed PP has also been used for car bumpers. Arpro expanded PP bead (of Arco Chemical) is claimed to offer high impact absorption and resilience. It has been used for the front and rear bumper core. Arpro offers a substantial cost saving and a weight reduction of 50% in replacing the steel bumper system.

Interior Components

There is a lot of competition among plastics manufacturers for car interior components although, however, this area is not expected to show much growth. ABS plastics continue to dominate the instrument panel market, while new materials like Dow Chemical's Pulse PC/ABS are being introduced. Bayer's Bayblend PC/ABS alloys are for use in dashboard housings, interior mirror surrounds, and steering column covers. The alloy's high heat stability allows pigmentation in pastel colors.

Arco Chemical's new line of Arpro expanded PP and Arpak expanded PE bead resins are competing for the interior components. Polyolefin foams provide a higher degree of chemical resistance, heat resistance, and cushioning as compared to foamed polystyrene. Potential applications include seat backs, head rests, and other critical energy-absorbing areas.

A poly(ethylene terephthalate) (PET) product named Curl Lock for car seats has been introduced by Tagaki. In the production of Curl Lock, PET fibers are crimped and curled and then interlocked at their crosspoints with adhesive. The material is formed into car seat upholstery by placing it in a mold.

BP Chemicals have developed a polyurethane (PU) system for the production of integrally-molded sand-insulated carpets. The carpet is laminated with a PU backing to improve shape retention. It is then placed in a mold and PU foam is injected to fill the space between the car floor contours and the carpet. A carpet made with the system has been in use in the Fiat Turbo Uno.

Load-Bearing Parts

Automobile structural components are expected to be a major growth area for advanced composites. Examples include steering column, main door frame, suspension system, and selected engine parts. An early commercial application was Ashland's Arimax 1000 structural RRIM resin, which was used to produce the spare tire recess cover or the trunk of a number of cars built by GM's Buick-Oldsmobile Cadillac group. Other possible applications of Arimax resins include door trim panels, floor pans, package shelves, radiator supports, and motor side compartments.

The first all-around automobile application of composite leaf-springs was made in the Sherpa 200 series van and mini buses built by Freight Rover. These fiber-reinforced plastic (FRP) components for high stress, high-temperature uses were developed by GKN. ICI's Verton, a long-fiber reinforced thermoplastic is also suitable for such applications.

Under-the-Bonnet [Hood]

The use of plastics in under-the-bonnet applications is expanding at a significant rate. Filled or reinforced polypropylene (PP) or nylon is being used for items such as heater housings. Stampable PP or polyacetal glass mat materials, also known as glass mat thermoplastics, are being used as acoustic shields for diesel engines, chain drives, and timing belts.

Glass-filled polyether ether ketone (PEEK) resins are used in the inlet manifold of a Rover racing engine, while polyether sulfone (Victrex) has been used for piston skirts. Fluoroplastics have found many applications in the sealing rings and bearings of automotive drives.

The Ford Polimotor, which has a virtually all-plastic engine that offers a weight saving of some 60% compared with conventional components, uses polyamide-imide for moving parts with fuel saving advantages.

Blends of polytetrafluoroethylene (PTFE) are replacing metals in applications such as gaskets, bearings, and piston rings in automotive engines. The Polycomp R range of LNP Corporation includes several grades of PTFE filled with polyphenylene sulfide, which show good mechanical properties. Polycomp E combines PTFE with a linear aromatic polyester. It is claimed to offer good electrical, mechanical, and tribological properties. Polycomp A contains a proprietary aromatic polymer developed for reinforcing PTFE.

In the electrical system, glass-reinforced PBTs and PETs are finding use because of their heat stability and good dimensional stability. An all-plastic headlamp with metallized polyester bulk molding compound or PBT for the housing and reflector is expected to find increasing use. Ford and GM light trucks are already using an abrasion-resistant coated polycarbonate in headlamps. A new high-intensity lamp introduced in Germany uses polyetherimide for the reflector, to resist the higher heat.

POLYMERS IN AEROSPACE APPLICATIONS

Advanced polymer composites, which are high-performance materials consisting of a polymer matrix resin reinforced with fibers such as carbon, graphite, aramid, boron, or S-glass, have their market in aerospace. This is also expected to be the fastest growing sector of plastics sales, with growth projected at 22% a year.

Much of the original impetus for a changeover from the use of aluminum materials to plastics in the aerospace industry came from the availability of light-weight polymer composites based on reinforcements such as Du Pont's aramid fiber (made under the tradename Kevlar) and honeycomb materials such as Nomex from Ciba-Geigy. By using these reinforcements through impregnation with resins such as epoxies, and with applications of precise tooling, components can be produced that involve complex shapes and meet difficult aerodynamic requirements. Weight reduction is the major advantage of using light-weight composites in place of metals. The incorporation of composite materials in Corcorde has afforded a saving of 500 kg in mass. The new Airbus 320 is said to have saved about one ton in mass by the use of advanced composites.

Within the aerospace industry, advanced composites are used to greatest extent in military aircraft. Commercial aircraft such as the Boeing 757 and 767 contain less than 5% advanced composites by weight, while military aircraft like the GA-18A contain 10–20%. Typical applications of composites include aircraft wings, tail sections, pressure vessels, and seat rails.

Advanced composites have been used most extensively in helicopters. Sikorsky's S-75 helicopter, for example, is about 25% composite by weight, mostly graphite-epoxy and aramid-epoxy composite materials. Composites are used in rotors, blades, and tail assemblies. Future military helicopters are likely to comprise up to 80% advanced composites by structural weight. Graphite-epoxy composites are likely to be used in the airframe, bulk-heads, tail bones, and vertical fins, while the less stiff glass-epoxy composites will be used in rotor systems.

Carbon Fibers

The fiber most commonly used in advanced polymer composites is carbon fiber. Research and development of carbon fiber components (CFCs) for aerospace applications gained impetus since 1970 when British Aerospace used them in the

Jet Provost rudder trim tabs. Since then CFCs have been used in the foreplanes, wings, access panels and doors, cockpit floor structures, and side panels of a number of aircraft including Jaguars, Tornados, and Harriers. Advantages of CFCs over aluminum materials include an increased strength-to-weight ratio; a reduction of up to 30% in the manufacturing cost, due to reduction in the number of detailed parts; a reduction in the volume of waste materials; and the ability to produce large, complex curved shapes required in advanced military aircraft.

The satellites ECS1 and AMSAT developed by Aerospatiale, France, feature two of the largest carbon-fiber reinforced honeycomb structures made in Europe. A sandwich structure reinforced with high modulus carbon fiber is used for the dual system capsule. The fiber, Grafil HM-S, provides exceptional dimensional rigidity for extremely low weight and matches very specific resonance characteristics required in the design. The fiber in epoxy resin is laminated to form composite skins using prepreg materials. The same grade of carbon fiber is also used in the manufacture of the actuator control struts for the second stage motor of the Ariane launcher.

Resins

There is a growing competition in the market for moderate-temperature materials for aircraft structures such as fuselages and cargo bays in response to the aerospace industry's endeavour to produce an all-composite aircraft, which is currently led by Boeing.

Shell Chemical has produced EPON HPT epoxy resins, which are used in large parts made of graphite-7 reinforced prepreg material for primary aircraft structures. EPON 9000 series of epoxy resins produced by the same company have been used for fabricating, by wet filament winding process, fuselages and other large components that have been used in the prototype of Beech Aircraft's experimental small business aircraft, the Starship 1.

Dow developed an experimental moderate-temperature resin, XV-71794, which is claimed to offer incredibly high toughness and low moisture absorption. Baltimore Aerospace has used Nextel and graphite fibers in a bismaleimide matrix for a hot air duct in its new jet engine. The duct is 75% lighter than one made of metal.

More recently, thermoplastics such as polyphenylene sulfide (PPS), polyether sulfone, and polyether ether ketone (PEEK) are entering the primary structures market in competition with thermosets.

PEEK, a relatively new thermoplastic material introduced in 1978 by ICI, is being used in aerospace applications. ICI Plastics have produced PEEK reinforced with 68% continuous carbon fiber in narrow tows suitable for braiding.

The braid is used to strengthen tubular structures without adding significantly to the weight. One application is in the repair of battle-damaged military equipment, particularly control rods and helicopter rotor shafts. The braid is slipped on where needed and built up in layers to strengthen the component. Another possible application of the braid is for the reinforcement of aluminum air frame struts in the Pegasus microlight hang-glider.

A 20% glass-filled PEEK introduced by ICI for injection molding was chosen by Boeing to replace aluminum fairing on its 757-200 aircraft. The substitution of PEEK for aluminum reduced weight by 30% and lowered costs considerably. Compared to other materials PEEK also showed greater heat and chemical resistance in addition to the required dimensional stability.

Carbon and glass-fiber reinforced PEEK has been used with advanced plastics technology in the redesigning of the helicopter blades profile, improving aerodynamic properties, and reducing overall weight. The use of PEEK composites in the RAF Linx helicopter enabled it to break the world helicopter speed record.

A good example of large-size fiber-reinforced components in aerospace application is the radome covering the underbelly radar on the Hercules transport aircraft. It is made from very thin polyethylene sulfide (PES) film interleaved with PES-impregnated glass fiber cloth, which is subsequently hot molded in a closed die. The composite radome, nearly 1 m in diameter and 6 mm thick, weighs only 10 kg.

Ryton PPS woven carbon fabric prepreg sheeting (AC31-60), supplied by Phillips Petroleum, is particularly suited for thermoforming into high-strength parts. Boeing Aerospace has designed and fabricated a prototype access door using this composite. It has resulted in weight reduction of 25% compared with a metal counterpart and a 67% cost saving compared with thermoset composite construction.

To improve on Ryton PPS, Phillips introduced a family of high-temperature performance polyarylene sulfide (PAS) polymers. The crystalline form, PAS1, is said to be suitable for structural components at temperatures up to 120°C and for nonstructural components at temperatures up to 230°C. PAS2, an amorphous form, has temperature limits of 160–200°C for structural components and 270°C for nonstructural components.

Raychem developed heat-shrinkable woven fabrics that incorporate heat shrink fibers in the weft and nonshrink protective fibers in the warp. Slipped over a cable harness, for example, they shrink down to a quarter of their size to form a tough outer tube. The problem of damage caused to aircraft radar covers, landing gear wiring, and hydraulic lines by stones and debris thrown up off runways, is being solved by the use of heat shrinkable woven fabric tubes containing high wear and impact resistant fibers such as Kevlar, polyether, and glass.

The British Aerospace 146 jet has a Kevlar-based heat shrink fabric protecting its undercarriage cables. Radar umbilical cables on RAF fighter aircraft are similarly protected.

A major problem relating to passenger safety has been the use of epoxy paste, which can emit toxic gas in the event of fire. The epoxy paste compounds are used to fill the honeycomb cells of sandwich panels, such as those used in aircraft wings and flooring, in the region of joints and attachments. A phenolic equivalent, Corfil 792, has been developed by Cyanamid Fothergill. It is a high strength, low density, spreadable thixotropic paste formulated for single stage curing of sandwich constructions and has been accepted by the aerospace industry worldwide.

Cellular materials are used in aircraft for a number of applications including cockpit insulation, duct insulation, and vibration damping for aircraft fuselage. Monsanto's Skybond polyimide foams have been used for sound deadening of jet engines. Solimide polyimide foams produced by Imi-Tech Corporation are said to be lightweight, fire resistant, and show good thermal and acoustic properties. They have been specified by Boeing for several aerospace applications. One space application for Solimide is cryogenic insulation for fuel tanks on major rocket propulsion systems.

POLYMERS IN ELECTRICAL AND ELECTRONIC APPLICATIONS

Polymeric materials are used in a wide range of electrical and electronic applications, including wire and cable coverings, housing adhesives, printed circuit boards, and insulations. Newer developments in electrical and electronic industries place greater demands on electrical, mechanical, chemical, and heat resistant requirements of plastics.

Wire and Cable Insulation

Du Pont's Neoprene synthetic rubber has been used for many years in the cable industry as it provides good low voltage insulation and mechanical toughness as well as good resistance to oil, high temperature, and oxidation. Alcryn, also from Du Pont, is a melt-processable rubber cable sheathing for signal and power transmission. It is claimed to have excellent oil resistance even at high temperatures, and good resistance to severe mechanical treatment. Another key product from Du Pont is Hypalon, a chlorosulfonated polyethylene that exhibits high resistance to oils and a host of chemicals. One application of this product is the jacketing of cables in chemical plants.

Fluoropolymers such as Teflon and Tefzel that have high mechanical strength and dielectric properties with heat resistance enable thin wall insulation for high current densities. Du Pont's fluoroelastomer Viton is used for insulation and heat shrinkable sleeving. In addition to excellent heat and chemical resistance it has high resistance to fuels, oils, and lubricants. Fluon, a polytetrafluoroethylene (PTFE) product from Fothergill Cables, which specializes in electronic cables for aerospace and defense industries, is applied to thin wall insulation for equipment wires, multicore cables, thermocouple wires, and radio-frequency coaxial cables. Fluon has a service temperature range of -75 to $+260°C$, nonflammability, high chemical resistance, good dielectric properties, and resistance to damage by soldering.

Polyesters in wire and cable applications include Hytrel (Du Pont), a thermoplastic polyester elastomer for the offshore industry; Kapton (Du Pont), a polyimide film offering weight saving and chemical resistance; and Mylar (Du Pont), used in primary insulation, shield isolation, chemical barrier, and mechanical protection.

ICI has developed a large number of products to suit different applications in the wire and cable industry. While for general purpose insulation and sheathing, poly(vinyl chloride) (PVC) is used with chlorinated plasticizers such as the company's Cereclor compounds, its Victrex polyether ether ketone (PEEK) wire and cable covering is designed to operate in high-demand environments. PEEK has high-performance electrical properties that permit the use of smaller diameter cable covering without loss of electrical power or reduction in insulation. Because of the product's good inherent flame retardancy, low smoke emission, and low toxic and corrosive gas emissions it is used in underground mass transit wiring and cable insulation. Victrex has also been used for insulating a coaxial control cable operating in a high radiation environment at a nuclear installation in the United Kingdom.

BP Chemicals have introduced a new technology for silane cross-linking in low-voltage cable insulations to produce more uniform cross-linking and greater shelf-life. The new silane-linked polyethylene compounds can be run on conventional extrusion equipment. The company has also produced a strippable double layer cable shield in cross-linkable polyethylene for medium voltage cables. The shield has a strippable outer layer formed by a coextrusion of 20% strip material and 50% standard semi-conductive material.

Printed Circuit Boards

Developments in the electronics industry are focused on two main areas—miniaturization and improved cost/performance ratios—creating the need for polymers that can tolerate higher temperatures for longer periods with good dimensional stability and mechanical strength.

Though the majority of printed circuit boards (PCBs) at present are two-dimensional and made of glass-filled epoxy resins, increasing miniaturization is leading to more PCBs being multi-layered with several copper circuits sandwiched between insulating layers. In these applications, because of the added stress due to thermal expansion, specialty materials such as polyamides and fluoropolymers are being increasingly used.

The present trend is towards injection-molded thermoplastic boards in place of thermoset laminate PCBs. The two main reasons for this shift are the difficulty of etching on thermoset laminates, and the greater design flexibility and potential cost saving in high-volume production by injection molding. A further advantage of injection molding is the ability to produce circuit boards with three-dimensional features such as ribs, bosses, heat sinks, and holding devices for snap-parts. Engineering polymers that can be used for molded circuit boards include Du Pont's Rynite polyethylene terephthalate (PET) grades and glass-reinforced grades of ICI's Victrex polyethersulfone (PES).

In the electronics industry, one of the most successful uses of advanced composites has been in PCBs. General Electric's Ultem polyetherimide is being used in transport PCBs, especially aerospace and missile guidance systems. Polysulfones and other high-performance thermoplastics are being used in multilayer PCBs for mainframes and supercomputers operating at gigahertz frequencies.

Processes developed for the injection molding of circuit boards include PCK Technology's mold-n-plate process and an in-mold metallization process developed by Allen-Bradley. The mold-n-plate process uses two shots, the first being one of catalytic resins, to form paths on plastic parts that are then processed in a copper bath to form circuitry. Special grades of polysulfone, polyethersulfone, and polyarylsulfone are used for the process. In the in-mold metallization process, various conductive, resistive, capacitive, or inductive components are embedded directly into the plastic part in the tool.

Connectors

The miniaturization of PCBs has created the need for connectors to be made from special materials suitable for shaping into thinner parts that can be packed into higher-density spaces and that can withstand higher soldering temperatures. Such materials available on the market include tailored grades of polyphenylene sulfide (PPS), PEEK, PET, polyetherimide, sulfone-based alloys, and liquid crystal polymers. Well-known commercial products are Rynite FR-530 and FR-945 modified PET grades of Du Pont; Valox 9730 developmental alloy of General Electric; Victrex high-performance thermoplastics of ICI; Mindel alloys of Amoco Performance Products, which combine various grades of sulfone polymers; and Ryton PPS grades R-4 and R-7 of Phillips Petroleum Chemicals.

Enclosures

Plastics are increasingly being used for electrical/electronic enclosures. Those used for enclosures of circuit chips should, however, be of high purity and high flow. Since miniaturization reduces the surface area of heat dissipation, the materials used are required to have high heat resistance in addition to high purity. While epoxy resins such as Quatrex 3430 and 3450 of Dow Chemical dominate the market, other materials used include polyurethanes, polyimides, and silicones.

Plastics used in electrical housings are now required to provide protection against electromagnetic interference (EMI), radio frequencies, and electrostatic charges that could damage sensitive equipment circuits. To impart EMI shielding properties to a plastic component, a coating of conductive layer may be provided by post-molding or in-molding processes or polymers may be used that have been metal or carbon black filled to make them conductive (see Chapter 5).

The methods mostly used to produce EMI shielding on plastic moldings are conductive paints, zinc arc spraying, vacuum metallizing, electroplating, and foil application. There are also a range of filled plastic compounds on the market that are based on metal fillers such as aluminum flake, brass fibers, stainless steel fibers, graphite-coated fibers, and metal-coated graphite fibers. The most cost effective conductive compounds are, however, based on carbon black.

Optical Fibers

A major advance in communications has been the introduction of optical fiber transmission. In optical fiber transmission, voice and data signals are converted into beams of light that then travel along extremely thin glass fibers. Major advantages of this transmission over the conventional current carrying metal wire are higher message speed and density, lighter weight, and immunity to electromagnetic interference. The main growth in the use of optical fibers is expected to be in telephone networks.

Plastics find extensive use in several areas of fiber optic cables. Buffer tubes, usually extruded from high-performance plastics such as fluoropolymers, nylon, acetal resins, or polybutylene terephthalate (PBT) are used for sheathing optical fibers. A blend of PVC and ethylene vinyl acetate (EVA) polymer, such as Pantalast 1162 of Pantasote Incorporated, does not require a plasticizer, which helps the material maintain stability when in contact with water-proofing materials. PVC and elastomer blends, Carloy 6190 and 6178, of Cary Chemicals are also used for fiber optic applications. (Stiffening rods for fiber optics are either pultruded epoxy and glass or steel. Around these is the outer jacketing, which is similar to conventional cable.)

Information Storage Discs

In recent years, information has begun to be stored magnetically, on tape and on disc. Today their position is being challenged by optical discs, because an optical disc can store 1,000 times more information than its magnetic equivalent, has a longer guaranteed lifetime (usually ten years), and more easily accessible data.

Although magnetic storage discs are beginning to be regarded as archival as compared to optical discs, the advantage they have over most optical systems is that information can be added and erased. The two types of magnetic discs on the market are Winchester hard discs and stretched discs.

Winchester discs are molded as blanks in engineering plastics, usually General Electric's Ultem polyetherimide, then coated with an epoxy-ferric oxide layer. The main requirement in manufacturing these discs is absolute flatness because the magnetic read/write head moves only 8–20 microinches above the disc surface and is spinning at a high speed of 3,600 rpm. Any microscopic bump can thus cause damage to both the disc and head.

The stretched disc, developed by 3M Company as an alternative to the Winchester disc, is said to combine the performance characteristics of hard discs with the economy and environmental tolerance of a floppy disc. The discs are 5¼ inches in diameter and are able to store 12-Megabytes of computer data with storage capacity up to 100-Megabytes is thought possible. Unlike Winchester discs, stretched discs consist of a magnetically coated polyester film stretched over an injection-molded disc. The film is bonded to the raised edges of the disc, thus producing a compliant surface with a 250 μm gap between the film and substrate. Since information is stored on the film rather than on the disc, flatness requirements are not critical as with Winchester discs.

The first optical disc to appear on the market was the laser-read video disc, 8 or 12 inches in diameter, which is produced in acrylic resin. The disc is made of two halves sandwiched together, which reduces the problems of warpage and means that the disc can be played on both sides. One example is Phillips' Laservision, which dominates the European market.

The compact audio disc (CD) based on digital recording and playback technology is 4.75 inches in diameter and can store 16 million bits of information in the form of minute pits in the substrate. It is the presence or absence of the pits that is read by the laser. The pits are 0.1 microns deep, 0.5 microns wide, and between 0.833 and 3.56 microns long. Each "track" comprises a spiral of these pits. The track is laid in polycarbonate that is backed with reflective aluminum and coated with a protective acrylic lacquer. The CD can thus be played only on one side.

The CDs are manufactured either by injection compression molding or injection molding. Injection compression molding minimizes molded-in stress but requires longer cycle times than injection molding and produces flash. For both processes very clean room conditions, up to 1000 times cleaner than medical molding, are essential.

The processability of the polycarbonate, or any other material used as the substrate, is crucial in the manufacture of all optical discs. The material must have low melt viscosity and high melt flow to minimize internal stresses caused by molecular orientation of the polymer during filling. Also important are high long-term dimensional stability, resistance to moisture and gas permeability, low birefringence, and superior optical properties. Bayer's polycarbonate grade Makrolon CD-2000 has been specially developed to fit such requirements. The molecular weight of the polycarbonate has been reduced to give an exceptionally high melt flow rate compared to a standard polycarbonate. In addition to polycarbonate, a number of other materials including acrylic resins, polyacrylates, thermoset epoxies, and blends of copolymers have been tested for their suitability in optical discs.

The compact read-only memory disc (CD-ROM) has been developed from the compact audio disc. A 4.75 inch CD-ROM has a storage capacity equivalent to 1,500 floppy discs, or 250,000 typewritten pages. The purity of the material and the absence of internal stresses and composition variation in the discs is even more crucial than for CDs since any birefringence can cause significant loss of data. The main disadvantage of CD-ROM is that material cannot be added or erased.

Two products, DRAW (Digital Read After Write) and EDRAW (Erasable DRAW), provide the ability to add more information besides the advantages of CD-ROM, thus bridging the gap between optical and magnetic storage discs. While much of the information on these products is proprietary, Daicel in Japan have reportedly produced magneto-optical discs based on an polycarbonate injection-molded substrate involving the application of "opto-magnetic-sensitive" coatings and more complex metallizing procedures than other formats. To "write," a high-power laser is applied to the magnetic field of a disc to change its polarity in a microscopic area, and to "read," a low-power laser measures the differences in polarity.

POLYMERS IN AGRICULTURE AND HORTICULTURE

The application of polymeric materials in agriculture and horticulture has increased considerably in recent years, not only as replacement for traditional materials but also as a means of effecting improvement. For example, the application of plastics has led to a significant improvement in technological processes in the growing and storing of agricultural crops, construction of storages for fertilizers, vegetables and animals, and in agricultural equipment and drainage technology.

Plastic Film

The main uses of plastic films in agricultural and horticultural applications are for covering fodder and cultivation units such as seed-beds, for mulching and warming soil, for the building of canals and reservoirs, and as packaging material for storing and transporting food.

The most widely used plastic film is polyethylene (PE). PE film tunnels and perforated flat PE film allow better use of natural resources such as solar energy, water, and soil. Crops may be made to ripen early making cultivation possible all year round. In addition to covering produce, PE film has a number of other uses. Livestock feeding stuff such as green foliage grass and maize can be preserved by covering the forage in a silo with large gastight black or white PE films. Shrinkable PE films are used for sheet steaming in horticulture.

Polyethylene films for agricultural applications need to have high strength and elasticity, resistance to wind forces, and a long service life. They are mostly produced from low density PE combined with linear, low density PE and ethylene-vinyl acetate copolymer.

As PE films, like all synthetic polymers, suffer from decomposition by environmental influences such as light and atmospheric oxygen, special photoprotective systems are added to delay these effects. Systems that are commonly used are ultraviolet (UV) light absorbers, quenchers, radical scavengers, and hydroperoxide decomposing agents.

Benzophenone and benzotriazole are UV light absorbers that are frequently used in packaging film. While the addition of UV absorbants increase the service life of the film by 100–200%, a disadvantage of such absorbants is that their effectiveness is dependent on the thickness of the film to be protected. For films less than 100 μm thick, they can only offer a limited protective effect.

The mode of action of quenchers, however, is not dependent on the thickness of the film to be protected. Quenchers are photoprotective compounds that can take up and conduct away energy that has been absorbed by chromophores, such as hydroperoxide, which are present in PE film. Organo-nickel compounds are quenchers that also act as decomposing agents of hydroperoxide.

Hindered amine light stabilizers (HALS), which represent more recent developments in photoreactive compounds and are referred to as scavengers, absorb light above 250 nm and therefore do not act as UV absorbers or quenchers.

Several specialized types of polyethylene film are also available. These include heat-resistant film, heat-retaining film, water-absorbing antistatic film, and photodegradable film.

A heat-resistant PE film for warming the soil will stand heating for 200 hr at 120°C and at a pressure of 0.2–0.3 MPa. A heat-retaining PE film with enhanced absorption in the long wave region of the IR spectrum enables the temperature under the film to be 3–4°C higher than when under normal PE film.

Moisture-absorbing antistatic film with enhanced permeability to UV radiation is used primarily for seed beds, as it does not become dusty and therefore creates better conditions for growing plants inside hot-houses. The film also has a surface that prevents the deposition of condensed droplets, increasing the yield of vegetable crops by 15–20% compared to normal PE film.

Photodegradable PE film is used for mulching the soil in vegetable growing. It improves soil drainage and discourages weed growth. The film breaks down within two weeks to three months as a result of solar radiation, combines with the soil, and is broken down by microorganisms.

Building

Plastics have several advantages over metals in the construction of agricultural buildings. For example, in the construction of low-rise farm buildings, metal roof sheeting shows deterioration after 2–3 years due to condensation of water vapor produced by dairy cows. In contrast, glass-fiber reinforced plastic (GRP) polyester sheet has a service life of 15–20 years and is unaffected by condensation of water vapor as well as by most acids, alkalis, grease, and caustic soda solutions. In translucent form it allows a high level of natural light but cuts out glare and excessive solar heating.

GRP sheet has also been used as an overall cladding material. GRP polyester sheet, Filon, produced by BIP Chemicals has a tensile strength of 10^4 psi (69 MPa), better than most other cladding materials of equal thickness and thus allows the wall thickness to be reduced.

Pipes and Hoses

Plastic pipes are being increasingly used in the agricultural industry for drainage and irrigation. The improved strength and greater crack resistance of new grades of polyethylene have allowed the wall thickness of such pipes to be reduced and their diameters increased.

Polyethylene has been irradiated to produce cross-links thus giving it improved strength. Drip irrigation for raw crops uses such irradiated LDPE tubing. For vine and tree crops the drop emitter is made of PE molded in large multicavity molds.

Plastics are now being used in a variety of ways to replace metal components in different irrigation systems. Interestingly, one of the first uses of PVC pipe in the 1960s was in sprinkler irrigation systems as a replacement of brass, gunmetal, or zinc alloys.

Hoses used in irrigation are invariably made of PE. Crimped PE pipes are being more frequently used for drainage and irrigation of soils. Flexible hoses that can be rolled up are used in spraying and sprinkling machines. Hoses reinforced with a rigid spiral are being used with machines to introduce chemicals into the soil for the protection of plants.

Greenhouses

The use of plastics in horticultural industry has grown substantially in recent years. Several plastics, and PVC in particular, are used in applications such as hosepipes, glazing strips, paint, and wiring. However, it has come to light that as greenhouses become airtight, some chemical vapors released by plastics may build up to levels toxic to plants. The plasticizer dibutyl phthalate (DBP) used in PVC has thus been found to be phytotoxic in a greenhouse environment. Research has shown that a fire retardant used in polystyrene trays and granules reduced the growth of some seedlings. Similarly, a silicone joining compound used in greenhouses and fumes from a paint used on an electric boiler were also found to be harmful to test plants.

Responding to the toxicity problem of plastics in greenhouse applications, the Dutch horticultural industry developed a test for screening. Plastics that pass the test qualify as safe horticultural plastics can be used in a greenhouse environment.

POLYMERS IN DOMESTIC APPLIANCES AND BUSINESS MACHINES

As plastics are more lightweight than glass or metal, are generally impact resistant, lend themselves well to manufacturing techniques, and allow designers greater freedom in design, they are continuing to replace metals in domestic appliances and equipment housings. The use of plastics has been well exploited in both these areas.

Large Appliances

While the use of plastics in refrigerators is well established, product development has led to replacement of some of the conventional plastics with newer ones of improved properties. For insulation rigid polyurethane foam has, on the whole, replaced fiberglass in cabinets and doors. In this application, diphenylmethane diisocyanate (MDI), which has lower vapor pressure than toluene diisocyanate (TDI), is now preferred.

For refrigerator lines low-cost polystyrene (PS) was the first choice, but this was soon replaced by ABS, which has higher stress crack resistance to trichloromethane, a foaming agent that is used to blow polyurethane insulation behind the liners. However, because of the higher cost of ABS there has been a swing back to PS with special grades of PS now being available. Dow Chemical and BASF produce special crack-resistant grades of PS, although the sheets are not as rigid as ABS.

Mobil Chemical Company produces a composite structure consisting of the company's PSMX7100A and an extrusion-grade high-impact PS. The structure is claimed to offer a cost saving of 30–40% over ABS when equal thicknesses are compared. Mobil's translucent impact grade polystyrene PSMX7800A can be coextruded or laminated as a cap layer to yield a surface glass.

Responding to new PS products, Borg Warner has produced special ABS resins tradenamed CTB, which permit more uniform drawdown and thus allow the use of thinner gauze sheet without the risk of thinning of corners. For interior refrigerator components such as crispers and dairy drawers, ABS grades are commonly used. Lexan PC film has been used for refrigerator consoles.

Because of the limitation of temperature resistance, the use of plastic components in ovens has been mostly limited to exterior trim. Plastics however are gradually being used for semi-structural components. End-caps with a metallic finish are a typical example. General Electric have used a sputter-coated modified Noryl polyphenylene oxide for a shiny metallic appearance for the Tappan oven. The material provides flame resistance, electrical insulation, and resistance to heat, grease, and abrasion.

Celanex thermoplastic elastomer produced by Celanese Engineering Resins has been used for certain oven components. The 30% glass reinforced grade, offering faster molding cycles, improved impact resistance, and greater flexibility and color range has replaced phenolic resins. Belling and Company Limited used the material for the two oven handles and corner panel supports of its Format Cook Center.

General Electric produces several blow molding grades of polycarbonate for the vertical panels of ovens and dishwashers. Advantages of polycarbonate panels over metallic panels include low cost product, rigidity, and dimensional stability.

In microwave ovens, General Electric's glass-filled Ultem polyetherimide is used for shelf supports and stirrers. Dartco's Xydar liquid crystal polymer is used for turntables, shelf supports, stirrers, and meat probes. Membrane touch switches for microwave ovens are normally made from polyester film such as Du Pont's Mylar.

Plastics are used to make a number of dishwasher components, such as impeller and other pump parts, inner door, tubs, and consoles. Polyphenylene sulfide is used to make dishwasher impellers. The material shows processing ease, good durability, and resistance to high heat and detergents. For dishwasher pump parts, tubs, and door liners, glass-filled polypropylene has all but replaced diecast zinc. Noryl-modified polyphenylene oxide (PPO) is replacing stamped metal for the dryer for blower wheels. Both rigid PVC and PC are used for consoles. Rigid PVC is used because of its inherent flame retardancy, rigidity, and high heat distortion temperature. Polycarbonate (PC) is used for its high heat resistance, dimensional stability, and good electrical insulating properties.

Plastics are replacing metals in washing machines. Procom GS 30H254, a 30% glass reinforced structural foam grade of ICI's polypropylene, has been used by Hotpoint for the oven drums of its washing machine replacing enamelled steel. The Procom drums provide improved insulation and quieter operation. Nonfoamed grades of the product are used for the drum front moldings and detergent dispenser. An acetal copolymer spin gear has been used in place of the traditional belt drive in Whirlpool's Design 2000 washing machine. The acetal copolymer eliminates the need for secondary finishing operations and affords a significant reduction in weight.

Propathene random copolymer PXC 22406 PP of ICI is used by Electrolux in its vacuum cleaners. The material is used to produce impact-resistant side and rear bumpers to protect the machine during use.

Small Appliances

The market for plastic cookware that can be used in the freezer and microwave is set for a big expansion. Most cookware is made up of fiberglass reinforced thermoset resins. High-barrier multilayer containers that have gained popularity in the market are heat retortable, microwaveable, and provide a shelf life of up to two years without refrigeration for a variety of convenience foods. Liquid crystal polymers have also been used to make cookware as they are transparent to microwave, offer high heat resistance, and give an excellent finish.

The most notable development in small appliances has been that of the electric kettle. Nearly all new kettle bodies, lids, and spouts are molded from plastics such as Kematal acetal copolymer of Celanese Corporation. Kematal is chosen for its strength, resilience to impact, smooth surface finish, and wide colorability. Use of the material has also enabled a slimmer design of kettle, suited to the needs of the consumer.

Coffeemakers incorporate plastic in both the jug and housing. The Koffee King decanter from Bloomfield Industries uses for robustness and transparency of its main body Amoco Chemical's Udel polysulfone as a replacement to glass. Krups used a grade of polybutylene terephthalate from Mobay Chemical Company for the housing of their coffeemaker. The material has chemical resistance, good surface quality, high dimensional stability, and odor stability along with high heat deflection temperatures and electrical insulation properties. The same material has also been used for the housing of electric irons.

Business Equipment

Business machine and equipment housings represent a fast growing area for ABS polymers that are commonly described as tough, hard, and rigid—a combination that is unusual for thermoplastics. It's light weight and the ability to economically achieve a one-step finished-appearance part have contributed to large volume applications of ABS. A variety of special ABS grades are used for equipment and housings. These include high-temperature resistant grades (for instrument panels, power tool housings), fire-retardant grades (for appliance housings, business machines, television cabinets), electroplating grades (for exterior decorative trim), high-gloss, low-gloss, and matte-finish grades (for molding and extrusion applications), and structural foam grades (for molded parts with high strength-to-weight ratio).

Since the time of Noryl, the first commercial blend of two dissimilar polymers introduced by General Electric in the 1960s, a large number of different blends or alloys has been introduced, some of which have been developed specifically for office equipment and housings. A number of polyphenylene ether (PPE) alloys have been introduced by Borg-Warner, Hulls and BASF. The great advantage of these PPE blends is that they can be made flame retardant without halogens.

Borg-Warner has various grades of its Prevex PPE/high-impact polystyrene (HIPS) blend. Prevex offers high impact resistance and high heat distortion temperatures. Applications are for office equipment, business machines, and telecommunication equipment. BASF's Luranyl PPE/HIPS blend includes general purpose types, reinforced grades, impact modified types, and flame-proof types. BASF also has a PPE/nylon blend tradenamed Ultranyl developed for a similar market. Hulls produces a PPE/nylon blend called Vestoblend that uses nylon-12 and nylon-6,12. It is claimed to absorb less water than competing types.

Noryl AS alloys, which are blends of PPO/PS and special fillers, have been developed specifically for office equipment and housings. Their main features are high strength, maximum stiffness, and ability to maintain tight dimensional tolerances over a wide temperature range. They are expected to replace metals in such applications as structural chassis for typewriters and printers.

POLYMERS IN MEDICAL AND BIOMEDICAL APPLICATIONS

A wide variety of polymeric materials are used in various medical and biomedical applications, which range from medical packaging and products used in medical treatment and diagnosis equipment to items that are actually implanted in the body, e.g., pacemakers, hip joints, and artificial heart valves. These products have unique requirements from absolute nontoxicity and compatibility with human tissue to mechanical strength, resistance to sterilization techniques, and, in the case of disposable end products, mass production on a large scale. Due to differing requirements a wide variety of materials are used. While medical grades of poly(vinyl chloride) (PVC) and polystyrene (PS) are the most commonly used polymers in the medical sector, high-performance engineering resins are now finding an increasing number of uses in this field. The areas in which plastics find use can be divided broadly into the following: medical packaging, nontoxic sterilizable items, appliances, and disposables.

Medical Packaging

Significant advances have been made in recent years in the packaging of medical products requiring medium and high barrier protection. The demand for sterilizable packaging for hospitals and industry is being met by special paper, plastic films, and other nonwoven materials.

The chlorinated tetrafluoroethylene (CTFE) fluoropolymer Aclar from Allied Corporation is one of the highest barrier materials against moisture transmission. It is used in an antibiotic test kit where even trace amounts of moisture would violate the accuracy of the test.

A large number of small medical containers are now molded in poly(ethylene terephthalate) (PET). Its advantage over glass include no breakage of filling lines and higher output. Tamper-evident closures, droppers, and dispensers are usually molded in polypropylene and high-density polyethylene.

Nontoxic Sterilizable Items

Polymers used in nontoxic sterilizable items, such as tubing, artificial organs, and wound coverings, must be able to withstand sterilization by ethylene oxide, steam autoclave, or gamma radiation. For medical products sterilized by gamma irradiation, only plastics that do not degrade or discolor on exposure to radiation, such as polyester and polycarbonates, can be used.

The biological compatibility of materials used in the nontoxic category is vital and is the subject of continuing research. Since no material is completely inert, it is the level of interaction between an implant and the surrounding tissue that determines the acceptability of the material. Several other facts that are important in determining acceptability include mechanical properties of the polymer, e.g., wear resistance and fatigue, and bulk chemical properties such as resistance to degradation by hydrolysis, sensitivity to enzymes, and the way it reacts to the deposition of protein.

PVC is used for nearly all surgical tubing. Recent developments include flexible PVC compounds with greater resistance to body fluids, and X-ray-opaque PVC compounds that enable PVC catheters and tubing to be traced after insertion into the body. Esmedica-V, a plasticizer-free flexible PVC compound produced by Sekisui, Japan, reportedly meets all requirements for medical use, including ability to stand ethylene oxide sterilization. ICI has produced Welvic VK-2004 for coronary dilation catheters. It is reported to offer high flexibility, high-pressure resistance, and low elongation.

The development of artificial wound coverings has received much attention. The major requirements for such coverings are protection from microbial attack from the environment, optimal water permeability, capability of adhering well to the wound, and ready removability without causing tissue damage. Other important factors are prevention of excessive formation of granulation tissue and optimal elasticity to facilitate an intimate cover of the wound.

Sustained-release wound dressings capable of delivering antibiotics or other biomedical agents in small concentrations have been obtained by microcapsulating a variety of drugs into a UV curable urethane elastomer. Thermedics Incorporated has developed special UV-curable polyurethane elastomers for such applications.

Developments related to medical implants, artificial organs, and prosthesis have improved the quality of life and increased the life expectancy of many individuals.

Silicone rubber is a highly biocompatible thermosetting elastomer that has found applications in prosthesis for opthalmology, neurology, facial reconstruction, replacement of finger, toe and wrist joints, cardiovascular applications, such as pacemaker coatings and lead wires, and tendon replacements. It is also used in drug delivery systems and tubes for carrying blood, drugs, and nutrients.

The artificial heart, Jarvik-7, which has been successful in keeping a recipient alive for more than a year, has valves made from modified polypropylene and the two ventricles made of polyurethane supported on an aluminum base. For arterial replacements or bypass of clogged blood vessels, polyester fiber such as Du Pont's Dacron remains the preferred material.

Ultrahigh molecular weight polyethylene (UHMW PE) is most commonly used for articulation surfaces in joints. It is also used for prosthesis components in total hip replacement. In the latter application, it may be reinforced with carbon fibers to increase wear properties.

Expanded polytetrafluoroethylene (PTFE) grafts have gained increasing popularity as synthetic or nontextile grafts for reconstructive procedures, such as, above- or below-the-knee bypasses for limb salvage.

The ability to undergo biodegradation producing nontoxic by-products is a useful property for some medical applications. ICI offers a biodegradable plastic under the trade name Biopol. It is a polyester made from hydroxybutyric and hydroxyvaleric acids. Because of its much higher price than conventional mass polymers Biopol will not find wide use as long as purely economic considerations determine the use of plastics. A niche market exists, however, for the medical grade of the polymer. For example, fibers can be used for surgical sutures. The compound is absorbed in the body and does not invoke immune reactions. The molecular weight lies between 30,000 and 750,000. The mixture of the two hydroxyacids is produced by a bacterium of the type *alcalignes eutrophus*.

Appliances

Plastics components are now used in a number of appliances such as crutches, walking sticks, and wheelchairs. The comfort of the product is an important consideration in all of these applications. An example is BASF's "comfy crutch," which uses the company's Ultramid KR4412 nylon for the handle component.

Thermoformed high density polyethylene (PE) and low density PE foams that can be molded into a variety of complex shapes are being increasingly used in supporting devices in the treatment of orthopaedic and disabling conditions. Some examples are wheelchair seats, spinal jackets, medical shoe inserts, and splint supports.

Specialized high-temperature plastics such as polyether sulfone and polyetherimide are used for components for medical analysis and diagnosis equipment.

Disposables

Disposable medical products such as syringes, test tubes, petridishes, dialysis products, and drainage devices are proving to be high-volume markets for transparent plastics such as rigid PVC, polystyrene, styrene acrylonitrile (SAN) copolymer, acrylic resins, and polycarbonate. The possibility of mass production is an important consideration in these applications.

POLYMERS IN MARINE AND OFFSHORE APPLICATIONS

Polymers find a wide range of applications in offshore and marine industry. The properties of materials used in these applications invariably include resistance to marine fouling, sea water, and crude oil. The main applications are in marine cables, coatings, and adhesives.

Cables

Due to the damage caused by smoke-filled or halogen-based fumes from cables in the event of a fire, the offshore industry is showing increased interest in low smoke, zero toxic gas emission cable sheath. Research has centered on obtaining halogen-free polymers having fire retardance as well as oil, chemical, and weathering resistance. One example is a series of formulations called NONHAL-E (elastomeric) and NONHAL-T (thermoplastic) developed by Sterling Greengate specifically for the offshore industry. The elastomeric formulations are based on nitrile rubber and are resistant to oil and other commonly encountered fluids. Cables insulated with ethylene propylene diene terpolymer (EPDM) and sheathed with NONHAL-E are suitable for very tough applications. NONHAL-T was developed to replace PVC in cable sheathing. Halogen-free cables produced with a cross-linked polyethylene and bedded and sheathed with NONHAL-T gives excellent performance in fire conditions.

Polyether ether ketone (PEEK) has been considered for use as insulators and cable jackets. The use of PEEK as an insulator allows thinner overall diameters and permits a higher level of continuous operating temperature. The extreme hardness and toughness of PEEK may also eliminate the need of a metal sheath for cable jacket.

New materials and manufacturing techniques have been used to minimize the problem of drag and friction that occurs with heavy umbilicals. Tufflite, a surface diving umbilical, combines the strength of its rope-like cabled lay-up with kink resistance and neutral buoyancy that allows improved maneuverability and lower drag. Tufflite is based on textile braid reinforced polyurethane (PU) hoses for gas and air supply and has a color-coded outer sheath of abrasion-resistant PU.

Du Pont's thermoplastic polyester elastomer (TPE) Hytrel has been used in an underwater fiber optic cable produced by Shiplex Wire and Cable Corporation. The elastomer is used to position the fiber optics and is chosen because of its modulus properties that help prevent microbends in the fiber optics. The thermoplastic polyester elastomer is extruded over a central steel wire that imparts strength to the cable. Six fiber optic strands with a diameter of 5 mils each are then positioned and covered by a second Hytrel layer, which is added as part of a coextrusion with nylon to obtain greater abrasion resistance. Additional steel layers and a longitudinally formed copper tape are added, and the whole structure is then enclosed in a polyethylene jacket.

Coatings

A wide range of coating materials based on polyurethanes and epoxy resins is available to protect drilling rigs from corrosion. Prodoguard S and Dark Screed are two epoxy resin products available from Prodorite Limited. The products offer good adhesion in conditions requiring chemical and corrosion resistance and have the advantages of nonslip properties, high-impact resistance and noise deadening qualities. Prodoguard S offers a greater weight saving. Both products are used on walkways and other metal deck areas on offshore structures.

Zebron is a polyurethane corrosion-resistant coating material available from ACO-Coatings. It is a solvent-free product that cures in cold wet conditions and can be rapidly applied to a high film thickness in one operation. The material provides a thick abrasion and water-resistant coating to steel or concrete.

An extremely fire-resistant coating material designed for spray applications to onshore and offshore oil and gas facilities has been introduced by Hempel's Marine Paints. The product, called ContraFlam 3810, is easy to apply and is particularly well-suited to the protection of structural steel work, walls, and floors. Another coating of a surface hardener/weather barrier material, such as Contraflam Topclad 3811, can then be applied followed by a color coat.

Most anti-fouling paints incorporate a copper pigment that provides a toxic environment for biological growth. The cuprion anti-fouling system introduced by Cathodic Protection Company Limited, however, differs in that it releases copper into solution by passing direct current through a copper electrode. This produces soluble copper in minute amounts making a hostile environment for the growth of algae, slimes, and mussels. The Cuprion system is used by BP to keep its platforms on its Forties Field in the North Sea free from marine growth.

Other Applications

Hoses for offshore applications are in most cases made of abrasion-resistant materials such as poly(vinyl chloride) or polyurethane. Polytetrafluoroethylene hoses have been developed for use as interconnectors to oil or gas combustion chamber burners or turbines. The inert qualities, ability to withstand the hostile environment, and high-pressure capability suit them to this application.

Special underwater adhesives are available for applying to submerged metal substrates and organic-based substrates. A range of such adhesives are available from Wessex Resins and Adhesives Limited as products of WRA series that show great joint strength and durability. Products such as WR 4301, WR 4302, WRA 4303, WRA 4401, and WRA 4501 are basically two-part underwater adhesives based on epoxy resins and a sacrificial pretreatment solution. The pretreatment solution promotes spontaneous and uniform spreading of the adhesive formulation when applied to metal substrates under water. For applying to organic-based substrates, for example fiber-reinforced plastic, the formulations may however be used without the pretreatment solution.

Special polymeric material is needed to make heavy, finned pistons used in through-the-flow-line well servicing operations. Dowty Seals has developed, in collaboration with Shell, a new type of polymer for such critical service duties. The material produced under Dowty's material reference No. 2471 is able to withstand 130 hours of simulated deep-well travel at temperatures intermittently peaking to 130°C, without suffering damage.

Several methods are available for the treatment of oil spillages. The Rigidoil system developed by BP has two components, a polymer and a cross-linking agent. The two components are sprayed simultaneously onto the spill where they mix and rapidly cross-link to produce a rubbery solid in which the oil becomes physically entrapped. The solid assembly is removed easily.

POLYMERS IN SPORT

The properties of polymers—resilience, toughness, and good strength-to-weight ratio—ensured the quick entry of polymers into the sports industry. In most cases, the polymer has been used as a replacement for another material and, being a competitive business, the sports industry has also been very receptive to new developments in polymer formulations and advances in design that contribute to improved sporting performance. Uses of polymers in the sports industry fall into three main categories: synthetic surfaces, footwear, and equipment.

Synthetic Surfaces

In sports such as athletics, traditional materials have been almost completely replaced by synthetic materials. Sports surfaces in modern indoor stadiums are usually made from a porous structure of rubber crumb and a binder such as polyurethane. Other polymeric materials used include flexible PVC or rubber sheet, polyolefins, polypropylene (PP) grasses, and foam laminates.

Following the use of Monsanto's woven PP Astro Turf at the Montreal Olympics in 1976, artificial surfaces have been accepted at World Championships level in hockey. Cricket is often considered a traditional and conservative game, but in fact a great many synthetic surfaces are now used. The acceptance of synthetic surfaces in soccer has been somewhat slower. The first synthetic pitch installed inside a stadium in England was by Queen's Park Rangers who used a PP omniturf surface. This was made of tufted PP rather than woven or knitted, with a pile of about 20 mm. Once laid, graded sand was spread into the carpet to within 1 to 2 mm of the surface to completely fill the pile.

The requirements of ice-skating surfaces, which are very different from other synthetic surfaces, have been met very successfully by plastics. In 1982 High Density Plastics launched the first full-size synthetic skating floor. The surface is made of interlocking panels of high density polyethylene (tradenamed Hi-Den-Ice), which become an ice-rink when sprayed with a gliding fluid. One big advantage over ice is that the surface only needs to be cleaned off and resprayed once a month. However, in a dry form the panels can also be used for basketball, football, or badminton.

PN Structures, a leading producer of synthetic surfaces in England, produces a range of synthetic Neoturf surfaces for outdoor tennis in a choice of hardwearing Neolast and Recaflex T. For indoor tennis, PNS can adapt its Neolast surface to provide varying area elasticity and shock absorption. Nondirectional sand-filled Neoturf for synthetic golfing greens is also offered. Other new surfaces include Supreme Court from Paw Sport and Leisure designed specifically for tennis, and Uni-Turf from Dunlop, a PVC sheet material with a specially designed embossed "Eldorado" finish used for indoor cricket.

Footwear

There has been significant development in sports footwear in recent years, greatly aided by polymeric materials that allow for better shape, manufacturing techniques, and material performance.

In the design of running shoes, much emphasis has been placed on anatomical comfort leading to the development of toe bars and heel canters, wedges, and midsoles, usually molded in polyethylene foams. Greater attention has been focussed on the outsole of sport shoes in order to cope with the increased use of indoor facilities and artificial surfaces. Turntec's Apex 580 sports shoe offers a unique feature of interchangeable soles for different surfaces. The soles are held on by Scotchmate, a fastening system developed by 3M and can be interchanged.

Ski boots use a wide variety of plastics adapted for low temperatures. Thermoplastic polyurethane/ABS blends and polyamides are used for the outer shell of ski boots and modified polyethylene terephthalate for the binding. Polyurethane (PU) foam is often used to line the boots. Microsphere fillers may be incorporated into foam-lined boots to add further thermal insulation to the wax binder. The antivibrational characteristics of PU foams have also led to their use in ski fittings. A sandwich construction of polyurethane elastomer and aluminum alloy has been fitted between the ski and the binding to reduce shock and vibration.

Equipment

Developments in plastics sports equipment are mainly aimed at improving player performance and increasing player safety. Items that fall into the first category include balls, bicycles, tennis rackets, and windsurfing boards. Crash helmets belong in the second category.

Since the maximum performance of sports balls is governed by strict specifications, the scope of using plastics to improve them is rather limited. Improvements can however still be made. For example, microsphere fillers have been used in golf balls and bowling balls. The travel distance of golf balls is decreased by up to 30% by the incorporation of microspheres. In bowling balls, incorporating of microspheres reduces the density of the ball's core enabling lightweight balls to be made for children and women players.

Plastics are being used to make a number of bicycle parts replacing metal. Examples include polyamides for structural components such as the bike frames and wheels, and polyacetals for tire valves, brake levers, and gear change and power switches.

Windsurfing is a sport that may be said to have been created by the development of suitable plastics. The sailboards themselves are often made of coextruded ABS sheets as General Electric's Lexan. Sailboard accessories use thermoplastic blends; for example, boom fittings and mast steps that have to withstand very high dynamic loads are made of glass-fiber-reinforced polyamide blends. Polycarbonate/ABS blends are commonly used for daggerboard cases. The most popular type of sails are film sails, which are laminates of a plastic film, usually a polyester such as Du Pont's Mylar, bonded to woven fabric. They are claimed to offer superior tear resistance. A more recent development in windsurfers is a sliding mast molded in Du Pont's Delrin acetal homopolymer. The sliding mast permits adjustment of its position while surfing.

The tennis racket is another example of sports equipment that has been improved significantly due to the availability of high-performance plastics. The carbon fiber-reinforced nylon tennis racket frame first introduced by Dunlop has become the industry's standard for this equipment. An improved design incorporates Atochem's Pebax nylon ether block copolymer into the racket frame where maximum vibrations are concentrated, i.e., between the two profiles of the frame in the handle and across the throat of the frame. For racket strings ICI introduced a high performance yarn called Zyex made from Victrex polyether ether ketone. It is claimed that the Zyex strings have a good recovery and low dynamic modulus like the material originally used (natural gut), but unlike the latter they are not affected by exposure to sunlight or moisture and retain their tension better.

An item of sports equipment that has been improved greatly with regard to player safety is the crash helmet. American football helmets are polycarbonate or ABS shelled, while impact nylons are often used on motorbike crash helmets. Another item of sports equipment in the United States where safety has been improved is a Reduced Injury Factor baseball that has a novel polyurethane core material.

BIBLIOGRAPHY

L. C. E. Struik, *Plastics: A class of modern engineering materials, Int. J. Materials and Products Technology*, 4(4): 319 (1989).

J. Baker, *Recent Developments in Polymer Technology*, RAPRA Technology, Shawbury, England, (1987).

P. E. Cassidy, Overview of polymers for harsh environments: aerospace, geothermal and undersea. American Chemical Society Division of Polymeric Materials: Science and Engineering, Washington DC, 56:380 (1987).

H. Hinsken, Composite plastic film in the packaging sector—a survey, *Kunstoffe—German Plastics*, 77(5): 3 (1987).

P. A. Moskowitz and C. A. Kovac, Plastics: the sine qua non of electronics packaging, *Plastics Eng.* 46(6): 39 (June 1990).

S. A. Wood, Performance polymers are finding greater use in packaging markets, *Mod. Plastics*, 67(8): 62 (1990).

C. Vipulanandan and E. Paul, Performance of epoxy and polyester polymer concrete, *ACI Materials J.* (American Concrete Institute), 87(3): 241 (May–June 1990).

J. Davidovits, D. C. Comrie, H. H. Peterson, and D. J. Ritcey, Geopolymeric concretes for environmental protection, *Concrete Int.: Design and Construction*, 12(7): 30 (July 1990).

R. Weibner and J. Adler, Plastics for the interior trim of passenger cars—present situation and trends for the future, SAE Technical Paper Series, SAE, Warrendale, Pennsylvania, (1987).

J. Rabe, Plastic elements in and around the engine, *International J. of Vehicle Design,* *11*(3): 246 (1990).

N. G. Marks, Polymer composites for helicopter structures, *Metals and Materials* (Institute of Metals), *5*(8): 456 (Aug 1989).

B. J. Stevens, Three dimensional circuitisation of plastic moldings, *Circuit World, 13*(4): 24 (July 1987).

L. T. Nguyen, Issues in molding of electronic packages, Proceedings of Manufacturing International, ASME, New York, *4*: 119 (1988).

W. F. Beach and T. M. Austin, Dielectrics: a chemist's overview of today's alternatives, *Hybrid Circuit Technology 7*(1): 33 (Jan 1990).

C. N. Bowman, A. L. Carver, S. N. Kenneth, M. M. Williams, and N. A. Peppas, Polymers for information storage systems. III. Cross-linked structure of polydimethacrylates, *Polymer 31*(1): 135 (Jan 1990).

M. Anderman and J. T. Lundquist, *Conductive Polymer Batteries, Assessment of Technology*, Electrochemical Society Pennington, New Jersey (1987).

P. A. Lewin and M. E. Schafer, Wide-band piezoelectric polymer acoustic sources, *IEEE Transactions on Ultrasonics, Ferroelectrics and Frequency Control, 35*(2): 175 (March 1988).

G. K. Lewis, Chirped PVDF transducers for medical ultrasound imaging, Ultrasonics Symposium Proceedings, IEEE, New York, (1987).

G. Kaempf, Special polymers for data memories, *Polymer J., 19*(2): 257 (1987).

J. M. Pearson, Polymers in optical recording, Proceedings of the ACS Division of Polymeric Materials: science and engineering, ACS, Washington DC, *55*:4 (1986).

C. L. McCormick, Polymers in agriculture: an overview, *Polymer Preprints, 28*(2): 90 (Aug 1987). (Published by the Division of Polymer Chemistry Inc., Newark, NJ, USA).

D. Gilead, Photodegradable films for agriculture, *Polymer Degradation and Stability, 29*(1): 65 (1990).

C. G. Gebelein, C. Tai, and V. Yang, Cosmetic, pharmaceutical and medical polymers: a common theme, Polymeric Materials Science and Engineering, Proceedings of the ACS Division of Polymeric Materials Science and Engineering, *63*: 401 (1990).

K. Petrak and E. Tomlinson, Synthetic polymers for controlled drug release and directed drug delivery, *Chemistry and Industry* (London), 45 (1987).

D. F. Williams, A. McNamara, and R. M. Turner, Potential of polyether etherketone (PEEK) and carbon fiber-reinforced PEEK in medical applications, *J. Materials Science Letters, 6*(2): 188 (Feb 1987).

R. J. Wagner, Synthetic polymer body prosthetic materials, Symposium Notes of the Technical Association of the Pulp and Paper Industry, TAPPI Press, Atlanta, Georgia (1986).

J. R. Ellis, Plastics/packaging, PVC. New products keep coming, *Medical Device and Diagnostic Industry, 12*(3): 90 (1990).

S. W. Shalaby, A. S. Hoffman, B. B. Ratner, and T. A. Horbett (eds.), *Polymers as Bio-materials*, Plenum Press, New York, (1984).

C. Lodge, Tomorrow's plastics may add years to your life, *Plastics World*, *45*(10): 5 (1987).

T. Merchant, Composites at sea—the quiet revolution, *Engineering* (London), *227*(6): 7 (1987).

Appendix 1A

Trade Names for Some Industrial Polymers

Trade name	Company	Type of polymer
Absafil	Fiberfill	Acrylonitrile-butadiene-styrene terpolymer
Abson	B. F. Goodrich	Acrylonitrile-butadiene-styrene terpolymer
Acrylan	Monsanto	Acrylic fiber
Acrylite	Cyanamid/Rohm	Acrylic resin
Adiprene	Du Pont	Polyurethanes
Afcoryl	Pechiney-Saint-Gobain	Acrylonitrile-butadiene-styrene terpolymer
Aflas	Asahi Glass	Tetrafluoroethylene-propylene + cure site monomer terpolymer
Aflon	Asahi Glass	Tetrafluoroethylene-ethylene copolymer
Akulon	Akzo	Nylon-6
Alathon	Du Pont	Low-density polyethylene
Alkathene	ICI	Low-density polyethylene
Alkox	Meisei Chemical Works	Poly(ethylene oxide)
Alloprene	ICI	Chlorinated natural rubber
Amberlite	Rohm & Haas	Ion-exchange resin

This selection confers no priorities and is not exhaustive.

Trade name	Company	Type of polymer
Ameripol-CB	B. F. Goodrich	Polybutadiene
Amidel	Union Carbide	Transparent amorphous polyamide
Antron	Du Pont	Polyamide fiber
Araldite	Ciba-Geigy	Epoxy resin
Ardel	Union Carbide	Polyarylate
Arnel	Celanese	Cellulose triacetate
Arnite	Akzo	Poly(ethylene terephthalate)
Arnite PBTP	Akzo	Poly(butylene terephthalate)
Arnitel	Akzo	Thermoplastic polyester elastomer
Arylef	Solvay	Polyarylate
Astrel	3M	Polyarylsulfone
Baylon	Bayer	Polycarbonate
Baypren	Bayer	Polychloroprene
Beetle	British Industrial Plastics	Urea-formaldehyde resin
Benvic	Solvay	Poly(vinyl chloride)
Budene	Goodyear	Polybutadiene
Buna-N	Chem. Werke Hüls	Butadiene-acrylonitrile copolymer
Buna-S	Chem. Werke Hülls	Butadiene-styrene copolymer
Butacite	Du Pont	Poly(vinyl butyral)
Butakon	ICI	Butadiene copolymers
Butaprene	Firestone	Styrene-butadiene copolymers
Butvar	Shawinigan	Poly(vinyl butyral)
BXL	Union Carbide	Polysulfone
Caprolan	Allied	Polyamide fiber
Capron	Allied	Polyamide resin
Caradol	Shell	Polyhydroxy compound for isocyanate cross-linking
Carbopol	B. F. Goodrich	Acrylic polyelectrolyte
Carboset	B. F. Goodrich	Acrylic polyelectrolyte
Carbowax	Union Carbide	Poly(ethylene oxide)
Cariflex I	Shell	cis-1,4-Polyisoprene
Carina	Shell	Poly(vinyl chloride)
Carinex	Shell	Polystyrene
Celanex	Celanese	Poly(butylene terephthalate)
Celcon	Celanese	Polyacetal
Cellidor B	Bayer	Cellulose acetate-butyrate
Cellit	Bayer	Cellulose acetate
Cellon	Dynamit-Nobel	Cellulose acetate
Cellosize	Union Carbide	Hydroxyethylcellulose
Celluloid	Dynamit Nobel	Cellulose nitrate plasticized with camphor
Chemigum	Goodyear	Polyurethane rubber
Cibamin	Ciba-Geigy	Urea-formaldehyde, melamine-formaldehyde resins

Trade name	Company	Type of polymer
Cibanoid	Ciba-Geigy	Urea-formaldehyde resin
Cis-4	Phillips	*cis*-1,4-Polybutadiene
Clariflex TR	Shell	Styrene-diene-styrene triblock elastomer
Cobex	Bakelite Xylonite	Poly(vinyl chloride)
Coral	Firestone	*cis*-1,4-Polyisoprene
Cordura	Du Pont	Polyamide fiber
Corvic	ICI	Poly(vinyl chloride)
Courtelle	Courtaulds	Polyacrylonitrile
Crastin	Ciba-Geigy	Poly(butylene terephthalate)
Creslan	Cyanamid	Acrylic fiber
Crystic	Scott Bader	Polyester resins
Cycolac	Borg-Warner	Acrylonitrile-butadiene-styrene terpolymer
Cymel	Cyanamid	Melamine-formaldehyde resin
Dacron	Du Pont	Poly(ethylene terephthalate) fiber
Daltoflex 1	ICI	Polyurethane rubber
Dapon	FMC Corp.	Diallyl phthalate resins
Darvic	ICI	Poly(vinyl chloride)
De-acidite	Permutit Co.	Ion-exchange resin
Delrin	Du Pont	Polyacetal
D.E.R.	Dow	Epoxy resin
Desmopan	Bayer	Polyurethanes
Desmophen A	Bayer	Polyurethane rubber
Diakon	ICI	Poly(methyl methacrylate) molding powder
Diene	Firestone	Polybutadiene
Dorlastan	Bayer	Spandex fiber
Dowex	Dow	Ion-exchange resins
Duolite	Chemical Process Co.	Ion-exchange resins
Duracon	Polyplastic	Polyacetal
Durel	Hooker	Polyarylate
Durethan	Bayer	Nylon-6
Durethan U	Bayer	Thermoplastic polyurethanes
Dutral	Montecatini	Ethylene-propylene copolymer
Dycryl	Du Pont	Photopolymer system
Dyflor	Dynamit Nobel	Poly(vinylidene fluoride)
Dynel	Union Carbide	Vinyl chloride-acrylonitrile copolymer
Econol	Carborundum	Poly (*p*-hydroxybenzoic acid ester)
Ekkcel	Carborundum	Aromatic polyester
Ekonol	Carborundum	Aromatic polyester
Elvacet	Du Pont	Poly(vinyl acetate)
Elvanol	Du Pont	Poly(vinyl alcohol)

Trade name	Company	Type of polymer
Elvic	Solvay	Poly(vinyl chloride)
Encron	Akzo	Polyester fiber
Epicote	Dow and Shell	Epoxy resin
Epodite	Showa Highpolymer	Epoxy resin
Epon	Shell	Epoxy resin
Epi-Rez	Celanese	Epoxy resin
Estane	B. F. Goodrich	Polyurethane elastomer
Estar	Eastman Kodak	Polyester film
Ethocel	Dow	Ethylcellulose
Filabond	Reichhold	Unsaturated polyester
Flemion	Asahi Glass	Carboxylated fluoropolymer
Flovic	ICI	Poly(vinyl acetate)
Fluon	ICI	Polytetrafluoroethylene
Fluorel	3M	Vinylidene fluoride-hexafluoropropylene copolymer
Forticel	Celanese	Cellulose propionate
Formica	Cyanamid	Melamine-formaldehyde resin
Gelvatol	Shawinigan	Poly(vinyl alcohol)
Geon	B. F. Goodrich	Poly(vinyl chloride)
Grilamid TR	Emser Werke	Transparent amorphous polyamide
Halar	Allied	Ethylene-Chlorotrifluoroethylene copolymer
Halon	Allied	Polychlorotrifluoroethylene
Herculoid	Hercules	Cellulose nitrate
H-film	Du Pont	Polyamide from pyromellitic anhydride and 4,4-diaminodiphenyl ether
Hitalex	Hitachi	Polyethylene
Hitanol	Hitachi	Phenol-formaldehyde resin
Hostadur	Hoechst	Poly(ethylene terephthalate)
Hostaflon C2	Hoechst	Polychlorotrifluoroethylene
Hostaflon ET	Hoechst	Tetrafluoroethylene-ethylene copolymer
Hostaflon TF	Hoechst	Polytetrafluoroethylene
Hostaform	Hoechst	Polyoxymethylene
Hostalen	Hoechst	Polyethylene; polypropylene
Hostalit	Hoechst	Poly(vinyl chloride)
Hostamid	Hoechst	Transparent amorphous polyamide
Hostatec	Hoechst	Polyether ketone
Hycar	B. F. Goodrich	Polyacrylate
Hydrin	B. F. Goodrich	Epichlorohydrin rubber
Hylar	Du Pont	Poly(ethylene terephthalate)
Hypalon	Du Pont	Sulfochlorinated PE
Hytrel	Du Pont	Segmented aromatic polyester

Trade name	Company	Type of polymer
Icdal Ti40	Dynamit Nobel	Polyesterimide
Ionac	Ionac Co.	Ion-exchange resins
Irrathene	G.E.	PE, cross-linked by radiation
Ixef	Solvay	Aromatic polyamide
Jupilon	Mitsubishi	Polycarbonate
Kalrez	Du Pont	Fluoroelastomer
Kamax	Rohm & Haas	Acrylic resin
Kapton	Du Pont	Polyimide film
Kel-F	3M	Polychlorotrifluoroethylene
Kel-F elastomer	3M	Vinylidene fluoride-chlorotri-fluoroethylene copolymer
Kematal	ICI	Polyoxymethylene
Kermel	Rhone-Poulenc	Polyimide fiber
Kerimid	Rhone-Poulenc	Polybismaleimide
Kinel	Rhone-Poulenc	Polybismaleimide
Kodapak	Eastman Kodak	Poly(butylene terephthalate)
Kodar PETG	Eastman Kodak	Copolyester based on 1,4-cyclohexylene glycol and a mixture of terephthalic and isophthalic acids
Kodel	Eastman Kodak	Polyester fiber
Kralastic	Uniroyal	Acrylonitrile-butadiene-styrene copolymer
Kraton	Shell	Thermoplastic styrene block copolymer
Krynac	Doverstrand	Carboxylated nitrile rubber
Kynar	Pennsalt	Poly(vinylidene fluoride)
Laminac	Cyanamid	Polyester resins
Leguval	Bayer	Polyester resins
Lekutherm	Bayer	Epoxy resins
Levapren	Bayer	Ethylene-vinyl acetate copolymer
Lewatit	Bayer	Ion-exchange resins
Lexan	G.E.	Polycarbonate
Lucite	Du Pont	Poly(methyl methacrylate) and copolymers
Luranyl	BASF	Poly(phenylene oxide) blend
Lustran	Monsanto	Acrylonitrile-butadiene-styrene terpolymer
Lustrex	Monsanto	Polystyrene
Lycra	Du Pont	Spandex fiber
Makrolon	Bayer	Polycarbonate
Marlex	Phillips	Polyethylene; polypropylene
Melinex	ICI	Polyester film
Melolam	Ciba-Geigy	Melamine-formaldehyde

Trade name	Company	Type of polymer
Melopas	Ciba-Geigy	Melamine-formaldehyde
Merlon	Mobay	Polycarbonate
Methocel	Dow	Methyl cellulose
Moltopren	Bayer	Polyurethane foam
Moplen	Montedison	Polypropylene
Mowicoll	Hoechst	Poly(vinyl acetate) dispersions
Mowilith	Hoechst	Poly(vinyl acetate)
Mowiol	Hoechst	Poly(vinyl alcohol)
Mowital	Hoechst	Poly(vinyl butyral)
Mylar	Du Pont	Poly(ethylene terephthalate) film
Nafion	Du Pont	Persulfonated fluoropolymer
Nalcite	National Aluminate Corp.	Ion-exchange resins
Napryl	Pechiney-Saint-Gobain	Polypropylene
Natene	Pechiney-Saint-Gobain	Polyethylene
Natsyn	Goodyear	Polyisoprene
Neosepta F	Tokoyama Soda	Ionic membrane (based on fluoro-polymer)
Nipeon	Japanese Geon	Poly(vinyl chloride)
Nipoflex	Toyo Soda	Ethylene-vinyl acetate copolymer
Nipolon	Toyo Soda	Polyethylene
Nitron	Monsanto	Cellulose nitrate
Nomex	Du Pont	Poly(m-phenylene isophthalimide)
Nordel	Du Pont	Ethylene-propylene-diene terpolymer
Norsorex	Cyanamid	Polynorbornadiene
Noryl	G.E.	Poly(phenylene oxide)-polystyrene blend
Novodur	Bayer	Acrylonitrile-butadiene-styrene terpolymer
Oppanol B	BASF	Polyisobutylene
Oppanol C	BASF	Poly(vinyl isobutyl ether)
Oppanol O	BASF	Copolymer from 90% isobutylene and 10% styrene
Orlon	Du Pont	Acrylic fiber
Paraplex	Rohm & Haas	Polyester resins
Parlon	Hercules	Chlorinated rubber
Paxon	Allied	Polyethylene
Pebax	Rilsan	Polyether block amide (thermo-plastic elastomer)
Pelaspan	Dow	Polystyrene (expandable)
Pellethane	Upjohn	Polyurethane
Penton	Hercules	Poly-3,3-bis(chloromethyl)-oxacyclobutane
PEO	Seitetsu Kagaku	Poly(ethylene oxide)

Trade name	Company	Type of polymer
Perbunan N	Bayer	Butadiene-acrylonitrile copolymers
Perlenka	Akzo	Nylon-6
Permutit	Permutit Co.	Ion-exchange resins
Perspex	ICI	Poly(methyl methacrylate) sheet
Petron	Mobay	Poly(ethylene terephthalate)
Petrothene	USI Chemical	Low-density polyethylene
Pevalon	May & Baker	Poly(vinyl alcohol)
Plastylene	Pechiney-Saint-Gobain	Polyethylene
Plexidur	Rohm & Haas	Copolymer from acrylonitrile and methyl methacrylate
Plexiglass	Rohm & Haas	Poly(methyl methacrylate)
Plexigum	Rohm & Haas	Acrylate and methacrylate resins
Pliofilm	Goodyear	Rubber hydrochloride
Pliolite	Goodyear	Styrene-butadiene copolymers
Pocan	Bayer	Poly(butylene terephthalate)
Poly-Eth	Gulf Oil	Polyethylene
Polylite	Reichhold	Polyester resins
Polymin	BASF	Polyethyleneimine
Polyox	Union Carbide	Poly(ethylene oxide)
Polysizer	Showa Highpolymer	Poly(vinyl alcohol)
Polythene	Du Pont	Low-density polyethylene
Polyviol	Wacker Chemie	Poly(vinyl alcohol)
Prevex	Borg Warner	Poly(phenylene oxide) blend
Profax	Hercules	Polypropylene
Propathene	ICI	Polypropylene
Pyralin	Du Pont	Polyimide
Pyre ML	Du Pont	Polyimide coating
Q2	Eastman	Polyamide from 1,4-bis-aminomethyl cyclohexane and suberic acid
Qiana	Du Pont	Polyamide fiber from trans trans-diamino-dicyclohexylmethane and dodecanedioic acid
Radel	Union Carbide	Polyether sulfone
Ravinil	ANIC, S.p.A.	Poly(vinyl chloride)
Resolite	Ciba-Geigy	Urea-formaldehyde resin
Restirolo	Societa Italiana Resine	Polystyrene
Revinex	Doverstrand	Carboxylated rubber
Riblene	ANIC, S.p.A.	Polyethylene
Rigolac	Showa Highpolymer	Polyester resins
Rilsan	Ato Chimie	Polyamide resin
Roylar	Uniroyal	Polyurethanes
Royalene	Uniroyal	Ethylene-propylene-diene terpolymer
Rucon	Hooker	Poly(vinyl chloride)

Trade name	Company	Type of polymer
Rucothane	Hooker	Polyurethanes
Rynite	Du Pont	Glass-reinforced poly(ethylene terephthalate) resin
Ryton	Phillips	Poly(phenylene sulfide)
Saflex	Monsanto	Poly(vinyl butyral)
Saran	Dow	Copolymers of vinylidene chloride and vinyl chloride
Scotchpak	3M	Polyester film
Silastic	Dow	Silicones
Silastomer	Dow	Silicones
Silopren	Bayer	Silicone rubber
Sirfen	Societa Italiana Resine	Phenol-formaldehyde resin
Sirotherm	ICI	Ampholytic polyelectrolyte
Sirtene	Societa Italiana Resine	Polyethylene
Skybond	Monsanto	Polyimide
Sokalan CP2	BASF	Polyelectrolyte
Solprene	Phillips	Thermoplastic elastomer
Solef	Solvay	Poly(vinylidene fluoride)
Solvic	Solvay	Poly(vinyl chloride)
Soreflon	Rhone-Poulenc	Polytetrafluoroethylene
Spanzelle	Courtaulds	Spandex fiber
Standlite	Hitachi	Phenol-formaldehyde resin
Styrocell	Shell	Polystyrene foam
Styrofoam	Dow	Polystyrene foam
Styrol	Idemitsu	Polystyrene
Styron	Dow	Polystyrene
Sumikon	Sumitomo Bakelite	Epoxy resin
Surlyn	Du Pont	Ionomer
Tecnoflon	Montecatini	Fluoropolymer
Tedlar	Du Pont	Poly(vinyl fluoride)
Tedur	Bayer	Poly(phenylene sulfide)
Teflon FEP	Du Pont	Tetrafluoroethylene-hexafluoropropylene copolymer
Teflon TFE	Du Pont	Polytetrafluoroethylene
Tenite	Eastman Kodak	Polyethylene; polypropylene
Tenite Acetate	Eastman Kodak	Cellulose acetate
Tenite Butyrate	Eastman Kodak	Cellulose acetate-butyrate
Tenite Propionate	Eastman Kodak	Cellulose acetate-propionate
Tenite PTMT	Eastman Kodak	Poly(tetramethylene terephthalate)
Teracol	Du Pont	Polyoxytetramethylene glycol
Terlenka	Akzo	Poly(ethylene terephthalate)
Terluran	BASF	Acrylonitrile-butadiene-styrene terpolymer
Terylene	ICI	Poly(ethylene terephthalate)

Trade name	Company	Type of polymer
Texicote	Scott Bader	Poly(vinyl acetate)
Texigels	Scott Bader	Acrylic polyelectrolyte
Texin	Mobay	Polyurethane elastomer
Thiokol	Thiokol	Polysulfides
Thornel	Union Carbide	Graphite yarn
Torlon	Amoco	Polyamide-imide
Toyarac	Toyo	Acrylonitrile-butadiene-styrene terpolymer
Toyoflon	Toyo	Polytetrafluoroethylene
TPX	ICI	Poly-4-methylpent-1-ene
Trans-4	Phillips	*trans*-1,4-Polybutadiene
Trevira	Hoechst	Polyester fiber
Tricel	Bayer	Cellulose acetate
Trogamid	Dynamit Nobel	Aromatic polyamide
Trolen	Dynamit Nobel	Polyethylene
Trolit AE	Dynamit Nobel	Cellulose ether
Trolit F	Dynamit Nobel	Cellulose nitrate
Trolitan	Dynamit Nobel	Phenol-formaldehyde resin
Trolitul	Dynamit Nobel	Polystyrene
Trosiplast	Dynamit Nobel	Poly(vinyl chloride)
Trovidur	Dynamit Nobel	Poly(vinyl chloride)
Tybrene	Dow	Acrylonitrile-butadiene-styrene terpolymer
Tynex	Du Pont	Nylon-6,6
Tyril	Dow	Styrene-acrylonitrile copolymer
Udel	Union Carbide	Polysulfone
Ultem	G.E.	Polyetherimide
Ultradur	BASF	Poly(butylene terephthalate)
Ultramid A	BASF	Nylon-6,6
Ultramid B	BASF	Nylon-6
Ultramid S	BASF	Nylon-6,10
Ultrapas	Dynamit Nobel	Melamine-formaldehyde
Ultrapek	BASF	Polyether ketone
Ultryl	Phillips	Poly(vinyl chloride)
U-Polymer	Unitika	Polyarylate
Urecoll	BASF	Urea-formaldehyde
Urepan	Bayer	Polyurethane
Valox	G.E.	Poly(butylene terephthalate)
Vectra	Celanese	Aromatic copolyester (liquid crystalline polymer)
Vedril	Montecatini	Poly(methyl methacrylate) molding powder
Versamid	General Mills	Fatty polyamides
Versicol	Allied Colloids	Acrylic polyelectrolyte

Trade name	Company	Type of polymer
Vespel	Du Pont	Polyimide resin
Vestamid	Chem. Werke Hüls	Nylon-12
Vestolen A	Chem. Werke Hüls	Polyethylene (low pressure)
Vestolen BT	Chem. Werke Hüls	Polybut-1-ene
Vestolen P	Chem. Werke Hüls	Polypropylene
Vestolit	Chem. Werke Hüls	Poly(vinyl chloride)
Vestoran	Chem. Werke Hüls	Styrene-acrylonitrile copolymer
Vestyron	Chem. Werke Hüls	Polystyrene
Vibrathane	Uniroyal	Cast polyurethane elastomers
Viclan	ICI	Vinylidene chloride copolymers
Victrex	ICI	Polyethersulfone
Vinapas	Wacker Chemie	Poly(vinyl acetate)
Vinavil	Montecatini	Poly(vinyl acetate)
Vinaviol	Montecatini	Poly(vinyl alcohol)
Vinidur	BASF	Poly(vinyl chloride)
Vinnol	Wacker Chemie	Poly(vinyl chloride)
Vinoflex	BASF	Vinyl chloride homo and copolymers
Vipla	Montecatini	Poly(vinyl chloride)
Vistalon	Exxon	Ethylene-propylene-diene terpolymer
Vistanex	Exxon	Polyisobutylene
Vithane	Goodyear	Polyurethanes
Viton A	Du Pont	Vinylidene fluoride-hexafluoropropylene copolymer
Viton B	Du Pont	Vinylidene fluoride-hexafluoropropylene-tetrafluoroethylene terpolymer
Vulcaprene	ICI	Polyurethanes
Vulkollan	Bayer	Polyurethanes
Vybak	Bakelite Xylonite	Poly(vinyl chloride)
Vydyne	Monsanto	Polyamide resin
Vyrene	U.S. Rubber	Spandex fiber
Welvic	ICI	Poly(vinyl chloride)
Wofatit	VEB Farbenfabrik	Ion-exchange resins
Xydar	Dartco Mfg.	Wholly aromatic copolyester injection molding resin
Zeo-karb	Permutit Co.	Ion-exchange resins
Zetafin	Dow	Ethylene-vinyl acetate copolymer
Zytel	Du Pont	Nylon-6,10
Zytel 101	Du Pont	Nylon-6,6

Appendix 1B

Commonly Used Abbreviations for Industrial Polymers

ABS	Acrylonitrile-butadiene-styrene terpolymer
ACS	Acrylonitrile-chlorinated polyethylene-styrene terpolymer
BR	Butadiene rubber
CA	Cellulose acetate
CAB	Cellulose acetate butyrate
CAP	Cellulose acetate propionate
CMC	Carboxymethyl cellulose
CN	Cellulose nitrate
CPE	Chlorinated polyethylene
CPVC	Chlorinated poly(vinyl chloride)
CR	Polychloroprene
CTA	Cellulose triacetate
EC	Ethyl cellulose
ECTFE	Ethylene chlorotrifluoroethylene copolymer
EMA	Ethylene methacrylate copolymer
EP	Epoxy resin
E/P	Ethylene-propylene copolymer
EPDM	Ethylene-propylene-diene terpolymer
EPM	Ethylene-propylene copolymer
EPR	Elastomeric copolymer of ethylene, and propylene
EPT, EPTR	Elastomeric copolymer of ethylene, propylene, and a diene

ETFE	Ethylene tetrafluoroethylene copolymer
EVA	Copolymer from ethylene and vinyl acetate
FEP	Fluorinated ethylene propylene copolymer
HDPE	High-density polyethylene
HIPS	High-impact polystyrene
IIR	Butyl rubber (isobutylene-isoprene copolymer)
IPN	Interpenetrating polymer network
IR	1,4-*cis*-polyisoprene rubber
LDPE	Low-density polyethylene
LLDPE	Linear low-density polyethylene
MF	Melamine-formaldehyde resin
NBR	Acrylonitrile-butadiene rubber (nitrile rubber)
NC	Nitrocellulose (cellulose nitrate)
NR	Natural rubber
PA	Polyamide
PAA	Poly(acrylic acid)
PAI	Polyamide-imide
PAN	Polyacrylonitrile
PB	Polybutadiene
PC	Polycarbonate
PBT	Poly(butylene terephthalate)
PE	Polyethylene
PEG	Polyethylene glycol
PEO	Poly(ethylene oxide)
PET	Poly(ethylene terephthalate)
PF	Phenol-formaldehyde resin
PI	Polyimide
PIB	Polyisobutylene
PIR	Polyisocyanurate foam
PMMA	Poly(methyl methacrylate)
PP	Polypropylene
PPG	Polypropylene glycol
PPO	Poly(phenylene oxide)
PPS	Poly(phenylene sulfide)
PS	Polystyrene
PTFE	Polytetrafluoroethylene
PTMG	Polytetramethylene glycol
PTMT	Poly(tetramethylene terephthalate)
PU	Polyurethane
PVA	Poly(vinyl acetate)
PVAL	Poly(vinyl alcohol)
PVB	Poly(vinyl butyral)

PVC	Poly(vinyl chloride)
PVDC	Poly(vinylidene chloride)
PVDF	Poly(vinylidene fluoride)
PVF	Poly(vinyl fluoride)
PVFM	Poly(vinyl formal)
PVOH	Poly(vinyl alcohol)
RTV	Room-temperature vulcanizing silicone rubber
SAN	Styrene-acrylonitrile copolymer
SBR	Styrene-butadiene rubber
UF	Urea-formaldehyde resin

Appendix 2A

Index of Trade Names, Manufacturers, and Suppliers of Antioxidants

The following index of trade names, manufacturers, and suppliers is based on a choice of representative antioxidants for thermoplastics. No claim is made for completeness. Detailed lists can be found in the source cited.

The letters A to M symbolize chemical classes of compounds with the following meanings: A = alkylphenols; B = alkylidene-bisphenols (molecular weight 300 to 600); C = alkylidene-bisphenols (molecular weight >600); D = thiobisphenols (molecular weight 300 to 600); E = hydroxybenzyl compounds (molecular weight 300 to 600); F = hydroxybenzyl compounds (molecular weight >600); G = Acylaminophenols (4-hydroxyanilides); H = octadecyl-3-(3,5-di-*tert*-butyl-4-hydroxyphenyl)-propionate; I = poly-hydroxyphenylpropionates (molecular weight >600); J = amines; K = thioethers; L = phosphites and phosphonites; M = zinc dibutyldithiocarbamate.

Trade name	Chemical class													Manufacturer/supplier
	A	B	C	D	E	F	G	H	I	J	K	L	M	
Age Rite									X				X	R. T. Vanderbilt Norwalk,
Vanox		X				X			X	X	X	X		CT, USA
Anox	X							X	X	X				Ente Nazionale Idrocarburi, Rome, Italy

Trade name	Chemical class													Manufacturer/supplier
	A	B	C	D	E	F	G	H	I	J	K	L	M	
CAO	X	X		X										Ashland Chemical Ltd., Columbus, OH, USA
Carstab							X	X						Morten Thiokol Inc., Danvers, MA, USA
Cyanox		X				X				X				American Cyanamid Co., Wayne, NJ, USA
Ethanox		X		X		X		X						Ethyl Corporation, Baton Rouge, LA, USA
Good-rite				X			X	X						B. F. Goodrich Chemical Group, Cleveland, USA
Hostanox			X								X	X		Hoechst AG, Frankfurt, Germany; American Hoechst Corp., Somerville, NJ, USA
HPM Garbefix		X	X	X							X	X		Societe Francaise d'Organo-Synthese, Nevilly-sur-seine, France
Ionol Ionox	X					X								Shell Nederland Chemie BV, Gravenhage, Netherlands
Irgafos Irganox		X		X	X	X	X	X	X	X	X	X		Ciba-Geigy AG, Basel, Switzerland; Ciba-Geigy Corp., Hawthorne, NY, USA
Isonox	X	X												Schenectady Chemicals Inc., Schenectady, NY, USA
Keminox	X	X	X	X		X								Chemipro Kasei Ltd., Chuo-Ka, Kobe, Japan
Lowinox	X	X		X				X	X	X	X	X		Chemische Werke Lowi GmbH, Waldkraiburg, Germany
Mark									X		X			Argus Chemical S.A.-N.V., Brussels, Belgium; Witco Chem. Co., New York, NY, USA
Mixxim		X										X		Fairmount Chem. Co., Newark, NJ, USA

Trade name	A	B	C	D	E	F	G	H	I	J	K	L	M	Manufacturer/supplier
						Chemical class								
Naugard		X	X					X	X	X	X	X	X	Uniroyal Ltd., Bromsgrove,
Naugawhite		X												Worcs., UK; Uniroyal
Polygard												X		Chem., Naugatuck, CT, USA
Nonox	X	X							X					ICI Ltd., London, UK
Topanol	X	X							X					ICI Americas, Wilmington,
Negonox			X											DE, USA
Oxi-chek		X						X						Ferro Corp. Chem. Div.,
														Walton Hills, OH, USA
Perkanox	X													Akzo Chemie, Nederland
														NV., Amsterdam,
														Netherlands
Permanax	X	X												Vulanax International, Saint
														Cloud, France
Samilizer		X		X				X	X		X	X		Sumitomo Chemical Co.
Antigene										X				Ltd., Osaka, Japan
Santonox				X										Monsanto Europe SA,
Santowhite	X	X		X										Brussels, Belgium;
Santoflex									X					Monsanto Co., Akron,
Santicizer				X								X		OH, USA
Seenox		X		X	X					X				Shipro Kasei Ltd., Kita-ku,
														Osaka, Japan
Tominox								X	X					Yoshitomi Pharmaceutical
Yoshinox		X		X										Ind. Ltd., Osaka, Japan
Vulkanox	X	X							X	X				Mobay Chem. Corp.,
														Pittsburgh, PA, USA
Weston												X		Borg-Warner Chemicals,
Ultranox	X	X		X			X		X			X		Parkersburg, WV, USA
Wing Stay	X			X					X					Goodyear SA, Rueil-
														Malmaison, France;
														Goodyear, Akron, OH,
														USA
Wytox	X	X							X			X		Olin Corp., Stamford, CT,
														USA

Source: T. J. Henman, *World Index of Polyolefin Stabilizers*, Kogan Page, London, 1982.

Appendix 2B

Index of Trade Names, Manufacturers, and Suppliers of Metal Deactivators

The following index of trade names, manufacturers, and suppliers is based on a choice of representative metal deactivators for thermoplastics. No claim is made for completeness. Detailed lists can be found in the source cited.

Trade name	Chemical class	Manufacturer/Supplier
Cyanox	Not disclosed	Cyanamid GmbH, Wolfratshausen, Germany American Cyanamid Co., Wayne, NJ, USA
Eastman Inhibitor OABH	Dibenzalhydrazone	Eastman Chemical Products, Kingsport, TN, USA
Hostanox	Phosphorous acid ester	Hoechst AG, Augsburg, Germany
Irganox MD	Diacylhydrazine	Ciba-Geigy AG, Division KA, Basel, Switzerland Ciba-Geigy Corp., Plastics and Additives Div., Hawthorne, NY, USA
Mark	Not disclosed	Argus Chemical SA, Drogenbos, Belgium Argus Chemical Corp., Brooklyn, NY, USA

Trade name	Chemical class	Manufacturer/Supplier
Naugard	Oxalamide derivative	Uniroyal Chemical, Naugatuck, CT, USA

Source: T. J. Henman, *World Index of Polyolefin Stabilizers*, Kogan Page Ltd., London (1982).

Appendix 2C

Index of Trade Names, Manufacturers, and Suppliers of Light Stabilizers

The following index of trade names, manufacturers, and suppliers is based on a choice of representative light stabilizers for thermoplastics. No claim is made for completeness. Detailed lists can be found in the source cited.

The letters A to J symbolize chemical classes of compounds with the following meanings: A = benzophenones; B = benzotriazoles; C = nickel compounds; D = salicylates; E = cyanocinnamates; F = benzylidene malonates; G = benzoates; H = oxanilides; I = sterically hindered amines and J = polymeric sterically hindered amines.

Trade name	Compound type	Manufacturer
Aduvex	A	Ward Blenkinsop and Co. Ltd., Wembley (Middx.), England
Anti-UV	A, D	Societe Francaise D'Organo Synthese, Neuilly-sur-Seine, France
Carstab	A	Cincinnati Milacron Chemicals Inc., New Brunswick, NY, USA
Chimassorb	A, C, J	Chimosa S.p.A., Pontecchio Marconi, Italy
Cyasorb UV	A, B, C, E, F, I, J	American Cyanamid Co., Wayne, NJ, USA

Trade name	Compound type	Manufacturer
Eastman Inhibitor	A	Eastman Chemical Products Inc., Kingsport, TN, USA
UV-Chek AM	A, C, G	Ferro Corp., Cleveland, OH, USA
Goodrite UV	I	B. F. Goodrich, Chemical Group, Cleveland, OH, USA
Hostavin	A, I, K	Hoechst AG, Frankfurt/M, Germany
Interstab	G	Interstab Ltd., Liverpool, England
Irgastab	C	Ciba-Geigy AG, Basel Switzerland
Mark	A, I	Argus Chemical Corp., Brooklyn, NY, USA
Mixxim	I, J	Fairmount Chemical Co., Newark, NJ, USA
Rylex	A, C	E. I. Du Pont de Nemours and Co., Wilmington, DE, USA
Salol	A, D	Dow Chemical Co., Midland, MI, USA
Sanol LS	I	Sankyo Co. Ltd., Tokyo, Japan
Seesorb	A, B, C, D, E, G	Sun Chemical Corp., Tokyo, Japan
Sumisorb	A, D	Sumitomo Chemical Co., Ltd., Osaka, Japan
Tinuvin	B, H, I, J	Ciba-Geigy AG, Basel, Switzerland
UV-Absorber Bayer	A, E	Bayer AG, Leverkusen, Germany
Uvinul	A, E	BASF AG, Ludwigshafen, Germany
Viosorb	A, B, D	Kyodo Chemical Co. Ltd., Tokyo, Japan

Source: T. J. Henman, *World Index of Polyolefin Stabilizers*, Kogan Page Ltd., London, 1982.

Appendix 2D

Index of Trade Names and Suppliers of Flame Retardants

The following index of trade names and suppliers is based on a choice of representative flame retardants for thermoplastics. No claim is made for completeness. Detailed lists can be found in the source cited.

Trademark	Chemical class	Producer
Amgard	Phosphoric acid ester	Albright and Wilson, London, England
Chlorowax	Chlorinated paraffins	Diamond Shamrock Plastics Div., Cleveland, OH, USA
Cereclor	Chlorinated paraffins	ICI, London, England
Dechlorane	Chlorine containing cycloaliphatics	Occidental Chemical Corp., Niagara Falls, NY, USA
Disflamoll	Organic phosphates and halogenated organic phosphates	Bayer AG, Leverkusen, Germany
Exolit	Chlorinated paraffins, organic phosphates	Hoechst AG, Frankfurt/M, Germany
Firebrake	Inorganic flame retardants	US Borax & Chem. Corp., Los Angeles, CA, USA
Fire Fighters	Organic bromine compounds	Great Lakes Chem. Corp., West Lafayette, IN, USA
Fireguard	Organic bromine compounds	Teijin Chem. Co., Nishishinbasi, Tokyo, Japan

Trademark	Chemical class	Producer
Firemaster	Organic bromine compounds	Great Lakes Chem. Corp., West Lafayette, IN, USA
Fire-Shield	Organic bromine compounds, antimony trioxide	PPG Industries Inc., Chicago, IL, USA
Firex	Organic halogenated compound with antimony trioxide	Dr. Th. Bohme KG, Geretstried, Germany
Flammastik	Flame retardant on the basis of antimony	Chem. Fabrik Grunau GmbH, Illertissen, Germany
Hydral	Hydrated alumina	Aluminum Company of America, Pittsburgh, PA, USA
Ixol	Organic bromine and chlorine compounds	Kali-Chemie AG, Hannover, Germany
Martinal	Hydrated alumina	Martinswerk GmbH, Bergheim, Germany
Nonnen	Organic bromine compounds	Marubeni Corp., Osaka, Japan
Nyacol	Flame retardant on the basis of antimony	Nyacol Products Inc., Ashland, MA, USA
Phosflex	Organic phosphates	Stauffer Chem. Co., Specialty Chem. Div., Westport, CT, USA
Proban	Phosphoric acid esters, quaternary phosphonium compounds	Albright and Wilson, London, England
Pyro-Check	Organic bromine compounds	Ferro Corp., Chem. Div., Bedford, OH, USA
Pyrovatex	Organic phosphates	Ciba-Geigy AG, Basel, Switzerland
Sandoflam	Flame retardants based on organic bromine compounds	Sandoz AG, Basel, Switzerland
Santicizer	Phosphoric acid esters, halogenated organic compounds	Monsanto Co., St. Louis, MO, USA
Saytex	Organic bromine compounds	Ethyl Corp., Chemical Group, Baton Rouge, LA, USA
Tardex	Organic bromine compounds	ISC-Chemicals, Avonmouth, Bristol, England
Timonox	Antimony oxides	Cookson Ltd., Stoke-on-Trent, England
Zerogen	Hydrated alumina	Solem Industries, Inc., Norcross, GA, USA
—	Smoke reducer based on molybdenum	Climax Molybdenum Co. Ltd., London, England

Source: J. Troitzsch, *International Plastics Flammability Handbook*, Carl Hanser Verlag, Munchen (1983).

Appendix 3

Formulations of Flame-Retarded Selected Thermoplastics

In the following formulations the fire performance properties of flame-retarded thermoplastics are compared with those of the base resin, the properties of the latter being given in brackets. The test methods used are: ASTM-D 635-77, the value for average time of burning (ATB) given in seconds and that for average extent of burning (AEB) in millimeters; ASTM 2863-77, limiting oxygen index (LOI) in % O_2.

LOW-DENSITY POLYETHYLENE (LDPE)

   90.5%   LDPE
    6.0%   Chlorinated paraffin (70% chlorine)
    3.5%   Antimony trioxide

Properties

ASTM	ATB	<5	(160)
	AEB	18	(100)
	LOI	23.75	(17.5)

HIGH-DENSITY POLYETHYLENE (HDPE)

89.9%   HDPE
5.0%   Octabromodiphenyl
1.6%   Chlorinated paraffin (70% chlorine)
3.5%   Antimony trioxide

Properties

ASTM	ATB	<5	(210)
	AEB	20	(100)
	LOI	25.75	(16.75)

POLYPROPYLENE

94.5%   Polypropylene
3.5%   Bis-dibromopropylether of tetrabromobisphenol A
2.0%   Antimony trioxide

Properties

ASTM	ATB	5	(195)
	AEB	14	(100)
	LOI	27.5	(18.25)

POLYSTYRENE
a. Crystal Polystyrene

97%   Polystyrene
3%   Hexabromocyclododecane

Properties

ASTM-D 635	ATB	<5	(117)
	AEB	28	(100)

b. Impact Polystyrene

80%   Impact polystyrene
15%   Octabromodiphenyl
5%   Antimony trioxide

Properties

ASTM	ATB	<5	(150)
	AEB	17	(100)
	LOI	25.0	(16.5)

c. Polystyrene Foam

82% Polystyrene
13% Decabromodiphenyl ether
5% Antimony trioxide

Properties

ASTM	ATB	<5	(90)
	AEB	19	(100)

PLASTICIZED PVC

53.5% Suspension PVC
27.7% Dioctylphthalate
4.9% Epoxidized oleic acid ester
4.9% Pentabromodiphenyl ether
8.3% Tricresyl phosphate
0.5% Lubricant
0.4% Stabilizer (butyltin carboxylate)

[The formulation passes the FMVSS 302 (Federal Motor Vehicle Safety Standard) test used for the assessment of the burning behavior of materials used for vehicle interiors. A specimen of $35.5 \times 10.0 \times$ max. 1.3 cm is mounted horizontally and ignited for 15 sec. The rate of flame spread should not exceed 10 cm/min.]

ABS TERPOLYMER

82% ABS
13% Octabromodiphenyl ether
5% Antimony trioxide

Properties

ASTM	ATB	<5	(195)
	AEB	10	(100)
	LOI	24	(18)

SATURATED POLYESTER

a. Polyester

88%   Poly(butylene terephthalate) (PBTP)
 8%   Octabromodiphenyl
 4%   Antimony trioxide

Properties

ASTM	ATB	<5	(240)
	AEB	16	(100)
	LOI	28.5	(22.0)

b. Polyester with Glass Fiber Reinforcement

87%   PBTP (30% glass fiber)
 9%   Decabromodiphenyl
 4%   Antimony trioxide

Properties

LOI	33	(19)

POLYCARBONATES

Flame-retarded polycarbonates preferably contain tetrabromobisphenol-A as flame-retardant added before polycondensation. Untreated polycarbonate can be processed to a flame-retardant product as follows:

93.1%   Polycarbonate
 5.0%   Decabromodiphenyl ether
 1.9%   Antimony trioxide

Properties

ASTM	ATB	<5	(17)
	AEB	12	(24)
	LOI	40	(25)

POLYAMIDES

a. Nylon-6

84%   Nylon-6

10% Decabromodiphenyl ether
6% Antimony trioxide

Properties

ASTM	ATB	<5	(178)
	AEB	17	(60)
	LOI	22.75	(21.75)

b. *Fiberglass-Reinforced Nylon-6*

53% Nylon-6
22% Fiberglass
18% Dechlorane 602
7% Antimony trioxide

Properties

LOI   32   (21)

Appendix 4

Formulations of Selected Rubber Compounds

Representative formulations (in parts by weight) of several rubber compounds and their normal curing conditions are given below.

Tread Compound for Passenger Tires

Smoked sheet	50
SBR 1712	70
Zinc oxide	5
Stearic acid	2
ISAF black	60
Softener	2
Antioxidant (diphenylamine-acetone condensate)	1.5
Accelerator (CBS)	1
Sulfur	2.2

Cure: 40 min. at 150°C

Sidewall Compound for Passenger Tires

Smoked sheet	50
SBR 1712	70
Zinc oxide	5
Stearic acid	2
HAF black	45

Softener	2
Antioxidant	1.5
(diphenylamine-acetone	
condensate)	
Accelerator (CBS)	1
Sulfur	2.2

Cure: 30 min. at 150°C

Tube Compound for Car Tires

Butyl (Polysar 301)	100
p-Dinitroso benzene (Polyac)	0.15
FEF black	60
Mineral oil (butyl grade)	20
Zinc oxide	5
Stearic acid	2
Accelerators	
MBT	1
TMT	1
Sulfur	1.5

Cure: 30 min. at 160°C

Friction Compound for Conveyor Belts

Smoked sheet	100
Zinc oxide	5
Stearic acid	2
SRF black	10
Whiting/activated calcium carbonate	15
Tackifier/softener	5
Antioxidant	1
Accelerator (CBS)	0.6
Sulfur	2.5

Cure: 20 min. at 150°C

Cover Compound for Conveyor Belts

Smoked sheet	50
SBR 1500	50
Zinc oxide	5
Stearic acid	2
Tackifier/softener	5
ISAF black	40
Antioxidant	1.5

Accelerator (CBS)	1.0
Sulfur	2.0

Cure: 20 min. at 150°C

Insulation Compound for Cables

Smoked sheet	100
Zinc oxide	20
China clay	30
Precipitated calcium carbonate	45
Paraffin wax	2
Stearic acid	0.5
Antioxidants	1.0
Accelerators	
DPG	0.5
MBTS	1.0
Sulfur	1.5

Cure: 15 min. at 140°C

Translucent Shoe Soling Compound

Pale crepe	100
Zinc oxide	3
Stearic acid	1
Precipitated silica	50
Paraffin wax	1
Spindle oil	2
Diethylene glycol	2
Antioxidant (styrenated phenol)	1
Accelerators	
TMTM	0.5
ZDC	0.75
Sulfur	2.5

Cure: 7 min. at 150°C

Microcellular Shoe Soling

Smoked sheet	20
SBR 1500	20
SBR 1958	60
Peptizer	1
Microcellular crumb	60
Zinc oxide	5
Stearic acid	3–5

Paraffin wax	1
Mineral oil	10
Coumarone-indene resin	5
Styrenated phenol	1
Aluminum silicate	40
China clay	100
Blowing agent (DNPT)	5
Accelerators	
DPG	0.5
MBTS	1.0
Sulfur	2.5

Cure: 8 min. at 150°C; oven
stabilization at 100°C, 4 h.

Appendix 5

Formulations of Selected PVC Compounds

The formulations given below (in parts by weight) are examples of typical formulations of PVC compounds from the many available, which have been designed to meet specific requirements of different applications.

Glass-Clear Food Packaging Film

PVC	100
Octyltin mercaptide	1–1.5
Glycerol ester	0.5–1.0
Ester wax	0.1–0.4

Furniture Film

PVC	100
Plasticizer	15–25
Epoxidized soybean oil	2–3
Liquid Ba/Cd stabilizer	1.5–2.0
UV absorber	0.2–0.4
Bis-amide wax	0.2–0.4
Processing aid	1
Pigments	0.5–10.0

Flexible FIlm for Outdoor Application

PVC	100
Plasticizer	40–80
Epoxy plasticizer	2–3
Tin carboxylate stabilizer	1–2.5
UV absorber	0.2–0.5
Oxidized polyethylene wax	0.2–0.4

Transparent Sheets for Exterior Application

PVC	100
Butyltin mercaptide	2–2.5
Fatty alcohol	0.5–0.8
Fatty acid ester	0.5–0.8
Polyethylene wax	0.1–0.2
Processing aid	1–2
UV absorber (benzotriazole type)	0.3–0.5

Artificial Leathercloth

Base coat	
PVC	100
Plasticizer	50–100
Epoxidized soybean oil	2–3
Barium/zinc stabilizer	1.5–2.0
Filler	0–20
Top coat	
PVC	100
Plasticizer	40–60
Epoxidized soybean oil	3–5
Barium/zinc stabilizer	1.5–2.0
UV absorber	0.2–0.3

Bottles for Food Packaging

PVC	100
Processing aid	0.5–1.0
Methyl- or octyltin mercaptide	1.0–1.5
Lubricants	0.5–1.5
Impact modifier	5–10

Potable Water Pipes

PVC	100
Methyltin mercaptide*, octyltin mercaptide*, butyltin mercaptide*	0.3–0.4
Calcium stearate	0.5–0.8
Solid paraffin wax	0.6–0.8
Polyethylene wax	0.1–0.2

*In compliance with specific national regulations.

Pressure Pipes and Conduits

PVC	100
Tribasic lead sulfate	0.5–1.0
Dibasic lead stearate	0.5–1.0
Calcium stearate	0.2–0.4
Stearic acid and/or paraffin wax	0.2–0.5

Calendered Floor Tiles

PVC	100
Plasticizer	30–70
Epoxidized soybean oil	2–5
Solid Ba/Cd stabilizer	1.5–2.5
Stearic acid	0.1–0.4
Chalk	50–100

Wire and Cable Insulation

PVC	100
Plasticizer	30–60
Tribasic lead sulfate	2–4
Normal lead stearate	0.5–1.5
Chalk	20–40
Stearic acid and/or paraffin wax	0–0.2

Window Profiles

PVC	100
Impact modifier	6–12
Modified tin maleate stabilizer	2–2.5
Paraffin wax	0.6–1.2

Oxidized polyethylene wax	0.6–1.2
Antioxidant	0.1
UV absorber	0.2–0.4
Titanium dioxide	2–4
Chalk	10

Shoes

PVC	100
Plasticizer	50–80
Epoxidized soybean oil	2–3
Liquid Ba/Cd stabilizer	1.5–2.0
Stearic acid	0.2–0.4

Appendix 6

Formulations of Polyurethane Foams

The formulations given below (in parts by weight) are three examples of typical formulations from the many available, which have been designed to meet specific requirements. Noteworthy points regarding the formulations are explained. An example of formulation for reaction injection molding (RIM) and reinforced reaction injection molding (RRIM) is also given

In order to achieve the desired balance of hydroxyl to isocyanate groups in a formulation, the *isocyanate index* is specified:

$$\text{Isocyanate index} = \frac{\text{Number of mole equivalents of isocyanate}}{\text{Number of mole equivalents of polyols}} \times 100$$

If water is included in the formulation, this is also included in the mole equivalents of polyol.

Formulation A   A Typical Rigid Foam Formulation

Component A		
Polyol mixture containing a significant amount of triols and hex/octols to produce cross-linking		100
Catalyst 1	*N,N*-cyclohexylamine	0.3
Catalyst 2	*N,N*-dimethylethanolamine	0.3

Freon 11	($CFCl_3$)	50
Water		1
Surfactant	(A block copolymer of polyether and silicone)	1

Component B

Impure liquid MDI	Equal in volume to component A and to give an isocyanate index near to 100	

NOTES ON FORMULATION A

1. In the formulation of the foam, the equipment performs better with approximately equal volumes of components A and B.
2. The MDI is of functionality around 2.2. This functionality together with the aromatic nature of the MDI will tend to give rigid foams.
3. The isocyanate index is quoted as 100 to give conditions essentially producing urethane and urea groups only.
4. Freon 11 blows to give closed cell structures whereas water produces open cells through carbon dioxide. This formulation is typical for a rigid foam for thermal insulation purposes. The heat build-up due to the reaction exotherm (about 80 kJ/mole for formation of urethane groups) is sufficiently dissipated through the open cell to avoid thermal degradation.
5. The catalyst combination allows a balance of reactions since catalyst 2 is specific to the water reaction. The values given are notional and depend on the polyols used.
6. A surfactant is used to stabilize the bubble structure.
7. It is likely that the product foam will have a density of 30 kg/m^3 which is roughly composed of 97% gas and 3% matrix by volume.

Formulation B   A Typical Flexible Foam Formulation

Component A

Polyether polyol with long chains and overall low functionality	100
Water	4.5
Catalyst 1 (a tertiary amine)	0.15
Catalyst 2 (stannous octoate)	0.2
Freon 11	10
Silicone surfactant	1.3

Component B

TDI (to index 112)	58.4

NOTES ON FORMULATION B

1.   This is a standard foam grade formulation. Other flexible foam formulations are available for supersoft, high resilience and special grades.
2.   The formulation is designed to give open cells.
3.   The isocyanate has a functionality of 2.0 and hence will not itself give cross-linking.

Formulation C   A Polyisocyanurate Foam Formulation

Component A		
A Flame retardant polyol		100
Catalyst 1,	DMP(tris-2,4,6-dimethylaminomethyl phenol)	3
Catalyst 2,	Sodium acetate: potassium acetate (1:1) 33% w/w in ethane-1,2-diol	2
Freon 11		40
Surfactant	(silicone type)	1.5
Component B		
A liquid isocyanate based on MDI (to give index 200)		142.6

NOTES ON FORMULATION C

1.   The high isocyanate index of catalyst system will promote the formation of isocyanate structures.
2.   The higher heat stability of the isocyanate structures will be enhanced by a fire-retardant grade of polyol.

Formulation D   A Typical RIM/RRIM Formulation

Component A		
Polyol mixture	Overall functionality 2.5, and containing primary hydroxyl terminated polyethers	100
Diol hardener	(1,4-Butanediol)	5
Soluble	(Soluble tin salt)	2
Surfactant		1
Blowing agent	Freon 11	5
Milled glass fiber	(For RRIM)	5–30% of total
Component B		
Modified isocyanate based upon MDI		To match index 95–110

A NOTE ON RIM AND RRIM

RIM and RRIM require that a liquid isocyanate compound (component B) is effectively mixed with another liquid (component A) which contains polyols, catalysts, and other agents. Components A and B are metered and pumped into mixing heads where the two immiscible components are each turbulently broken down into small droplets surrounded by the other component phase. (The nozzles on the mixing head are usually 3 mm in diameter. This corresponds to a viscosity-pumping pressure relationship of 8 Pa.s and 25 MPa, for example.) As the flow emerges from the mixing heads, the flow becomes laminar with phase separation (striation thickness) of about 10 µm. Reaction takes place at the interface during flow into the mold and subsequently in the mold until the product has sufficient strength to allow demolding.

The mixed liquids from the mixing head are forced (by pressure of liquid entering the mixing head) through a runner and gate and into the mold. It is generally required that the flow of the mixture should be laminar in the form of a film of about 1 mm thickness at a flow rate of about 2 m/s. Turbulent flow may cause air entrapment. The flow should be directed to the lower part of the mold to allow upward filling and any small amount of foaming to compensate for shrinkage (5%). The mold design should be such that air can escape through the parting lines.

Because of the low pressures (0.3 MPa) in the mold as compared to those encountered in thermoplastic injection molding (150 MPa) this process is suitable for the production of thin parts with large surface areas. A significant application for this is in panel formation in the automotive industry, including fascias, door panels, spoilers, grills, and bumpers. While a rigid foam part with a flexural strength of 700 MPa would require a thickness of 7 mm, a RRIM part, because of the reinforcement, would allow a much smaller thickness (<3 mm) by virtue of the increase in flexural modulus.

Appendix 7

Conversion of Units

SI UNITS AND CONVERSION FACTORS

Physical quantity	Name of SI unit	Symbol for SI unit	Definition of SI unit
Length	Meter	m	Basic unit
Mass	Kilogram	kg	Basic unit
Time	Second	s	Basic unit
Force	Newton	N	$kg\text{-}m\text{-}s^{-2}\ (= J\text{-}m^{-1})$
Pressure	Pascal	Pa	$kg\text{-}m^{-1}\text{-}s^{-2}\ (= N\text{-}m^{-2})$
Energy	Joule	J	$kg\text{-}m^2\text{-}s^{-2}$
Power	Watt	W	$kg\text{-}m^2\text{-}s^{-3}\ (= J\text{-}s^{-1})$

Physical quantity	Customary unit	SI unit	To convert from customary unit to SI units multiply by
Length	in.	m	2.54×10^{-2}
Mass	lb	kg	$4.535\ 923\ 7 \times 10^{-1}$
Force	dyne	N	1×10^{-5}
	kgf	N	$9.806\ 65$
	lbf	N	$4.448\ 22$
Pressure	dyne/cm^2	Pa or N-m^{-2}	1×10^{-1}
	atm	Pa or N-m^{-2}	$1.013\ 25 \times 10^5$
	mm Hg	Pa or N-m^{-2}	$1.333\ 22 \times 10^2$
	lbf/in.^{2} or psi	Pa or N-m^{-2}	$6.894\ 76 \times 10^3$

SI UNITS AND CONVERSION FACTORS (CONTINUED)

Physical quantity	Customary unit	SI unit	To convert from customary unit to SI units multiply by
Energy	erg	J	1×10^{-7}
	Btu	J	$1.055\ 056 \times 10^3$
	ft-lbf	J	$1.355\ 82$
Area	in.^{2}	m^2	$6.451\ 6 \times 10^{-4}$
	ft^2	m^2	$9.290\ 304 \times 10^{-2}$
Density	lb/ft^3	kg-m^{-3}	$1.601\ 846\ 3 \times 10$
Viscosity	poise	kg-m^{-1}-s^{-1} or N-s-m^{-2}	1×10^{-1}

Appendix 8

Typical Properties of Polymers Used for
Molding and Extrusion

[handwritten: DENSITY .0329 #/in³]

	ASTM test method	Polyethylene		Polypropylene
		Low density	High density	
1. Specific gravity	D792	0.91–0.925	0.94–0.965	0.900–0.910
2. Tensile modulus (psi $\times 10^{-5}$)	D638	0.14–0.38	0.6–1.8	1.6–2.25
3. Compressive modulus (psi $\times 10^{-5}$)	D695	—	—	1.5–3.0
4. Flexural modulus (psi $\times 10^{-5}$)	D790	0.08–0.6	1.0–2.6	1.7–2.5
5. Tensile strength (psi $\times 10^{-3}$)	D638, D651	0.6–2.3	3.1–5.5	4.5–6.0
6. Elongation at break (%)	D638	90–800	20–130	100–600
7. Compressive strength (psi $\times 10^{-3}$)	D695	2.7–3.6	12–18	5.5–8.0
8. Flexural yield strength (psi $\times 10^{-3}$)	D790	—	1.0	6–8
9. Impact strength, notched Izod, (ft-lb/in.)	D256	No break	0.5–20	0.4–1.0
10. Hardness, Rockwell	D785	D40–51(Shore)	D60–70(Shore)	R80–102
11. Thermal conduct. (cal/s-cm-K $\times 10^{4}$)	C177	8.0	11–12	2.8

Property	ASTM			
12. Specific heat (cal/g-K)	—	0.55	0.55	0.46
13. Linear therm. exp. coeff. ($K^{-1} \times 10^5$)	D696	10–22	11–13	8.1–10.0 ✳
14. Continuous-use temperature (°C)	—	80–100	120	120–160
15. Deflection temp. (°C at 0.45 MPa)	D648	38–49	60–88	107–121
16. Volume resistivity, ohm·cm	D257	$>10^{16}$	$>10^{16}$	$>10^{16}$
17. Dielectric constant at 1 kHz	D150	2.25–2.35	2.30–2.35	2.2–2.6
18. Dielectric strength (kV/in.)	D149	450–1000	450–500	500–660
19. Dissipation factor at 1 kHz	D150	<0.0005	<0.0005	<0.0018
20. Deleterious media	D543	Oxidizing acids	Oxidizing acids	Strong oxidizing acids
21. Solvents (room temperature) (Cl.H. = chlorinated hydrocarbons)		None	None	None

✳ USE .00009 IN/IN/°C FOR UNFILLED PP.

	ASTM test method	Polystyrene		Poly(methyl Methacrylate)
		Gen. purpose	Impact-resistant	
1. Specific gravity	D792	1.04–1.05	1.03–1.06	1.17–1.20
2. Tensile modulus (psi $\times 10^{-5}$)	D638	3.5–4.85	2.6–4.65	3.8
3. Compressive modulus (psi $\times 10^{-5}$)	D695	—	—	3.7–4.6
4. Flexural modulus (psi $\times 10^{-5}$)	D790	4.3–4.7	3.3–4.0	4.2–4.6
5. Tensile strength (psi $\times 10^{-3}$)	D638, D651	5.3–7.9	3.2–4.9	7–11
6. Elongation at break (%)	D638	1–2	13–50	2–10
7. Compressive strength (psi $\times 10^{-3}$)	D695	11.5–16	4–9	12–18
8. Flexural yield strength (psi $\times 10^{-3}$)	D790	8.7–14	5–12	13–19
9. Impact strength, notched Izod, (ft-lb/in.)	D256	0.25–0.40	0.5–11	0.3–0.5
10. Hardness, Rockwell	D785	M65–80	M20–80	M85–105
11. Thermal conduct. (cal/s-cm-K $\times 10^{4}$)	C177	2.4–3.3	1.0–3.0	4–6

12. Specific heat (cal/g-K)	—	0.32	0.32–0.35	0.35
13. Linear therm. exp. coeff. ($K^{-1} \times 10^5$)	D696	6–8	3.4–21	5–9
14. Continuous-use temperature (°C)	—	66–77	60–79	60–88
15. Deflection temp. (°C at 0.45 MPa)	D648	75–100	75–95	80–107
16. Volume resistivity, ohm·cm	D257	10^{16}	10^{16}	$>10^{14}$
17. Dielectric constant at 1 kHz	D150	2.4–2.65	2.4–4.5	3.0–3.6
18. Dielectric strength (kV/in.)	D149	500–700	300–600	400
19. Dissipation factor at 1 kHz	D150	0.0001–0.0003	0.0004–0.002	0.03–0.05
20. Deleterious media	D543	Strong oxidizing acids	Strong, oxidizing acids	Strong bases and strong, oxidizing acids
21. Solvents (room temperature) (Cl.H. = chlorinated hydrocarbons)		Aromatic and Cl.H.	Aromatic and Cl.H.	Ketones, esters, aromatic, and Cl.H.

	ASTM test method	Poly(vinyl chloride)		ABS medium impact
		Rigid	Plasticized	
1. Specific gravity	D792	1.30–1.58	1.16–1.35	1.03–1.06
2. Tensile modulus (psi $\times 10^{-5}$)	D638	3.5–6	—	3–4
3. Compressive modulus (psi $\times 10^{-5}$)	D695	—	—	2.0–4.5
4. Flexural modulus (psi $\times 10^{-5}$)	D790	3–5	—	3.7–4.0
5. Tensile strength _yield_ (psi $\times 10^{-3}$)	D638, D651	6–7.5	1.5–3.5	6–7.5
6. Elongation at break (%)	D638	2–80	200–450	5–25
7. Compressive strength (psi $\times 10^{-3}$)	D695	8–13	0.9–1.7	10.5–12.5
8. Flexural yield strength (psi $\times 10^{-3}$)	D790	10–16	—	11–13
9. Impact strength, notched Izod, (ft-lb/in.)	D256	0.4–20	—	3–6
10. Hardness, Rockwell	D785	D65–85(Shore)	A40–100(Shore)	R107–115
11. Thermal conduct. (cal/s-cm-K $\times 10^4$)	C177	3.5–5.0	3.0–4.0	4.5–8.0

Property	ASTM test			
12. Specific heat (cal/g-K)	—	0.2–0.28	0.3–0.5	0.3–0.4
13. Linear therm. exp. coeff. ($K^{-1} \times 10^5$)	D696	5–10	7–25	8–10
14. Continuous-use temperature (°C)	—	65–80	65–80	71–93
15. Deflection temp. (°C at 0.45 MPa)	D648	57–82	—	102–107
16. Volume resistivity, ohm·cm	D257	$>10^{16}$	10^{11}–10^{15}	2.7×10^{16}
17. Dielectric constant at 1 kHz	D150	3.0–3.3	4–8	2.4–4.5
18. Dielectric strength (kV/in.)	D149	425–1300	300–1000	350–500
19. Dissipation factor at 1 kHz	D150	0.009–0.017	0.07–0.16	0.004–0.007
20. Deleterious media	D543	None	None	Conc. oxidizing acids, organic solvents
21. Solvents (room temperature) (Cl.H. = chlorinated hydrocarbons)		Ketones, esters, swelling in aromatic, and Cl.H.	Plasticizer may be extracted. Otherwise like rigid PVC	Ketones, esters, some Cl.H.

	ASTM test method	Cellulose acetate	Cellulose acetate-butyrate	Fluoropolymers	
				—CF$_2$—CF$_2$—	—CF$_2$—CHCl—
1. Specific gravity	D792	1.22–1.34	1.15–1.22	2.14–2.20	2.1–2.2
2. Tensile modulus (psi × 10^{-5})	D638	0.65–4.0	0.5–2.0	0.58	1.5–3.0
3. Compressive modulus (psi × 10^{-5})	D695	—	—	—	—
4. Flexural modulus (psi × 10^{-5})	D790	—	—	—	—
5. Tensile strength (psi × 10^{-3})	D638, D651	1.9–9.0	2.6–6.9	2–5	4.5–6.0
6. Elongation at break (%)	D638	6–70	40–88	200–400	80–250
7. Compressive strength (psi × 10^{-3})	D695	3–8	2.1–7.5	1.7	4.6–7.4
8. Flexural yield strength (psi × 10^{-3})	D790	2–16	1.8–9.3	—	7.4–9.3
9. Impact strength, notched Izod, (ft-lb/in.)	D256	1–7.8	1–11	3.0	2.5–2.7
10. Hardness, Rockwell	D785	R34–125	R31–116	D50–55(Shore)	R75–95
11. Thermal conduct. (cal/s-cm-K × 10^4)	C177	4–8	4–8	6.0	4.7–5.3

12. Specific heat (cal/g·K)	—	0.0–0.4	0.3–0.4	0.25	0.22
13. Linear therm. exp. coeff. ($K^{-1} \times 10^5$)	D696	8–18	11–17	10	4.5–7.0
14. Continuous-use temperature (°C)	—	60–105	60–105	290	175–200
15. Deflection temp. (°C at 0.45 MPa)	D648	50–100	54–108	121	126
16. Volume resistivity, ohm·cm	D257	10^{10}–10^{14}	10^{11}–10^{15}	10^{18}	1.2×10^{18}
17. Dielectric constant at 1 kHz	D150	3.4–7.0	3.4–6.4	2.1	2.3–2.7
18. Dielectric strength (kV/in.)	D149	250–500	250–400	480	500–600
19. Dissipation factor at 1 kHz	D150	0.01–0.07	0.01–0.04	0.0002	0.023–0.027
20. Deleterious media	D543	Strong acids and bases	Strong acids and bases	None	None
21. Solvents (room temperature) (Cl.H. = chlorinated hydrocarbons)		Ketones, esters, Cl.H.	Ketones, esters, Cl.H.	None	Swells in Cl.H.

	ASTM test method	Nylon-6,6 (moisture conditioned)	Nylon-6 (moisture conditioned)	Acetal	Polycarbonate
1. Specific gravity	D792	1.13–1.15	1.12–1.14	1.42	1.2
2. Tensile modulus (psi $\times 10^{-5}$)	D638	—	1.0	5.2	3.5
3. Compressive modulus (psi $\times 10^{-5}$)	D695	—	2.5	6.7	3.5
4. Flexural modulus (psi $\times 10^{-5}$)	D790	1.75–4.1	1.4	3.8–4.3	3.4
5. Tensile strength (psi $\times 10^{-3}$)	D638, D651	11	10	9.5–12	9.5
6. Elongation at break (%)	D638	300	300	25–75	110
7. Compressive strength (psi $\times 10^{-3}$)	D695	—	—	18	12.5
8. Flexural yield strength (psi $\times 10^{-3}$)	D790	6.1	5.0	14	13.5
9. Impact strength, notched Izod, (ft-lb/in.)	D256	2.1	3.0	1.3–2.3	16
10. Hardness, Rockwell	D785	R120	R119	M94 to R120	M70
11. Thermal conduct. (cal/s-cm-K $\times 10^4$)	C177	5.8	5.8	5.5	4.7

12. Specific heat (cal/g-K)	—	0.4	0.38	0.35	0.3
13. Linear therm. exp. coeff. ($K^{-1} \times 10^5$)	D696	8.0	8.0–8.3	10	6.8
14. Continuous-use temperature (°C)	—	80–150	80–120	90	121
15. Deflection temp. (°C at 0.45 MPa)	D648	180–240	150–185	124	138
16. Volume resistivity, ohm·cm	D257	10^{14}–10^{15}	10^{12}–10^{15}	1.0×10^{15}	2×10^{16}
17. Dielectric constant at 1 kHz	D150	3.9–4.5	4.0–4.9	3.7	3.02
18. Dielectric strength (kV/in.)	D149	385–470	440–510	500	400
19. Dissipation factor at 1 kHz	D150	0.02–0.04	0.011–0.06	0.004	0.0021
20. Deleterious media	D543	Strong acids	Strong acids	Strong acids, some other acids and bases	Bases and strong acids
21. Solvents (room temperature) (Cl.H. = chlorinated hydrocarbons)		Phenol and formic acid	Phenol and formic acid	None	Aromatic and Cl.H.

	ASTM test method	Ionomers	Poly(phenylene oxide)	Polysulfone
1. Specific gravity	D792	0.93–0.96	1.06	1.24
2. Tensile modulus (psi $\times 10^{-5}$)	D638	0.2–0.6	3.55	3.6
3. Compressive modulus (psi $\times 10^{-5}$)	D695	—	—	3.7
4. Flexural modulus (psi $\times 10^{-5}$)	D790	—	3.6–4.0	3.9
5. Tensile strength (psi $\times 10^{-3}$)	D638, D651	3.5–5.0	9.6	10.2 (yield)
6. Elongation at break (%)	D638	350–450	60	50–100
7. Compressive strength (psi $\times 10^{-3}$)	D695	—	16.4	13.9 (yield)
8. Flexural yield strength (psi $\times 10^{-3}$)	D790	—	13.5	15.4 (yield)
9. Impact strength, notched Izod, (ft-lb/in.)	D256	6.0–15	5.0	1.2
10. Hardness, Rockwell	D785	D50–65(Shore)	R119	M69, R120
11. Thermal conduct. (cal/s-cm-K $\times 10^4$)	C177	5.8	5.2	2.8

Property	ASTM			
12. Specific heat (cal/g-K)	—	0.55	—	0.31
13. Linear therm. exp. coeff. ($K^{-1} \times 10^5$)	D696	12	3.3–5.9	5.2–5.6
14. Continuous-use temperature (°C)	—	70–95	—	150–175
15. Deflection temp. (°C at 0.45 MPa)	D648	38	—	180
16. Volume resistivity, ohm·cm	D257	$>10^{16}$	10^{18}	5×10^{16}
17. Dielectric constant at 1 kHz	D150	2.4	2.6	3.13
18. Dielectric strength (kV/in.)	D149	900–1100	400–500	425
19. Dissipation factor at 1 kHz	D150	0.0015	0.00035	0.001
20. Deleterious media	D543	Acids, esp. strong oxidizing acids	None	None
21. Solvents (room temperature) (Cl.H = chlorinated hydrocarbons)		None	Aromatic and Cl.H.	Aromatic hydrocarbons

	ASTM test method	Phenol-formaldehyde (cellulose fill)	Melamine-formaldehyde (cellulose fill)	Cast epoxy glass fiber fill
1. Specific gravity	D792	1.37–1.46	1.47–1.52	1.6–2.0
2. Tensile modulus (psi × 10^{-5})	D638	8–17	11–14	30
3. Compressive modulus (psi × 10^{-5})	D695	—	—	—
4. Flexural modulus (psi × 10^{-5})	D790	10–12	—	20–45
5. Tensile strength (psi × 10^{-3})	D638, D651	5–9	5–13	5–20
6. Elongation at break (%)	D638	0.4–0.8	0.6–1.0	4
7. Compressive strength (psi × 10^{-3})	D695	25–31	33–45	18–40
8. Flexural yield strength (psi × 10^{-3})	D790	7–14	9–16	8–30
9. Impact strength, notched Izod, (ft-lb/in.)	D256	0.2–0.6	0.2–0.4	0.3–10
10. Hardness, Rockwell	D785	E64–95	M115–125	M100–112
11. Thermal conduct. (cal/s-cm-K × 10^4)	C177	4–8	6.5–10	4–10

	ASTM			
12. Specific heat (cal/g-K)	—	0.35–0.40	0.4	0.19
13. Linear therm. exp. coeff. ($K^{-1} \times 10^5$)	D696	3.0–4.5	4.0–4.5	1–5
14. Continuous-use temperature (°C)	—	150–175	99	150–260
15. Deflection temp. (°C at 0.45 MPa)	D648	—	43	—
16. Volume resistivity, ohm·cm	D257	10^9–10^{13}	10^{12}	$>10^{14}$
17. Dielectric constant at 1 kHz	D150	4.4–9.0	7.8–9.2	3.5–5.0
18. Dielectric strength (kV/in.)	D149	200–400	270–300	300–400
19. Dissipation factor at 1 kHz	D150	0.04–0.20	0.015–0.036	0.01
20. Deleterious media	D543	Strong bases and oxidizing acids	Strong acids and bases	None
21. Solvents (room temperature) (Cl.H. = chlorinated hydrocarbons)	—	None	None	None

Conversion factors: 1000 psi = 6.895 MPa; 1 ft.lb/in. = 53.4 J/m; 1 cal = 4.187 J; 1 kV/in. = 0.0394 MV/m.
Data collected from *Modern Plastics Encyclopedia*.

Appendix 9

Typical Properties of Cross-Linked
Rubber Compounds

A. Diene-based polymers and copolymers[a]

	Styrene-butadiene random copolymer, 25% (wt) styrene (SBR)	Styrene-butadiene block copolymer, about 25% styrene (YSBR)	cis-1,4-Polyisoprene (natural rubber NR, also made synthetically IR)
Gum stock (cross-linked, unfilled)			
Density (g/cm^3)	0.94	0.94–1.03	0.93
Tensile strength (psi)[b]	200–400	1,700–3,700	2,500–3,000
Resistivity (ohm-cm, log)	15	13	15–17
Dielect. const. at 1 kHz	3.0	3.4	2.3–3.0
Dielect. str. (kV/in.)[c]	—	485	—
Diss. factor, at 1 kHz	0.003	0.01	0.002–0.003
Reinforced stock			
Tensile strength (psi)[b]	2,000–3,500	1,000–3,000	3,000–4,000
Elong. at break (%)	300–700	500–1,000	300–700
Hardness, Shore A	40–100	40–85	20–100
Cont. high-temp. limit (°C)	110	65	100
Stiffening temp. (°C)	−20 to −45	−50 to −60	−30 to −45
Brittle temp. (°C)	−60	−70	−60
Resilience	Good	Excellent	Excellent
Resistance to			
acid	Good	Good	Good
alkali	Good	Good	Good
gasoline and oil	Poor	Poor	Poor
aromatic hydrocarbons	Poor	Poor	Poor
ketones	Good	Poor	Good
chlorinated solvents	Poor	Poor	Poor
oxidation	Good	Good	Good
ozone	Good	Good	Good

[a]Abbreviations are according to ASTM.
[b]1000 psi = 6.895 MPa.
[c]1 kV/in. = 0.0394 MV/m.

cis-1,4 Polybutadiene (BR)	Polychloroprene (CR), Neoprene	Butadiene-acrylonitrile random copolymer, variable % acrylonitrile (NBR)	Reclaimed rubber (whole tires) (mainly NR and SBR)
0.93	1.23	1.00	1.2 (compd'd)
200–1,000	3,000–4,000	500–1,000	
—	11	10	
2.3–3.0	9.0	13	
—	150–600		
0.002–0.003	0.03	0.055	
2,000–3,500	3,000–4,000	3,000–4,000	500–1,000
300–700	300–700	300–700	300–400
30–100	20–100	30–100	50–100
100	120	120	100
−35 to −50	−10 to −30	0 to −30	−20 to −45
−70	−40 to −55	−15 to −55	−60
Excellent	Good	Fair	Good
Good	Good	Good	Good
Good	Good	Good	Good
Poor	Good	Excellent	Poor
Poor	Fair	Good	Poor
Good	Poor	Poor	Good
Poor	Poor	Poor	Poor
Good	Excellent	Fair	Good
Good	Poor	Fair	Good

B. Saturated, carbon-chain polymers[a]

	Polyisobutylene (butyl rubber, copolymer with 0.5–2% isoprene) (IIR)	Chloro-sulfonated polyethylene (CSM)	Ethylene-propylene random copolymer, 50% ethylene (EPM)
Gum stock (cross-linked, unfilled)			
Density (g/cm^3)	0.92	1.12–1.28	0.86
Tensile strength (psi)[b]	2500–3000	2500	500
Resistivity (ohm-cm, log)	17	14	16
Dielect. const. at 1 kHz	2.1–2.4	7–10	3.0–3.5
Dielect. str. (kV/in.)[c]	600	500	900
Diss. factor, at 1 kHz	0.003	0.03–0.07	0.004–0.008
Reinforced stock			
Tensile strength (psi)[b]	2,000–3,000	3000	1,000–3,000
Elong. at break (%)	300–700	300–500	200–300
Hardness, Shore A	30–100	50–100	30–100
Cont. high-temp. limit (°C)	120	160	150
Stiffening temp. (°C)	−25 to −45	−10 to −30	−40
Brittle temp. (°C)	−60	−40 to −55	−50 to −75
Resilience	Fair	Good	Good
Resistance to			
acid	Excellent	Good	Excellent
alkali	Excellent	Good	Excellent
gasoline and oil	Poor	Good	Poor
aromatic hydrocarbons	Poor	Fair	Fair
ketones	Excellent	Poor	Good
chlorinated solvents	Poor	Poor	Poor
oxidation	Excellent	Excellent	Excellent
ozone	Excellent	Excellent	Excellent

[a]Abbreviations are according to ASTM.
[b]1000 psi = 6.895 MPa.
[c]1 kV/in. = 0.0394 MV/m.

Ethylene-propylene random terpolymer, 50% ethylene (EPDM)	Poly(ethyl acrylate), usually a copolymer (ACM)	Vinylidene-fluoride-chlorotrifluoro ethylene random copolymer (FKM)	Vinylidene-fluoride-hexafluoropropylene random copolymer (FKM)
0.86	1.10	1.85	1.85
200	200–400	200–2500	2000
16	—	14	13
3.0–3.5	—	6	—
900	—	600	250–750
0.004–0.008	—	0.05	0.03–0.04
1,000–3,500	1,500–2,500	1,500–2,500	1,500–2,500
200–300	250–350	300–400	300–400
30–100	40–100	50–90	50–90
150	175	200	250
−40	—	−35	—
−50 to −75	−30	−50	−45
Good	Fair	Fair	Fair
Excellent	Fair	Excellent	Excellent
Excellent	Poor	Good	Good
Poor	Good	Excellent	Excellent
Fair	Good	Good	Excellent
Good	Poor	Poor	Poor
Poor	Poor	Good	Good
Excellent	Excellent	Good	Excellent
Excellent	Excellent	Good	Excellent

C. Heterochain polymers[a]

	Poly(dimethyl siloxane) silicone rubber, usually copolymer with vinyl groups (VMQ)	Poly(dimethyl siloxane) copolymer with phenyl-bearing siloxane and vinyl groups (PVMQ)
Gum stock (cross-linked, unfilled)		
Density (g/cm^3)	0.98	0.98
Tensile strength (psi)[b]	50–100	50–100
Resistivity (ohm-cm, log)	11–17	11–17
Dielect. const. at 1 kHz	3.0–3.5	3.0–3.5
Dielect. str. (kV/in.)[c]	100–600	100–600
Diss. factor, at 1 kHz	0.001–0.010	0.001–0.010
Reinforced stock		
Tensile strength (psi)[b]	500–1,200	500–1,500
Elong. at break (%)	200–700	200–700
Hardness, Shore A	30–80	30–80
Cont. high-temp. limit (°C)	250	300
Stiffening temp. (°C)	−50	−100
Brittle temp. (°C)	−50	−120
Resilience	Fair	Fair
Resistance to		
acid	Fair	Fair
alkali	Fair	Fair
gasoline and oil	Poor	Poor
aromatic hydrocarbons	Poor	Poor
ketones	Excellent	Excellent
chlorinated solvents	Poor	Poor
oxidation	Excellent	Excellent
ozone	Excellent	Excellent

[a]Abbreviations are according to ASTM.
[b]1000 psi = 6.895 MPa.
[c]1 kV/in. = 0.0394 MV/m.

Room-temp. vulcanizing silicone	Polysulfide (ET and EOT)	Polyurethane (AU and EU)
1.0–1.3 (compd'd)	1.35	1.25
—	100–200	2000–4000
15	12	11–14
2.8	7–9.5	5–8
500	250–600	350–525
0.003	0.001–0.005	0.02–0.09
400–800	1,300–1,800	3,000–10,000
100–200	200–500	200–600
30–50	25–85	20–100
200–250	120	120
−50 to −100	−25	−25 to −35
−50 to −100	−50	−50 to −60
Fair	Fair	Poor
Fair	Fair	Fair
Fair	Good	Poor to fair
Poor	Excellent	Excellent
Poor	Good	Good
Excellent	Good	Poor
Poor	Good	Poor
Excellent	Excellent	Excellent
Excellent	Excellent	Excellent

Appendix 10

Typical Properties of Representative Textile Fibers

Fiber/chemical name	Specific gravity	Breaking tenacity[a] (g/denier)	
		Standard	Wet
1. Acetate/cellulose acetate			
(a) diacetate	1.32	1.2–1.4	0.8–1.0
(b) triacetate	1.3	1.1–1.3	0.8–1.0
2. Acrylic/polyacrylonitrile	1.17	2.0–2.7	1.6–2.2
3. Aramid/aromatic polyamide			
(a) Kevlar (Du Pont)	1.44	21.7	21.7
(b) Nomex (Du Pont)	1.38	4.0–5.3	3.0–4.1
4. Cotton/α-cellulose	1.54	3.0–4.9	3.0–5.4
5. Fluorocarbon/poly(tetra-fluoroethylene)	2.1	0.9–2.0	0.9–2.0
6. Glass/silica, silicates	2.49–2.55	9.6–19.9	6.7–19.9
7. Nylon/aliphatic polyamide			
(a) Nylon-6	1.14	4.0–9.0	3.7–8.2
(b) Nylon-6,6	1.14	3.0–9.5	2.6–8.0
8. Olefin			
(a) polyethylene (branched)	0.92	1.0–3.0	1.0–3.0
(b) polyethylene (linear)	0.95	3.5–7.0	3.5–7.0
(c) polypropylene	0.90	3.0–8.0	3.0–8.0
9. Polyester/poly(ethylene terephthalate)	1.38	2.2–9.5	2.2–9.5
10. Spandex/segmented polyurethane	1.21	0.7–0.9	—
11. Viscose rayon/regenerated cellulose			
(a) regular	1.46–1.54	0.7–3.2	0.7–1.8
(b) high tenacity		3.0–5.7	1.9–4.3
12. Wool/protein	1.32	1.0–2.0	0.8–1.8

[a]Tensile strength (MPa) = tenacity (g/denier) × density (g/cm^3) × 88.3.
Source: Textile World, 128(8), 57 (1978).

Elongation at break (%)		Water absorbed at 70°F, 65% rel.	
Standard	Wet	humidity (%)	Thermal stability
			(a) Sticks at 175–205°C; softens, 205–230°C;
25–45	35–50	6.4	melts, 260°C
26–40	30–40	3.2	(b) Melts at 300°C
34–50	34–60	1.5	Shrinks 5% at 253°C
2.5–4	2.5–4	4.5–7	(a) Decomposes at 500°C
22–32	20–30	6.5	(b) Decomposes at 370°C
3–10		7–8.5	Decomposes at 150°C
19–140	19–140	Nil	Melts at about 288°C
			Softens at 730–850°C;
3.1–5.3	2.2–5.3	Nil	does not burn
			(a) Melts at 216°C;
16–50	19–47	2.8–5.0	decomposes, 315°C
			(b) Sticks at 230°C;
16–66	18–70	4.2–4.5	melts, 250–260°C
			(a) Softens 105–115°C; melts 110–120°C;
20–80	20–80	Nil	shrinks 5% at 75°C
			(b) Softens 115–125°C; melts 125–138°C;
10–45	10–45	Nil	shrinks 5% at 75–80°C
			(c) Softens 140–175°C; melts 160–177°C;
14–80	14–80	0.01–0.1	shrinks 5% at 100–130°C
			Sticks at 230°C;
12–55	12–55	0.4–0.8	melts, 250°C
400–625	—	1.3	Sticks at 215°C
15–30	20–40	11–13	(a) Loses strength at 150°C
9–26	14–34	—	(b) Decomposes at 175–240°C
20–40		11–17	Decomposes at 130°C

Index